# Lecture Notes in Networks and Systems

Volume 83

The series "Lecture Notes in Networks and Systems" publishes the latest developments in Networks and Systems—quickly, informally and with high quality. Original research reported in proceedings and post-proceedings represents the core of LNNS.

Volumes published in LNNS embrace all aspects and subfields of, as well as new challenges in, Networks and Systems.

The series contains proceedings and edited volumes in systems and networks, spanning the areas of Cyber-Physical Systems, Autonomous Systems, Sensor Networks, Control Systems, Energy Systems, Automotive Systems, Biological Systems, Vehicular Networking and Connected Vehicles, Aerospace Systems, Automation, Manufacturing, Smart Grids, Nonlinear Systems, Power Systems, Robotics, Social Systems, Economic Systems and other. Of particular value to both the contributors and the readership are the short publication timeframe and the world-wide distribution and exposure which enable both a wide and rapid dissemination of research output.

The series covers the theory, applications, and perspectives on the state of the art and future developments relevant to systems and networks, decision making, control, complex processes and related areas, as embedded in the fields of interdisciplinary and applied sciences, engineering, computer science, physics, economics, social, and life sciences, as well as the paradigms and methodologies behind them.

**** Indexing: The books of this series are submitted to ISI Proceedings, SCOPUS, Google Scholar and Springerlink ****

More information about this series at http://www.springer.com/series/15179

Samir Avdaković · Aljo Mujčić ·
Adnan Mujezinović · Tarik Uzunović ·
Ismar Volić
Editors

# Advanced Technologies, Systems, and Applications IV - Proceedings of the International Symposium on Innovative and Interdisciplinary Applications of Advanced Technologies (IAT 2019)

Springer

*Editors*
Samir Avdaković
Faculty of Electrical Engineering
University of Sarajevo
Sarajevo, Bosnia and Herzegovina

Aljo Mujčić
Elektrotehnički Fakultet
Univerzitet u Tuzli
Tuzla, Bosnia and Herzegovina

Adnan Mujezinović
Elektrotehnički Fakultet
Univerzitet u Sarajevu
Sarajevo, Bosnia and Herzegovina

Tarik Uzunović
Elektrotehnički Fakultet
Univerzitet u Sarajevu
Sarajevo, Bosnia and Herzegovina

Ismar Volić
Department of Mathematics
Wellesley College
Wellesley, MA, USA

ISSN 2367-3370 ISSN 2367-3389 (electronic)
Lecture Notes in Networks and Systems
ISBN 978-3-030-24985-4 ISBN 978-3-030-24986-1 (eBook)
https://doi.org/10.1007/978-3-030-24986-1



This Springer imprint is published by the registered company Springer Nature Switzerland AG
The registered company address is: Gewerbestrasse 11, 6330 Cham, Switzerland

# About this Book

This book presents innovative and interdisciplinary applications of advanced technologies. It includes the scientific outcomes of the conference 11th Days of Bosnian-Herzegovinian American Academy of Arts and Sciences held in Sarajevo, Bosnia and Herzegovina, June 20–23, 2019. This book offers a uniquely comprehensive, multidisciplinary, and interdisciplinary overview of the latest developments in a broad section of technologies and methodologies, viewed through the prism of applications in computing, networking, information technology, robotics, complex systems, communications, energy, mechanical engineering, economics, and medicine, among others.

# Contents

# About the Authors

**Samir Avdaković** was born in 1974 in Doboj, Bosnia and Herzegovina. He received his M.Sc. and Ph.D. in electrical engineering at the Faculty of Electrical Engineering, University of Tuzla, in 2006 and 2012, respectively. Currently, he is working in the Department for Strategic Development in EPC Elektroprivreda B&H and Faculty of electrical engineering - Department of Power Engineering - University of Sarajevo. Since October 2014, he has been Assistant Professor at the Faculty of Electrical Engineering, University of Sarajevo, where he currently teaches courses in fundamentals of power system operations and control and power system planning. His research interests include power system analysis, power system dynamics and stability, WAMPCS, smart systems, signal processing, and biomedical engineering.

**Aljo Mujčić** was born in 1969 in Dubrave Gornje, Bosnia and Herzegovina. He received his B.S. in electrical engineering from the University of Belgrade, Belgrade, former Yugoslavia, in 1992; his M.S. in electrical engineering from the University of Tuzla, Tuzla, Bosnia and Herzegovina, in 1999; and his Ph.D. from the University of Ljubljana, Slovenia, in 2004. From 1993 to 2001, he was Teaching Assistant at the University of Tuzla. From 2001 to 2004, he was with the Faculty of Electrical Engineering at the University of Ljubljana, engaged in a research project on high-speed digital power line communication over high-voltage power lines. He is currently Full Professor in the Department of Telecommunications, Faculty of Electrical Engineering, University of Tuzla. He published five textbooks and more than 70 journal and conference research papers. He participated in more than ten research projects. He was involved in developing graduate and undergraduate courses in the areas of electronics and telecommunications. His research interests include embedded systems, signal processing, optical access networks, and modeling of nonlinear components.

**Adnan Mujezinović** received his M.Sc. and Ph.D. in electrical engineering from the Faculty of Electrical Engineering, University of Sarajevo (Bosnia and

Herzegovina), in 2011 and 2017, respectively. From 2012, he has been with the same faculty as a teaching assistant and currently as an assistant professor. His research interests include numerical calculations of electromagnetic fields, cathodic protection, and grounding systems.

**Taik Uzunović** received the B.Eng. and M.Eng. in electrical engineering from the University of Sarajevo, Sarajevo, Bosnia and Herzegovina, and Ph.D. in mechatronics from Sabanci University, Istanbul, Turkey, in 2008, 2010, and 2015, respectively. He is Assistant Professor with the Department of Automatic Control and Electronics, Faculty of Electrical Engineering, University of Sarajevo, Sarajevo, Bosnia and Herzegovina. His research interests include motion control, robotics, and mechatronics.

**Ismar Volić** is Professor of Mathematics at Wellesley College in Massachusetts. He received a bachelor's degree from Boston University in 1998 and a Ph.D. in mathematics from Brown University in 2003. He has held postdoctoral or visiting positions at the University of Virginia, Massachusetts Institute of Technology, and Louvain-la-Neuve University before joining the Wellesley faculty in 2006. In 2018, he spent a semester at the University of Sarajevo as a US Fulbright Scholar. His research interest includes algebraic topology, more specifically calculus of functors and embedding spaces. He is the author of thirty articles and two books and has held over two hundred lectures in some twenty countries. His work has been recognized by various grants from the National Science Foundation, American Mathematical Society, the Simons Foundation, and the Clay Foundation, among others.

# Toward Finite Models for the Stages of the Taylor Tower for Embeddings of the 2-Sphere

Adisa Bolić[1], Franjo Šarčević[1], and Ismar Volić[2(✉)]

[1] Department of Mathematics, University of Sarajevo, Sarajevo, Bosnia and Herzegovina
adisa.bolic@gmail.com, franjo.sarcevic@live.de
[2] Department of Mathematics, Wellesley College, Wellesley, MA, USA
ivolic@wellesley.edu

**Abstract.** We provide the beginning of the construction of a finite model for the stages of the Taylor tower for embeddings of the 2-sphere in a smooth manifold. We show how these stages can be described as iterated homotopy limits of punctured cubes of embedding spaces where the source manifolds are homotopy equivalent to unions of disks.

## 1 Introduction

This paper is concerned with the Taylor tower for the space of embeddings $\mathrm{Emb}(S^2, N)$, where $S^2$ is the 2-sphere and $N$ is a smooth manifold (precise definition of $\mathrm{Emb}(S^2, N)$ can be found in Sect. 2). Namely, this is a tower of "approximations" of $\mathrm{Emb}(S^2, N)$,

$$\mathrm{Emb}(S^2, N) \longrightarrow \left(T_\infty \mathrm{Emb}(S^2, N) \to \cdots \to T_k \mathrm{Emb}(S^2, N) \to \cdots \to T_0 \mathrm{Emb}(S^2, N)\right) \quad (1)$$

where $T_k \mathrm{Emb}(S^2, N)$ is the $k^{th}$ *stage of the tower* and $T_\infty \mathrm{Emb}(S^2, N)$ is its inverse (homotopy) limit. The theory of Taylor towers for spaces of embeddings (and other related functors) was set up by Goodwillie and Weiss [GW99, Wei99] and has proved to be extremely useful in understanding the topology of embedding spaces, mainly due to a deep theorem about the convergence of the tower (Theorem 2.3). Some particular references of where the theory has been successfully used are given in the discussion following the proof of Theorem 3.1.

The definition of the $k^{th}$ stage is not easy (Definition 2.2) and involves taking a homotopy limit of a large topological diagram. However, there is a lesser-known version of the definition that does not quite have all the properties of the original one, but is good enough for many purposes. It involves "punching holes" in $S^2$ and looking at the $k^{th}$ Taylor stages of embeddings of the resulting manifolds. The process continues until $S^2$ has been reduced to a union of 2-balls, thereby


S. Avdaković et al. (Eds.): IAT 2019, LNNS 83, pp. 1–13, 2020.
https://doi.org/10.1007/978-3-030-24986-1_1

providing an iterative way of defining the stages. The advantage is that all the diagrams involved in the procedure are finite.

More detail about the construction of the Taylor tower and the process of "punching holes" can be found in Sect. 2. The main result of this paper, Theorem 3.1, simply says that this can be done in an easy $(k+1)$-step process, and the proof provides an algorithm for doing so. The same procedure can be performed for a general space of embeddings of any smooth manifold $M$ in $N$, but the case of $S^2$ is particularly nice since one arrives at embeddings of balls fast and in a combinatorially clean way.

We emphasize that all that we are essentially providing is a fairly straightforward algorithm of reducing $S^2$ to subsets that are homotopy equivalent to unions of balls. What is still needed is a way of organizing these equivalences so that they are compatible with the maps in the diagrams involved in the procedure. We will say more about this in the comments following the the proof of Theorem 3.1. We will also discuss how our procedure mimics a similar process that led to the cosimplicial model for the space of long knots, which has in turn been used to great effect in furthering our understanding of the topology of spaces of knots. We hope that the constructions in this paper will also lead to such a model and provide new insight into $\mathrm{Emb}(S^2, N)$.

## 2 Taylor Tower for Embeddings of the 2-Sphere

In this section we briefly recall the construction of the Taylor tower for the space of embeddings. We will provide very few details since they abound in literature, such as in the foundational papers [GW99, Wei99], as well as in expository work [Mun10, ŠV], or [MV15, Sect. 10.2]. In particular, we will assume the reader is familiar with the language of (punctured) cubical diagrams and their homotopy limits [Goo92], an exposition of which can for example be found in [MV15, Sect. 5.3].

Let $\underline{k} = \{1, 2, ..., k\} \subseteq \mathbb{N}$ and let $\mathcal{P}_0(\underline{k})$ be the poset of nonempty subsets of $\underline{k}$. Let Top be the category of topological spaces with maps as morphisms.

**Definition 2.1.** A *punctured* $k$-cube, or a *punctured cubical diagram of dimension* $k$ is a (covariant) functor

$$\begin{aligned} \mathcal{X}\colon \mathcal{P}_0(\underline{k}) &\longrightarrow \mathrm{Top}, \\ S &\longmapsto X_S. \end{aligned}$$

For example, a punctured 3-cube is the diagram

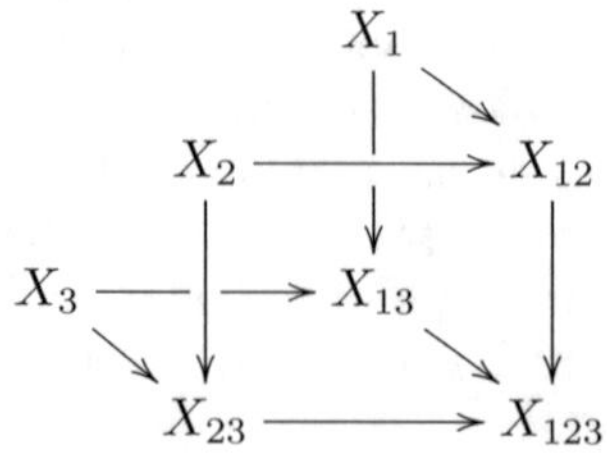

Of main interest here is the homotopy limit of a punctured $k$-cube, which is the subspace of the product of mapping spaces $\prod_{S\in\mathcal{P}_0(\underline{k})}\mathrm{Map}(\Delta^{|S|}\to X_S)$ given by certain compatibilities that naturally generalize the compatibilities in the ordinary limit in a way that endows the homotopy limit with homotopy invariance. The homotopy pullback is an example of such a homotopy limit where the punctured cube has dimension 2.

Now let $S^2$ be the 2-sphere, $N$ any smooth manifold of dimension $n$, and define $\mathrm{Emb}(S^2,N)$ to be the *space of embeddings of* $S^2$ in $N$, namely the set of smooth injective maps with injective derivative topologized using the Whitney $\mathcal{C}^\infty$ topology. Let $\mathrm{Imm}(S^2,N)$ be the *space of immersions of* $S^2$ *in* $N$, defined the same way except the maps need not be injective.

To define the Taylor tower for $\mathrm{Emb}(S^2,N)$, let $\mathcal{O}(S^2)$ be the poset of open subsets of $S^2$ and $\mathcal{O}_k(S^2)$ the subposet of $\mathcal{O}(S^2)$ consisting of open subsets that are diffeomorphic to at most $k$ open balls.

**Definition 2.2.** For $k\geq 0$, the $k^{th}$ *stage of the Taylor tower for* $\mathrm{Emb}(S^2,N)$ is defined as

$$T_k\,\mathrm{Emb}(S^2,N)=\operatorname*{holim}_{V\in\mathcal{O}_k(S^2)}\mathrm{Emb}(V,N) \tag{2}$$

The space $T_k\,\mathrm{Emb}(S^2,N)$ is "polynomial" of degree $k$ in a suitable sense and it can be thought of as trying to reconstruct $\mathrm{Emb}(S^2,N)$ from information about embeddings of balls in $N$. For details, see [Mun10, Sect. 4] or [MV15, Sect. 10.2]. The stage $T_k(-,N)$ is a functor so that $S^2$ can be replaced by any open subsets $U\in\mathcal{O}(S^2)$, and the definition would be the same, except the diagram would be indexed over the category $\mathcal{O}_k(U)$.

There are natural maps $\mathrm{Emb}(S^2,N)\to T_k\,\mathrm{Emb}(S^2,N)$ and $T_k\,\mathrm{Emb}(S^2,N)\to T_{k-1}\,\mathrm{Emb}(S^2,N)$. Putting these together gives the Taylor tower for $\mathrm{Emb}(S^2,N)$ as pictured in (1).

One well-known fact that we will use is that the linearization of the space of embeddings is the space of immersions, i.e.

$$T_1\,\mathrm{Emb}(S^2,N)\simeq\mathrm{Imm}(S^2,N). \tag{3}$$

For more on why this is true, see [Wei99, Example 2.3] or [Mun10, Examples 4.8 and 4.14].

The following result says that, under certain dimensional assumptions, the Taylor tower converges. First recall that a map $f\colon X\to Y$ is $k$-*connected* if it induces isomorphisms on homotopy groups up to degree $k-1$ and a surjection on the $k^{th}$ homotopy group.

**Theorem 2.3** [GW99]. *For $k\geq 1$, the map*

$$\mathrm{Emb}(S^2,N)\longrightarrow T_k\,\mathrm{Emb}(S^2,N)$$

*is*

$$(k(n-4)-1)\text{-}connected.$$

*Thus if $n>4$, the map $\mathrm{Emb}(S^2,N)\to T_\infty\,\mathrm{Emb}(S^2,N)$, where $T_\infty\,\mathrm{Emb}(S^2,N)$ is the inverse limit of the Taylor tower, is an equivalence.*

There is an alternative definition of $T_k \operatorname{Emb}(S^2, N)$ that is easier to understand but it has a defect in that it is not functorial. Nevertheless, it is useful since it provides a finite model for $T_k \operatorname{Emb}(S^2, N)$, as opposed to Definition 2.2 which involves a large topological category over which the homotopy limit is taken. Using this alternative definition, the $k^{th}$ stage can be expressed iteratively in terms of homotopy limits of punctured cubical diagrams.

Namely, let $C_i$, $1 \leq i \leq k+1$, be disjoint closed subsets of $S^2$. We then have a punctured cubical diagram of spaces given by

$$S \longmapsto \operatorname{Emb}(S^2 - \bigcup_{i \in S} C_i, N)$$

for $S \in \mathcal{P}_0(\underline{k+1})$. The maps in this diagram are restrictions of embeddings. We can apply $T_k(-)$ to each of these embedding spaces (as mentioned above, the definition is the same as Definition 2.2; just replace the domain $S^2$ by $S^2 - \bigcup_{i \in S} C_i$). We then have

**Proposition 2.4.** *There is an equivalence*

$$T_k \operatorname{Emb}(S^2, N) \longrightarrow \operatorname*{holim}_{S \in \mathcal{P}_0(\underline{k+1})} T_k \operatorname{Emb}(S^2 - \bigcup_{i \in S} C_i, N).$$

This proposition is true because of various polynomial properties of $T_k(-)$. For details, see [MV15, Example 10.2.18]. For more on how this map is defined, see the discussion following Example 3.4 in [ŠV].

But now we can apply Proposition 2.4 again to each $T_k \operatorname{Emb}(S^2 - \bigcup_{i \in S} C_i, N)$. That is, we can, for each $S$, "punch" up to $k+1$ new holes in $S^2 - \bigcup_{i \in S} C_i$ and form a punctured cubical diagram of $T_k$'s whose homotopy limit is $T_k \operatorname{Emb}(S^2 - \bigcup_{i \in S} C_i, N)$.

We can now iterate this process. If enough of these iterations involve holes that reduce the handle dimensions of the manifold from the previous stage, everything can be reduced to embeddings of unions of 2-balls, namely manifolds of handle dimension zero. (Handle dimension is the highest dimension of a handle required to build the manifold; the notion of handle dimension figures prominently in manifold calculus; see, for example, [Mun10].) We can stop here because of the following: Let $D^2$ denote the 2-dimensional open ball. Then

$$T_k \operatorname{Emb}\left(\bigcup_j D^2, N\right) \simeq T_k \operatorname{Emb}\left(\bigcup_k D^2, N\right), \quad \text{for } j > k; \tag{4}$$

$$T_k \operatorname{Emb}\left(\bigcup_j D^2, N\right) \simeq \operatorname{Emb}\left(\bigcup_j D^2, N\right), \quad \text{for } j \leq k. \tag{5}$$

The first equivalence is true because $T_k(-, N)$ is determined by what it does on $k$ balls (see proof of Theorem 10.2.14 in [MV15]). The second also follows from polynomial properties of $T_k(-, N)$ [Wei99, Theorem 6.1].

The goal of this paper is to systematically unravel this procedure, i.e. give a description of $T_k \operatorname{Emb}(S^2, N)$ as an iterated homotopy limit of diagrams of embeddings of 2-balls in $N$.

*Remark 2.5.* Everything that has been said so far, including the description of the process of "punching holes", still holds when $S^2$ is replaced by an arbitrary manifold $M$. The advantage of using $S^2$ is that in this case we can easily understand and control the way one arrives at a union of open balls by systematically removing closed sets from $S^2$.

## 3 "Punching Holes" Model for the Stages of the Taylor Tower for Embeddings of the 2-Sphere

In this section we carry out the "punching holes" procedure outlined in the previous section. Theorem 3.1 gives the statement of the main result, but its proof is the more interesting part since it gives an explicit algorithm for the iterative process. Following the proof, we provide some comments and examples illustrating the main construction.

**Theorem 3.1.** *The $k^{th}$ Taylor stage of the space of embeddings $\operatorname{Emb}(S^2, N)$ is given by an iterated construction involving homotopy limits of puctured cubes. After the $l^{th}$ iteration, all the spaces indexed by subsets $S \in \mathcal{P}_0(\underline{k+1})$ of cardinality $\leq l$ in the initial cube are given as iterated homotopy limits of punctured cubes of embeddings of 2-balls. The construction terminates after $k+1$ steps.*

To set some notation before the proof, let $A_j$ denote the open 2-ball with $j$ holes. More precisely, this is the open 2-ball $D^2$ with $j \geq 0$ subsets that are homeomorphic to closed 2-disks removed. Note that $A_1$ is the ordinary open annulus and $A_0$ is the open ball $D^2$. Let $N$ as usual be a smooth manifold.

*Proof.* Choose disjoint subsets $C_{i_1}$, $1 \leq i_1 \leq k+1$ of $S^2$ that are homeomorphic to closed disks. From Proposition 2.4, we get an equivalence

$$T_k \operatorname{Emb}(S^2, N) \longrightarrow \underset{S_1 \in \mathcal{P}_0(\underline{k+1})}{\operatorname{holim}} T_k \operatorname{Emb}(S^2 - \bigcup_{i_1 \in S_1} C_{i_1}, N). \tag{6}$$

The spaces $S^2 - \bigcup_{i_1 \in S_1} C_{i_1}$ are homotopy equivalent to $A_{|S_1|-1}$. In particular, for $|S_1| = 1$,

$$S^2 - C_{i_1} \simeq A_0 = D^2$$

and (6) gives an equivalence $T_k \operatorname{Emb}(D^2, N) \simeq \operatorname{Emb}(D^2, N)$.

We can now continue and, for each $S_1 \in \mathcal{P}_0(\underline{k+1})$ with $|S_1| > 1$, write $T_k \operatorname{Emb}(S^2 - \bigcup_{i_1 \in S_1} C_{i_1}, N)$ as a homotopy limit of a punctured cube of embeddings of spaces that are equivalent to unions of balls or unions of balls with a ball with fewer holes.

Namely, fix an $S_1$ with $|S_1| > 1$. Denote by $C_{i_1,1}, ..., C_{i_1,|S_1|}$ the closed sets removed from $S^2$ that correspond to elements of $S_1$. Then choose $k+1$ disjoint closed subsets $C_{S_1,i_2}$, $1 \leq i_2 \leq k+1$, of $S^2$ homeomorphic to closed disks in such a way that removing any of them removes a hole in the ball with holes $S^2 - \bigcup_{i_1 \in S_1} C_{i_1}$, as indicated in Fig. 1. (The "hole" corresponding to removal of $C_{i_1,|S_1|}$ is the space "around" the ball with holes.)

We then have

$$T_k \operatorname{Emb}\left(S^2 - \bigcup_{i \in S_1} C_i, N\right) \simeq \operatorname*{holim}_{S_2 \in \mathcal{P}_0(\underline{k+1})} T_k \operatorname{Emb}\left(S^2 - \bigcup_{i_1 \in S_1} C_{i_1} - \bigcup_{i_2 \in S_2} C_{S_1,i_2}, N\right).$$

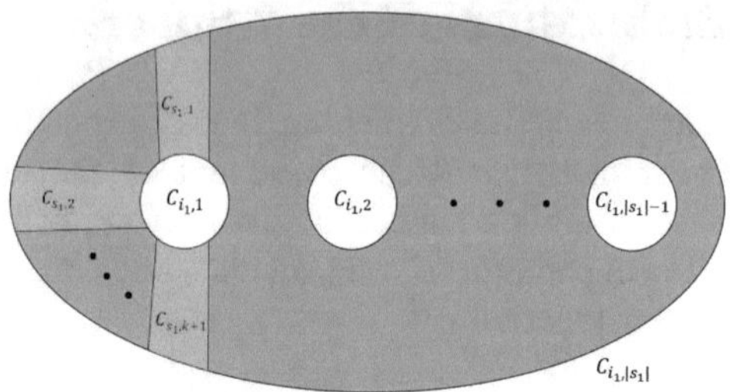


**Fig. 1.** A way of "punching holes" in $A_{|S_1|-1}$.

It is clear that

$$S^2 - \bigcup_{i_1 \in S_1} C_{i_1} - \bigcup_{i_2 \in S_2} C_{S_1,i_2} \simeq \left(\bigcup_{|S_2|-1} D^2\right) \cup A_{|S_1|-2}. \tag{7}$$

If $|S_1| = 2$, the space (7) is equivalent to $\bigcup_{|S_2|} D^2$. Then by (4) and (5), we have equivalences

$$T_k \operatorname{Emb}\left(S^2 - \bigcup_{i_1 \in S_1} C_{i_1} - \bigcup_{i_2 \in S_2} C_{S_1,i_2}, N\right) \simeq \begin{cases} \operatorname{Emb}\left(\bigcup_{|S_2|} D^2, N\right), & |S_2| \leq k; \\ \operatorname{Emb}\left(\bigcup_{k} D^2, N\right), & |S_2| > k. \end{cases} \tag{8}$$

If $|S_1| > 2$, then, for each such $S_1$ and for each $S_2$, we choose closed disjoint subsets $C_{S_1,S_2,i_3}$, $1 \leq i_3 \leq k+1$ of $S^2$ such that, when some of them are removed, the number of holes in $A_{|S_1|-2}$ is reduced by one and some disjoint 2-balls are created. More precisely, we have

$$T_k \operatorname{Emb}\left(S^2 - \bigcup_{i_1 \in S_1} C_{i_1} - \bigcup_{i_2 \in S_2} C_{S_1,i_2}, N\right)$$

$$\simeq \underset{S_3 \in \mathcal{P}_0(\underline{k+1})}{\operatorname{holim}} T_k \operatorname{Emb}\left(S^2 - \bigcup_{i_1 \in S_1} C_{i_1} - \bigcup_{i_2 \in S_2} C_{S_1,i_2} - \bigcup_{i_3 \in S_3} C_{S_1,S_2,i_3}, N\right)$$

with

$$S^2 - \bigcup_{i_1 \in S_1} C_{i_1} - \bigcup_{i_2 \in S_2} C_{S_1,i_2} - \bigcup_{i_3 \in S_2} C_{S_1,S_2,i_3} \simeq \left(\bigcup_{|S_2|+|S_3|-2} D^2\right) \cup A_{|S_1|-3} \quad (9)$$

Now if $|S_1| = 3$, this is equivalent to a union of $|S_2| + |S_3| - 1$ balls and we have

$$T_k \operatorname{Emb}\left(\bigcup_{|S_2|+|S_3|-1} D^2, N\right) \simeq \begin{cases} \operatorname{Emb}\left(\bigcup_{|S_2|+|S_3|-1} D^2, N\right), & |S_2| + |S_3| - 1 \leq k; \\ \operatorname{Emb}\left(\bigcup_k D^2, N\right), & |S_2| + |S_3| - 1 > k. \end{cases} \quad (10)$$

We continue iterating this procedure so that, given $l \leq k$ and fixing $S_1$, $S_2$, ..., $S_l$ with $|S_1| > l$, we can write

$$T_k \operatorname{Emb}\left(S^2 - \bigcup_{i_1 \in S_1} C_{i_1} - \bigcup_{i_2 \in S_2} C_{S_1,i_2} - \cdots - \bigcup_{i_l \in S_l} C_{S_1,\ldots,S_{l-1},i_l}, N\right)$$

$$\simeq \underset{S_{l+1} \in \mathcal{P}_0(\underline{k+1})}{\operatorname{holim}} T_k \operatorname{Emb}\left(S^2 - \bigcup_{i_1 \in S_1} C_{i_1} - \bigcup_{i_2 \in S_2} C_{S_1,i_2} - \cdots - \bigcup_{i_{l+1} \in S_{l+1}} C_{S_1,\ldots,S_l,i_{l+1}}, N\right)$$

with

$$S^2 - \bigcup_{i_1 \in S_1} C_{i_1} - \bigcup_{i_2 \in S_2} C_{S_1,i_2} - \cdots - \bigcup_{i_{l+1} \in S_{l+1}} C_{S_1,\ldots,S_l,i_{l+1}} \simeq \bigcup_{|S_2|+\cdots+|S_{l+1}|-l} D^2 \cup A_{|S_1|-(l+1)} \quad (11)$$

Then if $|S_1| = l + 1$, this is the union of $|S_2| + \cdots + |S_{l+1}| - l + 1$ balls and we have, again by (4) and (5),

$$T_k \operatorname{Emb}\left(\bigcup_{|S_2|+\cdots+|S_{l+1}|-l+1} D^2, N)\right)$$

$$\simeq \begin{cases} \operatorname{Emb}\left(\bigcup_{|S_2|+\cdots+|S_{l+1}|-l+1} D^2, N\right), & |S_2| + \cdots + |S_{l+1}| - l + 1 \leq k; \\ \operatorname{Emb}\left(\bigcup_k D^2, N\right), & |S_2| + \cdots + |S_{l+1}| - l + 1 > k. \end{cases}$$

Thus after $l$ iterations, all the spaces in the inital punctured cube indexed by $S_1$ with $|S_1| \leq l$ have been replaced by iterated homotopy limits of diagrams of embeddings where the spaces being embedded are homotopy equivalent to unions of at most $k$ balls.

The process terminates after $k+1$ steps with the case $|S_1| > k$, i.e. with $|S_1| = k+1$ At the end, all the spaces in the initial punctured cube defining $T_k \operatorname{Emb}(S^2, N)$ have been replaced by iterated homotopy limits of diagrams of embeddings of at most $k$ balls in $N$. □

*Remark 3.2.* The largest possible number of balls is when the cardinality of all $S_i$ is $k+1$, in which case $|S_2| + \cdots + |S_{k+1}| - k + 1 = k^2 + 1$ (but as usual we can replace $T_k \operatorname{Emb}(\bigcup_{k^2+1} D^2, N)$ by $T_k \operatorname{Emb}(\bigcup_k D^2, N)$).

Here are some comments on Theorem 3.1, its proof, and further directions.

- The main omission of the result is that, while we have replaced $S^2$ – (closed subsets) by unions of balls and balls with holes, we have not consistently defined the maps that would make the original diagrams of embedding spaces homotopy equivalent – as diagrams – to diagrams of embeddings of balls and balls with holes. What is required is to define maps from embeddings of balls to embeddings of more balls in a way that is consistent across the diagram. This leads to:
- Focusing on the case $N = \mathbb{R}^n$, one of the uses for this construction would be to come up with a cosimplicial model for $\operatorname{Emb}(S^2, \mathbb{R}^n)$ in the spirit of what was done for (long) knots $\operatorname{Emb}(S^1, \mathbb{R}^n)$ [Sin09]. The cosimplicial model for knots was a way to resolve the issue of "doubling" maps between embeddings of balls (1-balls in that case) and embeddings of more balls. This led to many exciting results and provided a deep understanding of the cohomology of $\operatorname{Emb}(S^1, \mathbb{R}^{\geq 4})$ [LTV10, ALTV08] and finite type invariants of $\operatorname{Emb}(S^1, \mathbb{R}^3)$ [BCKS14, BCSS05, Vol06].
The setup in the knot case was to replace each $\operatorname{Emb}(\bigcup_j D^1, \mathbb{R}^n)$ with the homotopy equivalent configuration space $\operatorname{Conf}(j, \mathbb{R}^n)$ (there would also be a Stiefel manifold as a factor here, but it can be "removed" by passing to the homotopy fiber of the inclusion of embedding to immersions, as is commonly done in this setup). Then the restriction maps $\operatorname{Emb}(\bigcup_j D^1, \mathbb{R}^n) \to \operatorname{Emb}(\bigcup_{j+1} D^1, \mathbb{R}^n)$ are replaced by doubling maps on compactifications of corresponding configuration space. Putting all these doubling maps together gives a cosimplicial space whose partial totalizations are precisely the stages of the Taylor tower for knots.
Like in the knot case, the goal in this setup would be to replace each $\operatorname{Emb}(\bigcup_j D^2, \mathbb{R}^n)$ with $\operatorname{Conf}(j, \mathbb{R}^n)$ (upon removing the immersions part) and again replace the restriction maps
$$\operatorname{Emb}(\bigcup_j D^2, \mathbb{R}^n) \to \operatorname{Emb}(\bigcup_{j+1} D^2, \mathbb{R}^n)$$
by doubling maps of compactified configuration spaces.
The difference is that, in the knot case, after punching holes once,

what remains are balls (handlebodies of handle dimension zero), and $T_k \operatorname{Emb}(S^1, \mathbb{R}^n)$ is thus equivalent to the homotopy limit of a single punctured cubical diagram. The situation is more complicated here since the punching of the holes has to happen iteratively in $k+1$ stages in order to arrive at embeddings of balls.
Making this precise, namely showing that such (iterated) cosimplicial diagrams have totalizations that are homotopy equivalent to the iterated homotopy limits considered in Theorem 3.1 would be a considerably nicer result. One would obtain a spectral sequence converging to $\mathrm{H}^*(\operatorname{Emb}(S^2, \mathbb{R}^n))$ for $n > m+2$ (this condition on dimensions comes from Theorem 2.3) that would be given by a graph complex arising from the cohomology of configuration spaces.

- In the case $n = 4$, the graph complexes mentioned in the previous item could be used to define finite type invariants of $\operatorname{Emb}(S^2, \mathbb{R}^4)$. Namely, following the situation from the classical knots case, one could declare the cohomology classes on the diagonal of the cohomology spectral sequence to be finite type invariants. This would be particularly interesting for $\operatorname{Emb}(S^2, \mathbb{R}^4)$, even if the Taylor tower is not known to converge in this case.
- There are of course different ways to choose where the holes are punched at each stage of the process. While the procedure presented here is combinatorially simple, it is not the most efficient. In particular, we could have punched holes so that the number of holes created in the previous step is always reduced in a maximal way. For example, in Fig. 1, $C_{S_1,j}$, $2 \le j \le |S_1|$, could have been chosen to connect holes $C_{i_1,j-1}$ and $C_{i_1,j}$. Subsequent $C_{S_1,j}$, $j > |S_1|$, would create more balls, but the total number of balls resulting from this process would be significantly reduced.
- It should be clear how to extend Theorem 3.1 to products of 2-spheres. In fact, one could use multivariable calculus of functors [MV12] to treat each sphere in the product as a variable an apply Theorem 3.1 in that slot. This would result in a multi-iterated model with more complicated combinatorics, but it is conceptually clear how this generalization would go.
- Another direction for generalization would be to replace $S^2$ by $S^m$, $m > 2$. This would require a more sophisticated method of "punching holes" because generalizing the picture from Fig. 1 in the naive way will not yield the same result. Namely, if $C_{s_1,1}$ and $C_{s_1,2}$ are thought of as neighborhoods of the paths from the edge of $C_{i_1,1}$ and the outer edge of the punctured annulus, then cutting them both out produces a disjoint $D^2$. For $S^{\ge 3}$, this would not produce a disjoint ball; the result would still be a path-connected space. This means that, starting with the second stage of the iterative process, the closed subsets would have to be chosen in a different way in order for the equivalences (7), (9), and, more generally, (11) to hold (if they hold at all).

We finish with two examples that illustrate the procedure from the proof of Theorem 3.1.

**Example 3.3.** The first stage of the Taylor tower for $T_1 \operatorname{Emb}(S^2, N)$ is

$$T_1 \operatorname{Emb}(S^2, N) \simeq \operatorname{holim} \left( \begin{array}{ccc} & & T_1 \operatorname{Emb}(D^2, N) \\ & & \downarrow \\ T_1 \operatorname{Emb}(D^2, N) & \longrightarrow & T_1 \operatorname{Emb}(A_1, N) \end{array} \right) \tag{12}$$

By (5) and the fact that both embeddings and immersions of balls are determined by where they send, say, the center point plus the framing information, $T_1 \operatorname{Emb}(D^2, N) \simeq \operatorname{Emb}(D^2, N) \simeq \operatorname{Imm}(D^2, N)$. The final space can be written as

$$T_1 \operatorname{Emb}(A_1, N) \simeq \operatorname{holim} \left( \begin{array}{ccc} & & \operatorname{Emb}(D^2, N) \\ & & \downarrow \\ \operatorname{Emb}(D^2, N) & \longrightarrow & T_1 \operatorname{Emb}(D^2 \cup D^2, N) \end{array} \right)$$

It should be understood that the maps are not some kind of "doubling" but are restrictions of an embedding of a ball to two subballs. One can think of the balls in the initial spaces as little more than two halves of the open annulus with the two balls in the final space as the intersection of those two "half-annuli".

One can also write $T_1 \operatorname{Emb}(D^2 \cup D^2, N) \simeq T_1 \operatorname{Emb}(D^2, N) \simeq \operatorname{Emb}(D^2, N)$, but that unfortunately obscures the maps in this diagram. Instead, we can punch holes again to say

$$\begin{aligned} T_1 \operatorname{Emb}(D^2 \cup D^2, N) &\simeq \operatorname{holim} \left( \begin{array}{ccc} & & T_1 \operatorname{Emb}(D^2, N) \\ & & \downarrow \\ T_1 \operatorname{Emb}(D^2, N) & \longrightarrow & T_1 \operatorname{Emb}(*, N) \end{array} \right) \\ &\simeq \operatorname{holim} \left( \begin{array}{ccc} & & \operatorname{Emb}(D^2, N) \\ & & \downarrow \\ \operatorname{Emb}(D^2, N) & \longrightarrow & * \end{array} \right) \\ &\simeq \operatorname{Emb}(D^2, N) \times \operatorname{Emb}(D^2, N) \\ &\simeq \operatorname{Imm}(D^2 \cup D^2, N). \end{aligned}$$

The last equivalence is true since $\operatorname{Emb}(D^2, N) \simeq \operatorname{Imm}(D^2, N)$ and map being an immersion is a local condition.

Putting all this together, (12) becomes

$$T_1 \operatorname{Emb}(S^2, N) \simeq \operatorname{holim} \left( \begin{array}{ccc} & & \operatorname{Imm}(D^2, N) \\ & & \downarrow \\ \operatorname{Imm}(D^2, N) & \longrightarrow & \operatorname{Imm}(D^2 \cup D^2, N) \end{array} \right) \tag{13}$$

The ordinary pullback of this diagram is precisely $\operatorname{Imm}(S^2, N)$ since, given immersions of spaces that agree on the intersection, we get an immersion of their union. According to the Smale-Hirsch Theorem, this limit is also the homotopy limit as the restriction maps in the diagram are fibrations. Thus $T_1 \operatorname{Emb}(S^2, N) \simeq \operatorname{Imm}(S^2, N)$, which provides a special case of (3).

**Example 3.4.** For the second polynomial approximation of the space of embeddings of the 2-sphere, we have

$$T_2 \operatorname{Emb}(S^2, N) \simeq \operatorname*{holim}_{S \in \mathcal{P}_0(\underline{3})} T_2 \operatorname{Emb}(S^2 - \bigcup_{i_1 \in S_1} C_{i_1}, N).$$

Then $S^2 - \bigcup_{i_1 \in S_1} C_{i_1}$ is equivalent to $D^2$ if $|S_1| = 1$, to $A_1$ if $|S_1| = 2$, and to $A_2$ if $|S_1| = 3$.

Using $T_2 \operatorname{Emb}(D^2, N) \simeq \operatorname{Emb}(D^2, N)$, we thus have that $T_2 \operatorname{Emb}(S^2, N)$ is given as

$$T_2 \operatorname{Emb}(S^2, N) \simeq \operatorname{holim} \left( \begin{array}{ccccc} & & \operatorname{Emb}(D^2, N) & \searrow & \\ & \operatorname{Emb}(D^2, N) & \longrightarrow & & T_2 \operatorname{Emb}(A_1, N) \\ & \downarrow & \downarrow & & \downarrow \\ \operatorname{Emb}(D^2, N) & \longrightarrow & T_2 \operatorname{Emb}(A_1, N) & \searrow & \\ \searrow & T_2 \operatorname{Emb}(A_1, N) & \longrightarrow & & T_2 \operatorname{Emb}(A_2, N) \end{array} \right)$$

We next have, for each $|S_1| = 2$,

$$T_2 \operatorname{Emb}(A_1, N) \simeq \operatorname*{holim}_{S_2 \in \mathcal{P}_0(\underline{3})} T_2 \operatorname{Emb}(A_1 - \bigcup_{i_2 \in S_2} C_{S_1, i_2}, N)$$

with $A_1 - \bigcup_{i_2 \in S_2} C_{S_1, i_2}$ equivalent to $D^2$ for $|S_2| = 1$, to $D^2 \cup D^2$ for $|S_2| = 2$, and to $D^2 \cup D^2 \cup D^2$ for $|S_2| = 3$.

Using $T_2 \operatorname{Emb}(D^2, N) \simeq \operatorname{Emb}(D^2, N)$ and $T_2 \operatorname{Emb}(D^2 \cup D^2, N) \simeq T_2 \operatorname{Emb}(D^2 \cup D^2 \cup D^2, N) \simeq \operatorname{Emb}(D^2 \cup D^2, N)$, we then have

$$T_2 \operatorname{Emb}(A_1, N) \simeq \operatorname{holim} \left( \begin{array}{ccccc} & & \operatorname{Emb}(D^2, N) & \searrow & \\ & \operatorname{Emb}(D^2, N) & \longrightarrow & & \operatorname{Emb}(D^2 \cup D^2, N) \\ & \downarrow & \downarrow & & \downarrow \\ \operatorname{Emb}(D^2, N) & \longrightarrow & \operatorname{Emb}(D^2 \cup D^2, N) & \searrow & \\ \searrow & \operatorname{Emb}(D^2 \cup D^2, N) & \longrightarrow & & \operatorname{Emb}(D^2 \cup D^2, N) \end{array} \right)$$

We again emphasize that the initial maps in the diagram are restrictions, while the maps to the final space are even more involved since the equivalence $T_2 \operatorname{Emb}(D^2 \cup D^2 \cup D^2, N) \simeq \operatorname{Emb}(D^2 \cup D^2, N)$ requires deeper understanding of the definition of polynomial functors.

Now, when $|S_1| = 3$, we have

$$T_2 \operatorname{Emb}(A_2, N) \simeq \operatorname*{holim}_{S_3 \in \mathcal{P}_0(\underline{3})} T_2 \operatorname{Emb}(A_2 - \bigcup_{i_3 \in S_3} C_{S_1, S_2, i_3}, N).$$

Here the space being embedded is equivalent to $A_1$, $D^2 \cup A_1$, or $D^2 \cup D^2 \cup A_1$ depending on whether the cardinality of $S_1$ is 1, 2, or 3. Hence

$$T_2 \operatorname{Emb}(A_2, N) \simeq \operatorname{holim} \left( \begin{array}{ccccc} & & T_2 \operatorname{Emb}(A_1, N) & & \\ & & \downarrow & \searrow & \\ & T_2 \operatorname{Emb}(A_1, N) & \longrightarrow & & T_2 \operatorname{Emb}(D^2 \cup A_1, N) \\ & \downarrow & T_2 \operatorname{Emb}(D^2 \cup A_1, N) & & \downarrow \\ T_2 \operatorname{Emb}(A_1, N) & \longrightarrow & & \searrow & \\ & \searrow & & & \\ & T_2 \operatorname{Emb}(D^2 \cup A_1, N) & \longrightarrow & & T_2 \operatorname{Emb}(D^2 \cup D^2 \cup A_1, N) \end{array} \right)$$

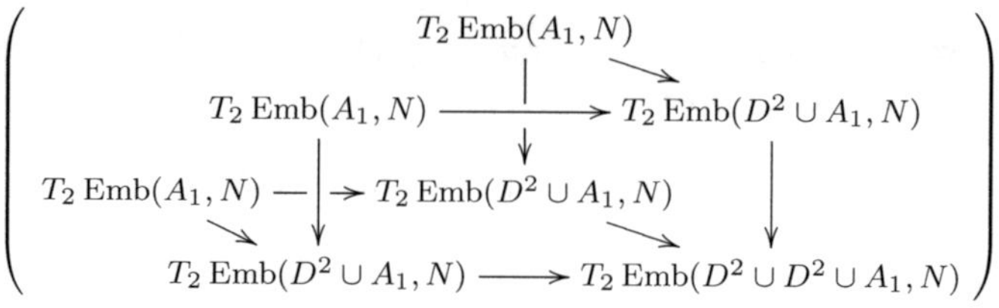

The initial space $T_2 \operatorname{Emb}(A_1, N)$ has been described above.

The space $T_2 \operatorname{Emb}(D^2 \cup A_1, N)$ is a punctured 3-cube with three initial vertices $T_2 \operatorname{Emb}(D^2 \cup D^2, N) \simeq \operatorname{Emb}(D^2 \cup D^2, N)$, three vertices $T_2 \operatorname{Emb}(\bigcup_{i=1}^3 D^2, N) \simeq \operatorname{Emb}(D^2 \cup D^2, N)$, and the final vertex $T_2 \operatorname{Emb}(\bigcup_{i=1}^4 D^2, N) \simeq \operatorname{Emb}(D^2 \cup D^2, N)$:

$$T_2 \operatorname{Emb}(D^2 \cup A_1, N) \simeq$$

$$\operatorname{holim} \left( \begin{array}{ccccc} & & \operatorname{Emb}(D^2 \cup D^2, N) & & \\ & & \downarrow & \searrow & \\ & \operatorname{Emb}(D^2 \cup D^2, N) & \longrightarrow & & \operatorname{Emb}(D^2 \cup D^2, N) \\ & \downarrow & \operatorname{Emb}(D^2 \cup D^2, N) & & \downarrow \\ \operatorname{Emb}(D^2 \cup D^2, N) & \longrightarrow & & \searrow & \\ & \searrow & & & \\ & \operatorname{Emb}(D^2 \cup D^2, N) & \longrightarrow & & \operatorname{Emb}(D^2 \cup D^2, N) \end{array} \right)$$

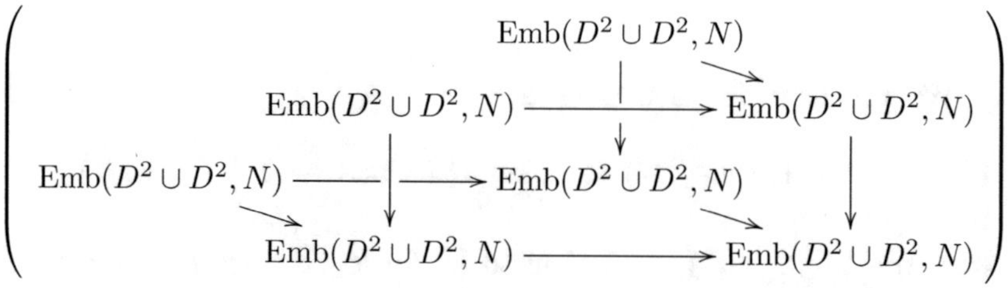

Again, we are not specifying what the maps in this diagram are.

Finally, $T_2 \operatorname{Emb}(D^2 \cup D^2 \cup A_1, N)$ can similarly be expressed as a homotopy limit of a punctured 3-cube of embeddings of $D^2 \cup D^2$.

## References


[ALTV08] Arone, G., Lambrechts, P., Turchin, V., Volić, I.: Coformality and rational homotopy groups of spaces of long knots. Math. Res. Lett. **15**(1), 1–14 (2008)

[BCKS14] Budney, R., Conant, J., Koytcheff, R., Sinha, D.: Embedding calculus knot invariants are of finite type. Algebr. Geom. Topol. **17**(3), 1701–1742 (2017). arXiv:1411.1832

[BCSS05] Budney, R., Conant, J., Scannell, K.P., Sinha, D.: New perspectives on self-linking. Adv. Math. **191**(1), 78–113 (2005)

[Goo92] Goodwillie, T.G.: Calculus II: analytic functors. K-Theory **5**(4), 295–332 (1991/1992)

[GW99] Goodwillie, T.G., Weiss, M.: Embeddings from the point of view of immersion theory II. Geom. Topol. **3**, 103–118 (1999)

[LTV10] Lambrechts, P., Turchin, V., Volić, I.: The rational homology of spaces of long knots in codimension $>2$. Geom. Topol. **14**, 2151–2187 (2010)

[Mun10] Munson, B.A.: Introduction to the manifold calculus of Goodwillie-Weiss. Morfismos **14**(1), 1–50 (2010)

[MV12] Munson, B.A., Volić, I.: Multivariable manifold calculus of functors. Forum Math. **24**(5), 1023–1066 (2012)

[MV15] Munson, B.A., Volić, I.: Cubical Homotopy Theory. New Mathematical Monographs, vol. 25. Cambridge University Press, Cambridge (2015)

[Sin09] Sinha, D.P.: The topology of spaces of knots: cosimplicial models. Amer. J. Math. **131**(4), 945–980 (2009)

[ŠV] Šarčević, F., Volić, I.: A streamlined proof of the convergence of the Taylor tower for embeddings in $\mathbb{R}^n$. Colloq. Math. **156**(1), 91–122 (2019)

[Vol06] Volić, I.: Finite type knot invariants and the calculus of functors. Compos. Math. **142**(1), 222–250 (2006)

[Wei99] Weiss, M.: Embeddings from the point of view of immersion theory I. Geom. Topol. 3, 67–101 (1999)

# The Macro-Political Economy of the Housing Market Through an Agent-Based Model

Faizan Khan and Zining Yang(✉)

Department of International Studies, Claremont Graduate University, Claremont, CA, USA
{faizan.khan, zining.yang}@cgu.edu

**Abstract.** Both, the housing bubble and financial crisis, are prime examples of complex events. Complex in the sense that there were several interconnected and interdependent root causes. This paper presents an agent-based model (ABM) to model the housing market from 1986 to 2017. We provide a unique approach to simulating the financial market along with analyzing the phenomenon of emergence resulting from the interactions among consumers, banks and the Federal Reserve. This paper specifically focuses on the emergence of "underwater mortgages" and the macroeconomics of the housing market. The market value of a property is heavily influence by the value of a neighboring property; therefore, individuals are able to gauge the probable value of a property that has not been developed yet. The blend of available financial products to consumers (i.e., ARM versus Fixed-Rate) certainly influences demand within the housing market given that ARM products are more affordable than fixed-rate products. Policymakers and financial institutions should work together to develop programs, which monitor the supply of these historically easy to access financial products and prevent the risk of underwater mortgages and crashes.

**Keywords:** Agent-based simulation · Housing market · Systemic risk · Complexity and emergence · Financial crisis

## 1 Introduction

For many people, the "American Dream" is defined as an opportunity to own a home. Home ownership allows for individuals to secure a potential appreciating asset. The real estate crash in 2007 started as a banking and securities crisis, which transformed into what is considered the worst crisis since the Great Depression. The diminishment of home prices influenced some individuals to strategically default on their mortgage payments because their mortgage was underwater (i.e. unpaid loan balance is greater than market value of property). Others defaulted without choice because of catastrophic events such as becoming unemployed. These delinquent payments inevitably led to foreclosed properties, which inevitably deteriorated the balance sheets of financial institutions. Researchers have attempted to understand the housing market by analyzing various underlying determinants, such as monetary policy implemented by the Federal Reserve, the role of sub-prime mortgages and mortgage-backed securities (MBS), lending standards by the financial services sector, and the list continues; however, there


S. Avdaković et al. (Eds.): IAT 2019, LNNS 83, pp. 14–26, 2020.
https://doi.org/10.1007/978-3-030-24986-1_2

is no consensus as to what caused the US housing market to tumble into this detrimental position. What if it was a "perfect storm" of all the above plus more? Both, the housing bubble and financial crisis, are prime examples of complex events. Complex in the sense that there were several interconnected and interdependent root causes.

The purpose of this paper is to share an agent-based model (ABM), which simulates the US housing market as a complex adaptive system. Through this approach, we are able to understand, visualize, and analyze the fluctuation in house prices and phenomenon of "underwater mortgages" due to variations in micro-level agent attributes and macro-level parameters such as interest rates. Once the world is generated, we perform stress testing and sensitivity analyses to assess varying outputs given a specific set of inputs. We demonstrate that the interdependent relationships in the market paired with varying input parameters such interest rates, percentage of adjustable rate versus fixed rate mortgages, and initial minimum and maximum house prices influence the average housing price over time. Substantial increases in house prices reduce the demand and ability for consumers to own a home. In the case of overvalued properties, this may further encourage delinquent payments and underwater mortgages, which may threaten a bank's balance sheet. Nonetheless, monetary policy implemented by the Federal Reserve plays a critical role in controlling the housing market. Policymakers and banks should work together to develop programs, which monitor and prevent the risk of underwater mortgages, overvalued properties, and crashes.

## 2 Literature Review

This section will briefly reference the various claims by researchers as to why and how the housing bubble occurred, and it will also mention literature that discusses the mechanics of the real estate market as a major component of the financial system. Additionally, we would like to begin this section of the paper by briefly addressing that the model presented in this paper was substantially enhanced from a Level 1 Axtell-Epstein (1994) agent-based model developed by a group of researchers at George Mason University (GMU) (McMahon et al. 2009). In addition, this paper focuses on similar methodologies implemented by Gangel et al. (2013a) to represent the contagion effect and contributes to real estate, foreclosure, and financial literature.

### 2.1 Literature and Background on the Real Estate Market

Many researchers have analyzed the housing market under different circumstances. For example, scholars have contributed to literature revolving around the foreclosure of properties (Khan and Yang 2019; Vernon-Bido et al. 2017; Seiler et al. 2013; Collins et al (2013); Gangel et al. 2013a and 2013b). With respect to the underwater mortgage literature, Archer and Smith (2010), underwater mortgages—loans with balances higher than the actual market-value of the property—were a significant driver for default of payment. Goldstein (2017) developed a series of ABMs to simulate the housing market in the Washington DC Metropolitan Statistical Area from 1997–2009. The author demonstrates that leverage (loan to value ratio or LTV) and expectations primarily contributed to the local bubble along with interest rates, income towards housing, and seller behavior. Goldstein claims that lending standards and refinance rules did not particularly influence the bubble; however, we will attempt to model these factors over time to assess whether they play significant roles in the nation's housing market. The financial sector, specifically banks, control the supply of loans in the market. Lenders typically do not hold onto mortgages and have the option to bundle volumes of them together. After the mortgages are bundled into a single product, they are sold to investors. These products are residential mortgage-backed securities (RMBS or MBS). A pair of researchers developed a hypothesis to assess the action by banks to created MBS products for investors called "originate to distribute" (Bord and Santos 2012). In other words, the more loans a bank can originate, the more MBS products they can sell to investors.

Additionally, small and mid-sized banks attempted to mimic their business strategies after large banks. Before the crisis, large banks had insurmountable MBS portfolios. How were small and mid-sized banks going to catch up with large banks? These banks evolved into mortgage originating machines, since this was a powerful strategy to generate high-revenue and remain competitive. During this time, a few factors were in place: (1) interest rates were at an all-time low—which was key for Adjustable Rate Mortgages, or ARMs, because interest rates were fixed for a certain amount of time (typically 5–10 years), and then home-owners would sell the property just before the teaser-rate expired to make a profit and avoid and adjustable rate, (2) Americans wanted to lock down low-interest mortgages and homes as soon as possible to retain affordable payments, (3) lending standards were loose, which may be correlated with originating to distribute (Mian and Sufi 2009); however, Gorton (2008) suggests securitizing mortgages did not lower these standards, and instead these loose standards became problematic for the securitization process. Essentially, once the initial fixed- rate (also known as teaser rate) for adjustable-rate mortgages ended, borrowers began defaulting on payments because they were too high while house prices did not appreciate. As we mentioned earlier, researchers have asserted that sub-prime mortgage borrowers, or high-risk borrowers with low credit scores, played a critical role in contributing to the housing bubble because they increased demand and prices (Mian and Sufi 2009). Given the three factors we previously listed, sub-prime mortgages were an easy solution for small and mid-sized banks to generate high volumes. The financial sector could innovate trading mechanisms for these products (i.e., credit default swaps), which contributed to the bubble and crash (Duca et al. 2011). Banks no longer had to be concerned with risk, since investors were only focused on pricing of securities and not how loans were originated. Ashcraft and Schuerman (2008) identify this as a principal-agent problem. Gorton (2008) and Khandani et al. (2009) claim refinancing and the appreciation of home prices contributed immensely to the real estate crash; however, Goldstein (2017) argues that refinance did not play a critical role. Eventually, investors understood that the pools of mortgages being purchased may contain sub-prime mortgages and it was too late. Researchers have also presented literature on how the international real estate markets operate and change over time by using agent-based simulations, such as the English housing market (Gilbert et al. 2009). We have included additional agents such as individual consumers and banks because their characteristics and behavior play critical roles in the market. In the U.S., ARMs typically rely on LIBOR, Treasury, and other financial indexes to represent “interest rate.” Although these rates do not drastically differ from the Fed funds rate, they are not identical values and do not represent the real components of adjustable-rate mortgages. We include attributes belonging to individual people such as employment, income, etc. We also argue that changes in home prices, whether or not the property has entered foreclosure, directly impact the property value of neighboring homes.

## 3 Model and Research Design

Several researchers have used this approach to simulate complex real-world problems. Schelling (1971) used the ABM approach to simulate racial segregation within housing in the United States. Abdollahian et al. (2013b) created an ABM to model SemPro, which simulates how competing interests and barriers impact siting outcomes and policy for sustainable energy infrastructure. Abdollahian et al. (2013a) also used ABM to simulate the Human Development theory and analyze the effects from interactions between economics, culture, society, and politics. As Haldane (2016) from Bank of England pointed out in 2016, ore and more scholars start to use ABM to study economic, fiscal and monetary policies, leveraging the model’s capability in studying crisis (Dosi et al. 2015; Geanakoplos et al. 2012). The model presented in this paper accounts for the housing construction market by initiating the ABM with more houses than people. The initial set up of the model plays an integral role in how the real estate market is simulated. The *initial density* parameter determines the density of houses in the world. The *rental house density* parameter determines how many properties are rental at tick 0, and the *percent occupied* parameter generates the initial number of people. Furthermore, the elements within each agent are then assigned using a random uniform distribution (i.e., income). People choose to live in affordable homes based on their available income to spend on a mortgage or rent. The tool outputs a variety of results, such as average house price, average mortgage cost, balance sheets of banks, real and natural unemployment rates, and more. Additionally, the “world” is a 32 $\times$ 32 grid in Netlogo with agents as people, houses, banks and mortgages. Each tick represents one month, and one patch is equivalent to one mile. Income and housing prices are in thousands of U.S. dollars. People can choose to rent or own one or more houses. Each house can either have a mortgage affiliated with a bank, or no mortgage at all.

Figure 1a includes pseudocode describing the consumer decision-making algorithm to determine whether or not the agent should purchase or rent a property. Figure 1b demonstrates the interdependent relationships among the agents and a high-level overview of their attributes. In this process, people own house(s), which are affiliated with mortgages that are owned by banks. We have chosen to incorporate LIBOR 12-month forward curves to more accurately depict how these mortgages are modeled. LIBOR 12-month assumes that interest rates will be adjusted every 12 months. For the sake of simplicity, we do not distinguish teaser and annual adjustable index rates pertaining to ARM loans for two reasons. First, there are several types of adjustable-rate mortgages (i.e., 5/1, 7/1, 10/1 etc.). The first number represents the number of years for the fixed teaser-rate; therefore, a 5/1 would represent a mortgage where the rate is fixed for five years, and then adjusts once every year for the life of the mortgage.

**(a)**

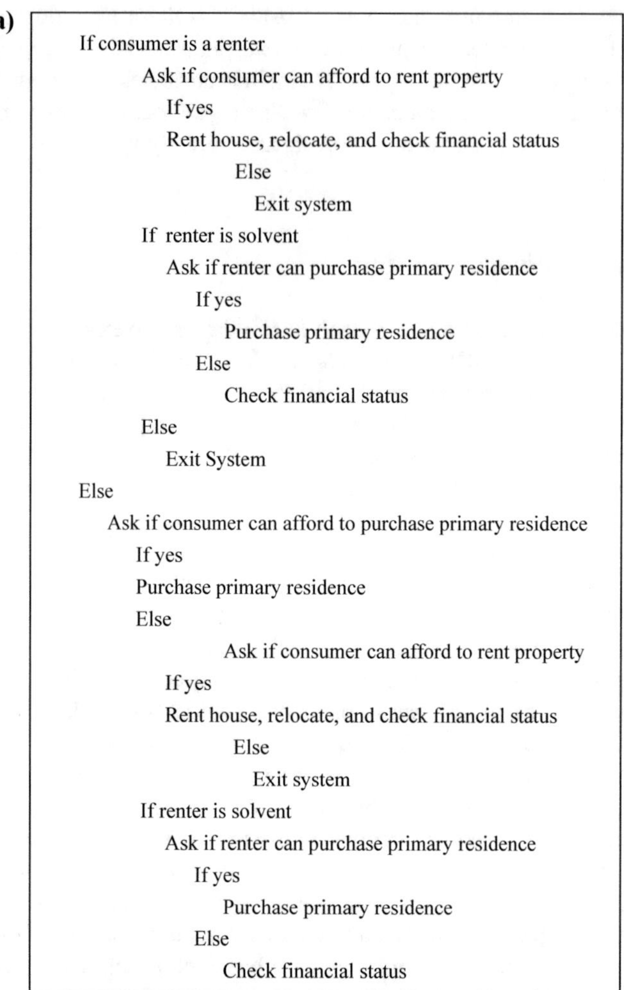

```
If consumer is a renter
    Ask if consumer can afford to rent property
      If yes
      Rent house, relocate, and check financial status
           Else
             Exit system
    If  renter is solvent
      Ask if renter can purchase primary residence
        If yes
          Purchase primary residence
        Else
          Check financial status
    Else
      Exit System
Else
  Ask if consumer can afford to purchase primary residence
    If yes
    Purchase primary residence
    Else
          Ask if consumer can afford to rent property
      If yes
      Rent house, relocate, and check financial status
           Else
             Exit system
    If renter is solvent
      Ask if renter can purchase primary residence
        If yes
          Purchase primary residence
        Else
          Check financial status
```

**(b)**

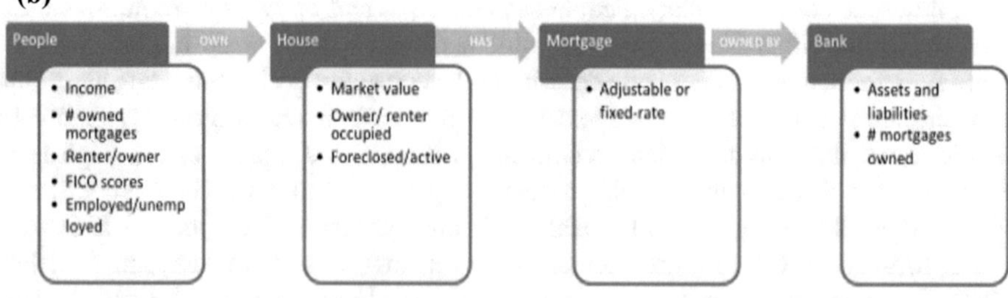


**Fig. 1.** (a) and (b). Pseudocode of consumer decision-making algorithm and flowchart of the model, which lists entities, relationships and attributes at a high level.

Our decision to not include these specific ARM products in the simulation allows us to observe patterns between fixed and adjustable rate mortgages more clearly. Secondly, the model already includes a mobility variable to account for the frequency and fraction of agents that move to different properties. This is a useful component because people typically sell their property just before the teaser-rate expires for ARMs. People choose to sell prior to the expiration of the rate, since ARMs tend to have lower initial rates than fixed-rate mortgages. Moreover, we have included both fixed and adjustable rate mortgages; however, the reader should note that majority of mortgages in the United States are fixed-rate products for 30 years.

## 3.1 Macro-level Agents and Variables

This section will summarize the different agents within the agent-based model, which are incorporated into the interface as displayed in Fig. 2 below.

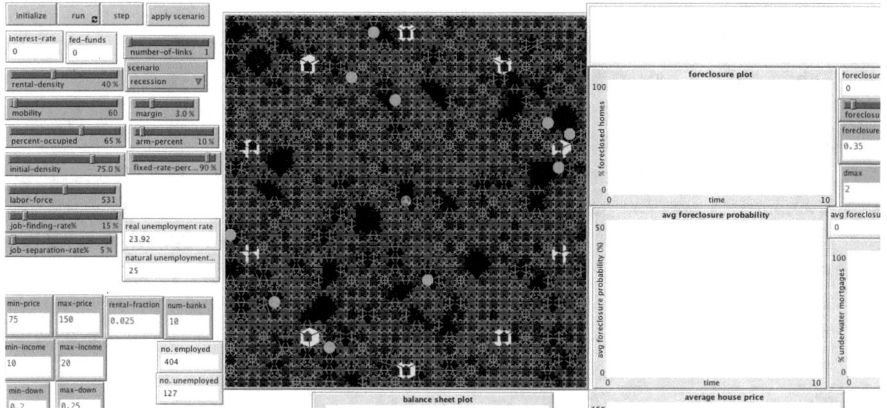


**Fig. 2.** NetLogo interface of agent-based model.

***Houses.*** When the model is initiated, houses are assigned a price, which follows a random uniform distribution of 75–150 K. These agents use color to depict whether a home is a rental (red), on the path to ownership (blue), foreclosed (pink), or empty (black). A darker gradient of red or blue indicates a higher mortgage cost or rent.

***Mortgages.*** Mortgages are represented as agents, which are owned and stored by banks; however, they are linked to a specific person and house. All loans include an interest rate, which is used to calculate individual amortization tables to determine monthly payments. A lag effect is introduced to account for people's reactions to changes in interest rates, which may lead to a refinance opportunity. Moreover, only a random percentage of adjustable rate mortgages will respond to changes in interest rates. This will allow the simulation to properly model adjustable rate mortgages, where the interest rate is fixed for *X* amount of time and then adjusts to the current offered rate. In a near-future version of the model, mortgages will be bundled together to create mortgage backed securities—a new agent. People will also be able to prepay mortgages, which will rely heavily on probabilities from a predictive econometric model as the foundation. Below is a table to explain how ARMs, fixed-rate mortgages, and rent is calculated (Table 1).

**Table 1.** Summary of equations for calculating mortgage amortization tables and rent.

| Product | Equation |
|---|---|
| ARM | $m = p\frac{r(1+(g+r))^n}{(1+(g+r))^n-1}$ |
| Fixed-rate | $m = p\frac{f(1+f)^n}{(1+f)^n-1}$ |
| Rent | $t = p * c$ |

where:

- ***m***: monthly payment
- ***p***: principal amount (in this case, house price after down payment has been subtracted for mortgages)
- ***r***: index of choice (in this case, LIBOR 12-Month), which is also divided by 12 to represent your annual interest-rate
- ***f***: Fed-funds rate, which is also divided by 12 to represent your annual interest- rate
- ***n***: number of expected payments you will make for the life of the loan
- ***g***: For ARMs, this represents the additional spread or margin applied to the index rate
- ***t***: rent
- ***c***: fractional constant

This model also uses Rama Cont's formula (2005) to represent the effect of aggregate demand on prices:

$$r_t = ln\frac{S_t}{S_{t-1}} = g(\frac{Z_t}{N}) \tag{1}$$

where:

- $r_t$: return on house price at time t
- $S_t$: house price at time t
- $Z_t$: excess demand for houses at time t
- $N$: number of agents
- $g(x) = x$: which is the appreciation or depreciation of house price, can then be represented as excess demand by solving for $S_t$.

We assume that properties undergo formal and informal appraisals of property value. Informal in the sense that an owner can determine an approximate appraisal value of his/her property through researching appraisal values of local and similar properties recently sold (Ling and Archer 2009). We also assume that neighboring properties have homogenous physical features, while the individuals who occupy each property may differ based on a distribution. The change in appraisal values of neighboring properties impacts an agent's property value through links. Every month, a random percentage of agents will assess the appraisal value of their property by observing the change in property values of neighboring homes with a maximum distance. The foreclosure discount is the negative percentage of price diminishment affiliated with foreclosed neighbors, and it is a function of change in price and distance from an appraised property. It is notated by $\mu$ as displayed in the contagion effect formulae. Within the model, there are a variety of reasons and equations to represent why a house may enter foreclosure such as equity, interest rate, investor, catastrophic foreclosures (Gangel et al. 2013a). The following equation is used to determine the appraisal value of each property when one of its neighbors have entered a foreclosure:

$$\Delta d_i = d_{max} - d_i, \quad \Delta p_i = p_{\substack{i \\ t+1}} - p_{\substack{i \\ t}}$$
$$Appraised\ Value_j = p_j - \sum_{i=1}^{n} \mu \cdot (p_j + \frac{\Delta p_i}{\Delta d_i}) \tag{2}$$

where:

**$\Delta d_i$**: difference between the i[th] property and maximum distance constant ($d_{max}$)
**$d_i$**: distance from i[th] property and appraisal property (j)
**$\mu$**: contagion effect severity for a single home
**$\Delta p_i$**: price change of i[th] property from t to t + 1
**$p_j$**: price of appraisal property

***Banks.*** Banks have balance sheets to monitor their assets and liabilities. In the current model, the entire mortgage value of homes is represented as a liability, while the monthly mortgage payments are represented as assets for the banks. If a house undergoes foreclosure, this will negatively impact the assets portion of the balance sheet.

## 4 Sensitivity Analysis and Results

We ran 288 different scenarios by using the BehaviorSpace (i.e., parameter sweeping) feature in NetLogo to perform behavioral experiments and run sensitivity analyses on housing prices using over 110,000 generated observations. Table 2 below summarizes the setup and randomization of parameters for the iteration to compile a total of 288 runs.

Three output variables are measured which may operate as dependent variables to measure the macro-political economy of the housing market. The variables include average mortgage and average house price. Below, Fig. 3 plots the ending average house price of the population against the initial minimum and maximum house prices

**Table 2.** Setup for the 288 simulation runs

| | | | |
|---|---|---|---|
| Percent-occupied 65 | Arm-percent 10 30 | Min-down 0.2 | Min-income 10 |
| Rental-density 40 | Mobility 60 | Rental-fraction 0.025 | Job-finding-rate% 10 |
| Max-down 0.25 | Max-income 0.20 | Margin 3 10 | Initial-density 75 |
| Min-price 75 100 125 | Job-separation-rate % 5 | Num-banks 5 15 | Fixed-rate-percent 90 70 |
| Max-price 150 175 200 | Foreclosure-constant 0.35 | Foreclosure-threshold" 0.4 | Contagion-effect 0.009 0.087 |

and average mortgages. The scatter plot in the bottom-left quadrant represents the phenomenon of underwater mortgages. Typically, you would expect a positive relationship between higher property values and mortgage costs; however, our sensitivity analysis demonstrates that many individuals at the end of each simulation run had very high mortgage costs for houses with low property values on average.

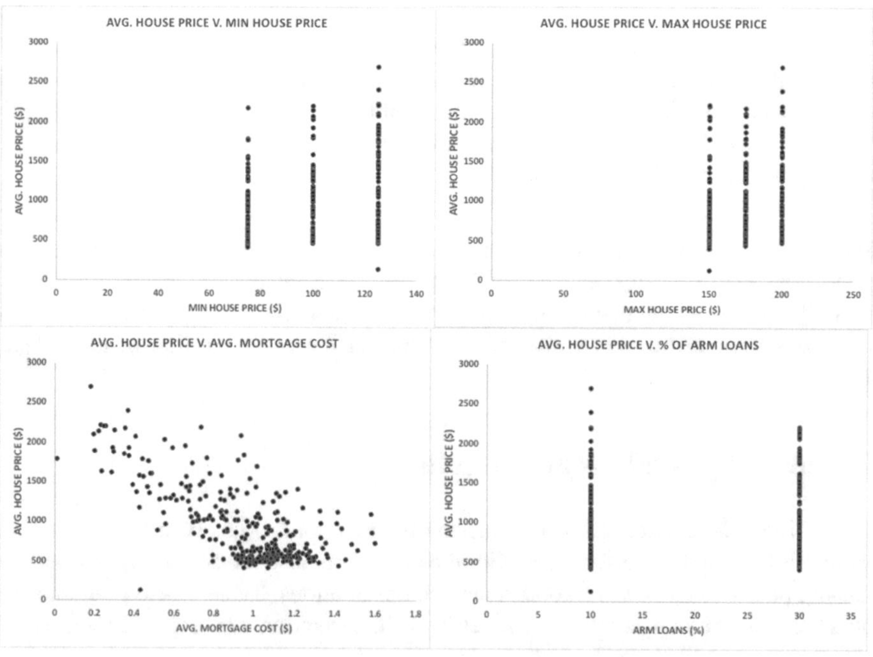


**Fig. 3.** Average foreclosures and house prices during sensitivity analysis.

Figure 4 shows a panel of time series plot of core outcome variables from the sensitivity tests. The price of occupied houses keeps increasing over time, though at different pace: fast at the beginning period and ending period of the simulation, but slower in the middle. Average mortgage cost shows a completely different pattern. It climbs at the beginning and reaches the peak very fast, fluctuates at the high level until the time period when house price starts to rapidly increase, then decreases with some fluctuation. The number of banks with income from houses stays at the same level across the entire simulation period after the very first few iterations when actions are taken place. The number of people with no investment capital has the largest fluctuation. However, as expected, this fluctuation follows the trend in average mortgage cost. After the mortgage cost peaks, when it goes high, the number of people with no investment capital also peaks.

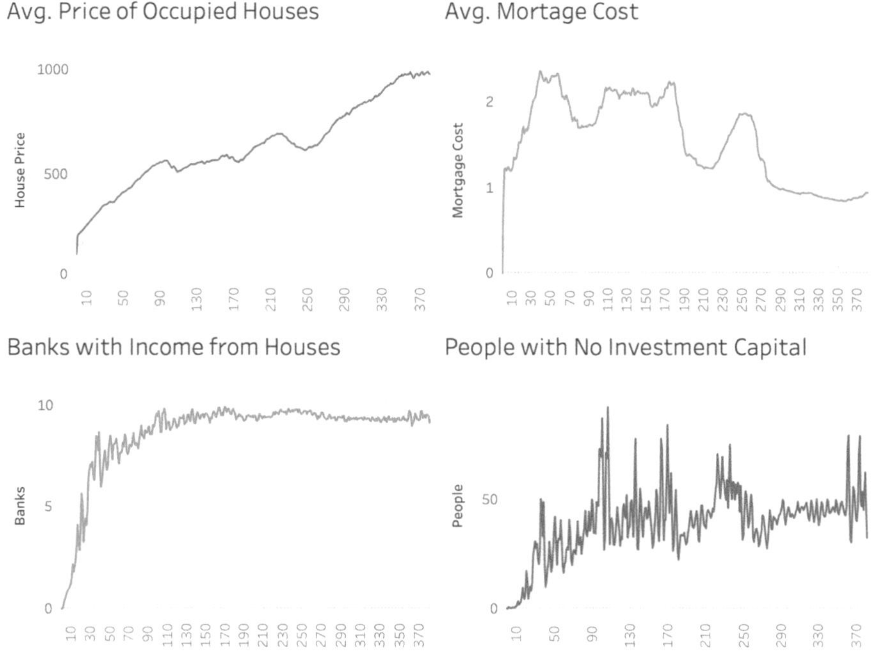


**Fig. 4.** Time series plot for core variables from sensitivity analysis.

Below, Table 3 quantifies the correlation between some of the important variables included in our agent-based model. The average house price and mortgage cost have a strong negative relationship, which further illustrates that higher property values do not necessarily imply a higher mortgage cost. The population of ARMs versus fixed rate mortgages was adjusted throughout the sensitivity analysis; therefore, exceedingly high mortgage costs may be attributed to an increase of the ARM population along with the spike in interest rates this product exhibits once the teaser rate expires. Additionally, this sensitivity run highlights the dilemma of underwater mortgages, which may lead to an increased probability of delinquent payments and eventually foreclosure.

**Table 3.** Correlation of variables.

| Dependent Var. | Average mortgage | Avg house price |
|---|---|---|
| Average mortgage cost | 1.000 | |
| Average house price | −0.4769 | 1.000 |
| Minimum house price | 0.0520 | 0.1554 |
| Maximum house price | 0.0657 | 0.1318 |

Table 4 shows the pooled regression results from our generated data. The results indicate that increases in initial minimum and maximum house prices lead to higher house prices on average regardless of how the agents behave and adjust their goal-oriented decisions. To reiterate, houses tend to be an asset that appreciates over time; therefore, an increase in the initial minimum and maximum house price will lead to a higher peak when assessing the property values in the world. This demonstrates the sensitivity of the initial input values for these variables, and the significance of understanding the property value distribution in a specific geography when we analyze the real world. Moreover, an increase in the percentage of adjustable rate mortgages tends to influence higher house prices on average, whereas an increase in fixed rate mortgages presents the converse. ARM products tend to be more affordable than fixed rate mortgages in the short run; therefore, an increase in the availability of these products to consumers will increase the ability and demand to purchase a home, which will increase the overall property values in the specific "neighborhoods" of the world. As displayed in the results, an increase in the population of fixed-rate mortgages tends to have a negative impact on house prices given that these products are more expensive; therefore, reducing demand.

**Table 4.** Pooled regression results after sensitivity runs.

| Variables | Average house price |
|---|---|
| Min. house price | 2.188*** |
| | (0.0386) |
| Max. house price | 1.856*** |
| | (0.0379) |
| ARM (%) | 0.388*** |
| | (0.0770) |
| Fixed rate (%) | −11.74*** |
| | (0.0770) |
| Constant | 1,013*** |
| | (8.775) |
| Observations | 110,299 |
| R-squared | 0.209 |

Robust standard errors in parentheses;
*** $p < 0.01$, ** $p < 0.05$, * $p < 0.1$

## 5 Conclusion and Future Work

This paper shows the power of simulation as an advanced approach to scientific inquiry and builds on previous agent-based models of the real estate market. Currently, the results of the example simulation along with the hundreds of experiments emphasize a tremendous influence from the Federal Reserve's monetary policy tool, the types of mortgage products available to consumers, and the initial minimum and maximum distribution of house prices when analyzing the housing market at a certain point in time. After full model setup, we conducted a number of sensitivity analyses and focused on variables that influence the average housing price at the end of each simulation run. Geographically, the price of a property is heavily influence by the market value of a neighboring property; therefore, homebuilders and future property owners are able to gauge the probable value of a property that has not been developed yet. The blend of available financial products to consumers (i.e., ARM versus Fixed-Rate) certainly influences demand within the housing market given that ARM products are more affordable than fixed-rate products. Most importantly, this ABM still has plenty of features that need to be included.

As we continue to build this model, we would like to place more emphasis on adding parameters and features to represent MBS trading, sub-prime mortgages, credibility and debt-to-income ratio of borrowers, credit default swaps, quantitative-easing, Dodd-Frank Act, lending standards and deregulation, etc. Additional agents will include rating agencies such as Standard & Poor, Fannie and Freddie, and construction companies to control the supply side and pricing of homes. Another necessary component for the model is the ability for borrowers to prepay and refinance mortgages. In sum, ABM is a valuable tool and can help us monitor and prevent systemic risk in the housing market, and it allows us to gain a better understanding of emergence and interconnectedness in the "world."

## References

Abdollahian, M., et al.: Human development dynamics: an agent based simulation of macro social systems and individual heterogeneous evolutionary games. Complex Adapt. Syst. Model. **1**(1), 18 (2013a)

Abdollahian, M., et al.: Techno-social energy infrastructure siting: sustainable energy modeling programming (SEMPro). J. Artif. Soc. Soc. Simul. **16**(3), 6 (2013b)

Archer, W.A., Smith, B.C.: Residential mortgage default: the roles of house price volatility, euphoria and the borrower's put option. Working paper, March 2010, No. 10-02 - Federal Reserve Bank of Richmond

Aschcraft, A.B., Schuerman, T.: Understanding the securitization of subprime mortgage credit. Federal Reserve Bank of New York staff reports, no. 318. March 2008

Axtell, R., Epstein, J.: Agent based modeling: understanding our creations. Bull. Santa Fe Institute **9**(4), 28–32 (1994)

Bord, V.M., Santos, J.A.C.: The rise of the originate-to-distribute model and the role of banks in financial intermediation. Econ. Policy Rev. **18**(2), 21–34 (2012)

Collins, A.J., Seiler, M.J., Gangel, M., Croll, M.: Applying Latin hypercube sampling to agent-based models: understanding foreclosure contagion effects. Int. J. Hous. Markets Anal. **6**(4), 422–437 (2013). ISSN: 1753-8270

Dosi, G., et al.: Fiscal and monetary policies in complex evolving economies. J. Econ. Dyn. Control **52**, 166–189 (2015)

Duca, J.V., Muellbauer, J., Murphy, A.: House prices and credit constraints: making sense of the US experience. Econ. J. **121**(552), 533–351 (2011)

Gangel, M., Seiler, M.J., Collins, A.: Exploring the foreclosure contagion effect using agent-based modeling. J. Real Estate Finance Econ. **46**(2), 339–354 (2013a)

Gangel, M., Seiler, M.J., Collins, A.J.: Latin hypercube sampling and the identification of the foreclosure contagion threshold. J Behav. Fin. **14**(2), 149–159 (2013b). ISSN: 1542-7560

Geanakoplos, J., et al.: Getting at systemic risk via an agent- based model of the housing market. Am. Econ. Rev. **1023**, 53–58 (2012)

Gilbert, N., Hawksworth, J.C., Swinney, P.A.: An agent-based model of the English housing market. In: AAAI Spring Symposium: Technosocial Predictive Analytics (2009)

Goldstein, J.: Rethinking housing with agent-based models: models of the housing bubble and crash in the Washington DC area 1997–2009. Dissertation, George Mason University (2017)

Gorton, G.B.: The panic of 2007. Working paper, National Bureau of Economic Research, September 2008

Haldane, A.G.: The dappled world. Speech (2016)

Khan, F., Yang, Z.: Simulation of financial systemic risk and contagion in the U.S. housing market. In: Cassenti, D., (ed.) Advances in Human Factors in Simulation and Modeling (AHFE 2018). Advances in Intelligent Systems and Computing, vol. 780. Springer, Cham (2019)

Khandani, A.E., Lo, A.W., Merton, R.C.: Systemic risk and the refinancing ratchet effect. Working paper, National Bureau of Economic Research, September 2009

Ling, D., Archer, W.: Real Estate Principles: A Value Approach. McGraw-Hill Irwin, Boston (2009)

McMahon, M., Berea, A., Osman, H.: An agent based model of the housing market. Housing. Market Rev. **8**, 3 (2009)

Mian, A., Sufi, A.: The consequences of mortgage credit expansion: evidence from the U.S. mortgage default crisis. Quart. J. Econ. **124**(4): 1449–1496 (2009). https://doi.org/10.1162/qjec.2009.124.4.1449

Vernon-Bido, D., Collins, A.J., Sokolowski, J.A., Seiler, M.J.: Using real property layouts to study the foreclosure contagion effect in real estate with agent-based modeling and simulation. J. Hous. Res. **26**(2): 137–155 (2017). ISSN: 1052-7001

Schelling, T.C.: Dynamic models of segregation. J. Math. Sociol. **1**(2), 143–186 (1971)

Seiler, M., Collins, A.J., Fefferman, N.H.: Strategic mortgage default in the context of a social network: an epidemiological approach. J. Real Estate Res. **35**(4), 445–475 (2013). ISSN: 0896-5803

Ghoulmie, F., Cont, R., Nadal, J.-P: Heterogeneity and feedback in an agent-based market model. J. Phys. condens. matter **17**(14), S1259 (2005)

# New Approach for Fault Identification and Classification in Microgrids

Tarik Hubana[1(✉)], Mirza Šarić[2], and Samir Avdaković[3]

[1] Technical University Graz, Graz, Austria
t.hubana@student.tugraz.at
[2] Public Enterprise Elektroprivreda of Bosnia and Herzegovina, Mostar, Bosnia and Herzegovina
m.saric@epbih.ba
[3] Faculty of Electrical Engineering, University of Sarajevo, Sarajevo, Bosnia and Herzegovina
s.avdakovic@epbih.ba

**Abstract.** Power system fault identification and classification continues to be one of the most important challenges faced by the power system operators. In spite of the dramatic improvements in this field, the existing protection devices are not able to successfully identify and classify all types of faults which occur power system. The situation in even more the complex in the case of microgrids due to their dynamic behavior and inherent peculiarities. This paper presents a novel method for identification and classification of faults in the microgrid. The proposed method is based on Descrete Wavelet Transform (DWT) and Artificial Neural Networks (ANN). The model is completely developed in MATLAB Simulink and is significant because it can be applied for practical identification and classification of faults in microgrids. The obtained results indicate that the proposed algorithm can be used as a promising foundation for the future implementation of the microgrid protection devices.

## 1 Introduction

The management and operation of modern power systems present numerous challenges for Distribution System Operators (DSO) throughout the world. One of the main challenges remains the detection and clearance of the power system faults. New energy paradigm stimulates development and operation of microgrids, which adds complexity to the existing protection schemes. There have been some significant improvements in this field, mainly owned to the advancement of signal processing, information and communication technologies, computational methods but also regulatory requirements. However, this remains and important area of research, especially when microgrids are considered.

This paper presents a new method for identification and classification of faults with particular attention to microgrid faults. This method is developed and tested in MATLAB Simulink and is based on Descrete Wavelet Transform (DWT) and Artificial Neural Networks (ANN). The contributions of this a papers are as follows:


S. Avdaković et al. (Eds.): IAT 2019, LNNS 83, pp. 27–39, 2020.
https://doi.org/10.1007/978-3-030-24986-1_3

(1) Design, test and validate a new method for identification and classification of the faults which can be applied in the realistic microgrid systems
(2) Provide a robust and flexible foundation for the future implementation of the microgrid protection devices.

There are numerous challenges in the area of microgrid protection, such as changes in the short-circuit level, false tripping signals, blindness of protection, prohibition of automatic reclosing, are unsynchronized reclosing. Microgrid protection depends on some key factors as: microgrid type, microgrid topology, type of DG resources, communication type, the time delay of communication links, a method of analysing data and detecting faults, relay type, fault type and method of grounding [1–4]. Existing methods of microgrid protection can be grouped in three categories: placement of special limitations on DGs, using external devices and protection system modification [5]. Protection system modification is the most promising method, especially the adaptive protection scheme.

Requirements for appropriate microgrid protection method, which can adapt dynamic changes of these networks and guarantee speed and selectivity of protection system, lead us to adaptive protection. Nowadays, with the advent of new technologies and digital relays and communication links, adaptive protection can play a pivotal role in protection of future networks [6]. Thus, researchers investigate this area using many methods, among them the artificial neural networks too [7–9]. However, the investments in the microgrid protection should be feasible, considering the fact that microgrids are usually set up on MV or LV networks [10].

Following the recent progress in the area of the advanced distribution system protection [11–13] and constant upgrades of the distribution grid, the power quality and reliability levels arrived at a point in which additional investments are just not feasible. This creates a need for an affordable and yet intelligent protection system. Since the currently available adaptive protection schemes most commonly use microprocessor based relays or industrial computers, an implementation of an advanced protection scheme that takes more parameters as input and provides better protection appears to be a promising direction for upgrades and future development. The use of signal processing and machine learning algorithms is a promising approach for global decision making which that lead to the successful fault isolation. Therefore, it is justified to further explore the area of a cost-effective, reliable and intelligent microgrid protection system that remains one of the highest priorities in this field.

## 2 Theoretical Background

### 2.1 Discrete Wavelet Transform

The wavelet transform is used in numerous engineering applications. It is regarded as a mathematical tool that has numerous advantages when compared with traditional methods in a stochastic signal-processing application, mainly because waveform analysis is performed in a time scale region [14]. WT of a signal $f(t) \in L^2(R)$, where $L^2$ is the Lebesgue vector space, is defined by the inner product between $\Psi_{ab}(t)$ and $f(t)$ as [14]:

$$WT(f, a, b) = \frac{1}{\sqrt{a}} \int_{-\infty}^{+\infty} f(t) \Psi\left(\frac{t-b}{a}\right) dt \tag{1}$$

where a and b are the scaling (dilation) and translation (time shift) constants, respectively, and $\Psi$ is the wavelet function which may not be real as assumed in the above equation for simplicity [14]. The Wavelet transform of the sampled waveforms is obtained by implementing DWT given by [14]:

$$DWT(f, m, n) = \frac{1}{\sqrt{a_0^m}} \sum_k f(t) \Psi\left(\frac{n - ka_0^m}{a_0^m}\right) \tag{2}$$

here a and b from Eq. (1) are replaced by $a_0^m$ and $ka_0^m$, k and m being integer variables. In a standard DWT, the coefficients are sampled from a continuous WT on a dyadic grid, $a_0 = 2$ and $b_0 = 1$, yielding $a_0^0 = 1$, $a_0^{-1} = 2^{-1}$, etc. [14].

## 2.2 Artificial Neural Networks

When it comes to artificial neural networks, the output is a feedback to the input to calculate the change in the values of the weights [15]. The weights of the back-error-propagation algorithm for the neural network are chosen randomly to prevent a bias toward any particular output. The first step in the algorithm is a forward propagation [15]:

$$a_j = \sum_i^m w_{ji}^{(1)} x_i \tag{3}$$

$$z_j = f(a_j) \tag{4}$$

$$y_j = \sum_i^M w_{kj}^{(2)} z_j \tag{5}$$

where $a_j$ represents the weighted sum of the inputs, $w_{ij}$ is the weight associated with the connection, $x_i$ are the inputs, $z_j$ is the activation unit of (input) that sends a connection to unit j and $y_i$ is the i-th output.

The second step is a calculation of the output difference [15]:

$$\delta_k = y_k - t_k \tag{6}$$

where $\delta_k$ represents the derivative of the error at a k-th neuron, $y_k$ is the activation output of unit k and $t_k$ is the corresponding target of the input.

The next step is back propagation for hidden layers [15]:

$$\delta_j = (1 - z_j^2) \sum_{k=1}^{K} w_{kj} \delta_k \tag{7}$$

where $\delta_j$ is the derivative of error $w_{kj}$ to $a_j$.

Afterwards, the gradient of the error with respect to the first- and the second-layer weights is calculated, and the previous weights are updated. The mean square error for each output in each iteration is calculated by [15]:

$$MSE = \frac{1}{N} \sum_{1}^{N} (E_i - E_o)^2 \tag{8}$$

where N is the number of iterations, $E_i$ is the actual output and $E_o$ is the output of the model.

After each step, the weights are updated with the new ones and the process is repeated for the entire set of input-output combinations available in the training-data set, and this process is repeated until the network converges for the given values of the targets for a predefined value of the error tolerance [15].

### 2.3 Microgrids

A microgrid is a group of interconnected loads and distributed energy resources within clearly defined electrical boundaries that acts as a single controllable entity with respect to the grid. Microgrid presents an active distribution network, which consists of distributed generation (DG) resources, different loads at the voltage level of distribution, and energy storage elements. A microgrid can connect and disconnect from the grid to enable it to operate in both grid-connected or island mode [16].

From a network perspective, a microgrid is advantageous because it is a controlled unit and can be exploited as a concentrated load. From the point of view of customers, a microgrid can be designed to meet their special needs such as higher local reliability, fewer feeder losses, better local voltages, increased efficiency, voltage sag correction, and uninterruptible power supply [10]. From an environmental standpoint, a microgrid reduces environmental pollution and global warming because it produces less pollutants [1].

The main components of the microgrid are DGs, distributed energy storage devices and critical/non-critical loads. Microgrid is connected to the public distribution network through Point of Common Coupling (PCC) and there is a separation device in the bus section, which can switch smoothly between the grid- connected mode and the islanded mode. The capacity of a microgrid is generally between kilowatts and megawatts, and it is interconnected to low or middle level distribution networks [17].

## 3 Test System

The 11-bus test system developed for the purpose of algorithm testing presents a part of a real medium voltage distribution system operating in the area of the City of Mostar (Bosnia and Herzegovina) which is similar to typical distribution systems used throughout Europe. However, this real power system is modified by adding the renewable resources that can generate enough power for the system to work in islanded mode. The DGs rated powers are the common rated powers of small hydro power plants (SHPP) and photovoltaic (PV) DGs that occur in the Bosnia and Herzegovina distribution network.

The microgrid is supplied from a 35/10 kV two parallel transformers (which presents a place of common coupling – PCC) via 10 kV feeders. In case of the fault in the main grid, the microgrid can be disconnected from the main grid via 10 kV circuit breakers (CBs) that are placed in the 35/10 kV substation. In that case, the DGs supply the 10 kV microgrid. The LV customers are supplied via 10/0.4 kV substations. The 11-bus test system supplies electricity in an urban/suburban area and mostly consists of underground cables. The simulation model is developed in MATLAB/Simulink software and presents a three-phase model of the previously described microgrid. A single line diagram of the test system is shown in Fig. 1.

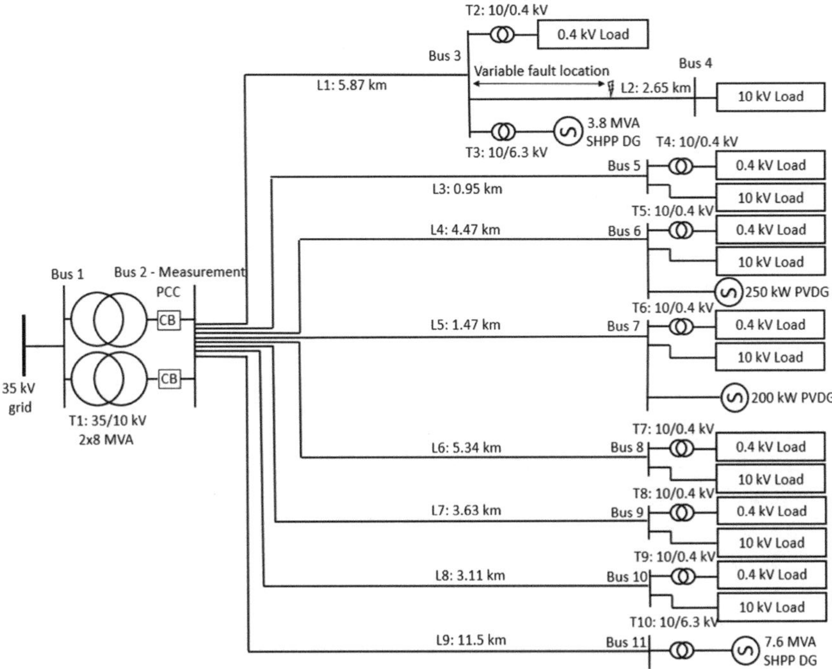


**Fig. 1.** 11-bus test system single line diagram

The faults are simulated on a 10 kV underground cable that feeds the entire consumption area. In order to test the algorithm behaviour in the challenging microgrid scenarios such as changes in the short-circuit level, false tripping signals, blindness of protection, the fault is simulated in the line between a SHPP and the rest of the consumers. Faults are simulated for different fault resistances (in the range from 0 Ω to 300 Ω) and at different fault locations. The simulated faults are phase A to the ground fault (AG), phase A to phase B to the ground fault (ABG) and phase A to phase B to phase C to the ground fault (ABCG). The measurement is performed at a 10 kV busbar in the 35/10 kV substation. The faults are simulated in both grid-connected and islanded mode of operation.

The sampling frequency of the current protection relays and measuring equipment in the Bosnia and Herzegovina EPDS is 3.2 kHz. Since the existing equipment already operates with this sampling frequency, a logical upgrade would be to use this equipment coupled with the proposed algorithm.

This fact is the reason for using the chosen sampling frequency in the measurement process. For the proposed DWT-ANN algorithm, the condition with no fault and the conditions with three types of the fault are simulated for various resistance values and fault locations, for both modes of microgrid operation giving a total of 89.384 fault scenarios.

## 4 Outline of the Computational Procedure

The proposed method for fault identification and classification is based on a combination of DWT and ANN. It is based on the previous research of the authors [12, 18]. It performs with details and approximation waveforms rather than with calculated coefficients. After simulating all the possible fault scenarios, for each fault and different values of the fault location and resistance, the voltage waveforms are generated. The fault is simulated during the entire simulation time, i.e. (0–0.08 s). When the voltage waveforms are generated, DWT is applied to these waveforms. The procedure of signature signal generation is shown in Fig. 2.

A Daubechies 4 wavelet is used at a 3.2 kHz voltage signal, therefore one approximation and four details are obtained for each voltage. Four levels of decomposition are used in this paper in order to get the following frequency bands [12]:

- First detail level - frequency band: [800, 1600] Hz,
- Second detail level - frequency band: [400, 800] Hz,
- Third detail level - frequency band: [200, 400] Hz,
- Fourth detail level - frequency band: [100, 200] Hz,
- Fourth approximation level - frequency band: [50, 100] Hz

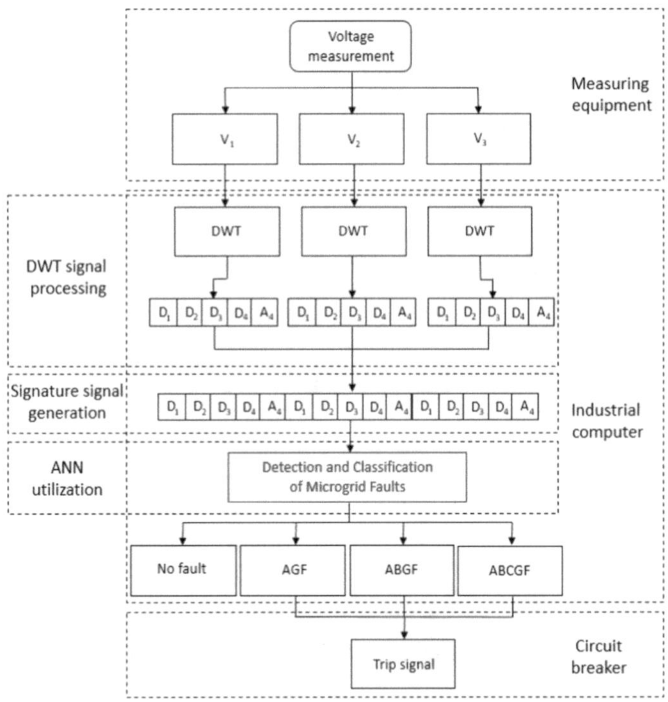


**Fig. 2.** Proposed method algorithm (signature signal generation)

The A4 waveform is a base sinusoidal wave and reflects the signal behavior during each fault. The rest of the DWT waveforms are higher harmonic components of the voltage signal, and therefore they reflect a distinctive voltage behavior during each fault type. The algorithm is also tested by the Symlet 4 and Biorthogonal 4.4 wavelet families, and the total algorithm output results do not differ from the Daubechies 4 wavelet family. Therefore, it is not necessary to be particularly cautious regarding the choice of the wavelet family. The DWT signal components give a good insight into the system behavior during fault conditions. For this reason, they are used as representative signals for each fault type. Afterwards, these DWT signals are combined and grouped and represent a unique "signature" for each fault, which represents the input to ANN. After that, ANN is trained with a large set of this data, becoming capable to detect and identify the EPDS faults.

Once trained, the ANN is capable of fault identification and classification, according to the algorithm shown in Fig. 3. With the measuring equipment installed in microgrid, the voltage waveforms can constantly monitored with the previously defined moving time frame, while the obtained signals are sent to an installed industrial computer with a DWT-ANN algorithm software. In the case of a fault identification, an appropriate trip signal can be sent to the circuit breaker.

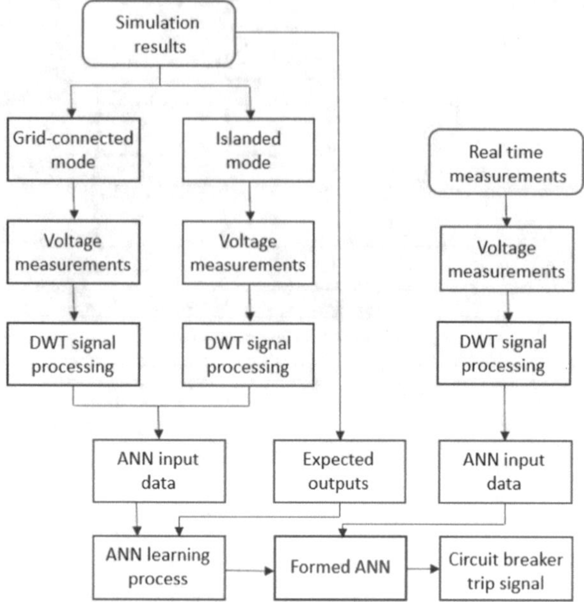


**Fig. 3.** Proposed algorithm learning and utilization procedure

## 5 Results and Discussion

In order to use the proposed algorithm in a real microgrid, the ANN needs to be trained to all the possible scenarios that may occur in the microgrid. Since the algorithm is planned to be used in the online mode to constantly monitor the system voltages, it can constantly improve and learn new possible microgrid operating scenarios, making the algorithm even better over time.

The algorithm consists of two phases. First one is the creation of fault scenario for the initial base of knowledge that ANN will learn from. Firstly, the input data for ANN needs to be created. In order to get a unique signature signal for every fault type, a signal that reflects each possible state needs to be created. For this purpose, 89.384 simulations for three types of the fault and normal operating conditions, with various fault resistances and fault locations, are carried out. For each fault scenario, each phase voltage is measured and preprocessed with the help of DWT. By combining the DWT signals of all phase voltages for each fault scenario, a signature signal that reflects the system behavior during each fault is created. This signature signal is a signal that is built simply by adding the start of the next signal to the end of the previous signal using the details and approximation waveforms for each fault scenario in the system, i.e. this signal is composed from the DWT waveforms of the currently measured voltage signal. A signature signal example for single line to ground fault is shown in Fig. 4.

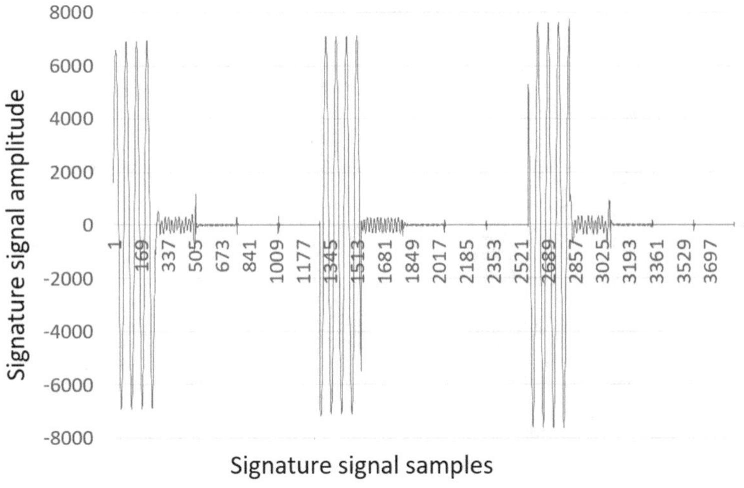


**Fig. 4.** Examples of signature signals for AG fault

The differences between the created signals for a particular fault type are apparently negligible, but ANN is capable to classify them correctly. Higher harmonic components, which are important in the identification process, can be efficiently identified in the DWT filters of the corresponding frequency range. Generally, DWT is widely used for the noise-removal applications [19, 20], which is one of the main reasons that it performs very well in this application. The noise with signal to noise ratio (SNR) value of 30 db is added to the signal measurements [21]. Moreover, since the proposed algorithm is paired with ANNs, which are known to have a high accuracy in the pattern classification and noise removal ability [18], this issue is addressed even more effectively.

After this unique signal for each fault scenario is created, the input set of data for ANN training is ready. The ANN takes the input set of 89.384 input vectors and 89.384 corresponding preferred outputs during a training process. After that, trained ANN has four possible outputs, where each output notes the normal operating condition and three types of microgrid faults.

As a result of the training process, the proposed DWT - ANN algorithm is capable to accurately identify the fault and distinguish between the three possible categories of faults for both modes of microgrid operation, regardless of the fault-resistance value and fault location, with the 85.74% accuracy in the 0–300 Ω fault resistance range, with the SNR value of 30 db. Of course, the accuracy improves with the noise reduction, thus having the accuracy of 96.92% with SNR value of 40 db (which is a common value in most 10 or 20 kV power networks), and having the 100% accuracy with SNR values over 50 db. The algorithm error for various SNR values is shown in Fig. 5.

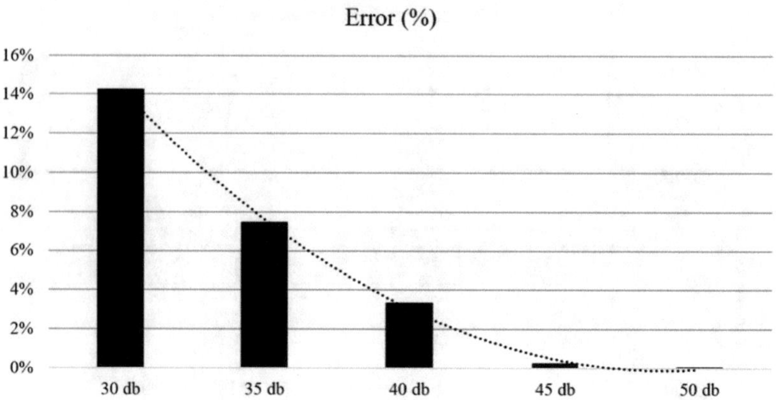


**Fig. 5.** Algorithm error for various signal to noise (SNR) values

After the training and testing process, the created ANN is perfectly capable to identify and classify faults in the both modes of microgrid operation. In order to get provide an insigtht into the algorithm efficiency, it is necessary to test it with new fault scenarios with new resistance values and different fault locations, i.e. fault scenarios that ANN is not trained to.

For this purpose, new simulations with new parameters are carried out. Table 1 shows classifier results to this fault scenarios, where column *Desired output* presents a simulated-fault type and column *Actual output* an evaluated fault type. The percentage values present the absence or presence of a specific fault. Column Actual output presents the ANN output for each fault scenario.

**Table 1.** Example of the DWT-ANN classifier output for different fault resistances, fault locations and fault types

| Resistance (Ω) | Fault location (% line length) | Desired output | | | | Actual output (ANN output) | | | |
|---|---|---|---|---|---|---|---|---|---|
| | | AGF | ABGF | ABCGF | No fault | AGF | ABGF | ABCGF | No fault |
| Grid-connected operation | | | | | | | | | |
| 20 | 16 | 1 | 0 | 0 | 0 | 0.9944 | 0.0022 | 0.0002 | 0.0031 |
| 110 | 31 | 0 | 0 | 0 | 1 | 0.0109 | 0.0001 | 0.0003 | 0.9885 |
| 200 | 46 | 0 | 1 | 0 | 0 | 0.0018 | 0.9952 | 0.0025 | 0.0003 |
| 290 | 78 | 0 | 0 | 1 | 0 | 0.0007 | 0.0022 | 0.9965 | 0.0004 |
| Islanded operation | | | | | | | | | |
| 20 | 16 | 1 | 0 | 0 | 0 | 0.9944 | 0.0022 | 0.0002 | 0.0031 |
| 110 | 31 | 0 | 0 | 0 | 1 | 0.0494 | 0.0005 | 0.0008 | 0.9491 |
| 200 | 46 | 0 | 1 | 0 | 0 | 0.0066 | 0.9925 | 0.0007 | 3.3e-06 |
| 290 | 78 | 0 | 0 | 1 | 0 | 0.0003 | 0.0021 | 0.9885 | 0.0089 |

The proposed algorithm is planned to be used in the online mode. The storage method is important since the grouped DWT components take a lot of storage. However, a present-day industrial computer will be enough for online monitoring. Few minutes old data can be deleted if no disturbances are recorded. However, voltage waveforms can be saved for a later analysis.

## 6 Future Research Directions

The list of the microgrid fault types is not exhausted by the faults included in this paper. The high level of noise can also lead to misclassificaiton. However, this algorithm is applicable to new scenarios since it can be easily extended by an additional training of the ANN. The proposed algorithm has a potential practical application in terms of its implementation in the microgrid protection system devices. In order to achieve this, it is necessary to improve the robustness of the proposed algorithm, which remains an important future research direction. Finally, an extension in the number of the system components and scenarios that can lead to a false tripping signal is proposed for future consideration.

## 7 Conclusion

Identification and classification of the power system faults, especially in the microgrid remains an important challenge faced by the DSO, because the existing protection devices are unable to detect nor classify such failures due to the complex and dynamic behavior of the microgrid. For these reasons, this topic presents an open research area. The paper proposes a method to improve the existing algorithms for identification and classification of microgrid faults based on DWT and ANN.

For algorithm testing, a test system based on a part of the real distribution network from Bosnia and Herzegovina is developed in MATLAB Simulink. The proposed method has a practical significance since it can be applied to a real microgrid and it accurately identifies and classifies faults in the 0–300 Ω range of the fault resistance for various fault locations. The proposed algorithm is believed to be a promising approach to the future implementation of the microgrid protection devices.

It is important to point out that the proposed algorithm is not immune to the large load and switching state changes, and in order to overcome this, further training with more operating scenarios is required. The possible upgrades of the existing algorithm are pointed out in the future research directions section. This study is a part of an ongoing research into advanced power system protection algorithms, whose ultimate goal is to develop a system that has satisfactory accuracy for detection and classification of the power distribution network faults.

## References


1. Hosseini, S.A., Abyaneh, H.A., Sadeghi, S.H.H., Razavi, F., Nasiri, A.: An overview of microgrid protection methods and the factors involved. Renew. Sustain. Energy Rev. **64**, 174–186 (2016)
2. Brearley, B.J., Prabu, R.R.: A review on issues and approaches for microgrid protection. Renew. Sustain. Energy Rev. **67**, 988–997 (2017)
3. Laaksonen, H., Hovila, P.: Enhanced MV microgrid protection scheme for detecting high-impedance faults. In: 2017 IEEE Manchester PowerTech, Manchester (2017)
4. Ustun, T.S., Ozansoy, C., Zayegh, A.: Recent developments in microgrids and example cases around the world—A review. Renew. Sustain. Energy Rev. **15**, 4030–4041 (2011)
5. Mohamed, N.A., Salama, M.M.A.: A review on the proposed solutions to microgrid protection problems. In: 2016 IEEE Canadian Conference on Electrical and Computer Engineering (CCECE), Vancouver (2016)
6. Memon, A.A., Kauhaniemi, K.: A critical review of AC Microgrid protection issues and available solutions. Electr. Power Syst. Res. **129**, 23–31 (2015)
7. Lin, H., Guerrero, J.M., Jia, C., Tan, Z.-h., Vasquez, J.C., Liu, C.: Adaptive overcurrent protection for microgrids in extensive distribution systems. In: IECON 2016 - 42nd Annual Conference of the IEEE Industrial Electronics Society, Florence (2016)
8. Yu, J.J.Q., Hou, Y., Lam, A.Y.S., Li, V.O.K.: Intelligent fault detection scheme for microgrids with wavelet-based deep neural networks. IEEE Trans. Smart Grid **10**(2), 1694–1703 (2019)
9. Zarrabian, S., Belkacemi, R., Babalola, A.A.: Intelligent mitigation of blackout in real-time microgrids: neural network approach. In: 2016 IEEE Power and Energy Conference at Illinois (PECI), Urbana (2016)
10. Hubana, T.: Coordination of the low voltage microgrid protection considering investment cost. In: 1. Conference BH BH K/O CIRED, Mostar (2018)
11. Hubana, T.: Transmission lines fault location estimation based on artificial neural networks and power quality monitoring data. In: 2018 IEEE PES Innovative Smart Grid Technologies Conference Europe (ISGT-Europe), Sarajevo (2018)
12. Hubana, T., Saric, M., Avdakovic, S.: Approach for identification and classification of HIFs in medium voltage distribution networks. IET Gener. Transm. Distrib. J. **12**(5), 1145–1152 (2018)
13. Hubana, T., Saric, M., Avdakovic, S.: Classification of distribution network faults using Hilbert-Huang transform and artificial neural network. In: IAT 2018: Advanced Technologies, Systems, and Applications III, pp. 114–131. Cham, Springer (2019)
14. Choudhury, M., Ganguly, A.: Transmission line fault classification using discrete wavelet transform. In: International Conference on Energy, Power and Environment: Towards Sustainable Growth (ICEPE), Shillong (2015)
15. Jamil, M., Sharma, S.K., Singh, R.: Fault detection and classification in electrical power transmission system using artificial neural network. SpringerPlus **4**, 334 (2015)
16. Hirsch, A., Parag, Y., Guerrero, J.: Microgrids: A review of technologies, key drivers, and outstanding issues. Renew. Sustain. Energy Rev. **90**, 402–411 (2018)
17. Chaudhary, N.K., Mohanty, S.R., Singh, R.K.: A review on microgrid protection. In: Proceedings of the International Electrical Engineering Congress 2014, Pattaya City, Thailand (2014)
18. Hubana, T., Saric, M., Avdakovic, S.: High-impedance fault identification and classification using a discrete wavelet transform and artificial neural networks. Elektrotehniški Vestn. **85**(3), 109–114 (2018)

19. Sarawale, R.K., Chougule, S.R.: Noise removal using double-density dual-tree complex DWT. In: 2013 IEEE Second International Conference on Image Information Processing (ICIIP-2013), Shimla (2013)
20. Haider Mohamad, A.R., Diduch, C.P., Biletskiy, Y., Shao, R., Chang, L.: Removal of measurement noise spikes in grid-connected power converters. In: 2013 4th IEEE International Symposium on Power Electronics for Distributed Generation Systems (PEDG), Shimla (2013)
21. Talebi, S.P., Mandic, D.P.: Frequency estimation in three-phase power systems with harmonic contamination: A multistage quaternion Kalman filtering approach. Imperial College London (2016)

# Analytical Solutions for Determination of Electrical Vehicle Starting Time and Corresponding Distance

Martin Ćalasan(✉) and Saša Mujović

Faculty of Electrical Engineering, University of Montenegro, Dzordza Vasingtona bb, 81000 Podgorica, Montenegro
{martinc, sasam}@ucg.ac.me

**Abstract.** The usage of the electrical vehicles (EVs) has a promising future. Therefore, different analyses related to the EVs characteristics are very popular. This paper proposes the novel-analytical approach in determination of the EV starting time and corresponding distance. The analytical solutions were obtained on the base of conducted mathematical analysis taking into account the inertial parameters of EVs, such as rolling resistance and the aerodynamic drag. It was concluded that the parameters have strong impact on the electrical vehicle starting time and corresponding distance, especially at higher values of vehicle speed. The proposed analytical approach is simpler and more convenient for using than the commonly used approach based on complex numerical methods.

## 1 Introduction

Electric vehicles (EVs) are considered as an effective solution for urban air pollution, fossil fuel depletion and global warming issues, and, therefore they are gaining in importance.

The EV uses one or more electric motors for propulsion [1–6]. The motor(s) may be powered from the off-vehicle sources, such as collector system, or EV may be equipped with a storage system (battery) or an electric generator [1, 2] both aimed at self-production of electricity. Therefore, there is a lot of potential variant for powering of electric vehicles.

On the other hand, the three main concepts of EV technologies are defined as follows:

- vehicle-to-home,
- vehicle-to-vehicle, and
- vehicle-to-grid.

Vehicle-to-home technology is based on onboard or offboard bidirectional charger. Hence, the vehicle is able to draw the energy from home or transfer its energy to home according to the control scheme. Vehicle-to-vehicle technology shows that vehicle can transfer their energy by bidirectional chargers through a local grid, and then distribute the energy among vehicles by a controller. Finally, vehicle-to-grid technology shows that vehicles can be connected to the power grid to obtain energy, as well as feed


S. Avdaković et al. (Eds.): IAT 2019, LNNS 83, pp. 40–51, 2020.
https://doi.org/10.1007/978-3-030-24986-1_4

energy back to the grid. The opportunities and challenges of vehicle-to-home, vehicle-to-vehicle, and vehicle-to-grid technologies are numerous [7].

In scientific view, electric vehicles are very interesting for research. They represent combination of electrical and mechanical components, but also, they are compact, complex and autonomous system. Therefore, in the available literature one may find papers addressing different topics related to EVs, such as their optimal control [8], vehicle battery settings [1], electric-vehicle stations [9], effect of speed and constant-power operation of electric drives on the vehicle design and performance [6], power electronics and motor drives for electric vehicle [1, 4], etc.

The main feature of EV is usually bounded with its acceleration time and distance covered from zero speed to a certain high speed [1]. This information is generally referred to testing at ground level. However, in available literature there is a very few information related to this problem - some papers deal with analytical method of EV velocity profile determination [10] or with influence of motor size and efficiency on acceleration [11], but without any discussion about the acceleration time calculation. In this paper will be presented analytical solutions for electric vehicle starting time and corresponding distance which take into account the rolling resistance and aerodynamic drag. Namely, the results of initial research related to this issue, published in [1], have confirmed necessity for involving of the mentioned parameters in the equations for the EV starting time and distance calculation. However, for calculation of acceleration time and corresponding distance, including rolling resistance and aerodynamic drag, in [1] a numerical integration technique is suggested.

This paper is organized as follows. In Sect. 2 a short description about basic equations for electrical vehicle starting time and corresponding distance is presented. In Sect. 3, a novel expression for electrical vehicle starting time and corresponding distance calculation, which take into rolling resistance and aerodynamic drag coefficients, is presented. The corresponding simulation results are presented in Sect. 4. The main paper benefits are highlighted in Conclusion section.

## 2 Electrical Vehicle Starting Time and Corresponding Distance

For a precise and correct mathematical modelling of an electric vehicle a sophisticated mechanical and mathematical knowledge is needed. The first step in vehicle performance modelling is to produce an equation for the tractive effort. According to Newton's second law, vehicle acceleration can be written

$$\frac{dV}{dt} = \frac{\sum F_t - \sum T_r}{\delta \cdot M}, \tag{1}$$

where $V$ is the speed of the vehicle, $\Sigma F_t$ the total tractive effort of the vehicle, $\Sigma F_r$ the total resistance, $M$ the total mass of the vehicle, and $\delta$ the mass factor [1]. Vehicle resistances, opposing its movement, include rolling resistance of the tires, aerodynamic drag and grading resistance.

The rolling resistance is primarily due to the friction of the vehicle tire on the road. Friction in bearings and the gearing system also play their part [1, 2]. The rolling resistance is approximately constant. The equation is

$$F_r = M \cdot g \cdot f_r \cdot \cos(\alpha), \tag{2}$$

where α is slope angle (see Fig. 1) and $f_r$ is the rolling resistance coefficient. The rolling resistance coefficient is a function of the tire material, tire structure, tire temperature, etc.

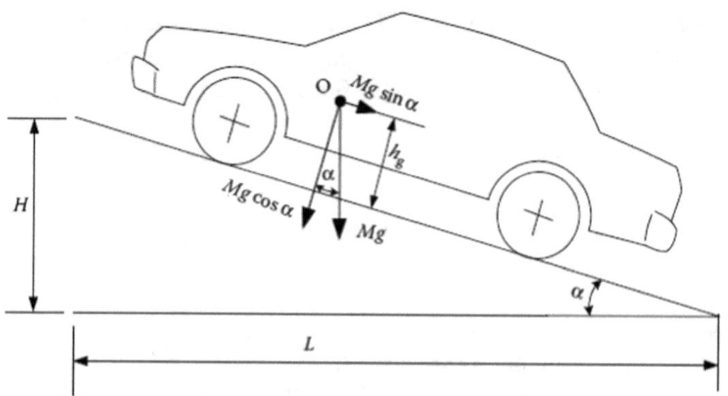


**Fig. 1.** Electric vehicle on sloped road [1]

The aerodynamic drag is a consequence of the force caused with friction of the vehicle body moving through the air. Aerodynamic drag is a function of vehicle speed, $V$, vehicle frontal area, $A_f$, shape of the vehicle body, and air density, $\rho$:

$$F_w = \frac{1}{2}\rho_a A_f C_D V^2, \tag{3}$$

where $C_D$ is the aerodynamic drag coefficient that characterizes the shape of the vehicle body. Finally, when a vehicle goes up or down a slope, its weight produces a component that is always directed in the downward direction (grading resistance). Therefore, this component either opposes the forward motion (grade climbing) or helps the forward motion (grade descending). Grading resistance, referring to Fig. 1, can be expressed as

$$F_g = M \cdot g \cdot \sin(\alpha). \tag{4}$$

In some literature, the tire rolling resistance and grading resistance together are called road resistance.

The acceleration performance of a vehicle is usually described by its acceleration time and distance covered from zero speed to a certain high speed on level ground.

Using Newton's second law the acceleration of the vehicle, on level ground, can be written as follows:

$$a = \frac{dV}{dt} = \frac{F_t - F_r - F_w}{M\delta} = \frac{\frac{P_t}{V} - M \cdot g \cdot f_r - \frac{1}{2} \cdot \rho_a \cdot C_D \cdot A_f \cdot V^2}{M\delta}, \tag{5}$$

where $P_t$ is the tractive power on the driven wheels transmitted from the traction motor corresponding the vehicle base speed [1].

Therefore, the acceleration time has the following form

$$t_a = \int_0^{Vf} \frac{M \cdot \delta}{\frac{P_t}{V} - M \cdot g \cdot f_r - \frac{1}{2} \cdot \rho_a \cdot C_D \cdot A_f \cdot V^2} \cdot dV, \tag{6}$$

where $V_f$ is final vehicle speed. Similarly, the acceleration distance can be calculated as follows:

$$S_a = \int_0^{V_f} \frac{M \cdot \delta \cdot V}{\frac{P_t}{V_b} - M \cdot g \cdot f_r - \frac{1}{2} \cdot \rho_a \cdot C_D \cdot A_f \cdot V^2} \cdot dV. \tag{7}$$

In [1] it is noted that it is difficult to obtain the analytical solution of Eq. (7). Also, in [1] the expression for accelerating time and distance are derived with neglecting rolling resistance and the aerodynamic drag, as follows [1]:

$$\mathrm{t}_{a\_simple} = \frac{\delta \cdot \mathrm{M}}{2 \cdot \mathrm{P_t}} \left( V_f^2 + V_b^2 \right), \tag{8}$$

and

$$S_{a\_simple} = \frac{\delta \cdot M}{6 \cdot P_t} \cdot (2 \cdot V_f^3 + V_b^3). \tag{9}$$

## 3 Novel Expressions for Electric Vehicle Accelerating Time and Distance

By observing Eq. (6), as well as taking into account machine mechanical characteristics, Eq. (6) can be rearranged in the following form:

$$t_a = \int_0^{Vb} \frac{M \cdot \delta}{\frac{P_t}{V_b} - M \cdot g \cdot f_r - \frac{1}{2} \cdot \rho_a \cdot C_D \cdot A_f \cdot V^2} \cdot dV + \int_{V_b}^{Vf} \frac{M \cdot \delta}{\frac{P_t}{V} - M \cdot g \cdot f_r - \frac{1}{2} \cdot \rho_a \cdot C_D \cdot A_f \cdot V^2} \cdot dV. \quad (10)$$

The most frequently used electric machine in electric vehicle applications is induction machine. The induction machine mechanical characteristics is presented in Fig. 2. On this characteristics one may differ two parts – constant torque zone (for speed form zero to base speed) and constant power zone (from base speed to final speed). The description about induction machine mechanical characteristics can be found in [12]. This is the reason for splitting of integral Eqs. (6) (or (7)) into two parts.

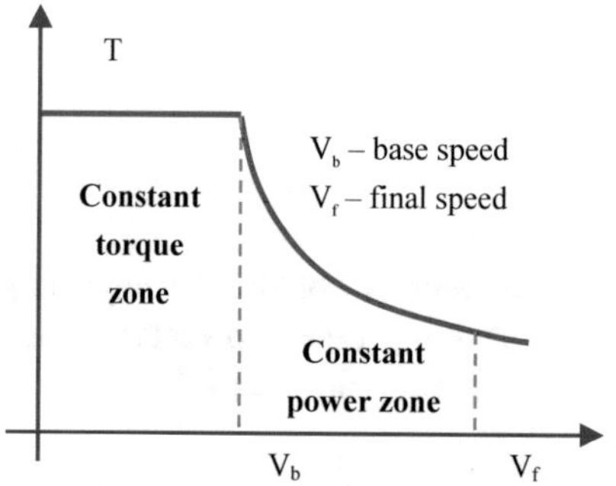


**Fig. 2.** Induction machine mechanical characteristics

Furthermore, Eq. (6) can be written as follows:

$$\begin{aligned} t_a &= \int_0^{Vb} \frac{a_1}{a_2 - a_3 \cdot V^2} \cdot dV + \int_{Vb}^{Vf} \frac{a \cdot V}{V^3 + c \cdot V + d} \cdot dV \\ &= \int_0^{Vb} \frac{a_1}{a_2 - a_3 \cdot V^2} \cdot dV + \int_{V_b}^{Vf} \frac{k_1}{V + x_1} \cdot dV + \int_{V_b}^{Vf} \frac{k_2 \cdot V + k_3}{R_1 \cdot V^2 + R_2 \cdot V + R_3} \cdot dV, \end{aligned} \quad (11)$$

where coefficients $a_1$, $a_2$, $a_3$, $a$, $c$, $d$, $x_1$, $R_1$, $R_2$, $R_3$, $k_1$, $k_2$ and $k_3$ can be simple calculated from Eq. (10): $a_1 = M \cdot \delta$, $a_2 = P_t/V_b - M \cdot g \cdot f_r$, $a_3 = \frac{1}{2} \cdot \rho_a \cdot C_D \cdot A_f$, $a = -\frac{2M \cdot \delta}{\rho_a \cdot C_D \cdot A_f}$, $c = \frac{2M \cdot g \cdot f_r}{\rho_a \cdot C_D \cdot A_f}$, $d = -\frac{2P_t}{\rho_a \cdot C_D \cdot A_f}$, $x_1$ is real root of equation $V^3 + c \cdot V + d = 0$, also $R_1 \cdot V^2 + R_2 \cdot V + R_3 = \frac{V^3 + c \cdot V + d}{V + x_1}$, and $k_1(R_1 \cdot V^2 + R_2 \cdot V + R_3) + (V + x_1)(k_2 \cdot V + k_3) = a \cdot V$.

Finally, an analytical solution for electric vehicle accelerating time has the following form:

$$\begin{aligned} t_a = & \frac{a_1}{a_3} \frac{1}{2\sqrt{\frac{a_2}{a_3}}} \log\left(\frac{\sqrt{a_2/a_3}+V_b}{\sqrt{a_2/a_3}-V_b}\right) + \mathrm{k_1} \log\left(\frac{\mathrm{V_f - x_1}}{\mathrm{V_b - x_1}}\right) \\ & + \frac{2k_3}{\sqrt{q}}\left(\mathrm{atan}\left(\frac{\mathrm{2R_1V_f + R_2}}{\sqrt{q}}\right) - \mathrm{atan}\left(\frac{\mathrm{2R_1V_b + R_2}}{\sqrt{q}}\right)\right) + \\ & k_2 \frac{1}{2 \cdot R_1} \left( \begin{aligned} & \log\left(\frac{R_1 \cdot V_f^2 + R_2 \cdot V_f + R_3}{R_1 \cdot V_b^2 + R_2 \cdot V_b + R_3}\right) \\ & - \frac{R_2}{R_1} \frac{k_3}{\sqrt{q}} \left(\mathrm{atan}\left(\frac{\mathrm{2R_1V_f + R_2}}{\sqrt{q}}\right) - \mathrm{atan}\left(\frac{\mathrm{2R_1V_b + R_2}}{\sqrt{q}}\right)\right) \end{aligned} \right), \end{aligned} \tag{12}$$

where $q = 4 \cdot R_3 \cdot R_1 - R_2^2$. Therefore, this equation is derived without ignoring the rolling resistance and the aerodynamic drag.

On the same way, we can also derive an analytical expression for starting distance. Namely, based on Eq. (7) we can write the following:

$$\begin{aligned} S_a = & \int_0^{V_b} \frac{M \cdot \delta \cdot V}{\frac{P_t}{V_b} - M \cdot g \cdot f_r - \frac{1}{2} \cdot \rho_a \cdot C_D \cdot A_f \cdot V^2} \cdot dV \\ & + \int_{Vb}^{V_f} \frac{M \cdot \delta \cdot V}{\frac{P_t}{V} - M \cdot g \cdot f_r - \frac{1}{2} \cdot \rho_a \cdot C_D \cdot A_f \cdot V^2} \cdot dV. \end{aligned} \tag{13}$$

Equation (13) can be rearranged as follows

$$\begin{aligned} S_a = & \int_0^{Vb} \frac{a_1 \cdot V}{a_2 - a_3 \cdot V^2} \cdot dV + \int_{Vb}^{Vf} \frac{a \cdot V^2}{V^3 + c \cdot V + d} \cdot dV \\ = & \int_0^{Vb} \frac{a_1}{a_2 - a_3 \cdot V^2} \cdot dV + \int_{V_b}^{Vf} \frac{k_{11}}{V + x_1} \cdot dV + \int_{V_b}^{Vf} \frac{k_{21} \cdot V + k_{31}}{R_{11} \cdot V^2 + R_{21} \cdot V + R_{31}} \cdot dV, \end{aligned} \tag{14}$$

where $k_{11}(R_{11} \cdot V^2 + R_{21} \cdot V + R_{31}) + (V + x_1)(k_{21} \cdot V + k_{31}) = a \cdot V^2$. The solution of Eq. (14) is as follows:

$$
\begin{aligned}
S_a = & -\frac{a_1}{2\cdot a_3}\log\left(\frac{a_2/a_3 - V_b^2}{a_2/a_3}\right) + \mathrm{k}_1\log\left(\frac{\mathrm{V_f - x_1}}{\mathrm{V_b - x_1}}\right) \\
& + \frac{2k_{31}}{\sqrt{q}}\left(\operatorname{atan}\left(\frac{2\mathrm{R}_{11}\mathrm{V_f} + \mathrm{R}_{21}}{\sqrt{q}}\right) - \operatorname{atan}\left(\frac{2\mathrm{R}_{11}\mathrm{V_b} + \mathrm{R}_{21}}{\sqrt{q}}\right)\right) \\
& + k_{21}\frac{1}{2\cdot R_{11}}\left(\begin{array}{l} \log\left(\dfrac{R_{11}\cdot V_f^2 + R_{21}\cdot V_f + R_{31}}{R_{11}\cdot V_b^2 + R_{21}\cdot V_b + R_{31}}\right) \\ -\dfrac{R_{21}}{R_1}\dfrac{k_{31}}{\sqrt{q}}\left(\operatorname{atan}\left(\dfrac{2\mathrm{R}_{11}\mathrm{V_f} + \mathrm{R}_{21}}{\sqrt{q}}\right) - \operatorname{atan}\left(\dfrac{2\mathrm{R}_{11}\mathrm{V_b} + \mathrm{R}_{21}}{\sqrt{q}}\right)\right) \end{array}\right).
\end{aligned}
\tag{15}
$$

Therefore, in this section, the analytical solutions for determination of electrical vehicle starting time and corresponding distance have been derived without ignoring the rolling resistance and the aerodynamic drag.

## 4 Simulation Results

In this section we have observed electric vehicle which parameters are as follows: 63 kW induction machine, mass of vehicle M = 1200 kg, rolling coefficient fr = 0.013, aerodynamic drag coefficient CD = 0.8, vehicle frontal area Af = 2, air density ρ = 1.205, the mass factor δ = 1.1 and Vb = Vf/4.

The acceleration time and distance versus final speed characteristics are presented in Fig. 3. It is evident that the starting time, as well as starting distance, depends on rolling resistance and aerodynamic drag. Furthermore, for higher value of the final speed, the impact of rolling resistance and aerodynamic drag is higher.

In order to check the impact of rolling resistance and aerodynamic drag on acceleration time and corresponding distance we have also observed the different values of rolling resistance coefficient (Fig. 4), the different values of aerodynamic drag coefficient (Fig. 5) and different values of vehicle frontal area (Fig. 6). It is evident that the higher value of rolling resistance coefficient (or higher value of aerodynamic drag coefficient, or higher value of vehicle frontal area) causes the higher value of the starting time and corresponding distance. Therefore, by using developed expression we can easily investigate the impact of different vehicle parameters or road conditions on starting time and corresponding distance.

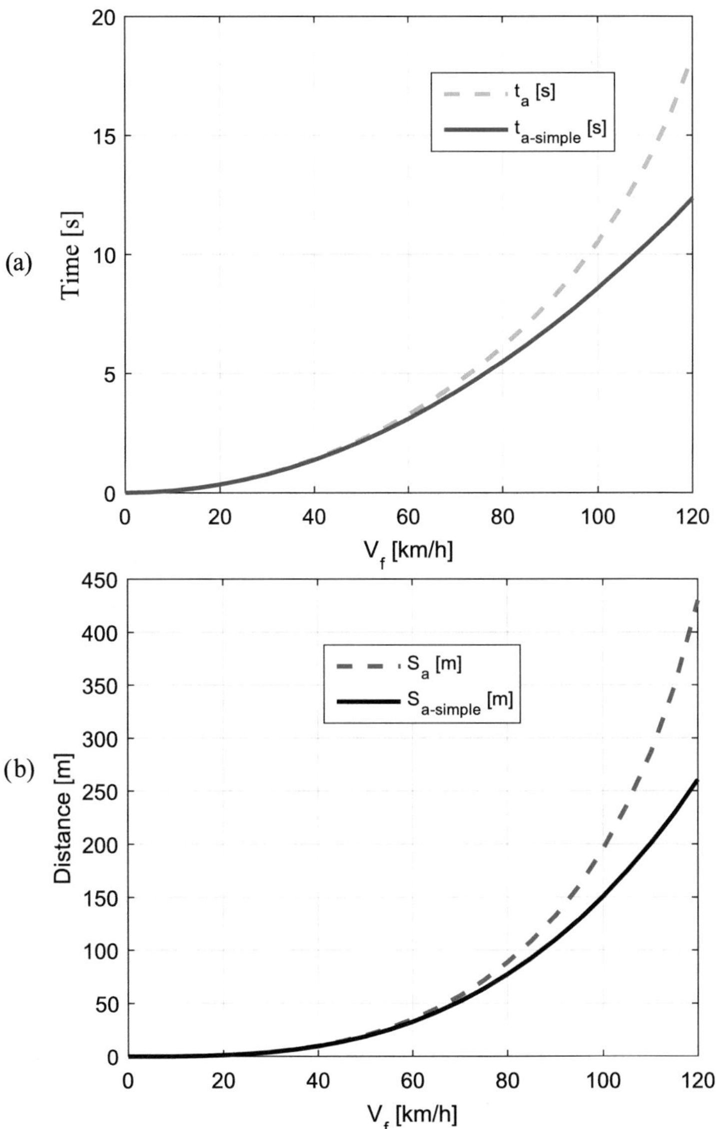


**Fig. 3.** Acceleration time (a) and distance (b) versus final speed

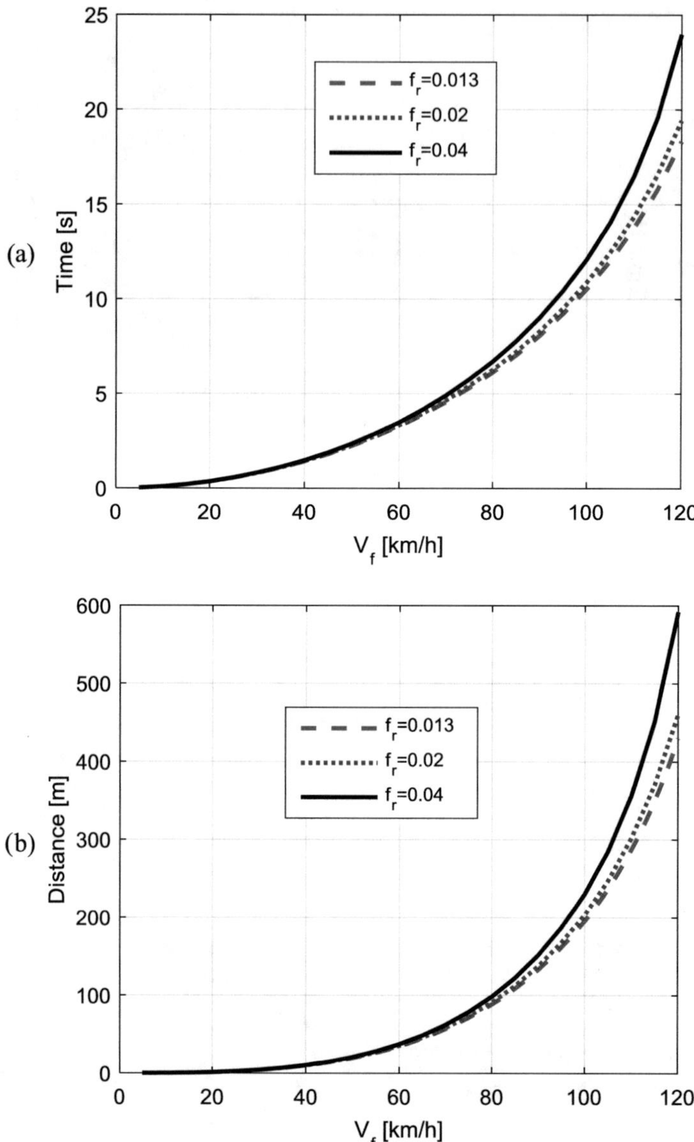


**Fig. 4.** Acceleration time (a) and distance (b) versus final speed for different value of rolling resistance coefficients

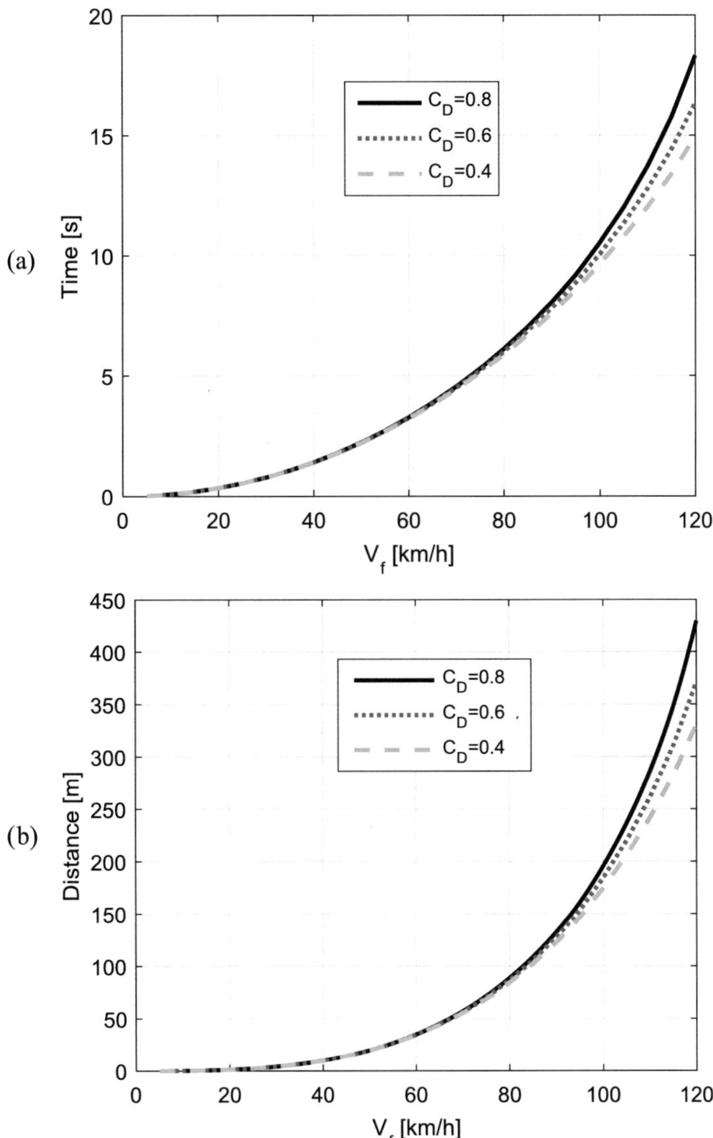


**Fig. 5.** Acceleration time (a) and distance (b) versus final speed for different value of aerodynamic drag coefficient

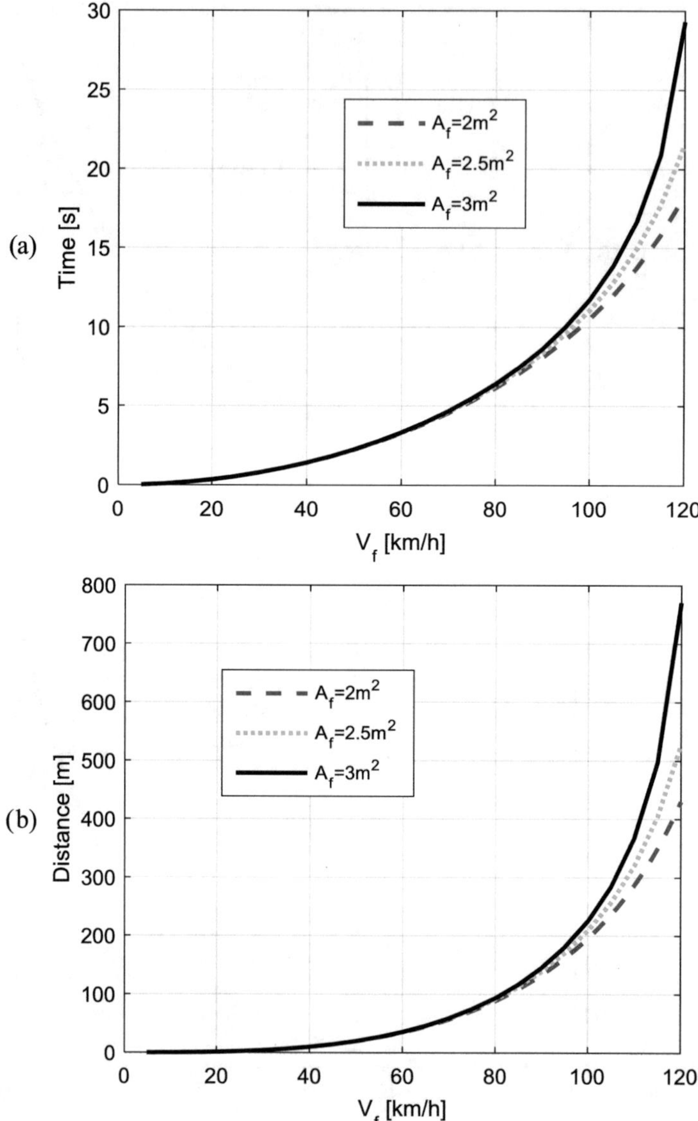


**Fig. 6.** Acceleration time (a) and distance (b) versus final speed for different value of vehicle frontal area

## 5 Conclusion

In this paper an analytical solution for starting time and corresponding distance of electric vehicle are derived. The expressions take into account the value of rolling resistance coefficient and the value of aerodynamic drag coefficient. The paper contains comparison between these results and the corresponding ones which ignore the values of aforementioned parameters. It was concluded that the parameters have strong impact on the electrical vehicle starting time and corresponding distance, i.e. their neglecting yields inaccurate results. Finally, it should be stated that the proposed solution, contrary to literature known techniques which are based on numerical integration, is very easy for implementation, does not require special software or choose of integration step, and therefore can be used in many vehicle studies (such as calculation of optimal drive cycle or similar).

## References


1. Ehsani, M., Gao, Y., Longo, S., Ebrahimi, K.M.: Modern Electric, Hybrid Electric, and Fuel Cell Vehicles, 3rd edn. CRC Press, Taylor & Francis Group (2018)
2. Chan, C.C., Chau, K.T.: Modern Electric Vehicle Technology. Oxford University Press, New York (2001)
3. Fan, J., et al.: Thermal analysis of permanent magnet motor for the electric vehicle application considering driving duty cycle. IEEE Trans. Magn. **46**(6), 2493–2496 (2010)
4. Emadi, A., Lee, Y.J., Rajashekara, K.: Power electronics and motor drives in electric, hybrid electric, and plug-in hybrid electric vehicles. IEEE Trans. Industr. Electron. **55**(6), 2237–2245 (2008)
5. Rahman, K.M., Ehsani, M.: Performance analysis of electric motor drives for electric and hybrid electric vehicle application. In: IEEE Power Electronic in Transportation, pp. 49–56 (1996)
6. Rahman, Z., Ehsani, M., Butler, K.: Effect of extended-speed, constant-power operation of electric drives on the design and performance of EV-HEV propulsion system. In: Society of Automotive Engineers (SAE) Journal, Warrendale, PA, 2003, Paper No. 2000-01-1557
7. Liu, C., Chau, K.T., Wu, D., Gao, S.: Opportunities and challenges of vehicle-to-home, vehicle-to-vehicle, and vehicle-to-grid technologies. Proc. IEEE **101**(11), 2409–2427 (2013)
8. Yu, H., Cheli, F., Castelli-Dezza, F.: Optimal design and control of 4-IWD electric vehicles based on a 14-DOF vehicle model. IEEE Trans. Veh. Technol. **67**(11), 10457–10469 (2018)
9. Etezadi-Amoli, M., Choma, K., Stefani, J.: Rapid-charge electric-vehicle stations. IEEE Trans. Power Deliv. **25**(3), 1883–1887 (2010)
10. Tanaka, D., Ashida, T., Minami, S.: An analytical method of EV velocity profile determination from the power consumption of electric vehicles. In: 2008 IEEE Vehicle Power and Propulsion Conference, pp. 1–3, Harbin (2008)
11. Zulkifli, S.A., Mohd, S., Saad, N., Aziz, A.R.A.: Influence of motor size and efficiency on acceleration, fuel economy and emissions of split-parallel hybrid electric vehicle. In: 2013 IEEE Symposium on Industrial Electronics & Applications, pp. 126–131, Kuching (2013)
12. Krishnan, R.: Electric Motor Drives – Modeling, Analysis and Control. Prentice Hall, Upper Saddle River, NJ (2001)

# Impact of Electric Vehicle Charging on Voltage Profiles and Unbalance on Low Voltage

## (Case Study of a Real Rural Low-Voltage Distribution Network)

Naida Nalo[1(✉)], Adnan Bosović[2], and Mustafa Musić[2]

[1] International Burch University, Sarajevo, Bosnia and Herzegovina
naidal497@gmail.com

[2] Department of Strategic Development, Public Electric Utility Elektroprivreda of Bosnia and Herzegovina, Sarajevo, Bosnia and Herzegovina

**Abstract.** Public interest in electric vehicles is rising with continuous development of this technology followed by global-warming awareness. Charging these vehicles introduces additional loading stress and could negatively impact the power quality parameters, especially if charged during daily peak-period. This paper investigates the effect of different load phasing and EV penetrations on voltage variations and unbalance. First part of analysis is based on examining the effects of different load distribution per phase on four base scenarios with no electric vehicles involved. The second part of analysis consists of nine scenarios with 10%, 20% and 50% of EV penetration in grid-to-vehicle mode. Analysis was performed using DIgSLENT Power Factory software. Seven scenarios remained within allowed limits from standards, while others violated at least one parameter. Results also showed that phase and point of connection are vital when it comes to impacting the voltage variation and unbalance.

## 1 Introduction

The conventional power systems were designed to be passive with a unidirectional power flow, where the central source feeds the customers at lower voltage levels, usually situated farther from the production point. However, an increase in society's interest in distributed generation is changing this concept of power systems. Public awareness of global warming effects helped renewable energy become widely accepted as a non-polluting source of energy [1].

Transportation sector is to be blamed for a huge portion of greenhouse gas emissions, since it was accounted for a quarter of energy-related emissions produced in year of 2009. Most promising solution was found to be Electric Vehicles (EVs). EVs do not produce tailpipe emissions and are therefore considered as clean and environment-friendly. Many countries are taking action in promoting EVs by offering incentives and policies that would help increase the implementation of EVs [2].

One must keep in mind that despite EVs themselves not being polluting, the type of source that feeds the network to which these vehicles are connected when being charged, impacts the emission production. If the network uses coal for most of its


S. Avdaković et al. (Eds.): IAT 2019, LNNS 83, pp. 52–73, 2020.
https://doi.org/10.1007/978-3-030-24986-1_5

production, or other polluting fuels, connection of EVs to such a network would introduce a higher emission of greenhouse gasses. To decrease these emissions, EV charging should be based on renewable energy sources [2].

Implementation of this technology also affects the power grid itself. In the case of a high penetration of EVs and uncontrolled charging, which gives the customers freedom to charge their vehicles whenever they want, this would cause a negative effect by increasing the daily peak load. Unregulated EV charging would amplify the power losses, equipment overloading and disturb power quality. However, the positive effects of EV implementation would be seen when EV charging is regulated or EVs act as small distributed generators, when operating in Vehicle-to-Grid mode [2].

Power quality represents a key component in the aspect of power systems operation that directly influences the efficiency, reliability and security. According to standard 61000-4-30, power quality is defined as: "The characteristics of the electricity at a given point of an electrical system, evaluated against a set of reference technical parameters." The ultimate measure of power quality is defined by the end-user's equipment performance [4].

Power quality parameters are categorized, according to EN 50160, as either a continuous phenomenon or single events. Under the term continuous phenomenon, the following parameters are addressed: power frequency, supply voltage variations, rapid voltage changes, supply voltage unbalance, harmonic voltages, interharmonic voltages and mains signaling voltages. Voltage events include interruptions of the supply voltage, supply voltage dips/swells and transient overvoltages [5].

For the purpose of this project, two parameters were observed: supply voltage variations and supply voltage unbalance. Supply voltage variations and supply voltage unbalance are both categorized as continuous phenomenon according to EN 50160 standard. Voltage variations should not exceed ±10% of the nominal voltage Un, while for voltage unbalance, values of the negative phase sequence component of the supply voltage must be within the range of 0%–2% of the positive phase sequence component [5].

Voltage unbalance that is less than 2% is tolerated by most equipment, especially motors however, if it gets greater than 2%, it causes overheat of motor and transformers. This happens due to the current unbalance that varies as a cube of the voltage unbalance applied to the terminals [6]. Voltage unbalance is caused by unbalanced single-phase loading in a three-phase system, malfunctioning capacitor banks and single phasing of equipment [3, 6].

This paper focuses on the impact of electric vehicles working in Grid-to-Vehicle mode, based on a part of one real low-voltage distribution network in Zavidovići. To ensure the analysis depicts all possible situations as close as possible, thirteen scenarios were created. Impact of unbalanced load on variations and unbalance was investigated in four scenarios while nine scenarios combine the analysis of unbalanced load and EV charging. Single phase charging was studied in all nine scenarios that involved EVs. The analysis was conducted using a real network topology and metering data of year 2016 for the analyzed transformer and loads.

Section 2 presents the review of related literature. Section 3 describes the modelled part of the network while Sect. 4 defines scenarios and limits from standards used throughout this research. Section 5 provides with the results and Sect. 6 gives conclusions and ideas for future research.

## 2 Literature Review

When in a G2V mode, EVs are considered as loads that additionally increase the loading of the network. In the case of large fleets of EVs charging simultaneously, system losses, voltage drops, equipment overloading, and phase unbalance could be expected. Number of studies have been considering these impacts in recent years [2]. Influence of electric vehicles depends on the charging strategies, therefore this topic was studied in most of the papers [4–8].

A study in [7] analyzed the impact of connecting EVs in the distribution network. Two cases were observed: when the vehicles are being charged and when they act as generators, providing the network with stored energy. Analysis was based on a typical winter and summer day and used both regulated and unregulated charging of EVs. The results of this study showed that the EV distribution and percentage of penetration in the network play a significant role on how they impact the voltage variations and unbalance.

An analysis performed in [8] did not consider vehicle-to-grid concept. The focus of this study was to analyze how EV impact the power losses and quality of the distribution grid of Belgium. Both regulated and unregulated charging was observed. Importance of smart meters was highlighted, which would allow better communication and make EVs a controllable load. The results of this study show that regulated charging would flatten the daily peak power curve and decrease losses. Main reason is that the charging would be done during off-peak period.

Study in [9] was focused on the impact of EVs on distribution network investment and energy losses. The concept of vehicle-to-grid was also analyzed in this research. Two network areas have been considered: residential (low-voltage) and a combination of industrial and residential area (medium-voltage and low-voltage). The results showed that using smart charging strategies would help decrease the investment costs, required to reinforce the network, by around 70%. The costs could be additionally decreased, by charging some of EVs during off-peak period, up to 5%–35%. Charging point and station costs have not been considered in this paper.

A study in [10] investigated how different connection points, charging and discharging levels of PEVs impact the voltage unbalance of a low-voltage distribution network. Sensitivity analysis of voltage unbalance when only one PEV is connected at the end of the feeder was performed. Results showed that voltage unbalance increases for about 6%, when compared to the network's unbalance without PEVs connected. Another important find of this paper was that voltage unbalance is significantly greater when a set of PEVs, with higher charging levels, are connected to only one phase and at the end of the feeder.

Work done in [11] focuses on maximizing the total energy that is to be delivered to EVs connected to the distribution network, making sure that no limits are exceeded due to their high penetration. Using linear programming, and varying the charging rate, which would be performed by the Distribution System Operator (DSO), the goal was to obtain an optimal charging strategy. Results showed that by controlling the charging rate, high penetration of EVs would be possible, with little or no upgrade needed, for the existing distribution network. Network reconfiguration would be required in the case of many EVs connected to the same phase.

A study in [12] uses the Monte Carlo method for analyzing the charging impact of EVs on a low-voltage UK Generic Distribution Network. The analysis considered types of EVs, locations of charging and plug-in time as well as distributed generation. Their results showed that distributed generation helps normalize the voltage profile even for high penetration of EVs. This paper also highlights that for a high penetration of EVs connected to the network, source of charging should be from renewables.

## 3 Analyzed Part of One Real Low-Voltage Distribution Network

The analysis performed in this paper was based on a case-study of a real low-voltage distribution network in Zavidovići, Bosnia and Herzegovina. This part of the network was a suitable choice for investigating voltage unbalance, since it represents an example of a typical rural low-voltage network in Bosnia and Herzegovina. The area is supplied by a 10/0.4 kV distribution transformer, with installed power of 160 kVA. The modelled part of the network involved 63 loads out of which 61 are households and the remaining two are of other type of consumers, distributed along the main feeder. Due to restriction of available educational DIgSILENT Power Factory version, 38 loads were modelled, while the remaining 25 were represented in a form of a lumped load, further in text referred to as equivalent. Georeferenced scheme of the analyzed part of the network is provided in Fig. 1.

Data used for modeling included meter reading of the transformer and smart meter readings for loads. Meter readings of the transformer were recorded on a 15-min interval. Based on the peak consumption of year 2016, $5^{th}$ of August was chosen as the day to be analyzed. Peak power of the transformer, for this day, was found to be 124.272 kW. Smart meters provided with records of daily consumption of energy for each load. Percentage of energy for each load was obtained by dividing their daily consumption of energy with the total energy of the transformer. This helped calculate the daily power consumption of loads, by multiplying the percentage of energy for each load with the peak power of the transformer. Around 32.83% of the peak power of transformer is consumed by the analyzed part of the network, having in total 40.79 kW of power distributed along the analyzed low-voltage feeder. Power factor was calculated to be $cos\varphi = 0.9999 \approx 1$. The 38 modelled loads had total power of 23.1508 kW, while the remaining 25 loads in equivalent had total power of 17.6431 kW.

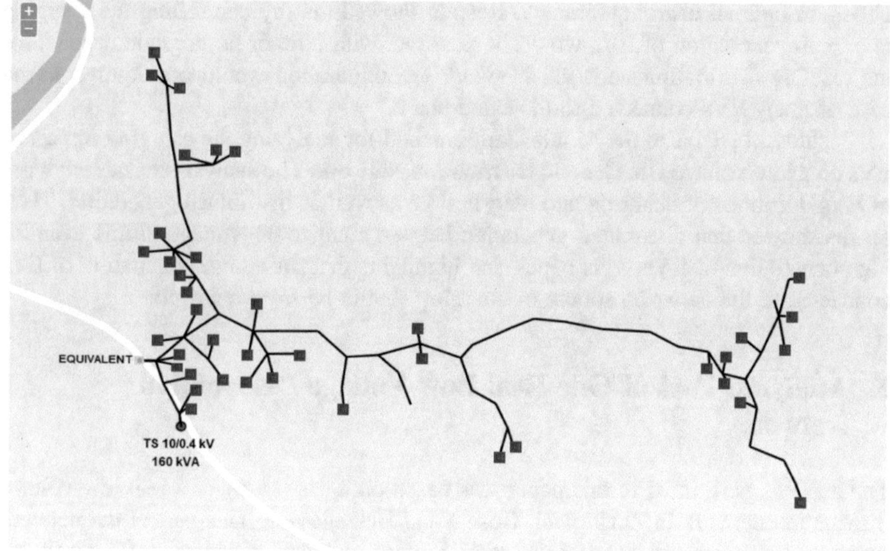


**Fig. 1.** Georeferenced scheme of modelled network

# 4 Methodology

Analysis performed in this paper was done using the DIgSILENT Power Factory software. Scenarios analyzed in this paper are defined by different loads distribution per phase. Variations option in DIgSILENT Power Factory software enabled changing the network topology by adding loads and analyzing the power flow for different scenarios, within one project. One should keep in mind that, in this project, different variations that were modelled are referred to as scenarios and have nothing to do with Operation Scenarios option, which allows only the modification of operational data.

Analysis performed in this paper can be split into two parts. First part of analysis included four scenarios with no electric vehicles connected. These scenarios were used as a 'base' for the second part of analysis that involved electric vehicles. The second part of analysis expanded three base scenarios with three additional ones based on EV penetration and their distribution per phase. Table 1 shows the distribution of loads per phase of the first part of analysis.

**Table 1.** Percentage of loads per phase for base scenarios

| Distribution of loads per phase | | | |
|---|---|---|---|
| First base scenario | Second base scenario | Third base scenario | Fourth base scenario* |
| 33.3%-33.3%-33.3% | 40%-30%-30% | 50%-25%-25% | 50%-25%-25% |

*Fourth scenario has the same distribution, but a set of loads near the end of the feeder is put on the same phase

Three scenarios that were chosen for the second part of analysis were: first, second and fourth. Third base scenario was excluded from the second part of analysis due its similarity to the fourth base scenario. Fourth base scenario was created as a modification of third, made worse by connecting the most remote set of loads to the same phase. The percentage defined for the first part of analysis was kept, however different percentage of electric vehicle penetration was added. Table 2 shows the distribution of EVs used in the second part of analysis.

All electric vehicles in this study were modelled as loads with power of 2.3 kW, which equals the household single-phase charging of EVs. This representation of vehicles in a form of load was suitable because it enables choosing a phase to connect it to and since only G2V mode was observed, therefore the aspect of battery giving energy back to the grid was not considered. Results of variations in base scenarios include all three phases, while for scenarios with EVs, only phase A is provided in graphs since the other two phases relatively kept the symmetry when compared to their base scenarios and have therefore been excluded from results.

Unbalanced three-phase AC Load Flow was applied to all scenarios and the resulting voltage variations and unbalance were observed. According to EN 50160, voltage variations should not exceed ±10% of the nominal voltage and voltage unbalance should be within range of 0%–2%. Nominal voltage of the analyzed network is 230 V, making the lower limit for allowed variation to be 207 V. Unbalance factor was included in the Flexible Data section of DIgSILENT Power Factory and used for further analysis of this paper. With respect to these limits from standards, evaluation of the obtained results is provided in the next section.

**Table 2.** Percentage of electrical vehicles per scenario of second part of analysis

| Penetration of EVs per scenario | | | | | | | | |
|---|---|---|---|---|---|---|---|---|
| First scenario | | | Second scenario | | | Fourth scenario | | |
| 10% EVs | 20% EVs | 50% EVs | 10% EVs | 20% EVs | 50% EVs | 10% EVs | 20% EVs | 50% EVs |

# 5 Results and Discussion

## 5.1 Scenarios Without EVs

**First base scenario**

First base scenario was defined to have an equal distribution of loads per phase and no electric vehicle connected. Out of 38 loads, 25 were three-phase loads, and the remaining 13 were single-phase loads. Total power of loads of 40.79 kW was distributed according to Table 3. Power of loads in equivalent and power of modelled loads is provided in Table 4. Obtained results regarding voltage variation are provided in Fig. 2.

**Table 3.** Power per phase of first base scenario

| Phase A | Phase B | Phase C |
|---|---|---|
| 13.61 kW | 13.63 kW | 13.55 kW |

**Table 4.** Power distribution of loads

| Modelled loads | | |
|---|---|---|
| Phase A | Phase B | Phase C |
| 7.729 kW | 7.749 kW | 7.669 kW |
| Loads in equivalent | | |
| Phase A | Phase B | Phase C |
| 5.881 kW | | |

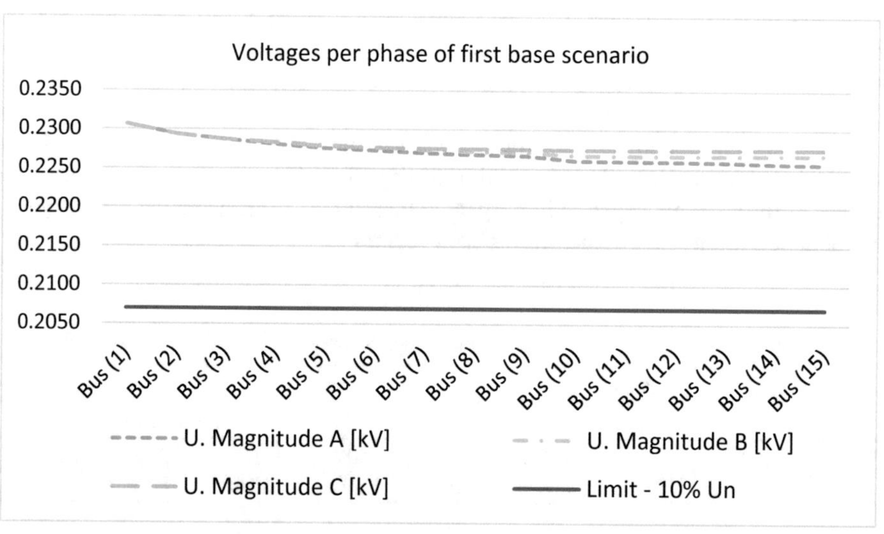


**Fig. 2** Voltage drop along the main feeder of first base scenario

From Fig. 2 we see that voltages of all phases are experiencing a certain drop while approaching the end of the feeder however, they are almost completely symmetrical and well within allowed limits of variation, presented as the red line.

**Second base scenario**

Second base scenario had distribution of 40% - 30% - 30% loads per phase. Total power of loads was distributed according to Table 5. Power of loads in equivalent and of modelled loads are presented in Table 6. Resulting voltage variation of second base scenario is presented in Fig. 3.

**Table 5.** Power per phase of second base scenario

| Phase A | Phase B | Phase C |
|---|---|---|
| 16.402 kW | 12.228 kW | 12.165 kW |

**Table 6.** Power distribution of loads

| Modelled loads | | |
|---|---|---|
| Phase A | Phase B | Phase C |
| 9.3448 kW | 6.9351 kW | 6.8721 kW |
| Loads in equivalent | | |
| Phase A | Phase B | Phase C |
| 7.0572 kW | 5.2929 kW | |

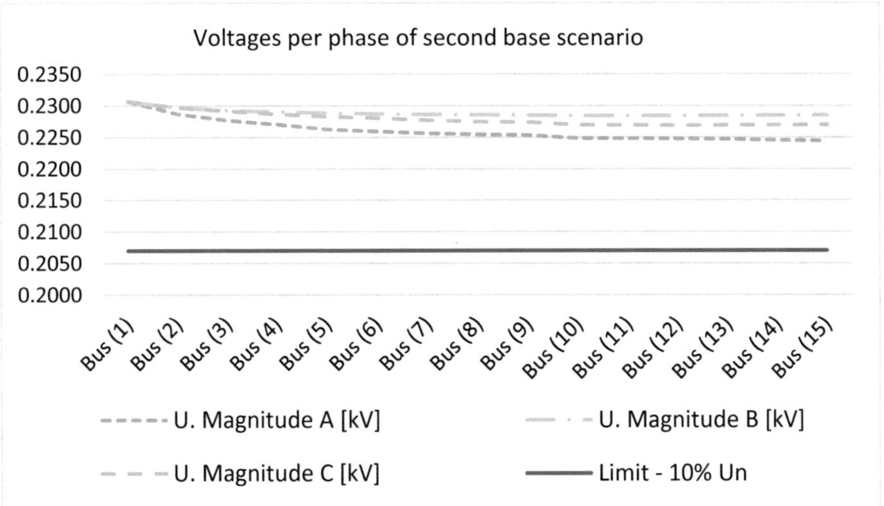


**Fig. 3** Voltage drop along the main feeder of second base scenario

Phase A experienced a bigger voltage drop when compared to previous scenario, which was expected according to the distribution of loads for this scenario. This affected the overall symmetry of phases, with phase B and C being more symmetrical when compared to phase A. However, this distribution did not influence negatively the variation limit, since all phases remain greater than -10% of Un, making this an acceptable scenario with regards to this limit from standards.

**Third base scenario**

Third base scenario was modelled to have distribution of 50% - 25% - 25% loads per phase. This scenario was one of the worse that were examined in this paper. Total power of loads was distributed according to Table 7. Power of loads in equivalent and of modelled loads is shown in Table 8. Voltage variation result of this scenario is provided in Fig. 4.

**Table 7.** Power per phase of third base scenario

| Phase A | Phase B | Phase C |
|---|---|---|
| 20.406 kW | 10.295 kW | 10.1923 kW |

**Table 8.** Power distribution of loads

| Modelled loads | | |
|---|---|---|
| Phase A | Phase B | Phase C |
| 11.584 kW | 5.884 kW | 5.782 kW |
| Loads in equivalent | | |
| Phase A | Phase B | Phase C |
| 8.8216 kW | 4.4108 kW | |

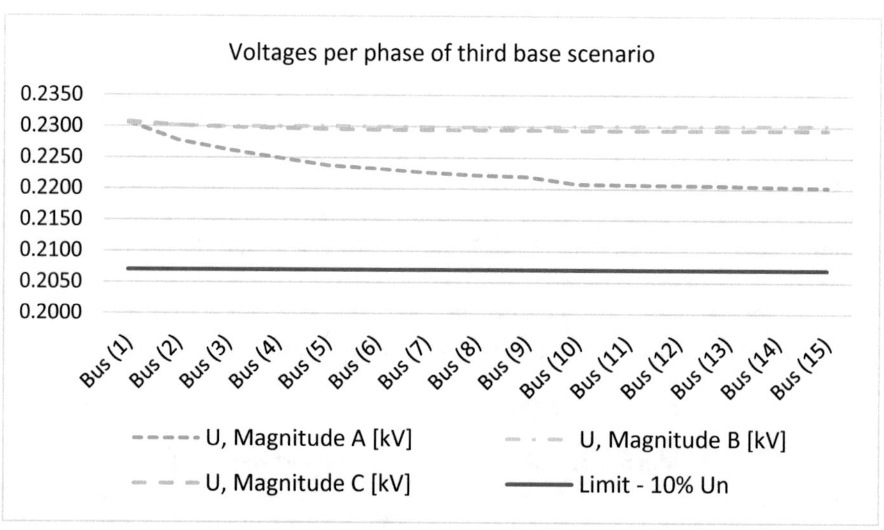


**Fig. 4** Voltage drop along the main feeder of third base scenario

With increase of loading on phase A, an increase in voltage drop along the main feeder is evident from the resulting graph. Phase B and C remain in symmetry. According to the limit from standards, all voltages are within allowed limit of variation, therefore this scenario is acceptable.

**Fourth base scenario**

Fourth base scenario was built as a modification of the third. Their difference lies in fourth base scenario having the most remote set of loads put on the same phase, which worsened the situation. Total power of loads was distributed according to Table 9. Power of loads in equivalent and of modelled loads is provided in Table 10.

**Table 9.** Power per phase of fourth base scenario

| Phase A | Phase B | Phase C |
|---|---|---|
| 20.452 kW | 10.295 kW | 10.047 kW |

**Table 10.** Power distribution of loads

| Modelled loads | | |
|---|---|---|
| Phase A | Phase B | Phase C |
| 11.63 kW | 5.884 kW | 5.636 kW |
| Loads in equivalent | | |
| Phase A | Phase B | Phase C |
| 8.8216 kW | 4.4108 kW | |

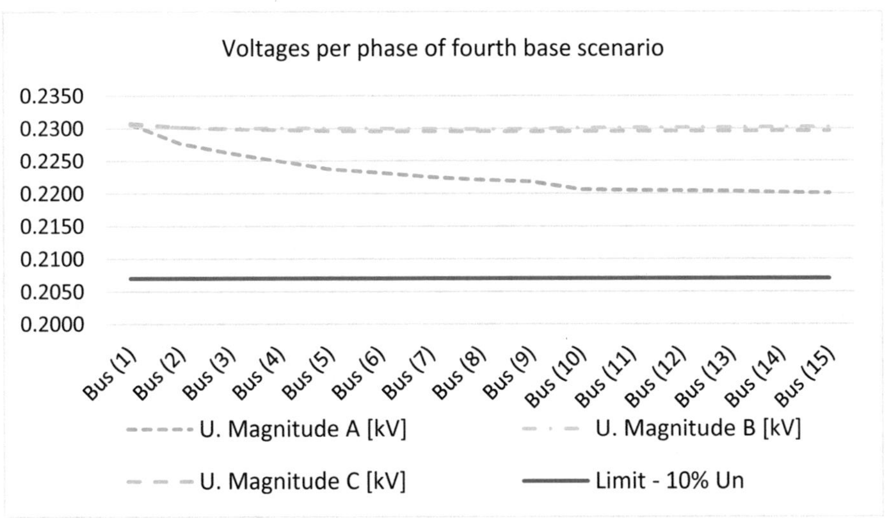


**Fig. 5** Voltage drop along the main feeder of fourth base scenario

Resulting voltage variation is provided in Fig. 5.

The result obtained for this load distribution is similar to the previous scenario. However, we see phase A experiencing a slight greater drop while approaching the end of the feeder, while phase B and C remain unchanged. This is due to loads that are connected at the end of the feeder being put on phase A. Voltage variation limit is not violated, making this scenario acceptable.

Resulting voltage unbalances along the main feeder for four base scenarios are combined in Fig. 6.

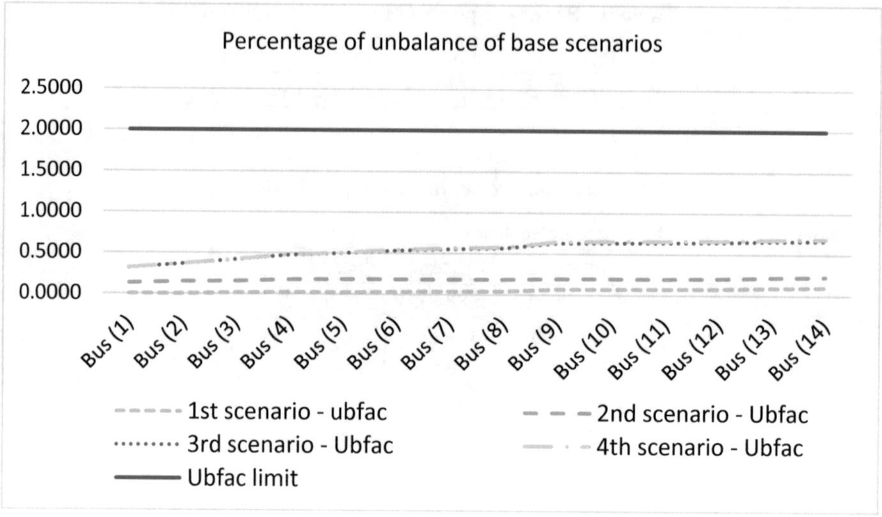


**Fig. 6** Voltage unbalance along the main feeder of base scenarios

With increase of loading on phase A, unbalance factor increases accordingly. Lowest unbalance factor at the end of the feeder was recorded for first scenario, being 0.1011%, while the highest unbalance factor occurred in fourth scenario, being 0.6865%. Figure 6 helps understand how only slightly different third and fourth scenario really are, with third scenario resulting in unbalance of 0.6684%, near the end of the feeder. All scenarios, investigated in the first part of analysis, remained within allowed limit from standards of voltage variation and unbalance. We conclude that all load distributions in scenarios without EVs are acceptable.

## 5.2 Scenarios with EVs

**First scenario with 10% of EVs**

First scenario with 10% of EVs involved 6 electric vehicles, distributed along the main feeder. Total power of loads was distributed according to Table 11. Power of loads in equivalent and of modelled loads are presented in Table 12. Distribution of EVs was done with respect to equal distribution of power of first base scenario.

**Table 11.** Power and EVs per phase of 1st scenario with 10% EVs

| Phase A | Phase B | Phase C |
|---|---|---|
| 18.21 kW | 18.23 kW | 18.16 kW |
| No. of EVs per phase | | |
| Phase A | Phase B | Phase C |
| 2 | 2 | 2 |

**Table 12.** Power distribution of loads

| Modelled loads | | |
|---|---|---|
| Phase A | Phase B | Phase C |
| 12.329 kW | 12.349 kW | 12.279 kW |
| Loads in equivalent | | |
| Phase A | Phase B | Phase C |
| 5.881 kW | | |

**First scenario with 20% of EVs**

First scenario with 20% of EVs included 12 electric vehicles, out of which 9 were modelled and 3 were put in equivalent. Total power of loads and number of EVs were distributed according to Table 13. Power of loads in equivalent and of modelled loads are provided in Table 14.

**Table 13.** Power and EVs per phase of 1st scenario with 20% EVs

| Phase A | Phase B | Phase C |
|---|---|---|
| 22.81 kW | 22.83 kW | 22.76 kW |
| No. of EVs per phase | | |
| Phase A | Phase B | Phase C |
| 4 | 4 | 4 |

**Table 14.** Power distribution of loads

| Modelled loads | | |
|---|---|---|
| Phase A | Phase B | Phase C |
| 16.929 kW | 16.949 kW | 16.879 kW |
| Loads in equivalent | | |
| Phase A | Phase B | Phase C |
| 5.881 kW | | |

**First scenario with 50% of EVs**

First scenario with 50% of EV*s* involved 30 electric vehicles, out of which 21 were modelled and 9 were put in equivalent. Total power of loads and no of electric vehicles per phase were distributed according to Tables 15 and 16. Power of loads in equivalent and of modelled loads are presented in Table 17.

**Table 15.** Power and EVs per phase of 1st scenario with 50% of EVs

| Phase A | Phase B | Phase C |
|---|---|---|
| 36.61 kW | 36.63 kW | 36.56 kW |

**Table 16.** Power and EVs per phase of 1st scenario with 50% of EVs (cont'd)

| No. of EVs per phase | | |
|---|---|---|
| Phase A | Phase B | Phase C |
| 10 | 10 | 10 |

**Table 17.** Power distribution of loads

| Modelled loads | | |
|---|---|---|
| Phase A | Phase B | Phase C |
| 30.729 kW | 30.749 kW | 30.679 kW |
| Loads in equivalent | | |
| Phase A | Phase B | Phase C |
| 5.881 kW | | |

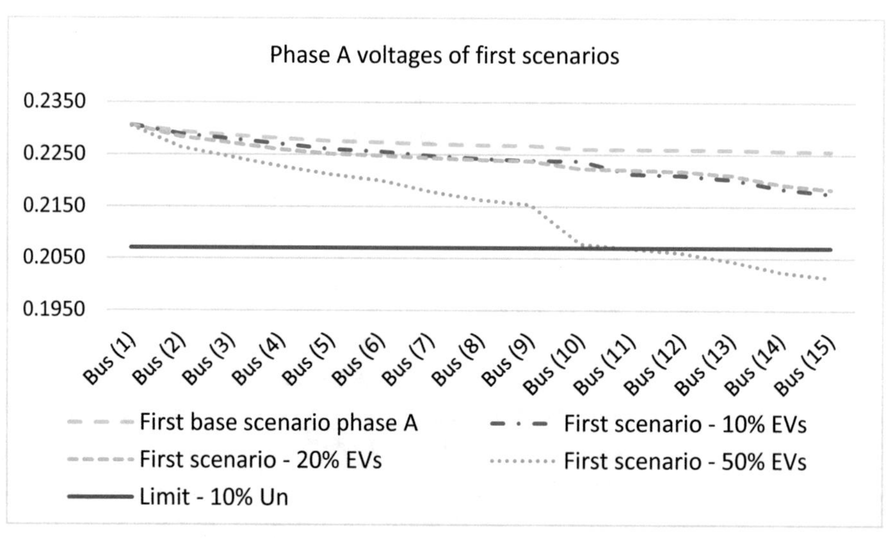


**Fig. 7** Voltage variation along the main feeder of phase A of first scenarios

Resulting voltage variations of phase A of first base scenario and first scenarios with 10%, 20% and 50% of EVs is provided in Fig. 7. Voltage unbalance along the main feeder is shown in Fig. 8.

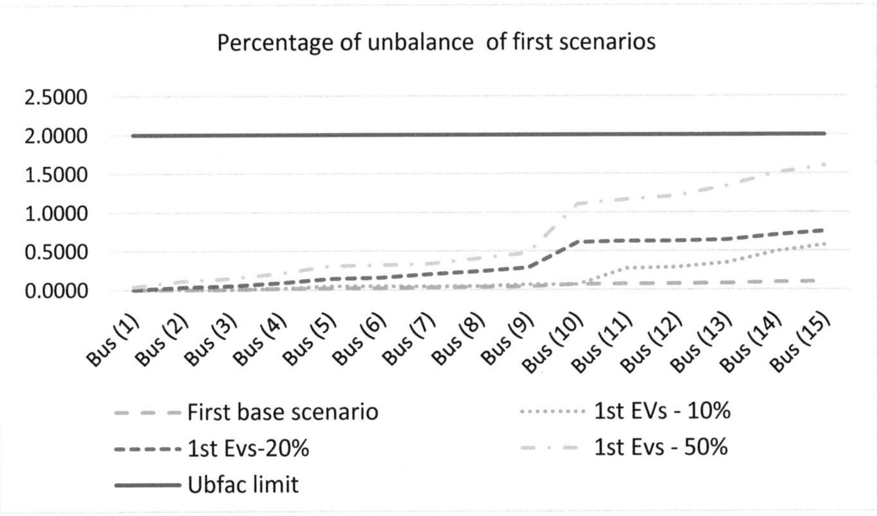


**Fig. 8** Voltage unbalance along the main feeder for the first base scenario and first base scenarios with 10%, 20% and 50% of EVs

All scenarios experienced an evident voltage drop, but only the scenario with 50% of EVs violated the limit from standards of voltage variation, making it unacceptable. Interesting find is that scenario with 10% of EVs had a greater voltage drop than the scenario with 20% of EVs. This was due to EV placement in the network. For a better understanding, EV placement of first scenario with 10% and 20% are provided in Tables 18 and 19, respectively. This distribution of vehicles along the feeder caused a greater drop of phase A in scenario with 10% EVs.

**Table 18.** EV placement of 1st scenario with 10% EVs

| Connection point of feeder | No. of EVs on phase A | No. of EVs on phase B | No. of EVs on phase C |
|---|---|---|---|
| Beginning | / | 1 | / |
| Middle | 1 | 1 | 1 |
| End | 1 | / | 1 |

**Table 19.** EV placement of 1st scenario with 20% EVs

| Connection point of feeder | No. of EVs on phase A | No. of EVs on phase B | No. of EVs on phase C |
|---|---|---|---|
| Beginning | 2 | 3 | 1 |
| Middle | 1 | 1 | 1 |
| End | 1 | / | 2 |

According to results, all scenarios remained within allowed limits of voltage unbalance. With increase of voltage variation, voltage unbalance increases accordingly. The highest unbalance of 1.6368% occurred in scenario with 50% of EVs.

**Second scenario with 10% of EVs**

Second scenario with 10% of EVs involved 7 electric vehicles, with 6 modelled and 1 put in equivalent. Total power of loads and number of EVs per phase are provided in Table 20. Power of loads in equivalent and of modelled loads are shown in Table 21.

**Table 20.** Power and EVs per phase of 2nd scenario with 10% of EVs

| Phase A | Phase B | Phase C |
|---|---|---|
| 23.302 kW | 16.83 kW | 16.77 kW |
| No. of EVs per phase | | |
| Phase A | Phase B | Phase C |
| 3 | 2 | 2 |

**Table 21.** Power distribution of loads

| Modelled loads | | |
|---|---|---|
| Phase A | Phase B | Phase C |
| 16.245 kW | 11.537 kW | 11.477 kW |
| Loads in equivalent | | |
| Phase A | Phase B | Phase C |
| 7.0572 kW | 5.2929 kW | |

**Second scenario with 20% of EVs**

Second scenario with 20% of EVs included 13 electric vehicles, out of which 9 were modelled and 4 put in equivalent. Total power of loads and number of EVs was distributed according to Table 22. Power of loads in equivalent and of modelled loads are provided in Table 23.

**Table 22.** Power and EVs per phase of 2nd scenario with 20% of EVs

| Phase A | Phase B | Phase C |
|---|---|---|
| 27.902 kW | 21.428 kW | 21.365 kW |
| No. of EVs per phase | | |
| Phase A | Phase B | Phase C |
| 5 | 4 | 4 |

**Table 23.** Power distribution of loads

| Modelled loads | | |
|---|---|---|
| Phase A | Phase B | Phase C |
| 20.8448 kW | 16.1351 kW | 16.0721 kW |
| Loads in equivalent | | |
| Phase A | Phase B | Phase C |
| 7.0572 kW | 5.2929 kW | |

**Second scenario with 50% of EVs**

Second scenario with 50% of EVs involved 31 electric vehicles, with 21 modelled and 10 put in equivalent. Total power of loads and number of EVs per phase was distributed according to Table 24. Power distribution of loads in equivalent and of modelled loads is shown in Table 25.

**Table 24.** Power and EVs per phase of 2nd scenario with 50% of EVs

| Phase A | Phase B | Phase C |
|---|---|---|
| 46.302 kW | 32.928 kW | 32.865 kW |
| No. of EVs per phase | | |
| Phase A | Phase B | Phase C |
| 13 | 9 | 9 |

**Table 25.** Power distribution of loads

| Modelled loads | | |
|---|---|---|
| Phase A | Phase B | Phase C |
| 39.245 kW | 27.635 kW | 27.572 kW |
| Loads in equivalent | | |
| Phase A | Phase B | Phase C |
| 7.0572 kW | 5.2929 kW | |

Resulting voltage variations of phase A for second scenarios are provided in Fig. 9. Voltage unbalance along the main feeder are presented in Fig. 10.

Additional loading of phase A caused all voltages to experience a larger voltage drop, when compared to previous scenarios. Results show that scenarios with 20% and 50% EV penetration are not acceptable, since they violate the limit from standards of voltage variation. Scenario with 10% of EVs is the only scenario involving electric vehicles which remained within allowed limit.

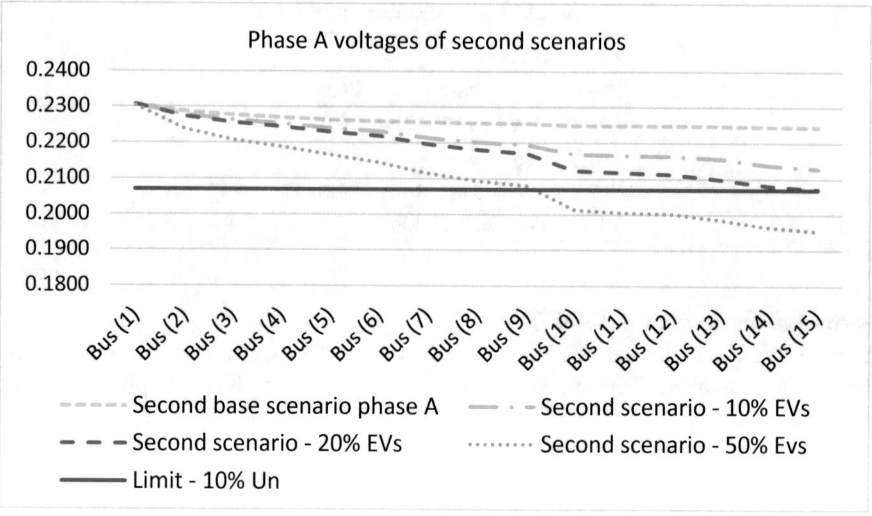


**Fig. 9** Voltage variation along the main feeder of phase A of second scenarios

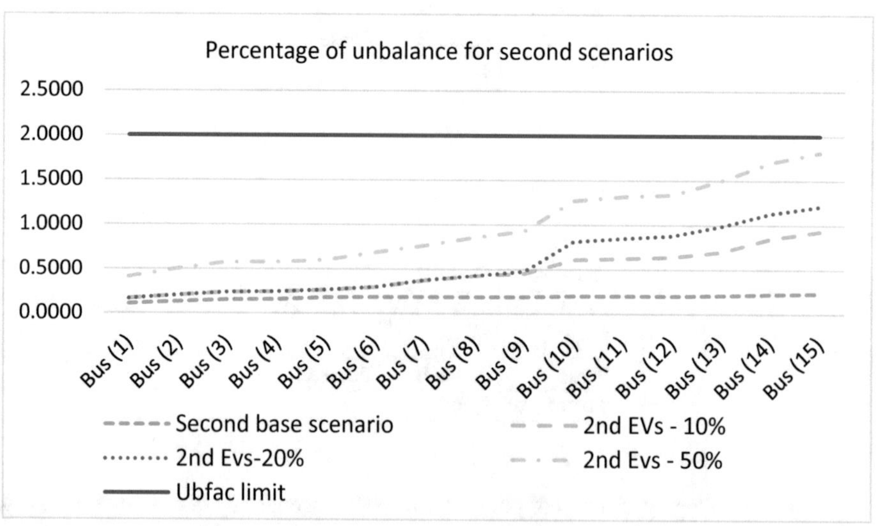


**Fig. 10** Unbalance factor along the main feeder of second scenarios

All scenarios experienced an increase of unbalance when compared to results of first scenarios. However, according to results presented in Fig. 10, all scenarios stayed within allowed limit of voltage unbalance. Scenario with 50% EV penetration resulted in highest unbalance of 1.8125%, near the end of the feeder.

**Fourth scenario with 10% of EVs**

Fourth scenario with 10% of EVs involved 8 electric vehicles, with 6 modelled and 2 put in equivalent. Total power of loads and number of electric vehicles per phase was distributed according to Table 26. Power of loads in equivalent and of modelled loads is presented in Table 27.

**Table 26.** Power and EVs per phase of 4th scenario with 10% of EVs

| Phase A | Phase B | Phase C |
|---|---|---|
| 29.652 kW | 14.895 kW | 14.645 kW |
| No. of EVs per phase | | |
| Phase A | Phase B | Phase C |
| 4 | 2 | 2 |

**Table 27.** Power distribution of loads

| Modelled loads | | |
|---|---|---|
| Phase A | Phase B | Phase C |
| 20.831 kW | 10.484 kW | 10.234 kW |
| Loads in equivalent | | |
| Phase A | Phase B | Phase C |
| 8.8215 kW | 4.4108 kW | |

**Fourth scenario with 20% of EVs**

Fourth scenario with 20% of EVs included 12 electric vehicles, with 8 modelled and 4 put in equivalent. Power of loads and number of electric vehicles per phase were distributed according to Table 28. Power of modelled loads and of loads in equivalent are shown in Table 29.

**Table 28.** Power and EVs per phase of 4th scenario with 20% of EVs

| Phase A | Phase B | Phase C |
|---|---|---|
| 34.252 kW | 17.195 kW | 16.947 kW |
| No. of EVs per phase | | |
| Phase A | Phase B | Phase C |
| 6 | 3 | 3 |

**Table 29.** Power distribution of loads

| Modelled loads | | |
|---|---|---|
| Phase A | Phase B | Phase C |
| 25.431 kW | 12.784 kW | 12.536 kW |
| Loads in equivalent | | |
| Phase A | Phase B | Phase C |
| 8.8215 kW | 4.4108 kW | |

**Fourth scenario with 50% of EVs**

Fourth scenario with 50% of EVs involved 32 electric vehicles, with 22 modelled and 10 put in equivalent. Total power of loads and number of electric vehicles per phase were distributed according to Table 30. Power of modelled loads and of loads in equivalent are shown in Table 31.

**Table 30.** Power and EVs per phase of 4th scenario with 50% of EVs

| Phase A | Phase B | Phase C |
|---|---|---|
| 57.252 kW | 28.695 kW | 28.447 kW |
| No. of EVs per phase | | |
| Phase A | Phase B | Phase C |
| 16 | 8 | 8 |

**Table 31.** Power distribution of loads

| Modelled loads | | |
|---|---|---|
| Phase A | Phase B | Phase C |
| 48.431 kW | 24.284 kW | 24.036 kW |
| Loads in equivalent | | |
| Phase A | Phase B | Phase C |
| 8.8215 kW | 4.4108 kW | |

Resulting voltage variations of phase A for second scenarios are provided in Fig. 11. Voltage unbalance along the main feeder are presented in Fig. 12.

Voltage variation limit was violated for all fourth scenarios that involved electric vehicles. The largest drop occurred in scenario with 50% EV penetration, resulting in 177.6 V. Accordingly, the same scenario experienced the largest unbalance factor of 3.79%, which is significantly higher than allowed 2% of unbalance. Fourth scenario with 50% of EVs was the only scenario of this study that violated both limit from standards.

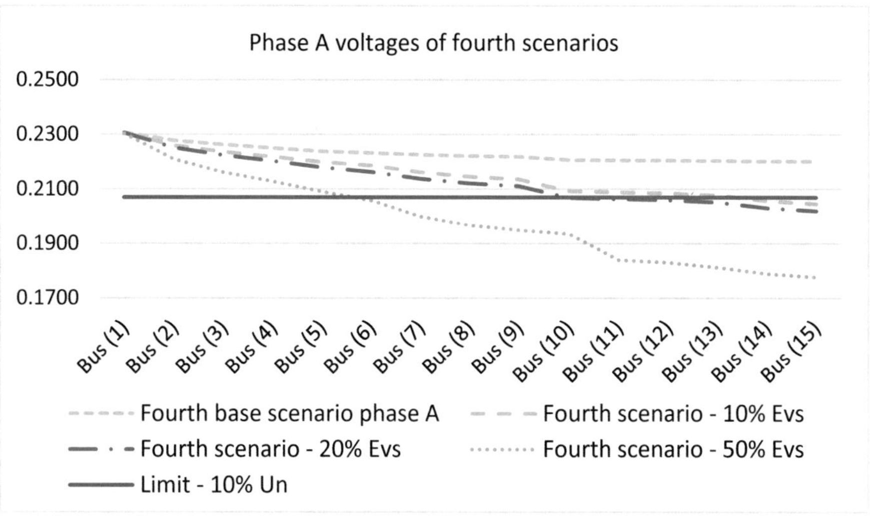


**Fig. 11** Voltage variation along the main feeder of phase A of fourth scenarios

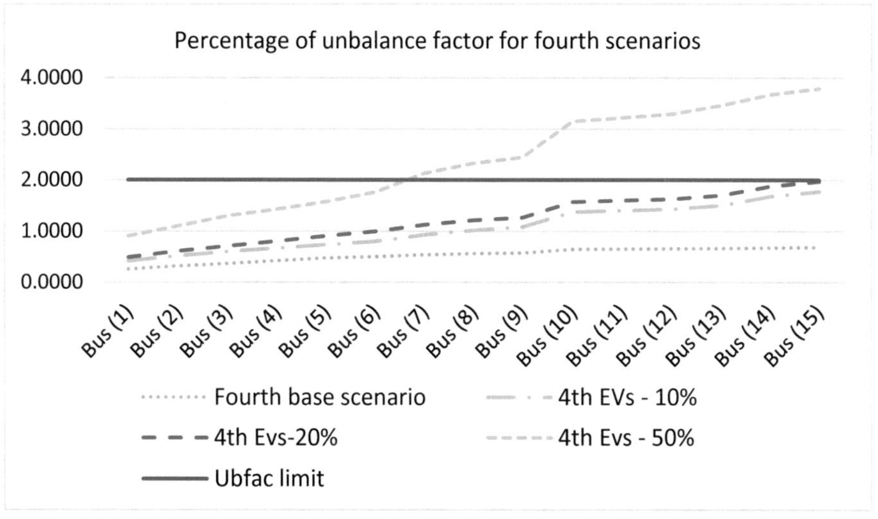


**Fig. 12** Voltage unbalance along the main feeder of fourth scenarios

## 6 Conclusion

Aim of this paper was to investigate the impact of different load distributions and EV penetrations per phase on voltage variation and unbalance. An important find was that connection point of electric vehicles and the phase to which it is connected to defines how EV influences the variation and unbalance of the distribution network.

Total of thirteen scenarios were modelled, with four base scenarios with no EVs involved and nine scenarios with different EV penetrations. In the case of the first scenario with 10% and 20% of EV penetration, results showed that despite having the same number of vehicles connected to each phase, their distance from the transformer highly impacts the outcoming voltage variation. This find confirms the results of study [10], highlighting the connection point of EVs in playing a key role in the effects of vehicle charging.

Out of thirteen scenarios, seven of them remained within allowed limits of both limits from standards examined in this paper and could therefore be declared as acceptable. Other six scenarios violated at least one boundary, that is, five violated the voltage variation limit while only the last scenario violated both variation and unbalance limit.

We conclude that charging electric vehicles impacts both parameters of power quality that were examined in this paper. All scenarios experienced a voltage drop which increased accordingly with higher penetrations of EVs, followed by increment of unbalance. For an already existing unbalance in the network, connection of EVs to such a network would inevitably contribute to increasing unbalance even more, since it decreases the network's capacity. Results showed that in the case of one phase being doubly loaded when compared to other two, not even 10% of electric vehicles could be connected to such a network.

Mass deployment of electric vehicles will force Distribution System Operators to pay more attention to unbalance in low-voltage networks and to balance the loads. Future work could include analyzing the vehicle-to-grid mode, regulated charging and optimal phasing of electric vehicles to ensure power quality parameters are not altered above allowed limits.

## References

1. Farhoodnea, M., Mohamed, A., Shareef, H., Zayandehroodi, H.: Power quality impact of renewable energy based generators and electric vehicles on distribution systems. Procedia Technol. **11**, 11–17 (2013)
2. Yong, J.Y., Ramachandaramurthy, V.K., Tan, K.M., Mithulananthan, N.: A review on the state-of-art technologies of electric vehicle, its impacts and prospects. Renew. Sustain. Energy Rev. **49**, 365–385 (2015)
3. Masoum, M.A.S., Fuchs, E.F.: Power Quality in Power Systems and Electrical Machines, 2nd edn. Elsevier, Boston (2008)
4. Dugan, R.C., McGranaghan, M.F., Santoso, S., Beaty, H.W.: Electrical Power Systems Quality, 2nd edn. McGraw-Hill, New York (2004)
5. European Standard: Voltage characteristic of electricity supplied by public electricity networks (EN 50160) (2011)
6. Kennedy, B.W.: Power Quality Primer. McGraw-Hill, New York (2000)
7. Putrus, G.A., Suwanapingkarl, P., et al.: Impact of electric vehicles on power distribution networks. In: IEEE Vehicle Power and Propulsion Conference (2009)
8. Clement-Nyns, K., Haesen, E., Driesen, J.: The impact of charging plug-in hybrid electric vehicles on a residential distribution grid. IEEE Trans. Power Syst. **25**(1), 371–380 (2010)

9. Fernandez, L.P., San Roman, T.G., Cossent, R., et al.: Assessment of the impact of plug-in electric vehicles on distribution networks. IEEE Trans. Power Sys. **26**(1), 206–213 (2011)
10. Shahnia, F., Ghosh, A., et al.: Voltage unbalance sensitivity analysis of plug-in electric vehicles in distribution networks. In: AUPEC. IEEE (2011)
11. Richardson, P., Flynn, D., Keane, A.: Optimal charging of electric vehicles in low-voltage distribution systems. IEEE Trans. Power Syst. 27(1), 268–279 (2012)
12. Shariff, N.B.M., Al Essa, M., Cipcigan, L.: Probabilistic analysis of electric vehicle charging load impact on residential distributions network. In: IEEE International Energy Conference (ENERGYCON) (2016)

# Determination of Effective Protection Distance in Front of Surge Protective Devices in Low Voltage Systems

Vladan Radulović(✉) and Zoran Miljanić

Faculty of Electrical Engineering, University of Montenegro, Podgorica, Montenegro
vladanra@ucg.ac.me

**Abstract.** Surge protective devices (SPDs) is widely used in low-voltage power systems for protection of devices and equipment against surges. The SPDs provide the optimized protection level only if it is installed directly across the terminals of the equipment to be protected. However, it is not always possible to locate SPD close to the protected equipment, and an inevitable separation distance along connecting cable exists. Maximum distance along cable from point of SPD connection for which desirable protective characteristics are still fulfilled represents effective protection distance of SPD. Effective protection distance is usually determined for cable length after location of SPDs toward protected equipment. However, it is known that SPD provides protective effects also on devices installed at certain distance before location of SPDs. Determination of effective protection distance in front of SPD installed in low-voltage power system has been given in the paper. Analyses are performed with extensive number of simulations in MATLAB and ATP/EMTP with different influencing parameters in wide ranges of values, and with application of standard surge testing waveforms.

## 1 Introduction

Modern electric and electronic devices (especially those based on advanced and sophisticated electronic components and microprocessors) in low-voltage systems (LVS) of different type of facilities usually have very low rated withstand impulse voltage level [1]. Therefore, these devices are susceptible to appearance and effects of surges in LVS, which can cause operational problems, degradation of equipment and/or insulation breakdown and failures.

Regardless of surge origin, it is necessary to provide adequate protection of equipment against surges in LVS. Surge protection devices (SPDs) are one of the main components of surge protection system. Their role is to provide significant reduction of electromagnetic impulse of surges by diverting surge energy and clamping surge voltage to value below of insulation withstand level of the equipment to be protected. Therefore, protection voltage of SPD as well as their energy absorption capability are important factors to be considered during selection process of SPD. On the other hand, different factors have influence on SPD protection performances. Effects of different


S. Avdaković et al. (Eds.): IAT 2019, LNNS 83, pp. 74–90, 2020.
https://doi.org/10.1007/978-3-030-24986-1_6

types and characteristics of load to be protected are investigated in [1–3]. Influence of incoming surges waveforms are described in [4, 5]. Arrangement of SPDs in protection system is analyzed in [6–8].

The SPD will offer the optimized protection level only if it is installed directly across the terminals of the equipment to be protected. Unfortunately, in practice, it is not always possible to locate SPD close to the equipment, and an inevitable separation distance along connecting cable exists [9]. This separation distance has significant influence on SPD protection performances. Therefore, it is necessary to determine effective protection distance of SPD, which represents maximal length along cable from point of SPD connection for which desirable protective characteristics are still fulfilled. In existing researches, effective protection distance is usually determined for cable length after location of SPDs in LVS toward protected equipment [3, 10–12].

However, it is known that SPD provides protective effects also on devices installed at certain distance before location of SPDs. This effect is partly investigated in [9], but only with limited values of influencing parameters. Therefore, aim of this paper is to provide comprehensive analysis and determination of effective protection distance in front of SPD arrangement in LVS. Obtained results should to provide possibilities to optimize number of installed SPDs, especially for administrative buildings which are characterized with large number of devices to be protected.

## 2 Surges in Low Voltage Systems

Surges that occur in LVS are originated by three types of event [13]:

- natural phenomena such as lightning which can strike the power system directly or induce overvoltages by striking other objects nearby;
- intentional actions on the power system, such as load or capacitor switching in the transmission or distribution systems by the utility, or in the low-voltage system by end-user operations;
- unintentional events such as power system faults and their elimination, or coupling between different systems such as interactions between power systems and signal/communication systems.

Lightning is a major electromagnetic interference in the electrical systems [14]. It is an unavoidable event, which affects LVS (power systems as well as signal/communication systems) through several mechanisms. The obvious interaction is a direct flash on the system, but other coupling mechanisms can also result in a system surges. Three types of coupling mechanisms are capable of producing surges in LVS [13].

- Direct flashes to the power system, which can occur on the primary side of the medium voltage (MV)/low voltage (LV) distribution transformer, on the LV distribution system (overhead as well as buried), and on the service entrance to individual buildings.
- Indirect flashes: flashes to nearby objects, which can produce surges in the LV distribution system by inductive coupling or by common-path coupling. While the

overvoltages and surge currents resulting from such flashes might be less severe than those associated with a direct flash, their frequency of occurrence is much greater.

- Direct flashes to the lightning protection system or to extraneous parts of the end-user building (structural steel, non-electrical components such as water lines, heating and air conditioning ducts, elevator shafts, etc.). Such flashes have two kinds of effect: inductive coupling from the lightning currents carried by the extraneous parts, and injection of lightning current from the building into the LV system made unavoidable by the necessary provision of surge protection devices between the LV system conductors and local earth, or the so-called equipotential bonding of the installation.

Mentioned mechanisms generate surges in LVS with different waveforms, steepness of wavefront, durations and amplitudes. The wide variety of surges that can be expected to occur in low-voltage AC power systems has been described in the database of IEEE Std. C62.41.1 [15]. There are no specific models that are representative of all surge environments; the complexities of the real world need to be simplified to produce a manageable set of standard surge tests.

As a first step toward a reduction of the complex database on surge occurrences, the concept of location categories is proposed in [15] to describe the scenario of surges impinging at the service entrance or generated within the building, exclusive of those associated with a direct lightning flash to the structure. The concept of location category rests on the considerations of dispersion and propagation of surge currents and surge voltages. For surge currents presented at the service entrance of a building, the increasing impedance opposing (impeding) the flow of surge currents further into the building (with or without the crowbar effect of a flashover that can occur at the meter or in the service-entrance equipment) reduces the surge current that can be delivered along the branch circuits. In contrast, a voltage surge, with an amplitude below the point of flashover of clearances and presented at the service entrance of a building, can propagate, practically un-attenuated, to the end of a branch circuit when no low-impedance load (equipment or local SPD) is present along the branch circuit. According to this concept, there are three location categories: Location Category A applies to the parts of the installation at some distance from the service entrance. Location Category C applies to the external part of the building, extending some distance into the building. Location Category B extends between Location Categories C and A [15].

In addition to concept of location categories, evaluation of the ability of equipment to withstand wide variety of surges, or of the performance of SPDs in dealing with surges, can be facilitated by a reduction of the database to a few representative stresses [16]. The reduction process should lead to selecting a few representative surges that will make subsequent laboratory tests uniform, meaningful, and reproducible. Since the environment is subject to change both for the better and the worse, it would be prudent to use these representative surges as a baseline environment.

Two types of representative surge test waveforms are proposed in [16]. The first type, standard waveforms, has a long history of successful application in industry and thus may be considered sufficient for most cases of surge immunity tests on the AC port of equipment. However, for special environments or difficult cases, there is a second type, additional waveforms. In this paper, only standard waveforms are taken into account.

The two recommended standard waveforms are the 0.5 μs–100 kHz Ring Wave and the 1.2/50 μs–8/20 μs Combination Wave.

The Combination Wave surge is delivered by a generator that applies a 1.2/50 μs voltage wave across an open circuit and an 8/20 μs current wave into a short circuit [16]. These waveforms are given in Fig. 1.

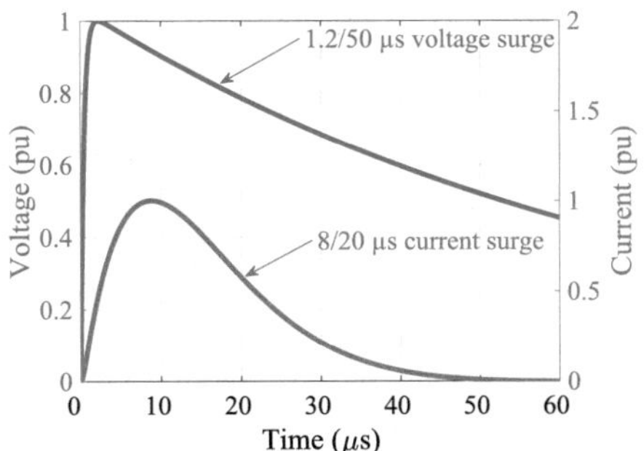


**Fig. 1.** Voltage and current surge waveforms for Combination Wave generator

A plot of the Ring Wave voltage waveform is shown in Fig. 2. No short-circuit current waveform is specified for the 0.5 μs–100 kHz Ring Wave [16].

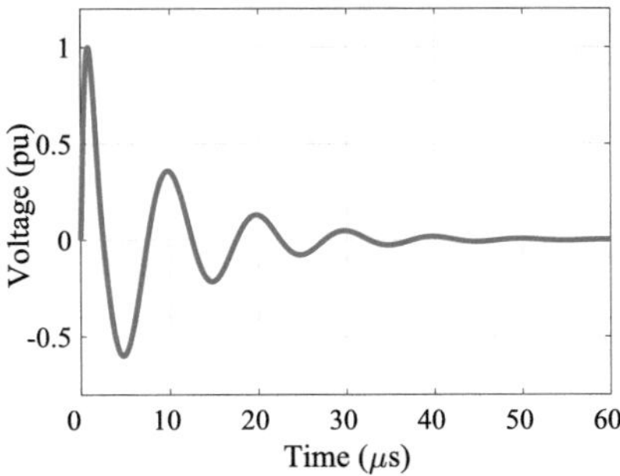


**Fig. 2.** Voltage surge waveform for Ring Wave generator

The exact waveforms that are delivered are determined by the generators and the impedance of the EUT to which the surge is applied [16].

The selection of the value of either the peak open-circuit voltage or the peak short-circuit current of these surges is based according to the location category. Definition of waveforms and acceptable tolerances are given in [17].

## 3 Surge Protection System

Simple model of the single phase TN-C-S system of low-voltage AC power circuits with nominal voltage of 230 V is used for the analysis of and determination of effective protection distance in front of SPD (Fig. 3). Model is consisting of surge generator (SG), SPD which is intentionally placed before a protected device (PD) and equipment under test (EUT). SG generates Combination Wave surge (CWS) or Ring Wave surge (RWS) as representation of surge environment in observed part of the LVS, which is assumed to belongs location category B according to IEEE C62.41.1 [15]. According to IEEE C62.45 [17], amplitudes of CWS voltage and current surge for location category B are 6 kV and 3 kA, respectively. Amplitude of RWS voltage waveform is 6 kV with generator effective impedance of 12 Ω. The observed protection system may correspond to cases with one-stage protection system (i.e. application of only one SPD within LVS) or to cases of multi-stage protection systems in which mentioned surges may appear. EUT represents load (with different type and values of active/reactive power) that has not intentionally protected with SPD, but which can be covered with protective effect of SPD if EUT fails within effective protection distance in front of SPD. Effective protection distance will be determined with assessment of length of cable EUT – SPD for which maximal voltages across EUT are below of its insulation withstand voltage.

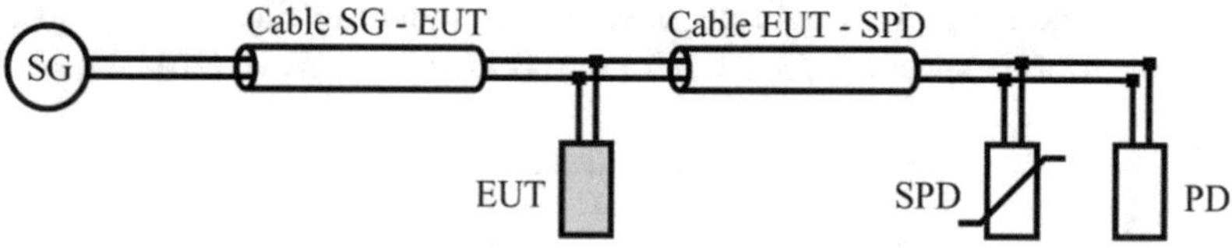


**Fig. 3.** Model of the observed protection system

It is assumed that analyzed EUT belongs to the equipment of the overvoltage category I according to IEC 60664-1 [18]. This overvoltage category involves equipment with withstands impulse voltage level of 1.5 kV and it is the most rigorous requirement for the protection effect of SPD.

SPD is voltage limiting type surge protective device of type 3 according to IEC 61643-11 [19], with protection voltage of 700 V and designed for mounting at power socket near PD. Value of 700 V for SPD's protection voltage is selected as value for which effective protection distance after SPD is infinite. Namely, according to IEC 62305-4 [20], SPD protective distance may be disregarded in following two conditions. One condition is that the connection cable length is shorter than 10 m. The second condition is that the value of protection voltage of SPD is smaller than half value of rated impulse withstanding voltage level of protected device. Since it is taken that rated impulse withstand voltage of devices in observed system is 1500 V, SPD's protection voltage of 700 V provides fulfilment of this condition.

Cables between arrester and suppressor (cable A-S) and between suppressor and EUT (cable S-EUT) are PVC-insulated cables 3 x 2.5 $mm^2$ with electric parameters: R = 0.00561 Ω/m, L = 0.324 μH/m, C = 0.1368 nF/m, G = 0 s/m.

## 4 Simulation Results for Effective Protection Distance

Determination of effective protection distance in front of SPD is analyzed by simulations of the circuit given in Fig. 3 in MATLAB Simulink and ATP/EMTP. Simulations are performed for three types of EUT load: resistive, inductive and capacitive, and for cases with different lengths of cable SG – EUT and cable EUT –SPD. Additional analyses are performed for different values of active/reactive power of EUT. Distance of cable EUT – SPD for which maximal voltages across EUT are bellow insulation withstand level of EUT (1500 V) is taken as value of effective protection distance in front of SPD.

In order to decrease number of influencing parameters, analysis with different types and values of protective device (PD) are performed with aim to investigate its effect on protective characteristics regarding maximal voltages across EUT. Observed results have shown that PD has no influence on maximal voltages across EUT. Example is given in Fig. 4 where voltage waveforms across EUT (with active power of P = 1000 W) is given for different types and values of PD power: PD with active power of P = 400 W, PD with active power of P = 1000 W, PD with inductive power $Q_{IND}$ = 100 VAr and PD with capacitive power $Q_{CAP}$ = –100 VAr, when Combination Wave surge (Fig. 4a) and Ring Wave surge (Fig. 4b) is applied.

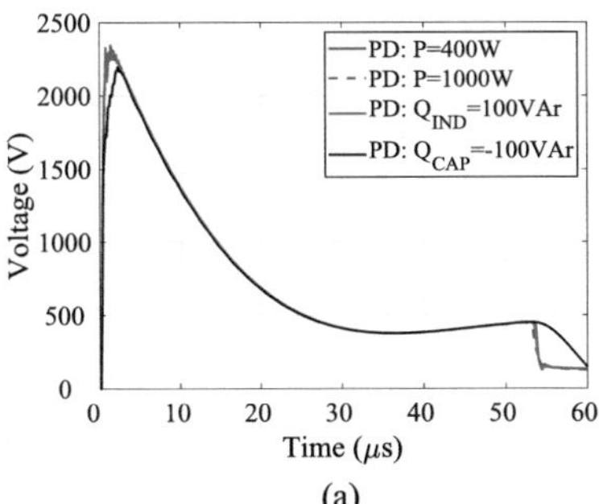


(a)

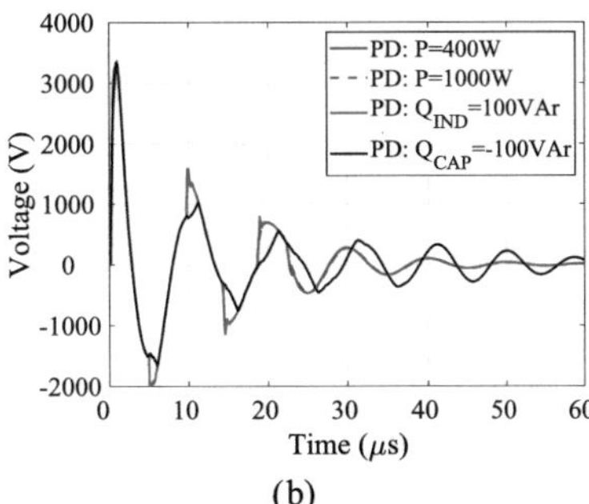


(b)

**Fig. 4.** Voltage waveform across EUT for different types and values of power of PD: (a) in case when Combination Wave surge is applied (b) in case when Ring Wave surge is applied

In following subchapters, simulation results for different types of EUT are presented.

### 4.1 Resistive Load

In case when EUT is purely resistive load, maximal voltages across EUT for different combinations of lengths of cable SG – EUT and cable EUT – SPD are given in Fig. 5 (in case when Combination Wave surge is applied) and in Fig. 6 (in case when Ring Wave surge is applied) for active power of EUT of P = 400 W. Cables lengths are taken in range of 0.5 m to 100 m. Figures 5b and 6b show zones in which maximal values of voltages across EUT are higher than value marked at zone boundaries.

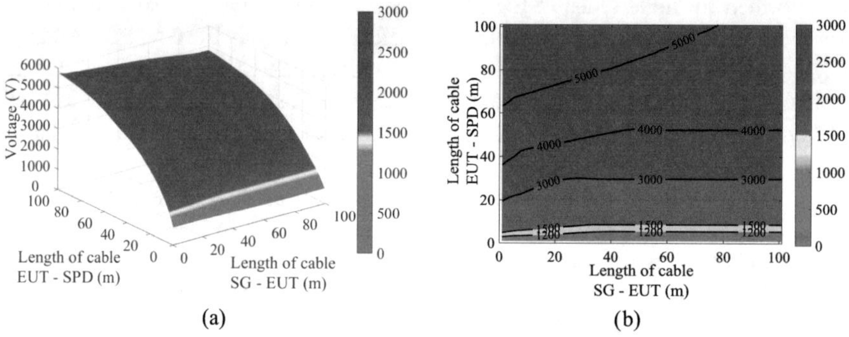


**Fig. 5.** Application of CWS: Maximal voltages across EUT with active power of P = 400 W for different combinations of lengths of cable SG – EUT and cable EUT – SPD in range of 0.5 m to 100 m: (a) 3D view (b) zone view

Obtained results show that maximal voltages across EUT are below of EUT's insulation withstand level only for narrow zone with relatively short cable EUT – SPD. Zoomed view of zones given in Figs. 5b and 6b are given in Figs. (7a and 7b. Numerical values of effective protection zones in front of SPD in case of resistive EUT with value of active power of P = 400 W are given in Table 1.

Obtained results show that effective protection distance in front of SPD depends on length of cable SG – EUT i.e. on the place of surge generation in relation to EUT. If it is assumed that surges arrive via supply power lines and distribution board, and if EUT is in relative vicinity of distribution board, observed SPD cannot provide protection effect for such EUT. Only SPDs installed relatively away from distribution board can provide certain effective protection distance in front of them.

In order to analyze effect of EUT's active power, maximal voltages across EUT for different combinations of lengths of cable SG – EUT and cable EUT – SPD are given in Fig. 8 (in case when Combination Wave surge is applied) and in Fig. 9 (in case when Ring Wave surge is applied) for active power of EUT of P = 1000 W.

By comparison of Figs. 5 and 8, as well as Figs. 6 and 9, it can be concluded that value of active power of EUT has a minor influence on effective protection distance in front of SPD. Effective protection distance is shorter in case of lower value of active power of EUT, and therefore, it should be used as more rigorous result. Example is given in Fig. 10 where voltage waveforms across EUT are given for EUT's active power of P = 400 W and P = 1000 W, for case of length of cable SG – EUT of 50 m,

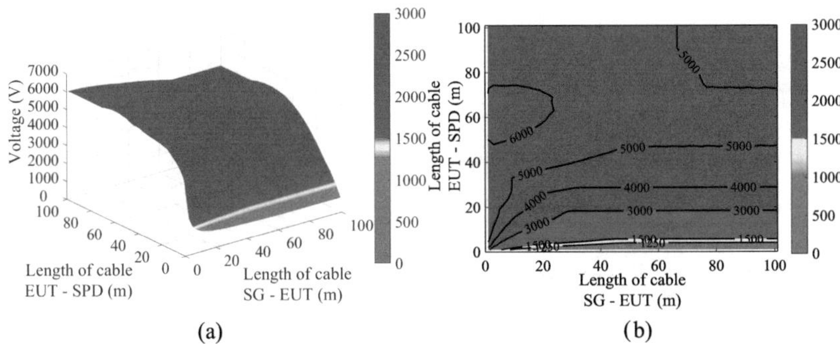


**Fig. 6.** Application of RWS: Maximal voltages across EUT with active power of P = 400 W for different combinations of lengths of cable SG – EUT and cable EUT – SPD in range of 0.5 m to 100 m: (a) 3D view (b) zone view

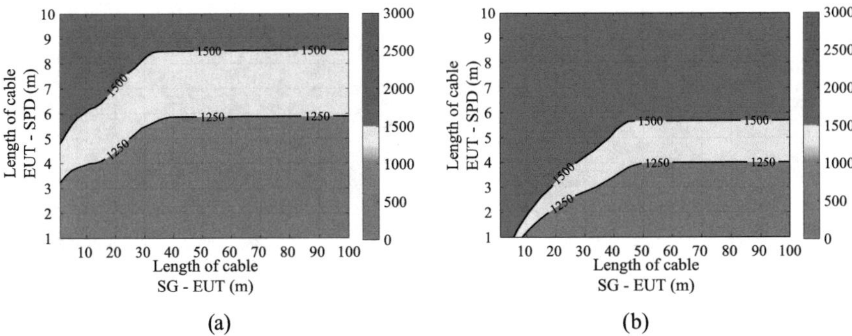


**Fig. 7.** Zoomed zone views for determination of effective protection distance for case of resistive EUT with P = 400 W: (a) application of CWS (b) application of RWS

**Table 1.** Effective protection distances in case of resistive EUT with P = 400 W

| Length of cable SG – EUT | Effective protection distance | |
|---|---|---|
| | CWS | RWS |
| 0–10 m | 5 m | 0 m |
| 10–20 m | 6 m | 2 m |
| 20–30 m | 7 m | 3 m |
| 30–40 m | 8 m | 4 m |
| 40–100 m | 8 m | 5.5 m |

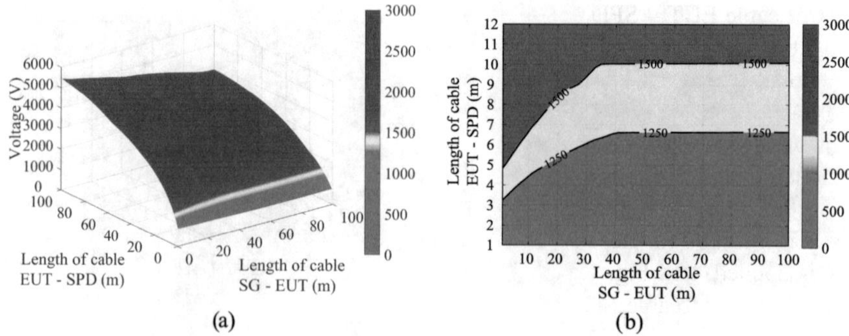


**Fig. 8.** Application of CWS: Maximal voltages across EUT with active power of P = 1000 W for different combinations of lengths of cable SG – EUT and cable EUT – SPD in range of 0.5 m to 100 m: (a) 3D view (b) zoomed zone view

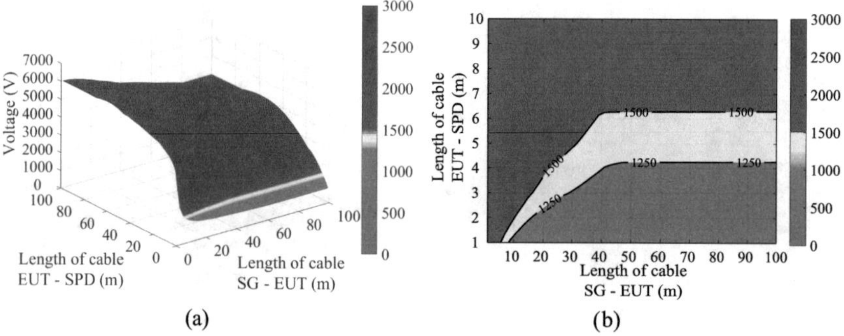


**Fig. 9.** Application of RWS: Maximal voltages across EUT with active power of P = 1000 W for different combinations of lengths of cable SG – EUT and cable EUT – SPD in range of 0.5 m to 100 m: (a) 3D view (b) zoomed zone view

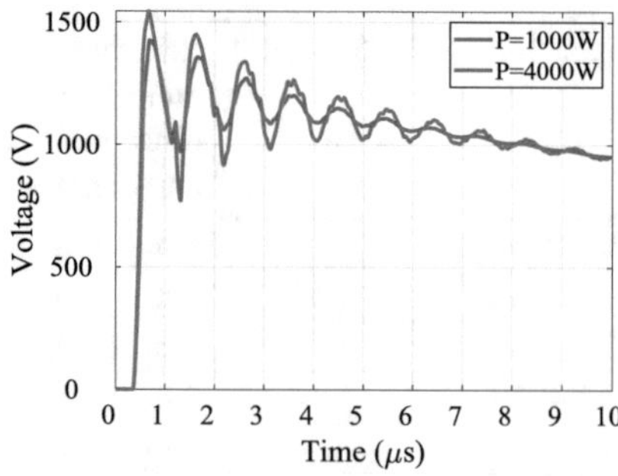


**Fig. 10.** Voltage waveforms across EUT for EUT's active power of P = 400 W and P = 1000 W, for case of length of cable SG – EUT of 50 m, length of cable EUT – SPD of 9 m, and application of CWS.

length of cable EUT – SPD of 9 m, and application of CWS. It is obvious that in case when EUT's active power is P = 1000 W, effective protection distance is longer than length of cable EUT – SPD (because maximal voltage across EUT is lower than 1500 V), which is not case when EUT's active power is P = 400 W.

In order to verify obtained results regarding influence values of active power, additional analyses are performed for different values of EUT's active power in wide range from 0.5 W to 2000 W, and for different values of cable EUT – SPD in range from 0.5 m to 100 m. Results are given in Fig. 11 (in case when Combination Wave surge is applied) and Fig. 12 (in case when Ring Wave surge is applied) for constant value of length of cable SG – EUT of 20 m.

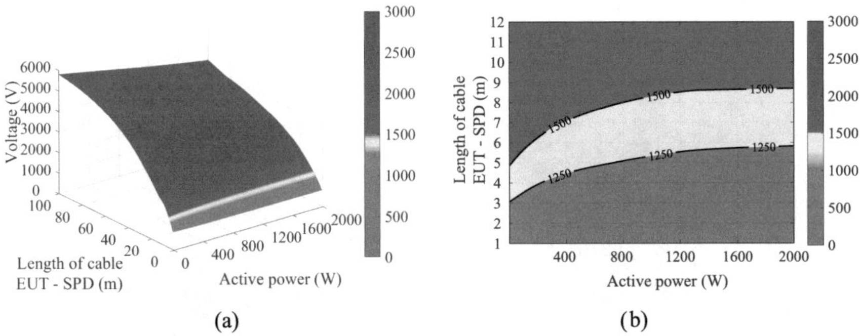


**Fig. 11.** Application of CWS: Maximal voltages across EUT for different combinations of EUT's active power in range from 0.5 W to 2000 W and length of cable EUT – SPD in range of 0.5 m to 100 m: (a) 3D view (b) zoomed zone view

Obtained results for different values of EUT's active power show that for constant value of cable SG – EUT length, effective protection distance in front of SPD is shorter for lower values of EUT's active power.

## 4.2 Inductive Load

In case when EUT is purely inductive load, maximal voltages across EUT for different combinations of lengths of cable SG – EUT and cable EUT – SPD are given in Fig. 13 (in case when Combination Wave surge is applied) and in Fig. 14 (in case when Ring Wave surge is applied) for reactive inductive power of EUT of $Q_{IND}$ = 10 VAr. Cables lengths are taken in range of 0.5 m to 100 m. Figures 13b and 14b show zones in which maximal values of voltages across EUT are higher than value marked at zone boundaries.

Numerical values for effective protection zones in case of inductive EUT with $Q_{IND}$ = 10 VAr are given in Table 2.

Additional analyses are performed for different values of EUT's reactive inductive power in wide range from 0.5 VAr to 100 VAr, and for different values of cable EUT – SPD in range from 0.5 m to 100 m. Results are given in Fig. 15 (in case when

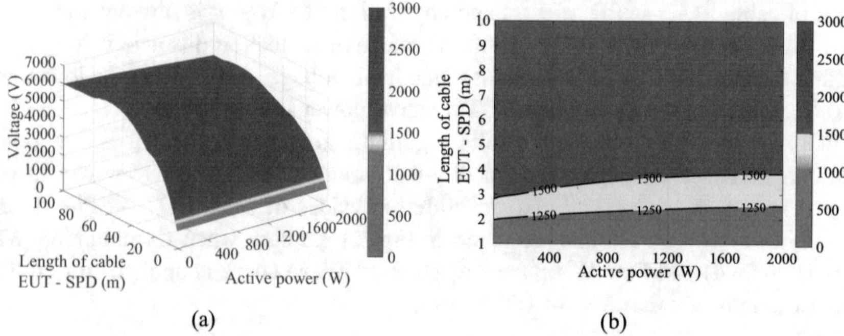


(a) (b)

**Fig. 12.** Application of RWS: Maximal voltages across EUT for different combinations of EUT's active power in range from 0.5 W to 2000 W and length of cable EUT – SPD in range of 0.5 m to 100 m: (a) 3D view (b) zoomed zone view

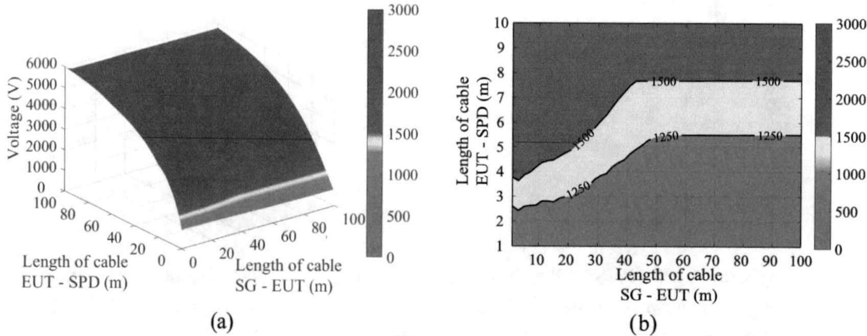


(a) (b)

**Fig. 13.** Application of CWS: Maximal voltages across EUT with reactive inductive power of $Q_{IND}$ = 10 VAr for different combinations of lengths of cable SG – EUT and cable EUT – SPD in range of 0.5 m to 100 m: (a) 3D view (b) zoomed zone view

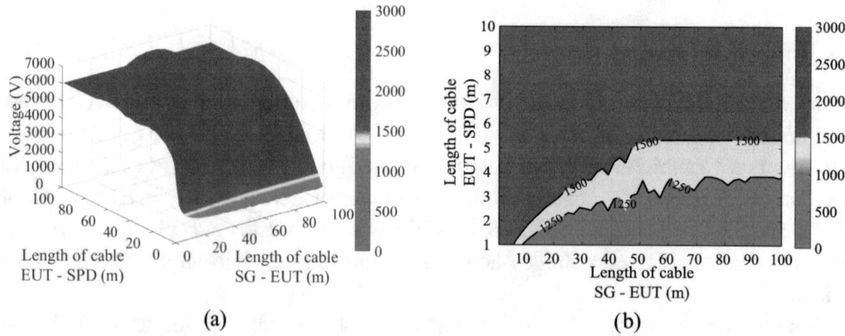


(a) (b)

**Fig. 14.** Application of RWS: Maximal voltages across EUT with reactive power of $Q_{IND}$ = 10 VAr for different combinations of lengths of cable SG – EUT and cable EUT – SPD in range of 0.5 m to 100 m: (a) 3D view (b) zoomed zone view

**Table 2.** Effective protection distances in case of inductive EUT with $Q_{IND}$ = 10 VAr

| Length of cable SG – EUT | Effective protection distance | |
|---|---|---|
| | CWS | RWS |
| 0–10 m | 3.5 m | 0 m |
| 10–20 m | 4 m | 2 m |
| 20–30 m | 5 m | 3 m |
| 30–40 m | 6 m | 3.5 m |
| 40–100 m | 7.5 m | 4–5 m |

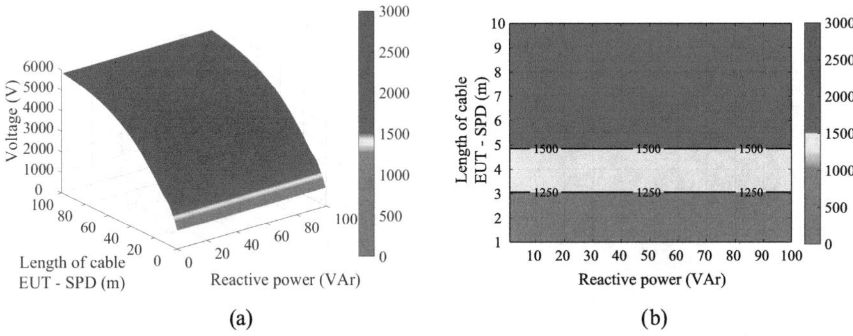


**Fig. 15.** Application of CWS: Maximal voltages across EUT for different combinations of EUT's reactive power in range from 0.5 VAr to 100 VAr and length of cable EUT – SPD in range of 0.5 m to 100 m: (a) 3D view (b) zoomed zone view

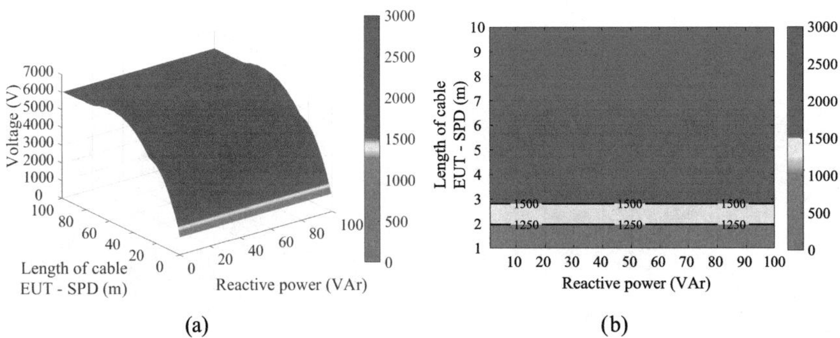


**Fig. 16.** Application of RWS: Maximal voltages across EUT for different combinations of EUT's reactive power in range from 0.5 VAr to 100 VAr and length of cable EUT – SPD in range of 0.5 m to 100 m: (a) 3D view (b) zoomed zone view

Combination Wave surge is applied) and Fig. 16 (in case when Ring Wave surge is applied) for constant value of length of cable SG – EUT of 20 m.

Obtained results show that effective protection distance in front of SPD in case of inductive EUT doesn't depend on value of EUT's inductive power.

## 4.3 Capacitive Load

In case when EUT is purely capacitive load, maximal voltages across EUT for different combinations of lengths of cable SG – EUT and cable EUT – SPD are given in Fig. 17 (in case when Combination Wave surge is applied) and in Fig. 18 (in case when Ring Wave surge is applied) for reactive capacitive power of EUT of $Q_{CAP} = -10$ VAr. Cables lengths are taken in range of 0.5 m to 100 m. Figures 17b and 18b show zones in which maximal values of voltages across EUT are higher than value marked at zone boundaries.

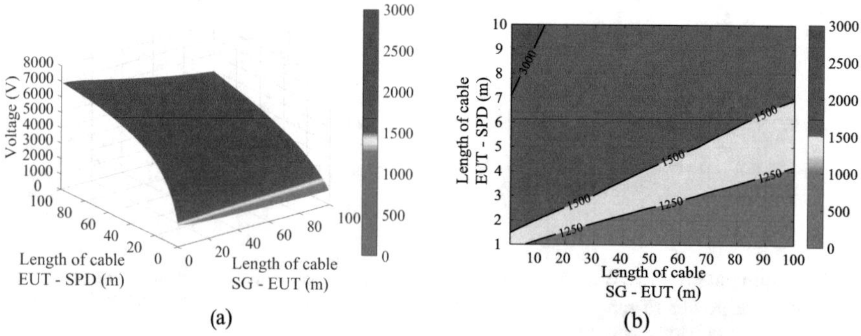


**Fig. 17.** Application of CWS: Maximal voltages across EUT with reactive capacitive power of $Q_{CAP} = -10$ VAr for different combinations of lengths of cable SG – EUT and cable EUT – SPD in range of 0.5 m to 100 m: (a) 3D view (b) zoomed zone view

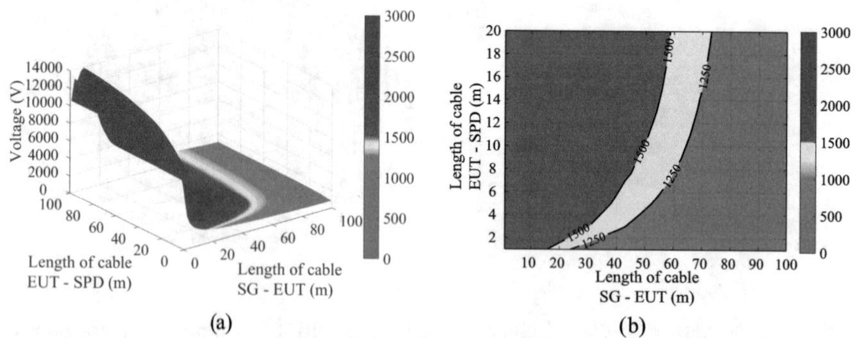


**Fig. 18.** Application of RWS: Maximal voltages across EUT with reactive capacitive power of $Q_{CAP} = -10$ VAr for different combinations of lengths of cable SG – EUT and cable EUT – SPD in range of 0.5 m to 100 m: (a) 3D view (b) zoomed zone view

Numerical values for effective protection zones in case of capacitive EUT with $Q_{CAP} = -10$ VAr are given in Table 3.

**Table 3.** Effective protection distances in case of capacitive EUT with $Q_{CAP} = -10$ VAr

| Length of cable SG – EUT | Effective protection distance | |
|---|---|---|
| | CWS | RWS |
| 0–10 m | 0.5 m | 0 m |
| 10–20 m | 2 m | 0 m |
| 20–30 m | 2.5 m | 2 m |
| 30–40 m | 3 m | 3 m |
| 40–100 m | 3.5–6 m | 6 m – 100 m |

Additional analyses are performed for different values of EUT's reactive capacitive power in wide range from –0.5 VAr to –100 VAr, and for different values of cable EUT – SPD in range from 0.5 m to 100 m. Results are given in Fig. 19 (in case when Combination Wave surge is applied) and Fig. 20 (in case when Ring Wave surge is applied) for constant value of length of cable SG – EUT of 20 m.

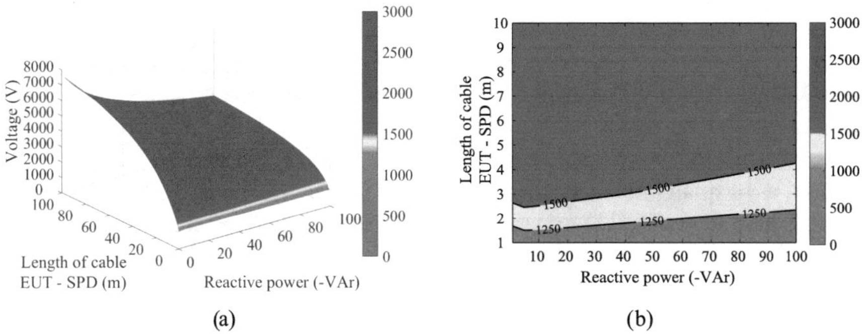


**Fig. 19.** Application of CWS: Maximal voltages across EUT for different combinations of EUT's reactive capacitive power in range from –0.5 VAr to –100 VAr and length of cable EUT – SPD in range of 0.5 m to 100 m: (a) 3D view (b) zoomed zone view

Obtained results show that in case of capacitive load effective protection distance in front of SPD is smaller than in case of EUT's resistive and inductive load in case of relatively short cable SG – EUT. However, for longer cable SG – EUT maximal voltages across EUT are below of its rated withstand insulation level for every length of cable EUT - SPD, i.e. effective protection distance in front of SPD is higher than 100 m. In addition, value of reactive capacitive power of EUT has significant influence on effective protection distance in front of SPD.

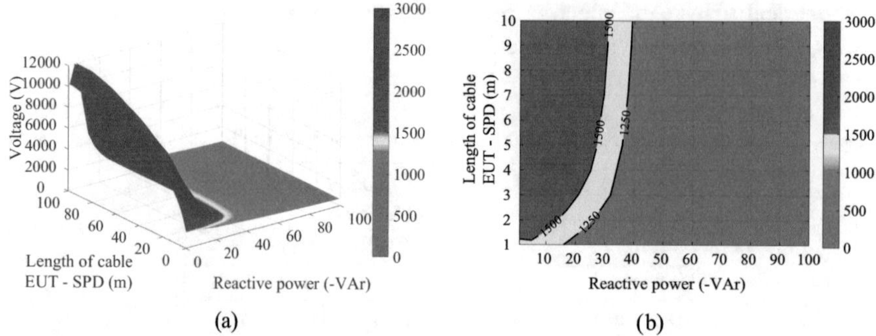


**Fig. 20.** Application of RWS: Maximal voltages across EUT for different combinations of EUT's reactive capacitive power in range from –0.5 VAr to –100 VAr and length of cable EUT – SPD in range of 0.5 m to 100 m: (a) 3D view (b) zoomed zone view

## 5 Discussion of the Obtained Results

Obtained results show that for applied standard representative surge waveforms (Combination Wave surge and Ring Wave surge) effective protection distance in front of selected SPD is finite and much shorter than effective protection distance after location of SPD.

By summarization of the presented results for different types of EUT's load, it can be concluded that effective protection distance in front of SPD depend on cable length between SG and EUT. For short cables SG – EUT effective protection distance in front of SPD is shorter (in some cases equals to zero) than in cases of longer cables SG – EUT.

Also, values of active/reactive power of EUT haven't influence (for inductive load of EUT), have small influence (for resistive load of EUT) or have significant influence (for capacitive load of EUT) on value of effective protection distance in front of SPD.

The main influencing parameters on obtained results regarding effective protection distance in front of SPD are amplitudes of surge voltage and current waveforms (which depend on location category of observed system) as well as value of SPD's protection voltage.

Taking into account quantitative values obtained for effective protection distance in front of SPD, it can be concluded that SPD can provide protection effect for very narrow zone in front of it or there is no protection effect at all. In many cases with standard design of low-voltage power installations in residential and administrative buildings, this effective protection distance may be insufficient. However, analysis for each case need to be conducted in order to confirm and to prove protection performances.

## 6 Conclusion

Protection of devices in low-voltage power systems against surges is provided with application of surge protective devices (SPDs). The SPD will offer the optimized protection level only if it is installed directly across the terminals of the equipment to be protected. Unfortunately, in practice, it is not always possible to locate SPD close to the equipment, and an inevitable separation distance along connecting cable exists [9]. Maximum distance along cable from point of SPD connection for which desirable protective characteristics are still fulfilled represents effective protection distance of SPD. Effective protection distance is usually determined for cable length after location of SPDs toward protected equipment. However, it is known that SPD provides protective effects also on devices installed at certain distance before location of SPDs.

Determination of effective protection distance in front of SPD installed in low-voltage power system has been given in the paper. Analyses are performed with extensive number of simulations in MATLAB and ATP/EMTP with different influencing parameters in wide ranges of values, and with application of standard surge testing waveforms.

Obtained results have shown that effective protection distance in front of SPD strongly depends on location of surge occurrence in low-voltage power system (i.e. length of cable between surge generator and equipment), amplitudes of surge voltage and current waveforms, as well as value of SPD's protection voltage. There are small or none dependence on type of load or value of load's active or reactive power. The quantitative value of effective protection distance may be insufficient for reliable expectation that SPD can provides adequate protection for devices connected in front of it.

## References


1. Radulovic, V., Durkovic, V.: Surge protection of resistive loads in low voltage power installations. In: 23rd International Scientific-Professional Conference on Information Technology (IT), Zabljak, Montenegro (2015)
2. Radulovic, V., Mujovic, S., Miljanic, Z.: Characteristics of overvoltage protection with cascade application of surge protective devices in low-voltage AC power circuits. Adv. Electr. Comput. Eng. **15**(3), 153–160 (2015)
3. He, J., Yuan, Y., Xu, J., Chen, S., Zou, J., Zeng, R.: Evaluation of the effective protection distance of low-voltage SPD to equipment. IEEE Trans. Power Del. **20**(1), 123–130 (2005)
4. Radulovic, V., Skuletic, S.: Influence of Combination Wave Generator's current undershoot on overvoltage protective characteristics. IEEE Trans. Power Del. **26**(1), 152–160 (2011)
5. Ziyu, H., Du, Y.: Influence of different impulse waveforms on coordination of two cascaded SPDs. In: 32nd International Conference on Lightning Protection, Shanghai (2014)
6. Lai, J.-S., Martzloff, F.: Coordinating cascaded surge protection devices: high-low versus low-high. IEEE Trans. Ind. Appl. **29**(4), 680–687 (1993)
7. She, C., Lei, S.: Analysis of two-stage cascade SPD coordination under the impact of lightning combination wave in 220 V low-voltage distribution system. In: 7th Asia-Pacific International Conference on Lightning, APL 2011, Chengdu (2011)

8. Skuletic, S., Radulovic, V.: Analysis of surge protection performance in low-voltage AC systems with capacitive load. In: 45th International Universities Power Engineering Conference, UPEC 2010, Cardiff, Wales, UK (2010)
9. Skuletic, S., Radulovic, V.: Optimization and protective distance of surge protective devices in low-voltage AC circuits. In: 6th International Universities' Power Engineering Conference, UPEC 2011, Soest, Germany (2011)
10. Hotchkiss, R.W.: Performance testing of surge protective devices for low-voltage AC power circuits: approach of the IEEE. In: IEEE Power and Energy Society General Meeting, Detroit, MI, USA (2011)
11. Shunchao, W., Jinliang, H.: Discussion on worst distance between SPD and protected device. IEEE Trans. Electromagn. Compat. **53**(4), 1081–1083 (2011)
12. Skuletic, S., Radulovic, V.: Effective protection distance from cascade coordinated surge protective devices to equipment in low–voltage AC power circuits. In: 43rd International Universities Power Engineering Conference, UPEC 2008, Padova, Italy (2008)
13. Low-voltage surge protective devices - Part 12: Surge protective devices connected to low-voltage power distribution systems - Selection and application principles, IEC 61643-12 standard (2008)
14. He, J.: Discussions on factors influencing the effective protection distance of SPD to loads. In: Asia-Pacific Symposium on Electromagnetic Compatibility (APEMC), Taipei (2015)
15. IEEE Guide on the Surge Environment in Low-Voltage (1000 V and Less) AC Power Circuits, IEEE C62.41.1-2002 Standard, April 2003
16. IEEE Recommended Practice on Characterization of Surges in Low-Voltage (1000 V and Less) AC Power Circuits, IEEE C62.41.2-2002 Standard, April 2003
17. IEEE Recommended Practice on Surge Testing for Equipment Connected to Low-Voltage (1000 V and Less) AC Power Circuits, IEEE C62.45-2002 Standard, April 2003
18. Insulation coordination for equipment within low-voltage systems –Part 1: Principles, requirements and tests, IEC Std. 60664-1 Standard (2002)
19. Low-voltage surge protective devices - Part 11: Surge protective devices connected to low-voltage power systems - Requirements and test methods, IEC 61643-11 Standard, March 2011
20. Protection againts lightning - Part 4: Electrical and electronic systems within structures, IEC Std. 62305-4 Standard (2006)

# Hybrid Power System Concepts for Two Different Consumption Categories – Industrial Consumers and Small Village

Ammar Arpadžić[1(✉)], Ermin Šunj[1], Ajla Merzić[2], Adnan Bosović[2], and Mustafa Musić[2]

[1] International Burch University, Sarajevo, Bosnia and Herzegovina
ammarbosna@gmail.com

[2] Department of Strategic Development, Public Enterprise Elektroprivreda of Bosnia and Herzegovina, Sarajevo, Bosnia and Herzegovina

**Abstract.** Hybrid power systems (HPS) allow remote regions that are far from grid to have power supply. The grid infrastructure is expensive for these regions and HPS is the ideal solution. HPS provides reliable and sustainable power supply for industry consumers. Some of the benefits from HPS are the reduction in energy losses, economic losses and emission of air pollutants as well. In this paper hybrid power system configuration and integration for two different consumer types has been performed, one located in Gračanica and the other in Radoč near Bužim, both in Bosnia and Herzegovina. The analysed load from Gračanica consisted of four industry consumers' power consumption throughout the year of 2018, while the type of the consumers analysed in Radoč is residential. The analysed consumers in Gračanica are supplied by 10 kV medium voltage (MV) distribution network, and the village Radoč is supplied by 0,4 kV low voltage (LV) distribution network. The modelling of the sustainable hybrid systems was done in HOMER software. The real data about the load was received from Public Enterprise Elektroprivreda of Bosnia and Herzegovina, while the Solar Global Horizontal Irradiance (GHI) resource, wind and temperature resource data were taken straight from the HOMER. Hybrid systems were designed with HOMER Optimizer™ option. One of the parameters for hybrid systems modelling was to contain solar panels, wind generator and diesel generator. One of the main challenges of this paper was the wind resource for Gračanica and Radoč II as well. Both locations do not have that great wind potential and our goal was to model hybrid systems with as much as possible renewable energy sources. The goal of this paper was to create the sustainable hybrid systems for two different type of consumers. The analysis conducted in this paper showed that both hybrid power systems are energy-efficient and cost-effective.

## 1 Introduction

Over the last few decades, we have witnessed the rapid development and progress of human civilization in every respect to the life around us. By discovering the resources that nature offers and using various resources, humanity through history seeks to ease


S. Avdaković et al. (Eds.): IAT 2019, LNNS 83, pp. 91–106, 2020.
https://doi.org/10.1007/978-3-030-24986-1_7

life for itself and others. Today, one cannot imagine life without some form of energy as it is electric. There are still many places in the world that do not have access to electricity. The reason for such a situation is usually the distance between certain areas of civilization and the poverty of individual countries. Industrial plants that are remote from the network also need independent systems.

In 2014, 1.06 billion, which is around 15 percent of global population, people still lived without access to electricity and about 3.04 billion still relied on solid fuels and kerosene for cooking and heating [1].

There will be something to change, energy is being spent more and more, and fossil fuels, apart from polluting the environment, are consumed. Therefore, renewable sources have a great future, so more and more countries are investing in the research and implementation of these sources. A lot of countries have already begun to encourage energy saving programs and switch to renewable energy sources.

Remote regions that are far from big consumer centers in the future will not be able to quickly develop grids, and it is questionable whether traditional grid expansion is profitable compared to nowadays available supply options. Electrification of rural areas is a big problem. The solution to this problem focused on the decentralization of the system in the rural areas. Great hope is in renewable energy sources, wind, sun, biomass and water, which are good for decentralized power generation. The problem with the exploitation of renewable energy sources is their intermittent nature, days will not be always sunny, and wind will not be always conducive for wind generator, because of that diesel generator is one of solution which will produce energy in unsuitable days for producing energy by renewable energy sources. System which is decentralized and formed of renewable energy sources with diesel generator is one part of hybrid system. The battery incorporation enhances the performance of the HPS and the consumers, both residential and industrial, will have power supply when weather conditions are not good enough for the electricity generation by RES.

## 2 Literature Review

Kalappan and Ponunusamy [2] did a project for Pichanur Village in Coimbatore, India. They utilized HOMER for hybrid system optimization. Their hybrid system consisted of photovoltaic (PV) system, wind generator (WG) and diesel generator (DG). The number of households in Pichanur Village is 33, which means that the village demand for electrical energy is not big. The daily consumption of electrical energy is 5 kWh. The average annual irradiation for the analyzed village is 4.87 kWh/m2/Day, while the annual average wind speed is 3.05 m/s. Their hybrid system had 1 kW DC PV, 2.7 kW DC WT, 1 kW AC DG, battery it500i 1505 Ah and converter with 90% of efficiency. The main objective of their paper was modelling a hybrid system in order to reduce the air pollutions since India imports around 80% of oils for the production of electrical energy.

Authors in [3] modelled the hybrid system with 80 kW PV20, 200 kW WG, 300 kW batteries and 30 kW invertor. The investment in hybrid system was $709,471. The lifetime of the project is considered to be 20 years. The annual average irradiance is 0.2 kW/m$^2$, while the average annual wind speed of this region is 3 m/s. The daily consumption of electrical energy for the analyzed location is 214 kWh and the peak is 30 kW. They used HOMER for hybrid system optimization for two cases with different capacity shortages. The capacity shortage in the first case was 0% and 5% in the second case. In both cases the representativeness of electrical energy produced by solar panels is significant. The percentage of capacity shortage has an impact on battery size which resulted with the hybrid system consisted of 90 kW PV, 100 kW WG and a battery whose capacity was lower for more than 100 kWh. In this paper they wanted to compare hybrid systems for different percentage of capacity shortages.

Elhassan et al. [4] designed a hybrid power system for Khartoum, Sudan. Sudan has an excellent potential for solar panels and the average wind speed is convenient for implementation of wind generators. Authors performed several analysis i.e. – analysis for one household, 20, 30, 40 and 50 households. Their system contained PV, WG, inverter and the battery with the controller. In the case with only one household they had 0.15 kW PV and two 360 Ah batteries. Their hybrid system for one household did not include WG because of the small energy consumption. Hybrid system for 50 households consisted of 3.5 kW PV, two 3 kW WG and eight 360 Ah batteries. They showed in their paper that it is not cost-effective to build WGs when energy consumption is small and that the cost of projects will decrease if the price of WGs in Khartoum decrease as well.

The objective of paper [5] is to show the optimal solution for energy crisis using HOMER. The project was done for Vathar, Kolhapur. The considered simulated hybrid renewable energy system consists of wind turbine, PV array with power converter, battery and Diesel generator. The energy consumption is 58 kWh per day and the peak is 5.5 kW. HOMER Optimizer made a hybrid system with 15 kW PV, 3 kW WT, 2.6 kW DG and 6 kW converter.

Maherchandani, Agarwal and Sahi [6] modelled a hybrid system with biomass generator, diesel generator, PV and WG for remote villages in Udaipur (Rajasthan). The electrical load in this area is mainly due to the household appliances and small industrial applications. The daily energy consumption of these consumers is 5826 MWh and the peak is 286 kW. They used 250 kW diesel generator ($25,700), 3 kW wind turbine ($5,950), 250 kW biomass generator ($68,750), 150 kW PV module ($225,000), batteries ($782) and 200 kW converter ($142,800). Total net present cost of this hybrid system was $1,017,651. The biomass generator was the main producer of electrical energy. It was selected to be the main producer due to its cost-effectiveness. Authors showed that this hybrid system would lower bills for electrical energy for 58.8%.

The case study of the paper [7] was the city of Brest in France. The city load has been scaled to 16 MWh/day. Seasons scale peak load is taken as 2 MW. PV modules should power the city and if excess electricity occurs it will be used for hydrogen electrolysis which will be stored in a tank. The hybrid system contained only PV (4.2 MW) and fuel cell (2 MW) system without batteries, converter 2 MW, electrolyzer 3.4 MW and Hydrogen tank 955000 kg. The investment in the project was $4,197,750, while the total net present cost was $8,942,636. The proposed optimal design study was focused on economic performance and was mainly based on the loss of the power supply probability concept.

Magarappanavar and Koti [8] designed a hybrid system with PV, WG and DG for BEC Campus. The energy consumption is 1516 kWh per day and the peak is 186 kW. Authors did not use HOMER optimization, but they offered several nominal values for components and it resulted in the hybrid system with 25 kW DG, 200 kW converter, 3000 Ah batteries, 280 kW PV and 30 kW WG. They showed the hybrid system can be designed for the specific location and that it is cost-effective.

Huang et al. [9] designed a hybrid system for Catalina Island in California. The energy consumption is 39 MWh, while the peak is 5.3 MW. The objective of this paper was to replace DG with renewable energy sources. Their hybrid system contained 5 MW PV, 15 MW WG and 15.20 MWh battery.

Authors in [10] modelled the optimal HPS using HOMER for Biret, Mauritania. Their HPS consisted of 61.44 kW solar field and 40 kW generating set with 2950 Ah storage system. The analysed load was of residential type with 58 kW peak and energy consumption 348 kWh/d. Optimized HPS showed to be cost-effective due to the Biret's Sun potential.

Study in [11] investigates HPS located in rural communities of Sokoto in Nigeria which consists of PV and WG. Authors used HOMER software for optimization of HPS components. Solar irradiation for analysed location is 6 kWh/m2/day and wind speed is 5 m/s at 10 m above height. HPS optimization resulted in a system with 35.21 kW PV, 3 x 25 kW WG, 12 x 24 V lead acid battery and 17.44 kW converter. The total capital cost for the system is $249,910.24, while its payback period is 5 years.

There is a lot of people who live in remote regions which do not have power supply. The objective of [12] is the efficiency of HPSs. The project was done for Sandip-para in Raujan upzila of Chittagong district. The HPS optimization was performed with HOMER software. The residential type of load had peak of 6.2 kW and daily consumption 31 kWh. The study showed that HPS is a better solution than implementation and construction of grid infrastructure.

## 3 Research Focus and Objective

The project refers to the problem related to the optimal hybrid energy system and to the investment cost of the system itself. The profitability of an application is considered of the wind turbine and photovoltaic system in combination with the diesel generator. The specificity of this paper, compared to papers mentioned in Chapter 2, is that it analyzes the optimization of HPSs for two different types of loads (residential and industrial) at two different locations. The difference in total cost and capacity of the proposed HPSs is due to the loads. The analyzed residential load does not need a huge power system because it is consisted of only 21 houses, while the analyzed industrial load needs since it is consisted of four large consumers.

Authors in [2] modelled HPS for a village in India in order to decrease emissions. Their objective is one of the objectives of this paper for Radoč village. Paper [4] proves that it is not cost-effective to incorporate WG in HPS when the wind potential is not great. Since wind potential for Radoč village is not that great, it would be better to install more solar panels instead of WG. The focus of [5] is the optimal solution for energy crisis using HOMER. This paper does not cover that topic, but it could be incorporated in the future work because industrial customers need constant power supply. The consumers of [6] are households and small industrial application. The analysis showed that the implementation of hybrid power systems lower electricity bills around 60%. The paper is similar to this paper, with the difference in the size of industrial consumers.

The main difference between residential and industrial customers is in the load curve. Residential load curve changes throughout the day. Its peak is between 16:00 and 18:00 when people start coming home from job. Industrial load curve is almost constant from 08:00 to 16:00 due to the work of the machines. Distribution system operator (DSO) in many countries has to pay a penalty to industry consumers if shortage of power supply occurs. Due to these two reasons HPS is the optimal solution for industrial consumers to assure constant power supply.

## 4 Analysis of a Real Distribution Networks at Location of Industrial Load Gračanica and Residential Network Radoč 2

### 4.1 Geographical Location of Gračanica Industrial Load

The first case-study of this project is placed in Gračanica, Federation of Bosnia and Herzegovina. It is located on 44°41.5'N, 18°17.8'E. The analysed network is a part of real medium-voltage (MV) distribution network. It is consisted of four industrial consumers (encircled on Fig. 1). The first consumer is the flooring company. The second is dealing with the final woodworking processes. The third consumer is the producer of the flexible packaging, while the fourth consumer is the producer of fiscal systems. A georeferenced scheme is presented in Fig. 1.

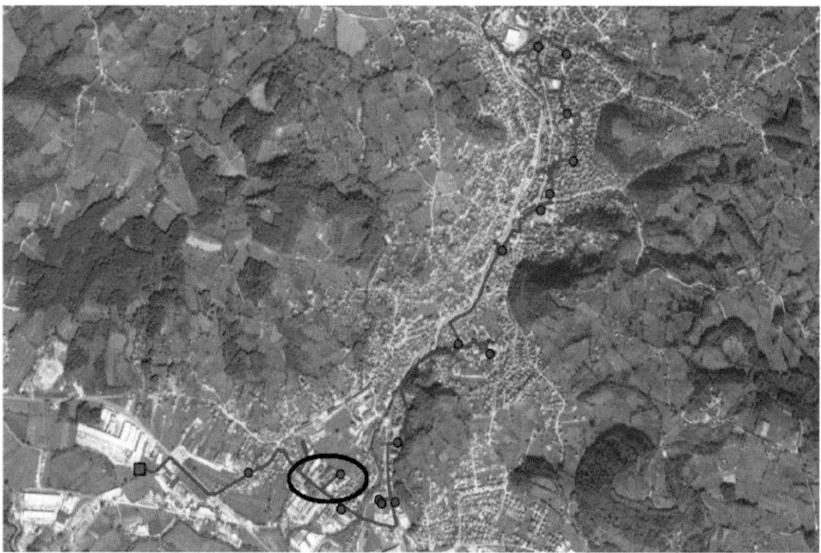

**Fig. 1.** Georeferenced scheme of the analyzed network in Gračanica

### 4.2 Geographical Location of Radoč 2 Residential Load

Second case-study of this project, low voltage residential network Radoč 2 is in the northwestern part of BiH, it is located on 45°04’08.9”N 16°06’10.8”E. Radoč consists of three villages, namely Šip, Mulalići and Šekići. The analysed network is low-voltage (LV) network. In these three places there are 21 consumers, most of them are residential. The total length of the network is about 3 km (Fig. 2).

**Fig. 2.** Georeferenced scheme of the analyzed network in Radoč 2

### 4.3 Consumption Profiles of Gračanica Industrial Load

The load consumption data of the analysed part of the grid was received from Public Enterprise Elektroprivreda of Bosnia and Herzegovina. The data contained 15-min interval readings (35040 readings) for the year 2018 of four consumers. All four consumers were of industrial type. Before inserting the data into the HOMER software, all readings were summed in order to get the complete consumption of the analysed part of the grid. Figure 3 illustrates the daily profile of power consumption. It can be seen that these firms start working between 07:00 and 08:00, but it is also visible that at least one firm works all 24 h. They finish with work between 16:00 and 17:00. The peak of the daily power consumption is 703.99 kW. The average energy consumption per day is 6261.59 kWh.

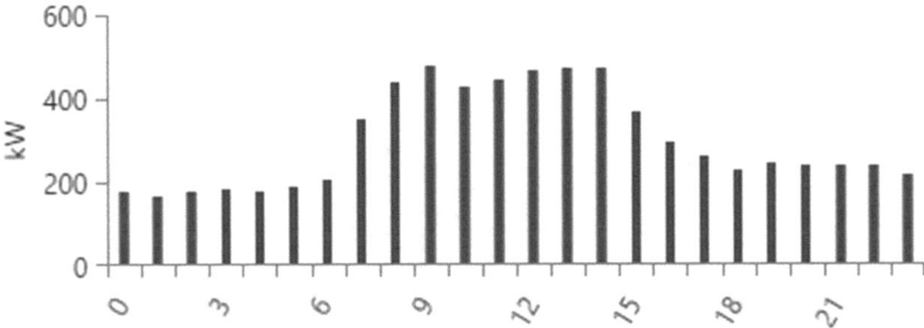


**Fig. 3.** Daily load profile of power consumption of analysed industrial customers in Gračanica

Figure 4 presents the seasonal profile of power consumption i.e.- throughout the year. The consumption is quite larger from January to March and from October to December.

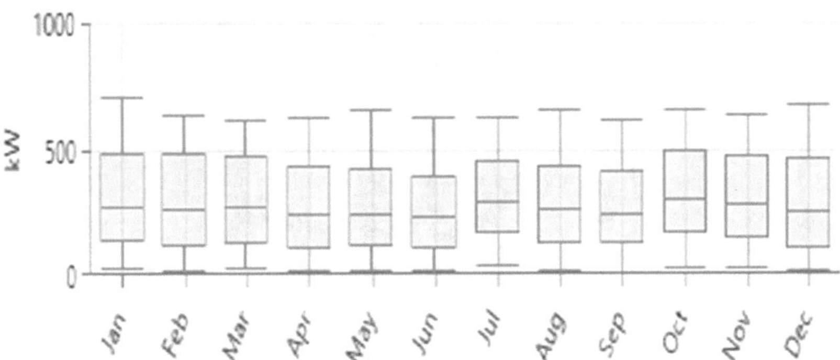


**Fig. 4.** The seasonal load profile of power consumption of analysed industrial customers in Gračanica

## 4.4 Consumption Profiles of Radoč 2 Residential Load

Daily consumption is small, because of residential consumers. HOMER simulates the operation of the system by calculating energy balance for each of 8760 h per year. Daily consumption is 133.62 kWh/day, and the annual maximum peak load is about 19.47 kW according to Fig. 5, which shows the daily load profile.

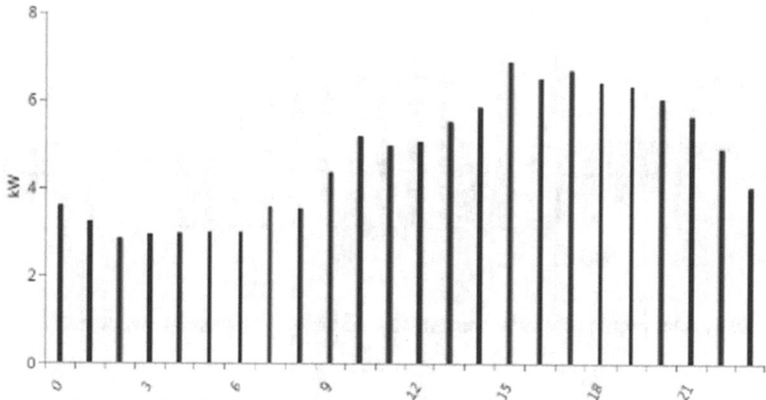


**Fig. 5.** Daily load profile of Radoč 2 power consumption

Figure 6 presents the seasonal profile of power consumption. The consumption is quite larger from June to September.

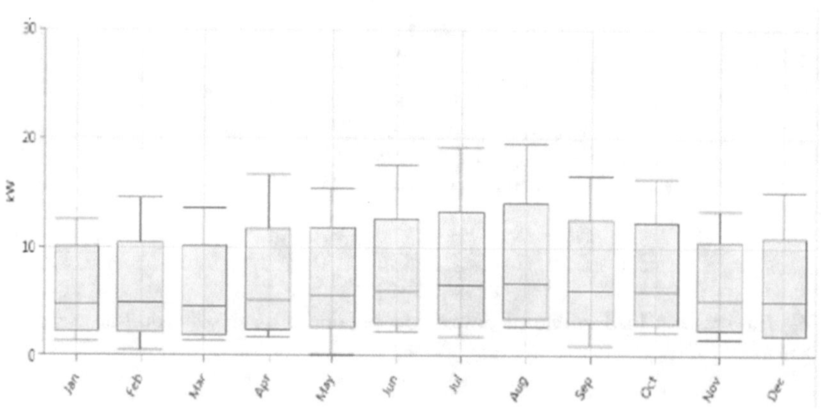

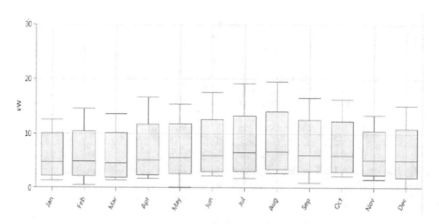

**Fig. 6.** Seasonal load profile of Radoč 2 power consumption

## 5 Methodology

Hybrid systems for both analysed locations have the same structure i.e. – they are consisted of PV, WG, DG, converters and batteries. Batteries are used to power the system in case of emergency or to store the excess of generated electrical energy. Diesel generator is used when renewable energy sources cannot produce electrical energy due to weather conditions. In both hybrid systems batteries are the most expensive components, especially in the case of Gračanica since the consumption is quite big due to four industrial consumers. The data about solar irradiation, wind speed and temperature are taken downloaded from HOMER. Besides the parameter that hybrid system has to contain five mentioned components, the other two parameters are that the minimal penetration of electrical energy produced by renewable energy sources is 50% and that the maximum capacity shortage is 1%. HOMER is used for the determination of optimal size of components and the operative strategy for the hybrid system of RES which is based on simulation, optimization and sensitivity analysis.

## 5.1 Natural Resources of Gračanica

Figures 7 and 8 show average solar irradiation per month and annual average wind speeds for Gračanica.

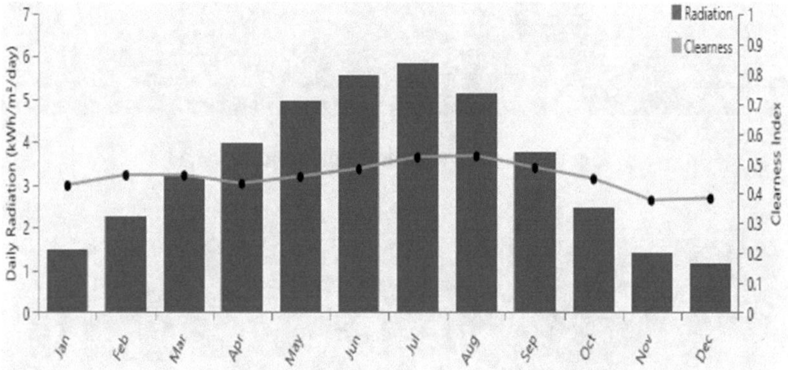


**Fig. 7.** Average solar irradiation for Gračanica

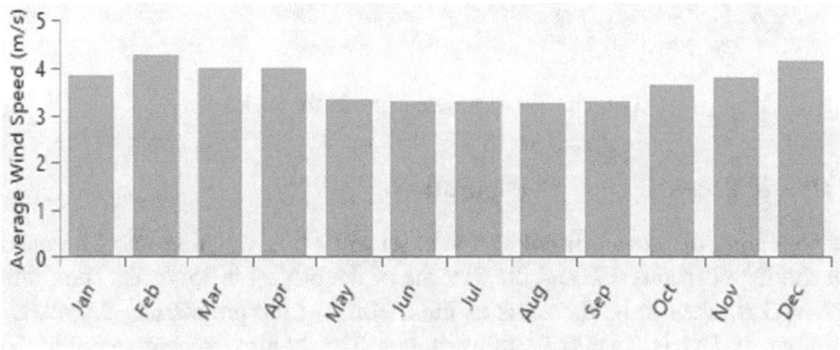


**Fig. 8.** Average wind speeds for Gračanica

## 5.2 Natural Resources of Radoč 2

Figures 9 and 10 presents average solar irradiation per month and annual average wind speeds for the second analysed location – Radoč 2.

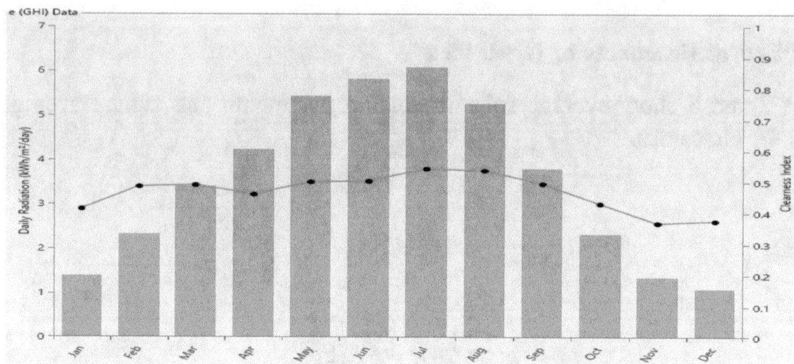


**Fig. 9.** Average solar irradiation for Radoč 2

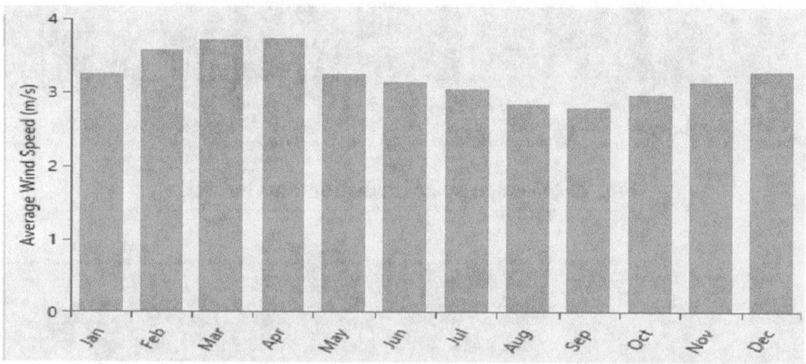


**Fig. 10.** Average wind speeds for Radoč 2

## 5.3 Hybrid Power System Configurations

At the beginning of project discount rate is set to be 6%, inflation rate 2%, maximum annual shortage capacity 1% and the lifetime of the project is 25 years. The lifetime of PV and WG is taken to be the same as the lifetime of the project i.e.- 25 years, while the lifetime of DG is 15,000 working hours. The lifetime of converter in HPS in Gračanica is set to be 12 years, while the lifetime of the converter in Radoč 2 HPS is 15 years. During the lifetime of the project replacement costs, operation and maintenance costs and fuel costs are taken into account as well.

Figures 11 and 12 presents the configuration of hybrid power systems at Gračanica and Radoč 2.

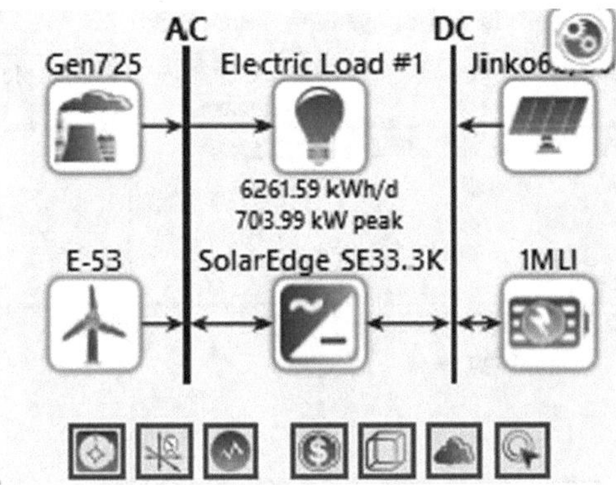


**Fig. 11.** Configuration of Gračanica hybrid system

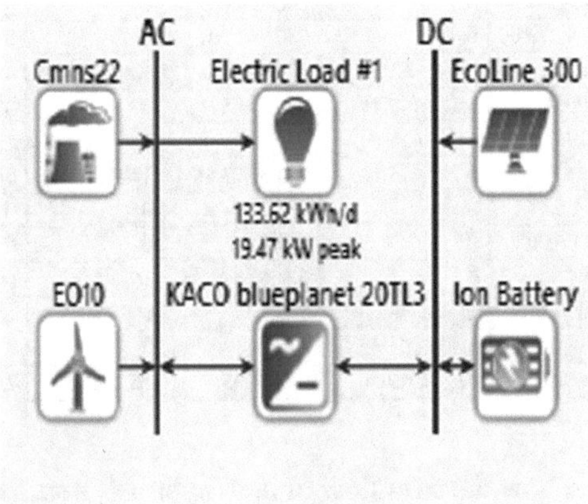


**Fig. 12.** Configuration of Radoč 2 hybrid system

# 6 Results and Discussion

After all parameters have been set for both locations calculation of all possible solutions was performed with HOMER Optimizer. Components and characteristics of both hybrid power systems are presented in Table 1.

From Table 1 it can be seen that the most feasible solution for Gračanica is consisted of 4040 kW of solar panels, one wind turbine E-53, 725 kW diesel generator, four 1MWh Li-Ion batteries and 721 kW of converter. Table 2 shows the cost summary of the most feasible solution of hybrid systems for chosen components with their parameters for both locations. It can be seen that the investment in Gračanica hybrid system is $5,452,701.05 and NPC is $10,783,000.36. Total investment in hybrid system located in Radoč 2 is $132,139.28 and NPC is $200,011.96. In both hybrid system the most expensive component is the storage, but it also has the biggest salvage value.

**Table 1.** Hybrid power system components – HOMER results

| Location | HPS configuration | Storage | Production kWh/year | Renewable percent | Capacity shortage %/year |
|---|---|---|---|---|---|
| Gračanica | PV: 4040 kW<br>WG: 800 kW<br>DG: 725 kW<br>Converter: 721 kW | Li-ion 1 MWh 4 strings | 2,285,479 | 95.3 | 0 |
| Radoč 2 | PV: 47.4 kW<br>WG: 10 kW<br>DG: 22 kW<br>Converter: 19.5 kW | Li-ion 182 kWh 1 string | 48,773 | 95.6 | 0 |

**Table 2.** Cost summary for HPSs

| Location | Component | Capital ($) | Replacement ($) | O&M costs ($) | Fuel ($) | Salvage ($) | Total ($) |
|---|---|---|---|---|---|---|---|
| Gračanica | PV | 1,072,237.64 | 0 | 636,409.68 | 0 | 0 | 1,708,647.32 |
| | WG | 1,234,273.00 | 0 | 830,393.90 | 0 | 0 | 2,064,666.90 |
| | DG | 290,000 | 117,243.23 | 87,024.12 | 1,845,187.25 | 69,762.26 | 2,269,692.34 |
| | Storage | 2,800,000.00 | 1,572,434.74 | 630,095.87 | 0 | 356,775.02 | 4,645,755.59 |
| | Converter | 56,190.41 | 57,737.15 | 0 | 0 | 19,689.35 | 94,238.21 |
| | Total | 5,452,701.05 | 1,747,415.12 | 2,183,923.56 | 1,845,187.25 | 446,226.62 | 10,783,000.36 |
| Radoč 2 | PV | 30,326.77 | 0 | 527.71 | 0 | 0 | 30,854.48 |
| | WG | 32,000.00 | 0 | 175.84 | 0 | 0 | 32,175.84 |
| | DG | 7,649.47 | 0 | 5,686.64 | 16,240.25 | 684.16 | 28,892.20 |
| | Storage | 60,000.00 | 41,425.49 | 5,275.17 | 0 | 1,828.68 | 104,871.98 |
| | Converter | 2,163.04 | 1,406.16 | 0 | 0 | 351.74 | 3,217.46 |
| | Total | 132,139.28 | 42,831.65 | 11,665.35 | 16,240.25 | 2,864.59 | 200,011.96 |

Figures 13 and 14 show the cash flow of both hybrid systems. Cash flow represents complete expenses throughout the lifetime of the project per each year.

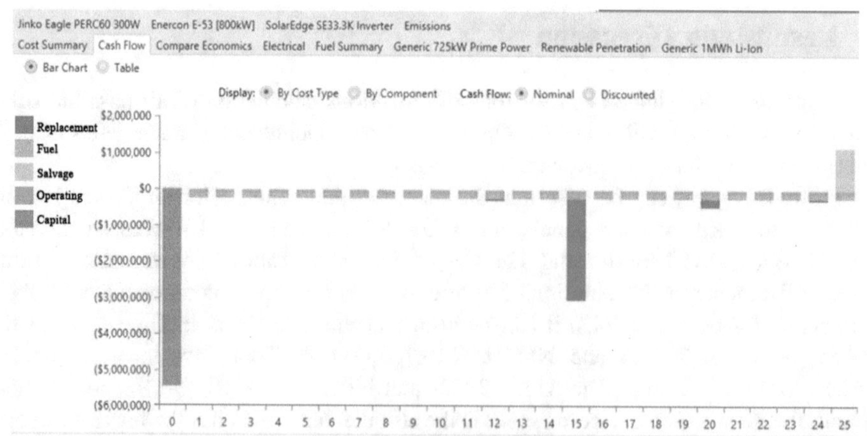


**Fig. 13.** Cash flow for Gračanica HPS

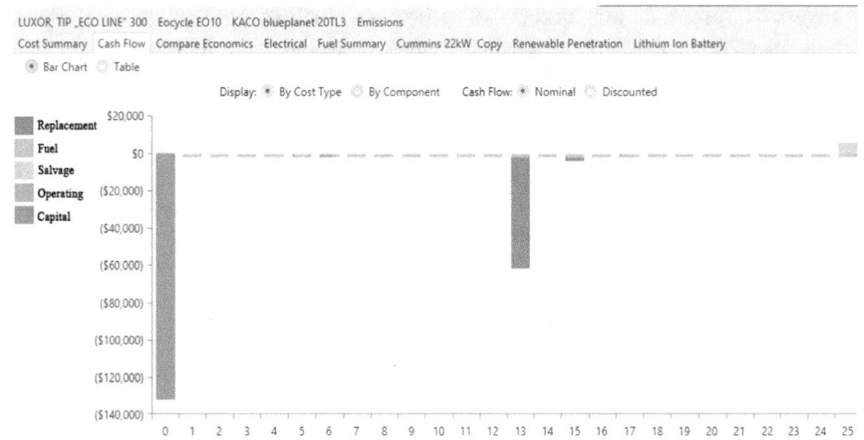


**Fig. 14.** Cash flow for Radoč 2 HPS

Figure 15 illustrates the amount of power production per source and monthly average electric production of Gračanica hybrid power system. On this figure it can be seen that the biggest producer of power are solar panels with 85% of total production. In this case wind turbine is the second largest producer of energy with 10.3% of production, but that is almost eight times less than production from solar panels. With the least percentage of energy production comes 725 kW diesel generator with 4.63% of total power production. With this arrangement of production, we got huge amount of excess electricity, i.e.- 60.1% of total produced electricity is excess. From the monthly average electric production graph, it is visible that solar panels have the biggest production in July and August, while the wind turbine has the biggest production in February and December. Diesel generator has almost no production in April, May and June. With these components we have achieved the goal of having capacity shortage below 1%. In this case it is 0%.

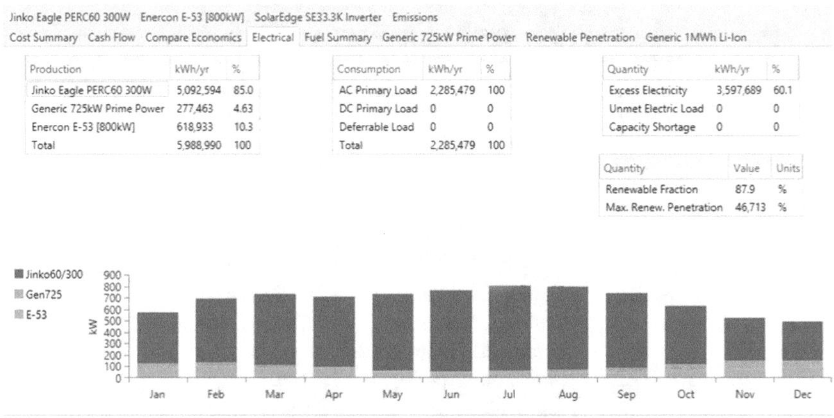


**Fig. 15.** Power production of Gračanica HPS

Figure 16 presents the amount of power production per source and monthly average electric production of Radoč 2 hybrid system. As it was the case with Gračanica HPS, the percentage of penetration of renewable energy resources is above 90%. The biggest power producer are solar panels with 79.2% of total production, while the wind generator had 16.4% of total production. The diesel generator had only 4.4% of total production and it did not work at all from March to July. Solar panels had the biggest production in July and August, while the wind generator had it in March and April. With this hybrid system there is no unmet electric load and the capacity shortage is below 1%.

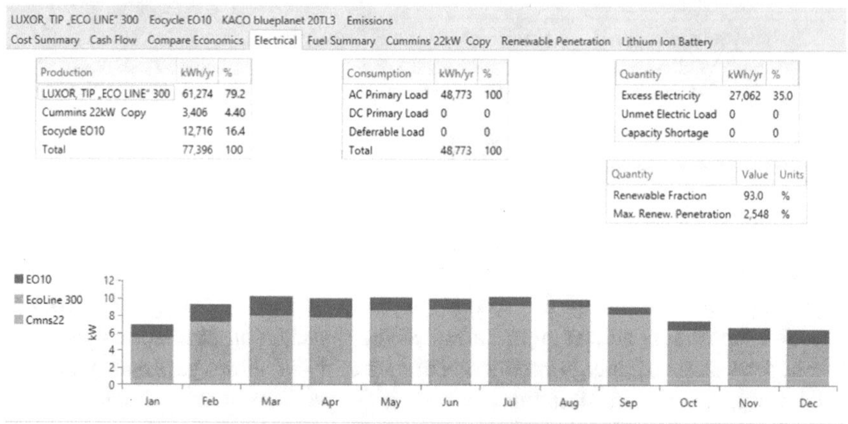


**Fig. 16.** Power production of Radoč 2 HPS

Table 3 shows the emissions of the hybrid power systems. It can be seen that in both hybrid power system emissions are very low when we take into consideration the lifetime of the project (25 years).

**Table 3.** Hybrid systems gas emissions

| Location | Quantity | Value | Units |
|---|---|---|---|
| Gračanica | Carbon Dioxide | 229,424 | kg/yr |
| | Carbon Monoxide | 962 | kg/yr |
| | Unburned Hydrocarbons | 110 | kg/yr |
| | Particulate Matter | 27.5 | kg/yr |
| | Sulfur Dioxide | 573 | kg/yr |
| | Nitrogen Oxides | 962 | kg/yr |
| Radoč 2 | Carbon Dioxide | 2,019 | kg/yr |
| | Carbon Monoxide | 0.489 | kg/yr |
| | Unburned Hydrocarbons | 0 | kg/yr |
| | Particulate Matter | 0.114 | kg/yr |
| | Sulfur Dioxide | 5.01 | kg/yr |
| | Nitrogen Oxides | 1.15 | kg/yr |

## 7 Conclusion

Hybrid power systems, slowly but surely, are increasingly occupying their place in the production of electricity from renewable sources. Composing more independent renewable sources to one unit has resulted in a higher degree of reliability of supply. Although fossil fuels still contribute most to energy production, the world is slowly turning towards new energy sources. New systems, or computer programs, are simulating the work of the system itself and allow for better exposure to the problem.

There are many factors that analyze the energy-efficiency of the system, that can be set during the system configuration phase. There is no universal method to quickly and easily obtain quality results. The power system that uses renewable energy sources can be implemented in many ways, at present one of the best softwares that combines energy and economic aspects is HOMER software. System configuration depends primarily on the load and location of the facility, based on the average load on the system, which is based on the conclusion of available power sources.

The analysis of this paper has been related to two different locations with two different group of consumers. HOMER software has enabled optimal analysis for both locations concerning their advantages and disadvantages from great solar irradiation to low wind speeds. The wind speeds are not measured with the proper instrumentation, rather they are downloaded from HOMER software. Larger wind speeds would enable WGs to generate more electrical energy for both analyzed locations. For places such as Radoč 2, who are far from the urban areas and whose number is large in Bosnia and Herzegovina, HPSs are more cost-effective and cheaper solutions for the generation of electrical energy than the development of the infrastructure in order to receive electrical energy from the main grid. The advantage of HPS for Gračanica, the place with huge number of industrial consumers, is the storage. The storage provides more reliable power supply with the help of DG if outages of power supply occur. Due to the large power consumption in Gračanica the storage should be of large capacity as well. Although the price for large capacity storage is high it would still be more cost-effective to include batteries in HPS than what would DSO have to pay to industry consumers in cases of power supply outages. By comparing the optimal power supply system, it comes to the conclusion that renewable energy is profitable in the case of long-term use. HOMER simulation software solutions can be used for conceptual projects.

## References

1. State of Report Electricity Access: International Bank for Reconstruction and Development / THE WORLD BANK, Washington DC 20433 (2017)
2. Kalappan, K.B., Ponnusamy, V.: Modeling, simulation and optimization of hybrid renewable energy systems in technical, environmental and economical aspects: case study Pichanur village, Coimbatore, India. Int. J. Appl. Environ. Sci. **8**, 2035–2042 (2013)
3. He, G.-X., Cheng, L., Xu, J., Chen, L., Tao, W.-Q.: Optimal configuration of a wind/PV/battery hybrid energy system using HOMER software. Chem. Eng. Trans. **61**, 1507–1512 (2017)

4. Elhassan, Z.A.M., Zain, M.F.M., Sopian, K., Awadalla, A.: Design of hybrid power system of renewable energy for domestic used in Khartoum. J. Appl. Sci. **11** (2011). https://doi.org/10.3923/jas.2011.2270.2275
5. Kumar, P., Deokar, S.: Optimal design configuration using HOMER. In: Konkani A., Bera R., Paul S. (eds.) Advances in Systems, Control and Automation. Lecture Notes in Electrical Engineering, vol 442. Springer, Singapore (2018)
6. Maherchandani, J.K., Agarwal, C., Sahi, M.: Economic feasibility of hybrid biomass/PV/wind system for remote villages using HOMER. Int. J. Adv. Res. Electr. Electron. Instrum. Eng. **1**, 49–53 (2012)
7. Mohammed, O.H., Amirat, Y., Benbouzid, M., Tang, T.: Optimal design of a PV/fuel cell hybrid power system for the city of Brest in France. In: 2014 First International Conference on Green Energy ICGE (2014)
8. Magarappanavar, U.S., Koti, S.: Optimization of wind-solar-diesel generator hybrid power system using HOMER optimization. Int. Res. J. Eng. Technol. **3**(6), 522–526 (2016)
9. Huang, R., et al.: Optimal design of hybrid energy system with pv/wind turbine/storage: a case study. In: 2011 IEEE International Conference on Smart Grid Communications (SmartGridComm). IEEE (2011)
10. Ramdhane, B., Ndiaye, D., Menou, M.M., Mahmoud, A.K., Yahya, A.M., Yahfdhou, A.: Optimization of electrical production of a hybrid system (solar, diesel and storage) pilot using HOMER in Biret, Southern Coast of Mauritania. Int. J. Phys. Sci. **12**(18), 211–223 (2017)
11. Mas'Ud, A.A.: The application of homer optimization software to investigate the prospects of hybrid renewable energy system in rural communities of Sokoto in Nigeria. Int. J. Electr. Comput. Eng. **7**, 596–603 (2017)
12. Bhuiyan, M.A.M., Deb, A., Nasir, A.: Optimum planning of hybrid energy system using HOMER for rural electrification. Int. J. Comput. Appl. **66**(13), 45–52 (2013)

# Willingness to Pay for Reliable Electricity: A Contingent Valuation Study in Bosnia and Herzegovina

Tarik Hubana[1(✉)] and Nerman Ljevo[2]

[1] Technical University Graz, Graz, Austria
t.hubana@student.tugraz.at
[2] Faculty of Management and Business Economy, Sarajevo, Bosnia and Herzegovina
nerman.ljevo@fmpe.edu.ba

**Abstract.** The expectation of electricity consumers to be served by electric utilities with higher levels of power quality and reliability becomes more stringent than ever, with the growth of total energy consumption worldwide, and mostly in developed areas and countries. Electricity supply interruptions less and less accepted by both consumers and regulators, imitated by the change of the consumer behavior and increased requirements by the regulator, making the requirements for the reliability of supply higher than ever. Meanwhile, distribution system operators struggle to achieve these requirements, mainly because of the infeasibility of the investments. However, if the consumers were willing to pay for the increased reliability of supply, these requirements can be achieved. Using a sample of 436 residential and business consumers in Sarajevo, Bosnia and Herzegovina, in this paper a willingness to pay (WTP) for an improved electricity service for both residential and business consumers is estimated by using the contingent valuation method (CVM). The average WTP of domestic consumers for avoiding one-hour interruption is estimated to be 3.02 BAM, while the average WTP of business consumers is 105.4 BAM. Information on the value of reliable service can be used to assess the economic efficiency of investments in distribution systems in order to strategically target investments to consumers segments that receive the most benefit from system improvements, and to numerically quantify the risk associated with different operating, planning and investment strategies.

## 1 Introduction

Electric power systems are subject to constant change in the last decade. New technologies, standards, and requirements from both consumers and regulators impose large investments in electric power system, in order to improve the power quality and reliability levels. In general, very severe outages and blackouts that occurred in the past decade in the in Europe clearly showed that, besides the price of electricity, continuity of service (in terms of reliability/continuity of supply) is also a very important issue for consumers and society as a whole. Hence, the power system faults are particularly unwelcome events because, apart from causing the system interruption, they create


S. Avdaković et al. (Eds.): IAT 2019, LNNS 83, pp. 107–116, 2020.
https://doi.org/10.1007/978-3-030-24986-1_8

numerous technical issues such as voltage sags [1]. Consequently, the management and operation of the electric-power distribution systems have also changed [2]. One of the guiding principles in evaluating investments designed to improve the reliability of electricity systems is that these investments should be economically efficient. That is, the cost of improving the reliability and power quality supplied by an electric system should not exceed the value of the economic loss to consumers that the system improvement is intended to prevent [3]. With the advances in the area of the advanced distribution system protection [4–6], and constant upgrades of the distribution grid, the power quality and reliability levels are brought to the margin where additional investments are just not feasible. On the other hand, to the distribution system operator the investments may be related to benefits related to rewards or penalties in accordance to the total non-delivered energy when individual standards of quality of service are not met [7]. This is a point where consumers need to take part in the process. In order to have a more reliable supply of energy, consumers can pay for an improved service. This paper estimates this value that presents a key parameter that can be used to assess the economic efficiency of investments in distribution systems in order to strategically target investments that will improve the reliability of that part of the distribution system.

There is a growing number of researchers that have focused on identifying the financial attributes that define the quality of service provided by an electricity supplier. In an attempt to provide guidance to regulators, and to assist distribution companies in designing service packages, Goett et al. [8] undertook a study on consumers' willingness to pay for various service attributes. Beenstock et al. [9] used a revealed preference approach (on business consumers and not domestic consumers) in which the cost of an outage may be inferred from the actions taken by consumers to mitigate losses induced by unsupplied electricity. They exploit data on investment in back-up generators that mitigate losses in the event of an outage [10].

Mansouri et al. [13] undertook a survey to identify environmental attitudes and beliefs, energy-use behavior and ownership levels for certain appliances and their utilization patterns, among household residents in the southeast of England. The results indicate that consumers are interested in receiving information concerning household energy use and the associated environmental impact, and are willing to modify their behavior in order to reduce household energy consumption and environmental damage [10].

Due to various economic, technical, and political problems, a low quality of electricity supplied by the grid can be expected in some developing countries. This issue can be characterized by the intermittent and unreliable supply of electricity to the consumers commonly through supply interruptions, either planned or unplanned [11]. A low reliability of electricity supply affects individuals and family life as well as being interrelated with the development and economic condition of a country. For instance, according to Murphy et al. [12], a reduction in the number of outages from 100 days per year to 10 days per year corresponds to more than a two-fold increase in GDP per person. Thus, an investment in the power system reliability, supported by the consumer's participation can present a driver for the development of certain societies.

## 2 Contingent Valuation Method

The contingent valuation method (CVM) is a widely used method for estimating economic values for all kinds of ecosystem services and environmental goods which are not traded in the market and hence have no market price. CVM is typically used to estimate the benefits (or costs) of a change in the level of provision (or in the level of quality) of a public good. The contingent valuation method is applied through conducting a survey in which people are directly asked how much they would be willing to pay (WTP) for a (change in) specific environmental service. It is also possible to ask people the amount of compensation that they would be willing to accept (WTA) to give up an environmental service. The first approach is called 'contingent' valuation, because people are asked to state their willingness to pay, contingent on a particular scenario and the environmental service described to the respondent [14].

The willingness to pay (WTP) is defined as the amount of money a household is willing to pay for a specific level of service, for example, in order to reduce the number and duration of outages from a base level to another level for the same duration. A willingness-to-pay study can provide evidence that the service provider can put before a regulator to support expenditure plans, by highlighting which aspects of service quality are important to consumers and, importantly, estimating the value consumers place on various service attributes. The estimated willingness to pay can then be compared with the incremental costs of achieving such improvements, as part of the service provider's business planning and the regulator's decision-making process. Consumers' willingness to pay provides an 'upper-bound' on the financial incentive or reimbursement that should be associated with service improvements. The 'lower-bound' is represented by the incremental cost to the business of improving that aspect of service quality [10].

The consumers' added value of service reliability can be quantified by the willingness of consumers to pay for service reliability, taking into account the resources (e.g., income) by the residential consumer or by a firm's expected net revenues associated with the added reliability. A system improvement is considered economically efficient if its marginal societal benefits (the economic value of the improvement in reliability) exceed the marginal societal costs (the cost of the investment, including direct as well as indirect (e.g., environmental costs). The cost of system improvements is usually estimated using engineering cost analysis. The economic value of the benefit to consumers is estimated as the avoided economic loss that would have occurred if the investment had not occurred [3].

Therefore, regulators and institutions are strongly promoting the improvement of continuity of service in the electricity sector. Then reliability of supply and its value are the key factors for the decision-making process underlying expansion plans not only of electricity generation systems but also of transmission and distribution networks. It is evident that low levels of investment can result in unreliable supply (unacceptable low continuity), while excessive investments can result in unnecessary expenditures with a resulting increase of the cost of electricity to consumers. The literature does not agree on whether or in which situations CA-based WTP measures outperform direct WTP measurements in terms of their accuracy in predicting actual buying/payment behaviors [15].

It is inconvenient to ask residential consumers to estimate the interruption costs because household respondents are unable to accurately estimate the costs. Even though business consumers can more accurately determine the interruption cost, both consumer groups are generally asked two questions: how much would you be willing to pay for electric service to avoid the power interruption in the case of this interruption (WTP)? and how much would you accept as a credit for a particular interruption scenario (willingness to accept or WTA)? In both cases they should be asked to assess these values in the real-life scenarios in order to make the estimation more accurate.

## 3 Reliability of Supply

Reliability of supply is defined as the ability of a unit (component or system) to perform a demanded function for an appointed time interval under preset conditions. To give a preliminary impression of the reliability of the electricity supply as a whole a calculation of the actual availability of the utility grid or the "mean time between failure (MTBF)" is usually conducted. Availability is defined as the percentage of time that a system is functional, or the time the system is up, divided by the total time at risk [12]. Also, MTBF can easily be calculated, since availability is MTBF divided by the total of MTBF + MTTR (mean time to repair) [12].

However, the almost universal measure of reliability is the average loss of supply (in minutes) per consumer over a year, or SAIDI (System Average Interruption Duration Index). SAIDI/SAIFI = CAIDI, where SAIFI (System Average Interruption Frequency Index) is the average number of interruptions to supply per consumer over a year; and CAIDI (Customer Average Interruption Duration Index) is the average duration of each supply interruption within a particular year. SAIDI may not be the best measure of reliability for the purposes of creating financial incentives for performance because it is a measure of the per consumer loss of supply, which implies that the distributor receives extra revenue by reducing minutes lost per consumer [10]. However, the self-reported reliability indices by the distribution system operators do not always represent the actual situation accurately. In this sense, knowledge about the experience of the grid users can be useful to evaluate the reality of the reliability indices of electricity service [11].

## 4 Methodology

The research is based on two samples: domestic and business consumers of electricity. A regression model is established based on the multiple regression for each of the samples. Using the inferential statistical method, it was shown which predictor has the most influence on the criterion variable (for both samples) and with the help of the descriptive statistics methods, it was possible to display and understand the sample characteristics. In the survey, in the sample of domestic consumers 364 respondents gave their views on the consumption of electricity, their properties, and personal life and consumer habits. Multiple linear regression represents the basic regression model and with it the corresponding transformation reduces the number of non-linear models. The model for willingness of paying domestic consumer for the reliability of electricity has eight variables:

$$Y = \alpha + \beta_1 X_1 + \beta_2 X_2 + \beta_3 X_3 + \varepsilon \qquad (1)$$

Where:
Y - dependent variable (for willingness of paying for the reliability of electricity – domestic consumers);
α - section on Y - axis (constant);
β1, β2, β3 - slope coefficients;
X1 - electric heating;
X2 - employment;
X3 - time spent at home;
ε - statistical member (standard error).

Within the research, in the sample of business consumers, the sample size was N = 72 electricity consumers - legal entities. In this part, WTP of the consumers for more reliability of electricity supply is considered a priori. The multi-regression equation for business users has four variables, according to Eq. (2).

$$Y = \alpha + \beta_1 X_1 + \beta_2 X_2 + \beta_3 X_3 + \beta_4 X_4 + \varepsilon \qquad (2)$$

Where:
Y - dependent variable (for willingness of paying for the reliability of electricity – business consumers);
α –section on Y – axis (constant);
β1, β2, β3, β4 - slope coefficients;
X1 - size of the enterprise;
X2 - sector of activity;
X3 - backup system;
X4 - monthly bill for electricity;
ε - statistical member (standard error).

Both models should anticipate the behavior of the criterion variable under the influence of the predictor, and it must be established which predictor has the greatest impact on the criterion variable in both samples.

## 5 Results and Discussion

Of the total number of respondents, 43% of men (156 respondents) and 57% of women (206 respondents) are in the sample of domestic consumers. The main goal of the research process in this part was to investigate the willingness of the users of electricity to pay for the improved reliability of electricity supply. The following table shows the willingness of respondents to pay for the improved reliability of electricity supply (Table 1).

**Table 1.** WTP for improved reliability: domestic consumers

| | | Frequency | Percent | Valid percent | Cumulative percent |
|---|---|---|---|---|---|
| Valid | NO | 91 | 25.0 | 25.6 | 25.6 |
| | YES | 265 | 72.8 | 74.4 | 100.0 |
| Missing | Total | 356 | 97.8 | 100.0 | |
| | System | 8 | 2.2 | | |
| Total | | 364 | 100.0 | | |

Of the total number of respondents, 74.4% are willing to pay more for the reliability of electricity supply, while 25.6% of them answered negatively on this issue (Table 2).

**Table 2.** Model summary – domestic consumers

| Model | R | R square | Adjusted R square | Std. error of the estimate | Change statistics | | | | |
|---|---|---|---|---|---|---|---|---|---|
| | | | | | R square change | F change | df1 | df2 | Sig. F change |
| 1 | .642[a] | .412 | .410 | .653 | .412 | 222.710 | 1 | 318 | .000 |
| 2 | .821[b] | .673 | .671 | .488 | .261 | 253.590 | 1 | 317 | .000 |
| 3 | .821[c] | .674 | .671 | .488 | .001 | .789 | 1 | 316 | .375 |

where:

a. Predictors: (Constant), Electric heating

b. Predictors: (Constant), Electric heating, Employment

c. Predictors: (Constant), Electric heating, Employment, Time spent in the house

In the previous table the sum of the multiple regression model is presented, which refers to domestic consumers. The R mark indicates the multi-correlation coefficient, while R Square is the multi-determination coefficient. What is significant in this analysis is that the adjusted R square coefficient, which represents the corrected multiple-determination coefficient, and a less biased score than the multi-determination coefficient. If taken as an example of variable one (electric heating), it can be concluded that there are 41% of individual differences in terms of the attitude towards increasing the payment for the reliability of electricity that we can explain in the population based on individual differences on the predictors taken together, i.e. based on their linear combination. The standard model error indicates that the model is not very accurate. In the context of testing hypotheses related to multiple regression, it can be set in two ways, i.e., to use ANOVA analysis, where the hypothesis of the multi-determination coefficient could be tested (which is also the general multi-regression hypothesis) and testing the hypothesis for each predictor variable. In addition, the statistical significance for each of the variables in the model can be determined, and this data can be obtained by inspecting the confidence interval. However, in this part of the research, two key questions are posed. First, it relates to the appearance of the model for the "WTP for improved reliability of electricity supply", and the second is: "Which predictor is the most important?" The model is constructed according to the Eq. 3.

$$WTP_{Domesticconsumers} = 110,9 + 31,2X_1 + 31,9X_2 - 2,18X_3 \tag{3}$$

The second question is which predictor is the most important, and in this part, it is important to note that it is necessary, in order to give an answer to this question, each new predictor to introduce gradually into the model, and see the contribution of each predictor to a criterion variable. The best approach is through the semi partial correlation of the predictor.

In this section, the R Square Change, or the F Change is considered in the order of the highest to the minimum: 222,71 (var 1) > 253,590 (var 2) > 0,789 (var 3).

According to the results, socio-economic parameters can be ranked in the following way (according to the effect of increased payment of reliability of electricity - from the highest to the least important parameter): electric heating, employment and time spent in the house. When it comes to business consumers, research has shown that even 97.2% of surveyed companies are willing to pay more for improved reliability of electricity supply, while only 2.8% companies have responded negatively to this issue. The conceptual multi-regression model that should explain the WTP for the reliability of electricity supply for business consumers has 4 predictors, which are: size of enterprise (micro, small, medium and large enterprises), business sector (agricultural, commercial, educational, health, service sector, other sector), back-up system possession (yes/no) and the electricity bills value. Within the research, 20.8% of micro enterprises, 37.5% of small companies, 30.6% of medium and 11.1% of large enterprises are included. When analyzing the sectors, the survey covered most of the enterprises from the commercial (37.5%) and service sector (30.6%), and the least from the education sector (6.9%). It is important to emphasize, in the context of considering back up the power supply system that 55% of enterprises use backup power, while 45% do not use this option. The large number of the analyzed companies have a monthly allocation for electricity over 1000 BAM, which is quite understandable (Table 3).

**Table 3.** Model summary – business consumers

| Model | R | R square | Adjusted R square | Std. error of the estimate | Change statistics | | | | |
|---|---|---|---|---|---|---|---|---|---|
| | | | | | R square change | F change | df1 | df2 | Sig. F change |
| 1 | .501[a] | .251 | .240 | .527 | .251 | 22.815 | 1 | 68 | .000 |
| 2 | .502[b] | .252 | .229 | .531 | .001 | .046 | 1 | 67 | .830 |
| 3 | .502[c] | .252 | .218 | .535 | .001 | .057 | 1 | 66 | .812 |
| 4 | .917[d] | .841 | .831 | .248 | .588 | 240.33 | 1 | 65 | .000 |

where:

a. Predictors: (Constant), Type of enterprise:

b. Predictors: (Constant), Type of enterprise: Activity sector;

c. Predictors: (Constant), Type of enterprise: Activity sector; Back – up system;

d. Predictors: (Constant), Type of enterprise: Activity sector; Back – up system; Electricity bill

The equation describing the behavior of the criterion variable is shown in Eq. 4.

$$\mathrm{WTP}_{\mathrm{Businessconsumers}} = 53,2 + 33,3X_1 + 2,4X_2 + 0,3X_3 + 29,7X_4 \quad (4)$$

When it comes to the statistical significance of each of the variables in its impact on the criterion variable, then we can conclude it by considering F Change statistics as follows: 240,330 (var 4) > 22,815 (var 1) > 0,057 (var 3) > 0,046 (var 2). It is shown that electricity bill is the most important predictor in the model. It is followed by the type of enterprise. The remaining two predictors in the model are much less significant, i.e. the backup system and sector which the company belongs.

In the comparative analysis, a comparison of two samples, domestic and business consumers, and their responses in terms of WTP will be made in addition to paying for the reliability of electricity. It is important to note that in the case of domestic users, several predictors have been considered that have an impact on the criterion variable (8) in relation to business users (4). In addition, the number of respondents within the "Domestic consumers" sample is significantly higher than the "Business consumers" sample. This is understandable because harder to approach and come to the process of research by the owners of companies/managers who are relevant to find themselves in the "Business Consumers" sample of those in the first mentioned sample. However, within these circumstances certain indications of the behavior of the population I (Domestic consumers) and population II (Business consumers) can be given. Thus, it can be concluded that business consumers are much more willing to additionally pay for higher returns (97.2% of analyzed companies), compared to domestic consumers (74.4%). The most important determinant of the readiness for additional payment for the reliability of electricity supply for business consumers is the size of the enterprise (type of enterprise, type of company: micro, small, medium, large enterprise), as well as the current amount of electricity bills. The most important determinant of WTP for the reliability of electricity delivery is electric heating, monthly income and employment.

Business consumers are those who are affected more by loss of electricity supply. This is primarily seen in their responses to the question of "fair compensation for electricity distribution" for the unannounced shutdown of electricity supply. In this segment, the largest number of respondents, 26.3% would accept 5 BAM for any unannounced interruption in electricity delivery (WTA) (Table 4).

**Table 4.** WTP and WTA values for domestic and business consumers

| Parameter | WTP | | WTA | |
|---|---|---|---|---|
| | Domestic | Business | Domestic | Business |
| Mean | 3.01 | 105.4 | 5.55 | 77.52 |
| Median | 2 | 20 | 5 | 10 |
| Standard deviation | 3.26 | 172.89 | 8.67 | 155.40 |
| Lower limit of confidence interval | 2.68 | 65.73 | 4.66 | 41.87 |
| Upper limit of confidence interval | 3.35 | 145.06 | 6.44 | 113.17 |
| Length of confidence interval | 0.67 | 79.32 | 1.78 | 71.29 |

Furthermore, 19.4% of them think that such an interruption should be refunded with 10 BAM, and 9.7% of them consider that such an interruption should be refunded with 20 BAM for every hour in which there will be no electricity deliveries. Domestic consumers are less sensitive to the cutoff of electricity supply, but of course, this problem affects them. However, this is considerably less than the business consumers are, which is seen through the answer to the same question, which was previously posted to business users. Thus, 27.5% of domestic consumers believe that every electricity cut per hour should be refunded from the distribution system operator (DSO) with a value of 3 BAM, 16.9% of them believe that this value should be 2 BAM, and 12.4% think that this should be only 1 BAM. This section shows significant differences in the importance of electricity supply to domestic and business consumers. Table 4 summarizes the WTA and WTP values for both domestic and business consumers.

In case of the WTP analysis, 18.1% of business consumers are willing to pay 10 BAM for the improved reliability of the electricity supply, 16.7% of them are willing to pay 5 BAM; 13.9% are willing to pay 50 BAM, and 12.5% are willing to pay 20 BAM. As stated in the previous paragraph, the importance of electricity for domestic and business consumers is confirmed here again, where 25.6% of domestic consumers don't want to pay for improved reliability of electricity at all. Significant data in this category is that 23% of respondents are willing to pay 5 BAM for increased reliability of electricity, while 12,9% are willing to pay 1 BAM. The average WTP of domestic consumers for avoiding one-hour interruption is estimated to be 3.02 BAM, while the average WTP of business consumers is 105.4 BAM.

## 6 Conclusions

In this paper, a research carried out in Bosnia and Herzegovina is presented. The results show that both residential and business consumers value reliability of the electricity service provided by DSOs. Residential consumers expressed an average willingness to pay of 3.02 BAM to avoid a one-hour outage. On the other hand, business consumers expressed an average willingness to pay of 105.4 BAM.

If an outage must occur, residential consumers did not express the difference between workdays and weekends. They are affected by the disappearance of electricity, and this disadvantage is most disturbed in the period between 18:00 h and 23:00 h. An important aspect of CVM is that there are biases that may affect the reliability and validity of the results. Some of these biases include respondents hypothetical WTP, while the real situation values are lower, different estimated amounts depending on the form of payment and non-response bias since individuals who do not participate in the survey are likely to have different values than individuals who do take part in it.

Information on the value of reliable electricity service can be used to assess the economic efficiency of investments in generation, transmission and distribution systems, to strategically target investments to consumer segments that receive the most benefit from system improvements, and to numerically quantify the risk associated with different operating, planning and investment strategies.

## References

1. Hubana, T., Begic, E., Saric, M.: Voltage sag propagation caused by faults in medium voltage distribution network. In: IAT 2017: Advanced Technologies, Systems, and Applications II, pp. 409–419. Springer International Publishing AG, Cham (2018)
2. Hubana, T., Saric, M., Avdakovic, S.: High-impedance fault identification and classification using a discrete wavelet transform and artificial neural networks. Elektrotehniški Vestnik **85** (3), 109–114 (2018)
3. Sullivan, M.J., Mercurio, M.G., Schellenberg, J.A.: Estimated Value of Service Reliability for Electric Utility Consumers in the United States. Ernest Orlando Lawrence Berkeley National Laboratory, Berkeley (2009)
4. Hubana, T.: Transmission lines fault location estimation based on artificial neural networks and power quality monitoring data. In: 2018 IEEE PES Innovative Smart Grid Technologies Conference Europe (ISGT-Europe), Sarajevo (2018)
5. Hubana, T., Saric, M., Avdakovic, S.: Approach for identification and classification of HIFs in medium voltage distribution networks. IET Gener. Transm. Distrib. J. **12**(5), 1145–1152 (2018)
6. Hubana, T., Saric, M., Avdakovic, S.: Classification of distribution network faults using Hilbert-Huang transform and artificial neural network. In: IAT 2018: Advanced Technologies, Systems, and Applications III, pp. 114–131. Springer, Cham (2019)
7. Hubana, T.: Coordination of the low voltage microgrid protection considering investment cost. In: 1. Conference BH BH K/O CIRED, Mostar (2018)
8. Goett, A., Hudson, K., Train, K.: Consumers' choice among retail energy suppliers: the willingness-to-pay for service attributes. Energy J **21**(4), 1–28 (2000)
9. Beenstock, M., Goldin, G., Haitovsky, Y.: The cost of power outages in the business and public sectors in Israel: revealed preference vs. subjective valuation. Energy J. **18**(2), 39–61 (1997)
10. Hensher, D.A., Shore, N., Train, K.: Willingness to pay for residential electricity supply quality and reliability. Appl. Energy **115**, 280–292 (2014)
11. Reinders, K., Reinders, A.: Perceived and reported reliability of the electricity supply at three urban locations in Indonesia. Energies **11**, 1–27 (2018)
12. Murphy, P., Twaha, S., Murphy, I.: Analysis of the cost of reliable electricity: a new method for analyzing grid connected solar, diesel and hybrid distributed electricity systems considering an unreliable electric grid with examples in Uganda. Energy **66**, 523–534 (2014)
13. Mansouri, I., Newborough, M., Probert, D.: Energy consumption in UK households: impact of domestic electrical appliances. Appl. Energy **54**(3), 211–285 (1996)
14. Sagoff, M.: The Economy of the Earth: Philosophy, Law, and the Environment. Cambridge University Press, Cambridge (1988)
15. Gerpott, T.J., Paukert, M.: Determinants of willingness to pay for smart meters: an empirical analysis of household consumers in Germany. Energy Policy **61**, 483–495 (2013)

# Auxiliary Power Systems of Advanced Thermal Power Plants

Azrina Avdić[1], Tatjana Konjić[1], and Nedis Dautbašić[2](✉)

[1] Faculty of Electrical Engineering, University of Tuzla, Tuzla, Bosnia and Herzegovina
[2] Faculty of Electrical Engineering, University of Sarajevo, Sarajevo, Bosnia and Herzegovina
nedis.dautbasic@etf.unsa.ba

**Abstract.** Large reserves and worldwide availability of coal indicate that this energy will play an important role for electricity generation in the future even if there is a significant increase in production from renewables. Besides that, the electricity market is open and the result is competition between the market actors. The goals of advanced thermal power plants are using as less as possible input energy to generate electricity as much as possible and reducing the $CO_2$ emission. In thermal power plants, 7–15% of the generated energy on the generator does not reach the power plant's threshold because it is geared back to pumps, fans and other auxiliary power systems. Given the fact that each MWh is important today, it is clear that auxiliary power systems of advanced thermal power plants must be energy efficient. In this paper contemporary regulated auxiliary power systems of advanced thermal power plant "Stanari" are presented. After an introduction, fresh air fan, flue gas fan and feed water pump analysis are presented. In addition, the paper analyzes the total power change of auxiliary power systems, depending on the change of the generator power. The most important findings are summarized at the end of the paper.

## 1 Introduction

Thermal power generation is a method of generating power by converting energy obtained from burning fuel into kinetic energy through the use of a turbine, which is then used to produce electrical energy by driving a generator. Because thermal power generation utilizes so-called fossil fuels there are many who say we should not use thermal power because it releases the greenhouse gas $CO_2$, especially after the adoption of the Paris Agreement at the 2015 [1]. Although there is increase in a usage of renewable energy sources such as solar energy and wind energy in the future, thermal power is still important as a back-up power source to these energy sources that their power generation depends on the weather conditions. Also, coal-fired thermal power in power system provides stabile operation of the system. Due to all mentioned reasons demand for thermal power is still high in the world [2–4]. Advanced thermal power plants must meet two basic requirements: high efficiency and low emissions of harmful pollutants. Achieving these goals is achieved through the application of high efficiency and low-emission technology, so-called HELE TECHNOLOGIES (High Efficient Low


S. Avdaković et al. (Eds.): IAT 2019, LNNS 83, pp. 117–125, 2020.
https://doi.org/10.1007/978-3-030-24986-1_9

Emission) [5, 6]. High-efficiency and low-emission technologies are groups of technologies developed to increase the efficiency of coal thermal power plants and the reduction of carbon dioxide and other greenhouse gases, as well as emissions of nitrogen oxides, sulfur dioxide and solid matter.

Advanced thermal power plants are expected to be flexible, due to renewable energy sources. In thermal power plants, auxiliary power systems allow the steam cycle to circulate securely and return to its thermodynamic starting point. Without auxiliary power systems, the steam-water cycle would suffer direct collapse or dangerous and unsustainable expansion. The main purpose of the auxiliary systems is to preserve the projected shape of this cycle in a wide range of operating conditions, using a minimum energy input and with maximum availability [7, 8]. If we take into consideration the above facts it can be concluded that it is extremely important the auxiliary power system be flexible. In other words, when generator power is decreased, power of auxiliary power systems should be reduce [9, 10].

## 2 Auxiliary Power Systems of TPP "Stanari"

The thermal power plant "Stanari" represents an advanced thermal power plant, located near the coal mine "Stanari" in the north of Bosnia and Herzegovina. TPP "Stanari" is the latest investment into a new thermal power plants in Bosnia and Herzegovina. Thermal efficiency of this thermal power plant is 0.34 [11]. Combustion in a circulating fluidized bed and air-cooled condenser are two technical aspects which TPP "Stanari" distinguishes from similar coal thermal power plants in the region. Larger consumers in auxiliary power systems are primary and secondary fresh air fans, flue gas fans, feed water pumps and condensate pumps. The feed water pumps and condensate pumps are controlled by a hydraulic coupler, and the regulation of other major consumers are done by cascading medium-voltage (6 kV) frequency regulators. Table 1 shows powers of major consumers in auxiliary power systems of TPP "Stanari".

**Table 1.** Major consumers in TPP "Stanari"

| Consumers | Total number of customers | Electric power (MW) |
|---|---|---|
| Feed water pump | 3 | 5.4 |
| Condensate pump | 2 | 1.12 |
| Flue gas fan | 2 | 4.2 |
| Primary fresh air | 2 | 3.2 |
| Secondary fresh air fan | 2 | 2.4 |

Below is an analysis of the primary fresh air fan, the flue gas fan and feed water pump.

## 2.1 Analysis of the Primary Fresh Air Fan

Table 2 shows the nominal motor parameters of the primary fresh air fan, and Table 3 the frequency regulator parameters.

**Table 2.** Nominal motor parameters of the PFAF

| | |
|---|---|
| Nominal power | 3.2 MW |
| Nominal voltage | 6 kV |
| Nominal current | 368 A |
| Nominal speed | 1490 r/min |
| Power factor | 0.88 |
| Number of pair of poles | 2 |
| Insulation class | F |
| Class of protection | IP54 |

**Table 3.** The frequency regulator parameters

| | |
|---|---|
| Nominal power | 4250 kVA |
| Output voltage | 0–6 kV |
| Output current | 0–408.96 A |
| Output frequency | 0–50 Hz |

Figure 1 shows DCS (Distributed Control System) data from March 7th, 2017 for the primary fresh air fan.

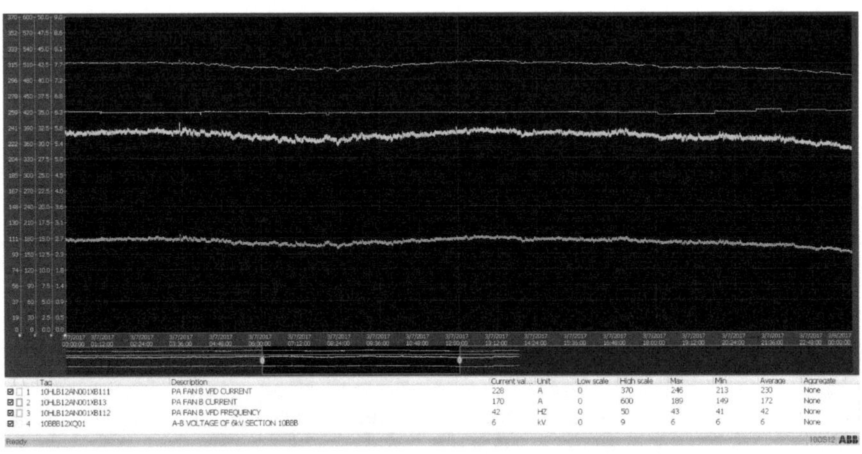


**Fig. 1.** DCS data for the primary fresh air fan

Figure 1 shows that for the primary fresh air fan, the following data are kept and stored:

- the frequency of the frequency regulator of the primary fresh air fan (first line from top to bottom),
- section voltage (second line from top to bottom),
- the primary fresh air fan frequency converter current (third line from top to bottom) and
- the primary fresh air fan current (fourth line from top to bottom).

The average current of the primary fresh air fan during the day was 180 A, which is considerably lower than current of the primary fresh air fan frequency regulator (the average value of the day was 235 A). The mean frequency for the primary fresh air fan frequency regulator was 42 Hz.

On the displays of protective relays, which are in the drive room, it is possible to see the following current values of major consumers:

- current by phases,
- power factor,
- active power,
- reactive power,
- apparent power,
- line voltage and
- phase voltage.

There is no system that collects and archives these data.

Current values from protective relays for primary fresh air fan are shown in Table 4 (March 5$^{th}$, 2017, generator power: 304 MW).

**Table 4.** Current values from protective relays for PFAF

| | |
|---|---|
| Current active power | 1722 kW |
| Current reactive power | 402 kVAr |
| Current apparent power | 1768 kVA |
| Current power factor | 0.97 ind. |

The primary fresh air fan is loaded with 53.8% $P_n$, and the power factor of this fan is 0.97. It can be concluded that the analyzed fan has a very good power factor, although it is overloaded. This is the result of frequency control.

## 2.2 Analysis of the Flue Gas Fan

Table 5 shows the motor nominal parameters of the flue gas fan, and Table 6 shows the frequency regulator parameters for flue gas fan.

**Table 5.** Nominal motor parameters of the FGF

| | |
|---|---|
| Nominal power | 4.1 MW |
| Nominal voltage | 6 kV |
| Nominal current | 475 A |
| Nominal speed | 995 r/min |
| Power factor | 0.86 |
| Number of pair of poles | 3 |
| Insulation class | F |
| Class of protection | IP54 |

**Table 6.** The frequency regulator parameters

| | |
|---|---|
| Nominal power | 5150 kVA |
| Output voltage | 0–6 kV |
| Output current | 0–495,56 A |
| Output frequency | 0–50 Hz |

Flue gas fan is monitored and stored the same data as for the primary fresh air fan. Table 7 displayed average values from DCS (March 7$^{th}$, 2017) for the flue gas fan.

**Table 7.** The average values from DCS for the FGF

| | |
|---|---|
| Average current of FGF | 265 A |
| Average current of FR | 353 A |
| Average frequency | 44 Hz |

Current values from protective relays for the flue gas fan are shown in Table 8 (March 5$^{th}$, 2017, generator power: 304 MW).

**Table 8.** Current values from protective relays for PFAF

| | |
|---|---|
| Current active power | 3038 kW |
| Current reactive power | 757 kVAr |
| Current apparent power | 3131 kVA |
| Current power factor | 0.97 ind. |

The flue gas fan is loaded with 74.1% $P_n$, and the power factor of this fan is 0.97. Owing to frequency control flue gas fan, as well as primary fresh air fan, has a very good power factor, although it is overloaded.

## 2.3 Analysis of the Feed Water Pump

Table 9 shows the motor nominal parameters of the feed water pump.

**Table 9.** Nominal motor parameters of the FWP

| | |
|---|---|
| Nominal power | 5.4 MW |
| Nominal voltage | 6 kV |
| Nominal current | 594 A |
| Nominal speed | 1493 r/min |
| Power factor | 0.899 |
| Number of pair of poles | 3 |
| Insulation class | F |
| Class of protection | IP54 |

The feed water pump is controlled by hydraulic coupling. Figure 2 shows DCS data for feed water pump.

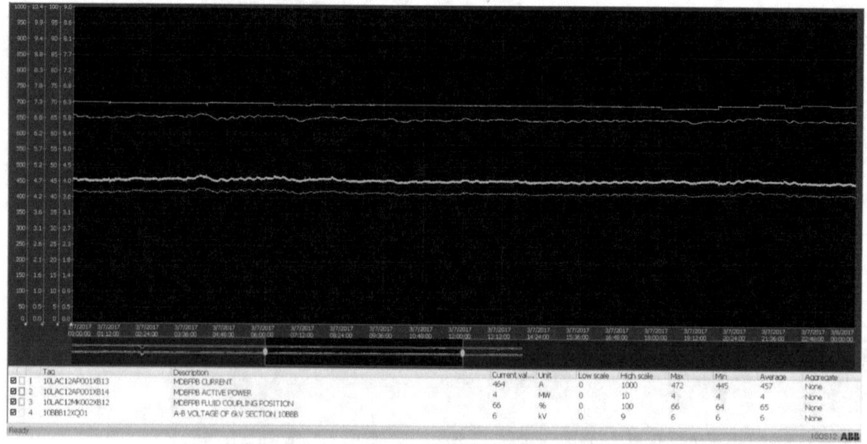


**Fig. 2.** DCS data for feed water pump

DCS collects and archives following data:

- voltage (first line from top to bottom),
- position of hydraulic coupling (second line from top to bottom),
- feed water pump current (third line from top to bottom) and
- active power (fourth line from top to bottom).

The average current of the feed water pump was 457 A and the average active power value was 4.4 MW (82% $P_n$). The average value of the hydraulic coupling position was 64%.

Current values from protective relays for feed water pump are shown in Table 10 (March 5th, 2017, generator power: 304 MW).

**Table 10.** Current values from protective relays for FWP

| | |
|---|---|
| Current active power | 4449.1 kW |
| Current reactive power | 2329.1 kVAr |
| Current apparent power | 5021.9 kVA |
| Current power factor | 0.89 ind. |

The feed water pump is loaded with 82.4% $P_n$ and the power factor is 0.89. This consumer has a lower power factor than primary fresh air fan and flue gas fan, because it is without frequency control.

## 2.4 Analysis of Total Power of Auxiliary Power Systems

Total projected power of auxiliary power systems is about 54 MW. Figure 3 shows the power at generator output in the period from March 1st to March 7th in 2017.

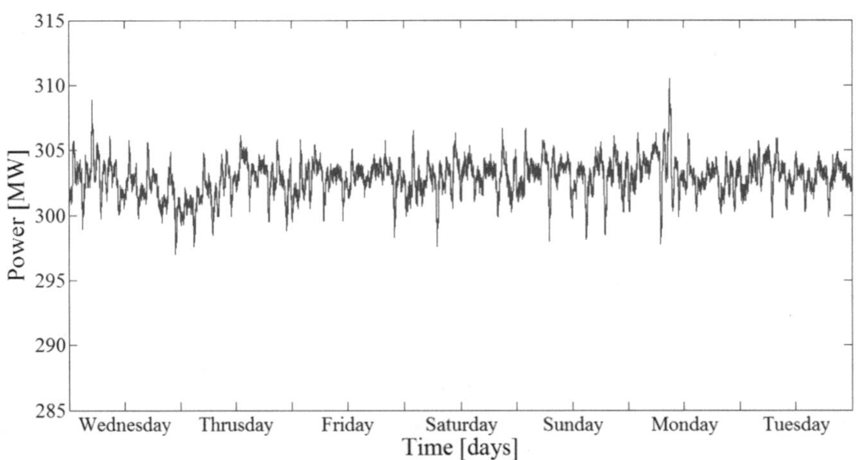


**Fig. 3.** The power at generator output

Figure 4 shows the total power of auxiliary power systems in the period from March 1st to March 7th in 2017.

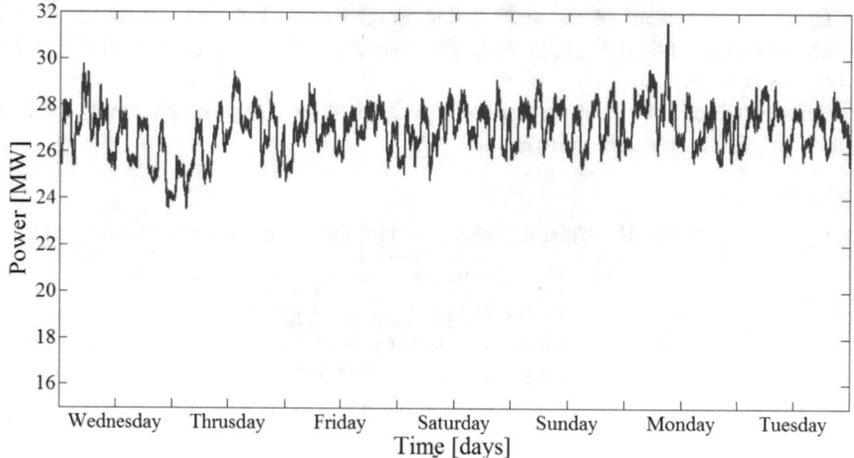


**Fig. 4.** The total power of auxiliary power systems

In the previous figures it is possible to notice that a small change of generator power is changing the total power of auxiliary power systems. This is the result of the frequent regulation of almost all major consumers in TPP "Stanari". The above Fig. 3 show that if power generators is about 300 MW (which is the nominal power of the unit) total power of auxiliary power systems is about 27 MW. Thus, using the frequency regulators for larger consumer control and other energy efficiency measures (lighting, etc.), the power of auxiliary power system is reduced by 50%. It should be noted that some consumers are spare, and that they have not been in operation at time in which the power of auxiliary power system is observed.

## 3 Conclusion

The paper presents the analysis of the major consumers of auxiliary power systems, as well as analysis of total power of auxiliary power systems in the TPP "Stanari", with combustion in a fluidized bed.

It can be concluded that analyzed frequency regulated fans have a very good power factor, although they are overloaded. Analyzed feed water pump has lower power factor, because hydraulic coupling regulation.

In TPP "Stanari" small changes of generator power are changing the total power of auxiliary power systems, which means that TPP "Stanari" have flexible auxiliary power systems. This is the consequence of the frequent regulation of almost all major consumers.

The implementation of frequency controlled drives in TPP "Stanari" increased power at the plant's threshold for the same amount of fuel, increased process efficiency, availability and flexibility through improved process control, reduced maintenance costs, as well as $CO_2$ emissions.

Based on all of the above, it can be concluded that it is essential that new blocks that are planned to be built are required to have a frequent regulation of their auxiliary power systems in order to adapt to market conditions and properly respond to the energy transition demands.

## References


1. Next-Generation Thermal Power Generation, focus NEDO No. 62 (2017)
2. Li,Y., Wen, Y.: Influence of plant steam system on thermal economy of thermal power plant and energy strategy. In: International Conference on Future Electrical Power and Energy Systems (2012)
3. Owusu, P.A., Asumadu Sarkodie, S.: A review of renewable energy sources, sustainability issues and climate change mitigation. Civil and Environmental Engineering (2016)
4. Islam, M.A., Hasanuzzaman, M., Rahim, N.A., Nahar, A., Hosenuzzaman, M.: Global renewable energy-based electricity generation and smart grid system for energy security. Sci. World J. (2014)
5. World coal association. https://www.worldcoal.org/. Accessed 20 Aug 2017
6. Alobaid, F., Mertens, N., Starkloff, R., Lanz, T., Heinze, C., Epple, B.: Progress in dynamic simulation of thermal power plants. Prog. Energy Combust. Sci. **59**, 79–162 (2017)
7. ABB, Power generation - energy efficient design of auxiliary systems in fossil-fuel power plants (2009)
8. Kuzle, I., Bošnjak, D., Pandži, H.: Auxiliary system load schemes in large thermal and nuclear power plants. In: 8th International Conference: Nuclear Option in Countries with Small and Medium Electricity Grids, Dubrovnik (2010)
9. Hesler, S.: Mitigating the effects of flexible operation on coal-fired power plants. Electric Power Research (2011)
10. Energiewende, A.: Flexibility in thermal power plants – with a focus on existing coal-fired power plants (2017)
11. Miljević, D., Mumović, M., Kopač, J.: Analysis of direct and selected indirect subsidies to coal electricity production in the energy community contracting parties (2019)

# Survey of Energy Poverty in Bosnia and Herzegovina

Majda Tešanović[1(✉)] and Jasna Hivziefendić[2]

[1] Faculty of Electrical Engineering, University of Tuzla, Tuzla, Bosnia and Herzegovina
majda.tesanovic@fet.ba

[2] International Burch University, Sarajevo, Bosnia and Herzegovina

**Abstract.** A large number of households facing difficulties in covering their energy costs and energy needs (around 11% in EU). The consequences of the energy poverty are financial problems, bad living conditions and health problems.

One of the main characteristics of power system in Bosnia and Herzegovina is that the same amount of energy achieves four times less than the gross national product then in the average EU countries, and double air pollution. Energy use in households and public sector is extremely inefficient. Low price of energy refused foreign investors in renewable energy sources.

Bosnia and Herzegovina is at the crossroads and must to decide where to go: continue the previous practice and ensure poverty to future generations or to change policies of energy management, and promote usage of RES and investments in energy sector.

The process of European integration represents a chance for B&H to join the EU trends and take a chance for energy reforms and guarantee sustainable development of country.

In this paper will be discussed a problem of energy poverty in Bosnia and Herzegovina on a national level. Energy poverty is very complex problem and its complexity will be shown by an overview of the current researchs.

The paper will present the results of realized researches on energy poverty in B&H.

The final part of the paper will be focused on possible means and mechanisms for prevention of energy poverty in the B&H.

## 1 Introduction

Energy poverty is one of the key problems today. The most famous definition of energy poverty (firewood poverty) adopted Brenda Bordman (1991/2009), which found that the some household is poor with energy if household must spend more than 10% of its income on firewood in order to maintain an adequate level of warmth. The current definition is focused on the home heating costs, but also includes the costs for water heating, lighting and the use of household appliances, as well as costs for cooking, …

The transition of the energy system in South-East Europe is an interesting example for the connection of social and technical developments and social and economical


S. Avdaković et al. (Eds.): IAT 2019, LNNS 83, pp. 126–136, 2020.
https://doi.org/10.1007/978-3-030-24986-1_10

changes. The transformation of a socio-technological infrastructure into liberal market structures and privatization process shocked vulnerable consumers.

The infrastructure of most utilities had to be restored. Energy consumption had to be billed and measured on the household level. There is still a lack of interaction between the development of public social systems, strategies of public housing and infrastructure and energy system and price developments (Buzar 2007) [1]. Most of the EU-countries do not consider the energy poverty problem as essential for their social or energy efficiency policies. There is a gap of comparative data. Until now, exists only one (quantitative) long-term study on the connection between the liberalization of the EU energy system and its consequences for the vulnerable consumers realized by Poggi and Florio in 2011 [2].

The European Commission has published a study on energy poverty in EU countries and ways of combating this kind of poverty on 2015. The study found that many countries of the European Union are implementing measures to protect vulnerable populations. Almost 11% of the population of the EU is energy poor, actually unable to adequately heat their homes at an affordable price.

It is estimated that this affected more than 54 million people in Europe. This problem is the result of rising energy prices, low income and the non-efficient energy homes. This problem is particularly present in the Central, East and South Europe, according to the study [1–3]. The situation in Bosnia and Herzegovina is almost 7 times worst in comparison with the EU countries.

Energy-poor households often do not have enough resources nor the knowledge to meet the basic energy needs. Energy poor households are socially isolated and without anyone to help them.

The conducted studies determined three main causes of energy poverty, which are:

- low household income,
- high energy prices and
- poor quality of accommodation (low level of energy efficiency).

Also, it is very important to recognize indicators of energy poverty. Many signs indicating that household has a problem with energy poverty, which are:

- low income (unemployment, pensions, social assistance);
- inadequately home heating;
- old and low energy efficient homes and buildings;
- delay or outstanding utility bills for energy;
- moisture in walls and floors;
- rotting window frames;
- lack of central heating;
- high housing costs relative to income.

## 2 Energy Poverty in Bosnia and Herzegovina

There is no official definition of energy poverty in B&H and specific quantitative data.

Problem of energy poverty is brought to the public mostly by NGOs activities. Subsidies for heating exist, but vary from canton to canton, Brčko District, Federation B&H and RS.

Due to the lack of an official definition of energy poverty in B&H the national statistical agency does not provide specific data on this issue.

Housing sector is the largest consumer of energy and the main source of greenhouse gas emissions, with about 58% of the initial net final consumption of energy for Bosnia and Herzegovina amounted to 145.54 PJ (3,476.1 ktoe), according to NEEAP (National energy efficiency action plan). Current Standard for construction in BH lag behind EU standards. The old buildings and houses are energy inefficient. Households are the biggest consumers of energy about 52%, industry and transport 40%, while services and agriculture alone consume 6% and 2%. The main source of energy for households is a wood with 57%, electric energy with 18.7% and coal with 10% [1–3]. About 83% of the population lives in family houses, with an average of 3.1 persons per household. According to statistics from the 2016, ratio of employees and pensioners was about 1.14:1, which is disastrous for B&H economy. The average for the EU is 4: 1. About 80% of pensioners is classified in the category of economically poor and on the brink of starvation. According to the NGO sector in the most difficult situation are single mothers. Even 58% of single mothers are not able to pay bills for rent and utilities in the previous period. 42% of single mothers said that they cannot even cover the costs for food.

The price of electricity in B&H is low in comparison with EU countries.

Low consumer electricity prices discourages private investment in renewable energy sources, and the because of this country is lateing in technological development.

Bosnia and Herzegovina has a development dilemma: to continue with the previous practice and guarantee poverty for future generations or change energy sector management policy, turning gradually towards to the rationalization of production and distribution, renewable energy and investments in energy efficiency.

There are indications for changes in the energy sector. First, the policy of subsidies and investments with old technology for energy production has reached a level of economic un-sustainability. Second, the process of European Integration and International Agreements represent a chance for B&H to join the trends in the European Union and get the chance for reforms that will ensure sustainable development.

The current EU trends in terms of investment in renewable energy sources are actually complementary with sustainable development. Investments in energy efficiency in Bosnia and Herzegovina would produce multiple benefits and ensure a return on investment in relatively short term.

At the national level there is no comprehensive energy strategy that could be used as a framework for analysis of energy policy. The responsibility for development of policies in the energy sector in B&H is not good defined.

In Bosnia and Herzegovina there is no definition of vulnerable customers on national level, but the legislations of its entities recognize a category of customers that have to be protected with respect to electrical energy consumption.

In certain cases all households in Bosnia and Herzegovina are protected against disconnection. The criteria for a vulnerable customer status typically are related to the level of income or health problems or disability of a person or its family member. Within the energy sector vulnerable customers are protected with respect to their electricity and gas consumption.

The general social welfare system protects vulnerable customers with respect to energy. Percentage of households qualified for economic support in B&H is 9.48% for electricity and 0.23% for gas. Customer categories receiving non-economic support within the energy sector are customers using medical devices necessary for help. The support scheme for vulnerable customers is seasonally based.

The analysis shows that there is a certain progress in terms of defining and protecting vulnerable customers.

The implementation of economic support schemes dominates over non-economic measures, but they are sometimes used in combination. Economic support systems are mainly based on direct budgetary subsidies. The lack of information might be due to the fact there is no institution following the practical implementation of support measures. Non-economic support within energy sector usually refers to protection against disconnection and it is in most cases provided for customers with poor health status.

District heating systems are still placed in major cities. Before the war most of urban population was connected to district heating systems. Today, most of these systems are in bad conditions and require considerable modernisation. Most of district heating sistems have a technical, market, and commercial problems. A Law on Consumer Protection states that the supplied energy is to be paid in accordance with consumption, rather than by square meter as is now the case. This showed that individual heat metering is a major priority for better quality of service and better management of district heating system.

Country support system should allow the customer groups to actively take part in the liberalized energy market in order to take advantage of cheaper offers and to be able to purchase energy for the best deal [8–10].

The energy production in B&H is based on technologies for thermal-power and hydro-power plants developed forty years ago. Old technologies caused big air pollution problems in the cities.

Experimental consumption metering is introduced locally as a basis for future consumption-based billing. The other district heating systems also consider this billing scheme.

Bosnia and Herzegovina has a major problem with air pollution. The Bosnian authorities haven't taken any concrete measures to decrease the pollution in the country. According to WHO's analysis (http://www.who.int/gho/publications/world_health_statistics/2017/en/) Bosnia and Herzegovina is on the top of the list of the most polluted countries in Europe, with the highest rate of deaths caused by air pollution, with other Balkan and Eastern Europe countries close behind. Bosnia and Herzegovina has particularly high concentrations of particulate matter (PM 2.5), a carcinogen and the main cause of respiratory diseases which kill 231 people per 100 000 across the country. Air pollution in Bosnia and Herzegovina is estimated to have caused over 3500 premature deaths in 2014. According to the World Health Organization (WHO), B&H is losing 21.5 of its GDP (US$ 7228 millions) due to heavy air pollution. The European Environment Agency (EEA), report for 2015. that 44,000 years of life are lost in B&H each year due to air pollution.

Official data published by Elektroprivreda for 2016 state that Tuzla power plant emitted 5861 tons of $NO_x$, 66 431 tons of $SO_2$ and 1017 tons of solid particles and 3 941 042 tons of $CO_2$. None of the power plants in B&H has desulphurization equipment installed.

Converting the pollution health impacts in to monetary equivalent shows that air pollution generates annual health externality expenses in the range of 99 million EUR/year.

UN Environment's Pollution (UNEP) has acted to improve air quality in Bosnia following Resolution 7 of the first United Nations Environment Assembly (UNEA-1), which mandated it to support governments through capacity building, data provision and assessments of progress.

Heating for homes and businesses is one of the biggest energy guzzlers in the country [16, 17].

B&H is without a law on energy efficiency at the State level.

Laws on renewable energy and cogeneration adopted at the Entity level are the main laws governing the use of renewable energy.

The state constitution proved to be a barrier for reforms in the energy sector.

Entities are responsible for energy policy creation and implementation but from the other side, the Energy Community recognized only the state as its partner for energy policy reforms [4–9].

## 3 Results and Discussion

Energy Statistics is not fully functional to ensure sufficient data in the BH renewable energy sector. In the sector of heating and cooling is planned the rise of the participation of renewable energy sources and the share of energy from renewable sources has to be increased from 43.3% to 52.4%. The objective in the sector of heating and cooling for Bosnia and Herzegovina is based on the parameters of the entity action plans for 2020.

In order to achieve action plan objectives it is necessary to use a biomass for heating in households and other forms of renewable energy with the aim of reduction of the share of energy from fossil fuels.

The planned increase in the thermal energy production from solid biomass during the planning period is based on replacement of heating systems in individual buildings and central heating systems from fossil fuels to biomass.

Biomass boilers will decrease carbon dioxide emissions by over 90 per cent while saving €1 million in fuel costs [11–13].

The average representation of flats in the overall structure of housing units surveyed households was 16.4%, while the family houses was 83.6% (Table 1) [14–16].

**Table 1.** The share of households in BiH

| | Share of housing units, % | |
|---|---|---|
| | Dwellings in buildings | Family houses |
| Bosnia and Herzegovina | 16,40 | 83,60 |
| Federation of B&H | 18,60 | 81,40 |

The average heated area of residential unit in Bosnia and Herzegovina is 51.2 $m^2$ and in the urban centers is about 13% higher than in the rural areas. The average size of housing units in Bosnia and Herzegovina which are cooled during the summer is 39.8 $m^2$ (Table 2.).

**Table 2.** The average size of housing units which is heated/cooled

| | The average size of housing units which is heated/cooled, $m^2$ | |
|---|---|---|
| | Heated during the winter | Cooled during the summer |
| Urban | 54.8 | 38.6 |
| Rural | 48.4 | 41.8 |
| Total | 51.2 | 39.8 |

The share of households in B&H, which predominantly heated one room is about 73%, with own central heating is 19% and with district central heating is 7.9%. The share of households without heating is 0.2%. Room heating means the heating with individual furnaces and "split systems" (Fig. 1).

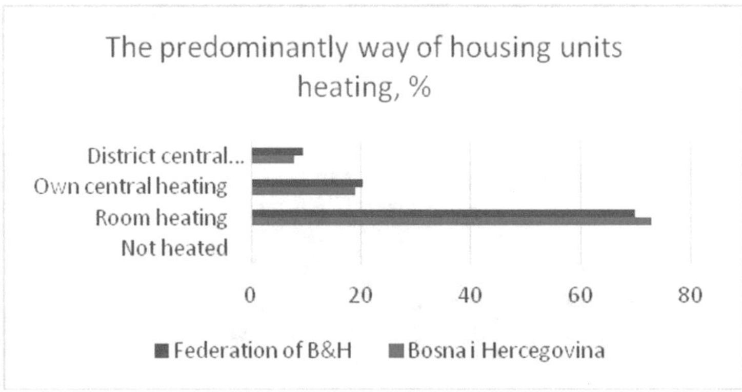


**Fig. 1.** The predominantly way of housing units heating, %

For households heating in BH dominant energy source is wood with 54.5%, then coal with 31.4%, natural gas with 9.1%, electricity with 3.4%, and fuel oil and other oil derivatives are insignificant with 1.6% (Fig. 2).

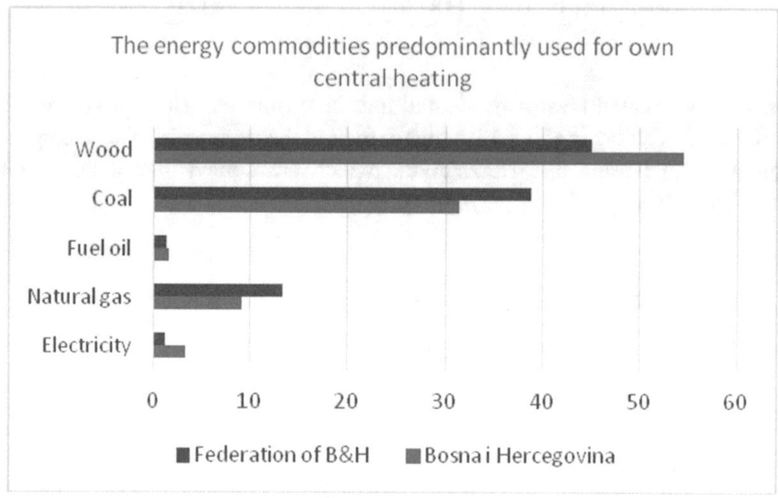


**Fig. 2.** The energy commodities predominantly used for own central heating

The most used energy source for cooking is electricity, 65.8%. The representation of wood for cooking is 27.2%. LPG is 4.5%, natural gas with 2.4%, and the smallest quantity of coal with only 0.1% (Fig. 3).

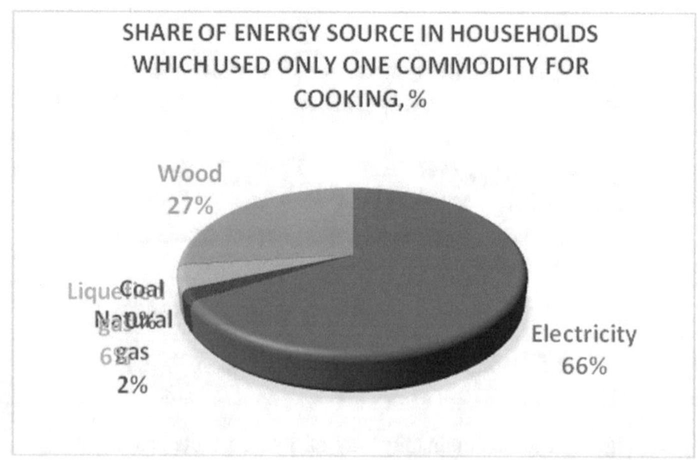


**Fig. 3.** Share of energy source in households for cooking

Electricity is the dominant in a structure of the energy use for hot water, almost 90%.

The average annual electricity consumption per household in B&H is about 4500 kW. Average consumption of firewood in households is around 7.7 $m^3$ per year.

The average coal consumption in B&H in households is about 4 tons per year. The average consumption of final heat in the BH in households is around 7900 kWh per year (Table 3.).

**Table 3.** The average annual energy consumption in households

| | The average annual energy consumption in households | | | | | |
|---|---|---|---|---|---|---|
| Electricity, kWh | Natural gas, $Sm^3$ | Liquefied gas, kg | Wood, spm | Wood and agricultural waste, kg | Coal, tons | Heat, kWh |
| 4568,20 | 871,70 | 67,40 | 10,80 | 3162,20 | 3,90 | 7909,50 |

The three European standardization bodies (CEN, CENELEC and ETSI) and relevant stakeholders established a Smart Metering Co-ordination Group (SM-CG) operating as joint advisory body and focal point for smart metering standardization, in 2009. An evaluation showed that the current level of penetration of smart meters is less than 1%.

The electrical equipment in B&H households is old, on average 10 to 15 years, which contributes to inefficient use of electricity. The research showed that households have about 39% of new TV sets, and 61% older than 10 years (Table 4.).

**Table 4.** The average annual energy consumption in households

| | Age of household appliances | | | |
|---|---|---|---|---|
| | 0–2 year | 3–5 year | 6–10 year | older |
| Freeze | 5,5 | 17,9 | 35,4 | 42,5 |
| Fridge without/with freezer | 7,8 | 25,3 | 38,4 | 32,3 |
| Washing machine | 8,2 | 27,9 | 38 | 26,8 |
| TV | 18,6 | 34,1 | 35,2 | 25,6 |
| Computer/Laptop | 21,1 | 51,1 | 27,8 | 5,5 |

Households have 41% of the new freezers and refrigerators, and 59% older than 10 years. Households have around 42% of new washing machines and about 58% older than 10 years.

All researchs were conducted within the projects realized by NGOs in Tuzla and Zenica-Doboj Canton. The aims of projects were to strength the energy poor households to take action in saving energy and changing their homes.

Also, one of the aims was establishment of energy poverty as a problem that requires customized policies and measures. Some of the projects were simultaneously implemented in Bosnia and Herzegovina, Serbia and Montenegro.

About 83% of monitored households use ordinary bulbs, 21% energy saving bulbs and 14% LEDs (Fig. 4).

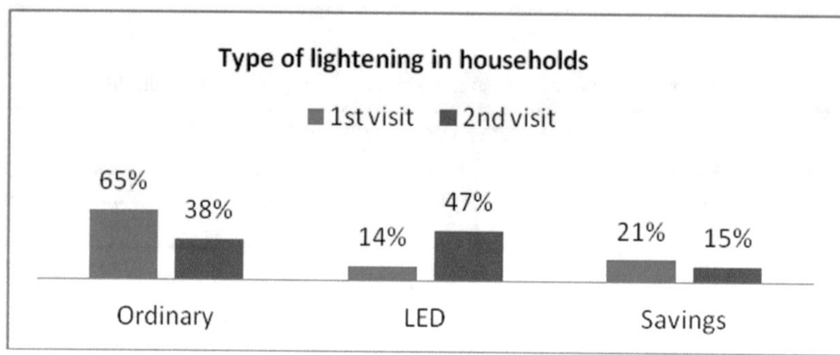


**Fig. 4.** Type of lightening in households

About 16% of monitored households did not know meaning of EE class and did not have knowledge about energy poverty (Table 5.). About 93% of monitored households did not have knowledge that old devices consume 50% more electricity than new devices (Table 6.).

**Table 5.** Check of the EE class during devices purchasing

| | Yes | No | Sometimes | I do not know meaning of EE class |
|---|---|---|---|---|
| 1st visit | 29% | 35% | 20% | 16% |
| 2nd visit | 71% | 6% | 18% | 5% |

**Table 6.** Knowledge that old devices consume 50% more electricity than new devices

| | Yes | No |
|---|---|---|
| 1st visit | 68% | 93% |
| 2nd visit | 32% | 7% |

It is encouraging that after visit of energy advisors, there have been seen a positive developments, such as:

- 90% of households give more attention on saving the electricity
- 78% of households decreased energy bill
- 70% of households start to think about replacement of old electric devices with new energy efficient

- 85% of households now have knowledge about energy poverty
- 79% of households passed on knowledge about energy saving to the others.

In monitored households electricity consumption is reduced by 6.2%, and at 66% of households who participated in the creation of savings, reduced electricity consumption by 20% [14–16].

The main tools for reducing energy poverty are:

- The energy efficiency rise
- Financial measures.

Energy poverty is a problem that relevant institutions should identify and put in the focus of its activities. It is necessary to establish a policies and measures at local, national and EU level for successful solution of energy poverty problem. Possible approach against energy poverty based on available data and analyzes could consist of:

- increase the amount of funds in the state budget for energy poverty problems
- determination of indicators for identification of energy-poor households
- identifying ways to increase energy efficiency and reduce energy poverty
- organization of a regular energy advising of vulnerable households
- development of programs for co-funding of energy rehabilitation of poor households (through installation of thermal insulation of facades, windows replacement, replacement of inefficient heating, the purchase of energy efficient appliances and efficient lighting)
- organization of campaigns and education about energy poverty problem.

## 4 Conclusion

Energy poverty is a very complex problem consist of:

- the financial inability of the energy poor households to invest in improvement of energy efficiency
- the environmental performance of these households and
- the social situation of a energy poor households.

This paper tries to give a new view to this complexity by an overview of the current researchs.

Energy poverty is wide spread in B&H and has several causes: rising share energy costs in the incomes, low incomes, inequality in income distribution, historic legacy, energy inefficient housing stocks, unsustainable energy sector, high energy losses and market and regulatory failures.

Regulatory or financial measures lead to the accomplishment of objectives such as energy market liberalization, reduction of pollution and greenhouse gas emissions, improvement in security of energy supply or reduction of the energy cost households.

From a technical point of view, new, market-ready technologies should be further fostered. Cost-effective insulation materials and heating solutions should be promoted and subsidized. The smart metering program should be further developed. The effectiveness of the measures depends on the funding and targeting processes. In B&H funding is inadequate and targeting process is very weak. It is necessary to provide funds for housing renovation and building retrofit programs and restrict access to private capital markets for low-income and energy vulnerable households.

Some characteristics of today's power energy sector are [17]: market oriented sector, geostrategicaly important sector, sector connected with ecology and climate changes and energy poverty, financial intensive sector.

Appropriate education within school programs and public campaigns will offer to consumer information about energy consumption, energy supply options and energy costs. The proposed measures will lead in the reduction of energy poverty.

## References

1. Bouzarovski, S., Petrova, S., Sarlamanov, R.: Energy poverty policies in the EU: a critical perspective. Energy Policy **49**, 76–82 (2012)
2. Poggi, A., Florio, M.: Energy deprivation dynamics and regulatory reforms in Europe: evidence from household panel data. Energy Policy **38**(1), 253–264 (2011)
3. Buzar, S.: The 'hidden' geographies of energy poverty in post-socialism: between institutions and households. Geoforum **38**(2), 224–240 (2007)
4. Dubois, U., Meier, H.: Energy affordability and energy inequality in Europe: implications for policymaking. Energy Res. Soc. Sci. **18**, 21–35 (2016)
5. SEERMAP: South East Europe Electricity Roadmap Country Report: Bosnia and Herzegovina (2017). ISBN 978-615-80814-9-8
6. Ruggeri Laderchi, C., Olivier, A., Trimble, C.: Cutting Energy Subsidies While Protecting Affordability. ISBN (paper): 978-0-8213-9789-3
7. In-Depth Review of Energy Efficiency Policies and Programmes. Energy Charter Secretariat (2012). ISBN: 978-905948-103-9 (English PDF)
8. Doleček, V., Karabegović, I.: Renewable energy sources in Bosnia and Herzegovina: situation and perspective. Contemp. Mater. (Renewable Energy Sources) **2**(IV), 152–163 (2013)
9. Granić, G., Željko, M.: Study of energy sector in BiH, Sarajevo (2008)
10. Armstrong, J., Hamrin, J.: What are "Renewable resources?" Chapter 1: Renewable Energy Overview, Cost-Effectiveness of Renewable Energy. Organization of American States (2013)
11. The International Energy Agency (IEA): Energy in the Western Balkans: The Path to Reform and Reconstruction (2008). ISBN: 978-92-64-04218-6
12. Jenko, J.: B.Sc. El. Eng: Power Sector Development background paper. In: World Bank – 4th Poverty Reduction Strategies Forum Athens, Greece, 26–27 June 2007
13. Council of Ministers of BIH Government of Federation of Bosnia and Herzegovina Government of Republika Srpska: Bosnia and Herzegovina: Poverty reduction strategy paper-mid-term development strategy, Bosnia and Herzegovina, March 2004
14. Agency for Statistics of Bosnia and Herzegovina, Survey on Household Energy Consumption in BIH (2015)
15. Centar za ekologiju i energiju, Energetsko siromaštvo u BiH, Tuzla, juni 2017
16. Centar za ekologiju i energiju PREGLED SITUACIJE U POGLEDU ENERGETSKOG SIROMAŠTVA U ZENIČKO-DOBOJSKOM KANTONU, mart, 2018
17. Tomšić, Ž.: Energetska sigurnost, niskougljični razvoj i investicije u elektroenergetici, 4. Energy Investment Forum, Hrvatska (2017)

# Electricity Consumption Forecasting in the Western Balkan Countries

Maja Muftić Dedović(✉), Emir Šaljić, Lejla Jugo, Zekira Harbaš, Azra Đelmo, Lejla Hasanbegović, and Samir Avdaković

Faculty of Electrical Engineering, University of Sarajevo, Sarajevo, Bosnia and Herzegovina
maja.muftic-dedovic@etf.unsa.ba

**Abstract.** This paper presents results of electricity consumption forecasting in the Western Balkan countries. Electricity consumption forecasting in the Western Balkan countries Bosnia and Herzegovina, Serbia, Montenegro, Albania, Kosovo and North Macedonia are done using conventional techniques based on extrapolation techniques using correlation between gross domestic product and electricity power consumption. Also, for electricity consumption forecasting is used advanced technique artificial neural intelligence. Inputs for training artificial neural network are for the first test case gross domestic product, population and decomposed time series of electricity consumption using Huang's Empirical Mode Decomposition. The forecasting results using time series of electricity consumption processed with Discrete Wavelet Transform as input for training artificial neural network are also presented. Results of all used techniques are also presented, compared and discussed. Forecasted results of electricity consumption in Western Balkan for 2035 year indicate increase of about 20% of electricity consumption compared to 2017 year.

## 1 Introduction

The prediction of electricity consumption guarantees reliable planning and management of distribution systems. It helps an electric utility to make important decisions such as purchase and generation of electric power, as well as development, planning and connection of new generation capacity to the transmission network. Forecasting of electricity consumption can be divided into three categories: short-term load forecasting (STLF) which is usually from one hour to one week, medium-term load forecasting (MTLF) which is usually from a week to a year, and long-term load forecasting (LTLF) which is required to be valid from 5 to 25 years. A LTLF is generally known as an annual peak load forecasting [1] and this paper addresses this type of prediction of electricity consumption. Historical load, time factors and weather data are factors which are considered for STLF and MTLF. LTLF is much more difficult and challenging problem and it is affected by the socio-economic and demographic data and their forecast such as gross domestic product (GDP) and population (POP). Other factors which can be considered are those that depend on the electric power system in the country like power losses in the power system (in MW) and load factor (LF).


S. Avdaković et al. (Eds.): IAT 2019, LNNS 83, pp. 137–147, 2020.
https://doi.org/10.1007/978-3-030-24986-1_11

The most commonly used techniques for performing accurate load forecasting include statistically based techniques like time series, regression techniques and computational intelligence method like fuzzy systems, ANNs, wavelets and neuro-fuzzy systems. Linear regression is a simple method for finding a relation between two or more variables. This technique is proposed in [2] and the error performances are obtained. Linear regression is usually insufficient as load data are highly nonlinear and nonstationary time series. Therefore, when variables are nonstationary it is better to obtain forecast on the basis of artificial neural networks [3]. The artificial neural network (ANN) is an artificial replica of the human brain which attempts to simulate the learning process and it splendidly copes with the problems of classification and forecasting. Generally, it copes well with all the problems in which a relationship between input and output variables is involved, irrespective of the high complexity of that link (nonlinearity) [4]. Using neural networks enhances the forecasting accuracy associated with other prediction methods due to the ease of use and the excellence in performance [5] and it is overall concluded that application of artificial intelligence techniques provides faster and more accurate results than conventional methods [6]. In order to increase the accuracy of electricity consumption prediction, in literature can be found hybrid models that are combination of ANN models and Huang's Empirical Mode Decomposition (EMD) [3]. Model based on EMD technique and ANN for long-term load forecast is proposed in the paper [7]. Simulation results indicate that accuracy of the proposed model is much higher than the traditional linear combination model. Discrete Wavelet Transform (DWT) has also been used to improve ANN forecasting accuracy [8, 9]. Simulation results obtained in [9] indicate that proposed model has shown much improvement in forecasting and it has shown the capacity to simulate non-linear behavior of the phenomenon more accurately than single ANN model.

This paper is organized as follow: In second section input data and methods for forecasting electricity consumption are presented. The third section results of electricity consumption forecasting for selected region are presented and discussed. The last section is reserved for conclusion.

## 2 Materials and Methods

In this section input data used for forecasting electricity consumption of Western Balkan countries are described. Also, materials and methods are shortly described.

### 2.1 Data

The main goal of this research is forecasting electricity consumption of Western Balkan countries. Unstable political situation and low economic growth are characterized for this region in the last 30 years. Western Balkan countries are Bosnia and Herzegovina, Serbia, Montenegro, Albania, Kosovo and North Macedonia (Fig. 1).

**Fig. 1.** Western Balkan countries

The Western Balkan annual electricity consumption for the period from 2000 to 2017 is plotted on Fig. 2, while Western Balkan GDP and population are plotted on Figs. 3 and 4.

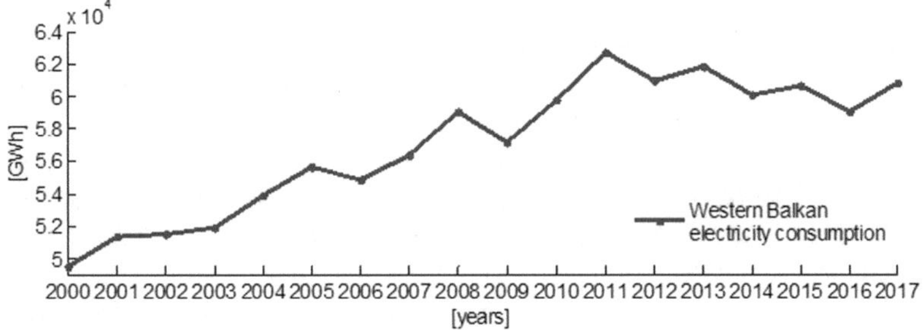


**Fig. 2.** Western Balkan annual electricity consumption

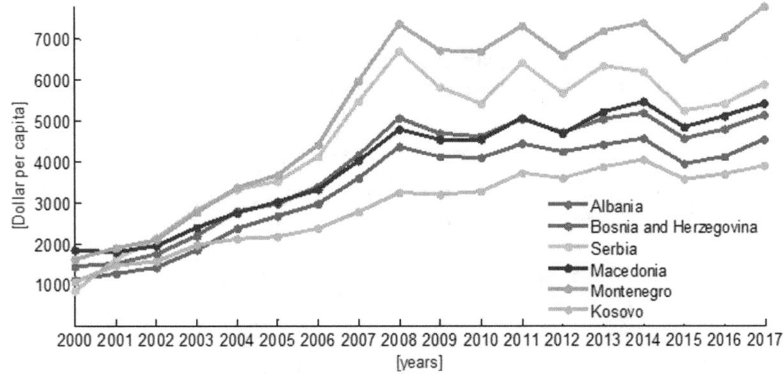


**Fig. 3.** Western Balkan countries GDP

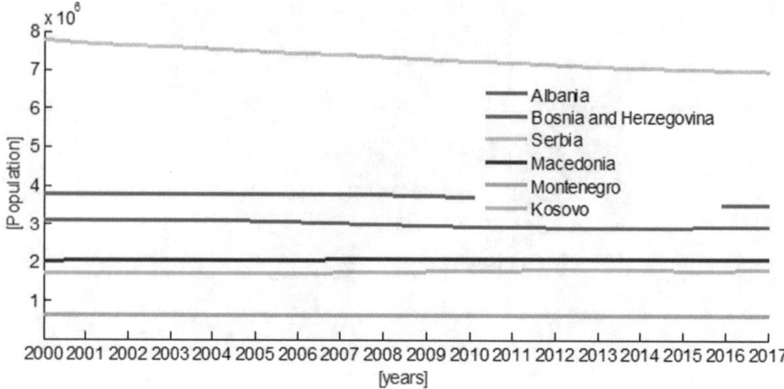


**Fig. 4.** Western Balkan countries population

From Fig. 3 can be concluded that GDP slightly increase for the all Western Balkan countries. On the other hand, for period 2000 to 2017 for the observed region population has been decreased for about million people (Fig. 3).

## 2.2 Linear Regression

Linear regression has been used for forecasting of electricity consumption in this paper. Linear regression represents a statistical technique used for finding a relation between two or more variables. It is called simple linear regression if the relation is found between two variables and if the relation is found for more variables it is called multi variable linear regression. After finding relation between the variables assumption is made that the parameters are varying with same relation and that relation is applied to the forthcoming parameters, which will give the values of dependent variable value for the corresponding forthcoming independent variable. Linear regression is quite simple method and it takes form $y = a + bx$, where b is the slope of the curve, a is the intercept, x is the independent variable and y is dependent variable [3]. In our case

$$W = a + b \cdot t \tag{1}$$

where W is electrical energy consumption and t is the time in years.

In order to calculate coefficient $a$ and $b$ the following formulas are used:

$$a \cdot n + b \cdot \sum_{i=1}^{n} t_i = \sum_{i=1}^{n} W_s(t_i) \tag{2}$$

$$a \cdot \sum_{i=1}^{n} t_i + b \cdot \sum_{i=1}^{n} t_i^2 = \sum_{i=1}^{n} t_i \cdot W_s(t_i) \tag{3}$$

$$n \cdot a + b \cdot \sum_{i=1}^{n} t_i = \sum_{i=1}^{n} W_s(t_i) \tag{4}$$

$$a \cdot \sum_{i=1}^{n} t_i + b \cdot \sum_{i=1}^{n} t_i^2 = \sum_{i=1}^{n} t_i \cdot W_s(t_i) \tag{5}$$

where $W_s(t_i)$ is real energy consumption in the past time.

According to correlation between two variables, in our paper variable $W(t)$ and GDP the following formulas are used:

If it is assumed that changing the x and y with time t is in form $x = a_x + b_x t$ and $y = a_y + b_y t$, the coefficients will be determined by the table procedure:

$$a_x = \frac{\sum_{i=1}^{n} x \cdot \sum_{i=1}^{n} t^2 - \sum_{i=1}^{n} t \cdot \sum_{i=1}^{n} tx}{N \cdot \sum_{i=1}^{n} t^2 - \left(\sum_{i=1}^{n} t\right)^2}; b_x = \frac{N \cdot \sum_{i=1}^{n} tx - \sum_{i=1}^{n} t \cdot \sum_{i=1}^{n} x}{N \cdot \sum_{i=1}^{n} t^2 - \left(\sum_{i=1}^{n} t\right)^2} \tag{6}$$

$$a_y = \frac{\sum_{i=1}^{n} y \cdot \sum_{i=1}^{n} t^2 - \sum_{i=1}^{n} t \cdot \sum_{i=1}^{n} ty}{N \cdot \sum_{i=1}^{n} t^2 - \left(\sum_{i=1}^{n} t\right)^2}; b_y = \frac{N \cdot \sum_{i=1}^{n} ty - \sum_{i=1}^{n} t \cdot \sum_{i=1}^{n} y}{N \cdot \sum_{i=1}^{n} t^2 - \left(\sum_{i=1}^{n} t\right)^2} \tag{7}$$

After finding coefficients $a_x$, $b_x$, $a_y$ and $b_y$ electricity consumption and GDP for the following period is predicted based on the expression:

$$GDP = a_x + b_x t \tag{8}$$

$$W(t) = a_y + b_y t \tag{9}$$

In order to set the correlation equation $y = a + bx$ coefficients a and b will be determined by formulas:

$$a = \frac{\sum_{i=1}^{n} y \cdot \sum_{i=1}^{n} x^2 - \sum_{i=1}^{n} x \cdot \sum_{i=1}^{n} xy}{N \cdot \sum_{i=1}^{n} x^2 - \left(\sum_{i=1}^{n} x\right)^2}; b = \frac{N \cdot \sum_{i=1}^{n} xy - \sum_{i=1}^{n} x \cdot \sum_{i=1}^{n} y}{N \cdot \sum_{i=1}^{n} x^2 - \left(\sum_{i=1}^{n} x\right)^2} \tag{10}$$

After finding coefficients a and b consumption for the following period is predicted based on the expression $W(GDP) = a + b \cdot GDP_i$, and correlation coefficient is determinate by expression:

$$r = \frac{\sum_{t=1}^{n} X_t Y_t}{\sqrt{\sum_{t=1}^{n} X_t^2 \cdot \sum_{t=1}^{n} Y_t^2}} \tag{11}$$

## 2.3 EMD Approach

The Empirical Mode Decomposition represents a technique used for decomposition given signal into a set of elemental signals called Intrinsic Mode Functions (IMF). This technique is the part of the Hilbert–Huang Transform. The Empirical Mode Decomposition comprises a Hilbert Spectral Analysis and an instantaneous frequency computations. Every IMF must satisfy two conditions. The first one is that the extreme number (min and max) of zero-crossing points should be equal or differ only by one. The second one is that the median value of envelopes at any point for IMF is equal to zero.

Algorithm of EMD can be explained in four steps. First one is finding all the local extrema and generating its upper and lower envelopes using a cubic spine line ($e_{up}(t)$ and $e_{low}(t)$, respectively). Then calculating the mean value of the $e_{up}(t)$ and $e_{low}(t)$: $m_1(t) = \frac{(e_{up}(t) + e_{low}(t))}{2}$, after that $h_1(t)$ is calculated by expression $h_1(t) = x(t) - m_1(t)$. These steps should be repeated until $h_1$ satisfies a set of predetermined stopping criteria for the IMF function: $c_1 = h_{1k}$. If $h_1(t)$ does not satisfy the criteria for the existence of IMF, the previous procedure is repeated and determined $h_{11}(t)$ as a new signal: $h_{11}(t) = x(t) - m_{11}(t)$, where $m_{11}(t)$ represents mean value of the upper and lower envelope of the signal $h_1(t)$. The last step is computing residue and repeating calculation of the mean value of the $e_{up}(t)$ and $e_{low}(t)$ 2 $k$ times to obtain $n$ IMFs and residual, then the new signal is:

$$x(t) = \sum_{i=1}^{n} c_i(t) + r_n(t) = \sum_{i=1}^{q} c_i(t) + \sum_{j=q+1}^{p} c_j(t) + \sum_{k=p+1}^{n} c_k(t) + r_n(t)$$

where $q<p<n$, $c_i(t)$ are *ith* IMF, $c_j(t)$ are the components representing properties of the series and $c_k(t)$ and $r_n(t)$ are the final residual, trend non-sinusoidal components [11–14].

## 2.4 DWT Approach

DWT is an algorithm that is being used for defining wavelet coefficients and scale functions in dyadic scales and dyadic points [15–17]. First, it's necessary to split approximation and discrete signal details to get two signals. Those two signals have the length of the original signal, which means that now we got double amount of the data. Using compression, the length of output signals is being split in half. The received approximation represents input signal in the following step. Digital signal $f(n)$ in the frequency range $0 - F_s/2$ ($F_s$ is the sampling frequency), passes through lowpass $h(n)$ and highpass $g(n)$ filter, so the each filter lets by just one half of the frequency range of the original signal. Then signals are subsampled with the goal to remove any other sample. $cA_1(k)$ and $cD_1(k)$ are outputs from $h(n)$ and $g(n)$ filter, respectively. Filtrating process and subsampling of input signal can be represented as: $cA_1(k) = \sum_n h(2k - n)$ and $cD_1(k) = \sum_n f(n)g(2k - n)$. Approximation of the first level of decomposition (coefficients $cA_1(k)$) represent input signal in frequency range $0 - F_s/4$, $cD_1(k)$ are coefficients of details in frequency range $F_s/4 - F_s/2$ Hz. Algorithm is being continued so the number of samples decreases which worsens time resolution, however, frequency resolution improves, because frequency ranges, for which the signal is observed, are getting narrower.

### 2.5 ANN Technique

Basic approach of ANN technique is presented and detail elaboration about ANN in this paper will be omitted. Based on [10], an artificial neuron needs to receive raw data or the output from the another neuron, data is being processed and one output is being produced. Network represents input layer, output layer with one or more hidden layers. Topology of the network may be different which depends on the problem to be solved, the type of input layer, output layers and other factors. Artificial neuron model is being created based on input variable to the ANN (for example GDP, load, temperature, etc.), what to forecast (yearly loads, peak load, etc.), neural network model (number of hidden layers or neurons in the hidden layer), training method and stopping criterion, size of the training and test data, etc. All methodological approach used in this paper and using ANN are implemented in Matlab. Input data used for training neural networks are presented in second section.

## 3 Result and Discussion

Based on methodology presented in second section results of forecasting electricity consumption using linear regression and correlations are plotted on Fig. 5. From Eq. (1) in our case characteristic equation is $W = 50481 + 692,1 \cdot t$, where W is electrical energy consumption for all countries of Western Balkan and t is the time in

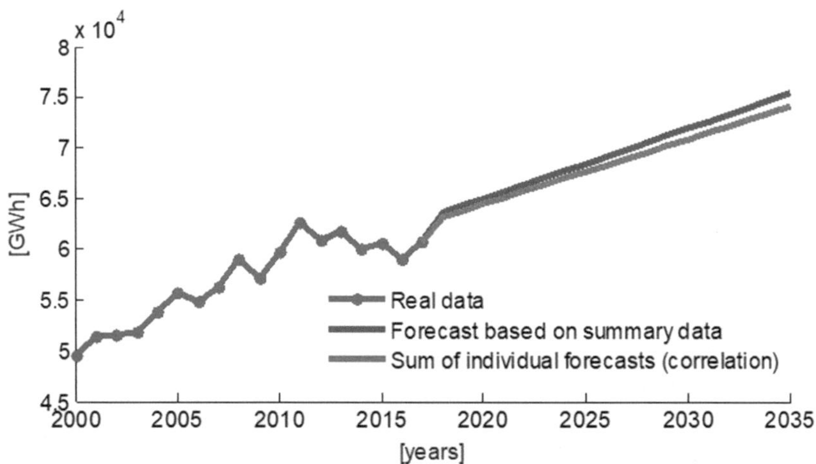


**Fig. 5.** Linear progression based on data from 2000 to 2017

**Table 1.** The characteristic coefficients for each country according to correlation technique

| | $a_x$ | $b_x$ | $a_y$ | $b_y$ | a | b |
|---|---|---|---|---|---|---|
| BiH | 1661.095 | 230.2707 | 9414.131 | 161.7874 | 8450.671 | 0.64969 |
| Montenegro | 1831.572 | 371.3126 | 3818.332 | −54.3131 | 3809.14 | −0.09457 |
| Serbia | 1947.058 | 280.8346 | 25033.28 | 235.7611 | 24155.68 | 0.67548 |
| Macedonia | 1734.147 | 232.1286 | 5687.761 | 81.15547 | 4800.492 | 0.420942 |
| Albania | 1366.523 | 208.042 | 3969.493 | 141.8159 | 3500.164 | 0.543411 |
| Kosovo | 1276.106 | 168.6055 | 2557.693 | 125.9349 | 1551.028 | 0.765516 |

years. Finally forecasting results using linear regression are presented on Fig. 5 as forecasted based on summary data. On the other hand, based on Eqs. (6–9) are calculated coefficients presented in Table 1 and represents characteristic coefficients for each country.

According to the characteristic coefficients forecasting electricity consumptions results for period 2018–2035 are obtained for each country and sum of all forecasting results is plotted on Fig. 5.

In order to increase the accuracy of electricity consumption prediction in this paper is used artificial intelligence techniques which provides faster and more accurate results than conventional methods. The test case denotes the long-term electricity consumption forecasting and is separated in two parts. In first case, the joint usage of the ANN and EMD approaches is introduced and the second test case is application of the two stage adaptive model based on joint usage of the ANN and DWT approach.

In Table 2 are presented parameters of the ANN structure for which are developed the best forecasting results. Feed-forward backprop neural networks is used. Then dataset are divided into train, train target, test, test target and network output. The ANN model has three layers. First layer are input data or train dataset, then there are hidden layer and output layer. The ANN are trained with following input data: GDP and population from 2000 to 2017, IMFs and residuals of electricity consumption time series (EMD method) and details and approximations of electricity consumption time series (DWT method). Each node in input layer is connected to nodes in hidden layer and every signal of nodes is obtained with activation function and then sent to the output layer reaching the target. The target in our test cases are forecasting electricity consumption results obtained by linear regression. This algorithm repeats until minimum error between outputs and targets are achieved.

The forecasting results according to input data case are plotted on Fig. 6. For the first test case the sum of individual forecast with GDP and population by ANN-DWT approach are obtained by DWT decomposition of the electricity consumption time series from 2000 to 2017 for each country of observed region and GDP and population of each country for period 2000 to 2017. After forecasting electricity consumption for each country the sum of the mentioned results is obtained. Furthermore, for the second test case the sum of individual forecast with GDP and population by ANN-EMD

**Table 2.** Parameters of the ANN model

| Case | Parameters of the ANN model | | | | |
|---|---|---|---|---|---|
| | ANN model | Network | Transfer function | Training function | Learning function |
| I | ANN-EMD (IMFs and residual, GDP and population of each country) | 5 × 10 × 1 | TANSIG | TRAINLM | LEARNGDM |
| II | ANN-DWT (Details and approximation, GDP and population of each country) | 7 × 10 × 1 | TANSIG | TRAINLM | LEARNGDM |
| III | ANN-EMD (IMFs and residual of Western Balkan electricity consumption and population of Western Balkan countries combined) | 3 × 10 × 1 | TANSIG | TRAINLM | LEARNGDM |
| IV | ANN-DWT (Details and approximation of Western Balkan electricity consumption and population of Western Balkan countries combined) | 6 × 10 × 1 | TANSIG | TRAINLM | LEARNGDM |

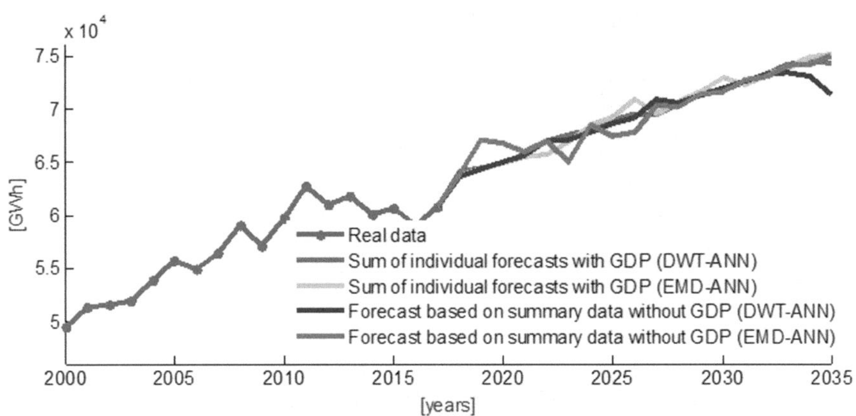


**Fig. 6.** The forecasting results according to input data case

approach are obtained by EMD decomposition of the electricity consumption time series from 2000 to 2017 for each country of observed region and GDP and population of each country for period 2000 to 2017.

In order to compare and validate results in the second test case forecasted results based on summary data are obtained. The third and fourth test cases are analyzed by ANN-DWT and ANN-EMD approaches respectively. Input data for those test cases are only population from 2000 to 2017 and DWT and EMD decomposition of summarized electricity consumption of Western Balkan.

In order to validate proposed ANN-DWT model and ANN-EMD model on training and testing data set, calculation of the coefficient of correlation *r* between actual and forecasted values is performed. The value of the coefficient of correlation *r* between target data and forecasted values obtained after application of ANN-DWT approach is 0,9595 and the value of the coefficient of correlation between target data and forecasted values obtained after application of ANN-EMD approach is 0,9564.

The values of the coefficients of correlations *r* between forecasted data with GDP as one of the input data for ANN and forecasted values obtained after application of ANN-DWT approach and ANN-EMD approach without the knowledge of GDP for ANN are 0,9689 and 0,9308, respectively. Those coefficients of correlations show accuracy of the model and association between output and target data. In addition, in both cases, the tested case results are satisfactory and presents a strong positive relationship.

## 4 Conclusion

In this paper are presented results of electricity consumption forecasting in the Western Balkan countries. First the results are obtained by linear regression. In order to achieve the more accurate results is used adaptive intelligent methodology with usage of advanced artificial intelligence with Huang's Empirical Mode Decomposition and Discrete Wavelet Transform. Results of all used techniques are promising and show the good performance of the applied methods. Forecasted results of electricity consumption in Western Balkan for 2035 year indicate increase of about 20% of electricity consumption compared to 2017 year although there is a decline in the number of inhabitants for about a million for the entire observed region. Also calculated coefficients of correlations for all test cases are indicators of strong correlation between target and forecasted variables.

## References

1. Feilat, E.A., Talal Al-Sha'abi, D., Momani, M.A.: Long-term load forecasting using neural network approach for Jordan's power system. Engineering Press (2017)
2. Dinesh Reddy, M., Valley, R.K., Vishali, N.: Load forecasting using linear regression analysis in time series model for RGUKT, R.K. valley campus HT feeder. Int. J. Eng. Res. Technol. **6**(5). ISSN: 2278-0181 (2017)
3. Muftić-Dedović, M., Dautbašić, N., Mujezinović, A.: Application of artificial neural network and empirical mode decomposition for predications of hourly values of active power consumption. In: 28th DAAAM International Symposium on Intelligent Manufacturing and Automation. https://doi.org/10.2507/28th.daaam.proceedings.xxx
4. Tačković, K., Nikolovski, S., Boras, V.: Kratkoročnoprognoziranjeopterećenjaprimjenom-modelaumjetneneuronskemreže, Energija 5, Zagreb (2008)
5. Mansouri, V., Akbari, M.E.: Neural networks in electric load forecasting: a comprehensive survey. J. Artif. Intell. Electr. Eng. **3**(10), 37–50 (2014)

6. Hadžić, A., Dautbašić, N., Konjić, T., Bećirović, E.: Prognozadnevnogdijagramaopterećenjaprimjenomneuronskihmrežai ARIMA modela, BosanskohercegovačkikomitetMeđunarodnogvijećazavelikeelektričnesisteme – BH K Cigre, Sarajevo (2017)
7. Bai, W., Liu, Z., Zhou, D., Wang, Q., Research of the load forecasting model base on HHT and combination of ANN. In: Power and Energy Engineering Conference, APPEEC (2009)
8. Khandelwal, I., Adhikari, R., Verma, G.: Time series forecasting using hybrid ARIMA and ANN models based on DWT decomposition. Procedia Comput. Sci. **48**, 173–179 (2015)
9. Deka, P.C., Haque, L., Banhatti, A.G.: Discrete wavelet-ANN approach in time series flow forecasting - a case study of Brahmaputra river. Int. J. Earth Sci. Eng. **5**(4), 673–685 (2012)
10. Zhang, W., Mu, G., Yan, G., An, J.: A power load forecast approach based on spatial-temporal clustering of load data. National Natural Science Foundation of China, Grant/Award Number: NSFC-51437003, 21 October 2017
11. Lee, K.Y., Park, J.H.: Short-term load forecasting using an artificial neural network. Trans. Power Syst. **7**(1), 124–132 (1992)
12. Huang, H.E., Shen, Z., Long, S.R., Wu, M.C., Shih, H.H., Zheng, Q., Yen, N.-C., Tung, C.C., Liu, H.H.: The empirical mode decomposition and the Hilbert spectrum for nonlinear and non-stationary time series analysis. Proc. R. Soc. Lond. A **454**, 903–995 (1998)
13. Dedovic, M.M., Avdakovic, S., Dautbasic, N.: Impact of air temperature on active and reactive power consumption - Sarajevo case study. Bosanskohercegovačkaelektrotehnika, under revision
14. Dedovic, M.M., Avdakovic, S., Turkovic, I., Dautbasic, N., Konjic, T.: Forecasting PM10 concentrations using neural networks and system for improving air quality. In: 2016 XI International Symposium on Telecommunications (BIHTEL), pp. 1–6 (2016)
15. Avdakovic, S., Nuhanovic, A., Kusljugic, M.: Wavelet theory and applications for estimation of active power unbalance in power system. Advances in Wavelet Theory and Their Applications in Engineering, Physics and Technology. IN-TECH. ISBN 979-953-307-538-8 (2012)
16. Avdaković, S., Čišija, N.: Wavelets as a tool for power system dynamic events analysis – state-of-the-art and future applications. J. Electr. Syst. Inf. Technol. **2**(1), 47–57 (2015)
17. Avdakovic, S., Ademovic, A., Nuhanovic, A.: Insight into the properties of the UK power consumption using a linear regression and wavelet transform approach. Elektrotehniški Vestnik/Electrotech. Rev. **79**(5), 278–283 (2012)

# ROCOF Estimation via EMD, MEMD and NA-MEMD

Maja Muftić Dedović(✉), Samir Avdaković, Nedis Dautbašić, and Adnan Mujezinović

Faculty of Electrical Engineering, University of Sarajevo, Sarajevo, Bosnia and Herzegovina
maja.muftic-dedovic@etf.unsa.ba

**Abstract.** This paper presents results of the rate of change of frequency (ROCOF) estimation using Huang's Empirical Mode Decomposition (EMD), Multivariate Empirical Mode Decomposition (MEMD) and Noise-Assisted Multivariate Empirical Mode Decomposition (NA-MEMD). On the generated test signals algorithms are performed and the obtained results are compared and discussed. The results are compared with actual values of the rate of change of frequency obtained by derivatization of the generated test signals as input data of the aforementioned algorithms. The performance of the algorithm are also tested using signals contaminated by zero-mean Gaussian noise. The results of rate of change of frequency estimation indicates that all three algorithms have great accuracy in the rate of change of frequency estimation.

## 1 Introduction

The empirical mode decomposition (EMD) is a data driven method and transform the signal into a finite set of intrinsic mode functions (IMFs) each of which contains a portion of the frequency spectrum and residual, leaving well-localized patterns at the instantons frequency levels. Addition of all IMFs with the residual after the application of the method reconstructs the original signal stay the same without loss of information or distortion of the signal. By using an EMD method, a complex signal can be decomposed to the IMFs, in which there is only one frequency component at a time so it is possible to calculate very well the estimation of the current frequency. It is data-adaptive, comprehensive and flexible method and can be nonlinear and nonstationary, having meaningful interpretation of nonlinear process [1]. EMD advantages are a compact decomposition, inherent ability to process nonstationary data and deterministic signals of oscillatory nature [2–5]. EMD disadvantages are that it is for signals with a sufficient number of local extrema and is designed for univariate data. But the most important disadvantages are mode-mixing phenomenon occurring in the EMD process at similar frequencies for different IMFs and aliasing, the overlapping of IMF due to sub-Nyquist extrema sampling [6, 7].

In order to overcome EMD disadvantages, the EEMD algorithm is developed and employs ensemble averaging of noisy signal realizations [8]. According to mode mixing EEMD has dyadic filterbank property and enhanced local mean estimation in noisy data yielding less prone to mode mixing IMFs [9]. As EEMD is a time–space


S. Avdaković et al. (Eds.): IAT 2019, LNNS 83, pp. 148–158, 2020.
https://doi.org/10.1007/978-3-030-24986-1_12

analysis method, the added white noise is averaged out with sufficient number of trials and the averaging process leave only parts or the components of the original signal treated as the true and more physical meaningful answer.

The MEMD algorithm was firstly developed for the bivariate signals. For multivariate data is developed extended standard EMD called multivariate EMD by [10]. Some of the properties of MEMD as coherence and data fusion algorithm are that data are analyzed in $p$-dimensional domain and estimates the local p-dimensional mean through averaging of multiple envelopes. Also, $p$-data channels have the same number of IMFs with scale integrity, however mode-mixing is still a problem because IMFs with same index have aligned composite scales.

The NA-MEMD is designed to add noise, actually $l$-independent WGN realizations to $n$-channel multivariate data. In the NA-MEMD n-variate IMFs corresponding to the original signals are extracted from the $n + l$-variate IMFs eliminating $l$-channel of the additive noise [11]. Using the MEMD method the multivariate signal with noise channel is processed and the IMFs are reconstructed to obtain the desired decomposition. According to this, in the NA-MEMD are prevented direct noise artifacts and alleviate the effects of noise interference in the univariate noise-assisted EMD [12].

Even those method have their advantages and disadvantages according to obtained IMFs, this paper has main goal to investigate behavior of residuals obtained by EMD algorithm and its noise assisted and multivariate extensions, multivariate EMD regarding estimation of the rate of change of frequency.

In this paper, second chapter is reserved for generation of input synthetic signals as frequency signals. Calculation of derivate of those frequency signals and application of the random noise to those signals. In third chapter results and discussion are outline and compared to the rate of change of frequency actual values. In the fourth chapter are given final concluding considerations.

## 2 Materials and Methods

In this section input data used for the rate of change of frequency estimation using Huang's Empirical Mode Decomposition (EMD), Multivariate Empirical Mode Decomposition (MEMD) and Noise-Assisted Multivariate Empirical Mode Decomposition (NA-MEMD) are presented. Furthermore, basic algorithms of the EMD, EEMD, MEMD and NA-MEMD are presented.

### 2.1 Test Signals

For the purpose of the rate of change of frequency estimation synthetic test signals are generated. This generated test signals represent the frequency signals plotted on Fig. 1 and 2. In order to obtain the actual values of the rate of change of frequency derivatization of the generated test signals is performed over frequency signals plotted on the Fig. 3. The actual values of the rate of change of frequency are obtained as the mean values of the derivatives of the frequency signals. Test signals are observed in time interval that lasted 2 s and sampling frequency is 666,66 Hz. First generated test signal has form:

$$f_1(t) = 50 - 20 \cdot t; \frac{df_1}{dt} = -20\left(\frac{Hz}{s}\right) \tag{1}$$

Second generated test signal is:

$$f_2(t) = 30 - 3 \cdot t - 3 \cdot t^2; \frac{df_2}{dt} = -3 - 6 \cdot t\left(\frac{Hz}{s}\right) \tag{2}$$

Third generated test signal is:

$$f_3(t) = 40 - 0.4 \cdot e^{-t} \cdot \sin(4 \cdot \pi \cdot t);$$

$$\frac{df_3}{dt} = e^{-t} \cdot (0.4 \cdot \sin(4 \cdot \pi \cdot t) - 5.02655 \cdot \cos(4 \cdot \pi \cdot t)\left(\frac{Hz}{s}\right) \tag{3}$$

Fourth generated test signal has form:

$$f_4(t) = 50 - 0.5 \cdot t + 0.5 \cdot e^{-t} \cdot \sin(4 \cdot \pi \cdot t);$$

$$\frac{df_4}{dt} = e^{-t} \cdot (-0.5 \cdot e^{-t} + 6.28319 \cdot \cos(4 \cdot \pi \cdot t) - 0.5 \cdot \sin(4 \cdot \pi \cdot t))\left(\frac{Hz}{s}\right) \tag{4}$$

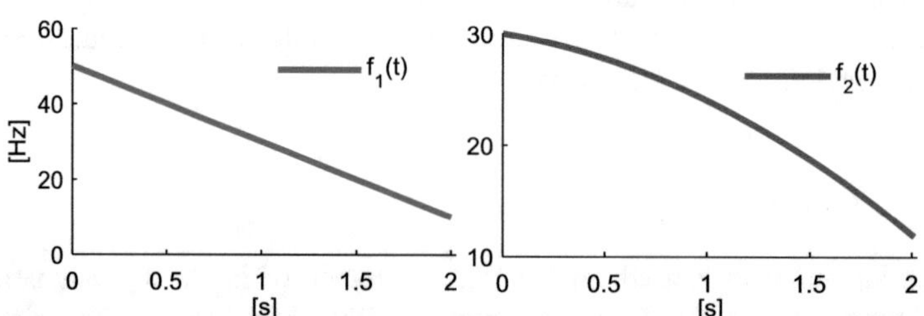


**Fig. 1.** Generated test signals $f_1(t)$ and $f_2(t)$

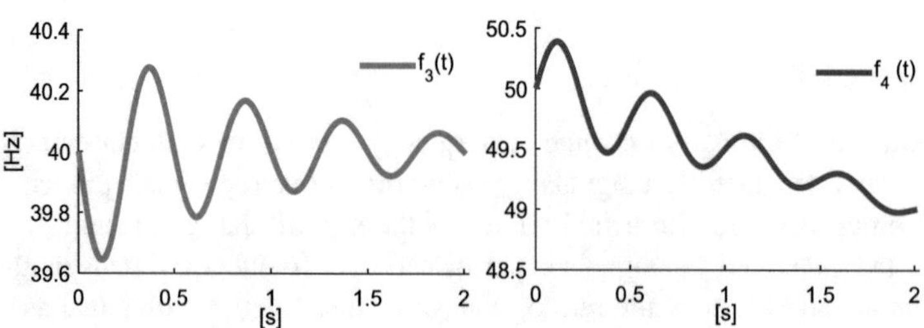


**Fig. 2.** Generated test signals $f_3(t)$ and $f_4(t)$

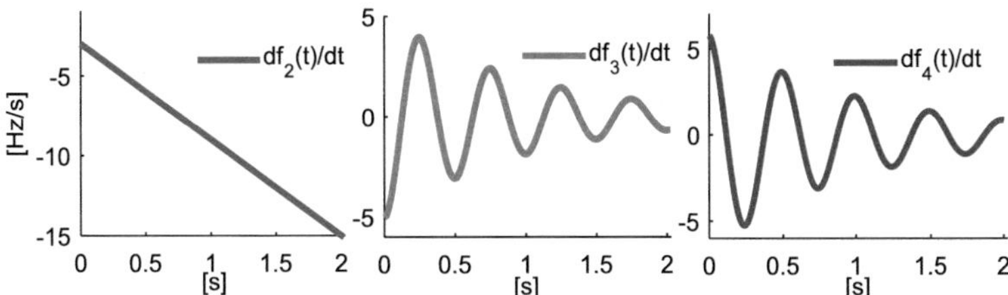


**Fig. 3.** Derivative of the generated test signals $f_2(t)$, $f_3(t)$ and $f_4(t)$

To investigate the influence of the noise, the test signals are contemned with additive noise SNR = 10 dB, 25 dB and 40 dB, respectively. On Fig. 4, 5, and 6 are plotted noise assisted input frequency signals according to aforementioned dB.

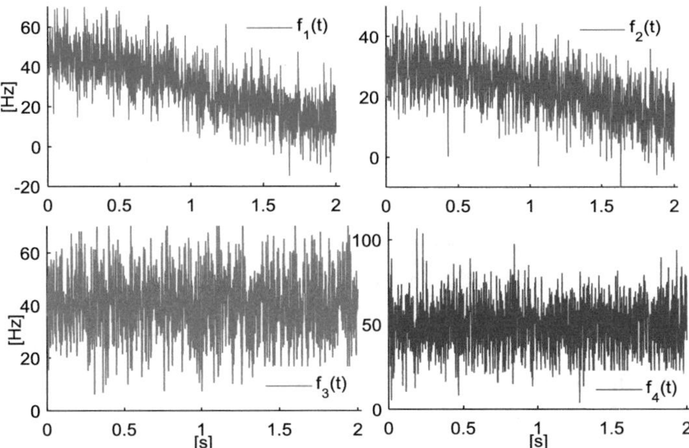


**Fig. 4.** Generated test signals $f_1(t)$, $f_2(t)$, $f_3(t)$ and $f_4(t)$ with additive noise SNR = 10 dB

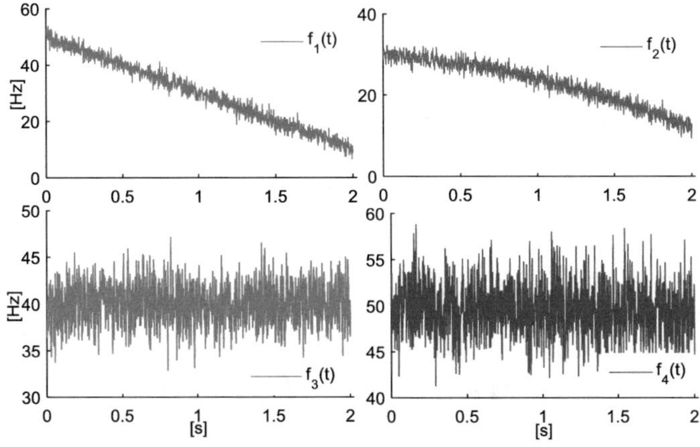


**Fig. 5.** Generated test signals $f_1(t)$, $f_2(t)$, $f_3(t)$ and $f_4(t)$ with additive noise SNR = 25 dB

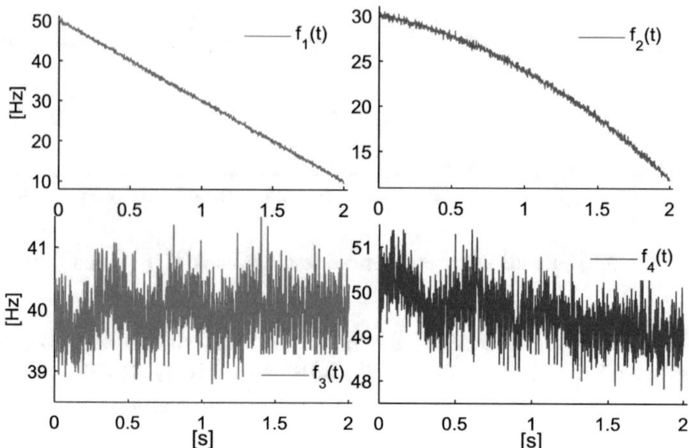


**Fig. 6.** Generated test signals $f_1(t), f_2(t), f_3(t)$ and $f_4(t)$ with additive noise SNR = 40 dB

## 2.2 Basics of EMD, EEMD, MEMD and NA-MEMD

In this section are shortly presented algorithms of EMD, EEMD, MEMD and NA-MEMD process.

Algorithm of EMD can be presented as follows:

Step 1. Find the local extrema and using a cubic spine line generate its upper and lower envelopes ($e_{up}(t)$ and $e_{low}(t)$, respectively) of the variable $x(t)$;
Step 2. Calculate the mean value of the $e_{up}(t)$ and $e_{low}(t)$: $m_1(t) = \left(e_{up}(t) + e_{low}(t)\right)/2$. And find $h_1(t)$ as follows: $h_1(t) = x(t) - m_1(t)$. Repeat steps until $h_1$ satisfies a set of predetermined stopping criteria for the IMF function as expressed: $c_1 = h_{1k}$
Step 3. If $h_1(t)$ does not meet the criteria for the existence of IMF, the previous procedure is repeated and determined $h_{11}(t)$ as a new signal: $h_{11}(t) = x(t) - m_{11}(t)$, where $m_{11}(t)$ is mean value of the upper and lower envelope of the signal $h_1(t)$.
Step 4. Compute residue and repeat step 2 $k$ times to obtain $n$ IMFs and residual, then the new signal is as follow: $x(t) = \sum_{i=1}^{n} c_i(t) + r_n(t) = \sum_{i=1}^{q} c_i(t) + \sum_{j=q+1}^{p} c_j(t) + \sum_{k=p+1}^{n} c_k(t) + r_n(t)$, where $q<p<n$, $c_i(t)$ are *ith* IMF, $c_j(t)$ are the components representing properties of the series and $c_k(t)$ and $r_n(t)$ are the final residual, trend non-sinusoidal components.

Algorithm of EEMD can be presented as follows:

Step 1: Generate $s_n(t) = x(t) + w_n(t)$ for n = 1, …, N; where $w_n(t)$ (n = 1, …, N) are N different realizations of WGN;
Step 2: Decompose the ensemble $s_n(t)$ (n = 1, …, N) by applying standard EMD to each realization $s_n(t)$ separately, obtaining $M_n$ IMFs for $s_n(t)$, denoted by $\{c_m^n(t)\}_{m=1}^{M_n}$;

Step 3: Average the corresponding IMFs from the whole ensemble to obtain the averaged (ensembled) IMFs; for instance, *mth* IMF can be obtained by using $\overline{c_m}(t) = \frac{1}{N}\sum_{n=1}^{N} c_m^n(t)$.

Algorithm of MEMD can be presented as follows:

Step 1. Generate the appropriate point-set-based for sampling over (n – 1) sphere.
Step 2. For the whole set of direction vectors $v$ = 1, … V, calculate the projections $q_{\theta_v}(t)$ of the input signal $s(t)$ along the direction vector $x_{\theta_v}$, to give $\{q_{\theta_v}(t)\}_{v=1}^{V}$ as a reflection set.
Step 3. The time constant $t_{\theta_v}^{j}$ corresponding to the maximum of the set of projected signals is found.
Step 4. Interpolation for all values of $v$ using $\left[t_{\theta_v}^{j}, s(t_{\theta_v}^{j})\right]$ is performed to obtain highly variable envelope curves $\{e_{\theta_v}(t)\}_{v=1}^{V}$.
Step 5. For a set of $v$ direction vectors, calculate the mean value as: $m(t) = \frac{1}{V}\sum_{v=1}^{V} e_{\theta_v}(t)$.
Step 6. $i$ is the order of the IMF, so 'detail' $c_i(t)$ is obtained by $c_i(t) = s(t) - m(t)$. If detail satisfies the stopping criterion for a multivariate IMF, apply the above procedure to $s(t) - c_i(t)$, if not, it is applied to $c_i(t)$.

Algorithm of NA-MEMD can be presented as follows:

Step 1: Check if the input signal fulfills the criteria of an IMF; if not so, proceed to the next steps otherwise stop the process;
Step 2: Create an uncorrelated Gaussian white noise time series ($l$-channel) of the same length as that of the input, with $l \geq 1$;
Step 3: Add the noise channels ($l$-channel) created in step 2 to the input multivariate ($n$-channel) signal $n \geq 1$, obtaining an $n + l$-channel multivariate signal;
Step 4: Process the resulting $(n + l)$-channel multivariate signal $n + l \geq 2$ using MEMD algorithm, to obtain multivariate IMFs;
Step 5: From the resulting $(n + l)$-variate IMFs, discard the $l$ channels corresponding to the noise, giving a set of $n$-channel IMFs corresponding to the original signal.

## 3 Result and Discussion

Based on methodology presented in second chapter, results of residuals obtained by application of EMD, MEMD and NA-MEMD on the generated frequency test signals are outline. On Fig. 7 are plotted residuals obtained after application of EMD algorithm over first and second generated test signals without additive noise and with noise 10 dB, 25 dB and 40 dB, respectively. On Fig. 8 are plotted residuals obtained after application of EMD algorithm over first and second generated test signals.

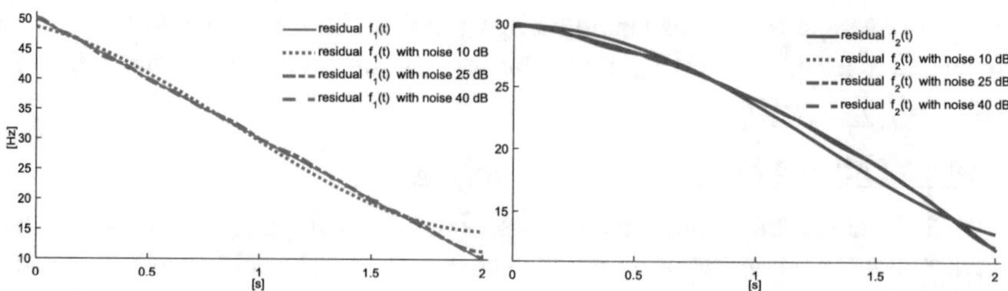


**Fig. 7.** EMD – $f_1(t)$ and $f_2(t)$ residuals

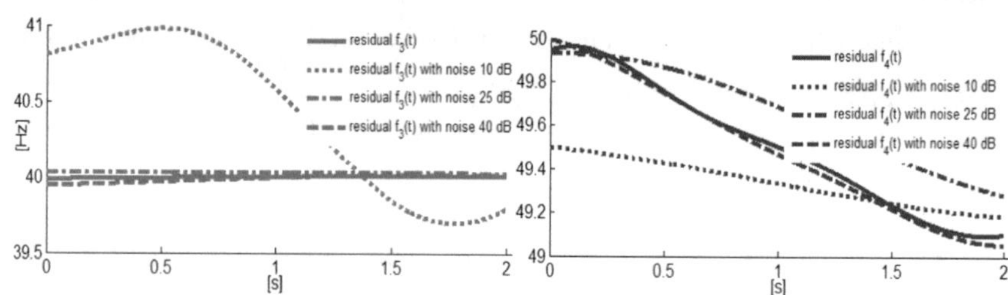


**Fig. 8.** EMD - $f_3(t)$ and $f_4(t)$ residuals

Residuals obtained after application of MEMD algorithm over the generated synthetic frequency signals are plotted on Fig. 9.

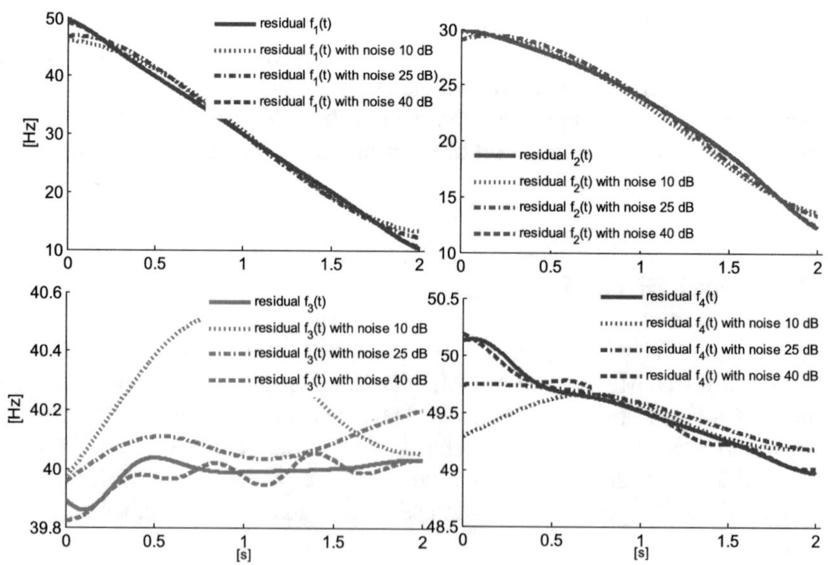


**Fig. 9.** MEMD - $f_1(t)$, $f_2(t)$, $f_3(t)$ and $f_4(t)$ residuals

Residuals obtained after application of NA-MEMD algorithm over the generated synthetic frequency signals $f_3(t)$ and $f_4(t)$ are plotted on Fig. 10.

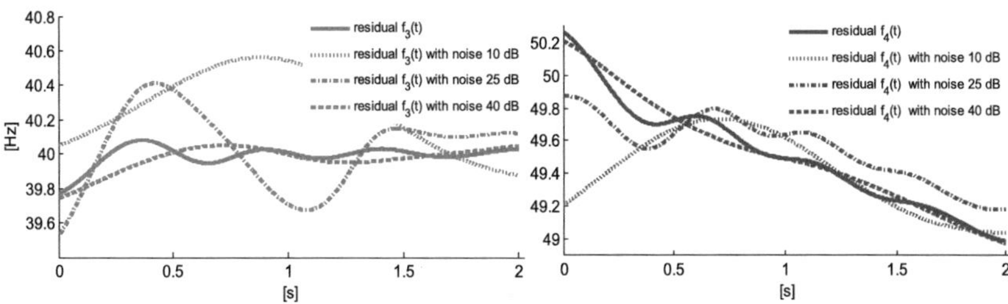


**Fig. 10.** NA-MEMD - $f_3(t)$ and $f_4(t)$ residuals

After obtaining residual, for the all test cases, according to four generated frequency test signals and additive noise, mean value is calculated over the observed time interval of 2 s. The forward difference method over the time series is introduced by expression: $(((x_{i+1} - x_i)/\Delta t), \Delta t = 0.0015\,\text{s})$. The mean value of the time series obtained after forward difference process over the test signal, represent value of the rate of change of frequency [13]. The actual value of the rate of change of frequency is already determinate with mean value of the derivative over the input frequency test signal. Furthermore, those values are compared and error is calculated with:

$$error = \left( \frac{df_i/dt - df/dt}{df/dt} \right) \cdot 100\% \tag{5}$$

where $df_i/dt$ is estimated value using one of the methods and according to additive noise and $df/dt$ is the mean value of the derivative over the input frequency test signal. Error is calculated for the all test cases, all applied methodologies (EMD, MEMD, NA-MEMD) and all additive noise (10 dB, 25 dB and 40 dB) and presented in Table 1. Moreover, on the figures below are plotted obtained estimated rate of change of frequency signals according to applied methodology. For the results representation two cases are chosen, first case denotes application of the EMD, MEMD and NA-MEMD over the second generated test signal $f_2(t)$, the second test case correspond to the application of the EMD, MEMD and NA-MEMD over the fourth generated test signal $f_4(t)$. On Fig. 11 are plotted actual ROCOF values obtained calculating mean value over the derivative of the $f_2(t)$ generated test signal time series. Also, $df_2(t)/dt$ representing ROCOF is calculated for the cases of additive noise 10 dB, 25 dB and 40 dB according to applied methodology, (a) EMD, (b) MEMD and (c) NA-MEMD.

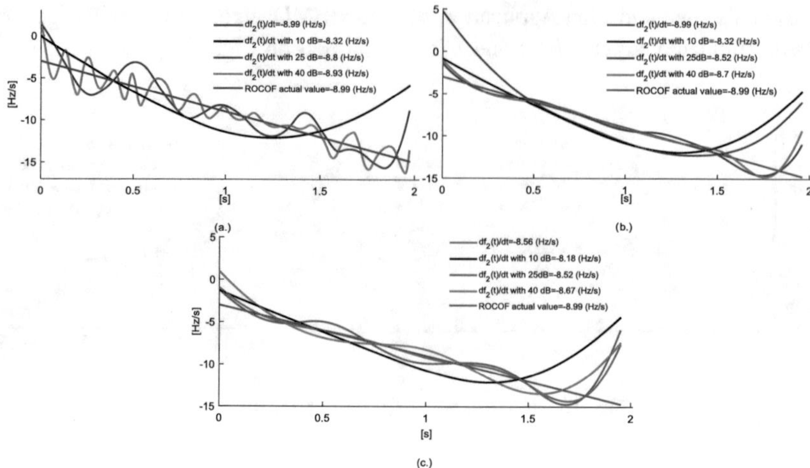


**Fig. 11.** ROCOF estimation - $f_2(t)$ generated test signal: (a) EMD, (b) MEMD and (c) NA-MEMD

On Fig. 12 are plotted actual ROCOF values obtained calculating the mean value over the derivative of the $f_4(t)$ generated test signal time series. In order to obtain the values of ROCOF, the mean value over the residuals derivatives of the input test signal $f_4(t)$ according to noise level is calculated. Those residuals are obtained after application of (a) EMD, (b) MEMD and (c) NA-MEMD over the $f_4(t)$ according to test case regarding additive noise.

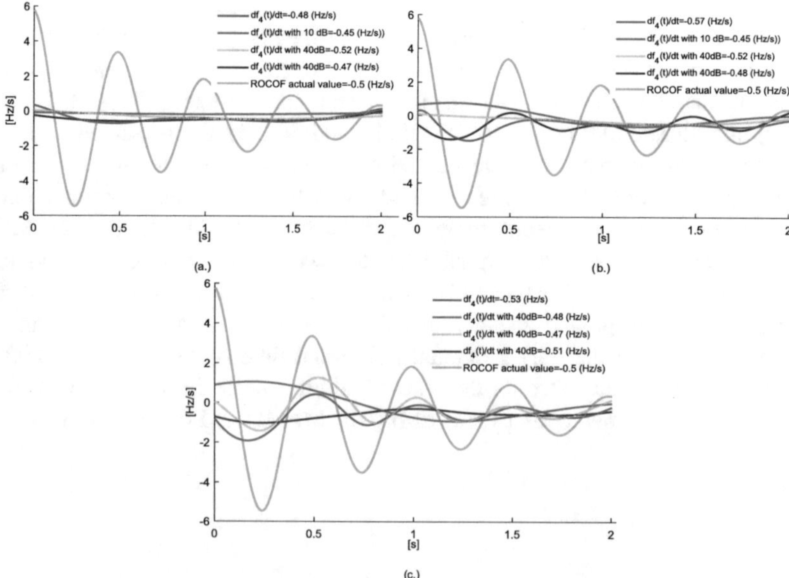


**Fig. 12.** ROCOF estimation - $f_4(t)$ generated test signal: (a) EMD, (b) MEMD and (c) NA-MEMD

From figures above can be concluded that residuals obtained by empirical mode decomposition based time–frequency analysis of multivariate generated test signal illustrates an excellent trend of behavior of the input test signals. The standard EMD show great result regarding obtained residuals compared to improve methods such as MEMD and NA-MEMD. From the Table 1 can be seen calculated error in percentage for all test cases. It can be concluded that the most satisfactory results, regarding ROCOF calculation belong to outcomes of NA-MEMD. Via NA-MEMD accuracy rates are between 91,9% and 100%, according to EMD accuracy rates are between 90% and 100% and regarding MEMD approach accuracy rates are between 86% and 100%. Also from Table 1 can be noted that the error decrease with the decrease of additive noise level in channels.

**Table 1.** Error (%) according to test case and additive noise in dB

| | Case | Rate of change of frequency | | | | |
|---|---|---|---|---|---|---|
| | | *Actual value* | *f(t)* | *10 dB* | *25 dB* | *40 dB* |
| EMD | I | −20 (Hz/s) | 0,2% | 1,5% | 0,8% | 0,2% |
| | II | −8.99 (Hz/s) | 0% | 6,5% | 1,1% | 0,3% |
| | III | −0,15 (Hz/s) | 1,4% | 5,5% | 1,1% | 1,3% |
| | IV | −0.5 (Hz/s) | 4% | 10% | 4% | 6% |
| MEMD | I | −20 (Hz/s) | 0,1% | 1,7% | 0,9% | 0,1% |
| | II | −8.99 (Hz/s) | 0% | 6,5% | 4,3% | 2,2% |
| | III | −0,15 (Hz/s) | 0,8% | 8,2% | 1,7% | 1,2% |
| | IV | −0.5 (Hz/s) | 14% | 10% | 4% | 4% |
| NA-MEMD | I | −20 (Hz/s) | 0% | 1,9% | 0,9% | 0,2% |
| | II | −8.99 (Hz/s) | 3,8% | 8,1% | 4,3% | 2,6% |
| | III | −0,15 (Hz/s) | 0,5% | 4,5% | 1% | 0,8% |
| | IV | −0.5 (Hz/s) | 6% | 4% | 6% | 2% |

## 4 Conclusion

In this paper is investigate behavior of residuals obtained over the generated test signals via EMD algorithm and its noise assisted and multivariate extensions, multivariate EMD regarding estimation. Although, NA-MEMD and MEMD approaches have better performance according mode-mixing and aliasing of obtained IMFs, results in terms of residuals are quite similar. On the obtained residuals from the decomposed test signals, are performed forward difference process and calculated mean value which represent result value of the rate of change of frequency (ROCOF). All methods show great accuracy, although NA-MEMD has the most satisfactory results. Also, as it was mentioned in discussion for all method can be seen better performances regarding additive noise of 40 dB over the additive noise level of 10 dB.

## References

1. Daubechies, I., Lu, J., Wu, H.-T.: Synchrosqueezing wavelet transforms: an empirical mode decomposition-like tool. Appl. Comput. Harmon. Anal. **30**(2), 243–261 (2011)
2. Huang, N.E., Shen, Z., Long, S.R., Wu, M.C., Shih, H.H., Zheng, Q., Yen, N.-C., Tung, C.C., Liu, H.H.: The empirical mode decomposition and the Hilbert spectrum for nonlinear and non-stationary time series analysis. Proc. R. Soc. London A **454**(1971), 903–995 (1998)
3. Hou, T.Y., Shi, Z.: Adaptive data analysis via sparse time–frequency representation. Adv. Adapt. Data Anal. **3**(1), 1–28 (2011)
4. Hou, T.Y., Shi, Z.: Data driven time frequency analysis. Appl. Comput. Harmon. Anal. **35** (2), 284–308 (2013)
5. Huang, N.E., Wu, Z., Long, S.R., Arnold, K.C., Chen, X., Blank, K.: On instantaneous frequency. Adv. Adapt. Data Anal. **1**(2), 177–229 (2009)
6. Meeson Jr., R.N.: HHT sifting and filtering. In: Huang, N.E., Shen, S.S.S. (eds.) Hilbert–Huang Transform and Its Applications, pp. 75–105. World Scientific, Singapore (2005)
7. Looney, D., Mandic, D.P.: Multi-scale image fusion using complex EMD. IEEE Trans. Sig. Process. **57**(4), 1626–1630 (2009)
8. Wu, Z., Huang, N.E.: Ensemble empirical mode decomposition: A noise-assisted data analysis method. Adv. Adapt. Data Anal. **1**(1), 1–41 (2009)
9. Mandic, D.P., ur Rehman, N., Wu, Z., Huang, N.E.: Empirical mode decomposition-based time-frequency analysis of multivariate. IEEE Sig. Process. Mag. **30**(6) (2013). https://doi.org/10.1109/msp.2013.2267931
10. Rehman, N., Mandic, D.P.: Multivariate empirical mode decomposition. Proc. Math. Phys. Eng. Sci. **466**(2117), 1291–1302 (2009)
11. Rehman, N.U., Park, C., Huang, N.E., Mandic, D.P.: EMD via MEMD: multivariate noise aided computation of standard EMD. Adv. Adapt. Data Anal. **5**(2), 1350007 (2013)
12. Rehman, N., Mandic, D.P.: Filterbank property of multivariate EMD. IEEE Trans. Sig. Process. **59**(5), 2421–2426 (2011)
13. Dedović, M.M., Avdaković, S.: A new approach for df/dt and active power imbalance in power system estimation using Huang's empirical mode decomposition. Int. J. Electr. Power Energ. Syst. **110**, 62–71

# The Influence of the Formation of a Regional Office for Coordinated Auctions of Cross Border Transmission Capacities in South East Europe

Omer Hadžić[1(✉)], Zijad Bajramović[2], and Irfan Turković[2]

[1] Independent System Operator in Bosnia and Herzegovina, Sarajevo, Bosnia and Herzegovina
o.hadzic@nosbih.ba

[2] Faculty of Electrical Engineering, University of Sarajevo, Sarajevo, Bosnia and Herzegovina

**Abstract.** In this paper, is describe coordinated auctions of cross-border transmission capacities between South East European Transmission system Operators through the operation of the regional company "Coordinated Auction Office in South East Europe" (SEE CAO). The main objective is to acquaint with the specific processes, functions, modules and contributions of the first regional company in the area of South-Eastern Europe. The importance of SEE CAO is reflected by upgrading previous unilateral auctions performed by each TSO on 50% of NTC (Net Transmission Capacity) following their national Auction Rules or bi-lateral auctions performed by two neighboring TSO (Transmission system operator) on 100% of NTC following bilateral Auction Rules to regionally coordinated auctions performed by SEE CAO, using one single set of the Auction Rules and the possibility for the traders to allocate cross-border capacity from Croatia to Turkey by using SEE CAO as one-stop-shop. At the end of the paper, the results of daily, monthly and annual auctions are presented via diagrams and give an overview of the results. Also, as a result, the auction data for 2018 and 2019 auctioned as well as the average prices for the same years are presented in the tables.

## 1 Introduction

Taking into account the facts related to formation SEE CAO [1], as well as by studying the available literature in this area [2–5] in the way that the comparison system has reached certain implicit conclusions.

SEE CAO [1], was established in 2014, after few years of intensive preparation, by its shareholders - TSOs from the region of South East Europe (HOPS - Croatia, NOSBiH - Bosnia and Herzegovina, CGES - Montenegro, OST - Albania, KOSTT - Kosovo, IPTO - Greece, MEPSO - North Macedonia, TEIAS - Turkey) - with the objective to perform the explicit allocation of cross-border transmission capacity in both directions between Control Areas of the Participating TSO's, through NTC based Auction Processes in accordance with the requirements of Regulation (EC) 714/2009


S. Avdaković et al. (Eds.): IAT 2019, LNNS 83, pp. 159–170, 2020.
https://doi.org/10.1007/978-3-030-24986-1_13

[6] which also became a part of Energy Community aquis in line with the stipulations of Energy Community Treaty [7].

Since the core service of the company is highly dependent on data provision, the technical framework for implementation of auction tool had to be met. Therefore, Auction Platform was developed and implemented based on SEE CAO technical specification and requirements from SEE CAO Auction Rules. The Auction Platform, as web based application, is implemented completely following the ECAN (ENTSO-E Capacity Allocation and Nomination, ENTSO-E European Network of Transmission System Operators for Electricity) standard [8], supporting the modules for Long Term and Daily Auctions, Secondary Market, Capacity Rights and Nominations (UIOSI), (UIOSI – "Use It Or Sell It"), Finance and Settlement. Users of the Auction Platform, besides SEE CAO Operators and SEE TSOs (shareholders which are at the same time service users), are numerous Auction Participants. Auction Platform enables their users to offer or allocate the cross-border capacities on yearly, monthly and daily level. Thus, the data exchange is enabled between Auction Platform and the users in accordance with the Electronic Data Interchange (EDI) library and standards, while the relevant data is also provided to ENTSO-E Transparency Platform.

## 2 Auction Process

SEE CAO performs in its own name but on behalf and for the account of the participating TSOs explicit allocation of available transmission capacities on the borders between participating TSOs. Auctions are NTC (Net Transmission Capacity) based and are conducted on three timeframes - yearly, monthly and daily using the Auction Platform [1, 5]. Simplified auction process diagram is shown on Fig. 1.

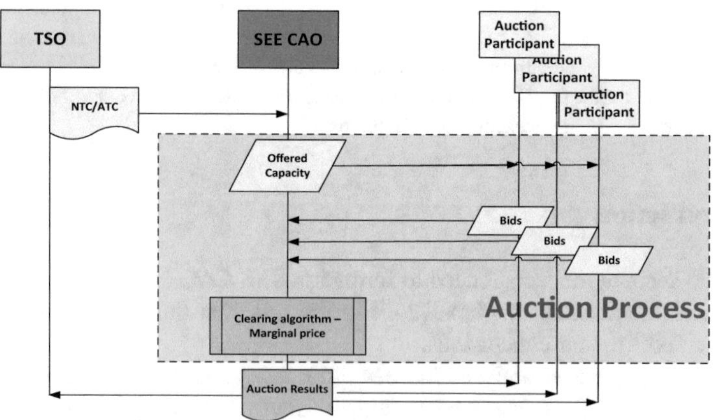


**Fig. 1.** Auction process [1, 5]

Based on the values of transmission capacities received from participating TSOs (NTCs, respectively ATCs- Available Transmission Capacity), SEE CAO schedules the auctions on the borders between participating TSOs. Auction Participants who are registered in SEE CAO and who are market participants eligible in at least one of participating TSO are bidding for the Offered Capacity during the bidding period [5]. After bidding period has expired, the auction algorithm calculates the auction results. Auction algorithm used is Marginal Price algorithm, where all successful Auction Participants pay the same price (last accepted) for the allocation of Physical Transmission Rights (PTRs). Auction results are distributed to Auction participants and participating TSOs, followed by Invoicing and Settlement procedures, Secondary market and finally, utilization of the allocated PTRs.

Since Marginal Price is expressed in €/MWh, the price of the allocated capacity per Auction Participants is calculated as product of marginal price, amount of allocated capacity and number of hours in timeframe which covers the auction. Total auction revenue collected from Auction Participants is distributed to the participating TSOs in accordance with the borders and auctions on which the congestion management revenue is obtained [5].

Prior to organizing the first auctions and launching the fully operational SEE CAO, most important preconditions and requirements which had to be met are as follows [5]:

(a) Auction Rules approved by all respective National Regulatory Authorities
(b) Agreement for Services signed with all respective TSOs
(c) Auction Platform delivered, installed, configured and tested
(d) Auction Operators employed and trained
(e) Auction Participants registered and trained
(f) All other requirements met by service user TSOs (legal, financial, technical)

On the side of TSOs, some requirements should also be met – TSO must assure that their NRA (National Regulatory Authority) has approved SEE CAO's Auction Rules, double taxation issue must be overcome either by TSO or their local legislative, technical interfaces between SEE CAO's Auction Platform and TSOs systems must be compatible, as well as neighbour participating TSO's system for the matching purposes.

To conclude, SEE CAO is performing yearly, monthly and daily auctions on 6 borders (HR – BA; BA - ME; ME - AL; AL - GR; GR – TR, GR - NM) with approximately 120 registered Auction Participants (HR -Croatia, BA - Bosnia and Herzegovina, ME - Montenegro, AL - Albania, TR - Turkey, NM - North Macedonia).

## 3 Auction Platform

Auction Platform was procured during the 2013 from the funds from EBRD and following the Bank's procurement procedure. Auction Platform was delivered and implemented in first half of 2014 and taken-over in June 2014 after successful testing. It was completely developed and configured in accordance with SEE CAO's technical specification from the tender procedure. Provider of the Auction Platform is Unicorn systems a.s., Czech Republic with their DAMAS system, which is well known among the TSOs and Auction Participants. Auction algorithm used is Marginal Price algorithm, where all successful Auction Participants pay the same price (last accepted) for the allocation of Physical Transmission Rights (PTRs). Auction results are distributed to Auction participants and participating TSOs, followed by Invoicing and Settlement procedures, Secondary market and finally, utilisation of the allocated PTRs. System is developed as a web application, with user friendly interfaces for its users – SEE CAO Auction Operators, Auction Participants and TSOs.

Auction Platform represents main tool which SEE CAO [1] uses to operate the auctions, and which is also used by Auction Participants to place a bid on the required auctions or TSOs to check the state and results of the auction.

Technical specification for the tender procurement was prepared following already mentioned ECAN standard [8], first draft of SEE CAO Auction Rules and their previous experience in TSOs.

Auction Platform is designed as a set of modules which perform specific functions from the Auction Rules and communicate between themselves to exchange the parameters and input/output information.

The most important section of the system is the System Administration, which enables the configuration of all crucial parameters and code tables – Auction Participant, TSO, Control Area, Border, Border Direction, Auction configuration, Auction timing, Communication channel, Process schedule, Data Flow, etc. [5]. Using the System Administration, particularities between borders, TSOs or auctions can be diverted and set as required by the stakeholders.

Communication is enabled towards the Auction Participants and TSOs, and additionally, the announcements of the auctions as well as results are delivered to SEE CAO [1] and ENTSO-E Transparency Platform Fig. 2.

Data exchange with Auction Participants and TSOs is enabled by using the format of documents following Electronic Data Interchange (EDI) library, ECAN and ESS (ENTSO-E Scheduling System) standard [1, 5]:

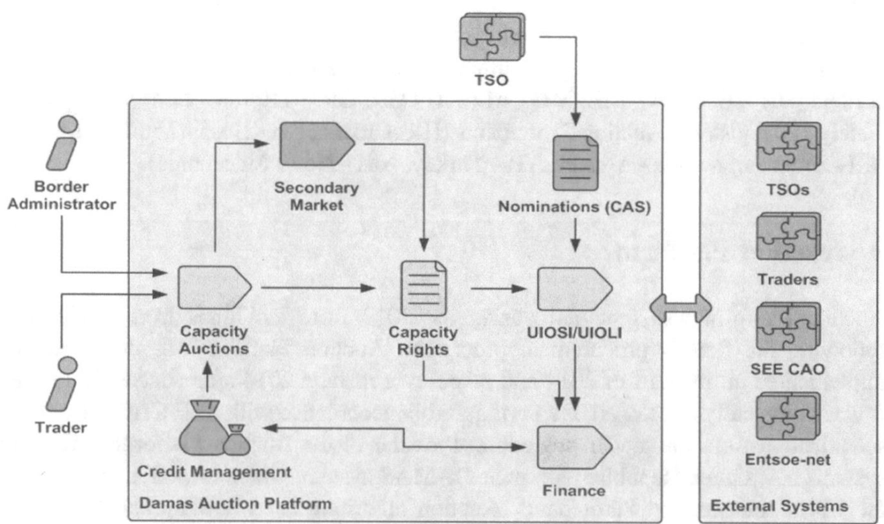


**Fig. 2.** Auction Platform communication with external systems [1, 5]

– Auction Specification Document,
– Bid Document,
– Total Allocation Result Document,
– Allocation Result Document,
– Rights Document,
– CAS Information for CAO (ESS – ENTSO Scheduling System),
– Acknowledgement.

Communication channels which are implemented as interfaces to and from auction platform system are manually insertion, upload or download using the web forms, receiving the notifications and/or documents by e-mail, as well as upload of particular files (for example, CIC file send by TSOs and Acknowledgement generated back by Auction platform), data exchange via SFTP (Secure File Transfer Protocol), and using of web services [5]. For data delivery to ETP (ENTSO-E Transparency Platform), SEE CAO is using the Energy Communication Platform - ECP [5]. It is important to note that SEE CAO was one of the first Data Providers who started reporting to ENTSO-E Transparency Platform, Fig. 3, as of January 5th, 2015, which has significantly contributed to the enhancement of the transparency in the SEE region, especially taking into account that prior to establishment of SEE CAO the auction results were not as transparent [5].

**Fig. 3.** ENTSO-E Transparency platform [1]

As it can be rated from the Fig. 3, daily auctions data (clearing price, offered, requested and allocated capacity) is published on ENTSO-E transparency platform. Same data is reported for long term auctions. Furthermore, SEE CAO reports to ETP data regarding allocation revenue, already allocated capacity, secondary market and auction configuration.

Moreover, all auction announcements and auction results are public and available on SEE CAO [1] for which the Auction Platform is using SFTP to deliver the information and web site is using web services to fetch the data.

## 4 Auction Results and Discussion

SEE CAO operate auction process of coordinated allocation the transmission capacities and auction participants place bids on the required auctions and check the state and results of the auctions. Yearly and Monthly auction processes are supported by functions: scheduling, opening, closing, evaluating, publishing results, allocated capacity and auction statistics. Also, daily auctions support complete process of scheduling, publishing the daily offered capacity, opening closing, evaluation, publishing results of daily auctions and auction statistics.

Some yearly, monthly and daily results from SEE region have been showed on following Figs. 4, 5, 6, 7, 8, 9, 10, and 11.

On Figs. 4, 5, 6 and 7 total allocated capacity is displayed with blue color while auction clearing price is marked by red color.

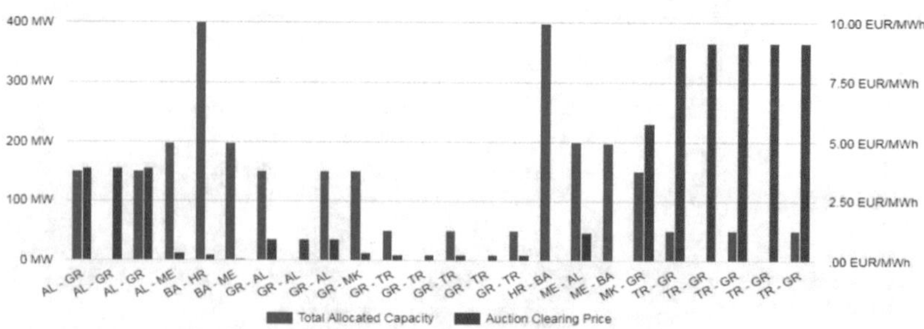


**Fig. 4.** Results of yearly auctions for 2018 [1].

In case shown on Fig. 4, lowest prices are achieved on border directions with highest offered capacity (BA-HR, HR-BA). Highest prices are achieved on Greek import border directions (AL-GR, MK-GR, TR-GR). Border directions AL-GR, GR-AL, GR-TR and TR-GR are displayed in multiple columns because they contain maintenance periods. Total auction income on yearly auctions for 2018 is 21 804 987.60 €.

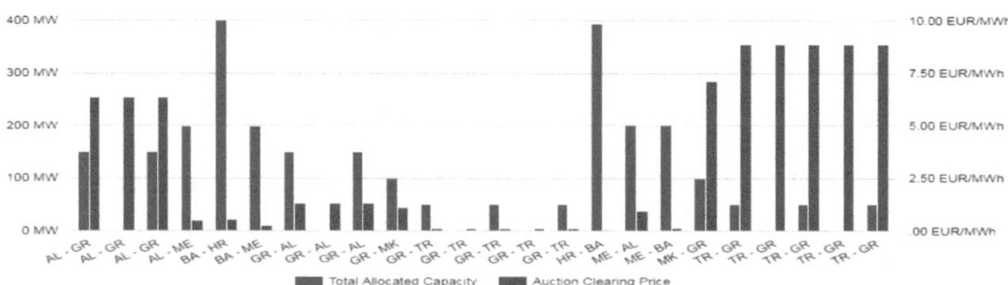


**Fig. 5.** Results of yearly auctions for 2019 [1].

In case shown on Fig. 5, lowest prices are achieved on border directions with highest offered capacity (BA-HR, HR-BA). Highest prices are achieved on Greek import border directions (AL-GR, MK-GR, TR-GR). Border directions AL-GR, GR-AL, GR-TR and TRGR are displayed in multiple columns because they contain maintenance periods. Total auction income on yearly auctions for 2019 is 25 938 963.60 €, which is increase for 19% comparing to 2018.

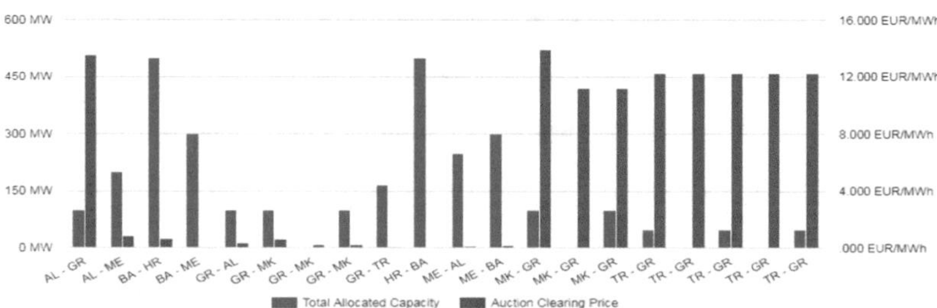


**Fig. 6.** Results of monthly auctions for January 2018 [1].

In case shown on Fig. 6, highest prices are achieved on Greek import border directions (TR-GR, MK-GR, AL-GR). Total auction income on monthly auctions for January 2018 is 1 649 410.80 €.

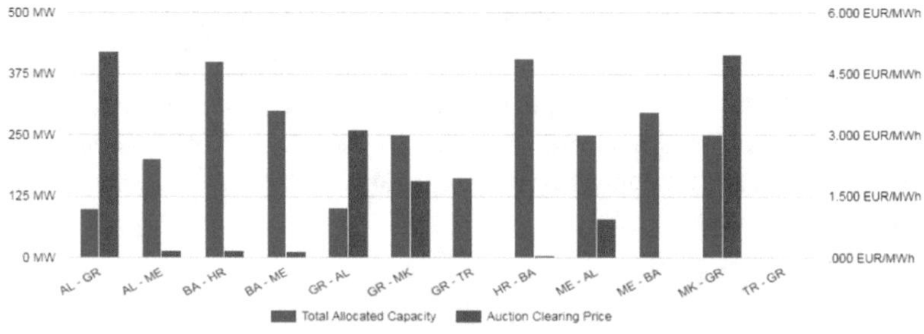


**Fig. 7.** Results of monthly auctions for January 2019 [1].

In case shown on Fig. 7, highest prices are achieved on border directions AL-GR and MK-GR, while there was no offered capacity on TR-GR border directions. Total auction income on monthly auctions for January 2019 is 2 190 209.52 €, which is increase for 33% comparing to January 2018.

From the Figs. 4, 5, 6 and 7 we can see that the highest offered capacities and lowest prices are on the border HR-BA, and highest prices are on the Greek borders. There are two reasons for this, Greece is an importer of electricity, and in addition, Greece and Italy are connected to a submarine cable through which electricity is often diverted to Italy (Italy is the largest importer of electricity in Europe)

On Figs. 8, 9, 10 and 11 daily auction results have been shown. Total allocated capacity per hour is displayed with blue color while auction clearing price per hour is marked as red.

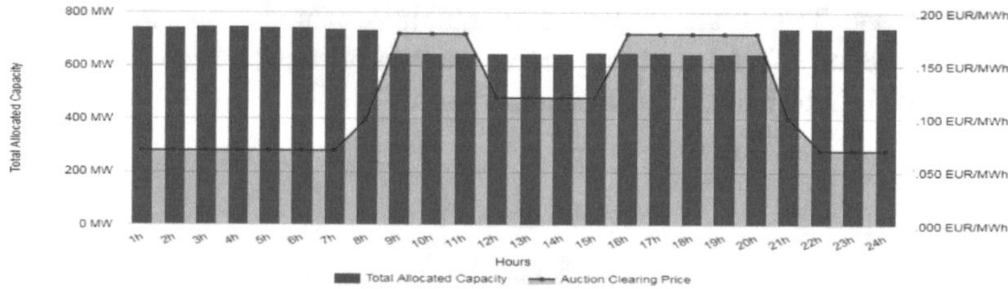


**Fig. 8.** Daily auctions BA – HR 31.12.2018 [1].

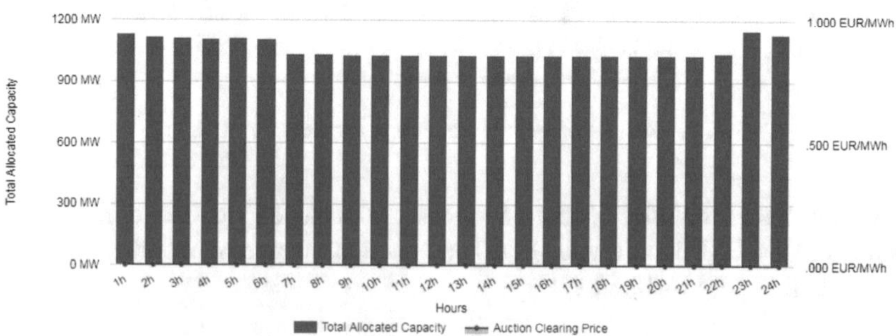


**Fig. 9.** Daily auctions HR - BA 31.12.2018 [1].

Figures 8 and 9 results of daily auctions for delivery day 31$^{st}$ January 2018 on BA-HR and HR-BA border directions, respectively. Congestion is reached on border direction BA-HR, while on opposite direction there is no congestion, which is common situation during the year.

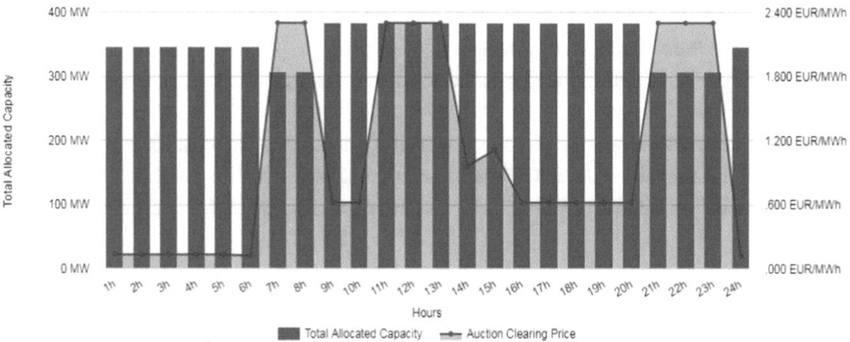


**Fig. 10.** Daily auctions BA – ME 01.03.2019 [1].

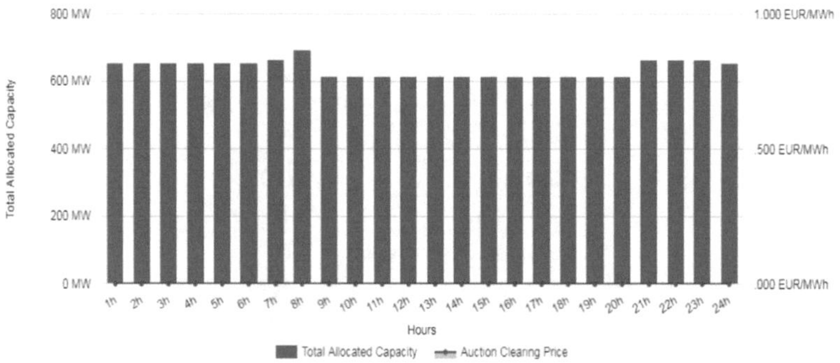


**Fig. 11.** Daily auctions ME - BA 01.03.2019 [1].

Figures 10 and 11 results of daily auctions for delivery day 1$^{st}$ March 2018 on BA-ME and ME-BA border directions, respectively. Congestion is reached on border direction BA-ME, while on opposite direction there is no congestion, which is common situation during the year.

On the Figs. 9 and 11, charts of daily auctions, we see that the price of capacity from Croatia towards BiH and from Montenegro towards BiH is zero. This is primarily because of fact that BiH is only country in the region that has electricity surpluses, so there is interest and pressure from traders to purchase and export of electricity from BiH across borders towards other countries and consequently there is great competition at the borders in the direction from BiH to Croatia and Montenegro. In opposite direction the situation is different, no competition, no congestion (the sum of all requests is within the available capacity), price is zero.

In the Tables 1 and 2 are showed comparable results for 2018 and 2019 (the first five months with certain observations).

Table 1 contains data (requested and offered capacity, clearing price, number of participants and number of successful participants) from yearly auctions for 2018 and 2019 on borders BA-ME and BA-HR. It can be noted from the table that number of participants increased in 2019 comparing to 2018. Since offered capacity for 2018 and 2019 was the same, increased interest of participants in 2019 resulted in increased amount of requested capacity and consequently, increased price and auction revenue in 2019.

Table 2 contains average prices on monthly auctions 2018 and 2019 so far, on borders BA-ME and BA-HR. Average price is increased on all border directions except BAHR. Average price for 2019 is calculated based on results for first five months of 2019.

**Table 1.** Yearly auctions on BH borders 2018 and 2019.

| | Year | 2018 | 2019 |
|---|---|---|---|
| BA-ME | Requested capacity (MW) | 461 | 927 |
| | Offered capacity (MW) | 200 | 200 |
| | Clearing price | 0.06 | 0.24 |
| | Participants | 11 | 18 |
| | Successful participants | 10 | 5 |
| ME-BA | Requested capacity (MW) | 446 | 690 |
| | Offered capacity (MW) | 200 | 200 |
| | Clearing price | 0.02 | 0.12 |
| | Participants | 10 | 17 |
| | Successful participants | 7 | 7 |
| BA-HR | Requested capacity (MW) | 1069 | 1376 |
| | Offered capacity (MW) | 400 | 400 |
| | Clearing price | 0.23 | 0.51 |
| | Participants | 13 | 18 |
| | Successful participants | 6 | 8 |
| HR-BA | Requested capacity (MW) | 774 | 929 |
| | Offered capacity (MW) | 400 | 400 |
| | Clearing price | 0.03 | 0.12 |
| | Participants | 9 | 17 |
| | Successful participants | 6 | 13 |

**Table 2.** Average prices on BH borders 2018 and 2019.

| Average price on monthly auctions (EUR/MW/h) | 2018 | 2019 |
|---|---|---|
| BA-ME | 0.04 | 0.18 |
| ME-BA | 0.06 | 0.07 |
| BA-HR | 0.39 | 0.34 |
| HR-BA | 0.01 | 0.11 |

## 5 Conclusions

There are numerous advantages and steps-forward which were recognised in the region after implementation of coordinated auctions. First and for sure most important, is the transition from unilateral auctions performed by most TSOs on their 50% of NTC to the coordinated auctions conducted by SEE CAO on 100% of NTC.

Hereby, the full harmonisation is reached by applying one single set of Auction Rules on 8 countries and single auction algorithm, synchronised timing of long term capacity rights nomination deadline and auctions on all borders, single registration and collaterals applied on all borders per Auction Participant, etc. The important thing to note is that in order for Auction Participants to be registered in SEE CAO, they must be on eligibility list of at least one participating TSO. Nevertheless, all Auction Participants are allowed to bid and allocate capacity on all auctions performed by SEE CAO, but when it comes to the usage of allocated PTR, the TSO eligibility list is applied and secondary market (transfer or UIOSI) is applicable for non-eligible TSOs on particular borders.

The most obvious step forward in electricity market integration is the complete appliance of ENTSO-E Capacity Allocation and Nomination (ECAN) standard in the region and introduction of new products. For example, UIOSI principles for the first time applied on all borders, which emphasizes the efficiency in capacity usage and reallocation. Regarding the products, for the first time daily auctions are performed on OST's and TEIAS' borders, while also on TEIAS' border even the yearly auction is introduced for the first time. The secondary market including transfer and resale is also introduced as a new product, which brought lot of advantages to the Auction Participant.

In the Figs. 4, 5, 6, 7, 8, 9, 10, and 11 where the results of yearly, monthly and daily auctions are shown, it can be concluded that by comparing 2018 and first five month of 2019, we have an increase in the number of participants in auctions and hence increased auction revenue. Also, we can conclude from the comparative Tables 1 and 2 that offered capacity for 2018 and 2019 was the same, increased interest of participants in 2019 resulted in increased amount of requested capacity and consequently, increased price and auction revenue in 2019, and also on monthly auctions 2019 compared to 2018 on BH borders average prices have increased.

## References

1. http://www.seecao.com/
2. Torriti, J.: Privatization and cross-border electricity trade: from internal market to European Supergrid? Energy **77**, 635–640 (2014)
3. Kristiansen, T.: Cross-border transmission capacity allocation mechanisms in South East Europe. Energy **35**(9), 4611–4622 (2007)
4. Adamsona, S., Noeb, T., Parkerc, G.: Efficiency of financial transmission rights markets in centrally coordinated periodic auctions. Energy, October 2011
5. Martinčić, D., Mijušković, A., Međimorec, D.: Regionally coordinated auctions of cross border transmission capacities between South East European transmission system operators. CIGRE, SEERC, Portorož (2016)

6. REGULATION (EC) No 714/2009 OF THE EUROPEAN PARLIAMENT AND OF THE COUNCIL of 13 July 2009 on conditions for access to the network for cross-border exchanges in electricity and repealing. Regulation (EC) No 1228/2003
7. Energy Community Treaty. www.energy-community.org
8. ENTSO-E Capacity Allocation and Nomination System (ECAN) Implementation Guide

# Determining the Applicability Area of Single-Lane Roundabout Depending on Traffic Flow Intensity

Ammar Saric(✉), Sanjin Albinovic, Mirza Pozder, Suada Dzebo, and Emira Muftić

Faculty of Civil Engineering, Department of Roads and Transportation, University of Sarajevo, Sarajevo, Bosnia and Herzegovina
ammar.saric@hotmail.com, sanjin.albinovic@gmail.com, pozder.mirza@hotmail.com, suada.dzebo.gf@gmail.com, emira.muftic@hotmail.com

**Abstract.** The main goal of this paper was to determine the area of single lane roundabout application depending on traffic flow intensity. Roundabouts are increasingly applied to solve traffic congestion problems, especially in big cities. However, the question arises whether they represent the optimal solution for each problem of such nature from different aspects. In this paper attention is focused on the aspect of capacity level that such intersection can provide depending on its design elements and traffic volume. Based on theoretical analysis of different traffic flow distribution patterns, applicability diagram was obtained for single-lane roundabouts and compared with similar diagrams published in different editions of the US Highway Capacity Manual.

## 1 Introduction

The movement of vehicles depends on a large number of factors, which is why a congestion description is a very complex process. Some of the most important factors influencing movement of vehicles in the traffic flow are: the size of the traffic flow, characteristics of the flow, vehicle dynamics, psychophysical characteristics and driver motivation, characteristics of traffic management and control system and environmental conditions (e.g. visibility). In this paper we have focused on one of these factors - the traffic management and traffic congestion system on interrupted flows, i.e. roundabouts.

Depending on the intensity of future traffic on main and minor approaches, it is possible to predict the most favorable type of intersection for actual traffic demand, based on diagrams suggested in the US Highway Capacity Manual 2000 (HCM 2000) (Fig. 1) [1]

According to the Fig. 1 roundabouts are a good solution if the traffic flow in the main and minor direction is approximately equal. The rest of the diagram (other combinations of traffic flow) do not particularly emphasize the use of roundabouts or it is undefined in which exact situations they are optimum solution, which may be a big drawback compared to real operating performances of built roundabouts.


S. Avdaković et al. (Eds.): IAT 2019, LNNS 83, pp. 171–186, 2020.
https://doi.org/10.1007/978-3-030-24986-1_14

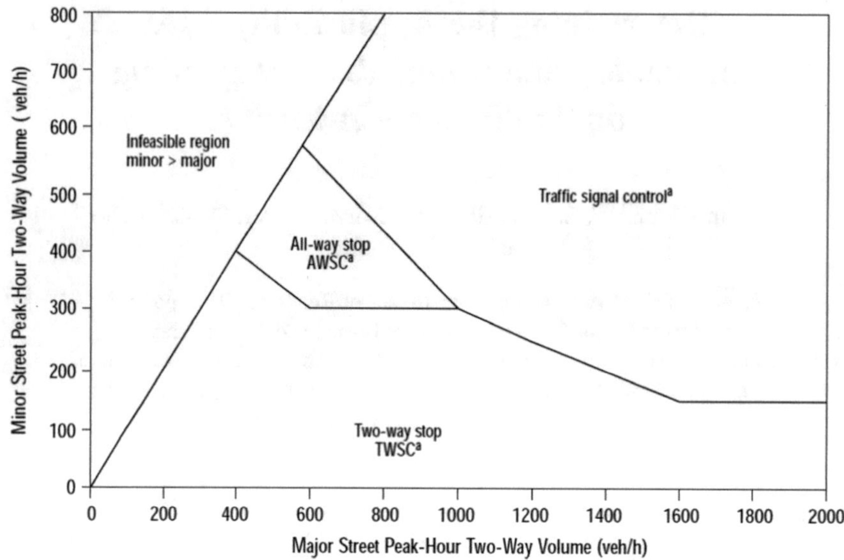


**Fig. 1.** Intersection control type and peak-hour volumes (a. Roundabouts may be appropriate within portion of these ranges) [1]

The lack of number of possible applications of roundabouts was overcome in a new edition of HCM called Planning and Preliminary Engineering Applications Guide to the Highway Capacity Manual 2016 [2]. Two new diagrams were presented.

The diagrams are quite different from the previous one because they have more restrictions and a more detailed overview of traffic load ranges for which certain types of intersections are being applied. As can be seen in the application of roundabouts, whether it is single-lane or multiple lane, it is possible in most cases. The diagram in Fig. 2 refers to the distribution of traffic volume 50/50 for the two-way traffic flow for both the main and minor approach, and the diagram in Fig. 3 refers to the distribution of traffic volume 67/33. For further information on the assumptions on which these diagrams are developed, refer to the reference [2].

Based on Fig. 2 single-lane roundabouts can be applied in area bounded with two dashed lines and traffic flow ranged from 490 veh/h to 1.900 veh/h for major approach, and 0 to 800 veh/h for minor approach. Two lane roundabouts are optimal solution when major traffic flow exceeds 1.400 veh/h in combination with minor traffic flow up to 800 veh/h.

Different traffic flow distribution in favour of major approach causes a little narrower area of single-lane roundabout application. While range of minor traffic flow remain the same, maximum value of traffic flow in major direction is 1.630 veh/h. At the same time, two-lane roundabouts can be considered as a good solution for major traffic flow between 1.060 veh/h and 2.00 veh/h.

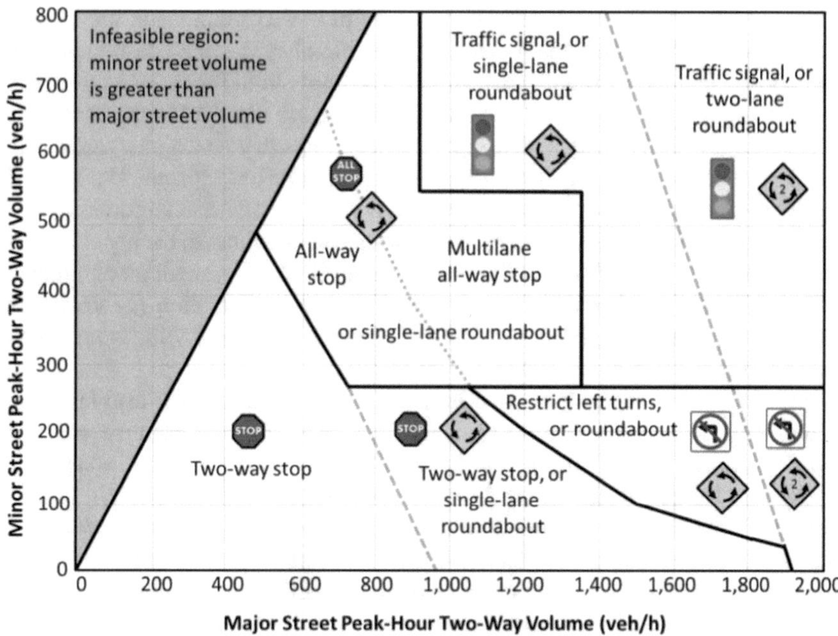


**Fig. 2.** Intersection control type by peak hour volume - 50/50 volume distribution on each street [2]

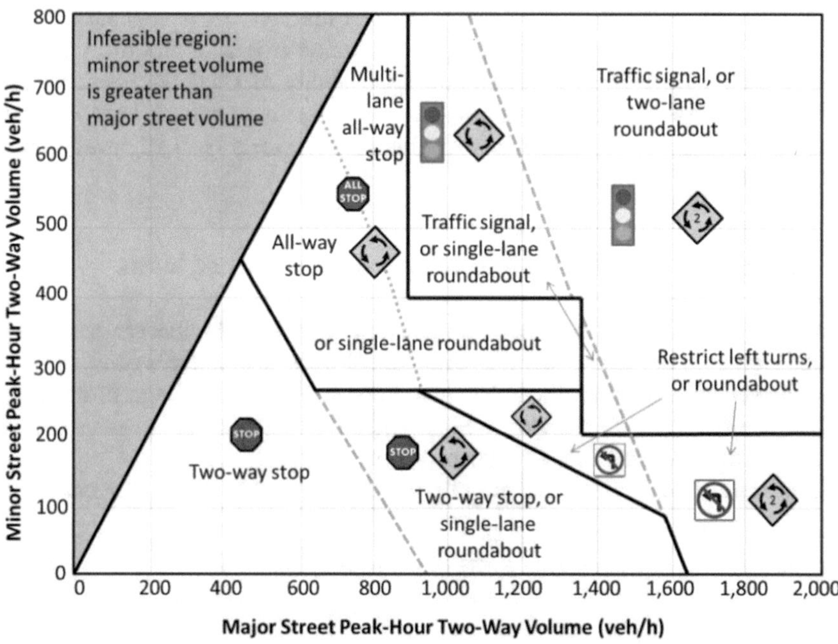


**Fig. 3.** Intersection control type by peak hour volume - 67/33 volume distribution on each street [2]

The application of these diagrams and use of proposed solutions in traffic analysis should be considered within the framework of local policy, i.e. the question is whether local policy is favored by roundabouts or not. The upper limit for applying the "STOP" sign on minor approaches occurs when all conditions are met for the use of a "STOP" sign on all approaches or when a demand on minor approach with a higher traffic volume exceeds the corresponding capacity, depending on what before happens. The lower limit for a single-lane roundabout is set for level of service C or D at the threshold of 25 s of average delay per vehicle for minor approach with a higher volume; the upper limit is set to 85% of the entry capacity of the single-lane roundabout. Also, as a limit of roundabout application or any other intersection, presents degree of saturation (x) whose value should be less than 0.90. When the saturation degree exceeds 0.90 it is necessary to improve the existing solution or find a new one [2].

Apart from the research presented in the HCMs, various authors also investigated this problem. Mauro and Branco [3] published results on comparative analysis of compact multilane roundabouts and turbo-roundabouts in terms of their capacity. They used capacity formula from "Kreisel" software developed by Brilon and Wu [4]. Based on five different traffic distribution scenarios, they provided diagrams of suitability domains for turbo-roundabouts and two-lane ("compact") roundabout. No capacity analysis for single-lane roundabout or field study verification provided in this study. Vasconselos et al. (2012) also investigated capacity of different roundabouts but with field results confirmation and based on Hagring's capacity formula. A number of studies have been published on different capacity models for roundabouts of which the following are listed in this paper: Çalişkanelli et al. [5], Qu et al. [6] and Hagring et al. [7].

For the planning purpose of future intersections or reconstruction of existing low-performance intersections, it is necessary to make the right decision about the appropriate type of intersection. In this respect, capacity is one of the most important decision factor. In this paper, research of applicability of single-lane roundabout in different traffic conditions based on capacity estimation is presented. Obtained results are compared with existing suitability diagrams from widely used Highway Capacity Manual, editions 2010 and 2016.

## 2 Capacity Determination of Single-Lane Roundabouts

The capacity of roundabouts is the maximum number of vehicles that can pass through a roundabout in a single time period, regardless of the length of the waiting time. This is the theoretical value that should be equal to or greater than the sum of intensity of traffic in the roundabout (at the each entry) and the intensity of the traffic at the entry joining the conflict point in the roundabout [8].

Geometric characteristics that are relevant to the capacity of the roundabout are:

- distance between entries and exits
- connection type of the access roads to the roundabout.

By increasing the distance between the entries and the exits, drivers get more time to estimate whether they can take advantage of the time gap and engage in the traffic flow in the roundabout. However, in practice it often happens that drivers leaving the roundabout do not give a right indicator to cause the vehicle at the entrance can not take advantage of time gap. This reduces the roundabout capacity [8].

Capacity calculation methods analyze roundabouts as a series of triangular T-links on a one-way circular road. Each T-link is individually analyzed and a special capacity is determined, depending on the method and degree of saturation, the waiting time and the length of the queue. If cyclists and/or pedestrians are present in the roundabout, the capacity must be reduced. In this paper the presence of pedestrians and cyclists is neglected. Also, the capacity of the roundabout depends on the intensity of circular flow (in roundabout), and is inversely proportional to the given intensity [8].

With respect to the model for determining the capacity, the main dilemma concerns whether to gap acceptance based method or empirical regression models [9].

The capacity calculation in this paper is determined on the basis of an analytical method based on the gap acceptance theory according to HCM methodology and Hagring method.

The HCM methodology for the calculation of the roundabout capacity is fully presented in the HCM 2010 in chapters: 21. Roundabouts and 33. Roundabouts Supplemental.

In the following sections these two methods are described in detail.

## 2.1 Gap Acceptance Based Methods

The basic concept of capacity models based on the gap acceptance theory can be briefly described as follows: a vehicle on the minor approach will entry on major road if the time gap t between the passage of two vehicles from the priority stream, i.e. major road traffic flow, greater than the critical value $t_c$ - critical time gap [10].

Namely, the application of models based on time gap acceptance theory requires two main parameters of the critical time gap and follow-up time. The distribution of follow-up time, i.e. the time of progression in the major floe, determines the value and the number of usable time gaps for the minor stream. Thus, critical time gap plays an essential role in implementing this theory in capacity estimation.

The critical time gap $t_c$ is defined as the minimum time interval that is long enough to allow at least one vehicle from the minor traffic stream to enter the intersection, in this case a roundabout.

The follow-up time $t_f$ is the average time interval between the departure of two successive vehicles from the minor stream into the conflicting cross-section of the priority flow in the conditions of the continuous queue in the minor approach, respectively the follow-up time is the average time interval between the vehicles in the queue entering the intersection during long enough time gap in the priority flow.

The critical time gap is not a constant value and is different from the driver to the driver and the traffic circumstances in which they are located. However, the theory of time gap acceptance implies that drivers are consistent and homogeneous driving because the practical application requires only one pair of critical time and follow-up time for the individual conflict flow, i.e. the values of these parameters must be constant in practical application. However, as the mentioned parameters depend on the traffic conditions, the type and dimensions of the intersection and the structure of the traffic flow, the above-mentioned assumptions are not realistic: drivers do not behave in the same way every time and often accept a shorter time gap than expected. However, simplifying the true nature of driving behavior does not have a significant impact on the final results of the capacity.

In 1998, Hagring [11] developed a universal formula for the capacity calculation that can be used for each multi-lane intersection. Each entry lane is characterized by a conflict with one or more circular lanes resulting in several parameters of gap accepting parameters that represent different behavior of the driver. Assuming that all major flows have Cowman's M3 distribution, the capacity can be calculated according to the following formula:

$$C = \frac{e^{\left(-\sum_{i \in I_k} \lambda_i \cdot \left(t_{c,i} - \Delta_i\right)\right) \cdot \sum_{i \in I_k} \lambda_i}}{1 - e^{\left(-\sum_{i \in I_k} t_{f,i} \cdot \lambda_i\right)}} \cdot \prod_{i \in I_k} \frac{\alpha_i}{\alpha_i + \lambda_i \cdot \Delta_i} \tag{1}$$

Where are:
C – capacity of entry lane (veh/h),
$\alpha$ - proportion of free vehicles,
$t_{c,i}$ - is the critical gap for each entry lane (s),
$t_{f,i}$ - follow-up time for each entry lane (s),
$\Delta$ - minimum headway of circulating vehicles (s),
$\lambda_i$ - Cowan's M3 parameter,
k - minor flow index,
$I_k$ – set of major flows i that conflict with minor flow k.

Research by Sullivan and Troutbeck (1994), Açelik and Chung (1994) and Hagring (1996) showed that the M3 distribution, introduced by Cowan (1975), provides a good consistency with the distribution of observed gaps between the vehicles in the circular stream.

The formula (1) can be presented more easily for the purpose of capacity calculation of single lane roundabout:

$$C = \frac{q \cdot \alpha \cdot e^{[-\lambda \cdot (t_c - \Delta)]}}{1 - e^{(-\lambda \cdot t_f)}} \tag{2}$$

Parameters $\alpha$, $\lambda$ and $\Delta$ are the parameters of the Cowan's distribution. Parameter $\lambda$ can be expressed by the following equation:

$$\lambda = \frac{\alpha \cdot q}{1 - \Delta \cdot q} \tag{3}$$

There are different expressions for parameters $\alpha$ and $\Delta$, according to several authors [12]. In this paper a Tanner's model was adopted for its simplicity and widespread use:

$$\alpha = 1 - \Delta \cdot q; \Delta = 2\,\text{sec} \tag{4}$$

The lane by lane approach based on the Hagring formula makes it convenient and easy to use when assessing the turbo roundabout capacity. This is a model based on theory and likelihood that describes very well the major traffic flow and expected drivers' and vehicles waiting at the roundabout entrances. It has been shown that the distribution of the major flow (circulating in the roundabout) to the circular lanes and the various critical time gaps for the vehicles in the circular lanes have a significant effect on the entry capacity [12]. It has also been shown that the uneven distribution of the total main flow to the circular lanes results in lower capacity. With the single lane roundabout, this is not a problem because they have only one circular lane and single lane entries. For this reason, the estimation of these parameters plays an important role in the capacity estimation. According to Miller and Pretty (1968) and Brilon et al. (1997), the method of maximum likelihood probability, besides the many other methods, proved to be the most efficient estimator of the required parameters.

### 2.2 Capacity Based on Highway Capacity Manual

The HCM 2010 [13] methodology is an analytical method based on the gap acceptance theory in roundabout. The procedures include the following three traffic participants: motor vehicles, pedestrians and cyclists. For the sake of simplicity in this paper, only motor vehicles have been considered.

For roundabout analysis, the following data is required:

1. Number and configuration of lanes of each approach,
2. One of the above:
   a. Vehicle volume for each entry maneuver (movement of vehicles entering the roundabout) and the number of pedestrians in peak 15 min, or
   b. Vehicle volume for each entry maneuver, number of pedestrians in peak hours, and peak hour factor for that hour,
3. Heavy vehicles percentage,
4. Distribution of traffic, i.e. volume of vehicles on lanes, for multi-lane approaches; and
5. The length of the analyzed time interval, usually the peak 15-min period. However, it is possible to analyze any 15-min period.

The three basic terms, i.e. flow of interest in this methodology are:

- entry traffic intensity = $v_e$
- conflict traffic intensity = $v_c$
- exit traffic intensity = $v_{ex}$

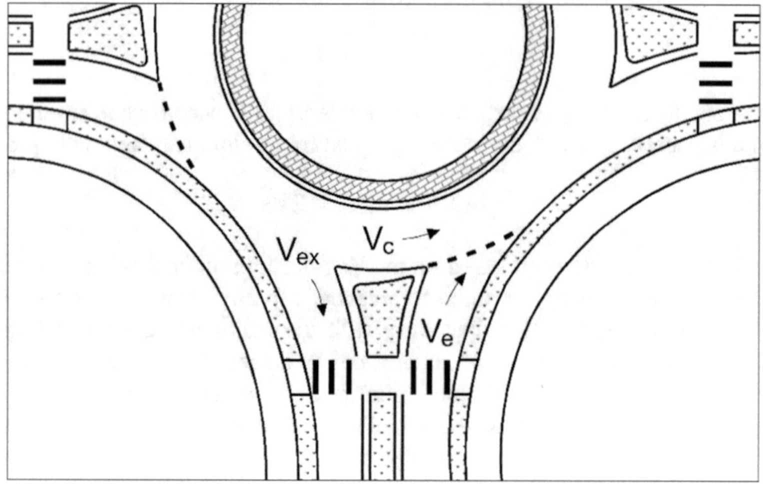


**Fig. 4.** Capacity analysis on single-lane roundabout (HCM 2010)

The above terms are shown on the following figure:

The approach capacity is reduced as the conflict flow increases. The conflict flow generally represents the circulating stream passing directly in front of the observed entry. The circulating flow is in direct conflict with the entry flow, but the exit flow may also affect the driver's decision to enter the roundabout. As long as drivers do not make their roundabout exit maneuver, especially if they have not signaled it, there is a certain uncertainty with the driver at the roundabout entry if the vehicle coming from their left side tries to get out of the roundabout or continue driving through the roundabout. At high values of both the flow, conflict and entry, can occur specific cases such as limited priority (circular vehicles adjust the reciprocal gaps at the entry to allow them to enter the roundabout) and reverse the priority (the traffic at the entrance compels circulating traffic to miss them). In these cases, more complex analytical or regression models give more realistic results.

The capacity of a single-lane entry that conflicts with the one circular lane (single-lane roundabout shown in Fig. 4) is based on conflict flow. Equation for capacity estimation is given by the following expression:

$$C_{e,pce} = 1130 \cdot e^{\left(-1{,}0 \cdot 10^{-3}\right) \cdot v_{c,pce}} \tag{5}$$

The calibration of the model for roundabout analysis is necessary to ensure realistic results, which are in line with the data collected in the field. The ideal scenario involves collecting data on a large number of different roundabouts, applying the methodology presented in the Highway Capacity Manual 2010, and finally calibration of the analysis model based on the collected data.

The capacity calculation model is calibrated by two parameters defined by field measurements, namely the critical gap $t_c$ and the follow-up time $t_f$ whose values are given in Table 1. Accordingly, the capacity formula can be expressed as follows:

$$C = A \cdot e^{(-B \cdot v_c)} \tag{6}$$

$$A = \frac{3600}{t_f} \tag{7}$$

$$B = \frac{t_c - \frac{t_f}{2}}{3600} \tag{8}$$

**Table 1.** Gap acceptance parameter for single lane roundabout

| $t_c$ (sec) | $t_f$ (sec) |
|---|---|
| 4,43 | 3,67 |

Gap acceptance parameters in Table 1 were obtained on local roundabouts in Sarajevo i Zenica, Bosnia and Herzegovina (see [14, 15]).

# 3 Estimation of Capacity Based on Theoretical Traffic Flow Distribution

Using the above-mentioned method for capacity evaluation of single-lane roundabout (Fig. 5), we have calculated capacity for different scenarios of traffic distribution.

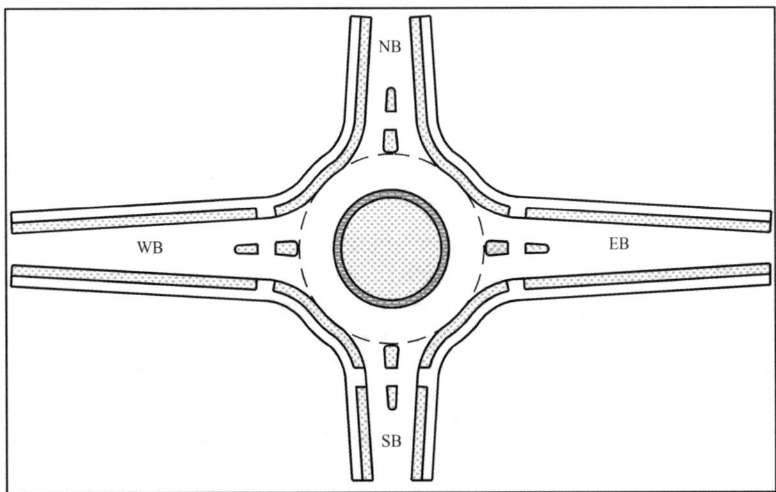


**Fig. 5.** Single lane roundabout

The traffic flows are assumed starting from 200 veh/h on minor road and 200 veh/h on major road. For each of the following combinations of 200 veh/h on minor road, it remained constant until the flow on the major entrances increased by 200 veh/h until it reached 1000 veh/h. When the value at the main entrances reaches 1000 veh/h, the process is repeated but the minor flow is increased by 200 veh/h (minor road = 400 veh/h, main road = 200, 400, 600, 800 and 1000 veh/h). The procedure was repeated each time in the same way until flor of 1000 veh/h is reached on both, minor and major, approaches. It should be noted that values for major, and also for minor approach, represent sum of flows on both major or minor approaches.

In order to estimate the roundabout capacity, seven theoretical traffic conditions were used which are presented as origin-destination matrices (Fig. 6). These matrices cover different traffic distributions such are: movements on all approaches are equal, dominant left or dominant right movement at roundabout, dominant through movement in major direction etc. U turns are neglected because they rarely have important influence in traffic distribution. Elements of these matrices are percentage of each traffic movement from one entry to another.

| | Scenario 1 | | | | Scenario 2 | | | | Scenario 3 | | | | Scenario 4 | | | |
|---|---|---|---|---|---|---|---|---|---|---|---|---|---|---|---|---|
| | **WB** | **EB** | **SB** | **NB** | **WB** | **EB** | **SB** | **NB** | **WB** | **EB** | **SB** | **NB** | **WB** | **EB** | **SB** | **NB** |
| **WB** | 0 | 0.33 | 0.33 | 0.33 | 0 | 0.7 | 0.15 | 0.15 | 0 | 0.8 | 0.1 | 0.1 | 0 | 0.80 | 0.10 | 0.10 |
| **EB** | 0.33 | 0 | 0.33 | 0.33 | 0.7 | 0 | 0.15 | 0.15 | 0.8 | 0 | 0.1 | 0.1 | 0.80 | 0 | 0.10 | 0.10 |
| **SB** | 0.33 | 0.33 | 0 | 0.33 | 0.15 | 0.15 | 0 | 0.7 | 0.1 | 0.3 | 0 | 0.6 | 0.00 | 0.50 | 0 | 0.50 |
| **NB** | 0.33 | 0.33 | 0.33 | 0 | 0.15 | 0.15 | 0.7 | 0 | 0.1 | 0.3 | 0.6 | 0 | 0.50 | 0.00 | 0.50 | 0 |

| | Scenario 5 | | | | Scenario 6 | | | | Scenario 7 | | | |
|---|---|---|---|---|---|---|---|---|---|---|---|---|
| | **WB** | **EB** | **SB** | **NB** | **WB** | **EB** | **SB** | **NB** | **WB** | **EB** | **SB** | **NB** |
| **WB** | 0 | 0.8 | 0.1 | 0.1 | 0 | 0.8 | 0.1 | 0.1 | 0 | 0.30 | 0.60 | 0.10 |
| **EB** | 0.8 | 0 | 0.1 | 0.1 | 0.8 | 0 | 0.1 | 0.1 | 0.30 | 0 | 0.10 | 0.60 |
| **SB** | 0 | 0.8 | 0 | 0.2 | 0.2 | 0.8 | 0 | 0 | 0.10 | 0.60 | 0 | 0.30 |
| **NB** | 0.8 | 0 | 0.2 | 0 | 0.2 | 0.8 | 0 | 0 | 0.60 | 0.10 | 0.30 | 0 |

**Fig. 6.** Traffic distribution scenarios

Detailed results of capacity calculation for each scenario will not be presented. It is important to note that capacity calculation was stopped at degree of saturation of 0,90. We used both methods described above with gap acceptance parameters from Table 1.

Capacity results obtained for theoretical traffic distributions are presented in following diagrams. The results are summarized in the form of applicability of single lane roundabout for different combinations of major and minor flows (Figs. 7, 8 and 9).

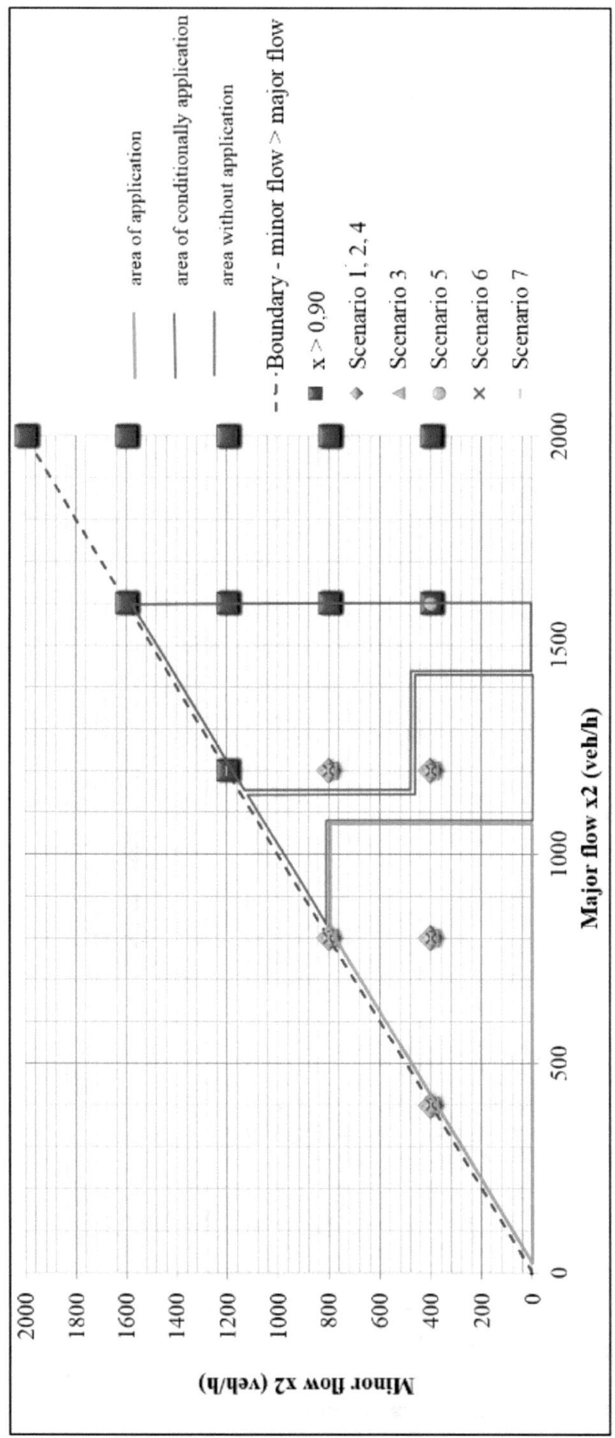


**Fig. 7.** Applicability of single lane roundabout based on HCM methodology

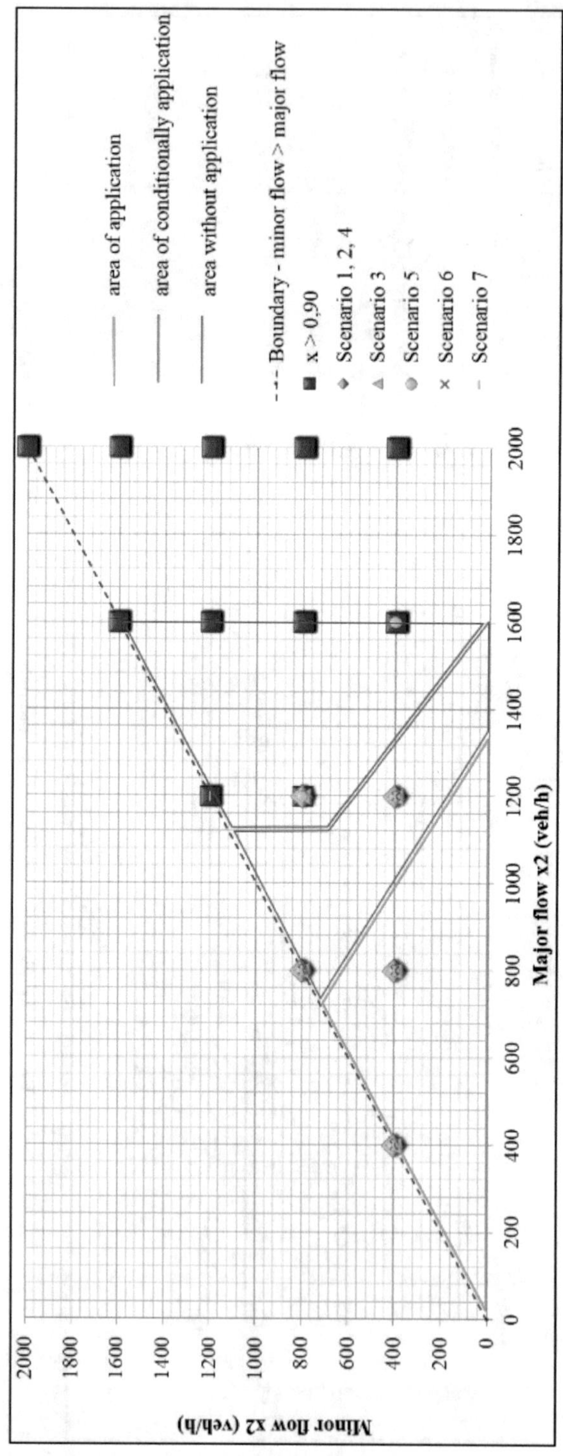


**Fig. 8.** Applicability of single lane roundabout based on Hagring methodology

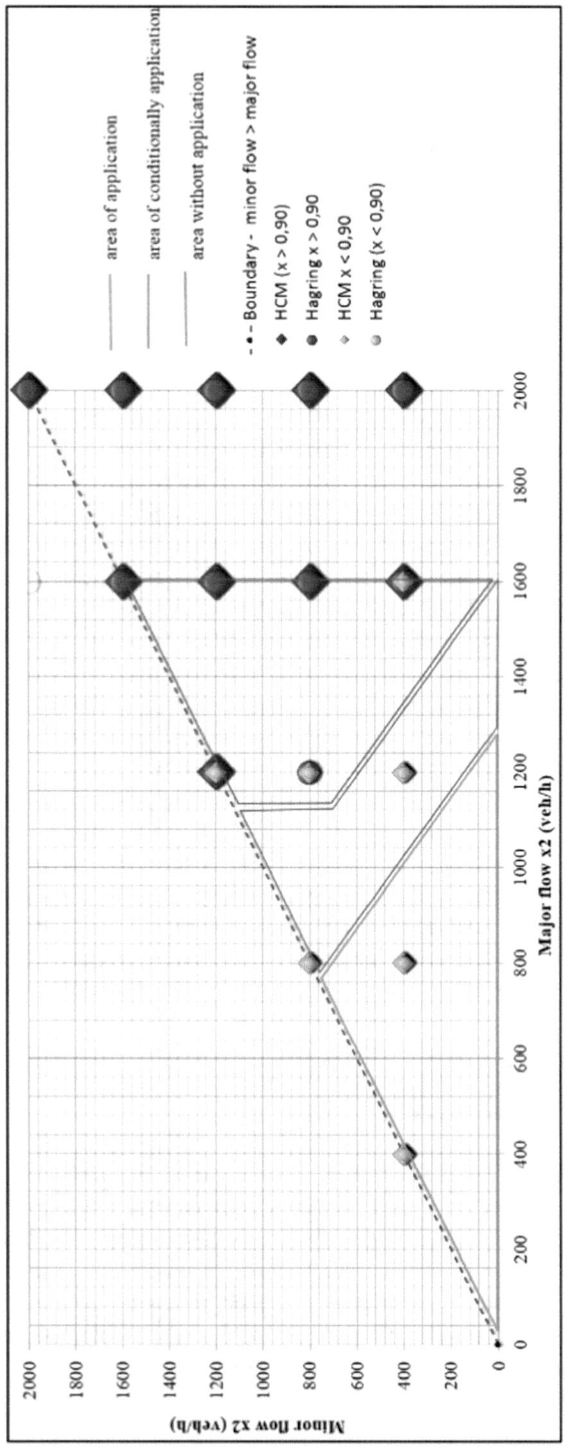


**Fig. 9.** Applicability of single lane roundabout based on both methodologies

Based on capacity, three different areas of single-lane roundabout applicability are visible:

1. Green area – single-lane roundabout can be applied,
2. Blue area – single-lane roundabout can be applied conditionally, which means that for some scenarios degree of saturation is greater than 0,90,
3. Red are – single-lane roundabout cannot be applied, which means that all scenarios have degree of saturation above 0,90.

It can be noted that Hagring method leads to smaller overall roundabout capacity compared to the HCM methodology. In terms of distribution scenarios, scenario 6 gives the lowest capacity.

The results obtained in this study can be compared to the diagrams in HCM 2000 and HCM 2016. Diagram in HCM 2000 indicates roundabout applicability in situations when traffic flow on major and minor roads are almost identical. As a consequence of this, applicability area is very narrow. Results from this study fully cover this area.

Two new diagrams in HCM 2016 have a much wider applicability area for single lane roundabouts. Although, those do not include a true traffic allocation on each approach, as we used in this research, those give a better overview of all situations in which single roundabouts are good solutions. The biggest difference between HCM 2016 recommendations and our results are in area with minor flow below 500 veh/h and major flow up to 1000 veh/h. In this area HCM 2016 suggests application of intersections with STOP signs on minor approach. This may be explained as a USA traffic culture and a widespread use of such traffic solutions. Based on our diagrams single lane roundabouts are very good solutions, regardless of traffic distribution, in area of low minor flow and major flow up to 1000 veh/h.

Second difference is for high value of both minor flow (above 500 veh/h) and major flow (above 1000 veh/h). For these values, HCM 2016 show single-lane roundabout as a very good solution together with signalized intersection and intersection regulated with STOP signs. Our results for good applicability area do not cover that values, which are quite high for real application. Also, combination of low minor flow and major flow higher than 1000 veh/h represent conditionally applicability area based on our area compare to the HCM 2016 where single lane roundabouts are favorable solution.

## 4 Discussion and Conclusion

Single-lane roundabouts are a widespread solution of traffic problems, especially in urban areas. Their application depend on numerous factors, such are: geometric elements, location, position in wider traffic system, traffic structure etc. One of the most important element based on which traffic engineers determine their applicability is capacity. They are two problems related to capacity: how to calculate it and how to determine applicability of single lane roundabouts based on obtained results.

First problem can be solved using gap acceptance theory with parameters of critical gap and follow up time which correspond to the field gathered data. Hagring's method seems the most appropriate for capacity calculation. HCM 2000 and 2010 method also can be used but with obligatory calibration of gap acceptance parameters.

Second problem is more difficult to overcome. Existing diagrams in HCM 2000 and HCM 2016 have major limitations. Those are not based on true distributions of traffic on each approach of single-lane roundabout. Also area of application of single-lane roundabout is or underestimate (HCM 2000) or overestimate (HCM 2016) compare to the results of this study. Although, diagram in HCM 2016 gives more situations in which these roundabouts are good solution, use of those are very questionable for very high values of minor or major flows.

Concept of capacity calculation which includes several different traffic distributions gives results which are more realistic and better described traffic operations on single-lane roundabouts. Final results highly depend on traffic distributions and must be an integral part of the applicability decision of single-lane roundabout. It is recommended that the theoretical scenarios be replaced or verified with micro simulations and with data on actual roundabouts.

## References


1. Highway Capacity Manual: Practical applications of research. U.S. Dept. of Commerce, Bureau of Public Roads (2000)
2. National Academies of Sciences, Engineering, and Medicine: Planning and preliminary engineering applications guide to the Highway Capacity Manual. The National Academies Press, Washington, DC (2016). https://doi.org/10.17226/23632
3. Mauro, R., Branco, F.: Comparative analysis of compact multilane roundabouts and turbo-roundabouts. J. Transp. Eng. **136**(4), 316–322 (2010)
4. Brilon, W., Wu, N.: Merkblatt für die Anlage von Kreisverkehren [Guideline for the Design of Roundabouts]. FGSV Verlag Gmbh, Cologne, Germany (2006)
5. Çalişkanelli, P., Özuysal, M., Tanyel, S., Yayla, N.: Comparison of different capacity models for traffic circles. Transport **24**(4), 257–264 (2009)
6. Qu, Z., Duan, Y., Hu, H., Song, X.: Capacity and delay estimation for roundabouts using conflict theory. Sci. World J. **2014**, Article ID 710938 (2014)
7. Hagring, O., Rouphail, N.M., Sørensen, H.A.: Comparison of capacity models for two-lane roundabouts. In: 82nd Annual Meeting of the Transportation Research Board, January 2003, Washington, DC (2003)
8. Kenjić, Z.: Roundabouts. Association of Consulting Engineers in Bosnia and Herzegovina, Sarajevo (2009)
9. Brilon, W.: Some remarks regarding the estimation of critical gaps. Transp. Res. Rec. J. Transp. Res. Board **2553**, 10–19 (2016)
10. Cvitanić, D., Lovrić, I., Breški, D.: Traffic flow theory (lectures on doctoral studies). Split: Faculty of Civil Engineering, Architecture and Geodesy, University of Split (2007)
11. Hagring, O.: Estimation of critical gaps in two major streams. Transportation Research Part B 34 (2000) (1999)
12. Giuffrè, O., Granà, A., Tumminello, M.L.: Gap-acceptance parameters for roundabouts: a systematic review. Eur. Transp. Res. Rev. **8**, 2 (2016)
13. Highway Capacity Manual: Practical applications of research. U.S. Dept. of Commerce, Bureau of Public Roads (2010)

14. Šarić, A., Lovrić, I.: Multi-lane roundabout capacity evaluation. Front. Built Environ. **3**, 42 (2017)
15. Saric, A., Albinovic, S., Pozder, M.: Capacity analysis of single and two-lane roundabouts: a case study in Bosnia and Herzegovina. Transport Research Arena – TRA 2018, 16–19 April 2018, Vienna, Austria (Poster presentation) (2018)
16. Vasconcelos, A.L.P., Silva, A.B., Seco, Á., Silva, J.: Estimating the parameters of Cowan's M3 headway distribution for roundabout capacity analyses. Balt. J. Road Bridge Eng. **7**, 261–268 (2012). https://doi.org/10.3846/bjrbe.2012.35
17. Sullivan, D.P., Troutbeck R.J.: The use of Cowan's M3 headway distribution for modelling urban traffic flow. Traffic Eng. Control **35**(7/8), 45450 (1994)
18. Akçelik, R., Chung, E.: Calibration of the bunched exponential distribution of arrival headways. Road Transp. Res. **3**(1), 42–59 (1994)
19. Hagring, O.: The use of the Cowan M3 distribution for modelling roundabout flow. Traffic Eng. Control **37**. https://trid.trb.org/view.aspx?id=463394
20. Cowan, R.J.: Useful headway models. Transp. Res. **9**, 371–375 (1975). https://doi.org/10.1016/0041-1647(75)90008-8
21. Miller, A.J., Pretty, R.L.: Overtaking on two-lane rural roads. Proc. Aust. Road Res. Board (ARRB) **4**, 582–591 (1968)
22. Brilon, W., Wu, N., Bondzio, L.: Unsignalized intersections in Germany-a state of the art. In: Proceedings of the Third International Symposium on Intersections Without Traffic Signals, Portland, Oregon, USA, July 21–23, pp. 61–70 (1997)

# Arch Bridge Quality Control Plans

Naida Ademović[1(✉)], Pavel Ryjáček[2], and Milan Petrik[2]

[1] Faculty of Civil Engineering, University of Sarajevo, Patriotske Lige 30, 71 000 Sarajevo, Bosnia and Herzegovina
naidadem@gmail.com

[2] Faculty of Civil Engineering, Department of Steel and Timber Structures, Czech Technical University in Prague, Prague, Czech Republic
pavel.ryjacek@fsv.cvut.cz

**Abstract.** The importance of infrastructure for economic development of countries and, it opens up the economy to grander, and bigger opportunities. Maintenance should be well-thought-out and be regarded as vital part any infrastructure development in a country. The main goal of infrastructure maintenance is life sustainability of major assets. The right planning of maintenance is necessary for right moment of investment, as its premature or late implementation will lead to unnecessary costs and it may compromise the asset's economic life duration. Different management systems developed across Europe leading to different choices and recommendations regarding maintenance actions. In order to overcome this issue a COST Action TU 1406 has proposed a guideline for standardization and development of quality control plans for roadway bridges. As in all management systems the starting points is collection of available data, defining element identification and grouping, recognition of vulnerable zones, damage processes and failure modes. On the basis of the identified damage processes identification and evaluation of performance indicators (PIs) is conducted, followed by the selection and identification of key performance indicators (KPIs), taking into account owners' demands, and finally creating Quality Control Plans (QCP) scenarios and comparing them by spider diagrams. The recommended procedure is applied on an arch concrete bridge in Czech Republic.

## 1 Introduction

Bridges due to their uniqueness, complexity and vital impact on the society were initiators for development of management systems. For a functional transport network suitable operation of bridges is required. In order to make investment decisions that would be cost-effective the asset management connects engineering and economic principles with complete business practice throughout the life-cycle of infrastructure assets [1]. To fulfill this requirement the bridge management system (BMS) has to be a part of the global network management system meaning that it is the network level which defines the required performance requirements [2]. The standards that are used in a vast number of European countries are based on ISO 2394 [3]. One needs to keep in mind that, since 2008, there has been a dramatic decrease of budget allocated for road maintenance in Europe. According to the data of European Commission grown of the


S. Avdaković et al. (Eds.): IAT 2019, LNNS 83, pp. 187–204, 2020.
https://doi.org/10.1007/978-3-030-24986-1_15

freight transport is continuing and it is foreseen to rise by around 40% by 2030 and by little over 80% by 2050. This will have a negative impact on the society from various aspects, like pollution, noise, congestion affecting economy, health and well-being of European citizens. In order for conduct correct maintenance measures a well management bridge system has to be established filled with all available data. According to the recommendations given in the COST TU 1406 once damages processes are identified they will be correlated with adequate performance indicators (PI), indicating the fitness for purpose of the bridge or its element. The next step is connection of PIs with identified Key PIs (Reliability, Availability, Safety and Cost (Economy)) [4].

## 2 Proposed Methodology

For defining and preparation of the maintenance and other intervention plans it is necessary to identify damage processes, PI and KPIs. This can be done by a proposed procedure which consists of static and dynamic tasks. Static tasks can be done in the office indicated by green color, or at the site colored by orange color. Once data regarding all state of the bridge are noted, dynamic tasks take place represented by blue color (from 12 to 15) defining various developments in the function of that foreseen scenarios. Figure 1 shows the proposed methodology [5].

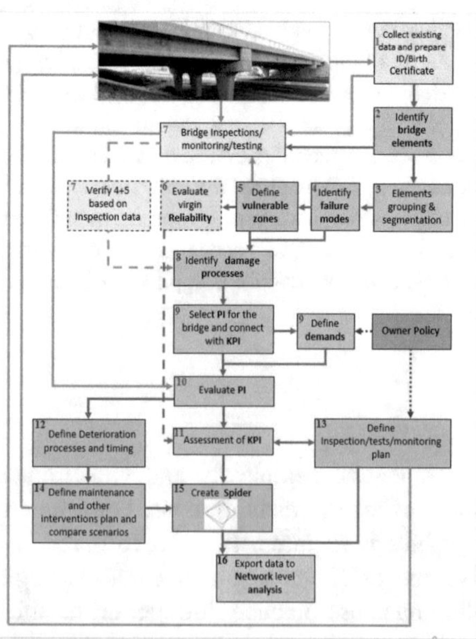


**Fig. 1.** Preparation process of case study [5]

All available data should be collected forming the basis for the database and making an ID/Birth certificate of the bridge. In order to have information regarding the current and actual stated of the bridge it is necessary to visit the site (geometric checks, element checks, etc.). Further steps would be on-site investigations (identification of different damages) and lab tests. Forth step in the static phase would be identification of the failure modes connected to the identified damages. Followed by the vulnerability zone identification and identification of damage processes. Damage of a bridge element is a physical disruption or a change in its condition, caused by external actions, such that some aspect of, either the current or future performance of the component (and perhaps consecutively a complete structure) is impaired [6]. Defining the PIs and KPIs forms the basis for calculation of the remaining service life of the bridge in respect to various possible foreseen scenarios. Knowledge of bridge deterioration rates is essential for cost-effective asset management and long-range transportation planning. The history of bridge deterioration in previous periods and the effects of individual factors are not explicitly considered. In order to predict changes in the individual bridge conditions over time one can used Markov Chain, however to variations in deterioration rates other methods have to be employed. Dependency of bridge conditions and age was developed by [7] with the application of a third-order polynomial function. Dunker and Rabbat [8] conducted a multi-stated study analyzing bridges that were constructed between 1950 and 1987. During their investigations they have realized that the choice of material for the bridge structure has an impact on structural deficiency. As the quality of the bridge decreases from concrete to steel to timber the structural deficiency increases. In their investigations age did not play a role. Once again, Kallen and Van Noortwijk [9] connected the bridge deterioration with age, indicating its high dependency, where the expected bridge condition and age are connected by a polynomial function. Confirmation that age represents the most important contributor to the structural safety was stated by Kim and Koon [10] in cold regions. The next two contributors are structural characteristics of the bridge and traffic volume.

Once the assessment and calculation of KPIs is done, planning of different maintenance and intervention plans with target reliability and safety over time which would take place. It is necessary to compare different scenarios in order to see their advantages and disadvantages. It is recommended to perform at least two scenarios, referenced and preventive. Referenced scenario is the most usual one, where no funds are available for any kind of repairs on the bridge. On the other hand, a preventive relates to a scenario with a 100-year service life, taking into account normal service life of bridges, with available budget for all the required inspections and maintenance/repair actions to take place. Spider diagrams could be utilized for the comparison purposes of net present KPI for certain instances in time or as a function of time giving a 3D shape. All KPIs are normalized in the range from one (the best) to five (the worst) conditions. Regarding the KPI Cost, the most expected financial resources are marked with five and other scenarios are scaled accordingly.

For the information to be efficient it should be transferred to the network level for further usage in the network domain.

## 3 Case Study- Road Concrete Arch Bridge Nerestce

### 3.1 General Data About the Bridge

A detailed inspection of a five-span concrete arch bridge (Fig. 2) build in 1953 was conducted. The bridge crosses the river Skalice between Čimelice and Osečany villages. As there are no original drawings and calculations some data could be only estimated according to the engineering judgment. In that respect according to the structure type and sketches from BMS it can be expected that the pad foundations are located on a rock structure.

The substructure consists of concrete abutments and massive foundation blocks of the arch. The loadbearing structure consists of three parts which are separated by dilatation. The two side parts are formed by the concrete frame consisting of two spans. The main arch is made of seven spans, meaning that a total of eleven spans makes the superstructure of the bridge. The deck is made of concrete and it is connected rigidly to the abutments and supported on two slab piers. In the central part the bridge deck is supported by a concrete arch of rectangular shape and by slab piers. The deck is fixed to the top of the arch by hinged connection (Fig. 3).

The carriage and pedestrian way are covered by asphalt. During the investigations an enormous thickness of the asphalt pavement on the carriage way was measured amounting to 200 mm. The handrailing is made of concrete. There is an open drainage system. The water is drained by an open drainage system with ten gullies and vertical pipes which are very close to the arch structure (cca 50 cm) causing additional damage to the concrete arch and piers (Fig. 2).

**Fig. 2.** Investigated bridge structure

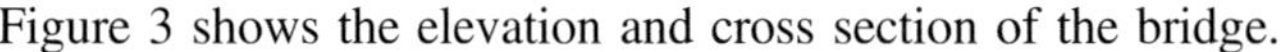

Figure 3 shows the elevation and cross section of the bridge.

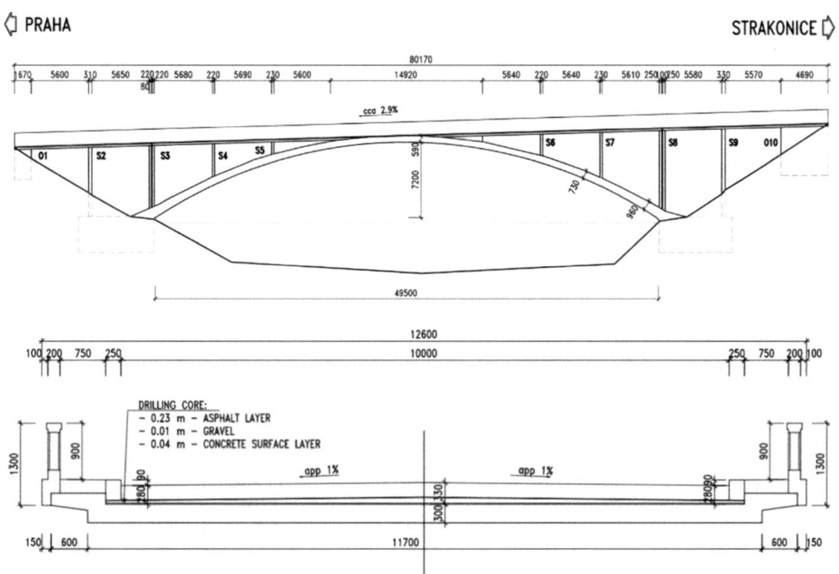


**Fig. 3.** Cross section of the bridge

According to the traffic information data from 2010, 9679 cars pass in 24 h, while number of heavy traffic is much smaller amounting to 1953 in 24 h. The traffic intensity is shown in Fig. 4.

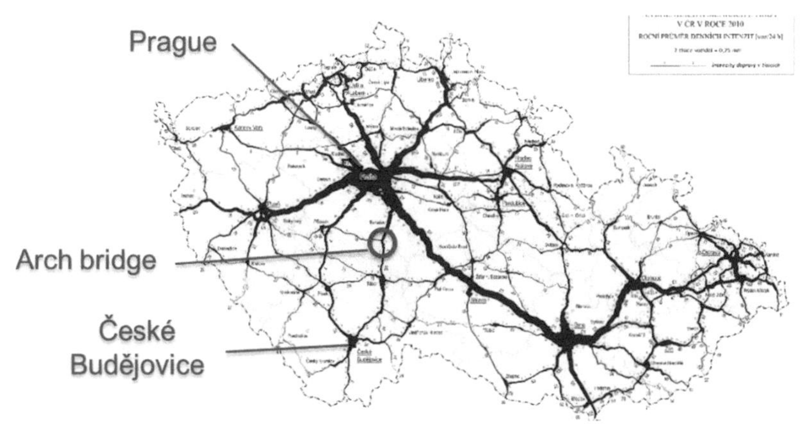


**Fig. 4.** Location of the bridge on the map of traffic intensity in 2010

### 3.2 Numerical Model

In order to check the load carrying capacity of the bridge a finite element model (FEM) calculation was made in SCIA Engineer software where three main cases were analyzed. The structure is modeled with shell elements and internal fixed links in the central part, where the arch and the deck are connected together (Fig. 5).

First case (case I) was taking into account an unlimited number of vehicles which represents a normal capacity of the bridge with the load of $V_n$ = 26.7 t. The second case (case II) elaborated the capacity of the one single vehicle on the bridge with $V_r$ = 66 t. Third case (case III) took into account the exceptional capacity for the heavy special transport of load $V_e$ = 175 t.

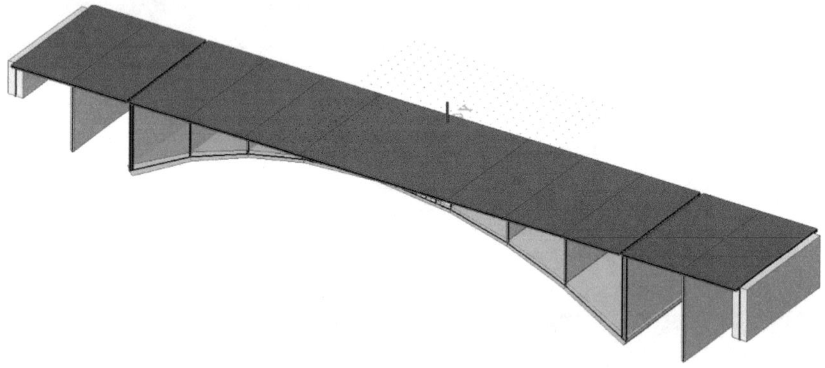

**Fig. 5.** FEM numerical model

The geometrically linear analysis was used, as the stiffness of the bridge and the deformations were small enough. The internal forces in the shell elements were taken with integration strips, that enabled to convert stresses to the moments and axial forces. The arch supports were considered to be fixed to the foundation block and also majority of connections were fixed as well. Some of the joints were hinged, as in the real structure (Fig. 6).

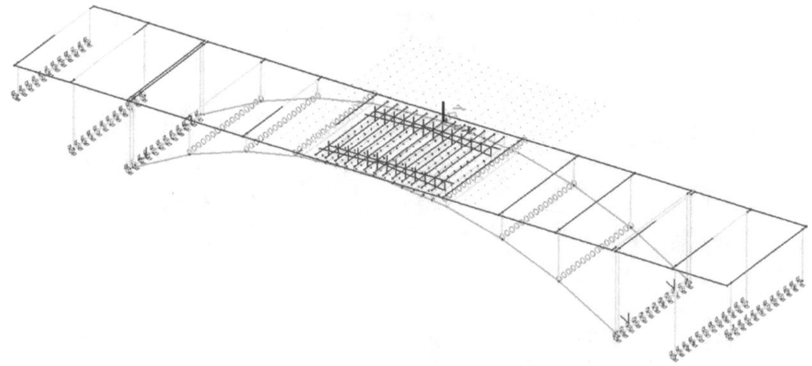

**Fig. 6.** FEM numerical model – supports and hinges

### 3.3 Defects and Vulnerable Zones of the Bridge

Identification of the vulnerable zones was done in accordance with the characteristic bridge typology. During the inspection of the bridge several defects were discovered and analyzed: fractures of spandrel walls, concrete deterioration and corrosion of reinforcement (mainly below the expansion joints on the piers and arch), visible cracks in the slab piers. Apart from these defects additional smaller defects were noticed as damage of waterproofing in the arch and spandrel walls, defects on the pavement, ineffectiveness of the drainage system and deterioration of the concrete railings.

Figures 7 and 8 shows the defects on the left and right view of the bridge respectably and Fig. 9 illustrates the defects on the lower surface of the arch structure and marked defects.

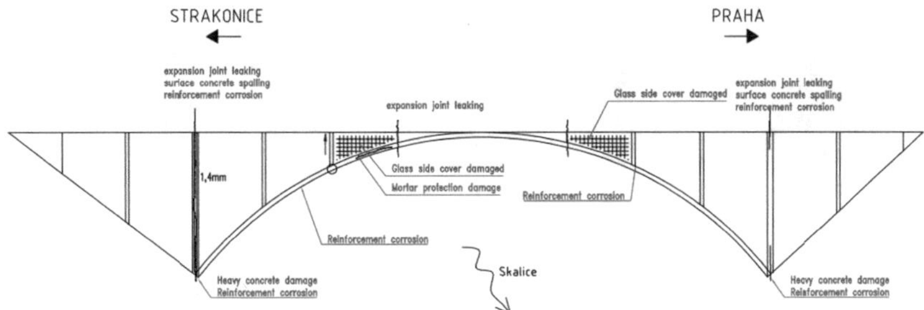


**Fig. 7.** Schematic left view on the bridge with the defects

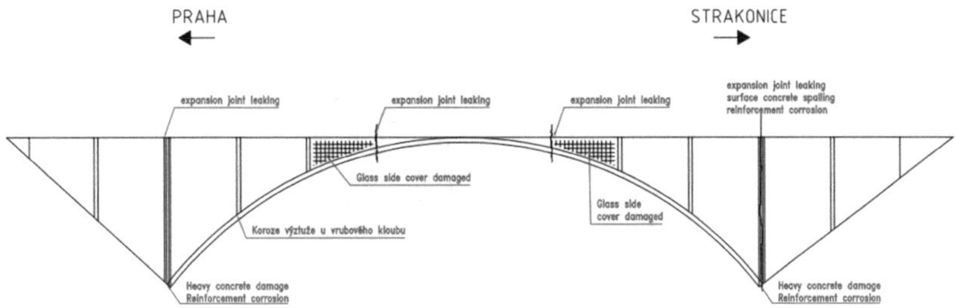


**Fig. 8.** Schematic right view on the bridge with the defects

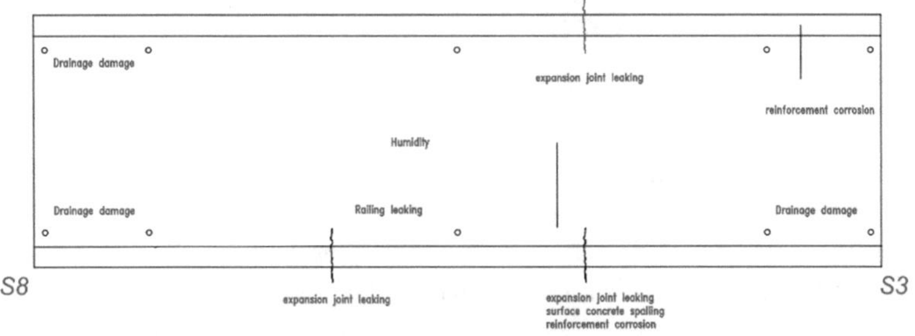


**Fig. 9.** The lower surface of the arch with the defects

### 3.4 Damages on the Bridge Structure

During the inspection which was conducted in October 2016 damage processes were identified on the bridge structure. A significant visible corrosion of the reinforcement on the piers S8-S9-O10 structure is noted (Fig. 10a), together with concrete deterioration (Fig. 10b). A major reinforcement corrosion and concrete deterioration was visible and identified on the pier S6 on the arch (Fig. 11).

a) b)

**Fig. 10.** Defects on piers S8-S9-O10

**Fig. 11.** The corrosion of the reinforcement, concrete deterioration – pier S6 on the arch

Inefficacy of the waterproofing system is seen on the abutment structure (Fig. 12). Poor waterproofing integrity can be related to several factors, from site procedure, workmanship, weather, to installation conditions, asphalt application temperatures, adhesive properties and similar. Additionally, the unusually high thickness of the asphalt layer together with the high traffic loading may contributed to its malfunctioning.

**Fig. 12.** Concrete defects

Visible cracks up to 1 cm are running perpendicular to the bridge structure. These cracks could be connected to thermal actions. It is not rare that large transverse cracks are caused by the pavement's contraction in cold weather. Possible causes for such defects could be shrinkage of the Hot Mix Asphalt (HMA) surface due to low temperatures or asphalt binder hardening or a formation of a reflective crack caused by cracks under the surface HMA layer (Fig. 13).

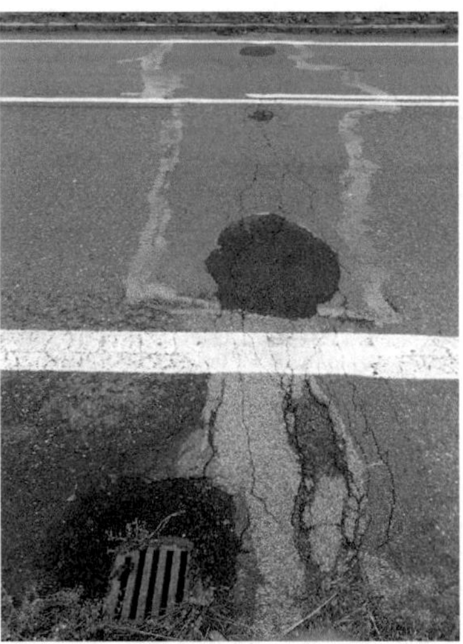

**Fig. 13.** Transverse cracks and asphalt thickness

### 3.5 Failure Modes

As a fourth step in the static stage failure modes were elaborated in the Ultimate limit state (ULS) and in the Serviceability limit state (SLS). Global bridge failure could be expected due to loss of stability of vertical walls under the expansion joint under live load. This is connected with the reinforcement corrosion and further concrete degradation under the leaking expansion joint. Additionally, a global bridge failure may be seen but due to main arch failure. There is a possibility for a top slab failure in the arch-slab joint. This can be expected in the weakest slab position due to water leakage which is further causing the concrete degradation and reinforcement corrosion. The fourth possible failure is connected to the loss of abutment stability. This failure can happen on the abutment 01 due to bad water management of pavement surface water (drainage system outlet). Regarding the SLS there is a possibility of pavement and parapet failure.

### 3.6 Minor Destructive Testing of the Bridge

In order to determine the compressive concrete strength of the bridge concrete samples were taken from characteristics locations (1x bridge deck, 3x arch, 2x piers) in the bridge structure. Figure 14 shows the prepared samples for testing and a sample during a testing procedure. The testing was conducted according to Czech standard ČSN 73 1317 [11]. Table 1 shows individual results of six concrete samples. According to the obtained data concrete can be considered as a C30/37.

**Fig. 14.** The samples for the concrete strength testing; Specimen no. NK1 during loading test

**Table 1.** Compressive strength of concrete

| Specimen | Height [mm] | Weight [kg] | Unit weight [kg.m$^{-3}$] | Force [kN] | Compressive strength [MPa] |
|---|---|---|---|---|---|
| NK1 | 97.0 | 0.955 | 2290 | 158.0 | 33.8 |
| 01_1 | 122.5 | 1.231 | 2338 | 142.0 | 31.7 |
| 01_2 | 116.0 | 1.210 | 2427 | 167.0 | 36.9 |
| 02_1 | 111.5 | 1.138 | 2374 | 135.0 | 29.8 |
| S1 | 154.5 | 1.511 | 2275 | 138.0 | 32.1 |
| S2 | 127.0 | 1.292 | 2367 | 141.0 | 31.8 |

It is well known that alkali–Silica reaction (ASR) is a main cause of premature degradation of concrete structures, mostly for highways, bridge decks, runways, and sidewalk. In order to determine the existence of the silica gel the Rhodamin b method, developed by Guthrie [12] was used. The reaction occurs between concrete pore water, some aggregate constituents (e.g., poorly crystalline silica phases), and alkali cations (Na and K) released by the cement and/or aggregate. Once the alkali content of the pore water rises, suspension of the silica phases increases, resulting in a release of silica to form the gel. This formation of gel will cause volume increase, causing internal pressure in the concrete and in due course forms cracks and fractures. This process can be intensified by freeze-thaw cycles and by salting of road surfaces. As shown on the Fig. 15, only cement changed colour to pink, not the aggregate meaning that ASR is not a problem here.

**Fig. 15.** Specimens after Rhodamin application

The potentiometric method according to the EN 196-21 [13] was used for chloride measurement. The results are given in Table 2. The chloride content in concrete varies between 1.4 mg/l and 14.7 mg/l. The chlorine ions content relative to the quantity of cement in tested samples in the worst-case amounts to 0.059%, which is lower that the limit value of 0,4% according to the EN 206-1 [14].

**Table 2.** Chloride measurement

| Specimen | Amount [mg/l liquid] | Amount [mg/g sample] | Max. amount of Cl- compared to the cement amount [%] |
|---|---|---|---|
| NK1 | 1.4 | 0.034 | 0.003 |
| 01_1 | 1.4 | 0.049 | 0.005 |
| 01_2 | 6.8 | 0.170 | 0.017 |
| 02_1 | 14.7 | 0.589 | 0.059 |
| S1 | 2.1 | 0.054 | 0.005 |
| S2 | 2.6 | 0.059 | 0.006 |

All samples were exposed to the 75 freezing cycles. All samples except one (Fig. 16) showed good Freeze-Thaw Resistance.

**Fig. 16.** Specimen no. 01_2 – after the freezing test

# 4 Key Performance Indicators and QC Plan

The expert judgement was the basis for determination of the key performance indicators. Figure 17 shows the identified failure mode, vulnerable zone and symptoms representing the damage processes that can be connected to the PI which are further correlated with the relevant KPIs, reliability and safety [15]. As it can be seen for the current state of the bridge the earlies failure of pavement can be expected, in the next five years, which is connected with the safety KPI. The structure safety, reliability, is directly connected to the load bearing elements (walls, arch, top slab or abutment). Most endangered element is the wall under the expansion joint which may fail within the next 20 years, followed by the failure of arch and top slab 15 years after and final failure of abutment.

| Structure | Component | Material | Failure mode | Vurnerable zone | Symptoms | KPI | Performance indicator | | Estimated failure time |
|---|---|---|---|---|---|---|---|---|---|
| Arch concrete bridge | Wall under E.J. | Reinforced concrete | Global failure | E.J. connection | E.J. leakage, reinforcement corrosion | Reliability (Structure safety) | 2 | 2 | **20 years** |
| | Arch | Reinforced concrete | Global failure | Arch under E.J. | E.J. leakage, reinforcement corrosion | | 2 | | **35 years** |
| | Top slab | Reinforced concrete | Local slab failure | Slab in hinge position | Hinge leakage and reinforcement corrosion | | 2 | | **35 years** |
| | Abutment 01 | Subsoil | Loss of stability | Abutment foundation | Undermined abutment | | 2 | | **40 years** |
| | Parapets | Reinforced concrete | Parapet collapse | Bottom section of parapet | Reinforcement corrosion | Safety | 2 | 2 | **10 years** |
| | Pavement | Asphalt concrete | Skid resistance | Top surface | Crack & sweating & deformation | | 2 | | **5 years** |

**Fig. 17.** Determination of the KPI values and estimated failure time

## 4.1 Verification of Reliability

The load baring capacity of the bridge was conducted taking into account the three load cases stated in previous chapters. The load capacity of the bridge was calculated taking into account the worst load case and taking into account the Eurocode regulations, meaning usage of partial safety factors for materials on one side ($\gamma_S$ = 1,15 – for prestressing steel,) and for actions on the other hand ($\gamma_G$ = 1,35 – dead load, $\gamma_Q$ = 1,35 –imposed load). As no information was available regarding the reinforcing steel, the safety factor was calculated according to the formulas in ČSN EN 13822 and its National annex [16]:

$$\gamma_s = \frac{f_{yk}}{f_{yd}} = \frac{e^{(-k_n V_{fy})}}{e^{(-\alpha_R \beta V_{Rx})}} \tag{1}$$

where:

$f_{yk}$ characteristic yield strength of steel,

$f_{yd}$ designed yield strength of steel,

$k_n$ factor for the evaluation of the required quantile of the characteristic resistance, depending on the number of tests,

$V_{fy}$ coefficient of variation for yield strength $fy$,

$\alpha_R$ sensitivity factor,

$\beta$ required reliability index,

$V_{Rx}$ variability index, including the uncertainty in the material, modelling and geometry.

Variability index can be written as:

$$V_{Rx} = \sqrt{V_{fx}^2 + V_{geo}^2 + V_{\xi}^2} \tag{2}$$

That variability for prestressing steel according to the literature [17] can be assumed to be equal to $V_{fy}$ = 0.05, variability of the geometry is medium high, as it is cast insitu member so $V_{geo}$ = 0.05, and the model uncertainty variability is assumed to be $V_{\xi}$ = 0.05.

Putting these values in Eq. 2 one obtains $V_{Ry} = 0.086$.

Equation 1 than reads:

$$\gamma_s = \frac{e^{(-1.64*0.05)}}{e^{(-0.8*3.8*0.086)}} = 1.19$$

The calculated partial safety value amounting to 1.19 is larger than the selected value of 1.15, however if value of $\beta$ is taken to be equal to 3.2 a value of the selected partial safety factor is obtained. The concrete strength is not important here, because the load capacity of the members depends mainly on the amount of steel.

For the dead load, as the geometry was not measured, it can be assumed for dominant load $\alpha_E$ = −0,7, and for variability $V_G$ = 0,1. For the factor of the model uncertainty we can assume $\gamma_{Sd}$ = 1,05. Than the partial safety factor for dead load reads according to [18]:

$$\gamma_G = (1 - \alpha_E * \beta * V_G) * \gamma_{Sd} \tag{3}$$

where:

$\gamma_G$ Partial factor for permanent actions, also accounting for model uncertainties and dimensional variations

$\alpha_E$ sensitivity factor,

$\beta$ required reliability index,

$V_G$ coefficient of variation for dead load,

$\gamma_{Sd}$ Partial factor associated with the uncertainty of the action and/or action effect model

$\beta$ Reliability index

$$\gamma_G = (1 + 0.7 * 3.8 * 0.1) * 1.05 = 1.329$$

In order to get a value of 1.35 for the dead load it is necessary to select the value of $\beta$ to be equal to 4.1.

In the other case if the live load is assumed and if the variability of the model uncertainties $V_{\Theta E}$ is assumed to be equal to $V_{\Theta E} = 0.1$ (Fig. 18), the value of $\beta = 3.5$ is obtained.

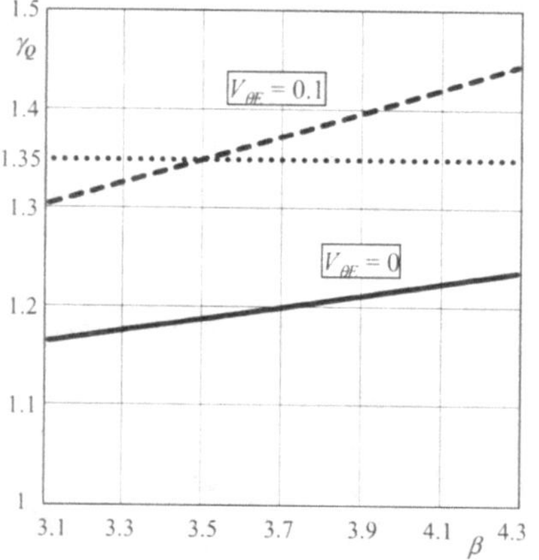


**Fig. 18.** The relation between β and $\gamma_Q$ [19]

Based on previous calculations and application of the standard safety factors, it can be concluded that the smallest value for $\beta$ was calculated for the live load impact and reinforcing steel. Taking the value of $\beta = 3.5$ for the whole bridge is conservative, leading to a safer scenario. For more precise load capacity verification (that also depends on the ratio between live and dead load), the slightly smaller load factors for the dead load can be taken, meaning if the value of $\beta = 3.5$ is selected there will be a very slightly increase the load capacity.

### 4.2 Different Approaches

In the case of unavailable funds, a reference approach is foreseen. This kind of approach is characterized by lack of any major repairs of superstructure and accessories except of basic pavement repairs leading to further defect development and at the end to bridge failure. In accordance with previous section the existing structure defects, development and estimated failure times are assumed as follows: pavement failure may be expected in five years due to crack development, sweating and deformation (as noted the pavement layer shall be repaired). During the process of pavement repair the availability will be temporarily decreased. Due to concrete parapets collapse there will be a decrease of availability and safety. New concrete barriers will partly increase availability, but not to the full extent as the carriage width will be decreased, and increase of safety will be evident. In the case that after 20 years failure of doubled wall under expansion joint failure happens this will lead to bridge failure and the necessity for construction of a new bridge. This will cause a significant drop in availability as the bridge will be closed. In order to extend the life of the new structure preventive measures as pavement replacement and bridge repair should be foreseen. Implications of this approach are presented in Fig. 19.

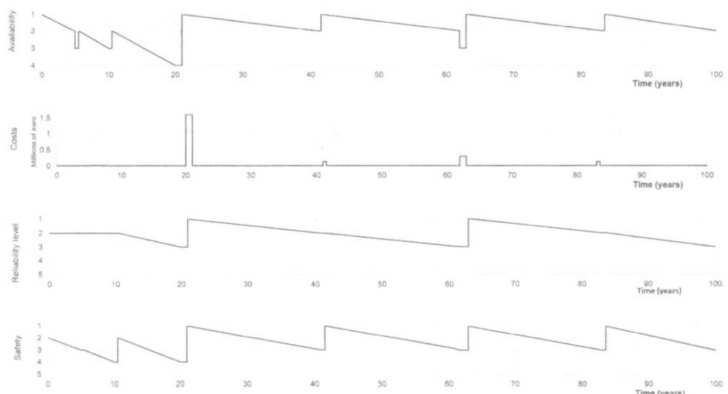


**Fig. 19.** Reference approach

The second approach is one of preventive nature which foresees a first major bridge major in the next 10 years. In the next five years there will be a need for major repair of pavement due to development of cracks, sweating and deformation. At the same time due to partial collapse of the concrete parapets their rehabilitation will take place. This will cause the drop of the availability, as the bridge most probably will have to be partly closed, leaving one lane in service. From this point on the same preventive measures will be considered as in the reference approach (Fig. 20).

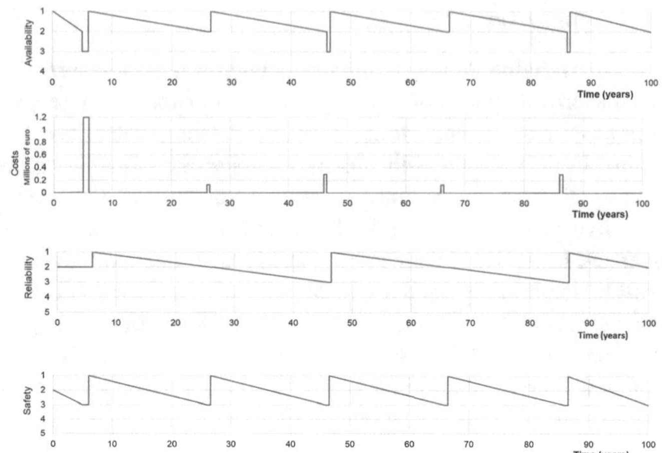


**Fig. 20.** Preventive approach

According to the performed analysis the preventative approach is more appropriate for the arch bridge - the indicators shows more favourable results for all aspects – safety, reliability, availability. Only the costs are almost comparable - the reason is the normalization of the costs based on the interest rate 2% (Fig. 21). A comparison of the two approaches can be done in a 3D spider graph, separately and in one image together for the whole period of 100 years (Fig. 22). Comparing the volumes of the two approaches, the preventive approach has a smaller volume for 22 $m^3$ making it more appropriate and cost effective.

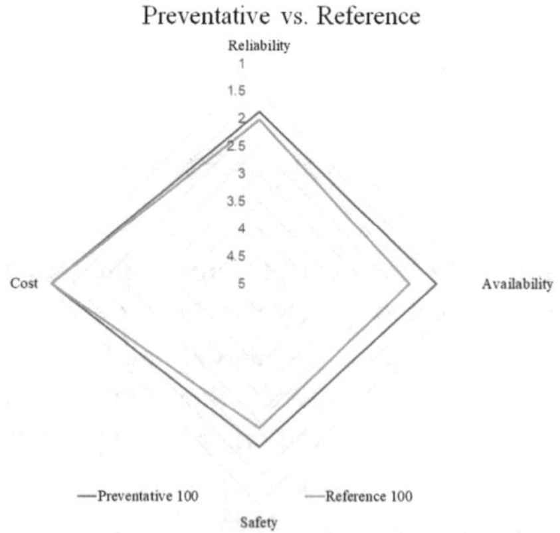


**Fig. 21.** Spider comparison

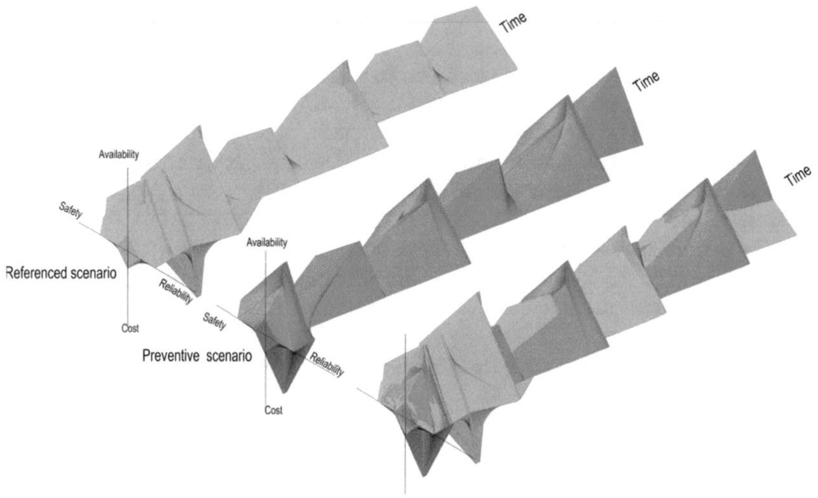


**Fig. 22.** The comparison of the safety, reliability, availability and cost in time and volume comparison

# 5 Conclusion

Bridge management systems are an excellent tool for better decision-making in the process of bridge management and maintenance. Throughout Europe different management systems exist which may lead to different scenarios and implementation. European globalization has grown due to advances in transportation and communication technology. In order to enable further unification, it is necessary to prepare guideline for standardization and development of quality control plans for roadway bridges. A proposal has been given as a result of the COST TU1406 Action. This is only a first step in this process which needs to be further upgraded and polished. The proposed methodology was applied on a concrete arch bridge, with the two scenarios indicating the benefits of the preventive measures through a 3D spider graph. Preventive maintenance is a strategy of extending service life by applying cost-effective procedure to bridge elements. It can be concluded that preventive measures can be seen as preservation of the main bridge components required for reaching the desired bridge state and sustaining it.

# References

1. Tao, Z., Zophy, F., Wiegmann, J.: Asset management model and systems integration approach. Transp. Res. Rec. **1719**(00–1162), 191–199 (2000)
2. Van der Velde, J., Klatter, L., Bakker, J.: A holistic approach to asset management in the Netherlands. Asset Manage. Civil Eng. Struct. Infrastruct. Eng. Maint. Manage. Life-Cycle Des. Perform. **9**(4), 340–348 (2013)

3. ISO (International Organization for Standardization): General principles on reliability of structures. ISO 2394 (2015)
4. Stipanovic, I., Chatzi, E., Limonggelli, M., Gavin, K., Allah Bukhsh, Z., Skaric Palic, S., Xenidis, Y., Imam, B., Anzlin, A., Zanini, M., Klanker, G., Hoj, N., Ademovic, N.: COST TU 1406, WG2 technical report – performance goals for roadway bridges (2017
5. Kedar, A., Sein, S., Panetsos, P., Ademović, N., Duke, A., Ryjacek, P.: COST TU 1406, WG4 technical report – preparation of a case study (2019)
6. Strauss, A., Mandić Ivanković, A., Matos, J.C., Casas, J.R.: COST TU 1406, WG1 technical report – performance indicators for roadway bridges (2016)
7. Jiang, M., Sinha, K.C.: Bridge service life prediction model using the markov chain. Transportation Research Records 1223. Transportation Research Board, Washington, DC (1989)
8. Dunker, K.F., Rabbat, B.G.: Highway bridge type and performance patterns. J. Perform. Constr. Facil. **4**(3), 161–173 (1990)
9. Kallen, M.J., Van Noortwijk, J.M.: Statistical inference for Markov deterioration models of bridge conditions in the Netherlands. In: Cruz, P.J., Frangopol, D.M., Neves, L.C. (eds.) Bridge Maintenance, Safety, Management, Life-Cycle Performance and Cost: Proceedings of the Third International Conference on Bridge Maintenance, Safety and Management, pp. 535–536. Taylor & Francis, London (2006)
10. Kim, Y.J., Yoon, D.K.: Identifying critical sources of bridge deterioration in cold regions through the constructed bridges in North Dakota. J. Bridg. Eng. **15**, 542–552 (2010)
11. ČSN 73 1317 (731317): Stanovení pevnosti betonu v tlaku. The Czech Office for Standards, Metrology and Testing (ÚNMZ) (2002)
12. Guthrie, G.D., Carey, J.W.: A simple environmentally friendly, and chemically specific method for the identification and evaluation of the alkali–silica reaction. Cem. Concr. Res. **27**(9), 1407–1417 (1997)
13. EN 196-21:2005: Methods of testing cement. Determination of the chloride, carbon dioxide and alkali content of cement. European Committee for Standardization, Brussels, Belgium (2005)
14. EN 206:2013+A1:2016: Concrete – Part 1: Specification, performance, production and conformity. European Committee for Standardization, Brussels, Belgium (2016)
15. Hajdin, R., Kušar, M., Mašović, S., Linneberg, P., Amado, J., Tanasić, N., Close collaborators: Ademović, N., Costa, C., Marić, M., Almeida, J., Galvão, N., Sein, S., Zanini, M., and Pendergast, L.: COST TU 1406, WG3 technical report – establishment of a quality control plan (2018)
16. ČSN ISO 13822:2005: Bases for design of structures – assessment of existing structures. The Czech Office for Standards, Metrology and Testing (ÚNMZ) (2005)
17. Partial factor methods for existing concrete structures: recommendation. Bulletin FIB –80 (2016)
18. Koteš, P., Vičan, J.: Reliability levels for existing bridges evaluation according to Eurocodes. Procedia Eng. **40**, 211–216 (2012)
19. Lenner, R., Sýkora, M., Keuser, M.: Partial factors for military loads on bridges. In: Proceedings of the ICMT13 International Conference on Military Technologies 2013, Brno, Czech Republic, pp. 409–419. University of Defense, Brno, 22–23 May 2013

# Accuracy Analysis of the Heights in Reference Networks in Bosnia and Herzegovina

Medžida Mulić[1(✉)], Amra Silnović[2], and Dževad Krdžalić[1]

[1] Faculty of Civil Engineering, University of Sarajevo, Sarajevo, Bosnia and Herzegovina
medzida_mulic@yahoo.com, medzida_mulic@gf.unsa.ba
[2] Office of Cadastre, Travnik, Bosnia and Herzegovina

**Abstract.** Modern Satellite geodesy methods allows the collection of geospatial information with high accuracy and resolutions that were not imaginable half a century ago. Thus, Global Navigation Satellite Systems ensure the point positioning with 3D accuracy below 1 cm, globally. On the other hand, satellite gravimetric missions have made global mapping of the Earth's gravity field, with incomparable accuracy and resolution. A large number of global geopotential models are estimated from the collected geospatial data.

Traditional geodetic methods used in the 19th and 20th centuries, such as the trigonometric networks, determined the point positions and heights applying optical method, with planned positioning accuracy of 10 cm. A classic terrestrial method of precise leveling still has unprecedented accuracy of 1–2 mm/km, but this method is expensive and time-consuming. The availability of precise modern data made it possible to analyze the accuracy of the conventional terrestrial methods. This paper is inspired by previous research, which showed that in the trigonometric networks of the first and second orders in Bosnia and Herzegovina blunders in trigonometric point's heights exist. This paper is a result of the research of the accuracy of the height of the trigonometric points of the third and fourth order in network of Bosnia and Herzegovina, which are still in the official use. Based on the points with coordinates available in both, modern global and classical local reference systems, investigation the two Global a Geopotential Models: Earth Gravitational Model - EGM2008 and A High-Resolution Global Gravity Combination Model Including GOCE Data - EIGEN6C presented. Total number of points whose deviations from official ones exceed the threshold of 30 cm is 18, or 4.52% of the total 398 points analyzed.

**Keywords:** Reference systems · Heights · Vertical datum · Global geopotential models · Geoid

## 1 Introduction

Satellite geodesy enables the acquisition of coordinates of geodetic points in the 3D Cartesian coordinate system, such as the ITRS (International Terrestrial Reference System), which is fixed to Earth's body and its origin is in the center of the Earth's mass - geocentar. The accuracy of the obtained 3D coordinates in ITRS is at centimeter


S. Avdaković et al. (Eds.): IAT 2019, LNNS 83, pp. 205–212, 2020.
https://doi.org/10.1007/978-3-030-24986-1_16

or even millimeter level, which was unthinkable and technically impossible to perform for some decades ago. Cartesian coordinates of any point P (X, Y, Z) or expressed as geodetic coordinates P ($\varphi$, $\lambda$, h) in ITRS can be used to calculate the heights of points $H_{GNSS}$, which represents the distance of the observed point P above the reference ellipsoid, what is a mathematical ideal representation of the shape of the Earth [1, 2]. However, geodetic applications require physical/orthometric heights H, i.e. vertical distances from the geoid surface.

Geoid is an equipotential surface [3] and, in accordance with definition, the direction of the tangent of the Earth's gravity vector in point of the observation, is perpendicular to the surface of the geoid at each point [4]. The shape of the geoid is geometrically irregular (Fig. 1) due to the uneven distribution of mass within the Earth's body, which causes the anomalies of gravity. So, Geoid is the reference surface for heights in geodesy [5, 6]. The basic methods of determining the height of the trigonometric points applied in the 19$^{th}$ and 20$^{th}$ centuries were the trigonometric leveling (decimeter accuracy) and the geometric leveling (millimeter accuracy). The latter method is most accurate, but it is expensive, and it is time consuming methods.

Geodetic works in Bosnia and Herzegovina were carried out for the first time in the territory of Bosnia and Herzegovina (B&H) during the reign of the Austro-Hungarian monarchy. However, basic trigonometric network in B&H was developed in the 19$^{th}$ century during the Ottoman rule, but the surveys were conducted by Austrian military officer, with the permission (*tur.* ferman) by Sarajevo's Governor (*tur.* pasha). Austrian officers performed geodetic and astronomical measurements in the trigonometric network, which represented the continuation of European trigonometric chains. These measurements are known as *gradus* measurements and were carried out to determine the precise dimensions and orientation of the Earth. As results, Bessel 1841 ellipsoid defined and used to be widely accepted in Europe as reference ellipsoid, approximating shape and dimension of the Earth during 19$^{th}$ and 20$^{th}$ century [2, 5]. The height of the trigonometric points was determined by the trigonometric method, and for the precise determination of the height, in the period from 1899 to 1906 a leveling network was developed. The heights of the benchmarks in the leveling network are calculated in relation to the mean level of the Adriatic Sea. The gauge on the Mole Santorio in Trieste was determined from registration of the sea level in 1875, and since that, this vertical geodetic datum is named Trieste. In the period of SFR Yugoslavia, the first leveling network of the high precision, known as NVT I (*bos. Nivelman visoke tačnosti I*) developed from 1945 to 1953. The leveling network NVT II was carried out in the period from 1970 to 1973 [7]. In 1986, after the adjustment of the leveling network NVT II, normal heights were adopted as the vertical datum Maglaj [8]. However, only the benchmarks included in these leveling lines have altitudes in Maglaj vertical datum.

However, the method for determining heights in the 21st century takes advantage of positioning using Globular Navigation Satellite Systems - GNSS, such as: GPS (Global Positioning System) in USA, Russian GLONASS, European Galileo, or Chinese BeiDou/Compass. Important and useful data were collected by satellite gravity missions, such as: CHAMP (Challenging Minisatellite Payload), GOCE (Gravity Field and Steady-State Ocean Circulation Explorer), and GRACE (Gravity Recovery and Climate Experiment [9]. These missions have collected valuable data for creating Global

Gravity Models and Global Geoid Models with few centimeters accuracy [10, 11]. Global geopotential models are mathematical functions that approximate the real potential of the Earth's gravity in the 3D space outside the Earth [12].

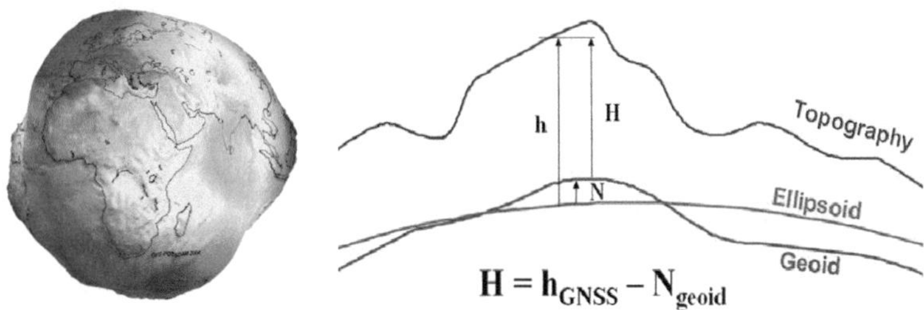


**Fig. 1.** The Global geoid model [13] left; Relations of ellipsoid, geoid and topographic surfaces, (right).

A global geoid allows calculation of the undulation of the geoid, N - height difference of the reference ellipsoid and geoid, as shown at Fig. 1, right (i.e. undulation is separation between red and green lines). Differences between the geoid and reference ellipsoid N range globally ± 100 m. The availability of 3D coordinates and undulations allow the determination of physical/orthometric heights H, with centimeter accuracy. This method is often referred to as GNSS leveling.

The accuracy of the points of trigonometric networks in B&H was investigated for the first time in [14]. The research showed that there was a disagreement of official heights in the trigonometric network of the first order, but the largest disagreement (0.7816 m) was shown at the trigonometric point of the second order in Livanjsko Polje, which was connected with the precise leveling network. The continuation of the research [15] confirmed the first-order trigonometric network disagreements with respect to GNSS leveling data. So, this paper shows results of extension of research to the network of lower orders.

## 2 Methods and Data

The subject of the study was trigonometric network of the third and fourth order. Geocentric coordinates (X, Y, Z) in ETRS89 (European Terrestrial Reference System) for the points of B&H trigonometric networks were provided by the Geodetic Administration of B&H, but global gravity models were used from web portal of the ICGEM (International Center for Global Earth Models).

An online calculator of the ICGEM service [16] was used to calculate the grid of geoid undulations for the territory of B&H. It serves to calculate the selected anomalies of the potential functions for the set of grid points on the reference ellipsoid. For the territory of B&H, long-term solutions of ten different global models were used and, based on them the selected functional (here, geoid undulation) were calculated for the territory of B&H (15.6°–19.7° E, 42.4°–45.5° N) in a grid of 0.0035° x 0.0035° geodetic latitude and longitude. The "mean-tide" model and the reference ellipsoid GRS80 (Global Reference System 1980) were used.

In this paper, only the results based on EGM2008 [17] and EIGEN6C [18] are shown, because they best approximate the area of Europe, and [19] shows that the standard deviations for Europe of these models are 0.125 m and 0.123 m respectively. When the deviation of the height calculated the heights by the models were subtracted from the reference height referring to the TRIESTE datum in accordance with relations:

$$\Delta h = h_{TRIESTE} - h_{GM} \tag{1}$$

Where $\Delta h$ is difference between the height expressed in vertical datum Trieste and calculated from global geoid model, $h_{TRIESTE}$ is height of point in Trieste vertical datum, $h_{GM}$ is height estimated from GNSS technics and respective geoid model.

The TRIESTE vertical datum differs from the vertical datum of the NAP (Normal Amsterdam's Peil) of the European Vertical Reference Frame 2007 in the amount of 0.31 m [20] and the (1) got the following form:

$$\Delta h = h_{TRIESTE} - h_{GM} - 0.31 \tag{2}$$

## 3 Result and Discussion

Total number of analyzed points is 398. They are classified in four different groups: **3 N**, **3-TMV**, **4-N**, and **4 – TMV**. Table 1 shows explanation of classification in accordance with order of the trigonometric network, method applied for heights determination, number of trigonometric points in specific group.

**Table 1.** Classification of the trigonometric points in accordance with order and method of leveling

| Group | Order | Marked as | Method | Points number |
|---|---|---|---|---|
| 3 N | 3rd | black spots | Precise leveling | 21 |
| 3-TMV | 3rd | red spot | Trigonometric leveling | 43 |
| 4 N | 4th | red circle | Precise leveling | 49 |
| 4-TMV | 4th | black circle | Trigonometric leveling | 285 |

Data analysis was carried out according to the mentioned groups, and the arrangement of points is shown in Fig. 2.

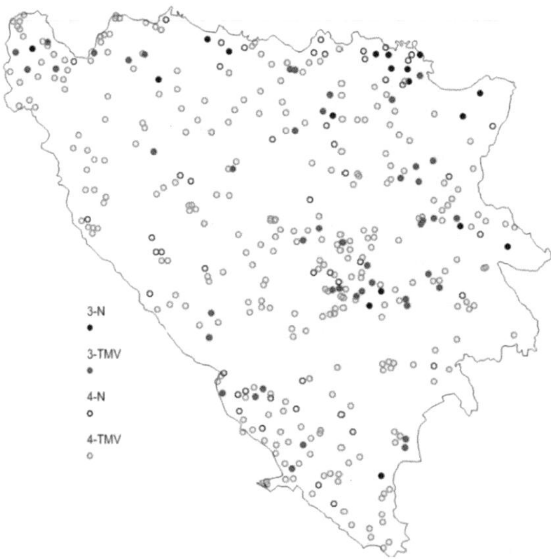


**Fig. 2.** Classification of points by groups in the territory of Bosnia and Herzegovina. [19]

Based on the analysis of all points, whose heights deviate from the threshold of 0.30 m and the average values of absolute differences for all groups, (calculated by all ten geoid models) it may be concluded that the minimum differences from the NAP datum show the models EGM2008 and EIGEN6C. Summary overview of the average values of the absolute and real differences are shown for these two models in Table 2 and Table 3, respectively.

Out of the total number of points 398, ten have deviations in heights exceeding threshold of 0.30 m in relation to the NAP datum, on the EGM2008 geoid model, representing 2.5%, and but each of ten points are determined by trigonometric measurements. Graphic interpretation of deviation of heights in B&H trigonometric network of third and fourth order is shown in Figs. 3 and 4.

**Table 2.** Average values of absolute differences of heights in groups in relation to NAP.

| Differences in heights (in m) by the geoid models in relation to the in Norman Amsterdam Pail (in meters) | | |
|---|---|---|
| Order/group | EIGEN6C2 | EGM2008 |
| 3N | 0.127 | 0.118 |
| 3T | 0.117 | 0.125 |
| 4N | 0.095 | 0.086 |
| 4T | 0.122 | 0.118 |

**Table 3.** The average value of the relative difference in the height of points in relation to NAP

| Differences in heights (in m) by the geoid models in relation to the in Norman Amsterdam Pail (in meters) | | |
|---|---|---|
| Order/group | EIGEN6C2 | EGM2008 |
| 3N | −0.018 | −0.012 |
| 3T | −0,004 | −0,016 |
| 4N | −0.028 | −0.030 |
| 4T | 0.122 | 0.118 |

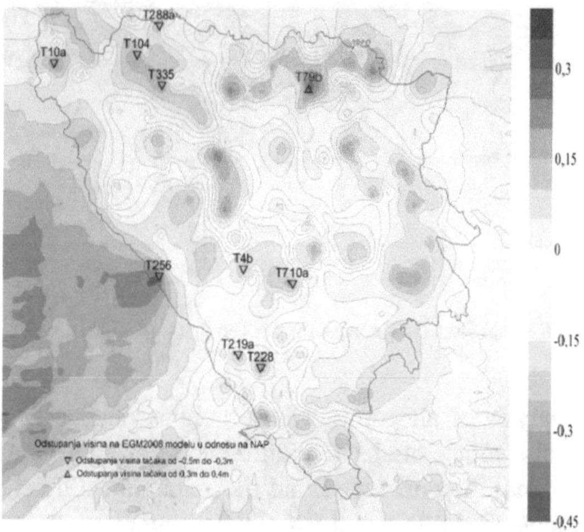


**Fig. 3.** Gradients of the difference in the heights of points in the B&H trigonometric networks of by the EGM2008 model in relation to the NAP datum. Blue triangles show deviations from −0.5 m to −0.3 m; the red triangle shows a deviation greater than +0.3 m and less than +0.4 m [19].

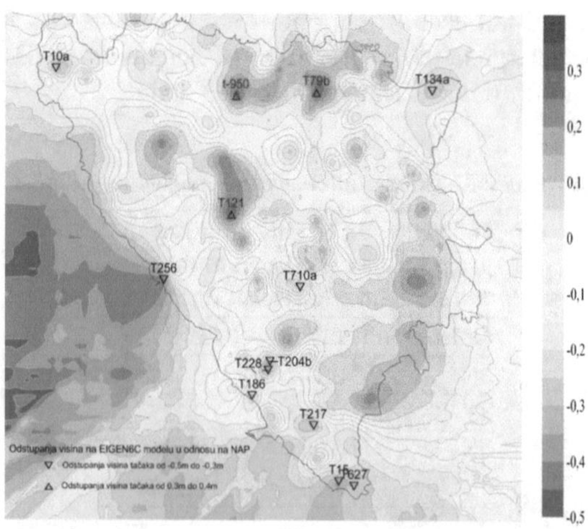


**Fig. 4.** Gradients of the difference in the heights of points in the B&H trigonometric networks by the EIGEN6C model in relation to the NAP datum. Blue triangles show deviations from −0.5 m to −0.3 m; the red triangle shows a deviation greater than +0.3 m and less than +0.4 m [19].

## 4 Conclusion

Statistical analysis of the heights of 398 trigonometric points, consisting the third and fourth order trigonometric networks in Bosnia and Herzegovina are presented. Beside 3D coordinates and station height, the analysis includes data of the satellite gravity missions and global geodetic geoid models. An online calculating service ICGEM was used. The analysis compared the height of the points obtained on the basis of ten different geopotential geoid models, and heights of networks referring to the datum Trieste and Amsterdam NAP. The results of calculating altitudes using global geoid models show the lowest average deviation for models EGM2008 and EIGEN6C. The total number of points on the EGM2008 and EIGEN6C models with deviations exceeding the 0.30 m threshold is 18 or 4.52%. The research should continue. In the next step the points of all four orders should analyzed together.

Generally, the number of trigonometric points with errors in heights in the B&H trigonometric network is not large if one takes into account the methods of determining the trigonometric points, and in particular that the lengths in the trigonometric level were calculated from the coordinates, which turned out to be less accurate than the planned 10 cm, due to the wrong orientation of the network originated in the 19th century.

## References

1. Moritz, H.: Geodetic reference system 1980. J Geod **74**(1), 128–162 (2000)
2. Mulić, M.: Geodetic reference systems. Faculty of Civil Engineering, University of Sarajevo (2018)
3. Heiskanen, W.A., Moritz, H.: Physical Geodesy. W.H. Freeman and Company, San Francisco (1967)
4. Blick, G.: ABLOS tutorial basic geodetic concepts. Office of the Surveyor-General Land Information New Zealand (2005)
5. Hofmman-Wellenhof, B., Moritz, H.: Physical Geodesy. Springer, Wien, New York (2006)
6. Kuhar, M., Mulić, M.: Fizikalna geodezija (skripta). Građevinski fakultet Univerziteta u Sarajevu (2009)
7. Krzyk, T.: Nivelmanske mreže viših redova i vertikalni datum na području Bosne i Hercegovine. Geodetski glasnik **35**, 22–34 (2001)
8. Bilajbegović, A., Mulić, M.: Selection of the optimal heights system on the example of Bosnia and Herzegovina's future leveling network. Geodetski glasnik **44**, 5–33 (2013)
9. Mulić, M., Đonlagić, E.: Gravitacijske satelitske misije. Geodetski glasnik **42**, 5–20 (2012)
10. Pavlis, N.K., Holmes, S.A., Kenyon, S.C., Factor, J.K.: An Earth gravitational model to degree 2160: EGM2008. Presented at the General Assembly of the European Geoscience Union, Vienna, Austria, April 13–18, 2008
11. Pavlis, N.K., Holmes, S.A., Kenyon, S.C., Factor, J.K.: The development and evaluation of the Earth Gravitational Model 2008 (EGM2008). J. Geophys. Res.: Solid Earth (1978–2012) 117(B4) (2012)
12. Barthelmes, F.: Global models. In: Grafarend, E. (ed.) Encyclopedia of Geodesy. Springer International Publishing, Cham (2014)

13. GFZ German Research Centar: Combined gravity field model EIGEN-CG01C-EIGEN-CG01C Geoid (2005). http://op.gfz-potsdam.de/grace/results/grav/g003_eigen-cg01c.html. 10 Apr 2019
14. Mulić, M.: Investigation of the impact of the ITRF realisation on the coordinates, their accuracy and the determination of the velocity vectors of the GPS points in B&H. Dissertation, Faculty of Civil Engineering, University of Sarajevo
15. Mulić, M., Đonlagić, E., Krdžalić, Dž., Bilajbegović, A.: Accuracy analyses of B&H trigonometric network heights by GPS/EGM Data. Geodetski glasnik **46**(42), 5–19 (2012)
16. GFZ ICGEM, at http://icgem.gfz-potsdam.de/home; calculation servise http://icgem.gfz-potsdam.de/calcgrid. 11 Apr 2019
17. Pail, R., Bruinsma, S., Migliaccio, F., Förste, C., Goiginger, H., Schuh, W.D., Höck, E., Reguzzoni, M., Brockmann, J.M., Abrikosov, O., Veicherts, M., Fecher, T., Mayrhofer, R., Krasbutter, I., Sansò, F., Tscherning, C.C.: First GOCE gravity field models derived by three different approaches. J. Geod. **85**, 819–843 (2011). https://doi.org/10.1007/s00190-011-0467-x. Springer
18. Shako, R., Förste, C., Abrykosov, O., Bruinsma, S., Marty, J.-C., Lemoine, J.-M., Flechtner, F., Neumayer, K.-H., Dahle, C.: EIGEN-6C: a high-resolution global gravity combination model including GOCE data. In Flechtner, F., Sneeuw, N., Schuh, W.-D. (eds.) Observation of the System Earth from Space - CHAMP, GRACE, GOCE and FUTURE MISSIONS (Geotechnologien Science Report; No. 20; Advanced Technologies in Earth Sciences), pp 155–161. Springer, Berlin [u.a.] (2014). http://doi.org/10.1007/978-3-642-32135-1_20
19. Silnović, A.: Analiza tačnosti visina referentih mreža u Bosni i Hercegovini, magistarski rad. Faculty of Civil Engineering, University of Sarajevo (2018)
20. Liebsch, G., Schwabe, J., Sacher, M., Rülke: A Unification of height reference frames in Europe. EUREF Tutorial. Leipzig, Germany, Federal Agency for Cartography and Geodesy (2015)

# Application of a Quality Control Plans on a Truss Bridge

Naida Ademović[1(✉)], Amir Keder[2], and Mor Mahlev[2]

[1] Faculty of Civil Engineering, University of Sarajevo, Patriotske Lige 30, 71 000 Sarajevo, Bosnia and Herzegovina
naidadem@gmail.com

[2] Kedmor Engineers Ltd, Hashachar Tower, 4 Ariel Sharon St, Givataim, Israel

**Abstract.** Infrastructure is vital to the prosperity and welfare of the citizens of a country. In order to take full advantage of its benefits to the society it is necessary to manage these valuable resources in a proper manner. A sophisticated computerized bridge management system (BMS) is a means for managing bridges throughout design, construction, operation and maintenance of bridges. Throughout Europe there are different management systems developed and adopted to the local requirements and needs, which may lead to different decisions on maintenance actions. In order to overcome this issue a COST Action TU 1406 has proposed a guideline for standardization and development of quality control plans for roadway bridges. In order to perform such analysis, first of all it is necessary to collect available data, conduct element identification and grouping, define vulnerable zones, damage processes and failure modes. This is followed by selection and evaluation of the performance indicators (PIs), calculation of key performance indicators (KPIs), establishing demands, and finally creating Quality Control Plan (QCP) scenarios and comparing them by spider diagrams. In this paper an example of the proposed procedure will be shown on a steel truss bridge in Israel.

## 1 Introduction

The starting point for development of management systems were actually bridges, thanks to their uniqueness, complexity, and the vital impact on society. Functionality of a bridge is vital for proper functioning of a society and their improper functionality may lead to *certain difficulties*. In order to acquire more value from assets and to minimize the use of threatened and scarce resources the infrastructure asset management (IAM) became a strategic tool. Aiming at the cost-effective investment decisions the asset management connects engineering and economic principles with comprehensive business practice throughout the life-cycle of infrastructure assets [1]. This indicates that bridge management system (BMS) has to be a part of the network management system as bridges are seen as physical assets, meaning that bridge performance levels *have* to satisfy the network performance requirements [2]. Most national design standards are based on ISO 2394 [3], which sets out the fundamental requirements for design, construction and maintenance of structures during the design working life. However, it is believed that the freight transportation is to be highly increase by 2030


S. Avdaković et al. (Eds.): IAT 2019, LNNS 83, pp. 213–228, 2020.
https://doi.org/10.1007/978-3-030-24986-1_17

posing a serious treat on the existing structures [4]. The infrastructure database can be helpful for IAM once the analyzed data has been linked in a proper way to the decision-making process which still has place for further improvement [5]. The obtained data has to be transferred into performance indicators (PI), which measures the fitness for purpose of the bridge or its element, which will them be correlated to Key Performance Indicators (KPIs) which in COST TU1406 have been limited to Reliability, Availability, Safety and Cost (Economy) [6], but need not to be. Through these multiple bridge performance goals different aspects of the bridge and network performance are taken into account. This gives a basis for creation of decision-making.

## 2 Proposed Methodology

The proposed methodology is conducted in several stages which generally can be divided into static and dynamic tasks. Static tasks run from point two to eleven and they are subdivided into work conducted on the site (orange color) and in the office (green color). Dynamic tasks are shown by a blue color (from 12 to 15) defining different processes that can change depending on the foreseen scenario. This is clearly presented in Fig. 1 [7].

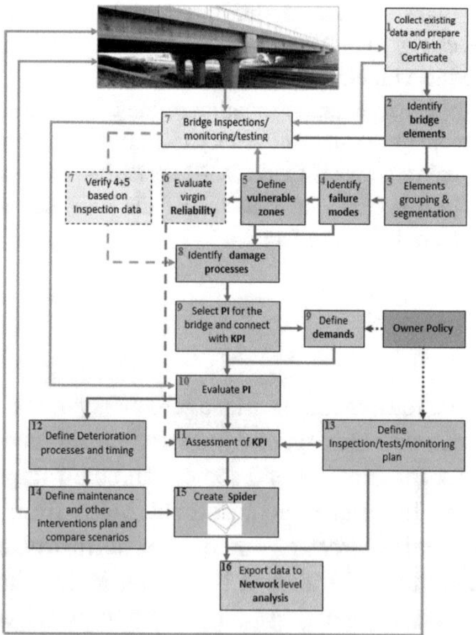


**Fig. 1.** Preparation process of case study [7]

At first, it is necessary to collect all existing data regarding the bridge in order to form an ID/Birth certificate of the bridge. Besides the information obtained from the design (if such data exists) and other documents, a valuable data could be obtained from the site, which is in some cases encouraged in order to identify the current bridge state (step 7). Beside the onsite investigations it is recommended to conduct lab investigations of different damage processes. Results of the data acquisition should be of the same form obtained from different methods for comparable purposes and for future usage of the data in the assessment procedures [8]. A taxonomy for different bridge types, due to their specific characterizes has been developed and they have been grouped according to different criteria (geometry, functionality, materials, exposure etc.) [8]. Depending on the typology, possible failure scenarios could be identified revealing the locations for detailed and comprehensive inspection on the site. The failure mode analysis is required for evaluation of the probability of a bridge failure, which is in a direct relationship with Reliability as a KPI. The failure modes are usually connected with the observed damages on the bridge. Throughout history it has been reported that different bridge typologies have their benefits and weaknesses which should be documented together with vulnerable zone which should be associated with the appropriate failure modes. The bridges are exposed to different types of actions from loadings, environmental exposure, aging, different processes which in most of the cases cannot be stopped, but may be prolonged if adequate maintenance measures are applied.

In order to determine the "current" reliability (structural safety and serviceability) of the bridge it is necessary to conduct the analysis of the undamaged (virgin) bridge taking into account all relevant load cases. The problem which arises here as with all other existing structures are the characteristics of the built-in material, load combinations that were taken into account in the original design and so on. Usually this information is not provided in the bridge database. The original finite element model (if exists) can be used for this analysis meaning that the resistance of the bridge in question for the analyzed limit states is in minimum as high as to sustain the internal forces due to the load model multiplied with the safety factor (Fig. 2) [9]. In order to obtain the degree of compliance, $n$, it is necessary to analyze the structure exposed to original design loading and current design loading.

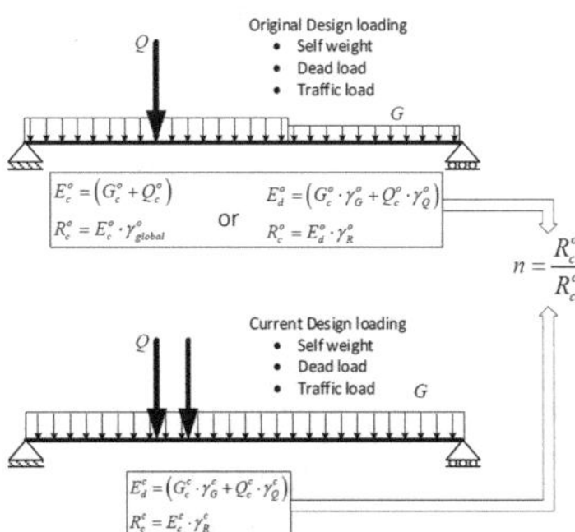


**Fig. 2.** Assessment of safety and serviceability margins [9]

For obtaining this value, it is necessary first of all to define the load effects used in the original design $E_c^o$. As in the previous codes global safety factor was implemented, this means that the above value was multiplied by this factor based on which the characteristic value of required resistance was calculated $R_c^o$. In the current codes the usage of partial safety factors is required meaning that the load effects $E_d^o$ are multiplied by the partial factor $\gamma_R^o$ and by this all the uncertainties related to the resistance model are covered giving $R_c^o$. It can be assumed that the resistance of the real structure will not have a value lower than $R_c^o$. On the other hand, the current traffic loading is reflected through the current design loadings which is multiplied by the partial safety factors for the actions as well. Multiplying $E_d^c$ with the adequate partial safety factor for resistance $\gamma_R^c$ one obtains the required resistance for the current design $R_d^c$. This indicates that the load effects should not be beyond the load effect of design loading $E_d^c$. Finally, if one divides $R_c^o$ by $R_d^c$ this gives the required compliance, $n$ [9]. Employing this procedure can lead to assessment of the KPIs without identifying the damage processes, selecting and identifying PIs. However, if this is not applicable it is necessary to conduct an investigation of the bridge on the site in order to identify the damage processes, select the PIs, and connect to the relevant KPIs in respect to the demands defined by the owners.

From this point on "dynamic" stages of the procedure are employed. Once different PIs and KPIs are determined and assessed, it is possible to calculate the remaining service life of the bridge as it stands without conducting any improvement measures on the bridge. This means that either Reliability or Safety as KPI will reach the defined threshold value (unacceptable return period for a failure) without any intervention. The calculation should be done taking into account the identified damage and deterioration processes (step 12) (physical/mechanical, chemical, and/or biological and organic), the speed of their progress and timing as well as damage forecast for which existing models could be used or calculation could be based on the expert judgement. As bridges are exposed to various effects from mechanical impact, harsh environment, natural disasters the performance of the brides deteriorates over time [10, 11].

Statistical models based on Markov property are mainly used for predictions [12, 13], which as an input take the conditions ratings obtained from visual inspections [14]. On the other hand, deterministic models can be divided into mathematical and physical deterministic deterioration models. Input for the mathematical models are statistical data of a large number of bridges [15], whereas data regarding physical and chemical phenomena (damage causing processes) [16], which cause the structure deterioration, is necessary for physical models being independent of the subjectivity of visual inspection. It is interesting to note that physical models in the form given in fib Bulletins [17–19], have not yet been implemented in any of the operational BMSs as much as the authors are aware. In the recent times there is a major use of the Bayesian optimization techniques and Artificial Neural Networks (ANNs) for prediction of the remaining service lives and estimation of serviceability conditions of bridges of different materials [20–22]. The next step would be planning different maintenance and intervention plans with target reliability and safety over a relevant period time based on the bridge age, forcast of trafic model or any other relevant criteria which would lead to different scenarios that ought to be compared in order to see the pros and cons of each scenario. Two maintenance scenarios are recommended: referenced and preventive. Referenced scenario represents a scenario when there are no available funds for necessary maintenance/repair actions and the bridge will be critically deteriorated and with traffic restrictions on year 100. Preventive refers to a scenario with a 100-year service life, according to the normal service life of bridges, for which there will be enough funds to undergo all the required inspections and maintenance/repair actions. Comparison can be done with the application of the Spider diagrams (Fig. 3a) of net present KPI for certain instances in time or as a function of time giving a 3D shape (Fig. 3b). All KPIs are normalized in the range from one (the best) to five (the worst) conditions. Regarding the KPI Cost, the most expected financial resources are marked with five and other scenarios are scaled accordingly.

Once this is finalized the data should be transferred to the network level analysis for further application in the network domain.

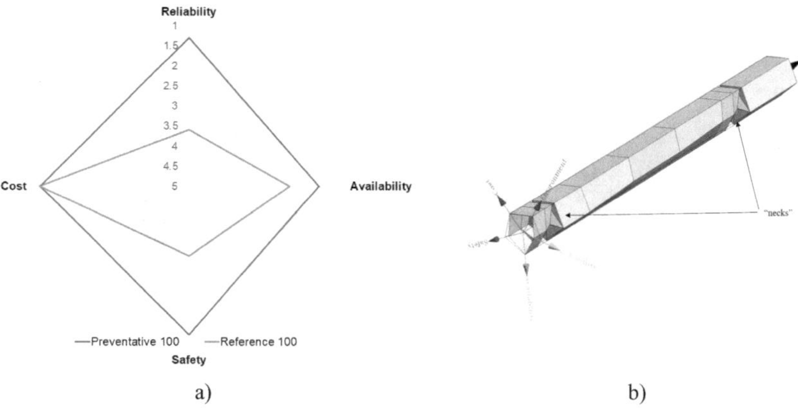


**Fig. 3.** Assessment of safety and serviceability margins

# 3 Case Study-Joseph Bridge

## 3.1 General Data About the Bridge

A detailed inspection performed on a 36-m single-span half-through steel truss bridge with reinforced concrete slab which was built in 1956. Figure 4 show the investigated bridge. The bridge is exposed to average annual daily traffic of 6800 with frequent heavy army vehicles crossing. The Original drawings were available so some data was extracted from these drawings. The superstructure is composed of 36 m long half through riveted steel truss divided into ten bays each 3.6-m long (Fig. 4).

**Fig. 4.** Investigated bridge structure

Figure 5 shows the cross section of the bridge showing the two parallel trusses being connected at the bottom cord by eleven rigid transvers cross girders forming a U shape rigid deck structure. The reinforced concrete deck is of variable depth (270 mm at the edge and 330 mm at the axis of the road). The bridge has two steel cantilever brackets. One cantilever is used for carrying high pressure sewage water pipe and the other carries a pedestrian concrete carriageway. An asphalt layer 60 mm thick is layered over the concrete slab.

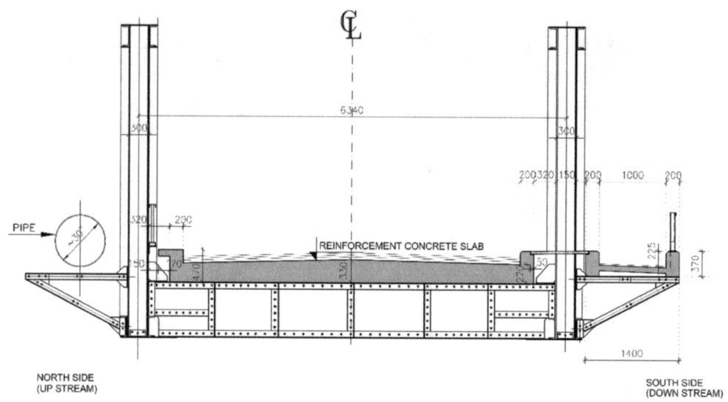


**Fig. 5.** Cross section of the bridge

## 3.2 Numerical Model

In order to check the load carrying capacity of the bridge a finite element model (FEM) calculation was made in LUSAS software. The theoretical capacity of each steel element composing the bridge was checked for the original design assuming that all elements and connections are in "as new" condition. Design checks were made in ULS according to the Israeli bridge code IS1227 [24] which is based on the old British code BS5400 [23] for HA, HB & HC loads and found to be satisfactory.

An inspection of the bridge conducted in 2011 revealed some weaknesses in the bridge structure (cross girder to truss connection) which were manifested in the excessive dynamic response of the bridge. As a consequence, and in order not to expose the bridge to excessive loading and further damage, the loading of the bridge was temporary reduced to 40 tons until further checks take place. Four main cases were analyzed for HC load case (1500KN) and for 600KN typical truck. Case A was the 'as designed' case which featured a monolithic connection between the transverse beams and the main truss. In order to take into account, the noted damages during the site visit on the structure, it was necessary to gradually demonstrate the release of the connections between the transverse girders and the main trusses (case B (red circles) – at two transverse girders; case C (yellow circles) -at four transverse girders; and finally case D (blue circles) was representing the theoretical case where all transverse girders connections with the main trusses are not functioning- Fig. 6).

**Fig. 6.** FEM numerical model

Since the connections between the cross girders and the truss controls the total stability of the main truss against global buckling, it was necessary to conduct the buckling analysis of the truss for the above cases. The global safety factor for truss buckling was calculated in all the defined cases using LUSAS buckling analysis module embedded in the software. The reduction in resistance was computed.

The load factor for the original design was 3.5 while it was drastically reduced by 7 times in the case if all connections were released between transverse girders and the main truss (Fig. 7).

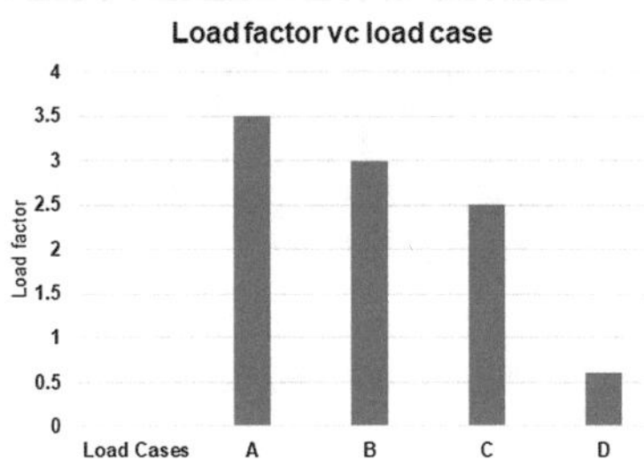


**Fig. 7.** Load factor in the function of the load case

The lateral sway of the upper chord was also investigated for the different loads (1500 kN and 600 kN) taking into account different levels of connectivity between the transverse girders and main truss girder. The obtained results are presented in Fig. 8.

The lateral sway of the upper chord when bridge was exposed to 1500 kN had values in the range from 8.25 mm to 65 mm (marked in red in Fig. 8). The change in the lateral movement when the bridge was exposed to a 600 kN truck load was not so drastic (marked in green in Fig. 8). This all indicates that the overall stability of the truss is directly related to the degree of the fixing of the lower cross girder connection with the truss.

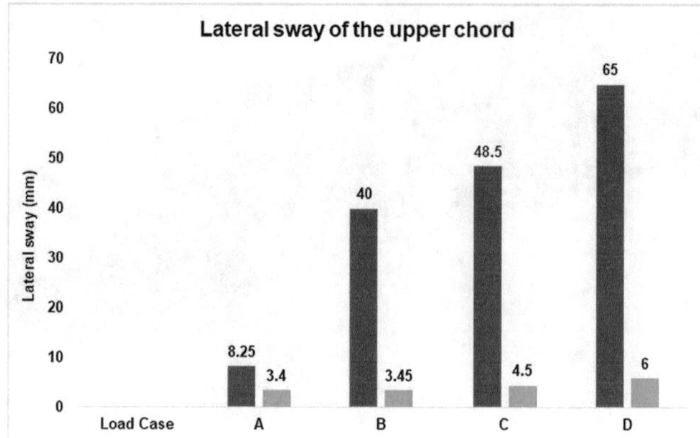


**Fig. 8.** Lateral sway of the upper chord

The first eigenmode was a vertical one and the calculated value was **3.91 Hz**, which is in good consistency with the measured data on the site (see Sect. 3.6) (Fig. 9).

**Fig. 9.** Eigen mode and eigen frequency

## 3.3 Vulnerable Zones

Identification of the vulnerable zones was done in accordance with the characteristic bridge typology. According to the load distribution and the characterizes of the single truss girder system vulnerability zones were noted as illustrated in Figs. 10 and 11.

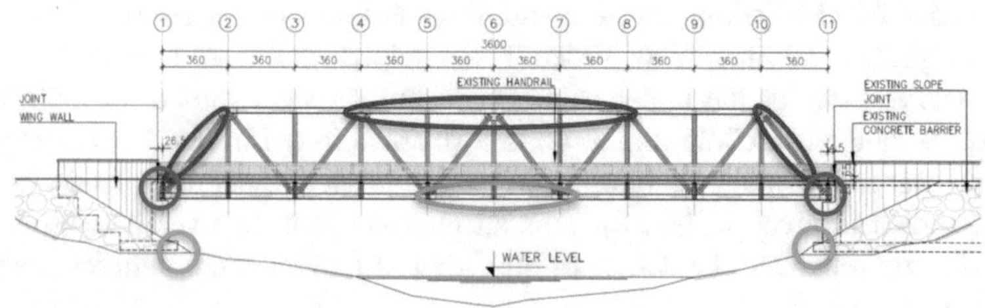


**Fig. 10.** Vulnerable zones – main truss
( high compression zones, high tension zone, bearing area, area possibly exposed to scour)

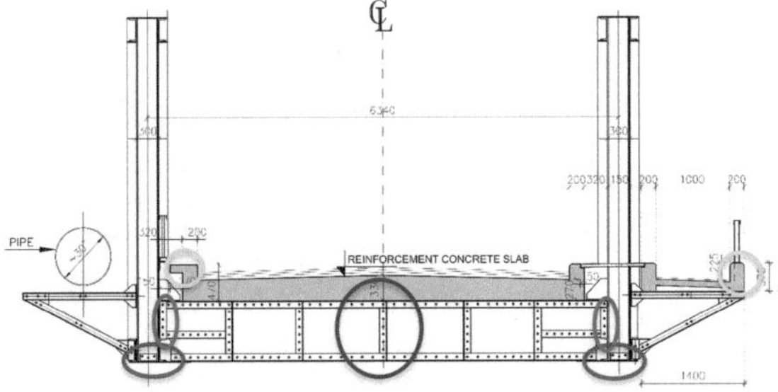


**Fig. 11.** Vulnerable zones – cross girder to deck connection
( compression zones, earing area, slab edge, cross girder sagging)

## 3.4 Damages on the Bridge Structure

Besides the increased vibrations of the bridge during the vehicle passing which was the basis for installation of the temporary structural monitoring system and dynamic bridge testing, several other defects were noted. Besides the mild corrosion of the steel structure (Fig. 12a), a visible corrosion at the connection between transverse girder and the deck slab with efflorescence due to water penetrating in between the girder upper chord was seen (Fig. 12b). In addition, corrosion can cause section loss in steel members and concrete reinforcement, which leads to strength degradation and increases the likelihood of bridge failure.

a) b)

**Fig. 12.** Corrosion of steel members

Due to the excessive dynamic loading an extreme relative movement of rivet heads was observed in many locations with visible shear deformations (Fig. 13).

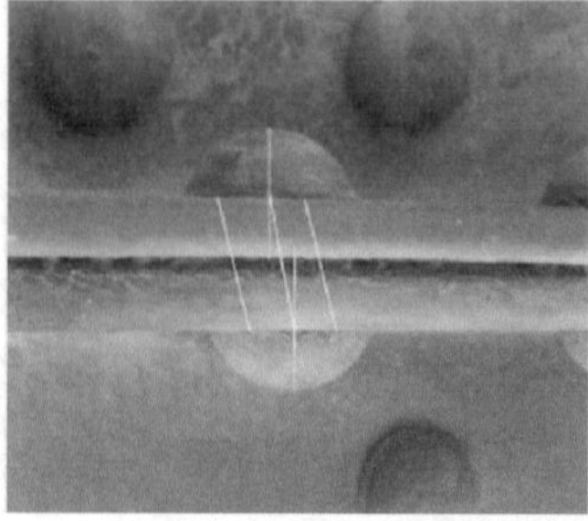

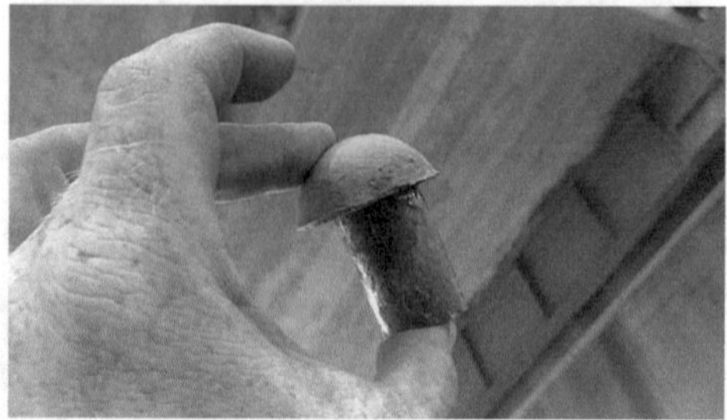

**Fig. 13.** Rivet movement and deformation

Deterioration of concrete part of the bridge structure was observed as well. Spalling and delamination of concrete along the bottom edge of the slab was observed and concrete surface abrasion at the massive abutments (Fig. 14).

**Fig. 14.** Concrete defects

Defects of pavement mainly near the expansion joints can be connected directly to the nonfunctionally of the roller bearing which was covered by large amount of debris and dust (Fig. 15).

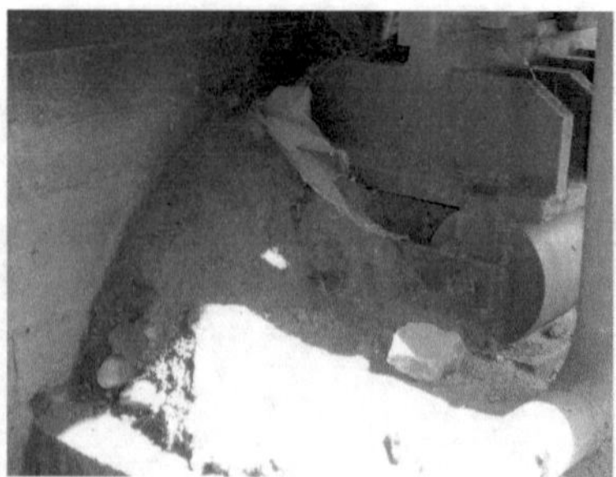

**Fig. 15.** Roller bearing

### 3.5 Failure Modes

As a fourth step, the static stage failure modes were elaborated in the Ultimate limit state (ULS) and in the Serviceability limit state (SLS). Truss failure could be caused due to several reasons. Local failure of truss members may occur as well as failure of disintegrated riveted section due to sheared rivets exposed to fatigue loading or local failure of vertical and diagonal truss members due to accidental load from heavy load vehicle as a result of nonfunctioning of safety barriers. This may lead to global failure of the vertical truss elements. Global failure of the bridge may be caused due to loss of stability of the truss structure and lateral buckling under heavy live load as a result of rivet failure connecting the transvers girder and the main truss. Additionally, due to excessive dynamic effect of heavy vehicles crossing the bridge a transverse girder bending/shear failure may occur. As the bridge is located in the high seismic zone failure due to seismic loading is possible as well, but it is not considered in the further analysis. Regarding the SLS there is a possibility for failure of the main safety barrier due to accidental load from heavy load transportation vehicle. Corrosion of the pedestrian safety handrail may lead to its failure. Evident limitation of the functioning of the roller bearings and rotation of the fixed bearings due to the corrosion and accumulation of debris. This is further reflected on the joints lowering their effectiveness and further causing asphalt pavement failure and abutment closing wall horizontal wide cracking.

### 3.6 Nondestructive Testing of the Bridge

In order to determine the fundamental frequency of the bridge a trailer truck of 600 kN passed over the bridge at various speeds (10 to 60 km/h) and additionally passed over a jumping rode which was placed at the location of the expansion joint in order to cause dynamic excitation of the bridge. The measured vertical frequency was **3.8 Hz ± 0.05**, while the lateral fundamental frequency of the truss in some cases was 10 Hz. The fraction of critical damping amounted to $\zeta = 0.012 \div 0.014$ (1.2%–1.4%). This is in quite a good agreement with the FEM value.

## 4 Key Performance Indicators and QC Plan

The determination of the key performance indicators was done on the basis of expert judgement. The performance indicators on the component level were determined on the basis of engineering practice and the failure time was estimated (Fig. 16). Failure mode in the form of the global buckling of truss upper chord may be caused due to fatigue of the shear rivets leaving only 15 years of bridge remaining life, or due to out of plane movement of lower connection plate which would extend the life for only 5 additional years. Truss bending failure mode and truss shear failure mode are mainly connected to the corrosion processes which due to the characteristic environmental aspect of this region can happen after 40 years. All of these segments are connected to the Reliability (structure safety) as the selected KPI. Equipment usually has a shorter life; this is a part of the bridge structure that has to be changed more frequently as its functionality directly affects safety as the selected KPI (Fig. 16).

| Structure type | Group | Component | Material | Design & Construction | Failure mode | Location/ Position | Damage /Observation | Damage process | KPI | Performance Indicator component level | | Performance value R | Performance value S | Estimated failure time [years] |
|---|---|---|---|---|---|---|---|---|---|---|---|---|---|---|
| TB | Structural elements | Main Trusses | Steel | 1954 | Truss Bending failure mode | Upper chord compression zone | Corroded plates | Corrosion | Reliability (Structure safety) | 2.3 | 4.1 | 4.1 | 2.1 | 40 |
| | | | | | | | Corroded rivet | Corrosion | | 2.3 | | | | 40 |
| | | | | | | Lower chord tension zone | Corroded plates | Corrosion | | 2.3 | | | | 40 |
| | | | | | | | Corroded rivet | Corrosion | | 2.3 | | | | 40 |
| | | | | | Truss Shear failure mode | Diagonals | Corroded plates | Corrosion | | 2.3 | | | | 40 |
| | | | | | | | Corroded rivet | Corrosion | | 2.3 | | | | 40 |
| | | | | | | | Accidental damage | Impact | | 2.0 | | | | 20 |
| | | | | | Global buckling of truss upper chord | Connection of truss verticals with deck cross girder | sheared rivet | Fatigue | | 4.1 | | | | 15 |
| | | | | | | | Out of plane movement of lower connection plate | Fatigue | | 4.1 | | | | 20 |
| | | Cross girders | Steel | 1954 | Bending | High sagging area | Shear connection with deck corroded | Corrosion | | 2.1 | 4.1 | | | 30 |
| | | | | | web plate buckling | Bearing area over main truss | Rivets are partially sheared | Fatigue | | 4.1 | | | | 20 |
| | | | | | Bending | Along the girder | Corroded rivet | Corrosion | | 2.1 | | | | 40 |
| | | Deck slab | Reinforced concrete | 1954 | Bending | HMS/bottom | delamination | Corrosion | Reliability | 2.1 | 2.1 | | | 30 |
| | | | | 1954 | Falling chunks | bottom | Spalling | Corrosion | Safety (Life and limb) | 2.1 | 2.1 | | | 30 |
| | | | | 1954 | Bending | HMH | Efflorescence | Leaching | (Symptom) | (2.1) | | | | |
| | | Bearings | Steel | 1954 | Bearing Failure | Abutment 1 (west) | Corrosion | Corrosion | Reliability | 2.0 | 4.0 | | | 40 |
| | | Bearings | Steel | 1954 | Bearing Failure | Abutment 1 (west) | Bearing restrained no movement due to corrosion and debris | Corrosion | Reliability | 4.0 | | | | 20 |
| | | Bearings | Steel | 1954 | Bearing Failure | Abutment 11 (east) | Loss of rotation ability due to Corrosion | Corrosion | Reliability | 3.0 | | | | 20 |
| | | Abutment | Reinforced concrete | 1954 | | Abutment 1 (west) | Spalling and delamination at closing wall | Joint leaking | Reliability | 3.0 | 3.0 | | | 20 |
| | | Abutment | Reinforced concrete | 1954 | Bearing Failure | Abutment 1 (west) | closing wall with horizontal crack | Closing of joint | Reliability | 3.0 | | | | 20 |
| | | Wing wall | Reinforced concrete | 1954 | | Wing wall | Horizontal cracking | | Reliability | 2.1 | 3.3 | | | - |
| | | Wing wall | Reinforced concrete | 1954 | | Wing wall | Spalling | Corrosion | Reliability | 3.3 | | | | - |
| | | Wing wall | Reinforced concrete | 1954 | | Wing wall | Surface abrasion | Abrasion | (Symptom) | 3.3 | - | | | |
| | | Expansion Joint | steel | 1954 | Closing | EJ 1 (west) | Closing of EJ | Deck movement | Reliability | 3.0 | 3.0 | | | |
| | | Pedestrian Deck slab | Reinforced concrete | 1954 | HMH | Over transvers supporting truss | Transvers cracks | Not active | Reliability | 2.3 | 2.3 | | | 20 |
| | | Pedestrian Deck slab | Reinforced concrete | 1954 | Falling chunks | South Edge | Spalling | Corrosion | Safety (Life and limb) | 3.3 | 3.3 | - | 3.3 | 20 |
| | Equipment | Safety barrier | Steel | 1954 | Falling of the deck | Safety barrier | Broken, missing parts | Impact | Safety (Life and limb) | 3.0 | 3.0 | | | 10 |
| | | Pedestrian Handrail | Steel | 1954 | Falling of the deck | Handrail anchoring | Corrosion of structural steel | Corrosion | Safety (Life and limb) | 2.7 | 2.7 | | | 30 |
| | | Curb | Reinforced concrete | 1954 | Falling chunks | Curb side | Spalling, delamination | Corrosion | Safety (Life and limb) | 3.3 | 3.3 | | | 20 |
| | | Pavement | Asphalt | Estimated 2005 | Sudden disturbance to driver | Expansion joints overlay | Open transvers cracks | Joint reflection cracking | Safety (Life and limb) | 3.3 | 3.3 | | | 5 |

**Fig. 16.** Determination of the KPI values and estimated failure time

In order to determine the life time costs, reliability, availability and safety of the bridge two life time cycles for the duration of 100 years were considered. The first approach named as "reference" approach envisages only some basic repair of the pavement, no other preventive measures were foreseen. Once a component or system failure is reached comprehensive interventions for this component are foreseen leaving the other components to gradually deteriorate and reach the failure level. In accordance with the current state of the pavement its failure is expected in five years mainly caused by the crack development around the expansion joints and formation of the potholes which additionally are affecting the safety of the drivers and possible accidental impact of the vehicles on the main truss elements. Due to existing defects on the rivets and further fatigue influence it is expected that the connection of the truss vertical members with the cross girders will fail in the next 15 to 20 years. Many steel elements are attacked by corrosion and according to the site climate and corrosion state it is expected that the peak of corrosion will be reached in the following 30 to 40 years (Fig. 17).

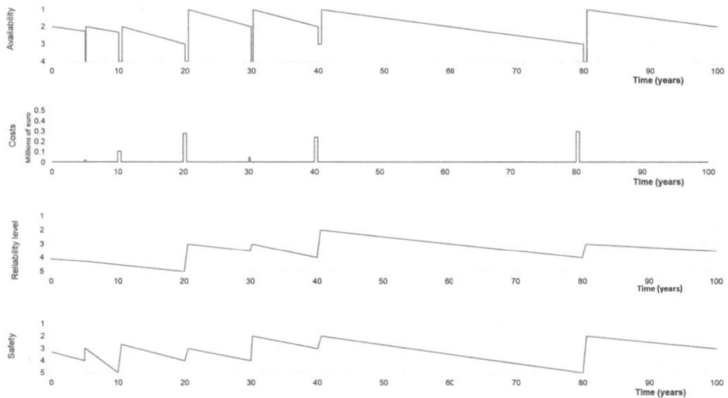


**Fig. 17.** Reference approach

The second approach, referred as "preventive" envisages a major rehabilitation of the bridge within the next two years bringing it to the state of an almost "new" bridge. From this point on, periodical interventions will take place preventing formation of large damages, deterioration and decay of the structure.

The rehabilitation works would include a complete repair of all the concrete and steel elements including the replacement of around 400 rivets and replacement of all corroded elements (trusses, pedestrian handrailing etc.), installation of new expansion joints, bearing rehabilitation, new waterproofing and asphalt overlay. It is envisaged that in the first intervention after the rehabilitation works (after 10 years) it would be necessary to rehabilitate the safety barriers and the upper layer of the asphalt paving. After 20 years from the first repair it will be necessary to conduct an overall concrete surface treatment, sandblasting and painting of all steel elements. It is believed that after 40 years replacement of rivets would be required, rehabilitation or replacement of the bearings and expansion joints as well as renewal of the waterproofing system (Fig. 18).

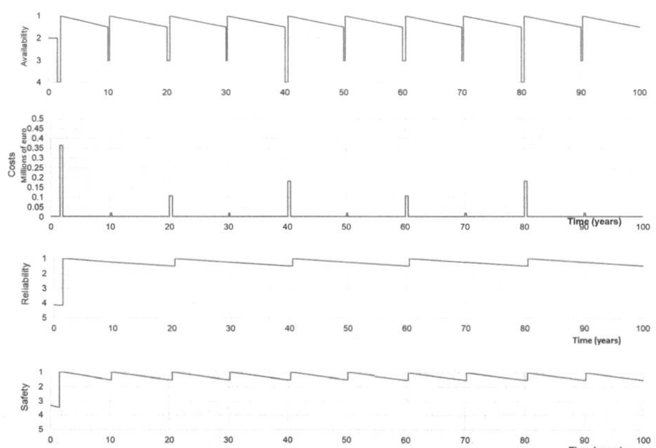


**Fig. 18.** Preventive approach

Comparing these two approaches after 100 years clearly shows the benefits of the preventive approach leading to higher safety, reliability and availability, while the cost value after 100 years would reach around the same amount (Fig. 19).

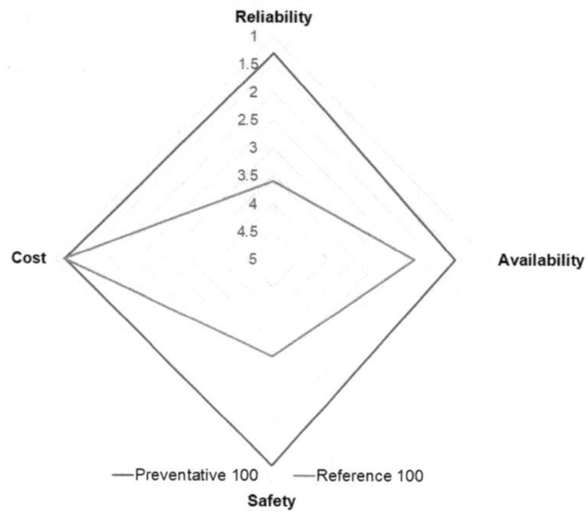


**Fig. 19.** Spider comparison

## 5 Conclusion

In order to conduct a better decision-making regarding bridge management it is necessary to use bridge management systems. It is well known that various European countries have different management systems. In order to overcome this issue a guideline for standardization and development of quality control plans for roadway bridges has been proposed by COST TU1406. The process is based on several stages with the foundation of performance indicators, performance goals and key performance indicators. A multi-criteria decision-making (MCDM) approach can thoroughly associate the inputs with cost-benefit models to rank existing decision options regarding bridges at component, system or network levels. The applicability of the method was shown on an example of a steel truss bridge. Further developments are seen in the standardization of the assessment procedures, collection of supplementary PIs and their connection with the KPIs and application of more robust maintenance tools in the practice.

## References

1. Tao, Z., Zophy, F., Wiegmann, J.: Asset management model and systems integration approach. Transp. Res. Rec. **1719**(00–1162), 191–199 (2000)
2. Van der Velde, J., Klatter, L., Bakker, J.: A holistic approach to asset management in the Netherlands. Struct. Infrastruct. Eng. **9**(4), 340–348. Maintenance, management, life-cycle design and performance, Asset management in civil engineering (2013)

3. ISO (International Organization for Standardization): General principles on reliability of structures. ISO 2394 (2015)
4. Capros, P., Mantzos, L., Papandreou, V., Tasios, N.: European energy and transport–trends to 2030–update 2007. European Commission, Directorate-General for Energy and Transport, Institute of Communication and Computer Systems of the National Technical University of Athens (2008)
5. Flintsch, G.W., Bryant, J.W.: Asset Management Data Collection for Supporting Decision Processes. US Department of Transportation, Federal Highway Administration, May 2006
6. Stipanovic, I., Chatzi, E., Limonggelli, M., Gavin, K., Allah Bukhsh, Z., Skaric Palic, S., Xenidis, Y., Imam, B., Anzlin, A., Zanini, M., Klanker, G., Hoj, N., Ademovic, N.: COST TU 1406, WG2 technical report – performance goals for roadway bridges (2017)
7. Kedar, A., Sein, S., Panetsos, P., Ademović, N., Duke, A., Ryjacek, P.: COST TU 1406, WG4 technical report – preparation of a case study (2019)
8. Rücker, W., Hille, F., Rohrmann, R.: Guideline for the assessment of existing structures. SAMCO Final Report (2006)
9. Hajdin, R., Kušar, M., Mašović, S., Linneberg, P., Amado, J., Tanasić, N. Close collaborators: Ademović, N., Costa, C., Marić, M., Almeida, J., Galvão, N., Sein, S., Zanini, M., Pendergast, L.: COST TU 1406, WG3 technical report – establishment of a quality control plan (2018)
10. Hakim, S.J., Abdul Razak, H.: Modal parameters based structural damage detection using artificial neural networks -a review. Smart Struct. Syst. **14**(2), 159–189 (2014). https://doi.org/10.12989/sss.2014.14.2.159
11. Frangopol, D.M., Soliman, M.: Life-cycle of structural systems: recent achievements and future directions. Struct. Infrastruct. Eng. **22**(1). http://dx.doi.org/10.1080/15732479.2014.999794
12. Yi, J., Kumares, S.C.: Bridge service life prediction model using the Markov chain. Transp. Res. Rec. **1223**, 24–30 (1989)
13. Ansell, A., Racutanu, G., Sundquist, H.: A Markov approach in estimating the service life of bridge elements in Sweden. In: 9th International Conference: Durability of Building Materials and Components, Paper 142, pp. 1–10 (2002)
14. Firouzi, A., Rahai, A.: Reliability assessment of concrete bridges subject to corrosion-induced cracks during life cycle using artificial neural networks. Comput. Concr. **12**(1), 91–107 (2013). https://doi.org/10.12989/cac.2013.12.1.091
15. West, H.H., McClure, R.M., Gannon, E.J., Riad, H.L., Siverling, B.E.: Nonlinear deterioration model for the estimation of bridge design life. Research project no. 86-07, Final report, Pennsylvania Transportation Institute (1989)
16. Puz, G., Radic, J.: Life-cycle performance model based on homogeneous Markov processes. Struct. Infrastruct. Eng. **7**(4), 285–294 (2011). https://doi.org/10.1080/15732470802532943
17. fib Bulletin 34: Model Code for Service Life Design of Concrete Structures. International Federation for Structural Concrete (fib), Lausanne (2006)
18. fib Bulletin 59: Condition Control and Assessment of Reinforced Concrete Structures Exposed to Corrosive Environment (Carbonation/Chlorides). International Federation for Structural Concrete (fib), Lausanne (2011)
19. fib Bulletin 76: Benchmarking of deemed-to-satisfy provisions in standards: durability of reinforced concrete structures exposed to chlorides. DCC Document Competence Center Siegmar Kästl e. K, Germany (2015)
20. Dissanayake, P.B.R., Narasinghe, S.B.: Service life prediction of masonry arch bridges using artificial neural networks. Eng. **XXXXI**(01), 13–16 (2008)

21. Hakim, S.J.S., Abdul Razak, H.: Structural damage detection of steel bridge girder using artificial neural networks and finite element models. Steel Compos. Struct. **14**(4):367–377 (2013)
22. Fathallaa, E., Tanakab, Y., Maekawa, K.: Remaining fatigue life assessment of in-service road bridge decks based upon artificial neural networks. Eng. Struct. **171**, 602–616 (2018)
23. BS 5400-3:2000 Steel, concrete and composite bridges. Code of practice for design of steel bridges, 15 October 2000, The British Standards Institution (BSI) (2000)
24. The standard Institution of Israel, Loads on Bridges: Highway Bridges: IS 1227, Part 1 (2002)

# Role and Efficiency of MBR Technology for Wastewater Treatment

Amra Serdarevic[1(✉)], Alma Dzubur[1], and Tarik Muhibic[2]

[1] Faculty of Civil Engineering, University of Sarajevo, Sarajevo, Bosnia and Herzegovina
amra.serdarevic@gf.unsa.ba
[2] Public Utility Company "JKP Vodovod i Kanalizacija", Konjic, Bosnia and Herzegovina

**Abstract.** The main objective of the operation of a wastewater treatment plant (WWTP) is to ensure effluent quality by meeting the parameter values determined by relevant legislation. In last decades, as application and improvement of performances of membranes grows, they are being used more often in wider range of the wastewater treatment process. The Membrane bioreactor (MBR) technology is applied in industrial wastewater treatment, leachate treatment and also for domestic wastewater treatment. In general, still major constraints for wider application of the MBR technology are significant operating costs and replacement of membrane cartridges. This approach is usually considered to be applied in the countries with low or medium income. Although the membrane technologies for wastewater treatmesnt are still not widely used in Bosnia and Herzegovina, the MBR technology has been selected for wastewater treatment of town Konjic, located in the south part of Bosnia and Herzegovina (BiH). Analyse of chosen technology, role and efficiency of the MBR technology, are based on full-scale experience. The MBR plant was designed and constructed in 2014. The experimental results of the plant operation has confirmed the viable long-term operation of membrane bioreactor for domestic wastewater treatment. A special emphasis was given to the role of pre-treatment. As an effluent of primary treatment contains less suspended solids and organic compounds than raw wastewater, operation of an MBR should be more efficient, reducing the overall water treatment costs. Technical details of the MBR Konjic, biotreatment and filtration, including design data and legal requirements, operation results are presented in this paper.

**Keywords:** Wastewater · Membrane · MBR · Effluent · Sludge · Pre-treatment

## 1 Introduction

Membrane bioreactor (MBR) is generally a term used to define wastewater treatment in combination of biological process (usually suspended growth bioreactor) with filtration through perm-selective membrane (micro or ultrafiltration). A membrane bioreactor is essentially a kind of conventional activated sludge system (CAS), where instead secondary clarifier, membranes are in use as a filter to separate the solids developed by biological processes. Therefore MBR is an excellent technology for obtaining high


S. Avdaković et al. (Eds.): IAT 2019, LNNS 83, pp. 229–237, 2020.
https://doi.org/10.1007/978-3-030-24986-1_18

quality standards for the effluent. The membranes are also barrier that retains colloids, bacteria and viruses, providing a complete disinfection of treated water. MBR can operate at high concentration of sludge (up to 10 g/L compared with usual 3 g/L) for conventional system, with significantly reduced sludge production [1].

Application of membrane technology in wastewater treatment processes, during the last decade, takes a significant place compared to the other wastewater treatment processes. The positive results of using MBR technology are recorded for treatment of municipal wastewater as well as industrial waste water and leachate [1].

Although the prices of membrane on the market are constantly dropping, the widespread implementation is limited still by its high costs, both capital and operation expenditure. High costs are mainly connected with membrane installation, replacement and high energy demand. Therefore, the choice of MBR technology is still considerably less common in countries with low income [6]. Intention of this paper is to present application of MBR technology of wastewater treatment in BiH, despite of the lack of relevant experience in operation of this type of technology.

MBR technology enables wastewater treatment in several basic phases (depending on the requirements for effluent) and a series of sub-phases, depending on the technical performance. The biological treatment process removes organic pollution and nutrients at minimal chemical input. After the biological treatment, water and activated sludge flows through the membranes that retain the microorganisms and all the residual organic and inorganic matter.

MBR technology was selected for the treatment of urban wastewater in the city of Konjic in BiH. Technical details, concept, operation and results are presented as follows.

## 2 Wastewater Issues of Town Konjic

Konjic is a municipality and town located in Herzegovina-Neretva Canton, Federation of Bosnia and Herzegovina, an entity of Bosnia and Herzegovina. It is located in northern part of Herzegovina, 50 km southwest from Sarajevo, capital of the Bosnia and Herzegovina. The whole area is a mountainous, heavily wooded, laid on area of 1.101 $km^2$ and includes more than half of the area of the upper watercourse of Neretva river. Konjic is 268 m above sea level [3].

According to the 2013 population census in the municipality there are 26 381 inhabitants, of which approximately 12,000 inhabitants live in the city. The industry is very little represented, and mostly with own discharge into the water stream. Sewerage system in Konjic is managed by the public utility company "JKP Vodovod i Kanalizacija d.o.o." and only about 30% of the population are connected to the sewerage system. The aged sewerage systems mainly collect combined sanitary and storm water drainage. This system causes problems such as, overloading of sewers and accumulation of large amount of grit and inert materials especially during heavy rains that could cause blockages of the pumps and screens. Heavy rains also cause a problem in manholes, pump stations and in operation of wastewater treatment plant [6, 8].

The central part of the city has built a new sewage system of a separate system of about 20 km in length, while in the rest of the city has not been solved the collection, and transport of domestic wastewater. Construction of the left and right collectors are in progress. The MBR plant is located downstream of the city, on the left bank of the river Neretva (Fig. 1) [2].

**Fig. 1.** Location of the MBR plant Konjic on the left embankment of Neretva river [2]

The MBR technology has been selected in accordance with several conditions which had been set up by the investor. The conditions were related to the effluent quality, the limited narrow space available for plant location, the proximity of the watercourses, the vicinity of the settlement, minimum of negative impacts to the environment (odors, aerosols, noise, dust, etc.) and simplicity for the operation and maintenance. The MBR technology was selected as the most favorable in regards to the environmental impacts and the quality of the effluent.

## 3 Design of MBR Konjic (BiH)

The MBR plant Konjic was built in 2014. The start-up of the plant was done with sanitary water, at the beginning of December 2015, and then a trial period with raw wastewater lasted until July 2016 when the MBR plant was officially commissioned.

Preliminary design of the WWTP Konjic was prepared in 2011 [4] and the MBR technology was selected as a most appropriate for requested requirements [4]. The main design, procurement and installation of the first stage of construction of wastewater treatment plant Konjic has been done by Aqualia Infraestructuras, S.A, Spain [3].

The MBR plant Konjic has been designed for the three phases, where phase I is for 5,000 PE (total capacity is 15,000 PE) for the tertiary treatment of urban wastewater using the membrane technology. The recipient is the accumulation of the HPP Jablanica on the Neretva River. The data of MBR Konjic design treatment capacity are presented in Table 1.

**Table 1.** Design treatment capacity of the MBR Konjic (BiH) [3, 4]

| Population Equivalent (PE) and daily flow [3, 4] | | | |
|---|---|---|---|
| Design phases | I (2016) | II (2023) | III (2038) |
| PE | 5000 | 10000 | 15000 |
| Average daily water consumption per capita (L/c/d) | 240,00 | 240,00 | 240,00 |
| Average dry weather flow (ADW) ($m^3/d$) | 1200,00 | 2400,00 | 3600,00 |
| Max. hourly flow ($m^3/h$) | 128,00 | 256,001 | 384,00 |
| **Designed pollutants loading [4]** | | | |
| (kg/day) | I (2016) | II (2023) | III (2038) |
| $BOD_5$ | 300 | 600 | 900 |
| COD | 600 | 1200 | 1800 |
| TSS | 350 | 700 | 1050 |
| TKN | 55 | 110 | 165 |
| TP | 9 | 18 | 27 |
| **Average concentrations of pollutants (according to the main design) [3]** | | | |
| (mg/L) | I (2016) | II (2023) | III (2038) |
| $BOD_5$ | 250 | 250 | 250 |
| TSS | 291,70 | 291,70 | 291,70 |
| TKN | 45,80 | 45,80 | 45,80 |
| TP | 7,50 | 7,50 | 7,50 |

Construction of the phase I of the WWTP Konjic started in 2014. Testing period started in December 2015 and continued in 2016. The MBR plant has a stabile operation since 2016 despite the facts that the situation with the collection of wastewaters still is not a satisfactory. Problems caused by combined sewage system resulted with flow oscillation, wastewater dilution (organic pollution) and overflow of the facilities of the WWTP. Those problems were tackled by solution provided with the pre-treatment and pre-screening facilities upstream of the raw water before the pumping station and inlet to secondary stage (biological treatment) of the MBR. These facilities consist of rectangular, horizontal flow grit channels to remove heavier grit particles and sets of coarse and medium screens.

Although the main project was calculated with approximately 60% population connected to the sewerage system, there is only about 30% of the population connected to the existing sewage system and the MBR plant. Tertiary treatment and disinfection is also included in the project due to the legal requirements [7].

The effluent discharge standards for MBR Konjic are compliant with legislation in FBIH which is harmonized with EU Municipal Wastewater Directive [7] (Table 2).

**Table 2.** MBR Konjic effluent discharge standards [7]

| Parameter | Unit | Maximum permissible concentration |
|---|---|---|
| $BOD_5$ | mg/L | 25 |
| COD | mg/L | 125 |
| TSS | mg/L | 35 |
| TN | mg/L | 10 |
| TP | mg/L | 1 |

Sanitary and industrial wastewater are collected partially by separate and combined sewerage system ($Q_{max}$ = 166 L/s) delivered by collector (500 mm diameter) at the inlet unit of WWTP Konjic. Water continue to flow by gravity through the coarse screen (s = 20 mm). The inlet channel has a lateral overflow of 2,0 meters in length to discharge the excess flow, or in the incidental situation to prevent damages of the pumps or other facilities of the plant (Fig. 2).

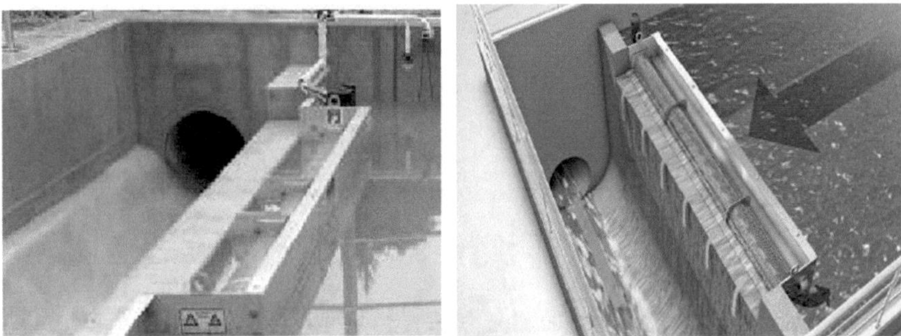

**Fig. 2.** Inlet with lateral overflow length of 2,00 meters for discharge of overflow [3].

From inlet channel, raw wastewater ($Q_{max}$ = 106,67 L/s) elevate by pumps for the raw wastewater 8 m above ground level to the pre-treatment compact station [3]. Pre-treatment compact station is dedicated to the MBR treatments that needs to receive wastewaters as much as possible without solids, grit and grease. The integrated pre-treatment station is able to carry out the filtration, washing and compaction of the screened material, oil, sand and grease removal, the washing and removal of sands. The compact station is completely covered to ensure the maximum safety and hygienic operation.

The raw wastewater entering the station passes through a filtering screen and deposits on screw filter the suspended solid matters. Water flows to micro-screening (3 mm) and requires daily cleaning.

Aeration system in grit chamber allows the oil and grease flotation and keep organic matters in suspension. The longitudinal conveyance screw carries over the sands to the screw classifier. Lack of good pre-treatment has been detected as a key technical problem for MBR plant operation. The operation of pre-treatment compacts station is completely automatic and controlled by the PLC. Protection against overload is secured by electronic load limiting devices.

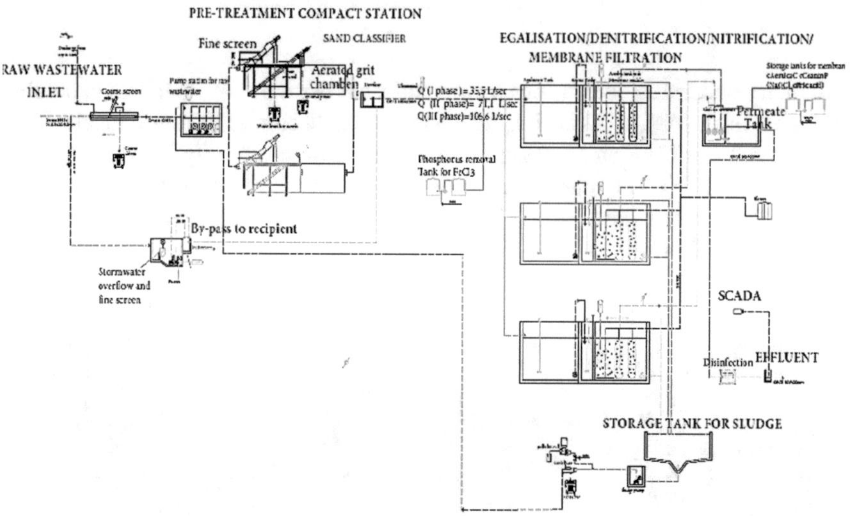


**Fig. 3.** Process flow scheme of WWTP Konjic [adapt. 4]

From pre-treatment compact station water flows to the egalisation tank (Fig. 3). The egalisation tank has the role of balancing wastewater inflow and frequency of filtration through membrane (flux). The volume of egalisation tank is 324,22 $m^3$.

The two pumps are installed in egalisation tank for the transport water into the bioreactor. The operating mode of the pump is controlled also through the SCADA system.

The volume of the bioreactor is designed in one line for the phase I, with a total volume of 825 $m^3$, divided into 3 parts: deoxidation zone (59,2 $m^3$), anoxic zone (180.03 $m^3$), aerobic zone I (450 $m^3$) and aerobic zone II with submerged membrane modules (135 $m^3$). The retention time of water in the bioreactor (HRT) is 25 h (approximately 1 day), with 30 $m^3$/h of inflow [3].

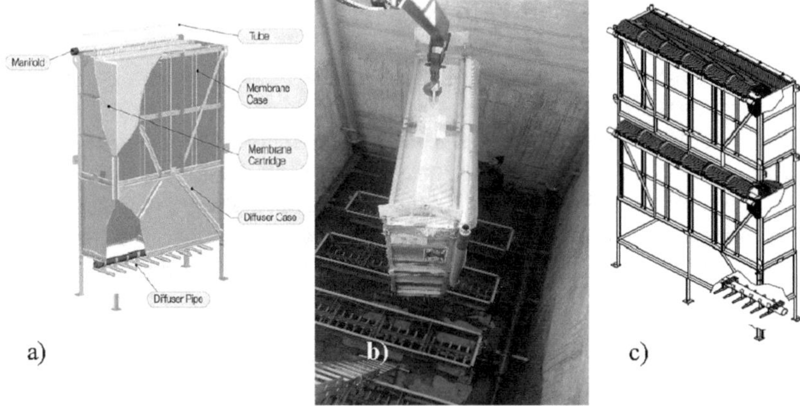


**Fig. 4.** Kubota (a) Single-deck and (c) Double-deck modules, (b) Installation of the flat sheet membrane module of MBR Konjic [1, 5]

From the bioreactor, the activated sludge was separated from water by membrane flat sheet (Kubota). Some of the biomass was recycled to the head of the aeration tank to provide designed concentration of the suspended solids (max 10 g/L) or to support process of denitrification (anoxic zone), while the remind biomass (excess sludge) is treated before disposal or reuse [3, 5, 8]. Iron chloride was added to water in egalisation tank for phosphorus elimination to the required threshold.

The aeration tank comprises four membrane units (modules), with 400 panels per unit, with (at) 1,45 $m^2$ membrane area per panel. The total surface area of the membrane per unit is 580 $m^2$, giving a total membrane panel area of 2320 $m^2$. The membranes modules are 0.5 m immersed from the bottom of the reactor while the depth of the reactor is 5,5 m. The installed membranes are flat sheet type, hydrophilicity, made by chlorinated polyethylene (PE), supported by a very robust nonwoven substrate (Fig. 4), which is welded on each side to the plate. The plate contains a number of narrow channels for even collection of permeate across the surface. The membrane pore size is 0.4 μm.

The average flux is generally below 20 L/$m^2$,h and the TMP usually between 0,02 and 0,11 bar. Permeate is extracted from a single point at the top of each membrane sheet. The membrane operation consist of sequences that includes 9 min of filtration, followed by 1 min of relaxation and 5 min of air scouring. The membrane is relaxed for 30–60 min daily as a result of low overnight flows [5].

The recommended mixed liquor suspended solids (MLSS) concentration range is 8 to 10 g/L, but it has been operated in concentration range from 2 to 3 g/L. The plant operates at SRT from 60 to 90 days, producing 0,10 to 0,3 kg sludge per kg of BOD. This condition is expected due to significant lower amounts of wastewater, lower load of biological unit and lower production of activated sludge [1, 3, 12].

After the treatment, the water is finally discharged into the Neretva River.

## 4 Operation Results and Comments

MBR Konjic has performed very well from the begging of operation until today. It needs to be pointed out as a positive example due to the lack of experiencewith this technology in Bosnia and Herzegovina. This is a simple and robust plant and it has generally proved very reliable. The failure for this plant during the first four years of operation was minimal. Permeate water quality has been always very good and measured parameters are under maximum permissible concentration. The plant produces very low odour levels. The quality of the effluent with yearly average concentration is presented in the Table 3.

**Table 3.** Average concentration of the influent and effluent of MBR (2016–2018) [5, 7]

| Parmeter | Unit | Influent | Effluent | Max. permissible conc. |
|---|---|---|---|---|
| $BOD_5$ | mg/L | 143,0 | 6,4 | 25 |
| COD | mg/L | 285,0 | 10,3 | 125 |
| TN | mg/L | 19,0 | 0,90 | 10 |
| TP | mg/L | 1,82 | 1,05 | 1 |
| TSS | mg/L | 232,0 | 0,0 | 35 |

The MBR plant was operated with a significantly lower load than the designed one. According to the activities on the sewage construction, approximately 60% of the population is expected to be connected in the near future, and then the MBR plant should be operated in accordance with the designed load.

Concentration of the mixed liquor suspended solid (MLSS) in bioreactor is in an average range from 2 to 3 g/L. This situation occurs due to the very low organic concentration and dilution by stormwater. That is also a reason of a very low amount of excess sludge. From the beginning of MBR Konjic operation, only 20 $m^3$ of sludge were drawing-off and disposed in the sludge storage tank. Sludge has been first dewatered by a centrifuge [3, 5].

Membranes are in operation approximately 3 years, with chemical cleaning in place with 0.5 wt% hypochlorite, according to the project documentation and the instructions of the contractor. TMP are in designed range. The long term operation of membrane without failure and membrane clogging is mainly result of the effective pre-treatment, proper operation and maintenance of the plant.

The local operator's staff (JKP ViK Konjic) are well trained to operate and maintain the MBR plant.

## 5 Conclusion

The MBR Konjic has been designed with the state-of-the-art technology, based on pre-treatment, activated sludge process and membrane filtration. The treated water is discharged into Neretva river. Selected technology removes suspended solids, organic material, nutrients and produces effluent fully compliant with the legislation.

Energy consumption is optimised by selection of the energy efficient aeration equipment and adequate membrane operation. The noise and odour emissions are controlled by covering the surfaces of process unit.

MBR plant Konjic has been successfully designed, constructed and taken into the operation by meeting all requirements and dealing with all constrains. This plant is a good practice example for further consideration of MBR technology in BiH.

## References

1. Judd, S.: The MBR Book: Principles and Applications of Membrane Bioreactors for Water and Wastewater Treatment. Elsevier Science, Oxford, UK (2011)
2. https://earth.google.com/web/@43.65800546,17.95151868,276.92173538a,2833.82865491d,35y,6.24891472h,7.47882354t,0r. Uploaded 10 Mar 2019
3. Aqualia Infraestructuras, S.A. Main design of the Wastewater treatement plant Konjic – first phase (I) (2014)
4. Zavod za vodoprivredu d.o.o., Preliminary design for WWTP Konjic, Sarajevo, BiH (2011)
5. Public Utility Company JKP Vodovod i kanalizacija Konjic, Technical documentation, operation results and photos, 2015–2019
6. Serdarević, A.: Razvoj i primjena MBR tehnologije u procesu preciščavanja otpadnihvoda/ Development and Application of MBR Technology in the Process of Wastewater Treatment. Vodoprivreda (2014). http://agris.fao.org/agris-search/search.do?recordID=RS2017000394
7. Uredba o uslovima ispuštanja otpadnih voda u okoliš i sisteme javne kanalizacije ("Službene novine FBiH", broj 101/15, 01/16 i 101/18)
8. Serdarevic, A., Dzubur, A.: Importance and practice of operation and maintenance of wastewater treatment plants. In: Avdakovic, S. (ed.) Advanced Technologies, Systems, and Applications III. IAT 2018. Lecture Notes in Networks and Systems, vol. 60. Springer, Cham, s.l. (2019)

# Qualitative and Quantitative Differences in the Calculations of Reinforced Concrete Frames

Marijana Hadzima-Nyarko[1(✉)], Naida Ademović[2], and Sanja Jović[3]

[1] Faculty of Civil Engineering and Architecture Osijek, University J.J. Strossmayer Osijek, Osijek, Croatia
mhadzima@gfos.hr

[2] Faculty of Civil Engineering in Sarajevo, University of Sarajevo, Patriotske lige 30, 71000 Sarajevo, Bosnia and Herzegovina
naidadem@gmail.com

[3] NWind GmbH, Haltenhoffstrasse 50 A, 30167 Hannover, Germany
sanja_jovic@web.de

**Abstract.** A comparison between the current codes and enforced codes at the time of construction of a real multi-storey reinforced concrete (RC) frame was made. This paper presents qualitative and quantitative differences in the calculation of reinforced concrete structures according to the Technical Standards for Concrete and Reinforced Concrete (PBAB87) and Eurocode 2 (EC 2). A model of the RC frame was created utilizing SAP2000 software to check the internal forces and moments and the basic dynamic properties of the structure. The value of the fundamental period of the modelled structure, as one of the most important dynamic properties of the structures, was compared with the values calculated according to the equations given in Eurocode 8 (EC 8), but also by other norms and according to several expressions given by the researchers. Large differences in the values of the fundamental period were found and it was from 3.4% up to 60%.

## 1 Introduction

Technical Standards for Concrete and Reinforced Concrete (RC) (PBAB87) which was adopted in 1987, was based on the Model Regulations of the European Committee of Concrete (CEB), the International Federation of Pre-Stressing (FIP) from 1978, the Swiss Standards Sia 162 and to a small extend on German DIN-1045 regulations and US ACI standards [1].

It is the limit state method that was the basis for the calculation assumptions stated in the PBAB87 [2], which is the basis for Eurocode 2 (EC2) [3], today as well. It was developed from the fracture theory which was established by Russian, then French and American scientists in the 30s of the twentieth century. The beginnings of its detailed explanation coincide with the development of thin-walled structures and structures made of pre-stressed concrete (the second stage of reinforced concrete development) in which the deficiencies of the classical theory were realized [4]. At that time, basis for the new method was to achieve an acceptable probability that a designed structure will


S. Avdaković et al. (Eds.): IAT 2019, LNNS 83, pp. 238–249, 2020.
https://doi.org/10.1007/978-3-030-24986-1_19

not be inappropriate for use. This means that all the technical and economic conditions applicable whenever and wherever, relevant to determining the safety, durability and functionality of the building, had to be considered. In this way, in comparison with classical theory, a reasonable safety is obtained, better utilization and better insight into the bearing capacity of the overall structure is obtained. More specifically: economic optimization of steel usage in centrally and eccentrically compressed elements, better utilization of concrete and steel in compressed elements, and the assessment of the structure reliability is located on a safer margin [4].

Eurocode as technical regulations were created in response to obstacles in the construction sector between individual EU Member States. In 1975, the European Commission launched the Treaty of Rome (The Treaty of Rome brought about the creation of the European Economic Community in 1957). In 1989, the concern for development and improvement of Eurocodes was taken over by the European Committee for Standardization (CEN) with the task of enforcing standards in all the EU Member States.

The final implementation of Eurocode in Croatia started in 2005, by the issuance of the Technical Regulations for Concrete Structures (NN 101/05) [5] with amendments and modification in 2006 and 2007. In 2009, these regulations were completely replaced (NN 139/2009) [6], and have already been amended and revised in 2010.

Since Croatia is located in a seismically active zone, the best preventive measures are in good planning, design and construction in accordance with the application of seismic regulations. In the last decade most of the structures were designed and constructed in accordance with the Eurocode 8 (EC8) [7] – provisions for earthquake-resistant design, however they count for only 4% of all buildings [8].

In order to determine the seismic behavior of this already constructed building, first of all it was necessary to conduct load bearing proofs regarding resistance and stability, dimensioning and calculation of reinforcement of the reinforced concrete structure according to EC2 in order to compare the obtained results. After that the structure was modelled in SAP2000® v16.1.0 Ultimate software package with the aim to determine the fundamental period as the input data for further analysis according to EC8. A correct and determination of the fundamental period characterizes a vital information during the seismic analysis of structures.

## 2 Description of the Structure

The office and residential reinforced concrete (RC) frame building, was built in Osijek, and consists of 3 dilatations. This article deals with the calculation of the structure which is separated by the first dilatation. The total area of the structure is 469.21 m$^2$. The structure consists of ground floor, first floor, II floor, III floor, IV floor and V floor (G + 5). The maximum height of the structure is 20.30 m, while the level of the roof is 18.56 m, measured from the ground clearance of ±0.00. The height of the ground floor is 4.30 m and the height of all other floors is 2.85 m. The roof structure is double wooded with a slight inclination of approx. 6° covered with metal sheets. In Fig. 1. the east façade of the building or the view from the street is shown, while in Fig. 2. the characteristic plan of the last floor is illustrated.

**Fig. 1.** East façade

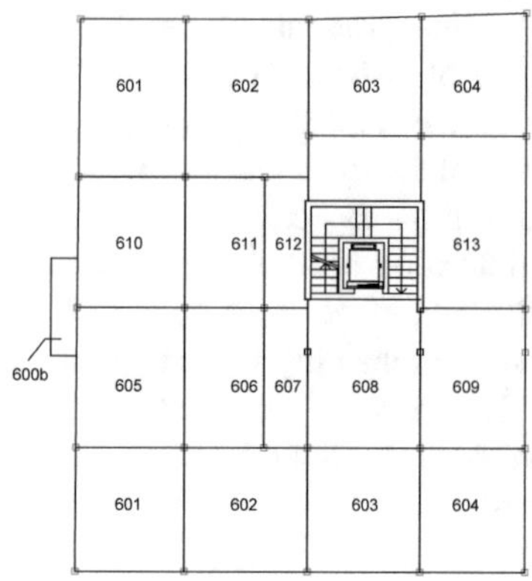


**Fig. 2.** Characteristic plan of the last floor

The loadbearing structure is a RC frame structure made of RC beams and RC columns. The infill in the RC frame is made of the brick block with thickness d = 25 cm. The inner partition walls are made of brick block 12 cm thick. Vertical communication is enabled by lifts located at almost the center of the structure (Fig. 2). The thickness of the reinforced concrete walls around the lift amounts to 20 cm.

The RC foundation slab has a height of d = 30 cm. The RC floor slabs are 16 cm thick. Beside the existence of the lift, the communication between the floors is done with the usage of RC stairs, with plate thickness d = 16 cm. The same thickness of the slab measured in the balconies. The structure is founded on a 60 cm slab plate with the effective foundation depth of $D_{f,approximately} = (-70)$cm measured from the ground level ± 0.00.

The selected concrete quality of all RC slabs and beams is C 25/30. The concrete grade C30/37 is selected for the columns in position 100, while the concrete grade C25/30 columns for positions 200, 300, 400, 500 and 600. The reinforcement is made of steel B 500.

## 3 Structure Model

The structure was modelled in SAP2000® v16.1.0 Ultimate software package. The 3D model consists of beam elements - beams and columns, and shell elements - walls and slabs. The model is loaded with constant (DEAD), imposed (LIVE) and earthquake (QUAKE) actions. Permanent and imposed loads are defined as continuous effects on beams and slabs. Earthquake action is defined by the response spectrum in two directions (x, y). The damping value is taken to be 5%. The selected soil category is ground type B, which corresponds to the location of the structure in the city of Osijek, with a maximum accelerated soil of 0.11 g for the 475-year return period [6]. The behavior factor for frame multi-storey building with more spans and medium ductility amounts to q = 3.9.

For the purposes of dimensioning, the following load combinations were used:
1,35•G + 1,50•Q
1,0•G + 0,3•Q + 1,0•EQ
where G: the dead load, Q: imposed load, EQ: earthquake action.

The model consists of 618 nodes, 669 beam and 271 shell elements (Figs. 3 and 4). The connection between the foundation slab and the columns is modeled as fixed. In the first step a linear static analysis was performed in order to check the reactions, stresses, strains, displacements due to self-weight of the structure and the overall behavior. By Modal Response Analysis the natural frequencies and mode shapes of the building are determined. On the basis of this analysis, the contribution of each mode can be obtained. This model gave the fundamental period of T = 0.58 s.

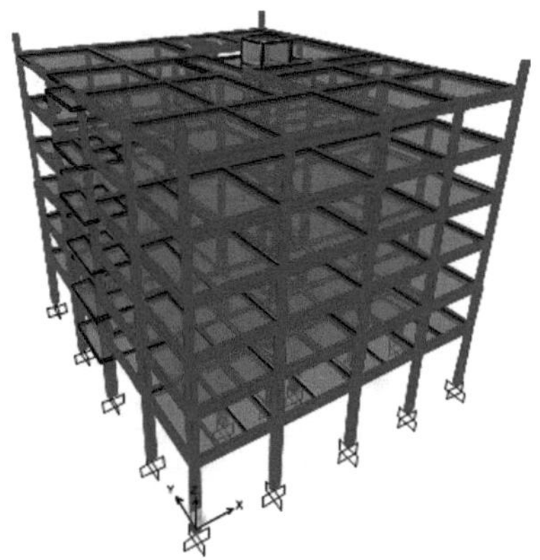

**Fig. 3.** Model of the structure in SAP2000

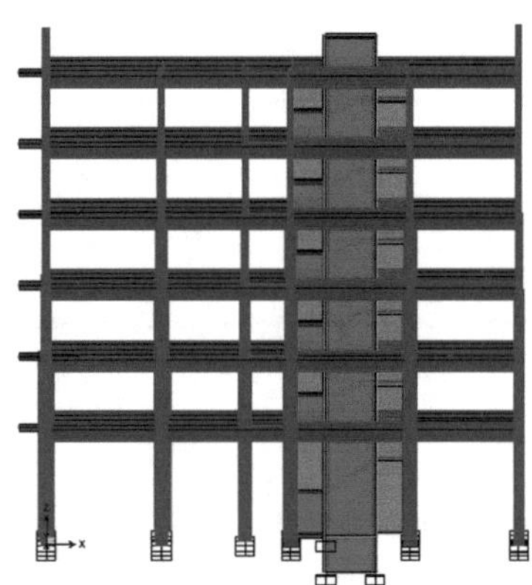

**Fig. 4.** Front view

# 4 Qualitative Comparison of the Calculations

Calculation criteria related to security and usage conditions are based on limit states; ultimate and serviceability limit states. The ultimate limit state in fact corresponds to the maximum capacity, and could be: loss of structural equilibrium or its element observed as a rigid body, ultimate fracture state or excessive deformation, loss of balance due to excessive deformation (theorem II), the ultimate fracture state caused by fatigue actions and transformation of the structure into the mechanism (calculation by plasticity theory) [2]. Determination of the static values, according to which dimensioning of the structure is done as well as designed capacity of the cross section is obtained, are determined depending on: the characteristic concrete strength, the yielding strength of steel, the dimension of the concrete section and the areas of the built-in reinforcement. They are calculated depending on the load-bearing mode of the observed element (bending, transverse force, longitudinal force, torsion, etc.) [7]. The differences in the mentioned approaches are shown in Table 1, where different application of the safety factor is evident. PBAB87 [4] used global safety factor for actions and materials, while EC2 [2] uses partial safety factors, one for actions and the other for materials.

**Table 1.** Differences in the calculations using PBAB87 and EC2

| | | PBAB87 | | EC2 | |
|---|---|---|---|---|---|
| Limit states | | $S_d \le R_d$ | | $S_d \le R_d$ | |
| Safety factor | | Global | | Partial | |
| | | Permanent | Imposed | Permanent | Imposed |
| | Favorable | 1.6 | 1.8 | 1.35 | 1.5 |
| | | 1.9 | 2.1 | | |
| | Unfavorable | 1.0 | 1.8 | 1.0 | 0.0 |
| | | 1.2 | 2.1 | | |
| | | | | Concrete | Steel |
| | | | | 1,50 | 1,15 |
| Combination of actions | | $S_u = 1{,}6 \times S_G + 1{,}8 \times S_Q, \varepsilon_s \ge 3‰$<br>$S_u = 1{,}9 \times S_G + 2{,}1 \times S_Q$, za $\varepsilon_s \le 0‰$ | | $\sum \gamma_{G,j} G_{k,j} \text{"+"} \gamma_{Q,1} Q_{k,1} \text{"+"} \sum_{i>1} \gamma_{Q,i} \psi_{0,i} Q_{k,i}$ | |
| | | $S_u = S_G + 1{,}8 \times S_Q, \varepsilon_s \ge 3‰$<br>$S_u = 1{,}2 \times S_G + 2{,}1 \times S_Q, \varepsilon_s \le 0‰$ | | $\sum \gamma_{GA,j} G_{k,j} \text{"+"} A_d \text{"+"} \psi_{1,1} Q_{k,1} \text{"+"} \sum_{i>1} \psi_{2,i} Q_{k,i}$ | |
| Concrete stress strain relation | | 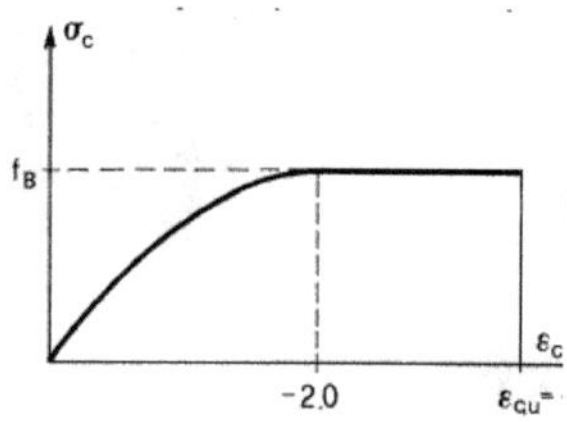 | | 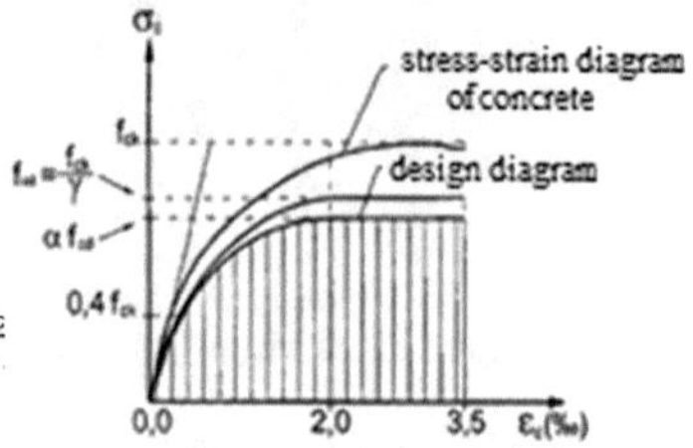 | |
| Steel stress strain relation | | 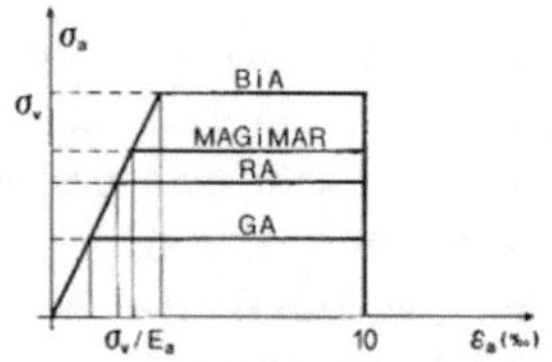 | | 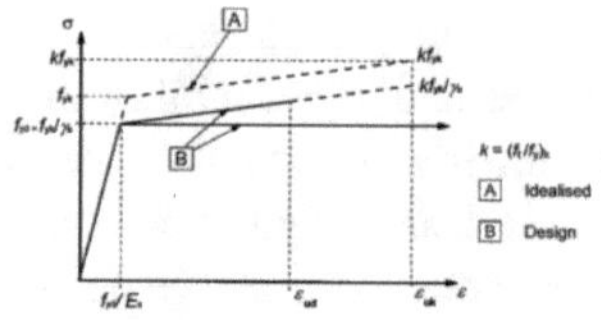 | |
| Load bearing capacity of a cross section | | $M_u = M_n$ | | $M_{Sd} = M_{Rd}$ | |
| Dimensionless bending moment | | $m_n = \dfrac{M_u}{b \cdot h^2 \cdot f_b}$ | | $\mu_{Sd} = \dfrac{M_{Sd}}{b \cdot d^2 \cdot f_{cd}}$ | |
| Coefficient of the cross section height | | $k_{hb} = \dfrac{h}{\sqrt{\dfrac{M_u}{b}}}$ | | $\zeta$<br>(taken from tables) | |
| Required reinforcement area | | $A_a = \dfrac{M_u}{k_z \cdot h \cdot \sigma_v}$ | | $A_{s1,req} = \dfrac{M_{Sd}}{\zeta \cdot d \cdot f_{yd}}$ | |

The fundamental period of the building has a major influence on its response. The ability to determine these characteristics is necessary for accurate calculation of the base shear. Additionally, the estimate of the fundamental period is important in order to detect the eventual possible resonance between the soil and the structure during the earthquake action [9].

The value of the building's fundamental period depends on many factors - structural irregularities, number of storeys, number of spans, dimensions of structural elements, filling elements, and load intensity [10, 11]. For the earthquake design of new RC structures, period of vibration is not known immediately, and because of that the simplified expressions are given in the construction rules, which usually link the base period with the height of the structure [12]. The majority of expressions for fundamental period given by various scientists and introduced in buildings codes are usually based on deterministic methods using experimental or analytical data. They do not take into account the presence of infill walls in the building (with or without opening), even though it is well known that the infill walls increase the stiffness and mass of the structure and, thus, lead to significant changes in the fundamental period [13–15].

Nevertheless, the so far available proposals in the literature for its estimation are often conflicting, making their use rather uncertain [16]. For this reason, it is not easy to accurately estimate the building fundamental period. Various investigators and researchers therefore give equations based on typological characteristics of the building, such as height, type of construction system and material, which can be used to estimate the fundamental period, as shown in Table 2.

**Table 2.** Expressions for the evaluation of fundamental period of vibration.

| Reference | **Expression** | **Equation No.** |
|---|---|---|
| EC 8 [7] | $T = 0.075 \cdot H^{0,75}$ | (1) |
| NBCC [17] | $T = 0,1 \cdot N$ | (2) |
| Hong and Hwang [9] | $T = 0.0294 \cdot H^{0.804}$ | (3) |
| Guler et al. [18] | $\mathrm{T}_A = 0.026 \cdot \mathrm{H}^{0.9}$ | (4) |
| Ditommaso et al. [19] | $\mathrm{T} = 0.026 \cdot \mathrm{H}$ | (5) |
| Navarro et al. [20] | $\mathrm{T} = (0.049 \pm 0.001) \cdot \mathrm{N}$ | (6) |
| Goel and Chopra lower bound [21] NEHRP-94 [22] | $\mathrm{T}_L = 0.0466 \cdot H^{0.9}$ | (7) |
| Goel and Chopra upper bound [21] | $\mathrm{T}_U = 0.067 \cdot H^{0.9}$ | (8) |
| HRN EN 1998-1:2011/NA [23] | $\mathrm{T} = 0.0016 \cdot \mathrm{H}$ | (9) |

## 5 Result and Discussion

In the following text an illustration of main reinforcement area in the slabs and beams is shown, as well as the value of the design internal forces in the columns. The differences in the values of the fundamental period obtained from the FEM model and the calculated periods according to the terms given in the norms and by different researchers are presented.

For only two slab positions, the adopted reinforcement according to PBAB87 is less than the reinforcement obtained for EC2. Minor differences have been observed in the reinforcement of the span, while a somewhat bigger difference is noted at the supports. The adopted amount of reinforcement is shown in Figs. 5 and 6.

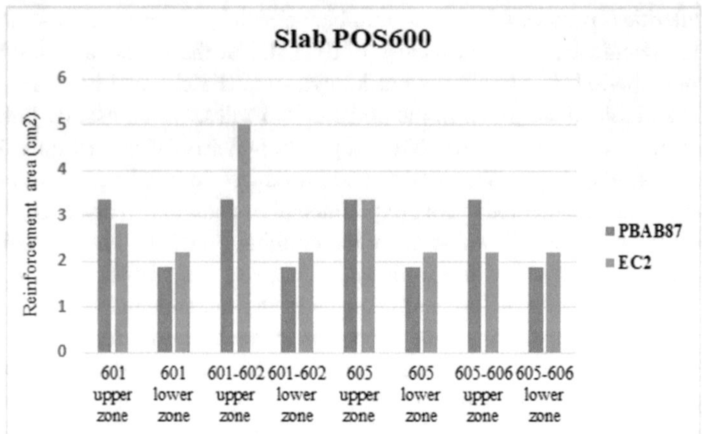


**Fig. 5.** Comparison of the area of longitudinal reinforcement in the slab of the last floor

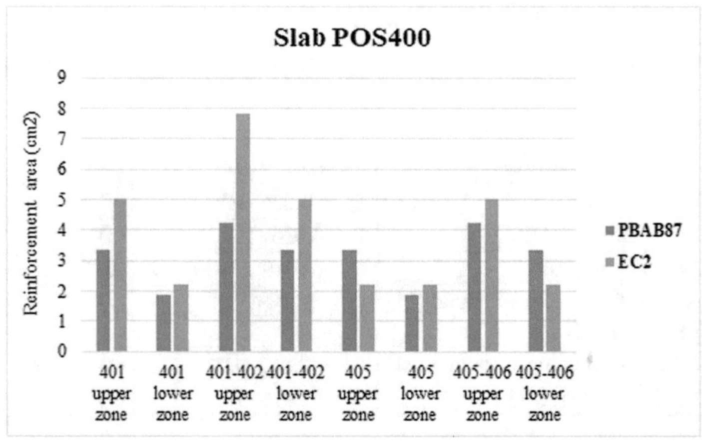


**Fig. 6.** Comparison of the area of longitudinal reinforcement in the characteristic slab

Graphical illustration of main beam reinforcement calculated according to PBAB87 and EC2 are shown in Figs. 7 and 8. The amount of reinforcement varies by up to 51.37% (for beam 6b located in G4).

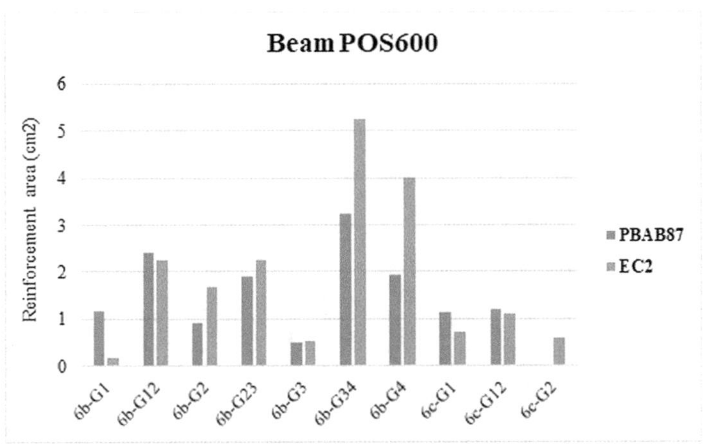


**Fig. 7.** Comparison of the area of longitudinal reinforcement in the beams of the last floor

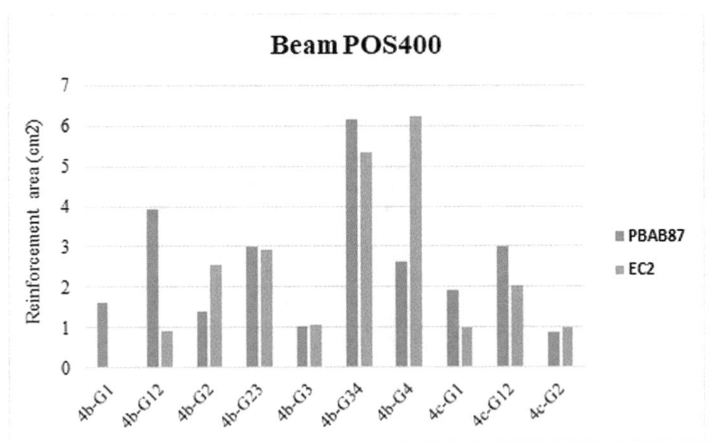


**Fig. 8.** Comparison of the area of longitudinal reinforcement in the beams of the characteristic floor

The axial forces in the columns obtained by the EC2 are generally larger than those obtained by PBAB87. The largest increase was observed on the column S43 POS600 in the amount of 42.21% (Figs. 9 and 10).

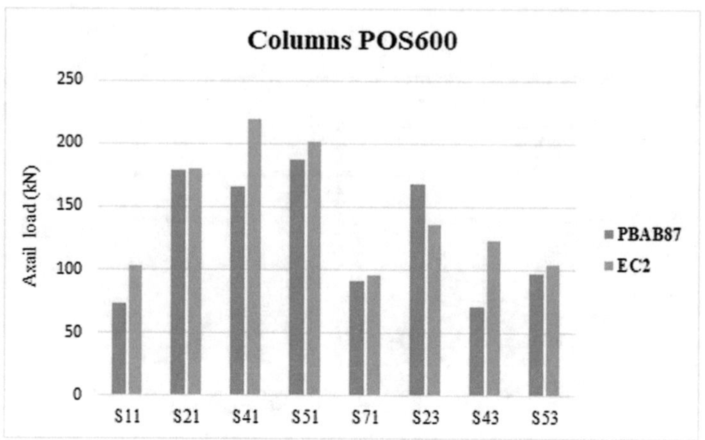


**Fig. 9.** Comparison of the axial forces in the columns in the last floor

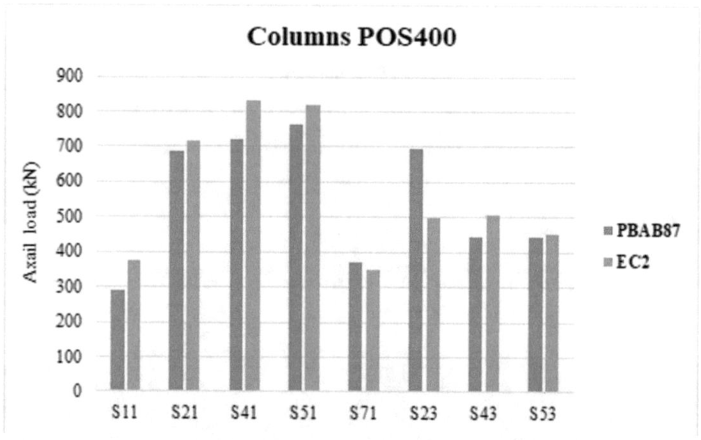


**Fig. 10.** Comparison of the axial forces in the columns in the characteristic floor

Comparison of the values of the fundamental period is shown in Fig. 11. Errors presented in a percentage value regarding the approximation of the fundamental period calculated according to the expression (Table 2) in relation to value obtained from the FEM model range from 3.4% to 60.1%. The difference between the fundamental period obtained from the SAP calculations and EC8 is 15.6%.

What do these differences mean, it is best presented in Design spectrum in Fig. 12. For the specific location and model of the structure the following parameters were used: ground type B, importance class II ($\gamma_I = 1$), Type 1 elastic response spectra (the expected surface-wave magnitude $M_s$ larger than 5.5) and viscous damping ratio (in percent) $\xi = 5\%$; the reference peak ground acceleration $a_{gR} = 0.1$ g and a behavior factor of $q = 3.9$ was taken into account for the DCH (Ductility Class High) structure.

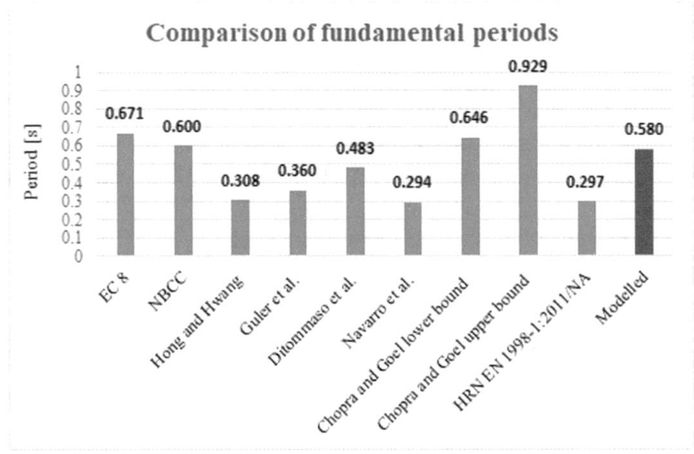


**Fig. 11.** Comparison of fundamental periods

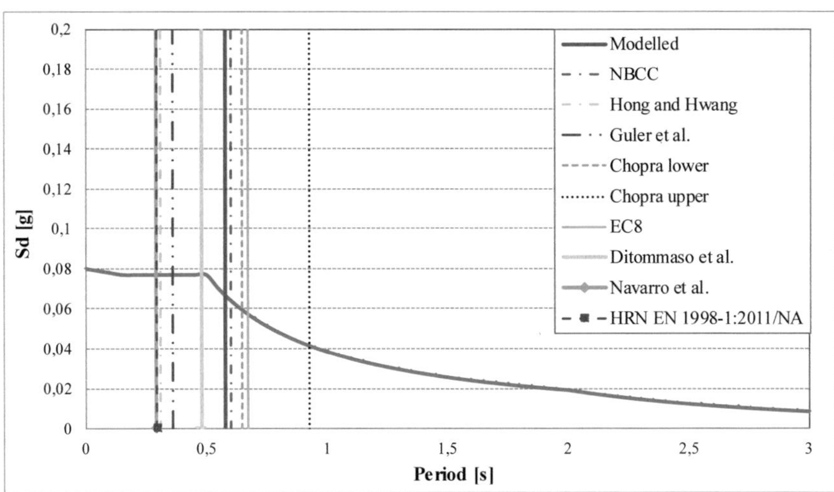


**Fig. 12.** Design spectrum with 5% viscous damping ratio for peak ground acceleration 0.1 g, ground type B and for behavior factor 3.9 with the value of fundamental periods obtained from different approaches

Figure 12 shows the corresponding design spectrum and all fundamental periods from empirical formulae (Table 2) and structural model reported on the graph. According to the selected method, the estimate of the period range is very wide from 0.294 to 0.929s, resulting in spectral values ranging between 0.077 g and 0.041 g. The period of vibration of the structural (numerical) model is 0.58s, resulting in spectral value of 0.066 g. Such a variation clearly indicates on the problem of accurate calculation of the base shear.

## 6 Conclusion

This paper presents qualitative and quantitative differences in the calculation of a multi-storey reinforced concrete frame building. The qualitative differences are shown by a comparative representation of the calculation of basic inputs for determining the load, strength and dimensioning of the elements exposed to bending according to PBAB87 and EC2.

Following the implementation of the dimensioning process of all structural elements: slabs, beams, columns and walls, according to the current (EC2 and EC8) regulations, a quantitative comparison of the obtained results with those from the main design of the structure, which was carried out using the PBAB87, and the seismic regulations valid at that time was conducted. The obtained values are shown in diagrams and it is concluded that there is no clear trend for the observed building, which would according to the regulations require larger amounts of necessary reinforcement depending on: static system, type and load distribution, in respect to slabs, beams and columns.

It should be noted that the comparisons were carried out using the values given in the main design of the structure, and for a further and more detailed comparison, the calculation according to PBAB87 should be done on a model of the structure in SAP2000 in order to confirm that the used assumptions were identical.

Another aim of this paper was to compare the fundamental period obtained from numerical analysis of a real RC frame building located in Croatian city Osijek with the periods obtained using the formulas in the European code EN1998-1:2004 (EC8) and other available literature. Numerical modal analysis was performed using SAP2000 software. Fundamental period of vibration calculated by currently available approximate equations show remarkable differences between "code-estimated". This opens a need for further evaluation and rechecking of the proposed equations and formulae in the codes.

## References

1. Tomičić, I.: Priručnik za proračun armiranobetonskih konstrukcija. Društvo hrvatskih građevinskih konstruktora, Zagreb (1993)
2. Pravilnih o tehničkim normativima za beton i armirani beton, Službeni list br. 11, Beograd (1987)
3. Eurocode 2. EN 1992-1-1:2004.: Design of Concrete Structures – Part 1-1: General Rules and Rules for Buildings. European Committee for Standardization, CEN, Bruxelles (2004)
4. Tomičić, I.: Betonske konstrukcije, 3. izmijenjeno i dopunjeno izdanje, Društvo hrvatskih građevinskih konstruktora, Zagreb (1996)
5. Tehnički propis za betonske konstrukcije (NN 101/05) (2005)
6. Tehnički propis za betonske konstrukcije (NN 139/2009) (2009)
7. Eurocode 8, EN 1998-1: 2004: Design of Structures for Earthquake Resistance – Part 1: General Rules, Seismic Actions and Rules for Buildings. European Committee for Standardization, CEN, Bruxelles (2004)
8. Kalman Šipoš, T., Hadzima-Nyarko, M.: Seismic risk of Croatian cities based on building's vulnerability. Tech. Gaz. **25**(4), 1088–1094 (2018)

9. Hong, L., Hwang, W.: Empirical formula for fundamental vibration periods of reinforced concrete buildings in Taiwan. Earthquake Eng. Struct. Dynam. **29**, 327–333 (2000)
10. Nikoo, M., Hadzima-Nyarko, M., Khademi, F., Mohasseb, S.: Estimation of fundamental period of reinforced concrete shear wall buildings using self organization feature map. Struct. Eng. Mech. **63**(2), 237–249 (2017)
11. Asteris, P.A., Repapis, C.C., Foskolos, F., Fotos, A., Tsaris, A.K.: Fundamental period of infilled RC frame structures with vertical irregularity. Struct. Eng. Mech. **61**(5), 663–674 (2017)
12. Draganić, H., Hadzima-Nyarko, M., Morić, D.: Comparison of RC frames periods with the empiric expressions given in Eurocode 8. Tech. Gaz. **17**(1), 93–100 (2010)
13. Asteris, P.G., Repapis, C.C., Tsaris, A.K., Di Trapani, F., Cavaleri, L.: Parameters affecting the fundamental period of infilled RC frame structures. Earthq. Struct. **9**(5), 999–1028 (2015)
14. Köse, M.M.: Parameters affecting the fundamental period of RC buildings with infill walls. Eng. Struct. **31**(1), 93–102 (2009)
15. Abo El-Wafa, W.M., Alsulami, B.T., Elamary, A.S.: Studying the effect of masonry infill walls on the natural period and lateral behavior of RC buildings in comparison with the SBC. Int. J. Sci. Eng. Res. **6**(6), 10–18 (2015)
16. Asteris, P.G., Nikoo, M.: Artificial bee colony-based neural network for the prediction of the fundamental period of infilled frame structures. Neural Comput. Appl. (2019). https://doi.org/10.1007/s00521-018-03965-1
17. NBCC, National Building Code of Canada: Canadian Commission on Building and Fire Codes. National Research Council of Canada (NRCC), Ottawa, Ontario (2005)
18. Guler, K., Yuksel, E., Kocak, A.: Estimation of the fundamental vibration period of existing RC buildings in Turkey utilizing ambient vibration records. J. Earthq. Eng. **12**, 140–150 (2008)
19. Ditommaso, R., Vona, M., Gallipoli, M.R., Mucciarelli, M.: Evaluation and considerations about fundamental periods of damaged reinforced concrete buildings. Nat. Hazards Earth Syst. Sci. **13**, 1903–1912 (2013)
20. Navarro, M., Sánchez, F.J., Feriche, M., Vidal, F., Enomoto, T., Iwatate, T., Matsuda, I., Maeda, T.: Statistical estimation for dynamic characteristics of existing buildings in Granada, Spain, using microtremors. Struct. Dyn., Eurodyn **1**, 807–812 (2002)
21. Goel, R.K., Chopra, A.K.: Period formulas for moment resisting frame buildings. J. Struct. Eng., ASCE **123**(11), 1454–1461 (1997)
22. NEHRP: Recommended provisions for the development of seismic regulations for new buildings. Building Seismic Safety Council, Washington, DC (1994)
23. HRN EN 1998-1:2011/NA. National Anex to EN 1998-1:2004: Design of Structures for Earthquake Resistance – Part 1: General Rules, Seismic Actions and Rules for Buildings. European Committee for Standardization, CEN, Bruxelles (2011)

# Behavior of Concrete Structures Under the Action of Elevated Temperatures

Samir Suljević, Senad Medić(✉), and Mustafa Hrasnica

Faculty of Civil Engineering, University of Sarajevo, Sarajevo, Bosnia and Herzegovina
samir.suljevic@gf.unsa.ba, senad_medic@yahoo.com, hrasnica@bih.net.ba

**Abstract.** Common building structures are designed to withstand the load that may occur during building service life. There are dead load and traffic loads of different characters, such as imposed load, wind load, snow load, earthquake load etc. However, special attention must be paid to ensure a satisfactory level of resistance in the case of fire action. The particular danger of fire damage exists in urban areas where low-rise and middle-high-rise residential buildings dominate, usually built of reinforced concrete. Because of all these factors, it is very important to analyse structural fire performance. Mechanical and thermal properties of reinforced concrete as well as other building materials, such as steel and wood, decrease at elevated temperatures. As an example, in this paper we present structural fire analysis of a concrete slab implementing analytical procedures given in EN 1992-1-2. Numerical analysis is performed using both the finite element method and the finite difference method taking into account nonlinear transient heat flow.

**Keywords:** Concrete structures · Fire · Heat analysis · Numerical modeling · Eurocode

## 1 Introduction

It is known that fire causes the most damages among all disasters. When we consider structural damages caused by fire, traditionally we have in mind steel or wooden structures. But, the fire could cause substantial damages to reinforced concrete structures, as well. Therefore, it is very important to analyse reinforced concrete structures exposed to fire.

Therefore, modern European regulations for analysis of various types of structures, Eurocode in section 1-2 of its specific codes (EN 1992, EN 1993, EN 1994, EN 1995, EN 1996) analyse effects of fire on structures. In the event of a fire, the structure is exposed to heat and mechanical action which produce thermal and mechanical response [1] (Fig. 1).

Since RC slabs are the most sensitive structural elements at elevated temperatures due to the fact that the main reinforcement in mid-span is located relatively close to the bottom surface of slab, its behavior has been studied by researchers for many years [8–13].


S. Avdaković et al. (Eds.): IAT 2019, LNNS 83, pp. 250–262, 2020.
https://doi.org/10.1007/978-3-030-24986-1_20

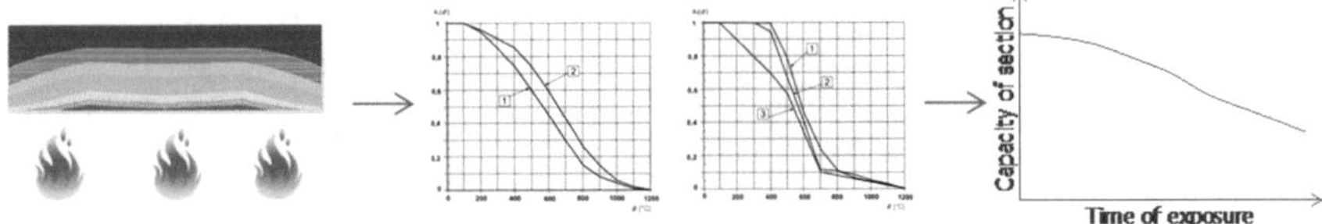


**Fig. 1.** Thermal actions cause reduction of mechanical properties of materials, and reduction in section capacity as well [1, 7, 10]

Various fire models, such as natural fire models and those based on standard ISO fire curve or parametric fire curve are used. Natural fire model usually represents fire more appropriately than models based on standard and parametric fire curve [10].

## 2 Thermo – Mechanical Analysis

The two-step coupling thermo-mechanical analysis is used for evaluating fire rating resistance. Thermal response of the structure is presented through the temperature field and based upon on these results, the mechanical calculation is carried out [11].

In this paper, two different numerical approaches are presented for obtaining temperature values in the slab: specially developed software based on the finite element method for spatial discretization and the finite difference method for time discretization, and another analysis which is performed using general purpose finite element program TNO DIANA [4].

The focus of this paper is on the one-way 220 mm-thick reinforced concrete slab, which is made of concrete with compressive strength equal to 25 N/mm$^2$, and reinforcement with yield stress equal to 500 N/mm$^2$. Welded ribbed meshes are used as a reinforcement. The concrete slab is assumed to be made of siliceous aggregate and to be restrained on its ends, providing appropriate top reinforcement near supports (Fig. 2).

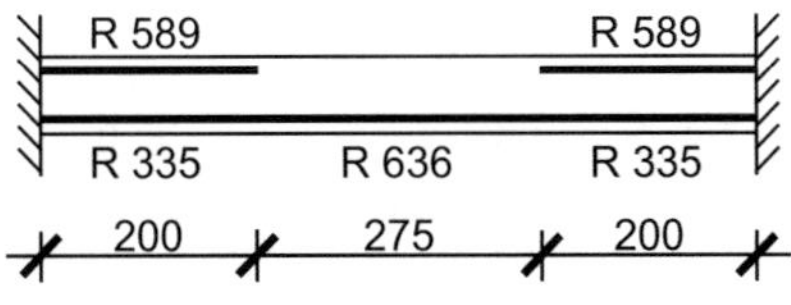


**Fig. 2.** Slab model

Thermal characteristics of concrete are taken from EN 1992-1-2 [1]. Temperature-dependent thermal conductivity, density, and specific heat are shown in Figs. 3, 4 and 5, respectively.

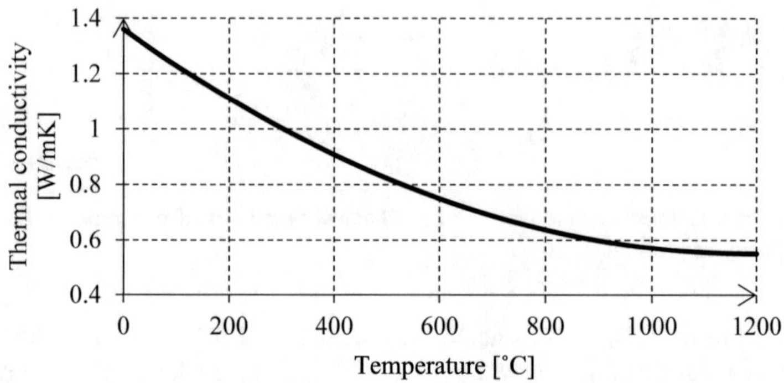


**Fig. 3.** Thermal conductivity with respect to temperature [1, 14]

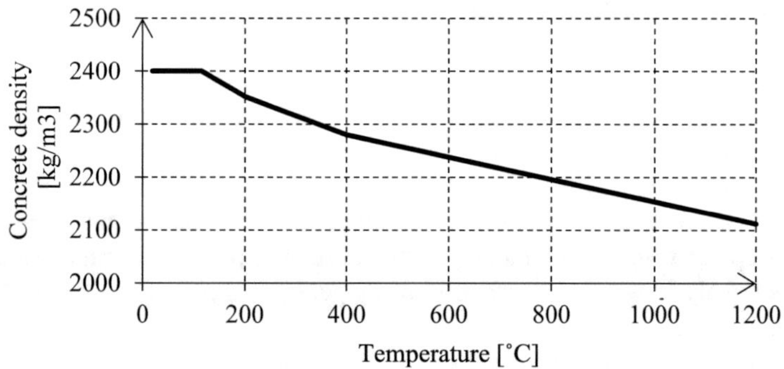


**Fig. 4.** Concrete density with respect to temperature [1, 14]

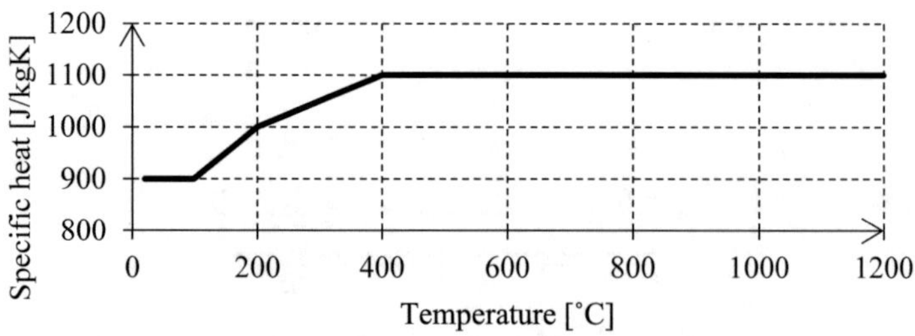


**Fig. 5.** Specific heat of concrete with respect to temperature [1, 14]

Standard ISO 834 temperature time-curve, which is defined by the expression:

$$\theta_g = 345 \cdot log_{10}(8t + 1) + 20 \tag{1}$$

is used as a fire model. $\theta_g$ is the gas temperature in [°C] and $t$ is the time of exposure in minutes. It is assumed one-side fire exposure (Fig. 6).

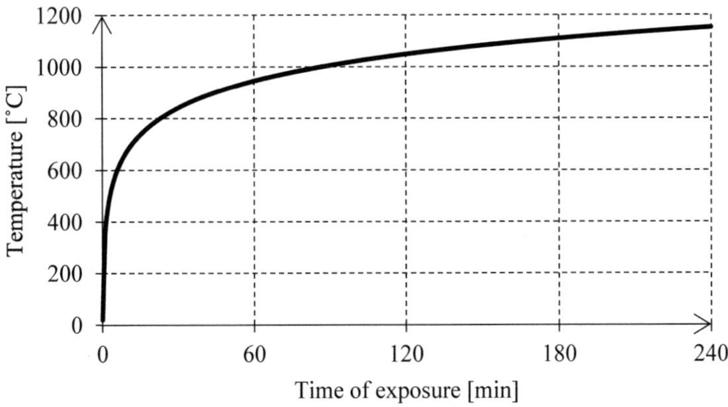


**Fig. 6.** Standard ISO 834 temperature-time curve [1, 12, 14]

Transient heat transfer is defined by a well-known partial differential equation:

$$\frac{\partial\theta}{\partial t} - \alpha \cdot \frac{\partial^2\theta}{\partial x^2} = 0 \ , \ \ \alpha = \frac{\lambda}{c \cdot \rho}, \tag{2}$$

where $\lambda$ is thermal conductivity, $c$ is specific heat, $\rho$ is density and $\alpha$ is thermal diffusivity. In case of fire, all 3 heat transfer mechanisms are presented: conduction, convection, and radiation. Robin's boundary condition involves thermal load in the form of convection and radiation, while heat transfer through the slab is described by conduction [1, 2, 6].

Heat transfer through slab is described by conduction, while thermal load is involved in a form of convection and radiation in Robin's boundary condition [1, 2].

Time discretization using the finite difference method is implemented, while spatial discretization is performed using FEM. The slab is divided into 22 equal 1D 2-noded finite elements assuming linear distribution of temperatures in the element (Fig. 7).

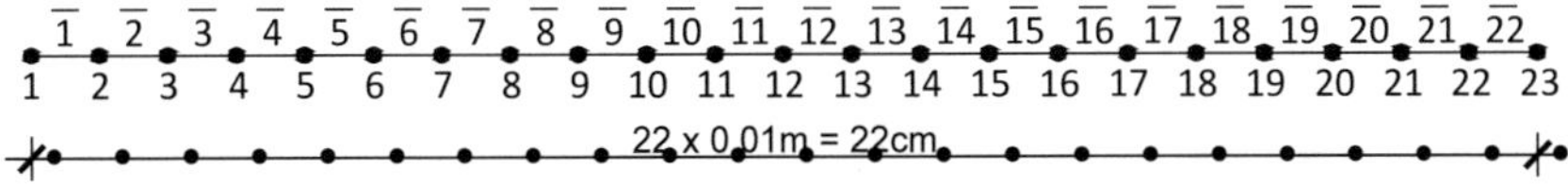


**Fig. 7.** Spatial discretization using 1D 2-noded finite elements [5]

The problem of transient heat flow in matrix form can be described by equation [3, 11]:

$$[M] \cdot \left\{\dot{\theta}\right\} + [K] \cdot \{\theta\} = \{F\}, \tag{3}$$

where $[M]$ is the global mass matrix, $[K]$ is the global thermal conductivity matrix, $\{\theta\}$ is the vector of nodal temperatures, the vector $\{\dot{\theta}\}$ contains time derivatives of nodal temperatures, and $\{F\}$ is the vector of net heat flow.

Numerical time integration is performed using the so-called $\alpha$- method. Using $\alpha = 0$, which corresponds to an explicit time integration scheme and with substitution:

$$\{\dot{\theta}\} \approx \frac{\Delta\{\theta\}}{\Delta t} = \frac{\{\theta\}^{j+1} - \{\theta\}^{j}}{\Delta t} \tag{4}$$

which is related to small time step, Eq. (4) takes a more convenient form [3]:

$$\frac{1}{\Delta t} \cdot [M] \cdot \{\theta\}^{j+1} = \left(\frac{1}{\Delta t} \cdot [M] - [K]\right) \cdot \{\theta\}^{j} + \{F\}^{j} \tag{5}$$

Since the temperature distribution in *j*-th time step is known, explicit relation for temperature distribution at *(j + 1)*-th time step is [2, 5, 6, 11]:

$$\{\theta\}^{j+1} = \{\theta\}^{j} + \Delta t \cdot [M]^{-1} \cdot \left(\{F\}^{j} - [K] \cdot \{\theta\}^{j}\right) \tag{6}$$

In this calculation approach, the time step $\Delta t$ is used as constant with the value of $2s$. This is smaller than a critical time step $\Delta t_{cr}$, so stability condition is satisfied [2, 5, 6].

$$\Delta t < \Delta t_{cr} = \frac{\rho \cdot c \cdot \Delta x^2}{\lambda + h/\Delta x}, \tag{7}$$

where $h$ is a convection coefficient equal to 25 $W/m^2K$ [1, 2, 15].

Thermal analysis results are shown in Fig. 8 and Table 1.

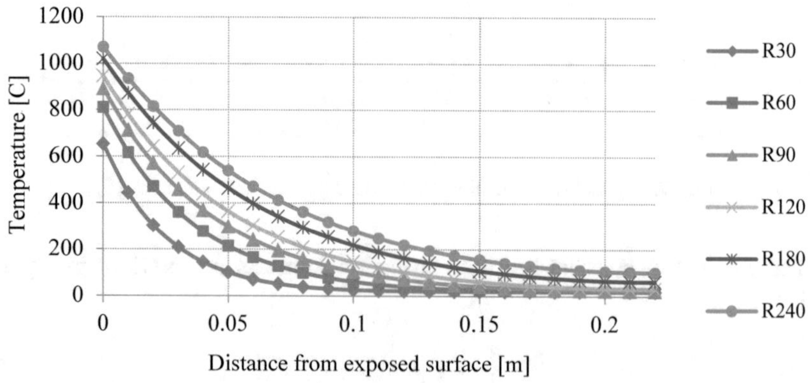


**Fig. 8.** Graphical interpretation of transient heat transfer analysis [5]

**Table 1.** Temperature of the exposed slab surface [5]

| Time step [min] | Temperature [°C] | Time step [min] | Temperature [°C] |
|---|---|---|---|
| 30 | 653 | 120 | 946 |
| 60 | 810 | 180 | 1020 |
| 90 | 891 | 240 | 1071 |

These results are well-in agreement with previous experimental and numerical research results [9, 10, 13] and with those given in building codes [1, 7].

## 3 Mechanical Analysis

It is aforementioned that mechanical analysis results are dependent on the results of thermal analysis, or precisely on the temperature distribution across the slab. The main objective of the mechanical analysis as well as the entire structural fire analysis is to determine fire resistance time. It is defined as a time in which structural element exposed to a standard furnace test elapsed before a fire limit is violated [12].

Obviously, the resistance of structural elements is related to, among others, mechanical properties of materials.

The stress-strain relationship for concrete and reinforcement is shown in Figs. 9 and 10, respectively.

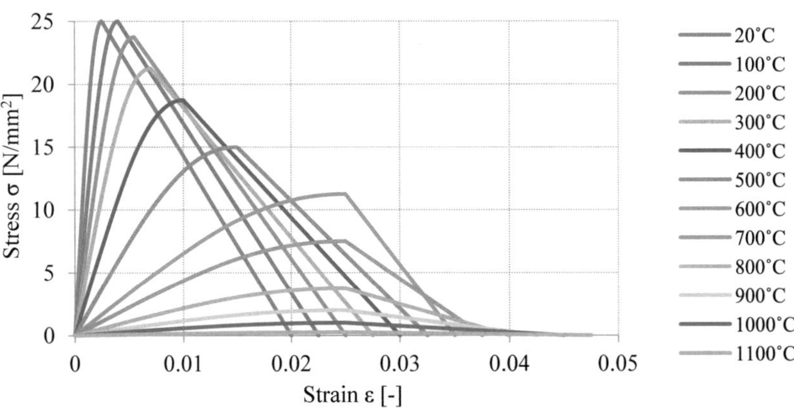


**Fig. 9.** Stress-strain relationship for concrete class C25/30 at elevated temperatures [1, 5, 14]

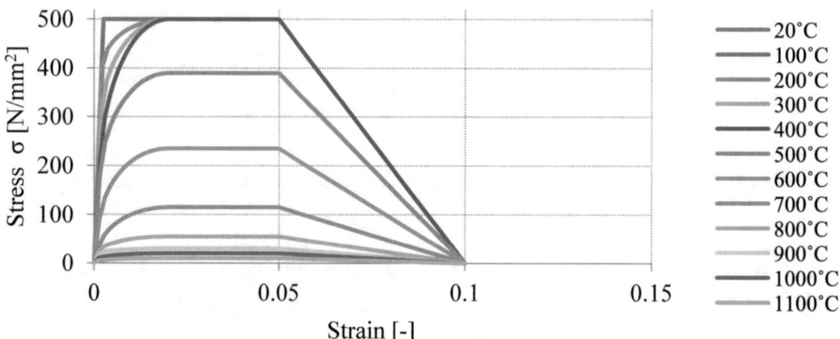


**Fig. 10.** Stress-strain relationship for reinforcement B500S at elevated temperatures [1, 5, 14]

According to Eurocode 2, there are following fire resistance rating: R30, R60, R90, R120, R180, R240. It means that structural element complies with criteria R (resistance) during 30, 60, 90, 120, 180, 240 min for exposure to a standard ISO 834 fire [12, 15].

Also, Eurocode 2 gives provision of treating one-way continuous slab as a simply supported slab if redistribution of bending moments at normal temperature design exceeds 15% [1]. Redistribution of moments is preferred because it exploits ductility and allows forming plastic hinges near supports.

For these reasons, mechanical analysis is carried out on a simply supported slab model. The analysis is done according to Eurocode 2 using 500 °C isotherm method [1] (Fig. 11).

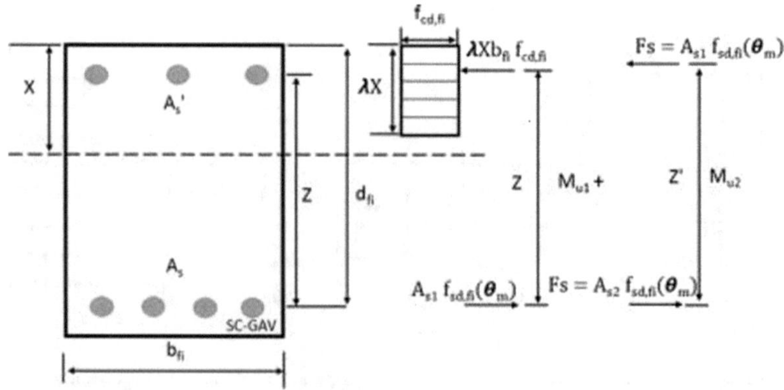


**Fig. 11.** Stress distribution at ultimate limit state for a rectangular cross section with compression reinforcement [1, 16]

It is known that most important factors which influence structural fire resistance are compressive strength of concrete, yield strength of steel, type of aggregate, concrete cover, heating regime, fire exposure conditions, etc. [7].

In this approach, different values of concrete cover are used to show its influence on the ultimate slab capacity. This parametric study includes concrete cover values of 15 mm, 25 mm, and 35 mm.

Design load, which include dead load and 30% of live load, in a fire situation is equal to 10 kN/m$^2$. It means that the design bending moment is equal to 56.95 kNm.

The fire rating is determined as the time when design flexural capacity of slab becomes less than the design bending moment in fire situation [13]. Calculation of structural fire resistance for different concrete cover is shown in the next tables (Tables 2, 3, 4).

**Table 2.** Calculation of structural fire resistance for concrete cover equal to 15 mm [5]

| Time t[h] | $T_{fire}$ [°C] | $T_{steel}$ [°C] | $F_{steel}$ [kN] | x [m] | $\lambda \cdot x$ [m] | $d-0,5 \cdot \lambda \cdot x$ [m] | $M_{Rd,fi}$ [kNm] |
|---|---|---|---|---|---|---|---|
| 0.5 | 841.8 | 303.2 | 318.0 | 0.016 | 0.013 | 0.200 | 63.58 |
| 1.0 | 945.3 | 470.4 | 254.4 | 0.013 | 0.010 | 0.200 | 50.87 |
| 1.5 | 1006.0 | 571.2 | 178.1 | 0.009 | 0.007 | 0.200 | 35.54 |
| 2.0 | 1049.0 | 642.9 | 117.7 | 0.006 | 0.005 | 0.200 | 23.53 |
| 3.0 | 1109.7 | 743.7 | 57.2 | 0.003 | 0.002 | 0.200 | 11.45 |
| 4.0 | 1152.8 | 815.0 | 35.0 | 0.002 | 0.001 | 0.200 | 7.00 |

**Table 3.** Calculation of structural fire resistance for concrete cover equal to 25 mm [5]

| Time t[h] | $T_{fire}$ [°C] | $T_{steel}$ [°C] | $F_{steel}$ [kN] | x [m] | $\lambda \cdot x$ [m] | $d-0,5 \cdot \lambda \cdot x$ [m] | $M_{Rd,fi}$ [kNm] |
|---|---|---|---|---|---|---|---|
| 0.5 | 841.8 | 209.3 | 318.0 | 0.016 | 0.013 | 0.190 | 60.40 |
| 1.0 | 945.3 | 360.7 | 318.0 | 0.016 | 0.013 | 0.190 | 60.40 |
| 1.5 | 1006.0 | 458.7 | 260.8 | 0.013 | 0.010 | 0.190 | 49.46 |
| 2.0 | 1049.0 | 530.7 | 216.2 | 0.011 | 0.009 | 0.190 | 41.08 |
| 3.0 | 1109.7 | 634.4 | 124.0 | 0.006 | 0.005 | 0.190 | 23.56 |
| 4.0 | 1152.8 | 709.3 | 70.0 | 0.003 | 0.003 | 0.190 | 13.29 |

**Table 4.** Calculation of structural fire resistance for concrete cover equal to 35 mm [5]

| Time t[h] | $T_{fire}$ [°C] | $T_{steel}$ [°C] | $F_{steel}$ [kN] | x [m] | $\lambda \cdot x$ [m] | $d-0,5 \cdot \lambda \cdot x$ [m] | $M_{Rd,fi}$ [kNm] |
|---|---|---|---|---|---|---|---|
| 0.5 | 841.8 | 144.9 | 318.0 | 0.016 | 0.013 | 0.180 | 57.22 |
| 1.0 | 945.3 | 277.6 | 318.0 | 0.016 | 0.013 | 0.180 | 57.22 |
| 1.5 | 1006.0 | 369.5 | 318.0 | 0.016 | 0.013 | 0.180 | 57.15 |
| 2.0 | 1049.0 | 439.2 | 267.1 | 0.013 | 0.011 | 0.180 | 48.07 |
| 3.0 | 1109.7 | 542.0 | 206.7 | 0.010 | 0.008 | 0.180 | 37.20 |
| 4.0 | 1152.8 | 618.0 | 136.7 | 0.007 | 0.005 | 0.180 | 24.61 |

Dependence of design flexural capacity with a time of exposure is shown in Fig. 12. Obviously, curves which describe flexural capacity are significantly different between each other. Therefore, concrete cover plays an important role in structural fire analysis of slab.

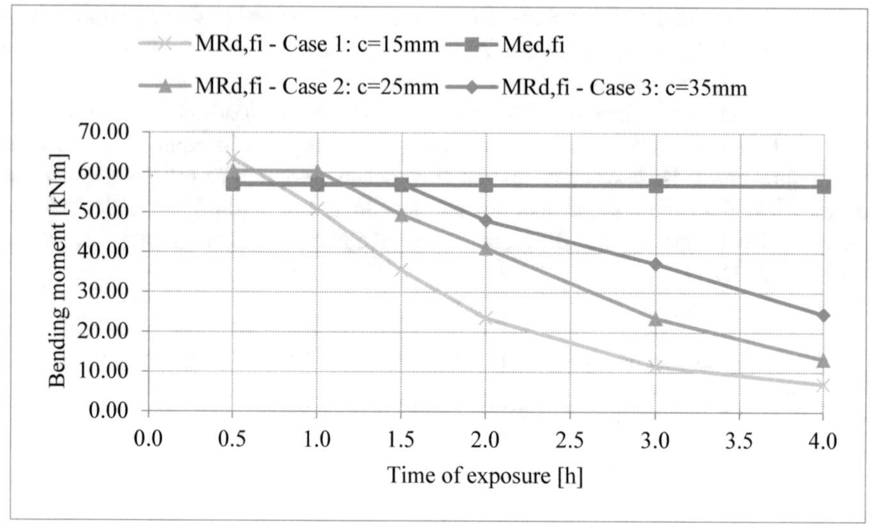


**Fig. 12.** Graphical interpretation of design flexural capacity of the slab in fire situation [5]

Hence it may be concluded that without any fire protection a slab of thickness 220 mm has fire rating resistance equal to R30, R60, R90 per Eurocode 2 provisions for the concrete cover of 15 mm, 25 mm, and 35 mm, respectively.

The same results would be obtained by applying a tabulated procedure, whereby concrete cover is the most important factor in determining fire rating.

Besides, these results are well-in agreement with previous research results [13, 15].

In addition to this calculation approach, another one-way coupling thermo-mechanical analysis of one-way slab is done using TNO DIANA software package [4]. The three-dimensional analysis is carried out on the slab model shown in Fig. 2. Temperature-dependent thermal and mechanical characteristics of concrete and reinforcement are used (Table 5).

**Table 5.** Details of analysed slab

| | |
|---|---|
| Width ($b$) | 1000 mm |
| Depth ($H$) | 220 mm |
| Load intensity ($q_{Ed,fi}$) | 10 kN/m$^2$ |
| Yield stress of reinforcement ($f_y$) | 500 N/mm$^2$ |
| Concrete compressive strength ($f_{ck}$) | 25 N/mm$^2$ |
| Reinforcement area - $A_{st}$ | 636 N/mm$^2$ |
| Concrete cover ($c$) | 15 mm |
| Fire exposure | One-sided |
| Fire model | Standard ISO 834 fire curve |

It is worth to mention that perfect adhesion between reinforcement and concrete is assumed. It means there is no slip between concrete and reinforcement. Many researchers investigate the behavior of bond between [7, 17, 18], and reliable anchorage of reinforcement within the support zone is specified as the main recommendation.

Reinforcement is modeled using 1D-truss elements, while concrete is modeled using 3D solid CHX60 elements [4, 5] (Fig. 13).

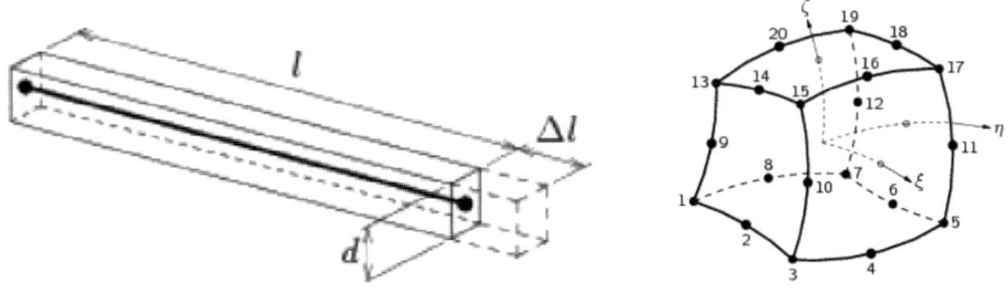


**Fig. 13.** Finite elements for modeling reinforcement and concrete [4, 5]

Mesh is divided by 11 layers of equal height, and mesh size in the layout is 250 × 250 mm. The analysis is performed using variable time steps values. Adopted time steps values are 60s for interval 0–300s, and 180s for other intervals [5]. Results of thermal analysis in the form of temperature distribution across the slab are shown below (Table 6).

**Table 6.** Temperature field of the concrete slab [4, 5]

| t=30min | t=60min |
|---|---|
| 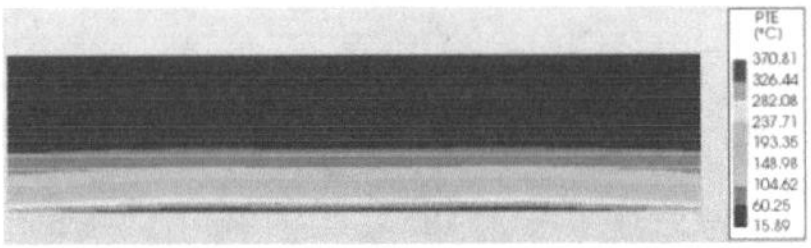 | 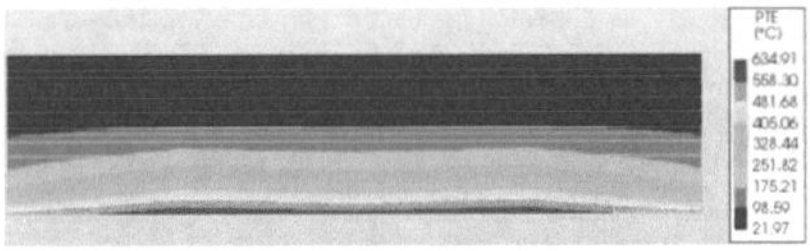 |
| t=90min | t=120min |
| 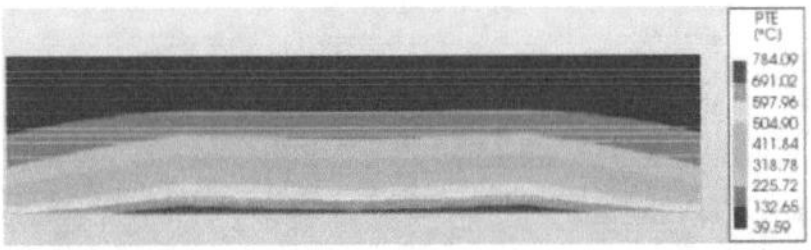 | 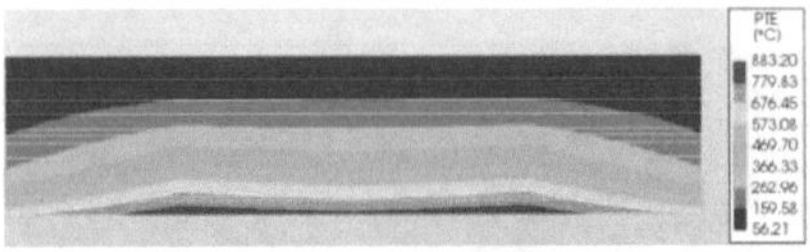 |
| t=180min | t=240min |
| 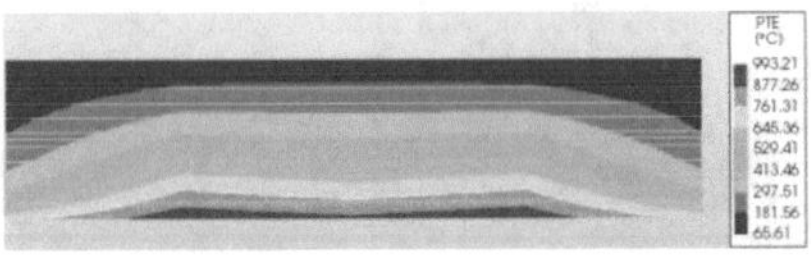 | 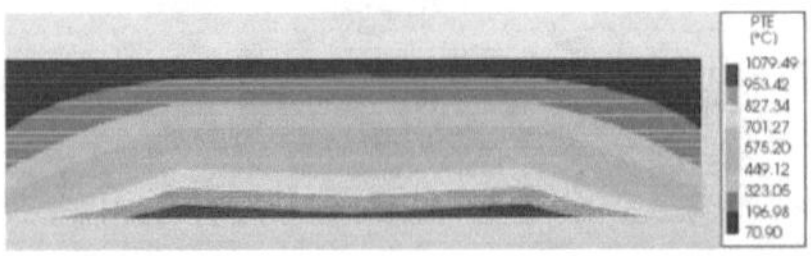 |

It can be observed that temperature varies predominantly along with the height of the slab, while temperature variety in the transverse direction is almost negligible. Since concrete is a material with noticeable thermal inertia, it takes a certain time for the heat penetrates into the interior of the slab. These temperature gradients are the most expressed at the 90 and 120 min of exposure to standard ISO 834 fire curve, in the areas close to the exposed surface.

Results of mechanical analysis in the form of bending moments, reinforcement stresses, and deflections are shown below (Figs 14, 15 and 16).

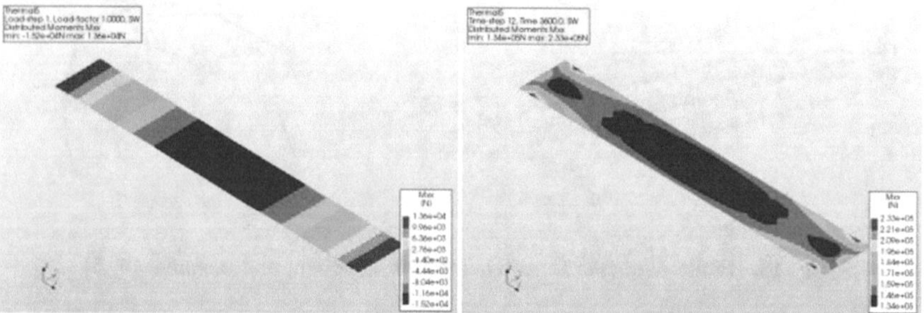

**Fig. 14.** Bending moments $M_{xx}$ at time step t = 0, t = 60 min [4, 5]

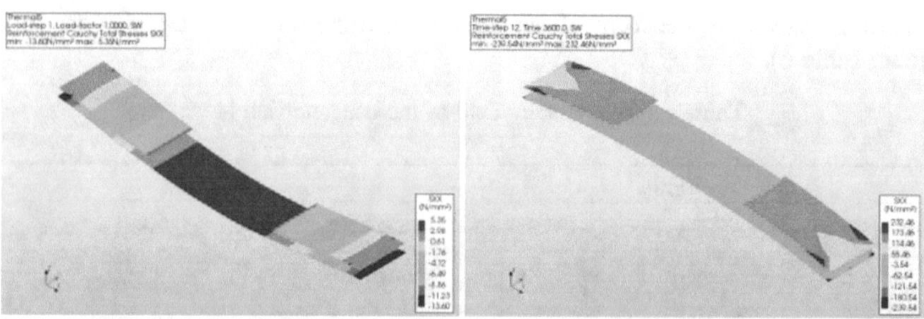

**Fig. 15.** Reinforcement stresses $\sigma_{xx}$ at time step t = 0, t = 60 min [4, 5]

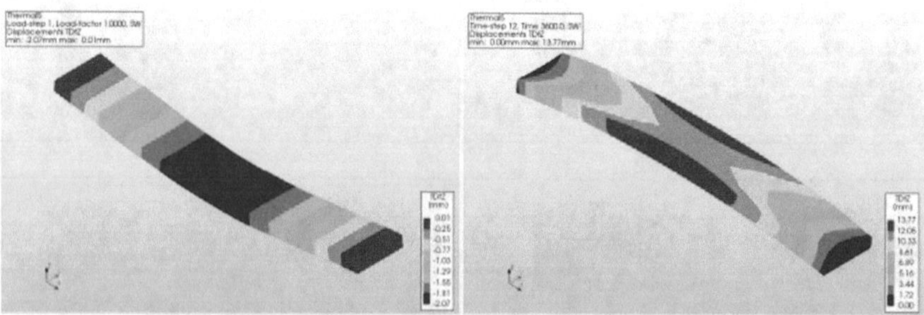

**Fig. 16.** Deflections at time step t = 0, t = 60 min [4, 5]

Thermal loads cause different stress state in comparison with standard gravity and variable loads. In most cases, stresses in constituent materials are several times higher than those at ambient temperatures under standard loads. It is noticeable from results that bending moments at ambient temperatures are uniformly distributed in the transverse direction, while bending moments in fire situation exist in both directions. The main consequence of this is both transverse and longitudinal curvature, as can be seen from the deflection diagram. Similarly, reinforcement stresses in a fire situation are substantially different from those in the calculation at ambient temperatures. Also, the stress distribution is non-uniform in both directions as a consequence of a complex stress state in the slab.

## 4 Concluding Remarks

In this paper, an analytical and numerical approach in structural fire analysis of concrete structures is presented. The emphasis is placed on the behavior of concrete slab in the fire, since the slab with one surface exposed to fire is the most common case when building sustains a fire accident.

Specific for the thermal load is its influence on both internal forces in structural member and properties of materials, which is encompassed through mechanical and thermal response. One-way coupling thermo-mechanical analysis is done, whereby one-dimensional temperature field across the slab with considerable temperature gradients substantially affects mechanical resistance of slab. It is validated in both analytical and numerical approach this work.

Importance of proper choice of the concrete cover of reinforcement is shown in analysis example. By decreasing the concrete cover, values of flexural slab capacity become slightly bigger at ambient temperatures, but slab performance in a fire situation is deteriorated. This leads to the opening of the cracks in the fire and increasing the risk of concrete spalling. As might be expected, the excessively thick concrete cover is not favorable from practical reasons and it can lead to potential increasing of crack width in serviceability state, which may affect the durability of the structure.

Certainly, it is worth to mention that reducing amplitude of the flexural capacity is almost the same to that of the yield strength of reinforcement, as a consequence of almost constant arm of internal forces in cross section. It is very important to know since the structural fire analysis is rarely performed in everyday engineering practice.

In this work, the behavior of bond between reinforcement and concrete, as well as end restraint conditions and potential spalling of concrete cover is not taken into account. In recent years, significant progress in determining of influence of all these factors on structural fire resistance is achieved. This opens up new opportunities and motives for further research in the field of structural fire engineering and fire protection engineering.

## References


1. EN 1992-1-2, Eurocode 2: Design of Concrete Structures. Part 1.2: General Rules – Structural Fire Design. Commission of the European Communities, Brussels (2004)
2. Wickström, U.: Temperature Calculation in Fire Safety Engineering. Springer, Switzerland (2016)
3. Logan, D.L.: A First Course in the Finite Element Method, 4th edn. CL-Engineering, USA (2011)
4. Diana, T.N.O.: User's Manual. DIANA FEA BV, Delft, The Netherlands (2016)
5. Suljevic, S.: Design of reinforced concrete structural elements exposed to fire. MSc thesis, University of Sarajevo (2018) (In Bosnian)
6. Hurley, M., et al.: SFPE Handbook of Fire Protection Engineering, 5th edn. Springer, USA (2016)
7. Guo, Z., Shi, X.: Experiment and Calculation of Reinforced Concrete at Elevated Temperatures. Elsevier, USA (2011)
8. Houang, Z.: The behaviour of concrete slabs in fire. Fire Saf. J. **45**, 271–282 (2010)
9. Sangluaia, C., Haridharan, M.K., Natarajan, C., Rajamaran, A.: Behaviour of reinforced concrete slab subjected to fire. Int. J. Comput. Eng. Res. **3**(1), 195–206 (2013)
10. Allam, S.M., Elbakry, H.M.F., Rabeai, A.G.: Behavior of one-way reinforced concrete slabs subjected to fire. Alex. Eng. J. **52**, 749–761 (2013). http://dx.doi.org/10.1016/j.aej.2013.09.004
11. Terro, M.J.: Numerical modeling of the behavior of concrete structures in fire. ACI Struct. J. **95**(2), 183–193 (1998)
12. Khoury, G.A.: Effect of fire on concrete and concrete structures. Prog. Struct. Mat. Eng. **2**, 429–447 (2000)
13. Balaji, A., Nagarajan, P., Madhavan Pillai, T.M.: Predicting the response of reinforced concrete slab exposed to fire and validation with IS456 (2000) and Eurocode 2 (2004) provisions. Alex. Eng. J. (2016). http://dx.doi.org/10.1016/j.aej.2016.06.005
14. fib Bulletin 46: Fire Design of Concrete Structures – Structural Behaviour and Assessment. International Federation for Structural Concrete, Lausanne, Switzerland (2008)
15. Purkiss, J.A., Li, L.: Fire Safety Engineering Design of Structures, 3rd edn. CRC Press, USA (2014)
16. Chandrasekaran, S., Srivastava, G.: Design Aids of Offshore Structures Under Special Environmental Loads Including Fire Resistance. Springer, Singapore (2018)
17. Khalaf, J., Huang, Z., Fan, M.: Analysis of bond-slip between concrete and steel bar in fire. Comput. Struct. **162**, 1–15 (2016)
18. Pothisiri, T., Panedpojaman, P.: Modeling of mechanical bond-slip for steel-reinforced concrete under thermal loads. Eng. Struct. **48**, 497–507 (2013)

# Use of GIS Viewer for Flood Risk Management on the Main Road Network in the Federation of Bosnia and Herzegovina

Slobodanka Ključanin[1(✉)], Suada Džebo[2(✉)], Meliha Drugovac[2(✉)], and Iris Bilalagić[3(✉)]

[1] University of Bihać, Bihać, Bosnia and Herzegovina
slobodanka63@yahoo.com
[2] University of Sarajevo, Sarajevo, Bosnia and Herzegovina
suada.dzebo.gf@gmail.com, meliha.musanovic@gmail.com
[3] Municipality Centar, Sarajevo, Bosnia and Herzegovina
irisbilalagic@gmail.com

**Abstract.** Planning for Sustainable Development and Risk Management from Natural Disasters requires a wide range of quality and reliable information. Based on such data it is possible to make timely and valid decisions. One of the most important forms of information is a variety of cartographic maps. Making such maps until recently was quite complex. Today, thanks to modern technology, it is relatively easy and quick to create the necessary thematic maps. This paper presents a method for creating/using a viewer and an example of creating a thematic map for managing flood risks in the Federation of Bosnia and Herzegovina.

**Keywords:** SDI · Natural disasters management · Floods · Road infrastructure · Spatial data · The quality of spatial data

## 1 Introduction

Floods represent an important threat to the roads and could lead to massive obstruction of traffic and road damage structures with possible long-term effects [1]. When the road network is interrupted by natural disasters, effects can be critical for emergency management. Transportation lifelines are generally considered the most important in an emergency because of their vital role in the restoration of all other lifelines [2].

In the past 20 years, the institutions for the maintenance of the road networks have increasingly been recognizing the impact of climate change on their networks and the need to better understand that risk in order to identify priority areas for improvement, resistance, and adjustment to climate change [3]. A better understanding of risks enables directing resources where they are the most useful and supports proactive rather than reactive approach to fighting the threats of climate change. Identification and management of risk are recognized as an integral part of asset management, it is included in ISO 55001:2014 [4]. The main steps for the assessment and management of risk are shown in the previous picture (Fig. 1).


S. Avdaković et al. (Eds.): IAT 2019, LNNS 83, pp. 263–275, 2020.
https://doi.org/10.1007/978-3-030-24986-1_21

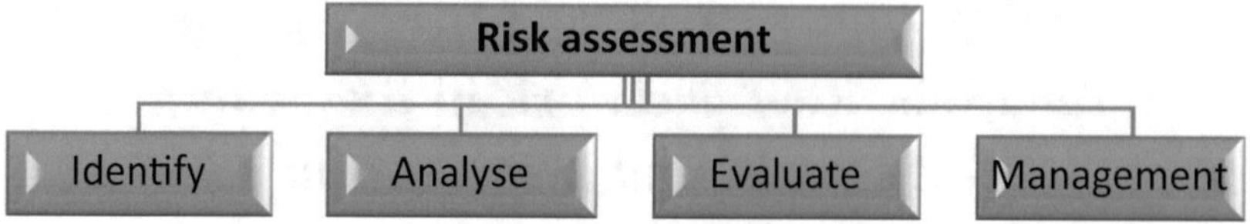


**Fig. 1.** Risk assessment steps

- Identification of hazards involves identifying and defining hazards and risk assessment can be focused on one of the hazards.
- Risk analysis involves collecting information on the risks and their analysis whereby analysis can be qualitative and quantitative.
- Risk evaluation presents a comparison of analyzed results with respect to levels of acceptable risk that is often used to inform risk management.

Risk assessment and mapping carried out in the broader context of disasters risk management. Risk assessment and mapping are central components of the overall process for identifying capacity and resources to reduce the level of identified risks, i.e. possible disaster effects (capacity analysis). It refers to the planning of measures to mitigate the corresponding risks, monitoring, and review of the hazards, risks, vulnerability to risks, as well as consultation and communication on the findings and results [5].

Due to lack of adequate databases in Bosnia and Herzegovina, there is a problem to use standardizing tools developed for updated datasets. The objective of this paper is to provide simplified tools for disaster risk management in accordance to available data and to emphasize the importance of application of international standards in B&H and INSPIRE Directive for establishing a database for the assessment of natural risk and mapping.

### 1.1 Floods in Bosnia and Herzegovina

The consequence of extreme weather events due to climate change are usually interruption of the road network and inaccessibility to urban and economic areas. In Bosnia and Herzegovina, climate change manifested itself by gradual changes in climate parameters in the last ten years [3].

It is well known that Bosnia and Herzegovina is exposed to a high risk of flooding. Significant floods in the recent past in Bosnia and Herzegovina occurred in 2001, 2010, and 2014 as well as the floods that hit the country in 2019. Floods have caused economic, infrastructure, environmental and social consequences. The flood damage cannot be completely avoided but can be mitigated. This implies a flood risk assessment, mapping of critical flood areas, property and people in these areas, and taking adequate and coordinating measures to mitigate the risk of floods.

The vulnerability of BiH's infrastructure to climate risks is exemplified by the impacts of the May 2014 floods and landslides. These natural hazards, which affected bridges, roads, homes, businesses, electricity distribution and flood protection infrastructure, highlight the need to include factor climate risks into decisions regarding investments in short-term and long-term infrastructure assets.

The devastating floods highlighted the need for increased investment in road infrastructure and policies that will help the country mitigate expected impacts from a shifting climate and shield the most vulnerable from natural disasters [3].

The climate change resilient road focuses on ensuring adequate service levels of the road network under extreme weather conditions. Innovation themes will address the adaptation of road operations and management to the effects of extreme weather to such extent that adequate service levels are ensured [6].

Data collation and collection are complementary processes with the aim to be as efficient as possible in the use of existing data while strategically collecting data where needed and possible with the available resources to enhance the level of detail, quality, and so on of the modeling result [7].

Based on the collected data and metadata and their previous analysis, in the case of natural disasters such as floods, certain preventive and preparedness measures may be taken to avoid or mitigate the consequences of such a disaster. The data read from the thematic map can be used through five steps of the integrated flood risk management cycle, namely:

- ***Prevention*** – activities to reduce flood risks and promote proper land use, agricultural land management, and forest management,
- ***Protection*** – activities to reduce the likelihood of flooding, i.e. to reduce the impact of flooding on a given site and increase the resistance to floods,
- ***Awareness*** – informing the population about flood risks and appropriate measures in case of a flood event,
- ***Preparedness*** – flood event activities, and
- ***Remediation*** – immediate recovery in the pre-flood state, analysis, and consideration of new facts.

Predicting floods and giving a timely warning about it caused by heavy rainfall is not easy, so the consequences mitigation process is limited. However, if constant monitoring is carried out and have up-to-date data, it can help determine the water motion and predict the spread of flood and its effect.

Protection measures for the purpose of flood prevention:

- Determination and consideration of floodplain areas
- Adjustment of land use in river basins
- Perform hydrological and meteorological monitoring
- Establish and maintain records for areas with flood risks
- Education and awareness raising about floods and risks
- Planning and implementation of flood protection building measures
- Implementation of individual (self-protection) flood protection measures
- Regularly check the effectiveness of existing buildings within the Flood Protection System
- Regular maintenance of river basins, water structures, and water resources
- Flood protection management

- Provision of financial resources for the implementation of the protection and rescue plan
- Designing a flood protection and rescue plan
- Flood forecasting
- Warning of floods
- Performing intervention measures in case of floods
- Estimates of damages and method of restoration measures after the flood
- Collection of documentation and analysis of flood events
- Systematic, normative, financial and other measures.

The practice has shown that in most cases it is impossible to completely eliminate the risk of flooding, i.e. to avoid the damage they inflict. Efforts are therefore directed towards reducing or mitigating damage through risk management and flood control activities. In order to achieve the effects of this process, it is primarily necessary to identify the risk of floods, to develop a strategy for its reduction and to create policies and programs that will enable the practical achievement of post-achievement goals. Risk management is an important activity that coordinates various initiatives.

## 1.2 Usage of the Viewer During the Floods

Sudden floods are a direct consequence of hydrological behavior of the river basin, and it is necessary to study the meteorological and physiographic characteristics of this area, for the monitoring of the formation and floods spreading. Significant contributions to the study of floodplains are based on the use of remote research, digital terrain model development, and GIS technology.

Based on the collected data and their analysis, it is possible to identify critical sections of the road network. In cases when sections of the main network are flooded, it is necessary to act urgently and under the leadership of the competent authorities, provide the following:

- Identify the roads that could be used for evacuation of the population and delivery of the necessary supplies;
- Identify which type of vehicles are able to be used on those alternative roads;
- Provide funds for purchase and equipment which is necessary for enabling road network.

Flood risk management in order to reduce the risk of floods is carried out through the implementation of the so-called **flood protection measures**, which can be divided into several types according to different criteria, but roughly are divided into **non – structural and structural flood protection measures**.

Non-structural measures include, for example, forecasting of floods; structural measures include, for example, the construction of high-water embankments or high-water containers.

Reducing the risk for a particular road network segment implies an impact on the occurrence, frequency or intensity of disasters. In case of a reduction in the natural disaster expectation, among other measures, monitoring is also used for timely warning, the regulation of river bed and afforestation is carried out.

Reducing the vulnerability of a road section may include the following:

1. **Increasing the resistance of the road section**. It is necessary to adapt the road section to the impact of natural disasters in order to improve its physical resistance. This involves increasing culverts, reinforcing retaining structures, increasing drainage structures, etc.
2. **Road network optimization**. It refers to the provision of the alternative road routes in the event of traffic interruption of the main roads due to natural disasters.
3. **Maintenance**. The management is obliged to ensure adequate inspections and maintenance of road infrastructure. Employees should have feedback from the field, on the basis of which they can assess whether a reconstruction of the existing drainage system is necessary and finally to define a specific action plan for inspection and maintenance.
4. **Monitoring**. Local meteorological stations and, for example, water level measurements in culverts can provide valid information on system response to specific weather conditions, such as heavy rain. Monitoring requires the installation of local meteorological stations, the setting of water level sensors in portholes, wells, underground pipes, gaps, watercourses and reservoirs, and the installation of video surveillance.
5. **Floods early warning systems**. The purpose of early warning is to provide certain time to authorized persons to carry out adequate actions before the actual problems arise.

That implies close cooperation with meteorological institutes and hydrological agencies. For example, it is possible to prohibit access to the affected areas timely or save people and goods from the affected area timely.

### 1.3 Guidelines and Recommendations in the Risk Assessments Process

A key role in the protection of people and their property is the management and control of floods. The practice has shown that in most cases it is impossible to eliminate completely the risk of flooding, i.e. avoid damages they cause. Therefore efforts are directed to reduce or mitigate damages, through flood protective damages, evacuation and rescue, also remediation of disasters consequences.

Protection and rescue act of FB&H document that concretizes the political and strategic idea of the preparation and implementation of protection and rescue in FB&H. It determines the resources, assets, appoints the authorities and legal entities as bearers of protection and rescue tasks.

Evacuation and rescue encompass a set of actions, processes, and rules that need to be implemented in the event of a natural flood disaster that can endanger the health and life of people. We can say that evacuation is a planned organized activity for the temporary transfer of citizens, material and cultural goods, animals, food, carried out by the state authorities and other institutions in accordance with the protection and rescue act.

The Civil Protection Plan prescribes that municipalities and cities for conducting evacuation shall determine:

- Procedures before, during and after evacuation
- The number and population census estimated for evacuation
- Concentrations of people
- The routes and directions of evacuation within the area of responsibility
- The routes and directions of evacuation outside the area of responsibility, to the reception centers
- Reception centers for evacuated population inside and outside the area of responsibility
- Transportation resources
- Participants and assets for evacuation
- Clearly defined protocol on informing people about evacuation
- System for documenting evacuees established
- The way of providing information about evacuees
- Evacuation routes within the area of responsibility
- Overview of reception centers and shelter sites capacities within the area of responsibility
- Logistic support for evacuation.

## 2 Methodology for the Development of a GIS Viewer for Flood Risks Management on the Main Network of FB&H

There are three components in risk analysis: hazards, vulnerability and element-at-risk [8]. They are characterized by both spatial and non-spatial attributes. Hazards are characterized by their temporal probability and intensity, derived from frequency-magnitude analysis. Intensity expresses the severity of the hazard, for example, water depth, flow velocity, and duration in the case of flooding [9]. Elements-at-risk or “assets” are the population, properties, economic activities, including public services, or any other defined values exposed to hazards in a given area [10]. The interaction of elements-at-risk and hazard defines the exposure and the vulnerability of the elements-at-risk. The spatial interaction between the element-at-risk and hazard footprints are depicted in a GIS by map overlaying of the hazard map with an elements-at-risk map [11].

In this paper hazard is flood with intensity expressed by water depth, and element-at-risk is the main road network in Federation of Bosnia and Herzegovina.

The methodology of creating a viewer and a thematic map involves the collection of relevant data sets, analyses of the quality of data collected, data processing and, finally, their visualization, alongside the visualization of identified critical sections of the road network in relation to the flow lines.

Based on the performed analyzes, guidelines and recommendations for managing the risks in case of natural disaster are given.

## 2.1 Spatial Data Collecting

The assessment of multi-hazards and the subsequent risk assessment is a very data intensive procedure. The availability of certain types of spatiotemporal data can be one of the main limitations for carrying out specific types of analysis [9].

The data needs to be collected for this purpose are administrative borders of B&H and FB&H, main road network, bridges, gaps, the hydrographic network of B&H, and flood areas for 100$^{th}$ annual and 500$^{th}$ annual water of FB&H.

After collecting data, an inventory of collected data was made with the indicated metadata they possess (Tables 1 and 2).

**Table 1.** Presentation of collected data

| Num. | Description | Administrative borders | Flood data | Traffic network |
|---|---|---|---|---|
| Data inventory | | | | |
| 1. | Type of data | Borders of B&H, BD B&H, entities, cities, municipalities | Data on floods for the Sava and Neretva river basin | Roads, bridges, gaps |
| 2. | Data description | | | |
| 3. | Data source | Federal Geodetic Administration of FB&H UoS | Agency for the Water Area of the Sava River and the Adriatic Sea | PC Roads of FB&H |
| 4. | Data format | ESRI Shapefile | | |
| 5. | Date of publication | | | |
| 6. | Document name | | | |
| 7. | Hazard | | | |
| 8. | Parameters | | | |
| 9. | Unit of measure | | | |
| 10. | Update data | | | yearly |

**Table 2.** Metadata of the road network

| Metadata of the road network | | |
|---|---|---|
| 1. | Georeferencing | y |
| 2. | Data format | ESRI Shapefile |
| 3. | Projection | USER PRJ FILE |
| 4. | Reference coordinate system | |
| 5. | Extension | No data |
| 6. | Description of provider | OGR data provider |
| 7. | Geometry type | Line, polygon, point |
| 8. | Number of data | |

According to Guptill and Morrison [12], there are seven elements of the quality of spatial data: the origin of data; positional accuracy; the accuracy of the attribute; completeness of data; logical consistency; semantic accuracy; time accuracy/information [13].

The data quality assessment for the "Road Network" in relation to the data element and quality has shown that the analyzed spatial data do not meet all the quality parameters. Thus, for example, data collection methods, conversion information, or transformation are not mentioned. Source data (persons, institutions or investors) are not listed. Data on the reference coordinate system used is available, but data related to placement accuracy is incomplete since all metadata (spatial data correction, scale, resolution, transformation used) are not available. Since it is the official digital data, it is assumed that the placement accuracy has been checked in accordance with national regulations. The logical consistency within the observed set exists, as the data within the analyzed "Road Network" theme is harmonized. Semantic accuracy of the analyzed data is questionable, given that the institutions or the persons authorized to create and manage the databases give their names voluntarily. Time information is available and relates to the date when the data was created or the date it was published. However, there is no data about control and no data of update dates.

After examining the data quality, it was concluded that the downloadable data can be used to create a GIS viewer because the data source is reliable and the data have placement accuracy i.e. the data is within the required accuracy limits.

## 2.2 Integration and Processing of Collected Data

The integration of the collected data is done in free GIS software, which enables the processing and visualization of spatial data.

Layers loaded into GIS software are grouped into three data sets, namely:

1. Administrative boundaries with layers:
   - BiH_polyline_border (border of Bosnia and Herzegovina)
   - FBH_region_border (border of Federation of B&H)
2. Road network
   - Main_Road_Network
   - Bridges
   - Culverts
3. Rivers and flood lines
   - Rivers
   - Lakes
   - Sava_Flood_100Y
   - Sava_Flood_500Y
   - Neretva_1_Flood_100Y

Different types of geometric data for the mentioned topics (raster, vector: point, line and polygon) and a series of attribute topic descriptions (Fig. 2) have been used. Topological rules were investigated and necessary repairs were made.

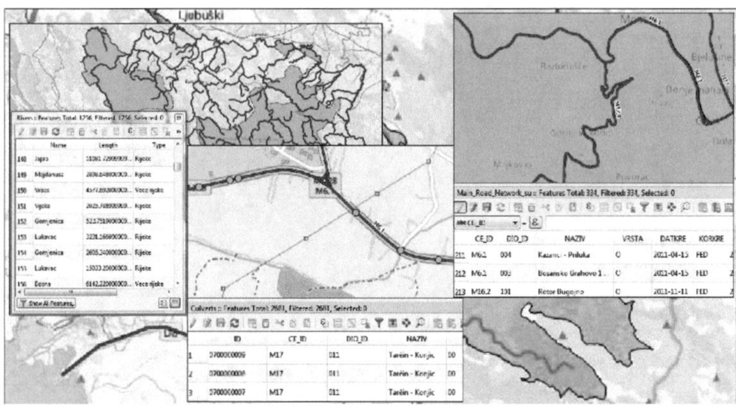


**Fig. 2.** Different types of geometric data

## 3 Creating a GIS Viewer for Flood Risk Management on the Main Road Network in FB&H – Case Study

The edited data are visualized on the GIS viewer (Fig. 3). This section demonstrates the utility of GIS interoperability for enhancing emergency management operations.

The case study covers the narrower area of the FB&H in the vicinity of the town of Doboj, which includes the Bosna River, Usora and Tešanjka, and the main roads M17 and M4.

After creating a GIS viewer, identification of critical areas was performed. These are areas overlapping overhead areas (100 year return period, Sava River Basin) with road infrastructure (M17 and M4) for flood risk management (using the "InaSAFE" software add-on). InaSAFE (Indonesian Scenario Assessment for Emergencies) is free software that enables disaster managers to study realistic scenarios of the impact of natural hazards for better planning, readiness, and response activities. It is a relatively simple tool for estimating exposure and losses from different hazards, using a Python plugin within the Open Source GIS Quantum – GIS. InaSAFE is not a hazard modeling tool, as hazard scenarios have to be provided as input in the software [9].

The first step is to define the layers of hazards and layers of risk. The layer "Sava_Flod_100Y" is defined as a layer representing a hazard, while the layer "Main_Road_Network" is defined as a layer representing elements exposed to hazard. Given that a complete road network is of the same (main) rank, it was not necessary to

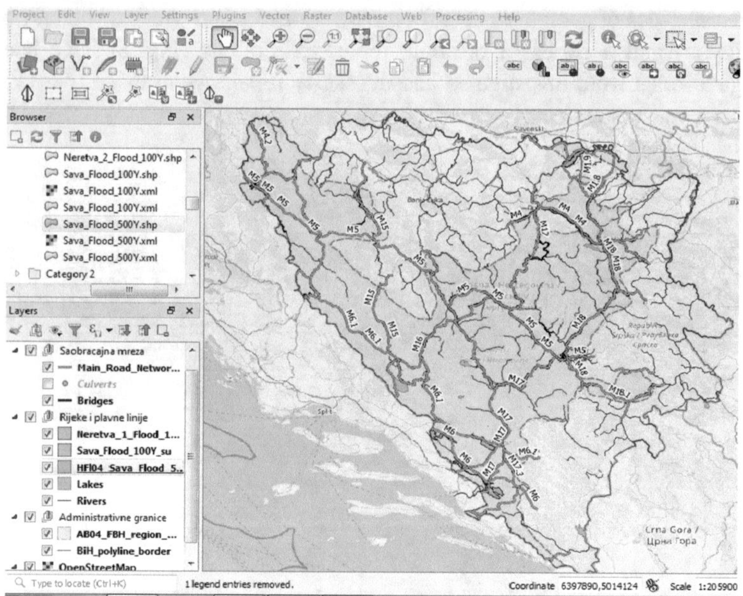


**Fig. 3.** The appearance of a created viewer

perform the ranking of the road network, while the flood lines were divided into three classes in relation to the water level:

Low 0–1000 mm
Medium 1001–4000 mm
High >4000 mm

After the analysis, the results are presented numerically and graphically (Figs. 4 and 5). Based on them, a thematic map was developed for flood risk management in the main road network (Fig. 6).

Analysis detail

Estimated length of roads (m) affected by road type

| Road type | Affected | | Not affected | | Total not exposed | Total |
|---|---|---|---|---|---|---|
| | High | Total affected | Low | Total not affected | | |
| M4 | 2,200 | 2,200 | 0 | 0 | 49,900 | 52,000 |
| M17 | 6 | 6 | 328 | 328 | 35,300 | 35,700 |
| Total | 2,200 | 2,200 | 328 | 328 | 85,200 | 87,600 |

Notes and assumptions

**Fig. 4.** The numerical results

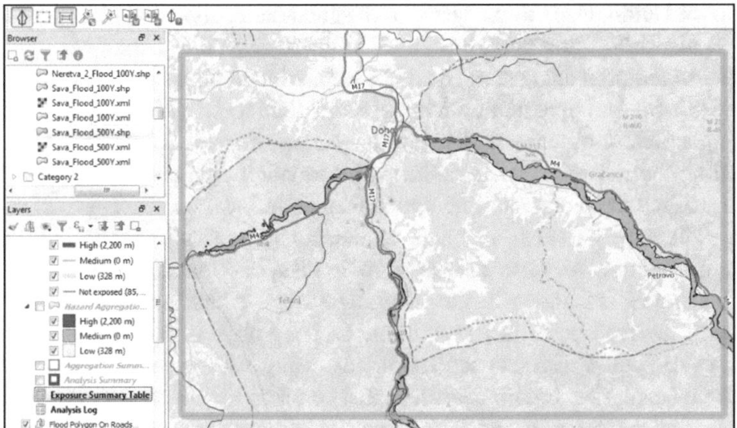


**Fig. 5.** Graphically display results obtained with InaSafe software for the case study area

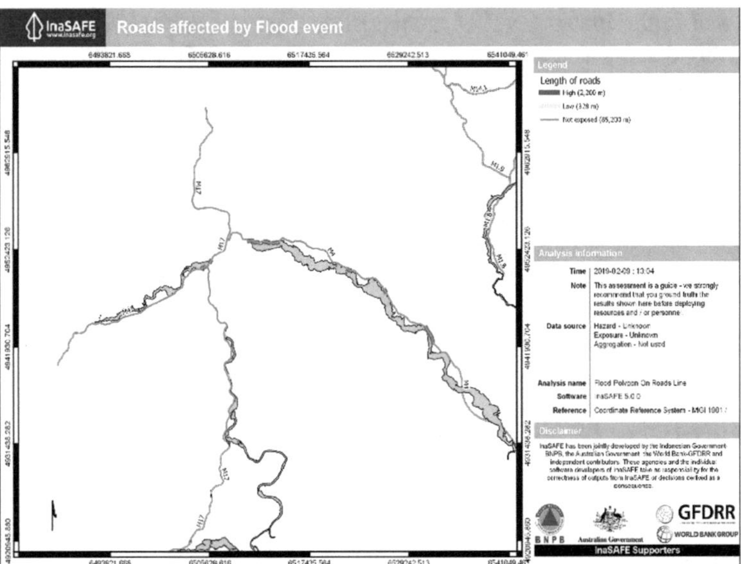


**Fig. 6.** Thematic maps for the case study

## 4 Conclusion and Recommendation

The functionality of geo-information systems depend on the possibilities of spatial data analyzing provided by the system, and the quality of data is essential for the quality of the system. Making the right decisions is only possible on the basis of valid data.

Based on the analysis of the elements of the quality of collected data, it can be concluded that acquisition, processing, storage and maintenance of geo-datasets with the use of information technologies in Bosnia and Herzegovina, which strives to join the European Union, leads to the need for their standardization and harmonization with the EU regulations in this field, first of all with the INSPIRE Directive.

In case all required data are not available or are incomplete, the obtained maps are a simplified way to find potential critical areas and then to provide a detailed analysis of every simple area in order to undertake the needed measures.

Based on the risk assessment collected information obtained from thematic maps, actions to reduce the risk should be undertaken as follows: increasing road resistance, providing alternative routes, adequate maintenance, constant monitoring, and early warning. In addition to these measures, it is also important for future projects, to include the impact of floods in the planning stages, which can greatly increase the resistance of the road network to such unwanted events.

In cases of natural disasters such as floods, using collected data and metadata and their previous analysis, certain measures must be taken to prevent flooding, evacuation and rescue, and recovery, in order to avoid or mitigate the consequences of potential road network disasters.

Large amounts of spatial data which are located in different institutions, beginning with the state government, administrative bodies of the BD B&H, across entities, cantons and to the lowest level of municipalities, are not useful if they are not comprehensively documented and accessible to interested users. Of particular importance is the fact that the documented data are comprehensible at the regional and global level, taking into account the complex state organization of B&H, it is necessary to initially harmonize SDI (Spatial Data Infrastructure) at all levels of government, from the municipality to the state structures. The above process is highly needed, and users and data producers would have the greatest benefit.

It is therefore of great importance to regulate this field by the laws in order to ensure continuous and expeditious work on the establishment and completion of the National Spatial Data Infrastructure (NSDI), which will further improve natural disasters management and the implementation of risk mitigation actions.

## References

1. Bles, T., Doef, M.V.D., Buren, R.V., Buma, J., Brolsma, R., Venmans, A., Meerten, J.V.: Investigation of the blue spots in the Netherlands National Highway Network. Deltares (2012)
2. Cova, T.J., Conger, S.: Transportation hazards. In: Handbook of Transportation Engineering, pp. 17.1–17.24. McGraw Hill, New York (2004)
3. Dzebo, S., Ljevo, Z., Saric, A.: Climate change impacts on roads in Bosnia and Herzegovina. In: 5th International Conference on Road and Rail Infrastructure CETRA, Zadar, Croatia (2018)
4. Britton, C., Elliot, J., Pohu, J., Correia, G., Gibb, M., Spong, C.: http://www.cedr.eu/download/Publications/2017/CEDR-Contractor-Report-2017-1-Implementation-Guide-for-an-ISO-55001-Managementt-System.pdf. 10 Jan 2019
5. EU, C.: Inspire Directive 2007/2/EC. https://eur-lex.europa.eu/legal-content/HR/TXT/PDF/?uri=CELEX:32007L0002&from=HR. 11 Jan 2019
6. Auerbach, M., Herrmann, C.: Adaptation of the road infrastructure to climate change. Transport Research Arena 2014, Paris

7. Prioritizing Climate Resilient Transport Investments in a Dedicated Environment, A Practitioners' Guide, The World Bank (2017). http://documents.worldbank.org/curated/en/818411499808641786/Prioritizing-climate-resilient-transport-investments-in-a-data-scarce-environment-a-practicioners-guide. Accessed 11 Jan 2019
8. Van Westen, C., Abella, C., Sekhar, L.K.: Spatial data for landslide susceptibility, hazards and vulnerability assessment: an overview. Eng. Geol. **102**(3–4), 112–131 (2008)
9. Van Westen, C.J., Damen, M., Feringa, W.: Theory book, National-scale multi-hazard risk assessment, Enschede, The Netherlands (2018)
10. UN-ISDR: Terminology of disaster risk reduction. United Nations, International Strategy for Disaster Reduction, Geneva, Switzerland (2004)
11. Van Westen, C.: Distance education course on the use of spatial information in multi-hazard risk assessment (2009). http://www.itc.nl/Pub/study/Courses/C11-AES-DE-01. Accessed 11 Jan 2019
12. Guptill, S.C., Morrison, J.L.: Elements of Spatial Data Quality. Elsevier, Oxford (1995)
13. Kljucanin, S., Posloncec-Petric, V., Bacic, Ž.: Osnove infrastrukture prostornih podataka. Dobra knjiga, Sarajevo (2018)

# Flood Impact and Risk Assessment on the Road Infrastructure in Federation of Bosnia and Herzegovina

Suada Džebo[1(✉)], Ammar Šarić[1(✉)], Sarah Reeves[2(✉)], Žanesa Ljevo[1(✉)], and Emina Hadžić[1(✉)]

[1] University of Sarajevo, Sarajevo, Bosnia and Herzegovina
suada.dzebo.gf@gmail.com, ammar.saric@hotmail.com, zanesalj@gmail.com, eminahd@gmail.com
[2] Transportation Research Laboratory, London, UK
sreeves@trl.co.uk

**Abstract.** The transport network is very important for the economic development of each country, enabling the movement of goods and people. Damage on road infrastructure poses a direct threat to people's safety, causing traffic disruption and economic and social impacts. The paper deals with the impact of floods on road infrastructure in the Federation of Bosnia and Herzegovina. The paper describes the interaction between the road network and flood as well as their vulnerability, hazards, risks and risk reduction. The main concept and methodological approach for estimating the flood risk for the road transport network have been presented. The proposed climate risk assessment is an indicator-based methodology adjusted to available data.

**Keywords:** Flood · Road network · Vulnerability · Hazard · Risk

## 1 Introduction

Floods as natural disasters are one of the most common causes of traffic crashes. In order to respond to such weather conditions, it is necessary to act adequately both before and after the disaster.

Flooding poses an important threat to roads as it may lead to massive obstruction of traffic and damage to the road structures, with possible long- term effects [1].

Transportation systems that are disrupted by hazardous event also play a critical role in emergency management. Road network disruptions can threaten the ability to provide medical care and other critical services [2]. Transportation lifelines are generally considered the most important in an emergency because of their vital role in the restoration of all other lifelines [3].

Over the past 20 years, the road network management institutions are increasingly recognizing the impact of climate changes on the transport network and the need for better understanding of risks in order to identify priorities for improving resilience and adapting to climate changes.


S. Avdaković et al. (Eds.): IAT 2019, LNNS 83, pp. 276–289, 2020.
https://doi.org/10.1007/978-3-030-24986-1_22

A better understanding of risk allows for targeting resources where the most useful approach in combating the threat of climate changes is proactive, rather than a reactive one. Risk identification and management are recognized as an integral part of asset management, which is included in ISO 55001: 2014 [4]. The main steps for assessment and risk management are shown in the following figure (Fig. 1).

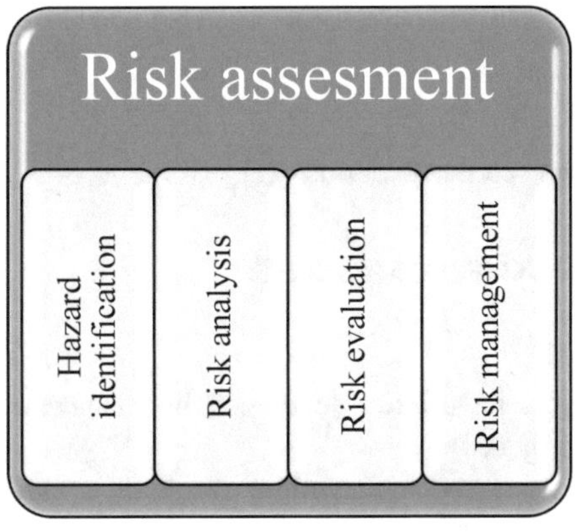


**Hazard identification** includes identifying and defining the hazard. Also includes a risk assessment that can be focused on one of the hazards.

**Risk analysis** includes the collection of risk information and their analysis, whereby analysis can be qualitative, semi-quantitative and quantitative.

**Risk evaluation** is a comparison of the analyzed results with respect to acceptable risk levels that are often used to inform **Risk management**.

**Fig. 1.** The steps for assessment and risk management

## 2 Interaction Between the Road Network and Floods

A road hit by natural phenomena induces two levels of consequences. On the one hand, direct consequences relate to humans and vehicles which can be respectively injured or destroyed. On the other hand, traffic disruption can have severe indirect consequences: closures of roads induce economic consequences (workers transportation, disruption of supply chains for factories, commercial units and stores, etc.), social consequences (loss of access to schools, universities…) or security-related consequences (loss of access to rescue, fire and police departments) which are difficult to assess [5].

There is also an influence of road infrastructure on the occurrence of floods. Road development in floodplains alters the floodplain hydraulics and affects the related aquatic ecosystems [6].

Construction of a road network often interferes with natural watercourses, both above and underground, and thus increases the risk of flooding certain area. The spatial position of the road route has a great influence on the type and frequency of interaction between the road and watercourses. A specific road segment may be a barrier, a source, a drain, or a water flow corridor (Fig. 2). The interaction between roads and streams may modify the magnitude and direction of water flows and debris, and water flows may transform into debris flows or vice versa [7].

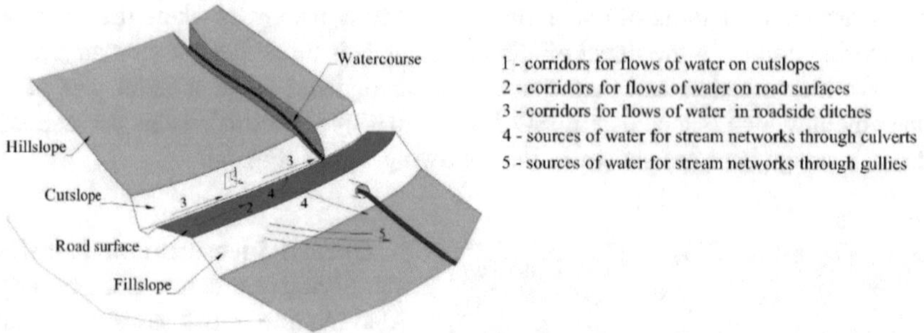


**Fig. 2.** Types of water flow on a road ***Source:*** [7]

## 3 Review of Road Infrastructure Risk Assessment Methodologies

The specified risk assessment methodologies have been developed by management structures or are the outcome of research projects.

There are differences in the approaches developed, including variation in assessment level, detail, scope and consideration of future uncertainty.

Consideration of the availability of data and resources and how the findings will be used are essential in selecting the most appropriate methodology for a particular application. All these methodologies have different use, benefits, complexity, and challenges.

The following chapters describe the two most current methods for climate changes risk assessment that is RIMAROCC and ROADAPT.

**RI**sk **MA**nagement for **RO**ads in a **C**hanging **C**limate – **RIMAROCC** has been developed through the project initiated by ERA-NET ROAD project of the European Commission with the aim of strengthening European road research by coordinating national and regional research programs and policies.

The RIMAROCC frame defines risk as a combination of hazards, vulnerabilities, and consequences. The methodology recommends different measures for assessment of territories, networks, sections and structures [8]. The general risk assessment framework is provided in seven steps and 22 sub-steps (Fig. 3) and it can be seen that it is very complex and difficult to follow.

The **ROADAPT** project was commissioned under the CEDR Call 2012 "Road owners adapting climate change". The ROADAPT project is based on RIMAROCC methodology in an effort to simplify complex procedures and numerous steps. The general stages of ROADAPT are given in Fig. 4.

The ROADAPT methodology developed a simplified approach, which involves steps from 1 to 4 of the RIMAROCC frame. After collecting existing data, knowledge and experiences related to the road network, evaluation is carried out at a series of stakeholder workshops with participants in the maintenance and management of the road network through semi-quantitative risk analysis. Semi-quantitative analysis using data and/or expert judgment to provide a numerical value to indicate the level of risk. A case study of the ROADAPT quick scan approach was carried out for the A24 motorway in Portugal [10].

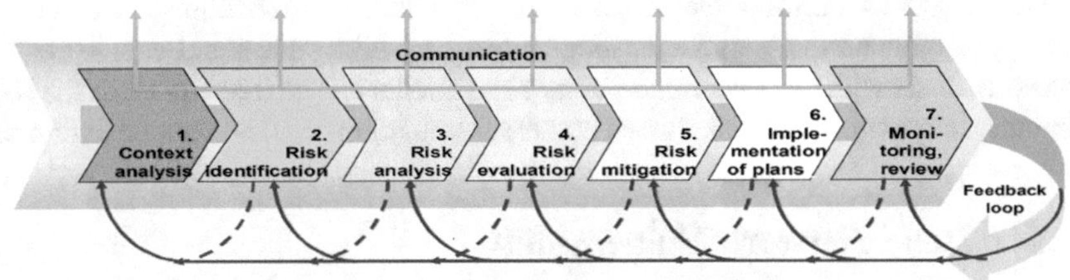


| Key steps | Sub-steps |
|---|---|
| 1. Context analysis | 1.1 Establish a general context<br>1.2 Establish a specific context for a particular scale of analysis<br>1.3 Establish risk criteria and indicators adapted to each particular scale of analysis |
| 2. Risk identification | 2.1 Identify risk sources<br>2.2 Identify vulnerabilities<br>2.3 Identify possible consequences |
| 3. Risk analysis | 3.1 Establish risk chronology and scenarios<br>3.2 Determine the impact of risk<br>3.3 Evaluate occurrences<br>3.4 Provide a risk overview |
| 4. Risk evaluation | 4.1 Evaluate quantitative aspects with appropriate analysis (CBA or others)<br>4.2 Compare climate risk to other kinds of risk<br>4.3 Determine which risks are acceptable |
| 5. Risk mitigation | 5.1 Identify options<br>5.2 Appraise options<br>5.3 Negotiation with funding agencies<br>5.4 Formulate an action plan |
| 6. Implementation of action plans | 6.1 Develop an action plan on each level of responsibility<br>6.2 Implement adaptation action plans |
| 7. Monitor, re-plan and capitalise | 7.1 Regular monitoring and review<br>7.2 Re-plan in the event of new data or a delay in implementation<br>7.3 Capitalisation on return of experience of both climatic events and progress of implementation |
| Communication and gathering of information | |

**Fig. 3.** RIMAROCC framework [9]

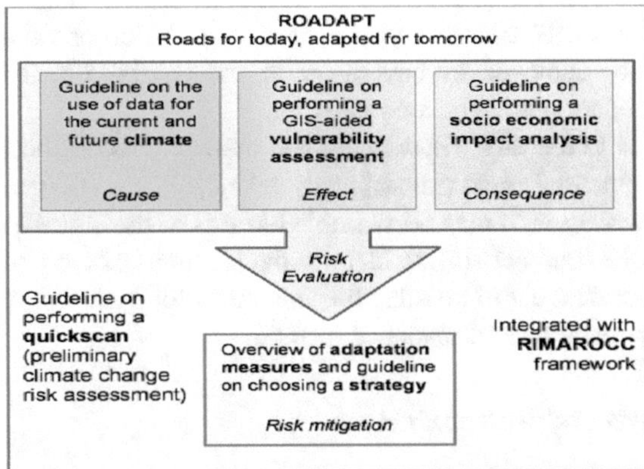


**Fig. 4.** Structure of the ROADAPT guidelines [8]

This approach narrows the scope of possible threats from climate changes and provides insights into vulnerable locations. Identified vulnerable sites become the focus of deep analysis with detailed vulnerability assessments and an assessment of the socio-economic impacts of high-risk threats and the formulation of an adaptation action plan.

# 4 The Road Network Vulnerability

The vulnerability of transport infrastructure can be viewed from different angles, i.e. peoples', vehicles, traffic, infrastructure, and the environment. The vulnerability can refer to the physical vulnerability of road users or the possibility of occurrence of an incident that would reduce the utilization of the transport system.

In the context of traffic flow, the vulnerability can be viewed in the context of network reliability, since a reliable network is less vulnerable [11]. An example of various vulnerabilities on the road network is road accident in a two-way tunnel that can temporarily disrupt regional transport system leading to significant delays, while the system with separate tunnel tubes in both directions would be less vulnerable [12].

People and environmental resources near the traffic corridor are also sensitive to harmful events. For example, when transporting hazardous substances along the inhabited area, the vulnerability of the corridor can vary significantly from one point to another, and two potential events over a distance of several kilometers can have very different outcomes depending on if there live people or not.

## 4.1 Hazard Analysis

The term "hazard" often refers to threats such as fog, wind, fire, poles, snow, and landslides. However, the term "transport hazard" exists at all levels from the curbstones on the sidewalk that could be stumbled by pedestrians to potential increase of the sea level that would flood the coastal road.

The focus of the hazard analysis is the identification of threats to the road transport system, its users, surrounding people and resources. In the simplest case, a list of possible threats that could affect transport systems in the observed area can be compiled. This could be achieved by creating a "threat matrix" indicating whether particular threat endangers particular road network.

The next step is to identify where and when these events can occur, as well as the probability of the potential event (impact analysis) calculation. This step can be done in two ways. The first way is to map all possible hazards in the entire observed area, and then overlap with the road network to identify the locations of the possible events. The second way is to create a list of possible hazards made for each section separately. The first approach requires a hazard mapping method.

## 4.2 Risk Analysis and Risk Reduction

The largest number of risk definitions include two elements: the possibility of an event occurring and its consequences. It is not possible to avoid all risks, but it is possible only to find a compromise between risks and benefits [11].

In the most general form, risk can be defined as [11]:

$$R = H \times C, \tag{1}$$

$$C = E \times V, \tag{2}$$

Where:
H – Hazard, i.e. probability of hazard event
C – Consequences related to the hazard. Consequences [10] are defined as
E – Element of risk, and
V – Vulnerability of risk element.

## 5 The Main Road Network of Bosnia and Herzegovina

According to the data from 2016, Bosnia and Herzegovina has 23,119 km of the roads of all categories. It is 6.65% more than in 1991 when the total road length was 21,677 km. The most important roads in Bosnia and Herzegovina connected to the European international road network are shown in Fig. 6:

**E-73** B. Šamac-Doboj-Zenica-Sarajevo-Mostar-Metković;
**E-761** Izacic-Bihac-Jajce-Travnik-Sarajevo-Višegrad-Vartište;
**E-762** Sarajevo-Foča-Hum;
**E-661** Gradiška-Banja Luka-Jajce.

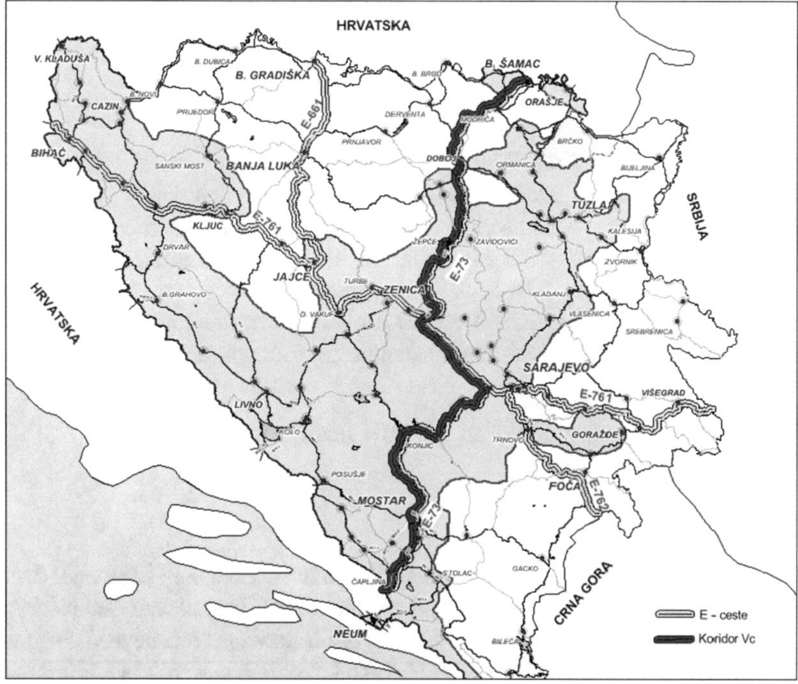


**Fig. 6.** The main routes in Bosnia and Herzegovina (source: author)

The aforementioned road sections represent major road directions that should be resilient to natural disasters, enabling crisis management for providing medical care and other critical services.

After floods in 2014 (Figs. 7 and 8), the overall assessed damage in the transport and communication sector was over 512 million BAM, out of which 221 million BAM was in the Federation of Bosnia and Herzegovina, 278 million BAM in the other entity of Bosnia and Herzegovina, i.e. Republika Srpska, and 12.6 million BAM in the Brcko District of Bosnia and Herzegovina [13].

**Fig. 7.** Floods 2014 Federation of B&H, location: Zenica

**Fig. 8.** Floods 2014 – Federation of B&H location: Prud

The recovery plan for the transport and communication sector is crucial, as many other sectors rely on this sector. The introduction of climate risks into the planning, design, and maintenance of existing roads is also very important.

## 5.1 The Methodology of Risk Assessment

Within the framework of the World Bank funded project entitled "Mainstreaming climate resilience risk in road management in Bosnia and Herzegovina", a risk assessment methodology was developed as adapted to the requirements of the public company "Roads of FB&H". It is similar to the CEDR (Conference of European Directors of Roads) project "The Risk Management for Roads in a Changing Climate" (RIMAROCC) and FHWA (US Department of Transportation, Federal Highway Administration) methods, but take into account the following:

- A network level assessment,
- The main hazards experienced on the FBH main road network,
- Data availability assessment,
- Use of the GIS platform, and
- Cooperation with stakeholders.

The proposed climate risk assessment approach is an indicator-based one. Based on stakeholder input, expert judgment, and data availability relevant indicators were identified (Table 1). A set of indicator scoring tables have been produced for flooding as a hazard. The involvement of local experts/stakeholders was key in addressing data gaps and refining the results of scoring. For each element of risk, a series of indicators were identified based on the main influencing factors and available data sources.

The methodology steps can be summarized as:

1. Identify H, E and V indicators for each type of hazard Establish thresholds for scoring each indicator (expert judgment - Table 1);
2. Consider if weightings should be applied;
3. Score the indicators for each road section based on data and expert judgment (1–10);
4. Normalize scores for H, E, and V (divided by the maximum potential score);
5. Multiply H, E, and V scores to produce an overall score for risk;
6. Map the risk score using the GIS platform – this will enable high scoring road sections for different types of hazard to be easily identified.

Not all influencing factors can be developed into indicators as the assessment will become too resource intensive to carry out. Careful consideration is required to decide which are the most significant indicators and also if there is data to support the indicator or if it can be scored using local knowledge from the field engineers or site observation. In some cases, data sets are not available for all road sections and therefore it may not be appropriate to include the indicator at this time as the assessment needs to be consistent across the network. The suggested initial indicators are given in the following table, but the model should be calibrated after feedback from the field.

Table 1 shows indicator types which include hazard, elements at risk and vulnerability. In brackets are the scores for defined thresholds.

**Table 1.** Flood hazard types and thresholds

| Hazard type: Fluvial flooding | | | | | |
|---|---|---|---|---|---|
| Indicator type | Description of indicator | Scoring method (Score)- thresholds | | | |
| | | 1 | 2 | 3 | 4 |
| Hazard | Number of times the monthly precipitation exceeded 120 mm over last 30 years ($P_n$) | $P_n \leq 50$ (1) | $50 > P_n \leq 100$ (4) | $100 > P_n \leq 150$ (7) | $P_n > 150$ (10) |
| Hazard | Average monthly precipitation over the last 30 years ($P_m$) | $P_m \leq 70$ mm (1) | 70 mm $> P_m \leq 90$ mm (4) | 90 mm $> P_m \leq 110$ mm (7) | $P_m > 110$ mm (10) |
| Hazard | Projected future annual precipitation (C) | $C \leq 600$ mm (1) | $600 > C \leq 800$ (4) | $800 > C \leq 1200$ (7) | $C > 1200$ mm (10) |
| Hazard | Location in flood zone | Road section not located in flood zone (1) | | Road section located in 500 year flood zone (7) | Road section located in 100 year flood zone (10) |

(*continued*)

**Table 1.** (*continued*)

| Hazard type: Fluvial flooding | | | | | |
|---|---|---|---|---|---|
| Indicator type | Description of indicator | Scoring method (Score)- thresholds | | | |
| | | 1 | 2 | 3 | 4 |
| Hazard | Flood defenses | Active flood defenses in good condition (1) | Active flood defenses in poor condition (4) | Natural vegetation (7) | No flood defenses (10) |
| Hazard | Elevation of road | High elevation (1) | Mostly high elevation (4) | Mostly low elevation (7) | Low elevation (10) |
| Hazard | Proximity of road to river (R) | R > 200 m (1) | 100 > R ≤ 200 (4) | 50 ≥ R ≤ 100 (7) | R < 50 m (10) |
| Hazard | Condition of the river | Minimal silt (1) | Small amount of silt (3) | Accumulation of silt is affecting the river capacity (5) | Accumulation of silt has significantly reduced the capacity of the river (10) |
| Hazard | Run-off coefficient | Rural (1) | Urban open spaces (e.g. sports field) (3) | Commercial and industrial area (7) | Residential area with a high number of houses per hectare (10) |
| Hazard | Culvert condition | Culverts clear and of appropriate capacity (1) | Culvert capacity slightly reduced (3) | Culvert capacity significantly reduced e.g. requires cleaning (7) | Completely blocked or broken culvert (10) |
| Elements at risk | Asset | Road without culvert (1) | Road with culvert (3) | Minor structure (5) | Major structure (10) |
| Vulnerability | Traffic in AADT | AADT ≤ 1,500 (1) | 1,500 < AADT ≤ 3,000 (4) | 3,000 < AADT ≤ 9,000 (7) | >9,000 AADT (10) |
| Vulnerability | Redundancy | Viable diversion exists (1) | Diversion of medium length (e.g. 30 km) (4) | Diversion long and/or on poor quality roads (7) | No viable diversion (10) |
| Vulnerability | Strategic importance | Low importance (1) | Provides access for a community (4) | Provides access to an essential service such as a hospital or a major town (7) | The road is of Federal or national importance (10) |

The explanation of these indicators will provide on the case study.

## 6 Case Study

A case study is used to illustrate how to employ climate risk assessment methodology. The results obtained should not be taken as a true assessment of the road section, the example provided is purely to illustrate the approach.

The case study used is the M15 section 004, Ključ to Tomina (Fig. 9). Each indicator is discussed in terms of the case study location and data available, and the overall score is provided at the end (Table 2).

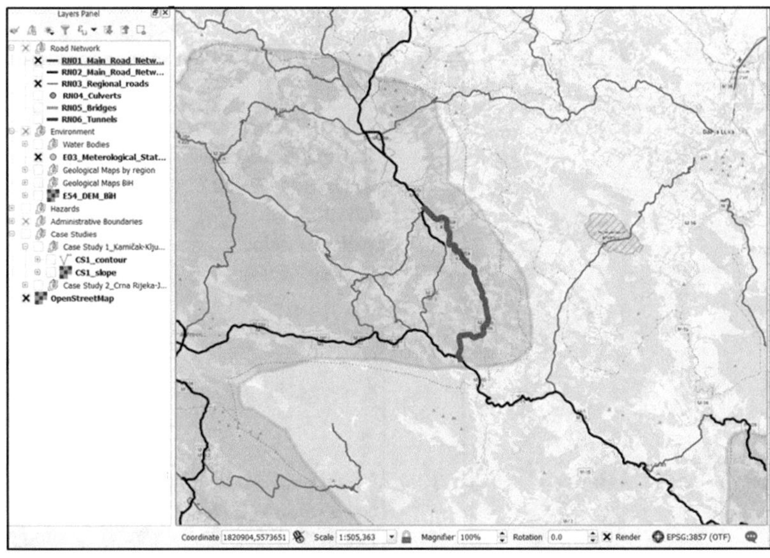


**Fig. 9.** Case study example M15 -004

**Table 2.** Hazard - Flooding indicator scores

| Indicator | Indicator type | Risk score | Data source used | Comments |
|---|---|---|---|---|
| Number of times the monthly precipitation exceeded 120 mm over the last 30 years | Hazard | 4 | Historical weather data from HydroMet (O19) and location of meteorological stations (E03) | The number of times the monthly precipitation exceeded 120 mm over the last 30 years at the nearest located meteorological station (Sanski Most) is 72 |
| Average monthly precipitation over the last 30 years | Hazard | 4 | Historical weather data from HydroMet (O19) and location of meteorological stations (E03) | The average monthly precipitation over the last 30 years at the nearest located meteorological station (Sanski Most) is 87 mm |

*(continued)*

**Table 2.** (*continued*)

| Indicator | Indicator type | Risk score | Data source used | Comments |
|---|---|---|---|---|
| Projected future annual precipitation | Hazard | 4 | Climate scenarios for meteorological stations (O42) and locations of meteorological stations (E03) | The projected change in annual precipitation at the nearest meteorological station (Sanski Most) is for a decrease of 40% for the 2050 s. The precipitation during the reference period is 1023.5 mm |
| Location in flood zone | Hazard | 10 | Flood extent for Sava region (HFl03 and HFl04) | The main road passing through Zgon is within the 100 year flood level |
| Flood Defenses | Hazard | 7 | Google Maps | It appears there are no flood defenses from Google Maps (these are low resolution). There is vegetation next to the banks of the river |
| Elevation of road | Hazard | 7 | Slope and contour analysis for case study site 1 (CS1) | Some sections appear to be at high elevation based on the contour map. There are many sections where the road is at low elevation next to the river |
| The proximity of the road to the river | Hazard | 7 | Google Maps | The proximity of the road to the river varies through the section. The river is less than 20 m away from the river and as far as 600 m in some places |
| Condition of the river | Hazard | | | This cannot accurately be done without field observation |

(*continued*)

**Table 2.** (*continued*)

| Indicator | Indicator type | Risk score | Data source used | Comments |
|---|---|---|---|---|
| Run-off coefficient | Hazard | 1 | Land use data (E55) and Google Maps | Located in a rural area with one small community at Zgon |
| Culvert condition | Hazard | | | This cannot accurately be done without PC Roads records or field engineer knowledge |
| Asset | Elements at risk | 5 | Culverts (RN04), bridges (RN05) and tunnels (RN06) | There are 3 bridges and 1 tunnel in this section. The tunnel is classed as a small structure because it is less than 100 m |
| Traffic | Vulnerability | 4 | AADT data (RN02) | The 2016 AADT (Annual average daily traffic) for this section was 2195 |
| Redundancy | Vulnerability | 1 | Google Maps | A diversion exists via the R410a from start to finish of the site. To get somewhere along the route would be a diversion of >25 km |
| Strategic importance | Vulnerability | 4 | Google Maps | Small community at Zgon |

The score assigned to the hazard, vulnerability, and element-of-risk indicators are combined and normalized (divided by the maximum potential score) to provide a value for hazard between 1 and 10. This provides a score from 1 to 10 for H, E, and V, which are multiplied to give an overall score out of 1000. A higher score indicates a higher risk. The case study scores for flooding are summarized in Table 3.

**Table 3.** H, E, V and risk scores for each hazard for the case study site

| Hazard | Indicator type | Total score | Normalized score (1–10) |
|---|---|---|---|
| Flooding | Hazard | 47 | 5.9 |
| | Elements at risk | 5 | 5 |
| | Vulnerability | 9 | 3 |
| | Overall risk score | | 88 (1–1000) Medium |

To make the scores more meaningful they can be categorized into low, medium, high and very high as shown in Table 4. Following this categorization flooding are medium risk. On the same way can be provided a risk assessment for landslides, rock falls, snow and wildfire.

**Table 4.** Risk categories (expert judgment)

| Risk category | Score (R) |
|---|---|
| Low | $R \leq 64$ |
| Medium | $216 \geq R > 64$ |
| High | $216 < R < 512$ |
| Very high | $R \geq 512$ |

## 7 Conclusion

The purpose of this paper is to present a methodology for carrying out a flood risk assessment on the road network in Bosnia and Herzegovina. While the floods are not the only natural disaster that can damage the road network, the flooded road network can increase the hazardous conditions. It is therefore very important to take into account the management of flood risks in the road network planning process.

There are currently several approaches to assessing the vulnerability of the road network. The choice of method depends on the purpose, scope and available data. However, the main purpose of all methods is to identify where and how disruption of road networks can cause negative consequences for the population and goods.

Risk mitigation measures need to be taken on the basis of assessed risks, such as increasing road resistance, providing alternative routes, adequate maintenance, constant monitoring, and early warning. In addition to these measures, it is also important for future projects to include the impact of floods in the planning stages, which can greatly increase the resistance of the road network to natural disasters in general.

The developed risk assessment methodology is similar to the CEDR (Conference of European Directors of Roads) project "The Risk Management for Roads in a Changing Climate" (RIMAROCC) and FHWA (US Department of Transportation, Federal Highway Administration) methods. The basic difference is in the availability of risk assessment data, which can sometimes be a limiting factor since Bosnia and Herzegovina does not yet have a well-established database according to the INSPIRE Directive, and this is an aggravating circumstance for assessing the risk of natural hazards on the BIH road network.

The example of the case study shows up employ developed climate risk assessment methodology for a flood event on the road section 004 on the main road M15. The result of risk assessment is medium risk.

This type of network wide risk assessment is a good way to identify areas which are at high risk or will be high risk in the future (e.g. as a result of the changing climate). The level of risk which is acceptable very much depends on the company responsible for the management and maintenance of the road network and will be influenced by the political context and available resources.

The risk assessment will also need to be reviewed at regular intervals. The level of risk will change over time as climate, traffic, condition, urbanization, etc. change.

## References

1. Bles, T., Doef, M.v.d., Buren, R.v., Buma, J., Brolsma, R., Venmans, A., Meerten, J.v.: Investigation of the blue spots in the Netherlands National Highway Network. Deltares (2012)
2. Jenelius, E., Mattsson, L.G.: Road network vulnerability analysis: conceptualization, implementation and application. Comput. Environ. Urban Syst. **49**, 136–147 (2014)
3. Cova, T.J., Conger, S.: Transportation hazards. In: Kutz, M. (ed.) Handbook of Transportation Engineering, pp. 17.1–17.24. McGraw Hill, New York (2004)
4. ISO office, International standard ISO 55001, Asset management - Management systems - Requirements, Published in Switzerland (2014)
5. Tacnet, J., Mermet, E.: Analysis of importance of road networks exposed to natural hazards. In: Proceedings of the AGILE 2012 International Conference on Geographic Information Science (2012)
6. Douven, W., Goichot, M., Verheij, H.: Best practice guidelines for the integrated planning and design of economically sound and environmentally friendly roads in the Mekong Floodplains of Cambodia and Viet Nam: synthesis report. Delft Cluster (2009)
7. Jones, J.A., Swanson, F.J., Wemple, B.C., Snyder, K.U.U.: Effects of roads on hydrology, geomorphology, and disturbance patches in stream networks. Conserv. Biol. **14**, 76–85 (2000)
8. Bles, T., Bessembinder, J., Chevreuil, M., Danielsson, P., Falemo, S., Venmans, A.: ROADAPT Roads for today, adapted for tomorrow. CEDR Transnational Road Research Programme (2015)
9. Bles, T., Ennesser, Y., Fadeuilhe, J.-J., Falemo, S., Lind, B., Mens, M., Ray, M., Sandersen, F.: Risk management for roads in a changing climate (RIMAROCC). Sweden, France, Netherlands, Norway: ERA-NET ROAD (2010)
10. Ennesser, Y.: Quickscan report: A24 motorway in Portugal. CEDR Call 2012: Road owners adapting to climate change, ROADAPT Roads for today, adapted for tomorrow, final version, March 2014
11. Chauncey, S.: Social benefit versus technological risk. Science **165**(3899), 1232–1238 (1969)
12. Rogelis, M.C.: Flood Risk in Road Networks. World Bank Group, Washington (2016)
13. Ministry of Foreign Affairs of Bosnia and Herzegovina, 4 Juli 2014. http://mvp.gov.ba/Baneri-Linkovi/4%20juli%202014%20Procjena%20potreba%20za%20oporavak%20i%20obnovu%20BiH%20-%20Sazetak.pdf

# The Concept and Application of the 3D Model Based on the Satellite Images

Mirko Borisov[1(✉)], Nikolina Mijić[2], Tanja Bugarin[1], Vladimir M. Petrović[3], and Filip Sabo[4]

[1] Faculty of Technical Sciences, University of Novi Sad, Trg Dositej Obradović 6, 21 000 Novi Sad, Serbia
mborisov@uns.ac.rs, tanjabugarin@gmail.com

[2] Faculty of Earth Science and Engineering, University of Miskolc, Miskolc Egyetemvaros, Miskolc 3515, Hungary
nikolinamijic7@gmail.com

[3] Institute of Chemistry, Technology and Metallurgy, Department for Ecology and Technoeconomics, University of Belgrade, Njegoševa 12, 11000 Belgrade, Serbia
vladimirpetrovic.gis@gmail.com

[4] European Commission (ARHS Italia S.R.L), Via E. Fermi 2749, TP 267, 21027 Ispra, VA, Italy
filip.sabo@ext.ec.europa.eu

**Abstract.** Users of digital elevation models must be aware of their main characteristics in order to judge better their suitability for a specific application. This paper describes concept of creating and applying of the *3D* model based on the satellite images. The quality and loyalty of the elevation model of the terrain depends of the data which are collected, i.e. on the type and resolution of the satellite images, horizontal and vertical datum, but also on the way of interpretation and visualization of the *3D* model. On the other hand, the organization and structure of data are influencing on the creation of the *3D* model.

**Keywords:** *3D* model · Satellite image · Erdas · Analysis · Application

## 1 Introduction

In recent years, we have witnessed great progress in improving existing and developing new methods and technologies for collecting and processing large amounts of data on objects and phenomena on the Earth's surface.

This is primarily the result of progress in the field of geoinformatics, i.e. the development of photogrammetry and remote detection, radar and shooting a laser from the air or from the Earth. In particular, the detection of a large number of quality sensors (passive and active) and the launch of satellites brought by sensors into Earth's orbit enabled the coverage of both the larger and the entire surface of the planet with data, relatively high accuracy and resolution.

The subject of this paper is the *3D* terrain model based on satellite images. The basic goal is creating and applying *3D* models in the selected area. The proof of this


S. Avdaković et al. (Eds.): IAT 2019, LNNS 83, pp. 290–304, 2020.
https://doi.org/10.1007/978-3-030-24986-1_23

can be seen clearly in the growing interest in *3D* applications, such as *Google Earth*. This fact is clear, because the world around us is unequivocally three-dimensional and most people no longer satisfy only *2D* views, whether in business or their personal needs [1]. Users and the whole society face an important change in the spatial data paradigm from *2D* to *3D*.

## 2 Digital Modelling of the Terrain and 3D Data Models

The two US engineers at the Massachusetts Institute of Technology introduced the term of digital terrain modeling. Definition which they gave fifties read [2]: "The digital terrain model - DTM (*Digital Terrain Model* - DTM) is a statistical performance continuous surface soil over a large number of selected points with known *X, Y* and *Z* coordinates arbitrary coordinate system".

Digital terrain models provide basic quantitative information on the Earth's surface. However, in the case of data users and professional users, the term *3D* models are used for both the digital terrain model and the digital surface model. In addition, the digital model of the terrain most often refers to the physical surface of the Earth (the height of the actual surface). While the digital surface model describes the upper surface that includes the height of the vegetation, the built objects of infrastructure and other surface objects, and the heights of the physical surfaces are described only in areas where no objects listed [3].

High-level data in *3D* models are stored, distributed and used in the form of a regular network, an irregular network *triangulated network (TIN*) or contour lines [4]. In 3D models give height as the basic data, and from them can get different geomorphological parameters such as pitch, brightness (side orientation), curvature of the surface, soil moisture index, etc. [5].

The altitudes and geomorphologic parameters, and a great number of applications in GIS applications such as hydrological analyses, land use planning, climate studies, relief mapping, terrain mapping (topographic mapping) modeling the acceleration of gravity and many other areas [6].

Other references to the *3D* data model [7] can be found in the literature. One of them is DTED *(Digital Terrain Elevation Data)*, which uses for its products National Geospatial Intelligence Agency of the United States (*National Geospatial-Intelligence Agency* - NGA, and *National Imaging and Mapping Agency* - NIMA).

## 3 Source Data and Test Model Defining

In the practical part of this work for creating *3D* terrain models, the original data from satellite images of the *WorldView1 were used*. They are a product of *Digital Globe*. For the test case, selected was the city of Belgrade. The area is representative of all types of terrain, from the lowlands, the hilly wilderness terrain to the much-razed relief. About chin and interpretation of images (digital processing and visualization) was done in the software environment of the *Earth Resources Data Analysis System* (*ERDAS*). Basic characteristics of satellite images are described (*multispectral, geometric, radiometric and temporal resolution*). In addition, in the digital processing of images, all steps are analyzed in detail, from pre-treatment (removal of sensor errors, geometric corrections, and calibration), improvements (filtering, contrast enhancement), and transformations (arithmetic operations, indexes, component analysis) to point classification.

The company *Digital Globe* QuickBird has recordings that have better resolution than many other available commercial satellite products (http://www.firstbasesolutions.com/digitalglobe.php). The resolution of the panchromatic sensor is 0.61 m, and 2.4 m multispectral. This system was launched in the early 21st century. It has a bandwidth of 16.5 km in the top and the largest capacity for archiving images on the platform itself. The orbit's height is 450 km, and the base scene is 272 km$^2$ (16.5 × 16.5 km, 27424 × 27552 pixels for panchromatic and 6856 × 6888 for multispectral mode) (Fig. 1).

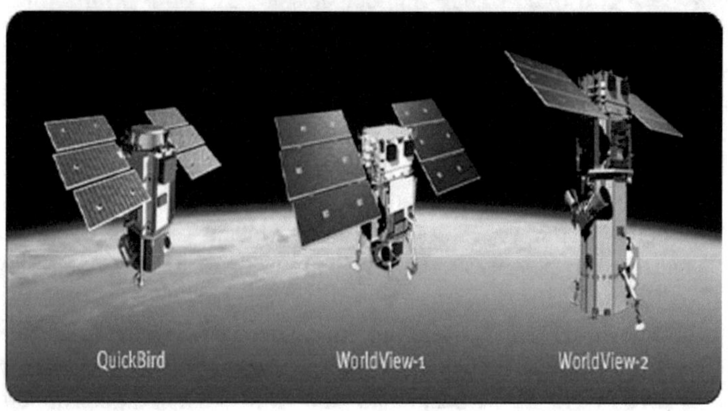


**Fig. 1.** Satellite configurations in the orbit of Digital Globe. Source: http://www.firstbasesolutions.com/digitalglobe.php (24.05.2017)

With the launch of *World View 2* satellites, *Digital Globe* has implemented new data collection capabilities with high-resolution multispectral images with eight channels (bands). In addition, *World View 2* is the second-generation satellite of the next generation of *Digital Globe*. Like *World View 1*, the *World View 2* is equipped with modern geolocation capabilities and represents the next commercial space *launcher* after *WorldView 1*. It is equipped with a gyroscope control, which allows increased performance, fast targeting and an effective stereo collection in the direction of motion. The enhanced agility combined with an operational elevation of 770 km provides a collection of approximately 1 million km$^2$ per day, a high-resolution shot, and after an average revision of 1.1 days around the Earth.

### 3.1 Basic Characteristics of the *WorldView* Satellite

Satellite *WorldView1 was* launched in September 2007. It was the first one in the *WorldView* satellite *series*. Panhro's recordings are 0,5 m in size. The satellite is at an altitude of 496 km. In addition, the *WorldView1* satellite has an average re-visit time of 1.7 days and can collect data from an area of 13 00 000 km$^2$ during one day at a resolution of 0,5 m. The recording width is 17,7 km. Its accuracy, without checkpoints, is about 4 m. Modern instruments that allow for high accuracy of geolocation, very rapid targeting of the location of the recording and efficient stereo data collection on the area, so for many needs no additional georeferencing is required overwhelm the satellite.

Satellite *WorldView2* is also a commercial satellite for observing the Earth, owned by *Digital Globe* and it was launched in October 2009. This is the first satellite mission of the high resolution of the 8 - channel multispectral commercial satellite. The satellite is at an altitude of 770 km. It provides commercially available panchromatic images of a resolution of 0.46 m, or multispectral images with 8 bands (channels), resolution of 1.85 m. Satellite *WorldView2* has an average re-visit time of 1.1 days and can collect surface data up to 1000 000 km$^2$, and 8 - shots for one day.

Keep in mind that the WorldView2 satellite belongs to the latest generation of fast satellite systems that could rotate around their own axis, creating stereo images on the same orbit. Because of the new mode of operation, the time difference between the two stereo images is in the range of only a few seconds or minutes. The advantage of such a short time interval is that the lighting is almost constant and changes to two stereo images are only visible in moving objects, such as, for example, vehicles. Thanks to state-of-the-art technology, a new generation of high-resolution sensors (*VHR-Very High Resolution*) allows modeling of surface with a level of detail that includes infrastructure objects and buildings, both in *2D* and *3D* views.

Recently, satellite *WorldView3* is also current (Fig. 2). It is the first multipurpose, super spectral, high-resolution commercial satellite. Its operational altitude is 617 km. The *WorldView3 satellite* has a resolution of 0,31 m in the panchromatic part of the spectrum and 1,24 m in the multispectral part of the spectrum, as well as a 3,7 m resolution in the short infrared portion of the spectrum. The average visit time is just under one day and can collect data up to 680 000 km$^2$ during the day. In addition, new and enhanced applications include opportunities of rapid mapping or field mapping, soil classification, monitoring and prediction of the activities around the weather and disasters, analysis of land and vegetation, as well as many other research and observation of objects and phenomena in space.

### 3.2 Software Environment Which Has Been Developed

For the digital processing and modeling of terrain data, the *Erdas* software environment were used.

More precisely *Erdas Imagine 2014*, its *Leica Photogrammetry* module. *Erdas Imagine* is a commercial software and it is leading in the field of processing optical satellite images, as well as radar recordings.

It also allows processing and support for many commercial and open source videos such as: *WordlView-1*, *WorldView-2*, *QuickBird*, *Iconos*, *Landsat 5*, *Landsat 7,* etc.

In addition to *Erdas*, *MicroStation (MS)* software, or its modules *TerraScan* and *TerraModeler*, were used. These modules serve to manually edit (edit) clouds and create *3D* terrain models. It was used a Software *System for Automated Geoscientific Analyzes (SAGA)*. An open source software SAGA develops GIS projects. The software itself is comfortable and easy to use. Its architecture is modular, where it also

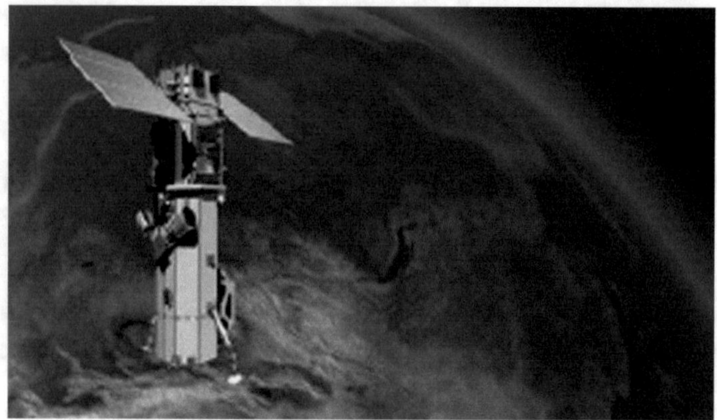

**Fig. 2.** *WorldView3* satellite

contains a DMT/DMP analysis module. It was also used to filter the *3D* terrain model in the Grid structure.

For some cases, the *ArcGIS* software product of the US Company ESRI (*Environmental Systems Research Institute*) was used. It belongs to the category of software for geoformation systems that allows data to work on a multitude of devices, as well as through the Internet browser. It is also used to create a *3D* model with a few features that enable geodata processing, spatial analysis, and statistics, as well as other complex processes that are performed over data.

### 3.3 Creating and Forming a 3D Model of the Terrain

Creation and formation of *3D* terrain models is performed based on an automatically extracted and calculated cloud point. Namely, *3D* terrain models are created differently, depending on the need and the chosen method of structuring and interpolating data. As the starting data for the formation of *3D* terrain models, i.e. the development of height models, terrain models and surface models, *WorldView-1* satellite images were used covering the territory of the city of Belgrade.

Original snowflakes are spatial resolution of 0.5 m. They are monochromatic, that is, they contain only one range, and this panchromatic spectrum. The used recordings are stereo recordings, that is, a certain part of the territory is taken from two angles, which enables the creation of *3D* terrain models. Due to time constraints and faster processing of images, cloud spots and terrain models, only one stereo shot of 15 stereo recordings (30 images) was used.

Namely, a satellite image with the mark *12MAR10093828-P2AS_R3C2-0529510 15010_01_P001 was used.* It is a third-order clip and another column. The satellite image is given in Fig. 3. The *Universal Transversal Mercator* (UTM) and the Projection Cartographic System Zone No. 34 were used for the cartographic projection of the image. Also, for the geodetic date, the ellipsoid *World Geodetic System 1984* (WGS84) was applied.

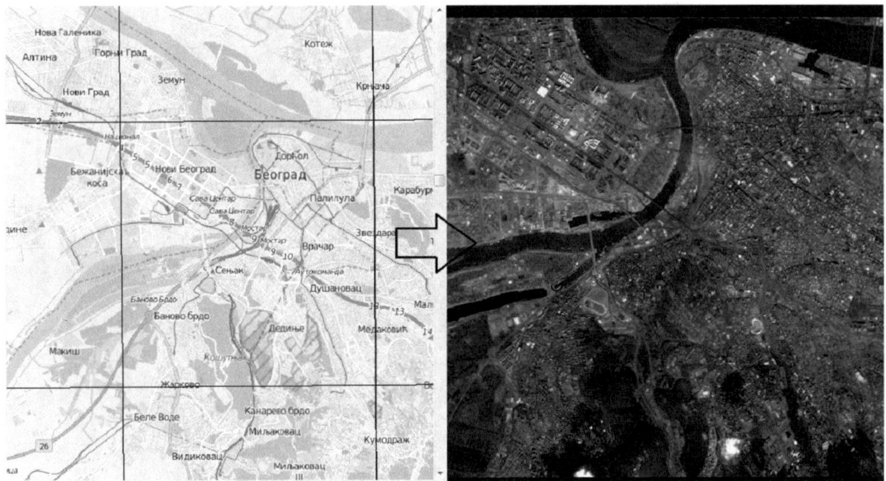


**Fig. 3.** The map and satellite image of the Belgrade area

The process of creating and creating *3D* terrain models, that is, the process of *3D* modeling in the *Erdas* software environment, proceeded at certain stages. At the beginning, it creates a block file and defines the model of the file itself. Namely, this is the *Worldview* platform, i.e. defining the *RPC* (*Remote Procedure Call*). After that, it was defined a cartographic projection and geodetic datum of the block file. It was used RPC files with the load of the stereo recordings and definition of the internal and external orientation. Finally, activates n is a program that is initiated nut tool for generating binding points. In addition, the algorithm recognizes the binding points on the recordings; generated points need to be checked (on the record or in the report). Software *LPS* (*Leica Photogrammetry System*) supports the automatic generation of clip-link points.

In our case, the number of automatically generated binding points is about 29. In this case, no checkpoints or orientation points were used. These points were not visible on the panchromatic record, i.e. the relative height model was obtained.

In addition, the layout and the number of connecting points did not have a major impact on the output accuracy of the block, as even with an enormous increase in the number of points, an accuracy increase just only 25%. Its impact is much more significant for reliability, so the scheme of the layout of the connecting points, during design, is usually determined by satisfying the criterion of reliability above all. The effect of improving the layout of the connection points on the reliability of the block is particularly expressed on the edge area of the block, as well as at the corners itself.

The process of forming the *3D* terrain model was begun in *Erdas*, starting the automatic extinction process using the classic ATE (*Automatic Terrain Extracting*) tool. This tool allows you to create a model from two shots. Spatial resolution is 5 m. The output parameters of the *3D* terrain model, i.e. DEM of Belgrade, are shown in Fig. 4.

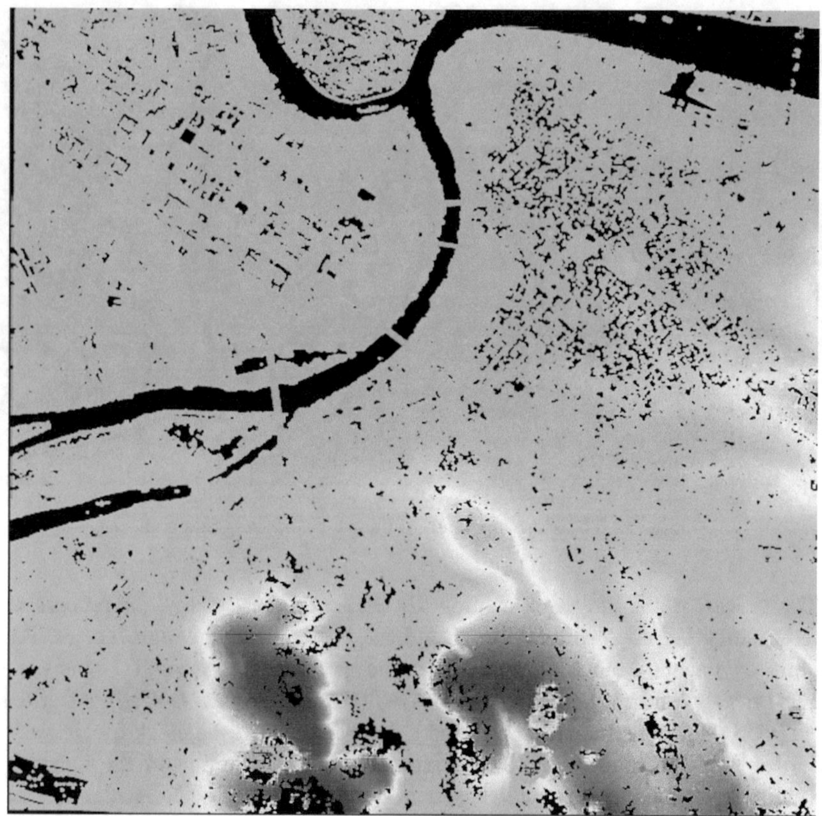

**Fig. 4.** 3D model of the terrain in the Grid structure of 5 m resolution

In addition to the classic ATE algorithm, *Erdas*'s eATE algorithm has been used to generate, a scene of terrain based on stereo images is shown on the Fig. 5. The given cloud of the dots is illustrated with certain colors based on relative heights. In doing so, no classification was performed with eATE, or the *Erdas* algorithm had no ability to allocate specific classes with points.

Further work included the modeling of the terrain model based on the cloud points obtained by the eATE. In order to create a *3D* terrain model and the classification of the point clouds must be done. The work will be continued in *MicroStation.*

After using the integrated *TerraScan* software, a point cloud representing fused clouds of the same resolution, obtained by combining two recordings, began the process of point classification. The process of point classification in *MicroStation* can be implemented in two ways:

- Manual - using the appropriate application software, systematically, or area by area. The processing process requires respecting the priorities that are classified, such as relief, objects, roads, rivers, and so on.

**Fig. 5.** Point cloud that is created using eATE algorithm

- Automatically or partially automatically - the process of automatic classification can contribute to the classification of points to some extent, but still details that are classified, must return to the manual classification step.

By automatic classification, all points are sorted in the *Default* class, and then the classification of details representing points on the surface of the ground or in the *Ground* class is done. Having in mind that the basic task is to form *3D* terrain models, this means the creation of a *3D* model based on points belonging to the *Ground* class.

In addition, one should bear in mind that the process of data modeling is one of the most demanding forms of computer graphics and visualization. Especially because it is difficult to achieve realistic views of the objects, or complexity in the work due to the large number of parameters that affect the appearance of the final *3D* terrain model. What arises from the process of digital data modeling is nothing more than a process of creating a mathematical representation of a three-dimensional object and is called a *3D* model.

In *MicroStation (MS)* software, the application solution *TerraModeler* is integrated, which is used for geoprocessing and the formation of *3D* terrain models. In Fig. 6, an editable terrain model is displayed in the TIN structure and cloud points in *MS*.

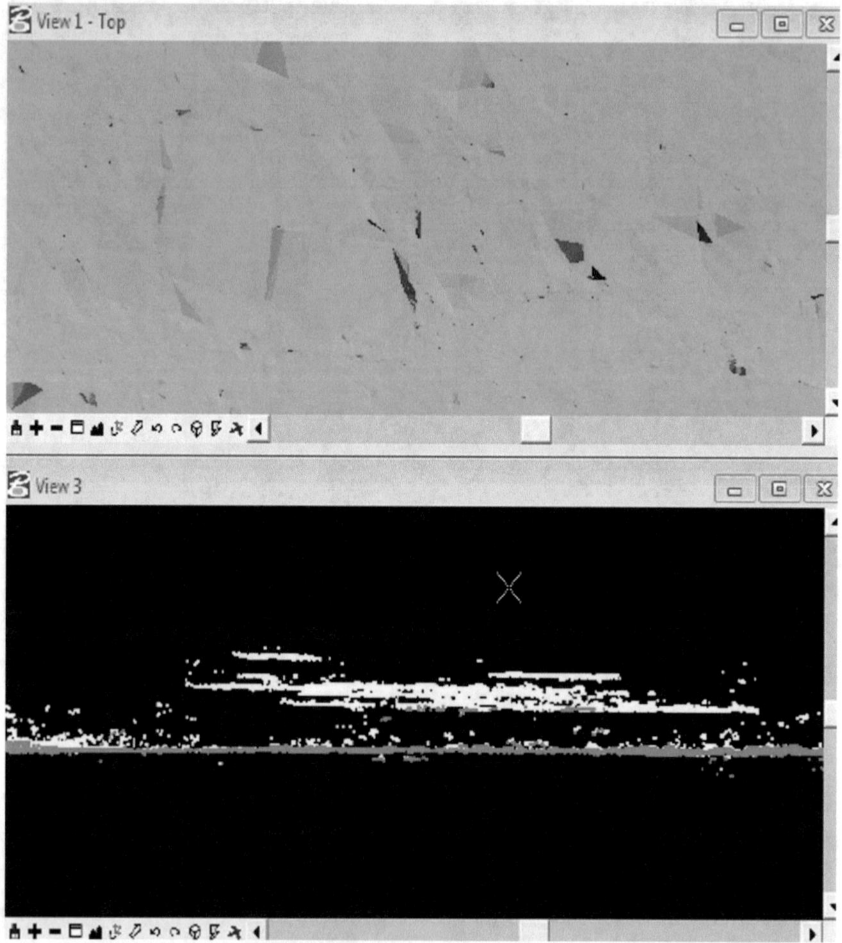


**Fig. 6.** Creation of the editable model in TIN structure (above) and point cloud in MicroStation

In doing so, the first step in creating a *3D* model is the creation of a special model (*Create Editable Model*) of surfaces of points belonging to the *Ground* class. This model provides a more realistic view and look at the terrain being processed. In addition, it is easier to identify possible errors during the classification process. The *3D* terrain model obtained from the points belonging to the *Ground* class contains certain errors and defects that are resolved by manual classification. In the process of manual classification, points located above the surface of the field and representing possible low objects, i.e. which are incorrectly included in the *3D* terrain model, are placed in the *Default* class. In addition, points that are possibly below the surface of the field, they are manually classified in the *Low Points* class, and because of that, they are excluded from the terrain model.

## 4 Results and Discussion

The next activities in the work involved quality analysis and application of *3D* terrain model. Analysis and comparison of terrain model *3D* terrain model using *TerraScan* and *TerraModeler* was performed after manual classification and data model cleaning. Based on satellite images, a point cloud is created which can serve primarily for the creation of DEM, and later with certain DMT and DMP algorithms. In addition to the terrain model in the Grid structure obtained from the *MicroStation* software, the terrain model is obtained directly from *Erdas* using the classic ATE. Namely, in the classic ATE, DEM was created. In doing so, it was necessary to approach the creation of a *3D* terrain model in the Grid structure, and for this procedure, the *Saga* software was used.

There are also two procedures for creating a *3D* terrain model based on DEM in *Saga*. The first step is to separate the DEM into *bare earth* (the future input for creating *3D* terrain models) and *removed objects*, where the *Saga's* DTM filter is used. Filtering parameters are the slope of the terrain as well as the search window itself. They are left at values 30 and 10 respectively. In addition, two grids, removed objects and a terrain model ("bare" land or *pale earth*) are shown in Fig. 7. As can be seen in the Fig. 7, the results of DTM filtering are shown. On the left side is displayed a raster of the future terrain model, while the raster of the removed objects from the DEM is displayed right.

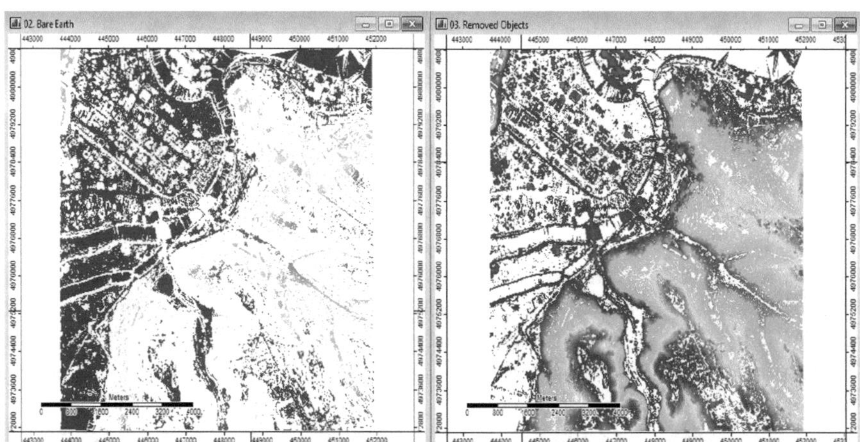


**Fig. 7.** Results of the DTM filtering

Namely, the *3D* terrain model was created based on the remaining pixels of the *bare-earth* raster. At the same time, most of the pixels came from the *removed objects*, especially in the hilly part of the Belgrade area.

In the area of New Belgrade (Novi Beograd), the situation is completely different, i.e. many pixels were added to the bare earth raster.

Then, based on the obtained partial terrain model (*bare earth*), the interpolation of the raster was started, so that the *3D* model of the terrain would cover the entire area of Belgrade. For this purpose, the *B* - spline method was used, which was also integrated into *Saga*. Finally, a *3D* model of the terrain with 5 m spatial resolution was created and shown in the Fig. 8.

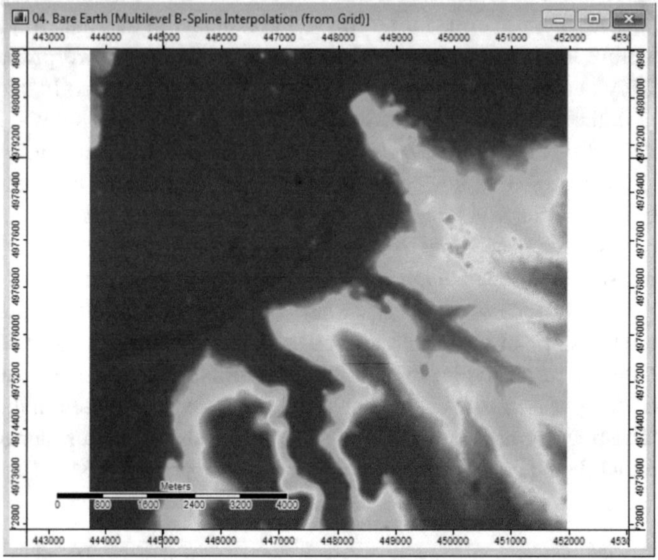


**Fig. 8.** The result of the *B*-plane interpolation. The ultimate look of the 3D model in Saga

It should be kept in mind that the used algorithms in *Saga are* fully automated and do not require much time to filter or interpolate. In addition, the user only adjusts the parameters necessary for processing the raster.

Based on the obtained *3D* models using *MicroStation* and *Saga*, their comparison was done at different locations in the city of Belgrade. Namely, the obtained *3D* model using *MicroStation* was used as a reference model, since the process of creating *3D* terrain model in *Saga* is more complex and takes much longer. On the other hand, the *3D* terrain model obtained by filtration and interpolation does not require much effort or much time.

Statistical indicators of differences these two models are shown in Table 1, the height of the lift of the ***MS*** into the model obtained using a *MicroStation* and **Z** *Saga* represents the height of the terrain model obtained by *Saga*. In addition, **ΔZ** is the absolute difference between these two models.

From Table 1, it can be noticed that 7 points have altitude value difference below one meter, while only one point drastically deviates, i.e. the difference in height is greater than 5 m. It is probably the result of a horizontal disagreement of the model and the processing of measurement data. In view of the experience of the other experiments, it may be considered that the remaining results were obtained in a satisfactory manner. In addition, the raster obtained by *Saga* can be further improved if the filtering and interpolation parameters are changed. In addition, you should keep in mind that the *3D* terrain model obtained from *MicroStation* software ***(MS)*** is not sufficiently accurate or not is one of the high-precision data models.

**Table 1.** Comparison of 3D terrain models for different point locations

| **Point description** - location | **X** | **Y** | **Z** *MS* [*m*] | **Z** *Saga* [*m*] | **ΔZ** [*m*] |
|---|---|---|---|---|---|
| Railway | 445643,24 | 4978134,96 | 123,950 | 118,348 | 5,602 |
| Highway | 445809,85 | 4978775,26 | 115,150 | 115,455 | 0,305 |
| Urban part | 446646,61 | 4978711,83 | 117,480 | 117,684 | 0,204 |
| Urban part | 445564,17 | 4979580,24 | 118,200 | 121,043 | 2,843 |
| The stadium C. Zvezda | 449481,30 | 4975380,71 | 143,620 | 143,952 | 0,332 |
| The stadium Partizan | 449025,21 | 4976016,32 | 161,350 | 162,135 | 0,785 |
| Forest area | 446263,81 | 4974429,70 | 229,179 | 228,755 | 0,424 |
| Urban part of the city | 449831,77 | 4976148,08 | 133,959 | 134,224 | 0,265 |
| Vegetation | 447553,95 | 4979492,45 | 117,390 | 118,207 | 0,817 |
| Urban part of the city | 449126,36 | 4979137,85 | 157,015 | 155,124 | 1,891 |

In this work, the ArcGIS software environment for geoprocessing and visualization of *3D* terrain models was also used. In Fig. 9, an example was illustrated where from the selected position (points) an observation was made, that is the visibility of certain parts of the territory marked with brighter colors and parts where there is no sight is marked with a darker color.

In Fig. 10, an exposure display is given, i.e. how the terrain can be oriented based on inclination in certain directions: north, northeast, east, southeast, south, south-west, west, north-west.

In addition, it is possible to identify areas with flat terrain, which could serve, for example, for landing aircraft (helicopter, aircraft) in emergencies.

The terrain exposure today is a very important natural factor and its study in recent times is increasingly gaining weight when it comes to evaluating the morphometric characteristics of the terrain. Namely, the exposition of the terrain represents the orientation of the inclination of the terrain in relation to the world. The exposure in DMT is calculated for each triangle in the TIN or for each grid cell when it comes to *Grid*. The exposition of the terrain can have values of 0° (north direction) to 360° (again north direction), which can be seen in Fig. 10. The value of each *Grid's* exposure cell points us to the orientation of the terrain surface depending on the angle of inclination. If the pitch is flat, it means that it is unclassified, and its value is taken (−1).

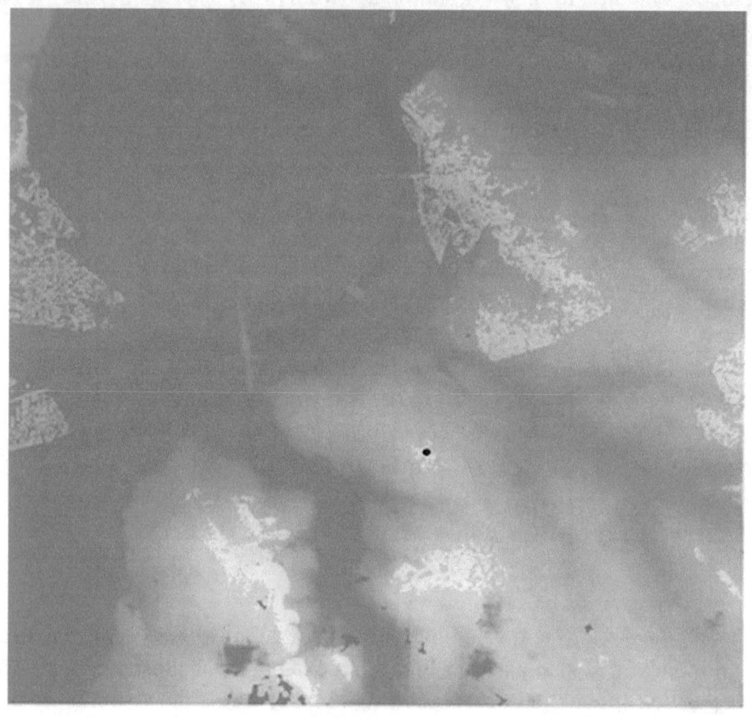


**Fig. 9.** Parts of the terrain that are seen or not from the certain position

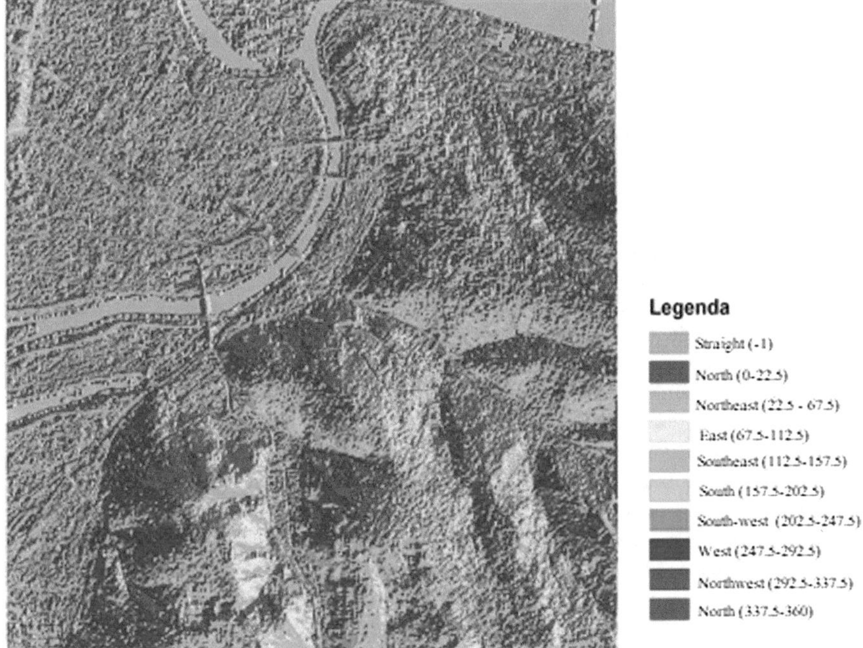


**Fig. 10.** Exposure (aspect) and orientation of the inclination of the terrain

## 5 Conclusion

Qualitative 3D data models are required in almost all scientific and business applications and applications. In this work, *3D* terrain models are created in different forms of display for the area of Belgrade based on the available satellite imagery *WorldView1*. For the processing and modeling of data, the software environment *Erdas Imagine 2014*, and its *Leica Photogrammetry* module were used. In addition to *Erdas*, *MicroStation* (MS) and its modules for manually editing cloud clouds and creating *3D* models were used.

In addition, *Saga* and *ArcGIS* software were used.

Based on the *3D* models created using *MicroStation* and *Saga,* a comparison of the results and interpretation of the models at different locations in Belgrade was performed. In addition, the obtained *3D* model using MS was used as a reference model, since the process of creating a terrain model in *Saga is* longer and more complex. On the other hand, the *3D* terrain model obtained by filtration and interpolation does not require more effort and time. By comparing the quality of these two models, it can be noticed that the higher number of points has relatively low altitude differences (below one meter), while only one point in the given m merges more (about 5 m).

Namely, it can be concluded that the results are satisfactory, and that the *grids* generated by *Sage* can be further improved if the filtering and interpolation parameters are changed. For a complete quality assessment, it is necessary to compare the height of 3D models with heights of geodetic points.

## References

1. Peckham, R., Jordan, G.: Digital Terrain Modeling. Springer, Berlin (2007)
2. Environmental Systems Research Institute/ESRI (2010): Using Arc GIS 3D Analyst, User Guide, Redlands, USA
3. Maune, D.F.: Digital Elevation Model Technologies and Applications: The DEM Users Manual. American Society for Photogrammetry and Remote Sensing Bethesda, North Bethesda (2007)
4. Li, Z., Zhu, Q., Gold, C.: Digital Terrain Modeling - Principles and Methodology. CRC Press, Boca Raton (2005)
5. Hirt, C., Filmer, M., Feathersstone, W.: Comparison and validation of the recent freely available ASTER GDEM ver1, SRTM ver.4.1 and GEODATA DEM9S ver3 digital elevation models over Australia. Aust. J. Earth Sci. **57**(3), 337–347 (2010). https://doi.org/10.1080/08120091003677553
6. Mukherjee, S., Joshi, P.K., Mukherjee, S., Ghosh, A., Garg, R.D., Mukhopadhyay, A.: Evaluation of vertical accuracy of open source digital elevation model (DEM). Int. J. Appl. Earth Obs. Geoinf. **21**, 205–217 (2013). https://doi.org/10.1016/j.jag.2012.09.004
7. de Smith, M.J., Goodchild, M.F., Longley, P.A.: Geospatial Analysis: A Comprehensive Guide to Principles, Techniques and Software Tools. Matador, Leicester, UK (2009)
8. Leica Geosystems Geospatial Imaging: LPS - Automatic Terrain Extraction, User′s Guide, USA (2006)
9. URL 1: Republic Geodetic Authority. http://www.rgz.gov.rs. Accessed 18 Apr 2019
10. URL 2: DEM explorer. http://ws.csiss.gmu.edu/DEMExplorer/. Accessed 18 May 2017
11. URL 3: Faculty of Technical Sciences University in Novi Sad, Geodesy and Geomatics. http://geo.ftn.uns.ac.rs/course/view.php?id=116. Accessed 19 May 2019
12. URL 4: Erdas. http://www.tips.osmre.gov/Software/remotesensing/Erdas.shtml. Accessed 20 May 2017

# Estimation of Longitudinal Dispersion Coefficient Using Field Experimental Data and 1D Numerical Model of Solute Transport

Hata Milišić(✉), Emina Hadžić, and Suvada Jusić

Faculty of Civil Engineering, Department of Water Resources and Environmental Engineering, University of Sarajevo, Sarajevo, Bosnia and Herzegovina
hata.milisic@gmail.com, eminahd@gmail.com, suvadajusic@yahoo.com

**Abstract.** The use of water quality models in natural environments is a very useful tool for the management of water resources. In the case of the transport of pollutants into natural watercourses, the advection-dispersion equation is widely used in its one-dimensional form to predict the spatial and temporal distribution of the dissolved substance, whether the release has occurred intentionally or accidentally. Among the important parameters of these models is the longitudinal dispersion coefficient. The objectives of this paper are: (1) the evaluation of dispersion coefficients using salt dilution method experiment and (2) the development, calibration and evaluation of numerical model for an instantaneous pollutant release in the Neretva River. In this study, field techniques are used to determine the longitudinal dispersion coefficient in the Neretva River (Bosnia and Herzegovina) using salt tracer test. Experiments are performed in order to corroborate the numerical predictions of the spatial and temporal distribution of the dissolved substance. A one-dimensional numerical model MIKE 11 is used for numerical simulation in this study. Using salt tracer data and hydrodynamic data collected from ADCP measurements for the Neretva River a dispersion coefficient was determined.

**Keywords:** Transport processes · Longitudinal dispersion coefficient · Salt tracer test · MIKE 11

## 1 Introduction

Determination of dispersion characteristics is the key task for solving problem of pollutant transport in streams and for modeling of water quality. In the applications with models of water quality in which longitudinal dispersion coefficient should be known, the most important are the studies focused on the evaluation of the environmental impact caused by the release of industrial or domestic waste. Dispersion in rivers has been studied since the 1960s, and has included observations of the processes, development of theories and derivation of mathematical models. As a result numerous tools have evolved that enable the impact of pollution incidents on rivers to be estimated. Pollutants undergo several processes when they are injected into rivers. Among these are advection


S. Avdaković et al. (Eds.): IAT 2019, LNNS 83, pp. 305–323, 2020.
https://doi.org/10.1007/978-3-030-24986-1_24

and diffusion processes. Pollutant transport processes are in essence three-dimensional, but it has been argued by many researchers [1–3] that they can be adequately represented or analyzed by one-dimensional process in a longitudinal direction. The advection dispersion equation (ADE) model is one of the most widely used model for dealing with solute transport processes. The argument is also owed to the limiting assumptions of the ADE model and estimation difficulties of its parameters [1–3].

In addition, there are numerous water quality models available: MIKE 11, QUALs, HEC-RAS and many others. The knowledge acquired from these models can be used to equip water managers with proper tools that will assist them to make reasonable water quality predictions and prevent further contamination in our rivers [4]. River dispersion models can also be applied in the development of early warning or alarm systems in order to determine the fate and the delay of a pollutant spill, accidentally discharged into a water body [4].

## 2 Theoretical Considerations

### 2.1 Mathematical Models of Solute Transport

Mathematical description of the distribution of wastewater in space and time in the water course is solved using a system of partial differential equations [5, 6]. The best solution would be to obtain a simultaneous analytical solution of the given equation system. However, such analytical solutions are virtually impossible, and solutions to individual, very simple cases are usually not of practical significance. The system of equations is solved numerically, using the methods known today, giving the results closest to the real state [5, 6]. In this way, starting from the given boundaries and initial conditions, a solution is obtained in the selected number of points in space and time. However, to solve this system, equations along the entire watercourse require a lot of computing time. In order to avoid this, in mathematical models, it is common to divide the watercourse into characteristic zones [7].

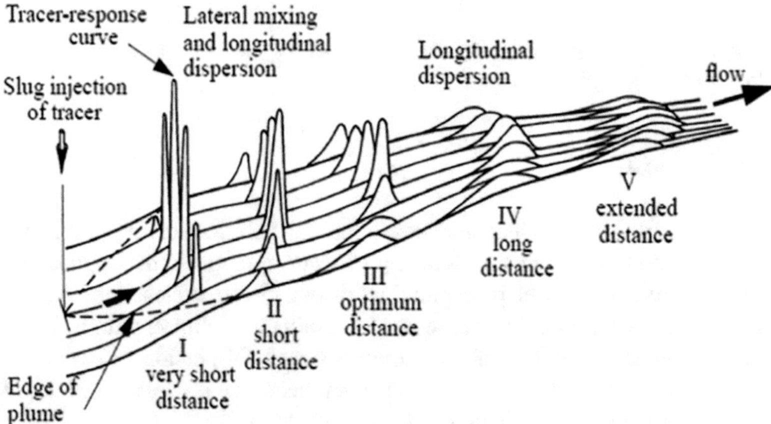


**Fig. 1.** Mixing zones - influence of transverse mixing and longitudinal dispersion on changes in concentration downstream of the discharging tracer [7]

This facilitates the simplification of the given equations, the reduction of the dimension of the problem, or the use of an approximate methods for solving the problem of distribution in particular zones. Bearing in mind how to solve the problem and the watercourse zone (Fig. 1) to which they can be applied, mathematical models can be divided into [2, 7]:

- Integral models developed primarily for the calculation of a near field (zone I, II, and III) and some cases of a distant field;
- Diffusion models for the final part of the zone III of the close field, and the middle and far field (zones IV, V and VI);
- Field models for completely solving the problem of propagation (all I–VI zones or just some selected zones).

After the comparison of the properties of the mentioned mathematical models (application area, accuracy of forecasting, economy and justification of use), diffusion models are considered most suitable for open watercourses.

Considering the turbulent characteristics of our watercourses, which are predominantly mountainous, and the existing relations of the order of the size of the discharge flow of the watercourses, a close field of relatively low length can be expected. Therefore, the middle and far field (zones IV, V, and VI according to Fig. 1) occupy the largest part of the watercourse, and the diffusion models are the most suitable for these areas.

Diffusion models are particularly simple in zone VI, where the solution of a one-dimensional transport equation is considered - the equation of a one-dimensional longitudinal dispersion. In rivers and canals, longitudinal dispersion is mainly due to vertical and transverse velocity gradients, while molecular diffusion and turbulent diffusion are generally negligible [2].

## 2.2 One-Dimensional Longitudinal Turbulent Dispersion

In the early stages of the transport process after the pollutant is discharged into the river, advection plays an important role in the transport of pollutants. In later phases, when the process of transverse mixing is complete, the longitudinal dispersion becomes important and the problem can be studied one dimensionally. Knowing the coefficient of longitudinal dispersion is required for the application of the theory of longitudinal turbulent dispersion [1, 2, 7]. The coefficient of longitudinal turbulent dispersion is also used to estimate the coefficient of reaeration and, in addition to its significance for determining the distribution of stable pollution, is needed for the analysis of the oxygen balance in the water stream (distribution of biochemical oxygen consumption and dissolved oxygen). The equation of longitudinal dispersion is made for flows that are primary in one direction (i.e. ignoring second currents), which we encounter in most practical problems (flows in pipes, channels, and natural watercourses). It is valid for a far field (zone IV according to Fig. 1) where the tracer is almost uniformly distributed across the entire cross-section [1, 2, 7].

The variations of the concentration centered by the cross-section are expressed in the direction of the flow, and it is possible to describe the distribution of the concentration with a one-dimensional equation, based on the sizes averaged by the cross-section. The insertion of the marker bar will be observed, uniformly at the cross-section at time t = 0, in the turbulent three-dimensional flow in a uniformly open channel. In the course of time, we have given a schedule of the average velocity in time in the longitudinal direction x, while the other average velocities in time, in the other two directions y and z, are $\bar{v} = 0$ i $\bar{w} = 0$. As seen in the figure (Fig. 2), in a short period of time $t = t_1$, we will have the spread of the marker strip due to an uneven distribution of the velocity at the cross-section [1, 2, 7].

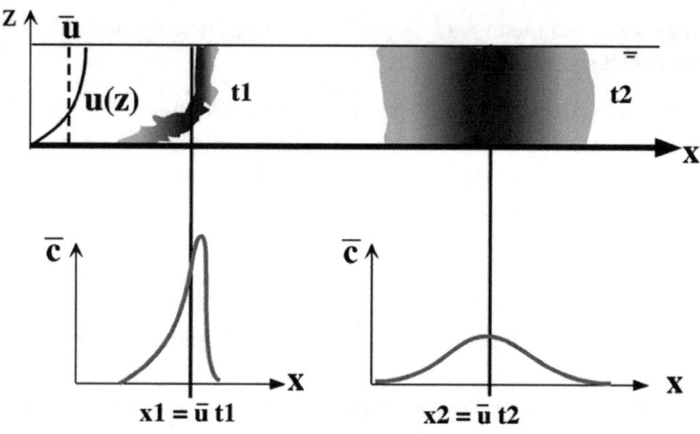


**Fig. 2.** Longitudinal dispersion - the concentration distribution centered by the cross-section at $t = t_1$ and $t = t_2$ [7]

In addition, the spread of tracers will act and turbulent diffusion in all three directions. The arrangement of the averaged concentration along the cross-section will be asymmetrical as shown in the Fig. 2. After a long time $t = t_2$, the marker will spread over the entire cross section. The variations of the concentration centered by the cross-section will be expressed in the direction of the flow, i.e. x, while in the other two directions they will be far less pronounced. Due to this arrangement of concentration, it will be possible to present the spreader of the marker with a one-dimensional equation, which will be based on the sizes averaged by the cross-section (velocity U and concentration $\bar{\bar{c}}$).

For the derivation of this equation for the unit volume (flow through one point), the equation of the turbulent diffusion will be used for this case $\bar{v} = 0; \bar{w} = 0$ [1, 2]:

$$\frac{\partial \bar{c}}{\partial t} + \bar{u}\frac{\partial \bar{c}}{\partial x} = \frac{\partial}{\partial x}\left(D_{Tx}\frac{\partial \bar{c}}{\partial x}\right) + \frac{\partial}{\partial y}\left(D_{Ty}\frac{\partial \bar{c}}{\partial y}\right) + \frac{\partial}{\partial z}\left(D_{Tz}\frac{\partial \bar{c}}{\partial z}\right) \tag{1}$$

If the variations of the averaged sizes over time are shown over their spatial variations, for this our case in a cross-section, taking the expression for turbulent diffusion in the direction x negligible in relation to the longitudinal dispersion in this direction, we obtain [1, 2]: $\bar{c} = \bar{\bar{c}} + c''$, $\bar{u} = U + u''$

$$\frac{\partial(\bar{\bar{c}}+c'')}{\partial t} + (U+u'')\frac{\partial(\bar{\bar{c}}+c'')}{\partial x} = \frac{\partial}{\partial y}\left[D_y\frac{\partial(\bar{\bar{c}}+c'')}{\partial y}\right] + \frac{\partial}{\partial z}\left[D_z\frac{\partial(\bar{\bar{c}}+c'')}{\partial z}\right] \quad (2)$$

Where $c''$ and $u''$ represent a variation with respect to the spatial average values of velocity U and concentration. If the Eq. (2) is averaged by the cross-section of the stream flow

$$\frac{\partial\bar{\bar{c}}}{\partial t} + U\frac{\partial\bar{\bar{c}}}{\partial x} + \overline{u''\frac{\partial c''}{\partial x}} = 0 \quad (3)$$

By replacing the expression $\frac{\partial u''}{\partial \mathrm{X}} = 0$, the following equation is obtained

$$\frac{\partial\bar{\bar{c}}}{\partial t} + U\frac{\partial\bar{\bar{c}}}{\partial x} + \frac{\partial}{\partial x}\overline{u''c''} = 0 \quad (4)$$

As we see, the last expression on the left-hand side of the Eq. (4), $\frac{\partial}{\partial x}\overline{u''c''}$, called the longitudinal turbulent dispersion, is the fraction of the advection, since the total convection in the stream flow is not accurately represented by averaging convection in the space in the second expression/$U\frac{\partial\bar{\bar{c}}}{\partial x}$/on the left side of the Eq. (4). From the expression of longitudinal dispersion it can be concluded that it depends on the velocity distribution $\bar{u}$ which affects the $u''$ and from the diffusion in the direction of the direction, which affects the size of $c''$. However, in order to obtain a one-dimensional equation, it is assumed that the term $\overline{u''c''}$ can be expressed, analogously to the molecular and turbulent diffusion, as $\overline{u''c''} = -D\frac{\partial\bar{\bar{c}}}{\partial x}$, which represents the size of the transport of the marked fractions of the fluid through the cross-section unit [1, 2].

At the end of the equation of the one-dimensional-longitudinal turbulent dispersion has the form [1, 2]:

$$\frac{\partial\bar{\bar{c}}}{\partial t} + U\frac{\partial\bar{\bar{c}}}{\partial x} = D\frac{\partial^2\bar{\bar{c}}}{\partial x^2} \quad (5)$$

Where are $\bar{\bar{c}}$, U and D are the sizes averaged by the cross-section.

For coefficient D, without detailed performances for the observed type of turbulent flow, where the mean value of the turbulence is sufficiently small compared to the variations in the cross section of the mean velocity in time, is the expression [1, 2]:

$$D = \overline{u''^2} \cdot T_L \quad (6)$$

Where are: $u'' = u - \bar{u}$; $T_L$ - Lagrange integral or diffusion time scale.

Taylor was the first to see [1] that the transport of mass in a single shear flow can, besides the existence of a change in velocity at a cross-section, be described by a one-dimensional diffusion equation. He carried out the Eq. (5) for a uniform flow in a real round tube and gave its solution.

Therefore, the equation solution (5), for the method described above, is given by the Eq. (7) under the same conditions for *U* and *D* (U and D are independent of the x constant). If we have a more complex case, let's say, the insertion of the marker bar, uniformly at the cross section $u_x = 0$, continuously for a longer period of time $0 < \tau < t_1$ in equal or variable wars over time, we can use this as a linear system properties of the superposition, and the solution obtained through the convolution integral. The resulting convolution integral can be solved by simple numerical methods. Accordingly, the solution would be for [1, 2]:

Case (a)

$$\bar{c}(x,t) = \frac{M}{A}\int_0^{t_1} \frac{1}{\sqrt{4\pi D(t-\tau)}} e^{-\frac{[x-U(t-\tau)]^2}{4D(t-\tau)}} d\tau \tag{7}$$

$\frac{M}{A} = c_0$; M - The mass of the marker is discharged in the unit of time

Case (b)

$$\bar{c}(x,t) = \int_0^{t_1} \frac{c_0(\tau, x=0)}{\sqrt{4\pi D(t-\tau)}} e^{-\frac{[x-U(t-\tau)]^2}{4D(t-\tau)}} d\tau \tag{8}$$

In the case of (a), in the limit case, when the discharge of the marker is continuous and when t is significantly greater than the time at which the complete passage of the first superposition response curve is realized, the time-concentration for each given distance x from the insertion point downstream from the total mixing profile, more longitudinal dispersion comes to the fore as a stationary state of concentration is established. In this way a known dilution formula is obtained [2]:

$$\bar{c} = \frac{c_0}{n} = \frac{c_0 q}{Q+q} \tag{9}$$

Where are:

n - The degree of dilution;

$c_0$ - Concentration of wastewater that flows into the water stream;

$\bar{c}$ - The concentration of the water of the watercourse after complete mixing, i.e. dilution;

Q - Stream flow- discharge;

q - Waste water flow, assuming that the water stream is not pre-polluted.

Since further exposition will only apply to dispersion, for simplicity, only one line or comma above the corresponding symbol will denote, in the future, averaging over the cross-sectional distribution and the corresponding fluctuations as already incorporated in the Eqs. (7), (8) and (9).

The Eq. (5) can be written in a different form, as well as other diffusion equations, assuming a moving coordinate system moving at the mean velocity of the "U" $\xi = x - Ut$

$$\frac{\partial \bar{c}}{\partial t} = D\frac{\partial^2 \bar{c}}{\partial \xi^2} \tag{10}$$

It is important to note that the Eq. (5) can be used only after one initial - convective - period, in the diffuse period where it is valid $c' \ll \bar{c}$. This is in agreement with the requirement $t \gg t^* \, or \, t \gg I_L$ for the application of equation of turbulent diffusion.

If this is not the case, we have to use the Eq. (1). In the diffusion period of the motion of the marker downstream and transversely to it, it is done according to the Eqs. (5) and (7) and describes Gauss's normal distribution curve at a given moment of time. It is important to note that, similar to the theory of turbulent diffusion, Taylor's theory of turbulent dispersion is based on Reynolds's analogy as well as its further application [1, 2]. As far as it is known, theoretical interpretation of the influence of the molecular diffusion of a scalar on a turbulent dispersion, analogous to that given for turbulent diffusion, has not yet been given.

After dealing with Taylor, the problem of longitudinal turbulent dispersion, Fischer made a significant contribution (1966–1968) in determining the duration of the convective and the formation of the diffuse period, and its application to natural watercourses, encouraging others to continue their work in that direction [1, 2]. The Eq. (5) derived for uniform flows can be used in approximately uniform and non uniform natural watercourses [1]. In this case, in the dispersion equation in addition $\bar{c}$ and $\bar{u}$, coefficient of longitudinal dispersion D along the watercourse, from the cross-section to the cross-section is also changed.

In order to solve the dispersion equation, we need to insert the mean value for D in the Eq. (5) in order to obtain the following equation [1, 2]:

$$\frac{\partial \bar{c}}{\partial t} + \bar{u}\frac{\partial \bar{c}}{\partial x} = D\frac{\partial^2 \bar{c}}{\partial x^2} \tag{11}$$

The basic problem, when applying the Eqs. (5) and (11) is the determination of the coefficient of longitudinal dispersion D. Analytical, empirical and experimental methods can be used for determining it. Experimental determination gives the total value of the coefficient $D_{ukupno} = D + D_T + D_M$.

However, taking into account the neglect of the turbulent $D_T$ and the molecular diffusion $D_M$, adopted in the derivation of the Eq. (5), we consider it that the coefficient of longitudinal turbulent dispersion is obtained D. The coefficient of longitudinal dispersion is generally a scalar which depends on the velocity distribution in the watercourse, that is, depends on the velocity change, as well as the geometry of the troughs if it significantly changes along the river. For one uniform flow at given flow conditions, the dispersion coefficient is constant.

## 2.3 Longitudinal Dispersion Coefficient

Dispersion in rivers has been studied since the 1960s, and has included observations of the processes, development of theories and derivation of mathematical models. As a result numerous tools have evolved that enable the impact of pollution incidents on rivers to be estimated.

Despite many tens of studies, the reliable prediction of dispersion coefficients in rivers is a difficult task [7, 8]. Although the ultimate goal would be to have a way of estimating dispersion coefficients that will apply to all rivers, this remains elusive and is probably unrealistic because of the great variability between rivers of important factors such as bed slope, structure and roughness; sinuosity and channel shape; flow rate and vegetation, both of which vary during the year [9]. The prediction of dispersion coefficients in longitudinal, lateral and vertical directions of flow are of utmost importance when evaluating the time concentration distribution at any point in a stream. To evaluate these coefficients, it is required to measure field data (velocity, width, depth, etc.) at various locations along the cross section of the river. Dispersion coefficients represent all the mixing processes in the flow [9, 10]. Longitudinal dispersion coefficient can be estimated directly using tracer test technique [1–3, 7–9]. Several empirical and analytical equations for computing the longitudinal dispersion coefficient have been recommended by various investigations [1, 11–13]. These equations produce values of longitudinal dispersion coefficient which vary widely for the same flow conditions.

According to the literature, the longitudinal dispersion varies within a range of $10^{-1}$ to $10^{7}$ $m^2/s$, while the molecular and turbulent diffusion are in the order of $10^{-9}$ and $10^{-2}$ $m^2/s$, respectively [6].

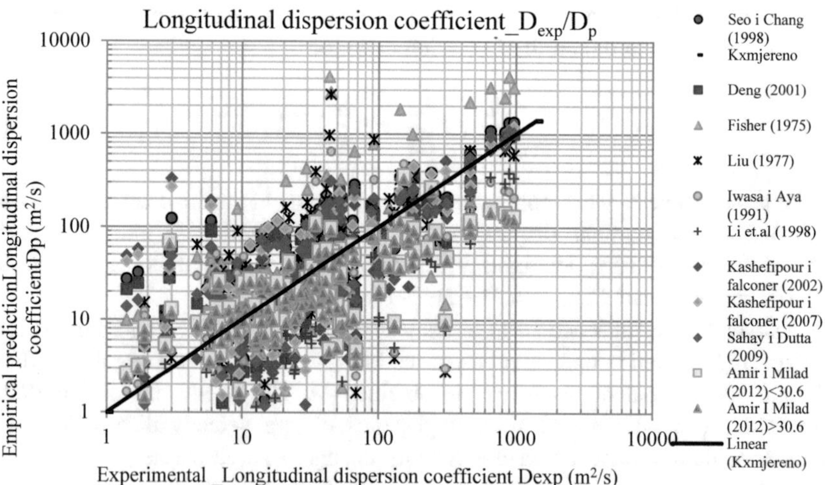


**Fig. 3.** Variation of dispersion coefficient - comparative review of measured and calculated dispersion coefficients by various authors

### 2.4 Estimation of Longitudinal Dispersion Coefficient by Tracer Dilution Methods

One of the most reliable ways to calculate the dispersion coefficient is by applying a tracer experiments - dye tracer test (or salt dilution method) as shown in the example of the next chapter. The longitudinal dispersion coefficient can be estimated by using tracer experiments in which the amount of conservative, soluble color or some other stable tracer that is injected into the river is known [2, 3, 7, 8]. The dye concentration profiles or some other stable tracker must be measured at minimum two locations downstream of the discharge into the watercourse, at a distance far enough so that the concentration at the cross-section is approximately uniform. In other words, the concentration profile should be measured downstream at least as much as Zone III (Fig. 1) [7]. If salt solutions are used as tracers, then chemical or conductivity measurement methods are used for detection and concentration measurements. Finely ground salt should be purchased for ease in mixing the solution if selected as the tracer. If dyes are used, then visual color intensity comparison standards may be used. The color-dilution method may be used for measuring small, medium, or large flows because the cost of the dye is relatively low. The salt-dilution method is applicable to measuring discharges in turbulent streams of moderate or small size where other methods are impracticable. Excessive quantities of salt are required on large streams. Tracer methods require special equipment and experienced personnel, and its use is relatively expensive [2, 3, 7, 8].

#### 2.4.1 Fischer Variance Method

To calculate the dispersion coefficient requires information on velocity and turbulent mixing that is often not known; it is therefore more common to estimate the coefficient from empirical equations or use tracer experiments by the second moment's [1, 2, 3, 4] demonstrated how the change of variance of measured concentration profiles could be used to calculate a longitudinal dispersion coefficient. This technique, called also the method of moment's, is valid as long as the rate of change of variance is linear.

In the method concentration distributions of a tracer material are measured at two (or more) points along the channel and the dispersion coefficient is calculated from the rate of change of variance of the distributions, as below.

The estimate of the longitudinal dispersion coefficient, based on the results of the tracing, uses similar expressions as for the transversal coefficient. Thus, the theoretical relation of the coefficient of dispersion is given by the expression [1–3]:

$$D = \frac{1}{2}\frac{d\sigma_x^2}{dt} \tag{12}$$

Where is $\sigma_x^2(t)$ the spatial longitudinal variance of the cloud of the tracer, and from the graph the concentration for the measurement times $t_1$ and $t_2$ follows the estimation by the expression:

$$D = \frac{1}{2}\frac{\sigma_x^2(t_2) - \sigma_x^2(t_1)}{t_2 - t_1} \tag{13}$$

In order to facilitate the practical application of the above expression, which requires measurement on a large number of transverse profiles to obtain a graph of concentration for a fixed $t$, transformation of the expression (13) is introduced. A spatial variance for fixed $t$ is transformed into a time variation for a fixed $x$, so that the estimate of the dispersion coefficient is obtained from the expression [1, 2]:

$$D = \frac{1}{2} U^2 \frac{\sigma_t^2(x_2) - \sigma_t^2(x_1)}{\bar{t}_2 - \bar{t}_1} \tag{14}$$

Where is:

$$\sigma_t^2(x_i) = \frac{\int\limits_{t=-\infty}^{\infty} (t - \bar{t}_i)^2 C(x_i, t) dt}{\int\limits_{t=-\infty}^{\infty} C(x_i, t) dt} \tag{15}$$

$$\bar{t}_i = \frac{\int\limits_{t=-\infty}^{\infty} t C(x_i, t) dt}{\int\limits_{t=-\infty}^{\infty} C(x_i, t) dt} \tag{16}$$

Average velocity in the experiment segment of the river is given by:

$$U = \frac{x_2 - x_1}{\bar{t}_2 - \bar{t}_1} \tag{17}$$

When using the dispersion coefficient calculation method described above, one should pay attention to the possible problem of long-term measurement of the concentration profile - time at two cross sections. When choosing a cross-sectional location, care must be taken to ensure that cross-sectional mixing is performed and that the cross-sections are far enough from the source to assume the validity of Fick's law [1–3]. This equation is applied in equilibrium zone where the longitudinal variance of the cross-sectional average tracer concentration increases linearly with time.

## 3 Materials and Methods

Numerical modeling has become an indispensable tool for solving various physical process [14–16]. One-dimensional (1-D) numerical models of solute transport in streams rely on the advection–dispersion equation, in which the longitudinal dispersion coefficient is an unknown parameter to be calibrated [17, 18]. In this context, we present a model of pollutant dispersion in natural streams for the far field case where dispersion is considered longitudinal and one-dimensional in the flow direction. The MIKE 11, which has earned a reputation as powerful and efficient numerical model, is used in this study [17, 18]. To validate our model, the results are compared with observations and experimental data from the Neretva River. The results show a good agreement with experimental data. The model can be used to predict the spatiotemporal changes of a pollutant in natural streams for effective and rapid decision making in a case of emergency, such as accidental discharges in a stream with a dynamic similar to that of the river Neretva (B&H) [17].

## 3.1 Study Area

The study area occupies the middle part of the Neretva river basin, located in the north region of Bosnia and Herzegovina. The Neretva River is the largest river in the catchment of the Adriatic Sea. The river length and basin is about 225 km and total area in FB&H of approximately 5745 km$^2$ [17, 18].

Together with its tributaries, it represents an individual environmental unit and a unique ecological system in this part of world. The river water is intensively used for hydroelectric power generation, domestic and industrial water supply and agricultural irrigation. River traversing nears the industrial municipalities and city of Mostar, is largely used for constant disposal of untreated effluents. Currently in Mostar, there is no waste water treatment plant. The major sources of pollution are industrial and domestic waste water and agricultural drainage. Consequently water quality of the river degrades particularly in the low flow months. The river reach considered in this work begins downstream from the Hydropower station Mostar to the Bačevići hydrometric station (Fig. 4), with a length of approximately 11 km.

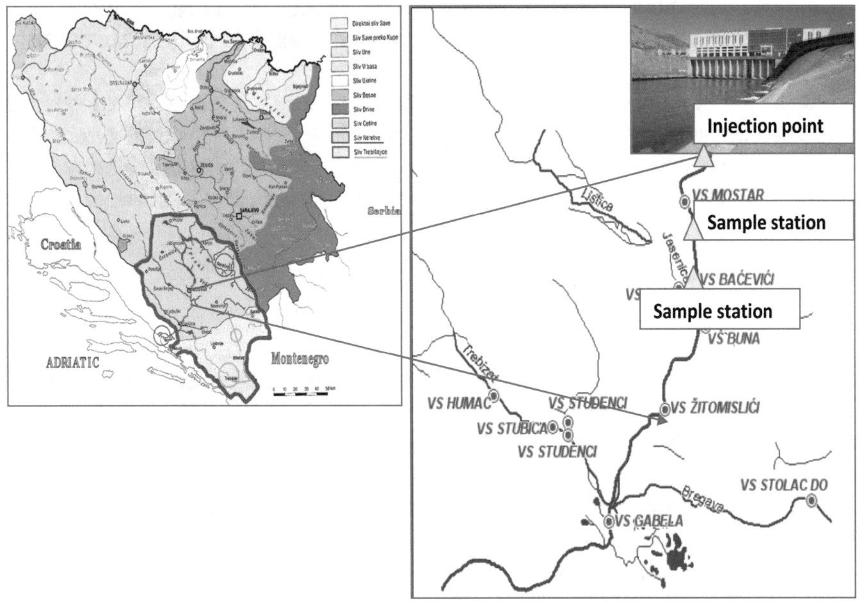


**Fig. 4.** Study area - Neretva river basin and sampling stations [17]

Three measuring sites were considered, injection site P0 (Raštani – downstream of the Hydroelectric Power Plants - HPP Mostar), sampling site P1 (Begić bridge in

Mostar) and sampling site P2 (Bačević). The location of sampling stations was established according to their accessibility (bridges), mixing conditions, weirs location, logistics and human resources availability [17, 18].

## 3.2 Numerical Modelling of Pollutant Transport in the Neretva River

The analysis of mixing processes in natural streams is very important to understand and predict water contamination and pollutant dispersion. Pollutant behavior is usually modelled by a standard advection-diffusion equation for the concentration [14–16]. One-dimensional models, relatively simple and widely used for global evaluation of water quality are applicable only after a certain distance downstream from the source of pollution, called the mixing length, distance over which the pollutant dispersion is done completely on the whole cross-section, so that the local deviations from the mean concentration become negligible. Among the possible existing software packages, MIKE 11 was chosen which consists of several modules for specific classes of the water flow [14, 17, 18].

### 3.2.1 MIKE 11 Model

DHI's MIKE 11 model [19] is a professional engineering software package for the simulation of flows, water quality and sediment transport in estuaries, rivers, irrigation systems and other water bodies. MIKE 11 is a one-dimensional, fully dynamic modelling tool. The hydrodynamic (HD) module forms the basis for most add-on modules including the advection-dispersion module (AD). MIKE 11 HD solves the vertically integrated equations for the conservation of continuity and momentum, i.e. the Saint - Venant equations.

The model simulates water levels and flows in response to specified boundary conditions (water levels, flows or Q-h relations). The one-dimensional (vertical and lateral variation integrated) equation for the conservation of mass of a substance in solution, i.e., the one-dimensional advection-dispersion-reaction equation, reads as follows [19]:

$$\frac{\partial \mathrm{AC}}{\partial \mathrm{t}} + \frac{\partial \mathrm{QC}}{\partial \mathrm{x}} - \frac{\partial}{\partial \mathrm{x}}(\mathrm{AD}\frac{\partial \mathrm{C}}{\partial \mathrm{x}}) = -\mathrm{AKC} + \mathrm{C_2 q} \tag{18}$$

Where are: $C$ is the concentration (arbitrary unit), $D$ is the dispersion coefficient [$L^2$ $T^{-1}$], $A$ is the cross-sectional area [$L^2$], K is the linear decay coefficient [$T^{-1}$], $C_2$ is the source/sink concentration [$ML^{-3}$], $q$ is the lateral inflow [$L^2$ $T^{-1}$], $x$ is the space co-ordinate [L] and $t$ is the time co-ordinate [T].

For advection-dispersion module, it is important to specify the substances considered in simulation and the value of dispersion coefficient. This value can be specified according to the equation [19]:

$$D = aV^b \tag{19}$$

Where are: $a$ – dispersion factor, $b$ – dispersion exponent and $v$ is a flow velocity.

### 3.3 The Field Experiments in the Neretva River

After the mathematical modeling of solute transport, the tracer requirements for model calibration and verification can be formulated. Measurement techniques, generally, must be selected or developed to meet individual model requirements [14, 15]. To plan a tracer experiment, the tracer break-through times (travel times) at every observation point in the river will be calculated initially with the model. The model, in turn, requires estimates of the parameter mean flow velocity and the longitudinal dispersion coefficient for this calculation [20]. Before tracer injection takes place, careful planning to provide a time table for measurements, laboratory analysis, and data interpretation is necessary. In this study, the salt dilution method was performed to estimate the longitudinal dispersion coefficient. Tracer test was carried out in the Neretva River in October 2013. The experiment consisted in instantaneous (slug) injections of tracer. The tracer used in this study was salt (NaCl), recommended by its characteristics: not toxic, not reactive, good diffusivity, good detectability, low sorptive and acidity and inexpensive. The tracer solution was prepared in a barrels by adding 2000 kg of NaCl in 4000 L of water (4 barrels per 1000 l) and mixed thoroughly so that the tracer salt was completely dissolved (the concentration of the dissolved salt was about 500 g/l). The tracer solution was then pulse-injected into the Neretva River at Injection site P0 (Figs. 5 and 6). Conductivity concentrations were measured at two downstream sections with a sampling period of 60 s using portable field conduct meters (HACH CO150 and HACH HQ14d). Background concentration conductivity is taken in all sampling sites. The location of the injection and the measurement profiles was chosen so that the study reach could be considered as approximately uniform. The distance from injection allowed the tracer to be well mixed over the cross-section at the measurement profiles. A monitoring program was carried out after injecting tracer in order to characterize in situ the transport and dispersion behavior of the river under one hydrodynamic regime: frequent (150 $m^3s^{-1}$) flow. Table 1 presents the information concerning the tracer injection performed during the sampling program included in this study. Sampling sites location (Figs. 5, 6 and 7) were established according to the aims of this monitoring program, the sites accessibility (bridges), river physics characteristics, mixing conditions, weirs location, logistical means and human resources availability.

**Table 1.** Data of tracer injections/sampling information

| Injection/sampling site | Date | Hour | Discharge ($m^3/s$) | Distance m | Salt/NaCl mass (kg) |
|---|---|---|---|---|---|
| Injection site P0 | 20/10/2013 | 10:40 AM | 115 | 0 | 2000 |
| Sampling site P1 | 20/10/2013 | 12:05 PM | 147 | 5570 | |
| Sampling site P2 | 20/10/2013 | 13:10 PM | 152 | 10968 | |

**Fig. 5.** Location of injection site - Profile P0

**Fig. 6.** Pictures of the Neretva river illustrating typical study reaches/injection site P0

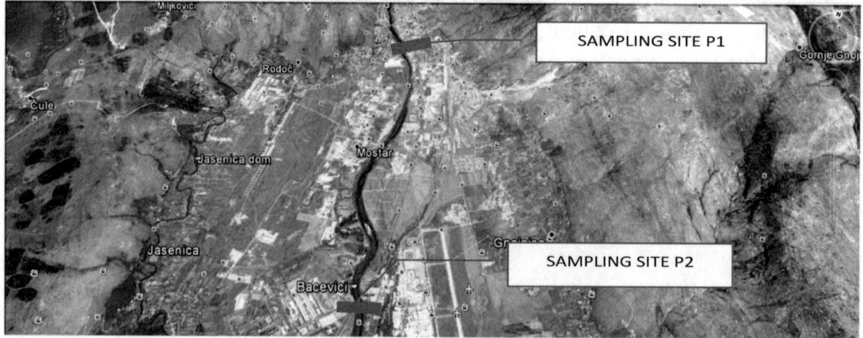


**Fig. 7.** Location of sampling site - Profile P1 and P2

The discharges and velocity values are obtained from conducted ADCP measurements of the velocity and discharge on October 20th, 2013 at three transects (see Fig. 8). Mean water velocity in reach was calculated with mean travel time and distances between sampling sites. Along with the ADCP measurements on the profiles P0 and P1, geodetic measurements of the profile distance and the absolute level of the water mirror with the total station and GPS devices were performed [17].

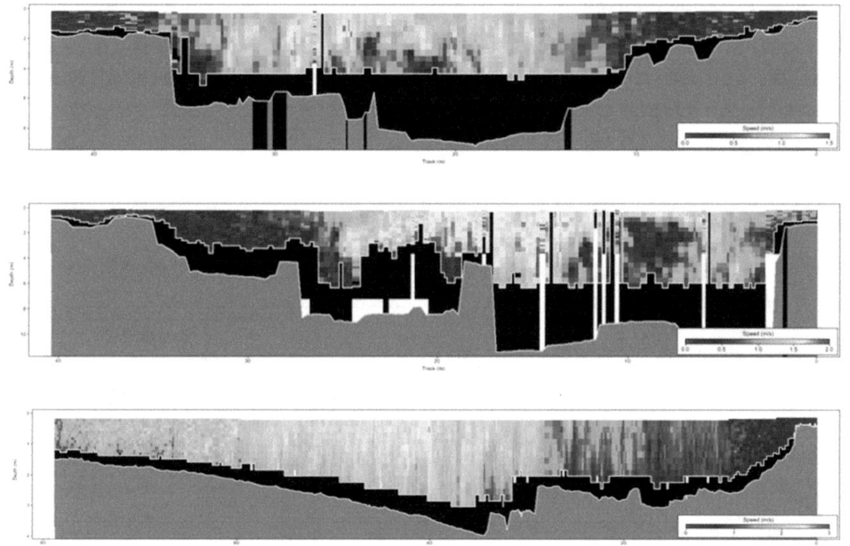

**Fig. 8.** ADCP measurements of the velocity and discharge on October 20th, 2013 at three transects (injection site P0, sampling site P1 and P2)

# 4 Results and Discussion

Breakthrough curves are the basis for determining the travel time of tracers and the dispersion characteristics of the watercourse. These travel times are used to calculate the mean speed of the tracker along the river, which is then used to estimate the longitudinal dispersion coefficient. Experimental longitudinal dispersion coefficient was calculated from concentration-time curve at consecutive sampling sites, using the methodology described in previous section for tracer studies.

The analysis of the measured break-through curves is performed using the moment method and a non-linear least squares fitting procedure. The calibrated MIKE 11-Neretva Model was verified by comparing calculated and measured concentration distributions from additional tracer experiments. These results are listed in Fig. 9 and Tables 2, 3 and 4.

Through the performed tests, in the field and in the laboratory, all the necessary data were obtained for the calculation of the dispersion coefficient for a given segment, or its individual segments. Calculation of the coefficient D was made according to the Fischer method of momentum.

The salt tracer injected mass recovered at each sites allows assessing the importance of physical and biochemical processes by quantification of precipitation, sorption, retention and assimilation losses.

The coefficient of longitudinal dispersion obtained by field measurements was used to calibrate (validate) the 1D model of the transport of pollution of the Neretva River. The model calibration procedure (Fig. 9) included the adjustment of the friction coefficients values and the longitudinal dispersion coefficients. Table 4 show compares the average velocity, travel time and dispersion results obtained from Neretva river model simulations with experimental tracer data. The model results show a good correlation with experimental data, accurately reproducing the tracer peak concentrations and the travel time between consecutive sampling stations. Data on water quality (conductivity), collected in 2013 was used for the verification of the AD - MIKE 11 model.

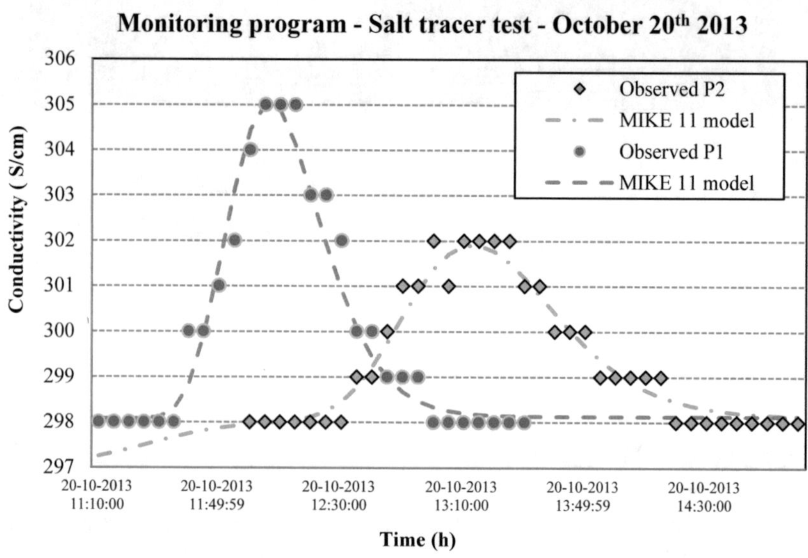


**Fig. 9.** The Neretva River model calibration with field data - comparison of MIKE 11 numerical model results and measurement of salt tracer experiment

**Table 2.** Total mass of salt on profiles P1 and P2

| **Sampling profile P1 - Begića most (distance from P0 = 5770 m)** | | | | |
|---|---|---|---|---|
| $S_1t^2$ | Q (l/s) | Amesurement (mg) | A injection (mg) | Am/Ainj (%) |
| 1029273,197 | 146000 | 1466585190,11 | 2000000000 | 73,33 |
| **Sampling profile P2 - Bačevići (distance from P0 = 10968 m)** | | | | |
| $S_1t^2$ | Q (l/s) | Amesurement (mg) | A injection (mg) | Am/Ainj (%) |
| 2105316,00 | 150500 | 1354499993,02 | 2000000000 | 67,73 |

**Table 3.** Determination of the coefficient of longitudinal dispersion by the moment method

| Profile $P_0$ | Determination of the coefficient of longitudinal dispersion by the moment method | | | | | | | | |
|---|---|---|---|---|---|---|---|---|---|
| $X_0$ | $X_1$ | $X_2$ | $P_1-P_2$ | $t_{sr}P_1$ | $t_{sr}P_2$ | Usr | $St^2$ $P_1$ | $St^2$ $P_2$ | $D_{mjer}$ |
| (m) | (km) | (km) | (km) | (sec) | (sec) | (m/s) | ($sec^2$) | ($sec^2$) | ($m^2/s$) |
| 0 + 00 | 5,8 | 11,0 | 5,2 | 5547,5 | 9678,0 | 1,26 | 1029273,20 | 2105316,00 | **206,28** |

Table 4 compares the average velocity, travel time and dispersion results obtained from Neretva river model simulations with experimental tracer data.

**Table 4.** The Neretva River model calibration results - experimental and numerical model results

| River - Reasch | Average velocity - V | | Travel time - T | | Long. dispersion coeff. - D | |
|---|---|---|---|---|---|---|
| | (m/s) | (m/s) | (h) | (h) | ($m^2$/s) | ($m^2$/s) |
| Neretva River | Tracer | MIKE 11 | Tracer | MIKE11 | Tracer | MIKE 11 |
| P1 – P2 | **1,26** | **1,16** | **1,15** | **0.98** | **206,28** | **225,00** |

## 5 Conclusions

Numerical modeling has become an indispensable tool for solving various physical processes. In this context, we present a model of pollutant dispersion in the Neretva River for the far field case where dispersion is considered longitudinal and one-dimensional in the flow direction.

Generally, the knowledge of transport processes in both large and small rivers are of increasing importance concerning the prediction of the pollutant distribution in aquatic systems. In principal, there are two approaches to calculate the transport of solutes in rivers. One is the more classical calculation based on exact river morphological input data and the other is the calculation based on estimation of transport parameters such as travel time and dispersion coefficients. Since exact morphological data are often unavailable, the parameter estimation technique is more promising. In both cases, tracer experiments are needed to calibrate and verify the calculation. Tracer experiments play a decisive role in the model development. Therefore, special measurement equipment is necessary. In the study described in this paper, MIKE 11 transport model based on one-dimensional dispersion equation are presented. Prior it application, river model should be calibrated and verified with consistent datasets, based on experimental or field data of high quality, and covering a representative range of river flow values.

The aim of this study was to predict the coefficient of longitudinal turbulent dispersion in natural open watercourse based on a comprehensive analysis of the results of field measurements and numerical modeling. The purpose of this paper was to describe the preliminary results of a study to determine the feasibility and accuracy estimation of longitudinal dispersion coefficient using the salt dilution method in the Neretva River for calibrated of numerical model MIKE 11.

For this purpose, field measurements were carried out, hydromorphologic with ADCP device (SonTek - RiverSurveyor S5) and field salt tracer experiment in the Neretva River on the selected water course segment. Through the performed tests, in the field and in the laboratory, all the necessary data for calculating the coefficient of longitudinal dispersion for a given share were obtained. Calculation of the coefficient D was made according to the Fischer method of momentum. Breakthrough curves are the basis for determining the travel time of tracers and the dispersion characteristics of the watercourse. These travel times are used to calculate the mean travel speed of the route along the river, which is then used to estimate the longitudinal dispersion coefficient. Description of the implementation of experiment with salt and all procedures for obtaining results as well as the budget itself that was carried out to determine the coefficient of longitudinal dispersion are described in detail in the previous research paper.

The coefficient of longitudinal dispersion obtained by field measurements was used to calibrate (validate) the 1D model of the transport of pollution of the Neretva river (numerical model MIKE11). The validated transport model of the Neretva River was further used to simulate suddenly discharged (incidental) pollution and its impact on the water quality in the Neretva River.

In general, models results showed a satisfactory agreement with experimental data. One-dimensional modeling, which was required to estimate short-term travel times typical of accidental spills, and the one point measurement technique are simple and reasonably reliable tools to be considered for estimating the distribution of solutes in large rivers. This procedure is of paramount interest in river basin management strategy for defining early warning or alarm systems, minimizing the effects from accidental pollutant spills, and to improve water sources protection practices. These capabilities of the MIKE 11 model therefore ensure its use for the present as well as future plans of the Neretva River. Simulation results can be used in engineering, water management, water quality management and planning applications. Most of the data explored and used in this calculation were obtained some years ago, and probably need updating.

## References

1. Fischer, H.B.: Longitudinal Dispersion in Laboratory and Natural Streams. Rept. H-4–12, California Institute of Technology, Keck Lab., Pasadena, CA, USA (1966)
2. Bajraktarević-Dobran, H.: Dispersion in mountainous natural streams. J. Environ. Eng. Div., ASCE **108**(EE3), 502–514 (1982)
3. Day, T.J.: Field procedures and evaluation of a slug dilution gauging method in mountain streams. J. Hydrol. (NZ) **16**(2), 113–133 (1977)
4. Heron, A.J.: Pollutant transport in rivers: estimating dispersion coefficients from tracer experiments. Master Thesis, School of Energy, Geoscience, Infrastructure and Society, Heriot-Watt University (2015)
5. Jobson, H.E., Sayre, W.W.: Predicting concentration profiles in open channel. Am. Soc. Civ. Eng. **96**(10), 1983–1986 (1970)
6. Rutherford, J.C.: River Mixing. Wiley, New York (1994)
7. Kilpatrick, F.A.: Techniques of water-resources investigations of the United States geological survey, Simulation of soluble waste transport and buildup in surface waters using tracers. USGS, Denver, CO (1993)
8. Singh, S.K., Beck, M.B.: Dispersion coefficient of streams from tracer experiment data. J. Environ. Eng. **129**(6), 539–546 (2003)
9. Socolofsky, S.A., Jirka, G.H.: Special topics in mixing and transport processes in the environment engineering - lectures. Coastal and ocean engineering division, Texas (2005)
10. Chapra, S.C.: Surface Water Quality Modelling. McGraw-Hill, New York (1997)

11. Deng, Z.Q., Singh, V.P., Bengtsson, L.: Longitudinal dispersion coefficient in straight rivers. J. Hydraul. Eng., ASCE **127**(11), 919–927 (2001)
12. Kashefipour, S.M., Falconer, R.A.: Longitudinal dispersion coefficients in natural channels. Water Res. **36**, 1596–1608 (2002)
13. Seo, I.W., Cheong, T.S.: Predicting longitudinal dispersion coefficient in natural streams. J. Hydraul. Eng., Am. Soc. Civ. Eng. **124**(1), 25–32 (1998)
14. Parsaie, A., Hamzeh, H.A.: Computational modeling of pollution transmission in rivers. Appl. Water Sci. **2017**(7), 1213–1222 (2017). https://doi.org/10.1007/s13201-015-0319-6
15. Duarte, A.A.L.S., Pinho, J.L.S., Vieira, J.M.P., Boaventura, R.A.R.: Comparison of numerical techniques solving longitudinal dispersion problems in the River Mondego. In: Bento, J., et al. (eds.) EPMESCVII: Computational Methods in Engineering and Science, vol. 2, pp. 1197–1206. Elsevier Science, Ltd., Oxford (1999)
16. Duarte, A.B.R.: Pollutant dispersion modelling for Portuguese river water uses section linked to tracer dye experimental data, vol. 4, ISSN: 1790–5079 1047, Issue 12 (2008)
17. Milišić, H.: Field and numerical investigations of the coefficient of longitudinal turbulent dispersion in the transport processes of open watercourses. Doctoral Thesis, Faculty of Civil Engineering, University of Sarajevo (2017)
18. Milišić, H., et al.: Mathematical modeling of surface water quality. Advanced technologies, systems and applications III. In: Proceedings of the International Symposium on Innovative and Interdisciplinary Applications of Advanced Technologies (IAT), vol. 2, Springer (2019)
19. MIKE bay DHI- MIKE 11: User Guide, A modelling system for Rivers and Channels (2008)
20. Launay, M., Le Coz, J., Camenen, B., Walter, C., Angot, H., et al.: Calibrating pollutant dispersion in 1-D hydraulic models of river networks. J. Hydro-Environ. Res., Elsevier, **9**(1), 120–132 (2015)

# Computation of the Fluid Flow Around Polyhedral Bodies at Low Reynolds Numbers

Muris Torlak(✉), Almin Halač, Mirela Alispahić, and Vahidin Hadžiabdić

Mechanical Engineering Faculty, University of Sarajevo, Vilsonovo šetalište 9, 71000 Sarajevo, Bosnia-Herzegovina
torlak@mef.unsa.ba

**Abstract.** The flow of incompressible fluid around the cube, octahedron and dodecahedron is investigated in the range of Reynolds numbers $\mathrm{Re}_d$ based on the largest polyhedron-width from 50 to 5000 using computational fluid dynamics (CFD) simulations. The velocity field as well as the drag coefficients obtained in the case of flow around the cube are compared to the results published in the literature by other authors, showing a reasonable agreement. In the steady flow regime the computed drag coefficients of cube agree well with the reference data. A certain discrepancy is found in the time-dependent, transitional flow regime, but the values stay between the computational results and the correlation formulae from the literature. Also beyond the transitional regime, the present results approach those from the literature.

## 1 Introduction

In a large majority of cases, the analysis of flow around particles or bluff bodies with more-or-less isometric shape is based on the models presuming spherical shape. In reality, the shape may be different, and typically, it is complex. It is widely accepted that particles formed through long lasting natural processes have smooth shapes with curved surfaces, while particles formed in artificial, technological processes typically have irregular shapes due to a number of mechanical deformations, wearing, fractures or other phenomena to which they are exposed. Despite the fact that the most analyses with spherical models of particles/bodies are accepted as satisfactory and valid, still it may be needed to apply more detailed models to describe behavior of particles or bodies with realistic shapes. The flow around some typical non-spherical, bluff bodies (cylinders, cubes or boxes, ellipsoids, discs, wedges) has been investigated since years, both experimentally and computationally, while the results of investigations of polyhedral shapes are much less frequently found in the literature.

This paper presents results of the simulation of incompressible flow around stationary bodies: a *cube*, an *octahedron*, and a *dodecahedron*, at low Reynolds numbers including laminar and transitional regime, with constant far-stream fluid velocity. The results of the flow around a cube are used to verify the model settings as well as the quality of discretization, which are used later for the simulation of flow around the octahedron and dodecahedron.


S. Avdaković et al. (Eds.): IAT 2019, LNNS 83, pp. 324–333, 2020.
https://doi.org/10.1007/978-3-030-24986-1_25

## 2 Problem Description

The solution domain extends from $-10d$ to $20d$ in the stream-wise direction ($x$-coordinate), and from $-5d$ to $5d$ in the cross directions ($y$- and $z$-coordinate), where $d$ is the largest width of the body – i.e. the diameter of the sphere circumscribed around the body considered. The body is placed $10d$ from the inlet boundary, in the middle of the channel cross section, see Fig. 1.

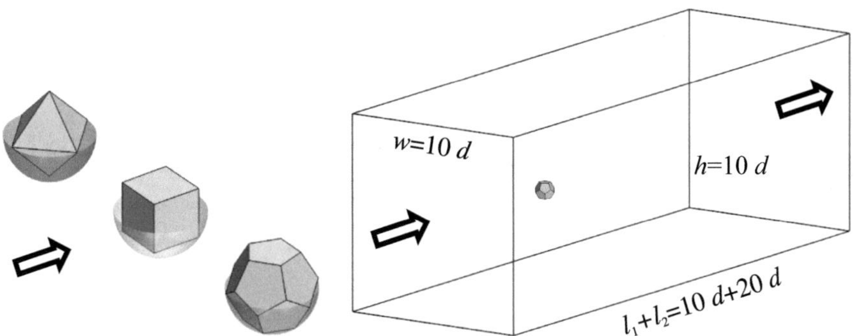


**Fig. 1.** The polyhedral shapes shown in orientation as investigated in this study (left) and the solution domain (right).

The body surface is defined as the no-slip wall, while the side boundaries are regarded as slip walls. The incoming velocity of the fluid, here air flow is adopted, is constant and uniformly distributed over the inlet boundary, while zero-gradient condition is assumed at the opposite outlet boundary. The inlet air velocity is varied from case to case in accordance with the Reynolds number considered.

The shapes of the investigated bodies require some geometric properties to be involved in evaluation and interpretation of the results. The basic geometric properties of the platonic polyhedra, to which the bodies investigated in this paper belong, are shown in Table 1.

**Table 1.** Geometric properties of the platonic polyhedra, with $a$ being the edge length.

| Polyhedron type | Number of faces | Volume | Area of the surface | Radius of the circumscribed sphere | Radius of the inscribed sphere | Diameter of the sphere with equivalent volume | Sphericity |
|---|---|---|---|---|---|---|---|
| tetrahedron | 4 | $\frac{1}{6\sqrt{2}}\,a^3$ | $a^2\sqrt{3}$ | $\sqrt{\frac{3}{8}}\,a$ | $\frac{1}{\sqrt{24}}\,a$ | $\frac{1}{\sqrt[6]{2}\sqrt[3]{\pi}}\,a$ | 0.671 |
| cube | 6 | $a^3$ | $6a^2$ | $\frac{\sqrt{3}}{2}\,a$ | $\frac{1}{2}\,a$ | $2\frac{\sqrt[3]{3}}{\sqrt[3]{4\pi}}\,a$ | 0.806 |
| octahedron | 8 | $\frac{1}{3}\sqrt{2}\,a^3$ | $2\sqrt{3}\,a^2$ | $\frac{\sqrt{2}}{2}\,a$ | $\frac{\sqrt{6}}{2}\,a$ | $\frac{2}{\sqrt{2}\sqrt[3]{\pi}}\,a$ | 0.846 |
| dodecahedron | 12 | $\frac{15+7\sqrt{5}}{4}\,a^3$ | $3\sqrt{25+10\sqrt{5}}\,a^2$ | $\frac{\sqrt{3}(1+\sqrt{5})}{4}\,a$ | $\frac{1}{2}\sqrt{\frac{5}{2}+\frac{11}{10}\sqrt{5}}\,a$ | $[\frac{3(15+7\sqrt{5})}{2\pi}]^{1/3}\,a$ | 0.910 |
| icosahedron | 20 | $\frac{5(3+\sqrt{5})}{12}\,a^3$ | $5\sqrt{3}\,a^2$ | $\frac{\sqrt{10+2\sqrt{5}}}{4}\,a$ | $\frac{\sqrt{3}(3+\sqrt{5})}{12}\,a$ | $[\frac{5(3+\sqrt{5})}{2\pi}]^{1/3}\,a$ | 0.939 |

In all simulations it was adopted that the bodies have the same width – the diameter of the circumscribed sphere ($d$ = 10 mm). The choice of this property as the common reference length is implied by use of the same settings in the mesh generation (for typical size of the mesh cells near the body), as well as by comparison of the results.

In order to analyze relations to much more frequently encountered spherical models of particles or spherical bodies, agreement of the real body shape with the ideal spherical one is usually described by the term of *sphericity*. Sphericity is defined as the ratio of the area of a sphere occupying the same volume as the body under consideration, and the area of the bounding surface of that body:

$$\phi = \frac{S_{\text{sphere}}}{S_{\text{body}}}.$$

For comparison with the data from literature, in this work the flow regime is described by three different Reynolds numbers, depending on the length scale applied:

$$Re = \frac{\rho v a}{\mu},$$

where $\rho$ is the density of the fluid, $v$ is the far-stream velocity, $\mu$ is the dynamic viscosity of the fluid, and $a$ is the polyhedron edge length (this is standard choice in the case of analysis of flow around a cube or a box),

$$Re_{\text{es}} = \frac{\rho v d_{\text{es}}}{\mu},$$

where $d_{\text{es}}$ is the diameter of an effective sphere of the same volume as the body considered (usually used in the analysis of non-spherical bodies and particles), and

$$Re_{d} = \frac{\rho v d}{\mu}.$$

Similarly, depending on the literature source, two different definitions of the force coefficients can be encountered. The mostly used definition is:

$$c_{D,A} = \frac{F_x}{\frac{\rho v^2}{2} A_{\text{front}}},$$

where $F_x$ is the drag force exerted on the body, $A_{\text{front}}$ is the frontal projected area of the body, exposed to the incoming flow direction, and the expression based on the projected area of the volume-equivalent sphere is:

$$c_{D,\text{es}} = \frac{F_x}{\frac{\rho v^2}{2} \frac{d_{\text{es}}^2 \pi}{4}}.$$

## 3 Simulation Method

Simulations are done using the program Simcenter STAR-CCM + v.13.06 (Siemens PLM 2018). As an integrated part of the simulation workflow, the solution domain is filled by a computational mesh, building thus a set of computational cells. The conservation laws of continuum mechanics (conservation of mass and conservation of linear momentum) with a corresponding constitutive relation for flow of incompressible, Newtonian fluid are applied to each cell in the mesh. These governing equations are solved using a finite-volume method, based on collocated variable arrangement (all solution variables refer to the computational points placed at the centroids of cells). For discretization of the convective term linear-upwind scheme is used, while the gradients are approximated using a variant of the Gauss method, yielding formally second-order accuracy of the spatial discretization. Anticipating time-dependent flow at higher Reynolds numbers (oscillating flow due to wake behavior, instabilities and transition to turbulence) the simulations are performed using integration in time. An implicit, second-order accurate scheme is applied for approximation of time-dependent properties. In the flow regimes considered here, turbulence is *not* modelled. Laminar flow solution approach is applied to all Re numbers, imitating direct numerical simulation.

The implemented method is capable of handling unstructured meshes with cells of polyhedral shape. In this work, the computational meshes are automatically generated using the trimmed cells – cubical cells in the core of the flow, while the cells in the vicinity of the boundaries are trimmed in order to adapt to the complex boundary shapes. Additionally, prism layers are created adjacent to the no-slip walls (polyhedron surface) in order to provide higher mesh resolution in direction normal to the walls. 10 layers of prisms with stretching of subsequent layers in direction from the wall toward the bulk flow are used. Downstream from the body additional refinement is applied in order to improve the resolution in the wake region.

For all polyhedra, three levels of mesh densities were used, with the cell width (in all three global Cartesian directions) of $d/20$, $d/40$ and $d/80$ near the wall surface. Thus, the finest mesh for the flow around cube contains 627.411 cells. The same template of the cell size distributions is applied also to the case of the flow around octahedron and flow around dodecahedron, which have the same diameter of the circumscribe sphere $d$ as the cube, yielding the meshes of similar size in terms of number of cells. Several illustrations of a typical mesh used here are shown in Fig. 2.

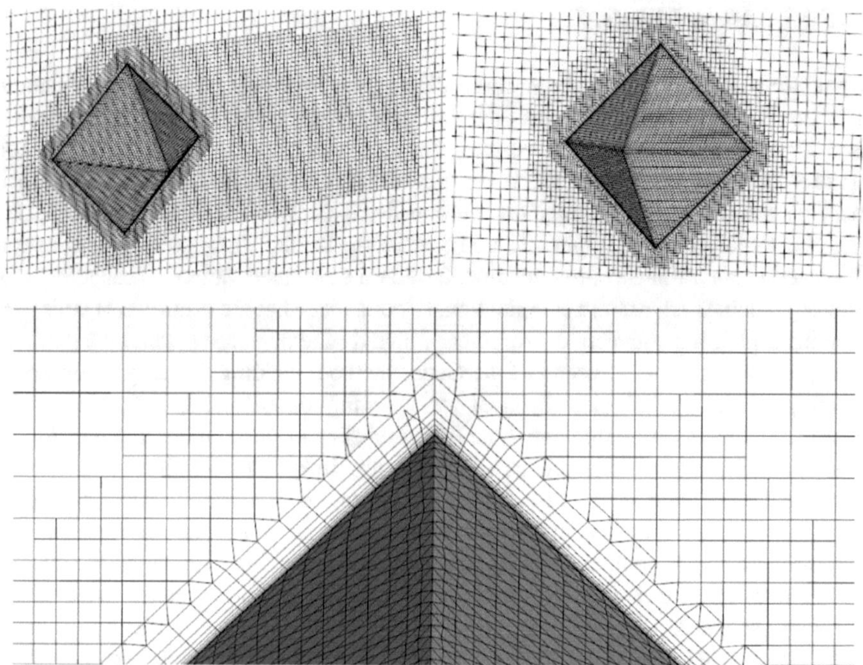

**Fig. 2.** The mesh around the octahedron: the wake refinement behind the body (top left), the frontal side of the body (top right), prism layers and the mesh on the body surface (bottom).

## 4 Results and Discussion

Figure 3a and 3b show the velocity field around the cube and in the wake behind it.

At the Reynolds-numbers up to $Re_d = 200$, Re = 116 the flow field is steady and symmetric. The primary separation develops from the rear edges of the cube. With the increase of Re, the recirculation region becomes longer and wider.

At $Re_d = 500$, Re = 289 the flow field is asymmetric and time dependent. Additional separation from the frontal edges is more apparent. These findings are in accordance with the results of Saha (2004), who found that time-dependent flow arises about Re = 270. At the higher Re values the transient phenomena and flow instabilities develop and become stronger. The separation from the frontal edges is dominant, while the number of vortical structures in the wake increases with further increase of Re. The flow disturbances triggered by the body, which are evidently stronger in this flow regime, are transported to the boundaries affecting the maintenance of the boundary conditions during the flow simulation and herewith the convergence rate of the calculation process.

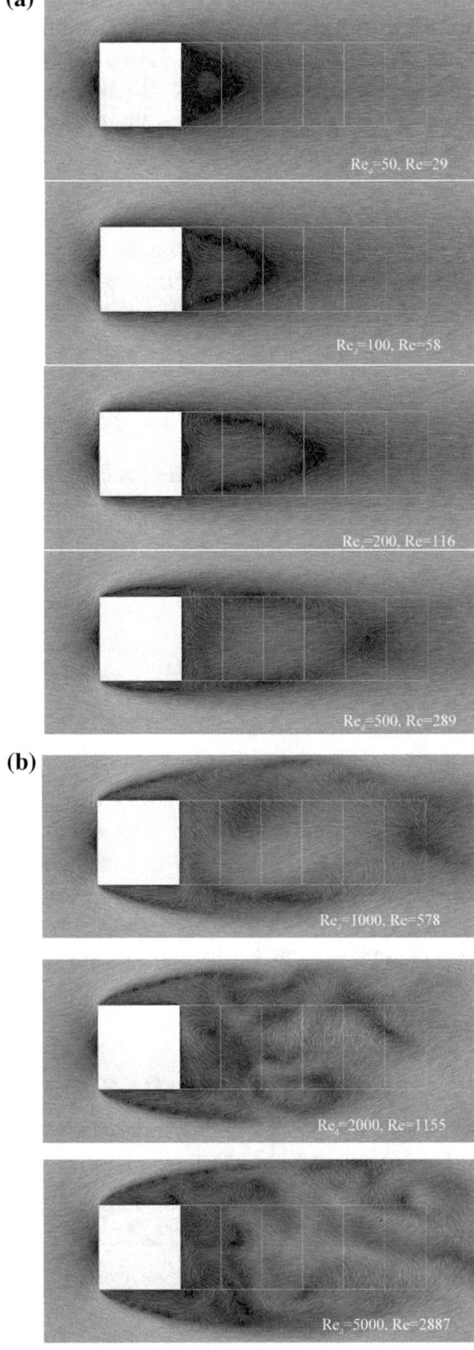


**Fig. 3.** (a) The flow velocity field around a cube at different Reynolds numbers (b) The flow velocity field around a cube at different Reynolds numbers.

The values of the drag coefficient calculated using three different meshes at two selected Reynolds numbers corresponding to the steady and the unsteady flow regime are shown in Tab. 2. Reduction of the spatial discretization error in the case of steady flow is stronger than in the transient case. Supposedly, flow stability in the steady case alleviates the calculation process and the error reduction. Calculation of transient flow involves additional, time-discretization error which degrades the convergence rate. The values in the unsteady regime are obtained by averaging the calculated drag coefficients in time, which was started after $t_{start} = 2.5\, t^{*}$, where $t^{*} = (l_1 + l_2)/v$ is a characteristic time at which the fluid particles pass the entire domain, or $t_{start} = 75\, t^{**}$, where $t^{**} = d/v$ is a characteristic time at which the fluid particles pass the polyhedral body. The averaging period is adopted to be of the same size: $T_{aver} = 2.5\, t^{*} = 75\, t^{**}$.

**Table 2.** Drag coefficient of the cube calculated on three different meshes.

| Refe-rence cell size | Cell width at the wall surface | Thickness of all prism layers *) | Total number of cells in the mesh | Drag coefficient $c_d$ | |
|---|---|---|---|---|---|
| | | | | $Re_d = 200$, Re = 116 | $Re_d = 5000$, Re = 2900 |
| 4 mm | $d/20 = 0.5$ mm | 1 mm | 52,856 | 1.313 | 1.173 |
| 2 mm | $d/40 = 0.25$ mm | 0.5 mm | 155,463 | 1.247 | 1.115 |
| 1 mm | $d/80 = 0.125$ mm | 0.25 mm | 627,411 | 1.2265 | 1.08 |

*) 10 layers with stretching factor of 1.5

Table 3 gives the information about the cell-center distance from the wall in the finest mesh used for the flow around a cube. In Fig. 4a distribution of the dimensionless wall distance $y^{+}$ at $Re_d = 5000$, Re = 2900 is shown for the same case. Figure 4b shows a comparison of the near-wall velocity distribution and the mesh used for the flow around an octahedron at $Re_d = 1000$, Re = 580. According to these data, the adopted mesh spacing seems to be sufficiently fine to capture the near-wall gradients properly.

**Table 3.** Near-wall mesh size.

| Reference cell size | Prism layers thickness | Flow around a cube, $Re_d = 5000$, Re = 2900 | | | |
|---|---|---|---|---|---|
| | | $y_P$ | $d/y_P$ | $a/y_P$ | $y^{+}_{max}$ |
| 1 mm | 0.25 mm | 0.0011 mm | ≈9,090 | ≈5,250 | ≈0.25 |

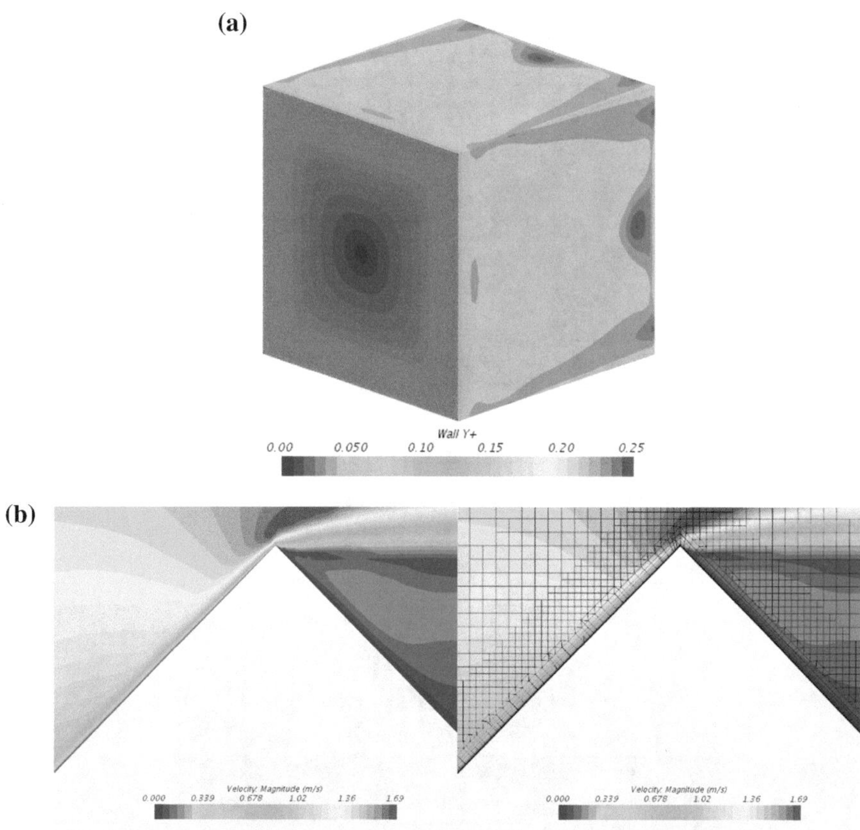


**Fig. 4.** (a) Dimensionless wall distance $y^+$ calculated for the flow around a cube at $\mathrm{Re}_d = 5000$, Re = 2900 (b) Comparison of the near-wall velocity distribution and the mesh for the flow around an octahedron at $\mathrm{Re}_d = 1000$, Re = 580.

The drag coefficients of all polyhedral shapes obtained in the present computer simulations are depicted in two variants. The first one, shown in the Fig. 5 (top) presents the drag coefficients based on the projected area of the volume-equivalent sphere, i.e. the value of $c_{\mathrm{D,es}}$, as function of Reynolds number based on the diameter of the volume-equivalent sphere $Re_{\mathrm{es}}$. The values obtained using the correlation expressions published by Haider and Levenspiel (1989), Saha (2004) and Hoelzer and Sommerfeld (2008) for the flow around the cube are shown for comparison, as well. The data agree very well in the regime of steady, symmetric flow, except for the expression by Hoelzer and Sommerfeld (2008), which yields higher drag coefficients for $Re_{\mathrm{es}} > 50$. However, approaching the turbulent regime (about $Re_{\mathrm{es}} = 4000$) the deviation of the present simulation from the Hoelzer-Sommerfeld equation is reduced. Agreement of the present computation with the equations of Haider and Levenspiel is qualitatively good, however, both in the transitional and the turbulent regime, the CFD results show underprediction.

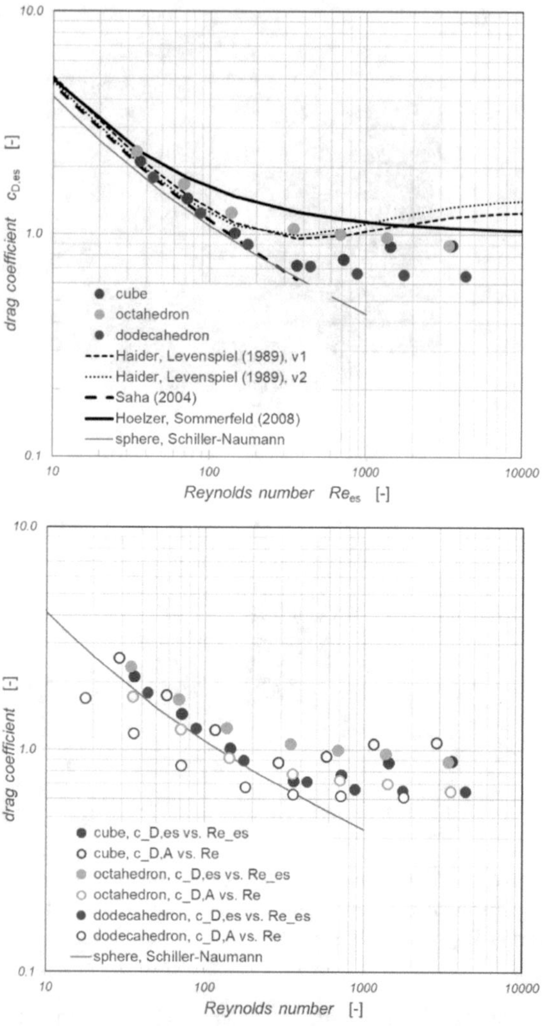


**Fig. 5.** Comparison of the drag coefficients obtained by simulation (colored symbols), and those obtained using the formulae from literature (lines).

According to the values given in the plots, the calculated drag coefficient of the octahedron is higher than for the other two shapes in the steady, and particularly in the transitional regime. This is not expected having in mind that the orientation of the octahedron applied in this study should be more convenient regarding the resistance to the flow (it should cause less deflection of the flow direction and the projected shape is the same as for the cube). Approaching the turbulent flow, the drag coefficient of the octahedron and the cube get closer, while the drag coefficient of the dodecahedron is lower, as expected due to its higher sphericity.

The second variant of drag coefficient illustration, given in Fig. 5 (bottom), shows a comparison of the $c_{D,es}$-$Re_{es}$ relation which is usually used in analysis of non-spherical bodies and the $c_{D,A}$-$Re$ relation which is typically used in aerodynamics of bluff bodies. Indeed, the interpretation of the results depends on the way how the non-dimensional values are defined in the case of non-spherical bodies. Both for octahedron and dodecahedron the $c_{D,A}$-$Re$ relation shows lower values of the drag coefficient (although for dodecahedron the difference vanishes with the increase of Re number). However, by careful comparison and appropriate use of non-dimensional parameters one can draw the same conclusions as previously discussed.

## 5 Conclusions

CFD simulation of the flow around bodies of polyhedral shapes provides detailed investigation of the flow properties. Comparison of the basic properties of the flow around cube with the results from literature shows acceptable agreements. However, further investigation is needed in the transitional flow regime in order to offer sufficient data for more reliable assessment of the drag force. It is clear that sphericity of the bodies affects the drag force, particularly in the transitional regime, while in the case of very low Reynolds numbers its effect can be neglected, provided the comparison is based on appropriately defined non-dimensional quantities.

**Acknowledgements.** This work is financially supported by Ministry of education, science and youth of Canton Sarajevo, Bosnia-Herzegovina in the period 2018/19, which is gratefully acknowledged.

## References

Siemens PLM: Simcenter STAR-CCM + v 13.06., documentation and user manual (2018)

Haider, A., Levenspiel, O.: Drag coefficient and terminal velocity of spherical and non-spherical particles. Powder Technol. **58**, 63–70 (1989)

Saha, A.K.: Three-dimensional numerical simulations of the transition of flow past a cube Phys. Fluids **16**(5), 1630–1646 (2004)

Hoelzer, A., Sommerfeld, M.: New simple correlation formula for the drag coefficient of non-spherical particles. Powder Technol. **184**, 361–365 (2008)

# Application of Console Methods in Experimental and Numerical Determination of Internal Stress of Varnish

Esed Azemović, Ibrahim Busuladžić(✉), and Izet Horman

Faculty of Mechanical Engineering,
University of Sarajevo, Sarajevo, Bosnia and Herzegovina
busuladzic@mef.unsa.ba

**Abstract.** The basic task and purpose of this paper is to demonstrate the use of console method in experimental and numerical studies. Materials commonly in use for surface treatment of wood were used in this paper. The results of this paper provide a clear picture and confirmation that the method is acceptable for this type of analysis.

The console method determining internal stresses in polymer films appeared in the sixties of the last century. This method is based on the principle that one end of the thin elastic narrow panel is fixed. Using the numerical finite element method, a console was modeled with known dimensions and basic technical characteristics of the experimental test.

**Keywords:** Internal stress · Varnish · Console method · Film thickness · Numerical · Experimental methods

## 1 Introduction

Console method was chosen here to determine internal stresses inside film [1]. Later, many researchers used this method and worked with [2]. Physically speaking, the cause for formation of internal stresses is difference in the changes in plate dimensions in the film - substrate system, if both, the film and the substrate are simply conceived as panels [3]. All polymer materials used in surface treatment are subject to volumetric change (curing) during the hardening period. In the material of film, applied on a solid substrate, the adhesion is possible only vertically on the substrate, while the change of dimensions parallel to the substrate is prevented by film and substrate adherence. That leads to the appearance of internal stresses in the stretched film and at the edge of the film-substrate shear stress appears [4, 5].

This paper shows that the console method can be used in experimental and numerical determination of internal stresses of varnished surfaces.

## 2 Theory of the Console Method

The influence of the internal stress of the film on the substrate bending can be replaced by attacking momentum (M) acting on the edges of the plate [6] as in Fig. 1.


S. Avdaković et al. (Eds.): IAT 2019, LNNS 83, pp. 334–344, 2020.
https://doi.org/10.1007/978-3-030-24986-1_26

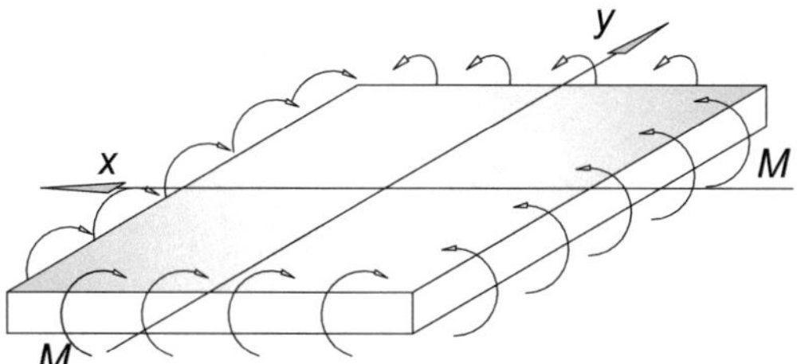


**Fig. 1.** Scheme of the influence on internal stresses in the varnish on the plate element

Bending moment, evenly distributed at the edge of a thin rectangular plate, which acts per unit length of the edge, is defined as:

$$M = \frac{E_1 S^3}{12\left(1 - \mu_1^2\right)} \left(\frac{\delta^2 f}{\delta y^2} + \mu \frac{\delta^2 f}{\delta x^2}\right) \tag{1}$$

where:

$E_1$- panel elasticity modulus, S - panel thickness, $\mu_1$ - Poisson coefficient for plate.

Adopting that:

$$\frac{\delta^2 f}{\delta y^2} = \frac{\delta^2 f}{\delta x^2} = \frac{1}{R} \tag{2}$$

where:

R- radius of curvature.

The term (2) goes into the form:

$$M = \frac{E_1 S^3}{12(1 - \mu_1)R} \tag{3}$$

With sufficient accuracy it can be assumed that the neutral part of the newly formed biplane passes through the middle of the system, so the connection between the attack moment and internal stress is given by the expression:

$$M = \sigma_1 t \frac{(S + t)}{2} \tag{4}$$

where:

$\sigma_1$ - internal stress in the film at focused thin substrates,

t - thickness of the film.

Based on the relation (3) and (4) the internal stresses can be determined in the film:

$$\sigma_1 = \frac{E_1 S^3}{6Rt(S + t)(1 - \mu_1)} \tag{5}$$

Internal stresses ($\sigma$) in films over rigid substrates are greater than internal stresses in the same films ($\sigma_1$) applied to a thin elastic substrate for additional strain:

$$\sigma_2 = \varepsilon \frac{E_2}{1 - \mu_2} \quad (6)$$

where:

$\sigma_2$ - additional internal stresses,
ε - dilation,
$E_2$ - elasticity modulus of the film,
$\mu_2$ - Poisson coefficient for the film.
Dilation can be determined from the expression:

$$\varepsilon = \frac{1}{R}\left(\frac{S+t}{2}\right) \quad (7)$$

From the relation (6) and (7) we get:

$$\sigma_2 = \frac{E_2(S+t)}{2R(1-\mu_2)} \quad (8)$$

Taking this into account, the overall internal stresses in surface of finishing film are defined by the term:

$$\sigma = \sigma_1 + \sigma_2 \quad (9)$$

or by incorporating (5) and (8) we obtain:

$$\sigma = \frac{E_1 S^3}{6Rt(S+t)(1-\mu_1)} + \frac{E_2(S+t)}{2R(1-\mu_2)} \quad (10)$$

In case E1 >> E2 and S >> t, the second member of the expression (10) is less than 1% of the first member and can be neglected. Internal stresses are particularly pronounced on unstable substrates, such as wood and wood souproducts. Unsaturated or poorly saturated massive wood or wood panels react dimensionally (swelling or rubbing) but with the least change in relative humidity [3].

Films that are continuously in the state of stress certainly have lower resistance to most external influences compared to uncompressed free films [4].

Summing up so far, it comes to the conclusion that the internal stress is a negative phenomenon because it tends to lower the level of mechanical characteristics of the film.

## 3 Console Method

Sanžarovski published a console method for determining internal stresses in polymer films. Later, this method worked well with many researchers such as Grigore and others [2].

The method consists of the following: On one side thin, narrow and long sheets of elastic material, whose dimensions and elastic properties are known, the appropriate amount of film forming material is applied and after the hardening process the film is formed.

During the hardening process, the film attempt to shorten, but the adhesion forces to the substrate obstruct it. The result of such a process within the film is the internal stresses in it. As the substrate is flexible in the case of a console method, it follows each dimensional change of the film by bending, Fig. 2.

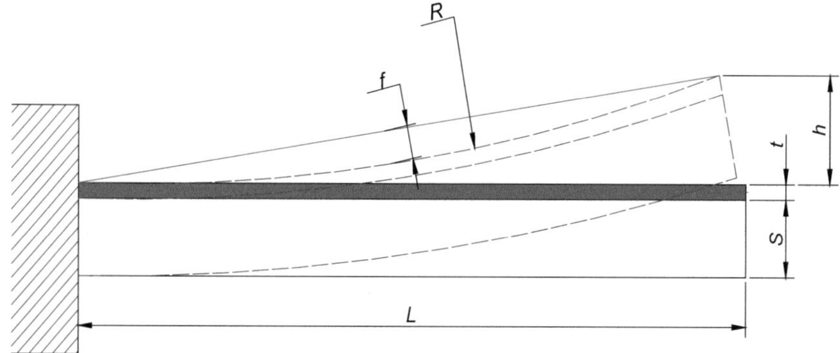


**Fig. 2.** Scheme of elasticity of the resilient plate with the applied film, before and after hardening

Where: S-plate thickness; t-film thickness; L-panel length; R-radius curve; f-sagging; h-deflection console.

For this following relationship is valid:

$$R = \frac{L^2}{8f} = \frac{L^2}{2h} \tag{11}$$

In accordance with the relations (5) and (11) the internal stresses in the film applied to the thin narrow elastic plate can be determined from the expression:

$$\sigma_1 = \frac{4fE_1S^3}{3L^2t(S+t)(1-\mu_1)} \tag{12}$$

respectively;

$$\sigma_1 = \frac{hE_1S^3}{3L^2t(S+t)(1-\mu_1)} \tag{13}$$

Based on the relation (9) and using the expression (13) we obtain:

$$\sigma = \frac{hE_1S^3}{3L^2t(S+t)(1-\mu_1)} + \frac{hE_2(S+t)}{L^2(1-\mu_2)} \tag{14}$$

In case where E1 >> E2 and S >> t, the second member of the expression (14) is less than 1% of the first member and can be neglected.

In the console method of polymeric testing films for wood surface treatment, the term (14) is usually used without the second member because it is effective for E1 >> E2 and S >> t.

As a substrate for testing this method, metal is commonly used, but for wood surface treatment of polymeric materials, it is better to use a wooden substrate because it is closer to realistic conditions.

# 4 Testing Method

## 4.1 Numerical Method

Numerical methods consist of translating the basic equations of mathematical physics into the system of linear algebraic equations [7]. In order to achieve this, time, space and equations must be discretized. In this study we use Finite Element Method for numerical determination of internal stress inside the film of varnish.

## 4.2 Finite Element Method

The Finite Element Method is based on the idea of building complex objects using simple blocks, i.e. blocks of the finite and known geometric shape.

The physical structure of any form that needs to be analyzed, i.e. to determine stresses and deformations, is replaced by a model composed of finite and known dimensions and shapes of blocks. This block known as the finite element allows its state of stress and deformation to be described by algebraic equations [7].

They are then used to describe the state of stress and deformation throughout the construction. This allows to the considered domain with infinitely many degrees of freedom to be replaced by a model consisting of finite elements with a finite number of degrees of freedom. The elements should be in such sizes to get accurate results and to avoid too much memory of computers (Fig. 3).

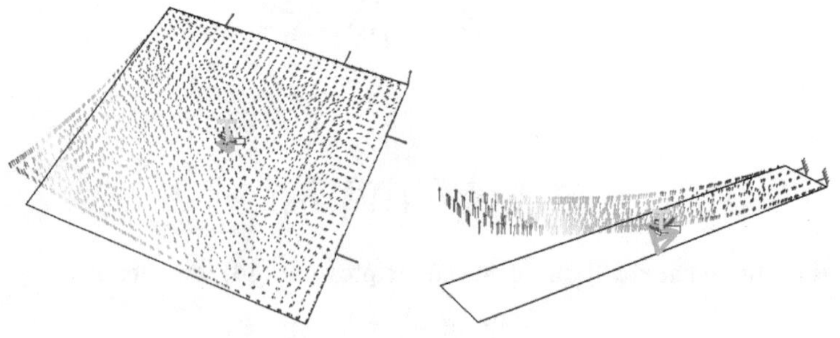

**Fig. 3.** Examples of numerical modeling using the FE method

### 4.3 Experimental Testing of the Internal Stress of Console Method

For studying internal stresses in polymer materials, the console beam method is used (Fig. 4). In the case of experimental analytical stress tests, using this method, it is necessary to provide:

- console,
- console bracket,
- glass shelf,
- pressure beam,
- clamping bolts.

Overall dimensions of the bracket are 600 × 60 × 60 [mm] with a slot having dimensions 600 × 40 × 30 [mm]. When making a slot on the console, it must be ensured that the depth and height of the slot form an angle 90° in order to have parallel position of the tube and glass shelves, on which the movable scale is supported.

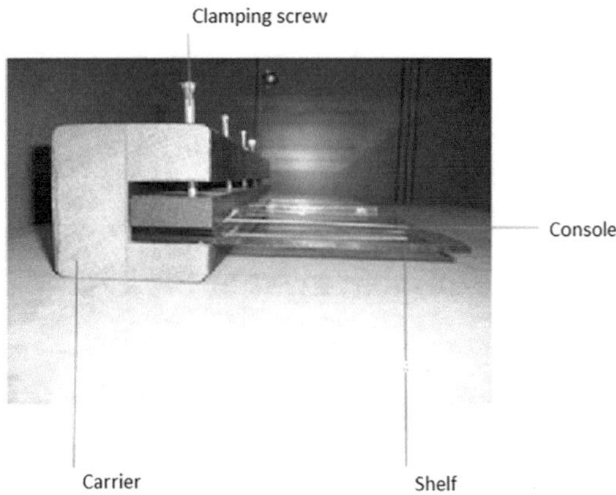


**Fig. 4.** Clamping the console in the bracket

When fixing the specimen in the carrier, the slightly concave side of the specimen is facing upwards. For each internal stress test, using this method, it is necessary to take five aluminum specimens. The dimensions of the specimens that are placed in the bracket are 140 × 13 × 0.18 [mm], Fig. 5. The total length of the free bracket is 100 [mm].

The basic properties and technical characteristics of aluminum forming the substrate are:

- Elasticity modulus: $7 \times 10^4$ [MPa]
- Poisson coefficient: 0.346
- Heat expansion coefficient: 0.0062 1/K

**Fig. 5.** Water and UV water specimens

All elements of the test set, except the specimen where the varnish is applied, need to be protected. The reason for the protection is the precision in the test and the correct measurement of the displacement of the console at time intervals [h]. By using a console method with defined substrate as in this case, the total internal strain in the film is calculated by the expression:

$$\sigma = \frac{hE_1S^3}{3L^2t(S+t)(1-\mu_1)} \tag{15}$$

## 5 Measurement of the Displacement of the Console When Drying Naturally

The basic feature of this method is to measure console deflection at different time intervals. With this drying method it is possible to follow the kinetics of the internal stress.

When measuring the deflection of the console for V-1 water varnish specimen, five time intervals has taken; 0; 0.5; 1; 24; 48 [h]. For measurement of the V-2 water varnish specimen, there are several time intervals (nine in total) so that the kinetics of internal stresses can be analyzed in more detail. Prior to applying the water varnish to the console, it is necessary to measure the deflection of the console $h$ as in Fig. 6. The meter is fixed to the console by means of a magnetic carrier that has a flexible hand that allows positioning of the movable scale at the desired location. The first measurement is called measurement zero, i.e. measurement before dispersing the varnish. Applying the varnish on the surface, using the same procedure, in the first, second, third, fourth intervals the instrument is used to measure the deflection of the console $h$. Fifty specimens were used for the test of water-based varnish.

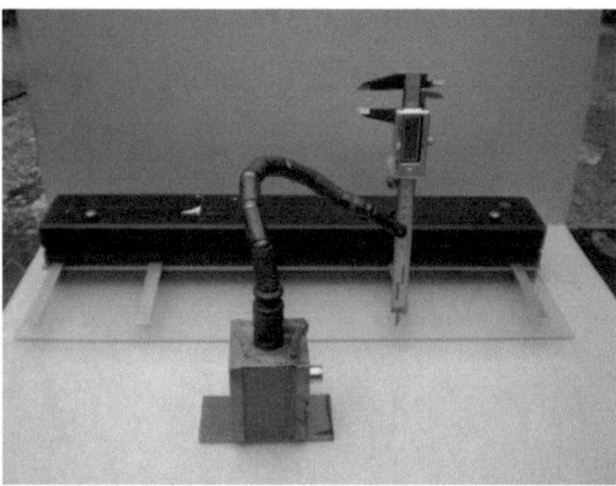

**Fig. 6.** Measurement of the displacement of the console

The zero measurement is subtracted from the remaining four intervals and real distortions of the console are obtained at time intervals.

# 6 Results and Discussions

## 6.1 Numerical Results - Stresses in the Varnish at the Console

Using Finite Element Method, a console was modeled with known dimensions and basic technical characteristics of the experimental model. An important point in modeling is the establishment of the relation between the substrate and the varnish deposit, i.e. the difference in the linear expansion coefficients. From the thickness of dry and wet film and the well-known analogy of evaporation of volatile components and reduction of film thickness, a linear expansion coefficient was determined for the film. When modeling the console, a constant temperature T = 21°C is set (Table 1).

**Table 1.** Dimensions and characteristics of the substrate and the film

| Dimensions and characteristics | Substrate | Film |
|---|---|---|
| Dimensions [mm] | 100 × 100 × 0,18 | 100 × 100 × 0,06 |
| Elasticity Module [MPa] | 7 × 104 | 3 × 104 |
| Poisson coefficient | 0,346 | 0,3 |
| Heat expansion coefficient 1/K | 0,0062 | 0,028 |

When computing stresses in the console, the maximum stresses in the film are calculated, the normal stresses as well as the shear stress, tangential stress.

## 6.2 Total Stresses in the Varnish on the Console

It can be clearly seen from Fig. 7 that the maximum vector of movement is at the free end of the console, or at the points that are most distant from the clamping itself. The smallest is in the bottom of the console, or in engraving. The maximum vector of displacement is: t = 1.13 [mm].

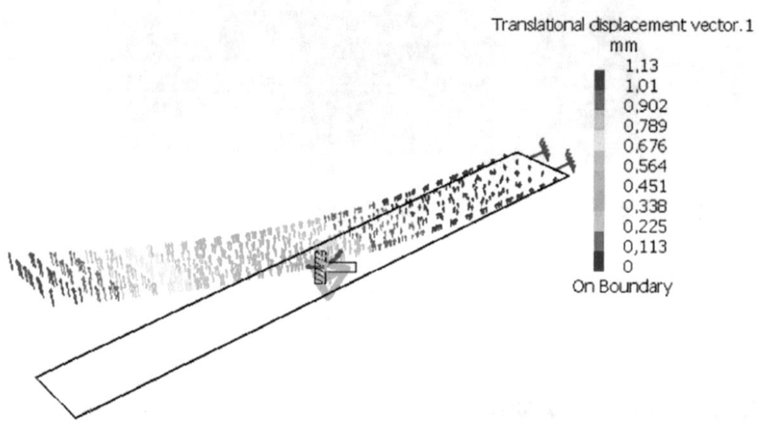


**Fig. 7.** Total displacement vectors of console points

Using the finite element method, the maximum total displacement of the console at the numeric calculation is h = 1.13 [mm] Fig. 8.

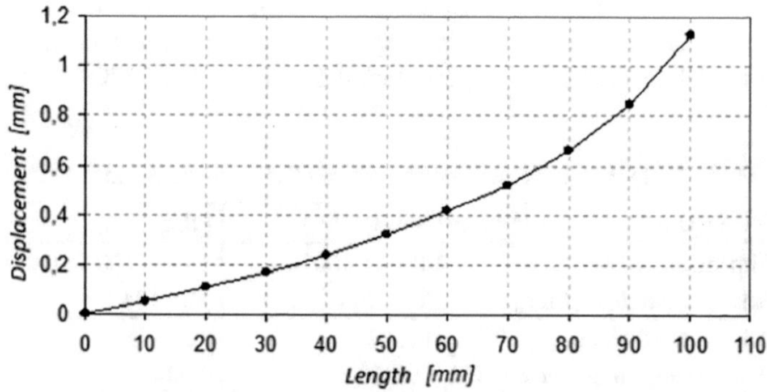


**Fig. 8.** Vertical displacement of the console at numerical calculation

To confirm the accuracy of the internal stress results, the vertical displacement of the console was analyzed in the experimental and numerical calculation in Fig. 9. The experimental result was expressed with the vertical displacement of water-based varnish, which was dried at room temperature 21 °C at 48 [h]. Numerical result are expressed also with vertical displacements at the console.

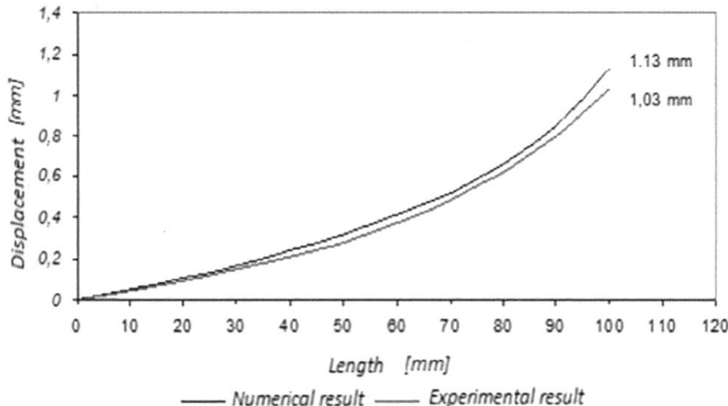


**Fig. 9.** Vertical displacement of the console in numerical and experimental calculations

The vertical experimental displacement of the console is $h_{s,e} = 1.03$ [mm], while in the numerical calculation the displacement is $h_{s,n} = 1.13$ [mm]. The vertical displacement difference for these two results is $h_r = 0.1$ [mm].

The comparison of these two results shows that there is no big difference between numerical and experimental results.

## 7 Conclusion

Using finite element methods, a console was modeled. When modeling substrate and film; elements, dimensions, basic properties and technical characteristics were taken from experimental analysis.

This method analyzes the stress in the film of the console. The results of the numerical and experimental analysis are approximate, i.e. maximum vertical displacement of the console for experimental results of water varnish is 1.03 mm, and the maximum vertical displacement of the console for numeric results is 1.13 mm.

These results confirm that the console method is applicable to the experimental and numerical analysis for the internal stresses of all cover materials.

## References

1. Sanzarovski, A.T.: Metodi opredelenija mehaničeskih i adgezionih svoistv polimerih pokriti. Moskva (1974)
2. Grigore, E., Ruset, C., Short, K., Hoeft, D., Dong, H., Li, X.Y., Bell, T.: In situ investigation of the internal stress within the nc-Ti2 N/nc–TiN nanocomposite coatings produced by a combined magnetron sputtering and ion implantation method. Surf. Coat. Technol. **200**(1–4), 744–747, ISSN 0257-8972 (2005)

3. Mrvoš, N.: Unutrašnja naprezanja u polimernim prevlakama. Drvna industrija, Zagreb (1984)
4. Alić, O.: Površinska obrada drveta. Mašinski fakultet, Sarajevo (1997)
5. Doleček, V., Karabegović, I., Martinović, D., Blagojević, D., Šimun, B., Vukojević, D., Kudumović, Dž., Uzunović Zaimović, N., Bjelonja, I.: Elastostatika. Tehnički fakultet, Bihać (2003)
6. ASTM D 6991 – 05, Standard test method for measurements of internal stresses in organic coatings by cantilever (beam) method
7. Ivanković, A., Demirdžić, I.: Finite volume stress analysis. University of London (1997)

# More Accurate 2D Algorithm for Magnetic Field Calculation Under Overhead Transmission Lines

Adnan Mujezinović(✉), Nedis Dautbašić, and Maja Muftić Dedović

Faculty of Electrical Engineering, University of Sarajevo, Sarajevo, Bosnia and Herzegovina
adnan.mujezinovic@etf.unsa.ba

**Abstract.** In this paper 2D algorithm for the calculation of the magnetic induction in vicinity of the three phase overhead transmission line is presented. The presented algorithm is composed of the two stages. First stage of the presented algorithm is determination of the current phasors in all system conductors including grounding wires and second stage is calculation of the magnetic induction based on results obtained in first stage of algorithm. The presented algorithm was verified by comparing with the measurement results.

**Keywords:** Magnetic field · Overhead transmission line · Magnetic induction · Phase conductors · Grounding wires

## 1 Introduction

The negative influences of electromagnetic fields on living beings have been subject of numerous studies around the world more than 70 years. The concerns of the public regarding the possible negative biological effects due to exposure to electromagnetic fields, and the relevant research on this topic begin after the Second World War [1]. The first studies of human exposure were focused to radio frequency electromagnetic fields. The first studies pointing to the link between the extremely low frequencies (ELF) magnetic fields $(50 - 60\,\text{Hz})$ and childhood leukaemia were published in the late seventies of the last century [2]. Nowadays, ELF magnetic fields are classified as potential carcinogenic for humans.

The sources of ELF electric and magnetic fields are most often elements of the electric power system, such as high voltage and distribution substations, overhead lines, underground cables etc. The basic requirement of protecting people from electromagnetic radiation is meeting the limit values of extreme low frequency electric and magnetic fields that people may be exposed to. Limit values of electric and magnetic fields are defined by appropriate international and national standards and recommendations [1]. The International Commission on Non-Ionizing Radiation Protection (ICNIRP) published the most widely accepted international recommendations on reference limit values. According to these recommendations, the reference magnetic induction allowed limit value is $1000\,\mu\text{T}$ for the working population, while for the general population the allowed limit is $200\,\mu\text{T}$ [3]. Therefore, the calculations of the ELF electric and magnetic


S. Avdaković et al. (Eds.): IAT 2019, LNNS 83, pp. 345–354, 2020.
https://doi.org/10.1007/978-3-030-24986-1_27

fields near electric power facilities are carried out both on facilities that are in operation and on objects that are in the design phase in order to determine whether the obtained values of the electric and magnetic fields are within the allowed limits.

Electromagnetic fields caused by electric power system objects are ELF fields therefore they can be approximated as quasi-static fields [4, 5]. Furthermore, in this case the Maxwell´s displacement current can be ignored therefore measurement and calculation of electric and magnetic fields can be carried out separately [1]. In this paper advanced 2D algorithm for the calculation of the magnetic induction in the vicinity of high voltage overhead lines is presented. The presented algorithm was used to calculate the magnetic induction of 400 kV overhead transmission lines. Validation of the presented algorithm was performed by comparing calculation results obtained by presented algorithm with measurement results.

## 2 Mathematical Model

For the calculation of the ELF magnetic induction under high voltage overhead transmission lines different numerical methods have been proposed. Most simple calculation method of the ELF magnetic field induction of the overhead transmission line is based on Biot-Savart law with numerous simplifications such as equivalence of conductors and/or bundle of conductors with one current point source and assumption that transmission line can be treated as infinite perfectly straight line. Also, some authors neglect the impact of the ground surface [6] and use the phase conductors as only source of field neglecting the current flow through other system conductors (shield wires and passive loops), i.e. assume that the system is symmetrical [7].

Within this paper 2D algorithm based on Biot-Savart law is presented. The presented algorithm is composed of two stages, first stage is calculation of the current intensity in all system conductors and second one is calculation of the ELF magnetic induction based on results obtained from first stage.

### 2.1 Pre Magnetic Induction Calculation Procedure

Pre magnetic induction calculation procedure is to determine current that flow in all system conductors including ground wires current and passive loop current if this loop is applied. A detailed implementation of this algorithm when taking into account passive loop can be found elsewhere [8, 9]. Within this paper the brief description of the used mathematical model is given with assumptions that each phase conductor is composed of two sub-conductors and there is no passive magnetic loop in the system. Current flow in all system conductors can be calculated from following matrix equation:

$$\begin{bmatrix} [\overline{Z}_{aa}] & [\overline{Z}_{ab}] & [\overline{Z}_{aG}] \\ [\overline{Z}_{ba}] & [\overline{Z}_{bb}] & [\overline{Z}_{bG}] \\ [\overline{Z}_{Ga}] & [\overline{Z}_{Gb}] & [\overline{Z}_{GG}] \end{bmatrix} \cdot \begin{Bmatrix} \{\overline{I}_a\} \\ \{\overline{I}_b\} \\ \{\overline{I}_G\} \end{Bmatrix} = \begin{Bmatrix} \{\Delta\overline{U}_a\} \\ \{\Delta\overline{U}_b\} \\ \{\Delta\overline{U}_G\} \end{Bmatrix} \tag{1}$$

where:

$\left[\overline{Z}\right]$ - square matrix of transmission line impedances,

$\left\{\overline{I}\right\}$ - matrix vector of sub-conductors current phasors,

$\left\{\Delta\overline{U}\right\}$ - matrix vector of voltage differences on sub-conductor on two ends of transmission line and indexes $a$, $b$ and $G$ represents the sub-conductors of the system and grounding wire.

The elements of $\left[\overline{Z}\right]$ impedance matrix can be determined from the per-unit-length impedances of the transmission lines equations. Per-unit-length transmission line impedances are defined by following analytical equations [10]:

$$\overline{Z}_{ii} = (R_{dc} + \Delta R_{ii}) + j\left(2\omega 10^{-4} ln\frac{h_{ii}}{R_i} + \Delta X_{ii}\right) \tag{2}$$

$$\overline{Z}_{ij} = \Delta R_{ij} + j\left(2\omega 10^{-4} ln\frac{h_{ij}}{d_{ij}} + \Delta X_{ij}\right) \tag{3}$$

where:

$R_{dc}$ - per-unit-length DC resistance of $i$-th sub-conductor,

$R_i$ - radius of $i$-th sub-conductor,

$h_{ii}$ - double height of the $i$-th sub-conductor,

$h_{ij}$ - distance of the $j$-th sub-conductor image and $i$-th sub-conductor,

$d_{ij}$ - distance between $i$-th and $j$-th sub-conductor,

$\Delta R_{ii}, \Delta R_{ij}$, $\Delta X_{ii}, \Delta X_{ij}$ - correction factors caused by earth surface and can be calculated by following equations [10]:

$$\Delta R_{ij} = 4\omega 10^{-4}\left(\frac{8}{\pi} - \frac{\sqrt{2}}{6} \cdot 0.00281 \cdot h_{ij} \cdot \sqrt{\frac{f}{\rho}}\cos\varphi\right) \tag{4}$$

$$\Delta X_{ij} = 4\omega 10^{-4}\left[\frac{1}{2}\left(0.6159315 - \text{In}\left(0.00281 \cdot h_{ij} \cdot \sqrt{\frac{f}{p}}\right)\right) + \frac{\sqrt{2}}{6} \cdot 0.00281 \cdot h_{ij} \cdot \sqrt{\frac{f}{p}}\varphi\right] \tag{5}$$

where:

$\omega$ - angular frequency,

$\rho$ - soil resistivity,

$f$ - system frequency

By using Eqs. (2) and (3) the per-unit-length transmission line impedances can be calculated. Therefore, obtained values need to be multiplied by transmission line length.

The graphical representation of the geometric parameters is given on Fig. 1.

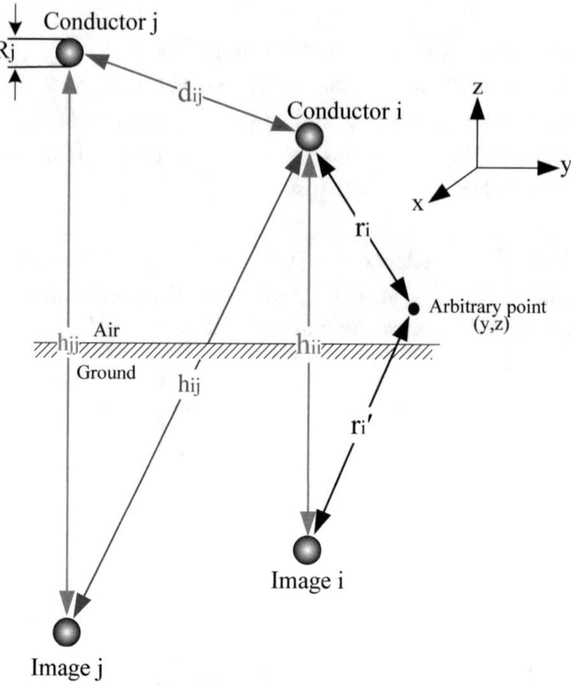


**Fig. 1.** Graphical interpretation of geometric parameters

In order to solve the matrix Eq. (1) the first Kirchhoff law must be applied, fact that voltage of each sub-conductor in the bundle is equal and voltage of grounding wire is equal to zero. Previous can be mathematically expressed as follows:

$$\{\bar{I}_a\} + \{\bar{I}_b\} = \{\bar{I}_P\} \tag{6}$$

$$\{\Delta\overline{U}_a\} = \{\Delta\overline{U}_b\} \tag{7}$$

$$\{\Delta\overline{U}_G\} = 0 \tag{8}$$

where:

$\{\bar{I}_P\}$ matrix vector of phase conductor currents

By introducing the Eqs. (6–8) in matrix Eq. (1), equations for the calculation of all phasors of all system conductors can be obtained as follow:

$$\{\bar{I}_G\} = [-\overline{Z}_{GG}]^{-1} \cdot ([\overline{Z}_{Ga}] \cdot \{\bar{I}_a\} + [\overline{Z}_{Gb}] \cdot \{\bar{I}_b\}) \tag{9}$$

$$\{\bar{I}_a\} = [\overline{Y}] \cdot (([\overline{Z}_{bb}] - [\overline{Z}_{ab}]) \cdot \{\bar{I}_P\} + ([\overline{Z}_{bG}] - [\overline{Z}_{aG}]) \cdot \{\bar{I}_G\}) \tag{10}$$

$$\{\bar{I}_b\} = [\overline{Y}] \cdot (([\overline{Z}_{aa}] - [\overline{Z}_{ba}]) \cdot \{\bar{I}_P\} + ([\overline{Z}_{aG}] - [\overline{Z}_{bG}]) \cdot \{\bar{I}_G\}) \tag{11}$$

where is:

$$[\overline{Y}] = ([\overline{Z}_{aa}] - [\overline{Z}_{ab}] - [\overline{Z}_{ba}] + [\overline{Z}_{bb}])^{-1} \tag{12}$$

By using the matrix Eq. (9–11) the phasors of all currents in the system conductors can be determined from measured phase currents.

Within this paper it was assumed that distance between two adjacent transmission line towers is same on whole rout between substations connected by analysed transmission line. The impact of conductor sag was taken into account by measuring conductors' heights at middle of span between two transmission lines towers since in this point the conductors are closest to the ground surface. In calculation it was assumed that height of the conductors of all spans of rout is same. The effect of the conductor sag is taken into account by hyperbolic functions [8, 9].

## 2.2 Calculation of the Magnetic Induction

When calculating the magnetic induction in the vicinity of high-voltage overhead transmission lines, it is possible to introduce appropriate approximations in order to simplify the mathematical model. For calculating the magnetic induction, the high-voltage transmission lines can be treated as infinitely long straight lines. This further enables the problem to be considered in the 2D space, which greatly simplifies the mathematical model and reduce the calculation time [11]. In presented algorithm all system conductors and sub-conductors have been equivalented by multiple current point sources as showed on Fig. 2. These current sources have been placed on the surface of the stand conductors of analysed system.

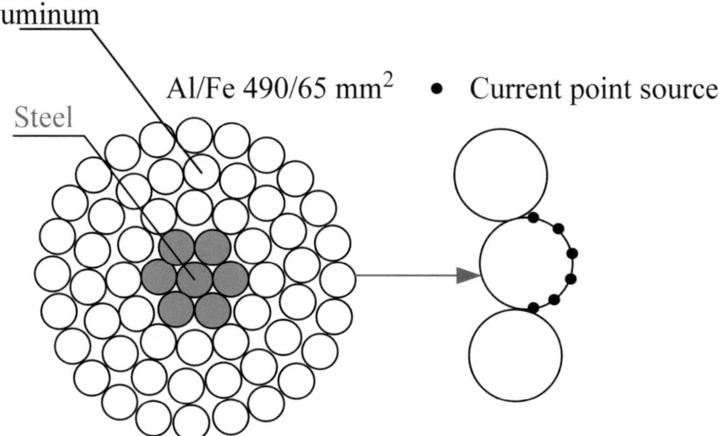


**Fig. 2.** Position of the current point sources on conductor

Current intensity of the current point sources can be calculated from the results obtained from the first stage, by using following equation:

$$|\bar{I}_{i,j}| = \frac{|\bar{I}_i|}{n_w} \tag{13}$$

where:

$|\bar{I}_{i,j}|$ - current intensity of the $j$-th current source on $i$-th sub-conductor,
$|\bar{I}_i|$ - current intensity of the $i$-th that flow in $i$-th conductor,
$n_w$ - number of the point current sources on $i$-th sub-conductor.

Magnetic induction vector in arbitrary point with coordinates $(y, z)$ can be calculated by using following equation:

$$\underline{\vec{B}}(y, z) = \frac{\mu_0}{2\pi} \sum_{i=1}^{n} (K_y \vec{y} + K_z \vec{z}) \bar{I}_i \tag{14}$$

where:

$$K_y = \left( -\frac{z - z_i}{r_i^2} + \frac{z + z_i + \bar{\alpha}}{r_i^{'2}} \right) \tag{15}$$

$$K_z = \left( \frac{y - y_i}{r_i^2} - \frac{y - y_i}{r_i^{'2}} \right) \tag{16}$$

where:

$\underline{\vec{B}}(y, z)$ - of magnetic induction phasor at arbitrary point with coordinates $(y, z)$,
$\mu_0$ - magnetic permeability of the air,
$n$ - total number of current point sources in the system,
$\bar{I}_i$ - phasor of current of the $i$-th current point source,
$(y, z)$ - coordinates of the point at which the magnetic induction is calculated,
$(y_i, z_i)$ - coordinates of $i$-th current point source,
$r_i$ - distance of the $i$-th current point source and the arbitrary point,
$r_i^{'}$ - is the distance of the image of the $i$-th current point source and the arbitrary point,
$\vec{y}, \vec{z}$ - unit vectors

Based on Eq. (14), individual components of the magnetic induction vector at the arbitrary point caused by $n$ current point sources can be calculated using the following equations:

$$\vec{\underline{B}}_y(y,z) = \frac{\mu_0}{2\pi}\sum_{i=1}^{n}\left(-\frac{z-z_i}{r_i^2} + \frac{z+z_i+\overline{\alpha}}{r_i'^2}\right)\overline{I}_i \tag{17}$$

$$\vec{\underline{B}}_y(y,z) = \frac{\mu_0}{2\pi}\sum_{i=1}^{n}\left(\frac{y-y_i}{r_i^2} - \frac{y-y_i}{r_i'^2}\right)\overline{I}_i \tag{18}$$

In the previous equations, the effect of ground surface is taken into account using the complex image theory, whereby image current point sources are placed at a distance below the ground surface which is equal to the distance of original point current sources above the ground surface, but increased for the complex distance $\overline{\alpha}$. The complementary distance $\overline{\alpha}$ can be calculated using the following equation [12]:

$$\overline{\alpha} = \frac{2}{\sqrt{-j\omega\mu_0(\sigma_{soil} - j\omega\varepsilon_{soil})}} \tag{19}$$

Finally, the effective value of magnetic induction at the arbitrary point can be determined using the following equation:

$$B = \sqrt{\left|\overline{B}_y\right|^2 + \left|\overline{B}_z\right|^2} \tag{20}$$

## 3 Case Study

Measurements of the magnetic induction were carried out below sections of high voltage overhead transmission line SS Tuzla 4 - SS Ugljevik. Analysed overhead transmission line on which measurements were carried out is horizontal configurations, with the standard dimensions. Measurements were made at a height of 1 m according to recommendations [13]. Measurements of the magnetic induction were carried out at the middle of the span between the two adjacent pillars. The reason for this lies in the fact that at these points the magnitude induction value is expected to be the reason that in that part the high voltage line is closest to the ground. In order to obtain a complete picture of the state in the considered section, measurement of the height of each phase conductor as well as the grounding wires were performed [14]. The measurement results of the height of the phase and grounding conductors are given in Fig. 3.

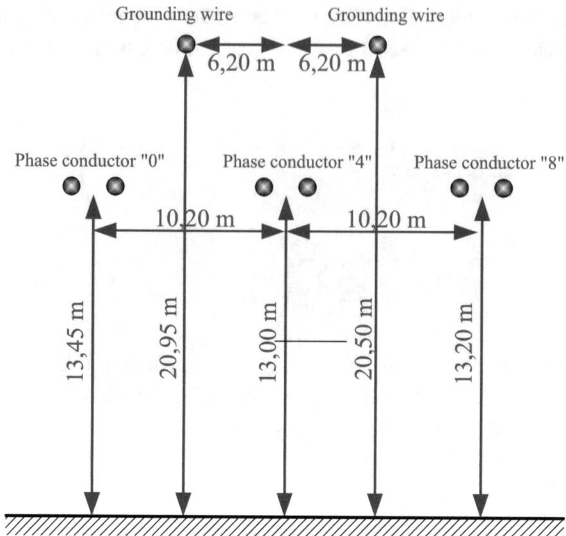


**Fig. 3.** Dimensions of the 400 kV overhead transmissions line SS Tuzla 4 - SS Ugljevik with measured heights

For the calculation of the current in the system conductors (first stage of the algorithm) due to simplicity, the mean value of conductor heights was used. For the phase conductors' height of 13.22 m while for the grounding wire 20.725 m were used. For the calculation of the magnetic field induction measured values of heights were used. The heights of the phase and grounding conductors on the transmission line tower were 20.5 and 27.7 m, respectively. Also, it was assumed that distance between two transmission towers was 350 m.

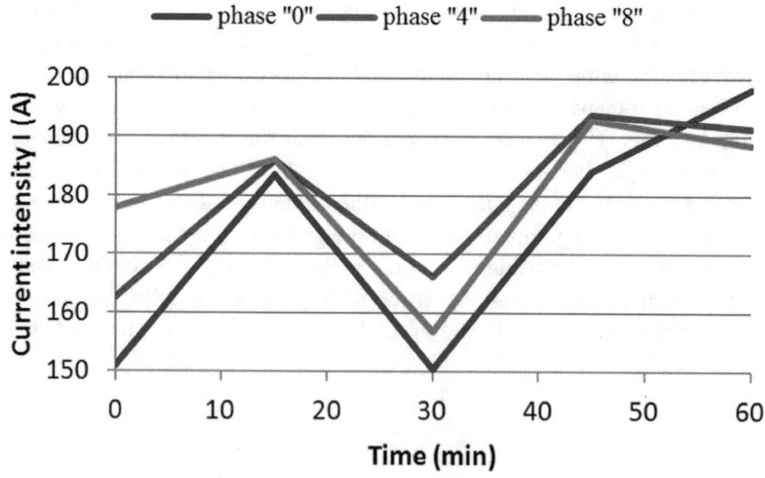


**Fig. 4.** Measured current intensities on phase conductors

Measurement of the magnetic induction was carried out at 1 m intervals from the axis of transmission line tower to a distance of 20 m in one direction. The measurements were performed only in one direction because the magnetic induction distribution of horizontal arrangement of the phase conductors is symmetric.

In order to calculate the magnetic induction in the vicinity of high-voltage overhead transmission lines, it is necessary to have the exact values of the phase currents. The measured values of phase currents at 15 min intervals during the magnetic induction measurement are given in Fig. 4. It was assumed that magnetic induction measurement was conducted when phase conductors currents was $I_{p1}$ = 183,5 A, $I_{p2}$ = 185,9 A and $I_{p3}$ = 186,1 A. Based on measured geometric parameters and phase conductor currents by using the previously presented algorithm the magnetic induction have been calculated and compared with the measured results. The comparison of the calculated and measured results is given on the Fig. 5.

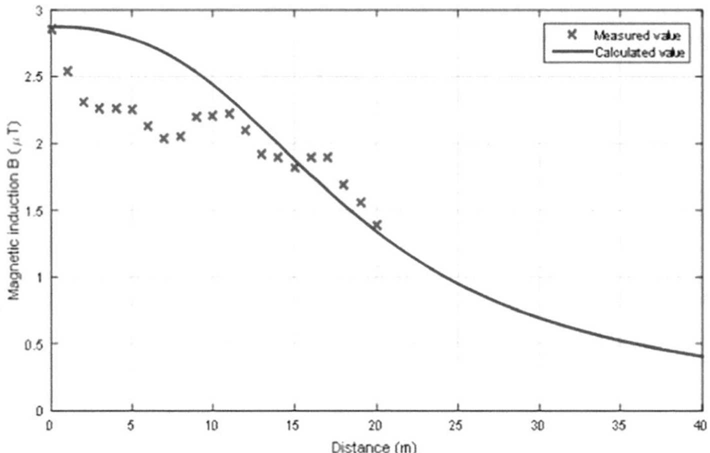


**Fig. 5.** Comparison of calculated and measured results

From results given on Fig. 5 it is noticeable that there is a certain discrepancy between the calculation and the measurement values of the magnetic induction. These deviations are caused by the phase conductor current intensity changes during the measurement of magnetic induction. Also, magnetic induction values are significantly lower than the reference limit values proposed by the International Commission for the Non-ionizing Radiation Protection (ICNIRP).

## 4 Conclusion

The paper presents a 2D algorithm for the calculation of magnetic induction in the vicinity of high voltage overhead transmission lines. Verification of the presented mathematical model was performed by comparing with magnetic induction measurement results carried out at 1 m height from the ground surface in vicinity of the high-voltage overhead transmission lines. The comparison was made on the transmission lines of the 400 kV rated voltage with horizontal configuration of phase conductors. It is important to emphasize that this deviations in results is mostly influenced by current intensity change over the time when measurement of the magnetic induction was performed.

## References


1. Poljak, D.: Human Exposure to Non-ionizing Radiation. Kigen, Zagreb (2006)
2. Modrić, T., Vujević, S., Lovrić, D.: 3D computation of the power lines magnetic field. Prog. Electromagn. Res. **41**, 1–9 (2015)
3. ICNIRP Safety Guideline: Guidelines for limiting exposure to time-varying electric and magnetic fields (1 Hz-1000 kHz). Health Phys. Soc. **99**(6), 818–836 (2010)
4. Trkulj, B., Štih, Ž.: Proračun kvazistatičkog elektromagnetskog polja složenih elektroenergetskih objekata. Energija, Časopis hrvatske elektroprivrede, **5**. Zagreb, 580–591 (2008)
5. Mujezinović, A., Čaršimamović, A., Čaršimamović, S., Muharemović, A., Turković, I.: Electric field calculation around of overhead transmission lines in Bosnia and Herzegovina. In: Proceedings of the 2014 International Symposium on Electromagnetic Compatibility (EMC Europe 2014), pp. 1001–1006. Gothenburg, Sweden, September 2014
6. Salkić, H., Madžarević, V., Muharemović, A., Hukić, E.: Numerical solving and experimental measuring of low frequency electromagnetic fields in aspect of exposure to non-ionizing electromagnetic radiation. In: The 14th International Symposium on Energy, Informatics and Cybernetics: ISAS (2008)
7. Madžarević, V., Muharemović, A., Mehinović, N., Salkić, H., Tešanović, M., Kasumović, M., Hadžimehmedović, A.: EMC of power facilities – calculations and measurement of the low frequency electric and magnetic field. Faculty of Electrical Engineering in Tuzla, Tuzla (2010)
8. Faria, J.A.B., Almeida, M.E.: Accurate calculation of magnetic-field intensity due to overhead power lines with or without mitigation loops with or without capacitor compensation. IEEE Trans. Power Delivery **22**(2), 951–959 (2007)
9. Gouda, O.E., El Dein, A.Z.: Mitigation of magnetic field under double-circuit overhead transmission line. Telkomnika **10**(8), 2272–2284 (2012)
10. Sadović, S.: Power System Analysis. Faculty of Electrical Engineering, University of Sarajevo, Sarajevo (2014)
11. Budnik, K., Machczyński, W.: Power line magnetic field mitigation using a passive loop conductor. Pozn. Univ. Technol. Acad. J.Electrical Eng. **73**, 137–145 (2013)
12. Salari, J.C., Mpalantinos, A., Silva, J.I.: Comparative analysis of 2-and 3-D methods for computing electric and magnetic fields generated by overhead transmission lines. IEEE Trans. Power Delivery **24**(1), 338–344 (2009)
13. CIGRE WG C4.203: Technical Guide for Measurement of Low Frequency Electric and Magnetic Fields near Overhead Power Lines. August 2008
14. Report on measurement of the electric and magnetic fields, ETF Sarajevo, 2 July 2014

# A Low-SWaP, Low-Cost Transceiver for Physically Secure UAV Communication with Visible Light

Burak Soner(✉) and Sinem Coleri Ergen

Electrical and Electronics Engineering, Koç University, Istanbul, Turkey
{bsoner16, sergen}@ku.edu.tr

**Abstract.** Unmanned aerial vehicles (UAV) are expected to utilize optical wireless technologies alongside radio frequency technologies for reliable, secure and high bandwidth communications. While terrestrial and atmospheric laser-based solutions in the past have achieved physically secure communication with very complex beam tracking/pointing mechanisms with large and costly telescopes, such systems are neither suitable nor necessary for medium-range (<100 m) commercial UAV communications. With the proliferation of low-cost solid-state lighting equipment and visible band photodetectors, visible light communications (VLC) offer a low-size-weight-and-power (SWaP) and low-cost solution. This paper presents a novel low-SWaP and low-cost transceiver for physically secure VLC in medium-range commercial UAV applications. Full implementation details for a proof-of-concept prototype built completely with off-the-shelf components are also reported.

## 1 Introduction

Unmanned aerial vehicles (UAV) require reliable, secure and high bandwidth wireless communication with other UAVs and ground stations, both for fulfilling a variety of mission requirements and for maintaining safe, collision-free flight [1]. In order to better fulfill these requirements for a wider variety of environmental and channel conditions, radio frequency (RF) communications for UAVs can be complemented by optical wireless communications (OWC) [2]. OWC refers to use of the optical band ($\sim$200–1600 nm) for communication purposes. Compared to its RF counterpart, OWC offers higher and unregulated bandwidth and higher robustness to electromagnetic interference. Predominantly a line-of-sight (LoS) technology, OWC additionally achieves highly directional, physically secure beams at much lower size-weight-and-power (SWaP) and cost figures compared to RF. A detailed introduction to OWC is provided in [3].

Physically secure communication systems require most or all the energy emanating from the transmitter (TX) to reach the intended receiver's (RX) antenna through a LoS path, leaving practically no energy for non-LoS eavesdroppers. Although this provides the ultimate security, these systems require very precise (<μrad [4]) acquisition, tracking and pointing (ATP) in order to keep the TX and RX within each other's LoS, especially for mobile TX and RX. Prior art has thoroughly investigated the ATP


S. Avdaković et al. (Eds.): IAT 2019, LNNS 83, pp. 355–364, 2020.
https://doi.org/10.1007/978-3-030-24986-1_28

problem for a subset of OWC which deals with very long-range terrestrial, aerial, satellite and lunar communications using visible (~390–750 nm) or near IR band (~750–1600 nm) lasers and highly complicated low-divergence optics: free-space optical communications (FSO) [4]. Despite many successful deployments, while still much lower than those of its RF counterpart, the SWaP and cost figures of FSO ATP are still too large for use in medium-range (<100 m) high volume commercial applications like communications for small-payload UAVs [5]. For such applications, another subset of OWC, visible light communications (VLC), can provide feasible low-SWaP and low-cost solutions.

VLC utilizes the very cheap and highly available visible band light emitting diodes (LED), laser diodes (LD), photodetectors (PD) and optics on the lighting equipment market for versatile design choices in performance, SWaP and cost [6]. Five main challenges have been identified for using VLC in UAV communication in [7]: (i) asymmetry of ground-facing and air-facing links in terms of ambient noise, (ii) high data rate, (iii) long range, (iv) tracking/pointing for high mobility, (v) energy efficiency. While [8] addresses the data rate, range and energy issues for a similar application (inter-satellite VLC) via simulations, the dynamic tracking/pointing issue is not explicitly addressed. The physically secure UAV VLC transceiver proposed in this paper presents a low-SWaP and low-cost solution to the high mobility tracking issue. Furthermore, since the % of the total TX energy reaching the RX increases with lower beam divergence, the proposed transceiver decreases path loss and increases signal-to-noise ratio and effective range [9], presenting also a solution to the long-range VLC challenge.

This paper initially discusses the system model for medium-range (<100 m) physically secure UAV communications with VLC in Sect. 2. Afterwards, it presents the design of a low-SWaP, low-cost VLC transceiver which is feasible for use in this system in Sect. 3. Full implementation details for a proof-of-concept prototype for this design realized with cheap, off-the-shelf components are provided in Sect. 4. While the performance of this prototype does not reflect the full potential of the proposed transceiver, it proves that a low-SWaP and low-cost VLC solution to the tracking/pointing problem exists. The paper is concluded in Sect. 5 after a summary of the presented work.

## 2 System Model

The system model, depicted in Fig. 1, comprises of two UAVs, <100 m apart, which carry VLC transceivers (TRX) with 360° rotation pan-tilt-type gimballed pointing mechanisms [10] capable of perfectly synchronous communication with each other as long as energy from the TX beam reaches the RX detector. TRXs contain LEDs with standard radially symmetric polar beam patterns (TX) [11], a photodetector (RX), associated drivers and optics and the pan-tilt gimbal mechanism that allows for them to point their LED towards the other's photodetector and track each other's LoS throughout their trajectories for sustained communication.

A tracker/pointer (TP) can be considered to have obtained a LoS link if the RX detector stays within the full width at half maximum (FWHM) of the TX beam footprint on the RX plane, and its accuracy can be measured by how well the beam central

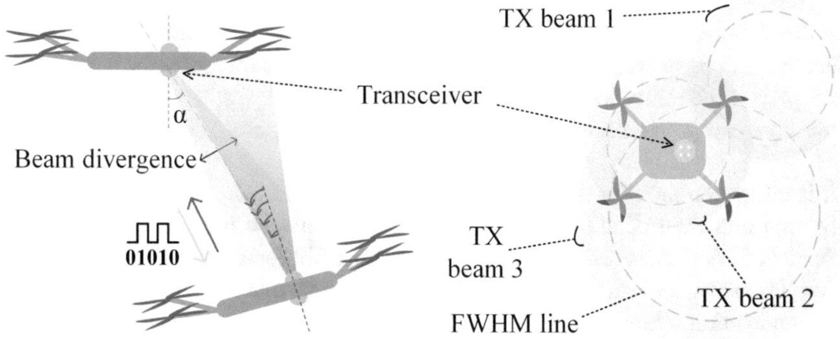


**Fig. 1.** Investigated system model for medium-range (<100 m) UAV communications with VLC

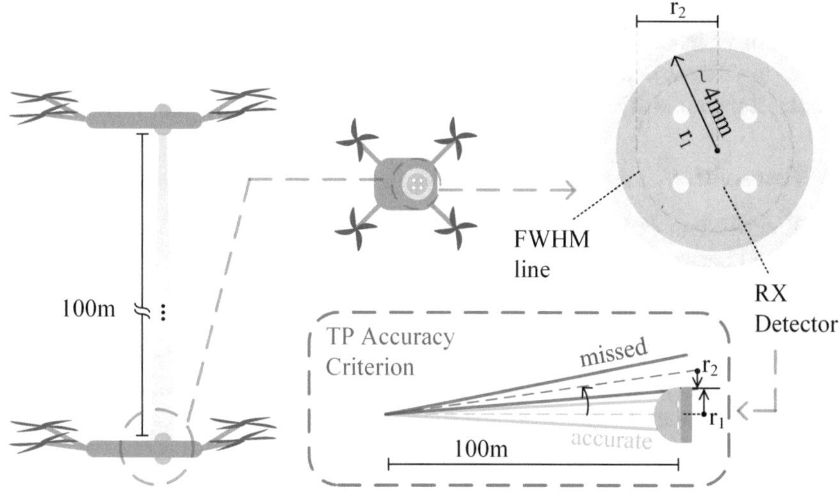


**Fig. 2.** Tracking/pointing (TP) accuracy requirements for complete physically security at 100 m

axis is aligned with the RX detector surface normal vector. A higher TP accuracy brings higher signal-to-noise-ratio and thus more reliable communication. Furthermore, the physical security level of a beam can be measured by what % of the total energy emanating from the TX falls onto the RX detector. These concepts are demonstrated in Fig. 1 where TX beam 1 is not in LoS, beam 3 is in LoS but not very accurate, and beam 2 has the highest accuracy. Beam 2 is also more secure than 3 since a much larger % of the total TX beam falls onto the RX detector, leaking out much less energy for potential eavesdroppers. The most stringent TP accuracy requirement for this system occurs when complete physical security (FWHM of TX beam is fully inside RX detector area) is required at the furthest distance (100 m). This spatial beam confinement requirement, depicted in Fig. 2, is common in long-range FSO applications and requires large and costly telescopes and ultra-high precision TP stages for >km ranges since the TP accuracy requirement becomes sub-μrad. The medium-range UAV VLC application can be treated with much lower SWaP and cost solutions since the required accuracy is inherently lower at smaller distances and furthermore the accuracy requirement can be traded for minor compromises in physical security, like shown in Fig. 1.

It is not uncommon for the two TRX in FSO and VLC links to lose LoS due to bad TP or environmental conditions. When LoS is lost between two TRX, an acquisition mechanism is necessary to regain LoS and resume TP for sustained communication. The requirements for acquisition mechanisms vary too much depending on the individual application requirements and are thus left out of this investigation. A summary on state-of-the-art acquisition methods can be found in [4].

## 3 Proposed Low-SWaP, Low-Cost Transceiver

This Section proposes a transceiver which can be used for physically secure VLC between UAVs in <100 m distances described in the previous Section. The low-SWaP and low-cost transceiver utilizes a simple standard VLC TX subsystem and a single photodetector on the RX subsystem for both TP and VLC purposes.

### 3.1 Transmitter

The TX subsystem of the proposed TRX simply consists of the TX LED, its driver circuitry and beam-shaping optics which is either composed of static lenses for a given beam divergence, or a set of moveable lenses in order to accommodate for different conditions with dynamic beam divergence control. The TX beam contains AC-coupled VLC signals modulated around a DC bias intensity level by methods like Manchester-coded on-off-keying (OOK), frequency shift keying (FSK) or variable-pulse-position (VPPM) to reduce message-content-dependent flicker [6].

### 3.2 Receiver

The RX subsystem of the proposed TRX uses a quad photodiode (QPD) which can infer the angle-of-arrival (AoA) of the incoming TX beam from the relative intensity difference on each of its cells. QPDs are widely used for laser beam tracking applications with accuracy requirements on the order of a few μrads [12]. Rather than sensing the AoA explicitly though, the QPD here is used simply as a differential sensor in the TRX controller which constantly tries to keep AoA at zero by moving the TRX with the pan-tilt and balancing the relative intensity on the QPD cells. To converge the incoming beam onto the QPD as a suitably sized spot [13], a hemispherical lens is used [20], and each QPD cell reading is amplified and fed to the TRX microcontroller for both TP and VLC purposes. The optics for the proposed RX subsystem is depicted in Fig. 3. Since the VLC signal content of the TX beam is AC-coupled, the QPD amplifier for each cell is also AC-coupled, rejecting the DC ambient noise and conveying the TX signal for that cell as clearly as possible to the microcontroller (MCU). The MCU then uses the QPD readings for both VLC and TP. This process is explained in detail in Sect. 3.3.

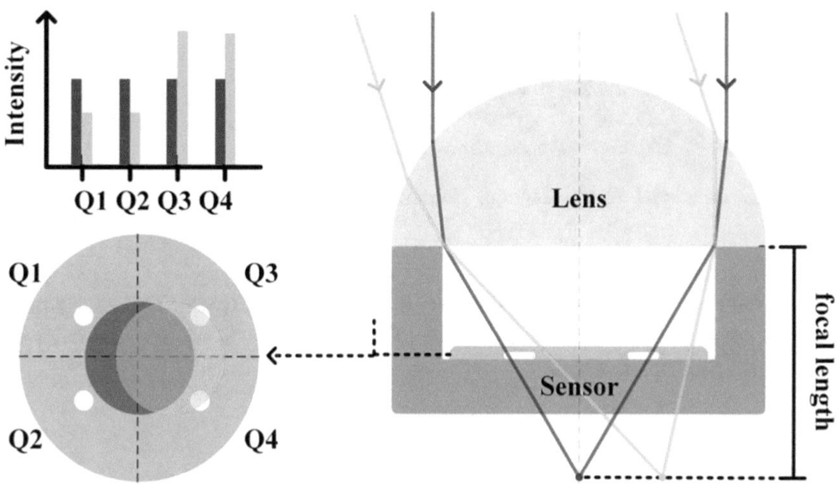


**Fig. 3.** Receiver optics for the proposed low-SWaP and low-cost transceiver.

## 3.3 Tracking and Pointing

Most long-range FSO ATP solutions separate the communication and ATP subsystems since the two require individual optical designs in order to meet the stringent sub-μrad TP requirements. While some previous works have used separate "beacon" beams for TP [14], others have used a single beam for both but separated it with beam splitters and processed the beam for the TP PD and the beam for the communication PD separately [15]. These designs address their application's needs, but they have resulted in high SWaP gimballed telescopes with high cost. The simple, single-PD-single-beam TRX design proposed in this paper meets the requirements for the application investigated in this paper and is low-SWaP and low-cost.

The QPD in the proposed TRX utilizes the TX beam shining on it for both VLC and TP purposes. Using these QPD cell readings, the microcontroller runs one task for VLC and one task for TP. The VLC task takes the sum of the cells, demodulates this to retrieve the VLC message and then responds to the message by modulating the TX LED. The TP task first calculates the signal energy on each QPD cell and deduces the position of the TX beam's spot on the QPD x-y frame. Then it commands the pan-tilt to turn the TRX accordingly to align the TX beam center with the QPD center, achieving accurate LoS tracking between the two TRXs. The functional block diagram for complete TRX operation is shown in Fig. 4.

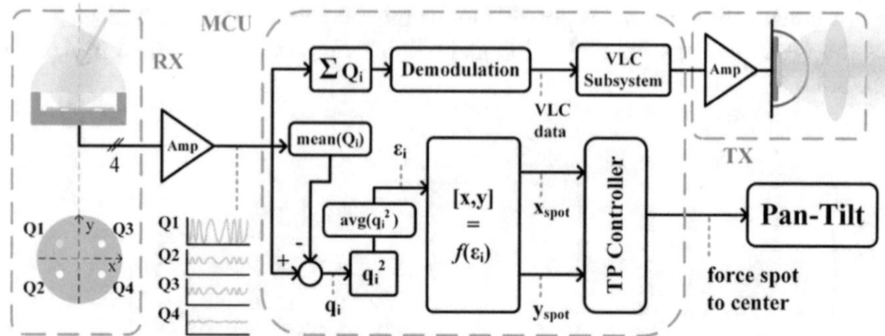


**Fig. 4.** TRX functional block diagram. Equation (2) is summarized by $[x,y] = f(\varepsilon_i)$. avg: average.

The signal energy for a certain duration ($\varepsilon_i$) is computed by the TP task by making each QPD reading for that duration ($Q_i$) zero-mean, taking its square, and calculating the average value of the result:

$$\varepsilon_i = average\left(\left(Q_i - \overline{Q_i}\right)^2\right) \tag{1}$$

While increasing the duration over which the signal energy is computed brings better robustness against noise, too large a duration slows the controller down. The signal energies are then used for computing the x-y position of the spot on the QPD using the well-known equation for these applications [12], referring to the cell indexing and QPD x-y axis convention shown in Fig. 4:

$$x_{spot} = \frac{(\varepsilon_3 + \varepsilon_4) - (\varepsilon_1 + \varepsilon_2)}{\varepsilon_1 + \varepsilon_2 + \varepsilon_3 + \varepsilon_4}, y_{spot} = \frac{(\varepsilon_1 + \varepsilon_3) - (\varepsilon_2 + \varepsilon_4)}{\varepsilon_1 + \varepsilon_2 + \varepsilon_3 + \varepsilon_4} \tag{2}$$

For accurate alignment of the TRX LoS, the spot needs to be exactly on the center of the QPD. The TP controller thus always has a zero reference for $x_{spot}$ and $y_{spot}$ while turning the TRX pan-tilt towards the incoming TX beam center.

## 4 Proof-of-Concept Prototype

This section presents a realization of the transceiver proposed in the previous section using completely off-the-shelf components and provides in-depth implementation details. The prototype is pictured in Fig. 5 and a demo video can be found in [18]. The prototype consists of one small-scale RX and TX unit built as testing platforms for further designs of complete TRXs. Due to the ultra-cheap components used and visible errors in mounting/build (e.g. Figure 5c, lens/QPD axis skew), performance and TP accuracy of this prototype does not reflect the limits of the proposed transceiver. It's merely an example which demonstrates the single-QPD VLC+TP concept, which is novel for commercial medium-range UAV communications.

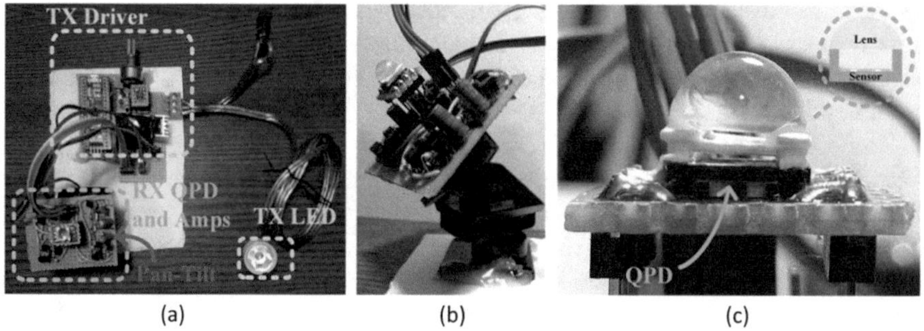


**Fig. 5.** Proof-of-concept prototype. TX+RX (a), Pan-tilt (b) and QPD+Lens RX (c).

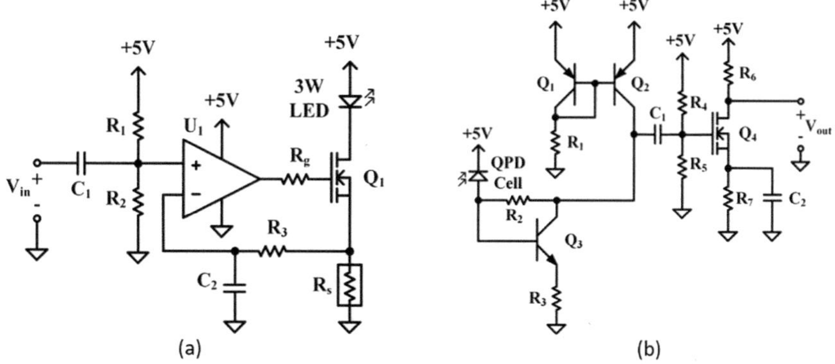


**Fig. 6.** Schematics for transmitter (a) and receiver electronics (b).

**Table 1.** Components for transmitter electronics

| $R_1$ | $R_2$ | $R_3$ | $R_g$ | $R_s$ | $C_1$ | $C_2$ | $Q_1$ | $U_1$ |
|---|---|---|---|---|---|---|---|---|
| 100 kΩ pot | | 100 Ω | 10 Ω | 4.2 Ω* | 10 uF | 100 pF | IRF530 | LMH6643 |

*$R_s$ is a sandstone resistor

**Table 2.** Components for receiver electronics

| $R_1$ | $R_{2,6}$ | $R_3$ | $R_{4,5}$ | $R_7$ | $C_1$ | $C_2$ | $Q_{1,2}$ | $Q_3$ | $Q_4$ |
|---|---|---|---|---|---|---|---|---|---|
| 560 Ω | 1 kΩ | 100 Ω | 100 kΩ | 200 Ω | 100 nF | 10 uF | 2N3906 | 2N3904 | 2N7000 |

## 4.1 Transmitter

The transmitter consists of a standard 3 W white TX power LED [8], a 5° divergence angle focusing lens [16] and the TX driver electronics. The 5° FWHM beam divergence from the source results in a moderate physical security level (beam radius is

8.7 cm at 1 m distance). While lower divergence beam-shaping optics are possible with more aggressive off-the-shelf lenses, the lens used in this prototype is a parabolic design with great form factor and holds the LED in place perfectly. The TX design resembles the TX used in [19].

The TX driver schematic and the components used are shown in Fig. 6a and Table 1 respectively. The driver is essentially a current servo where input voltage $V_{in}$ is the reference for the current flowing over the LED. Luminous flux generated by this LED increases linearly with current [11]. The op-amp linearizes the loop dynamics which include the current sense resistor $R_s$ and the MOSFET. The LED bias intensity, around which the modulation waveform swings, was set manually by a 100 kΩ potentiometer on $R_1$ and $R_2$ to fully utilize the LED dynamic range without turn-off or saturation/clipping. $V_{in}$ was fed from an AD9833 Digital Direct Synthesizer (DDS) which generated the binary FSK (BFSK) waveform that modulated the LED intensity for encoding VLC messages. The BFSK symbols used were 3 kHz -> 0 and 2 kHz -> 1 with a bit-rate of 1kbps, and the output voltage swing was 0.7 $V_{pk-pk}$. Since the prototype was focused on testing the TP principle, the carriers and bit-rate were kept moderately low for ease on the MCU side and random data was transmitted. With higher-end MCUs, higher bandwidth communication is possible.

### 4.2 Receiver

The receiver consists of a QPD with SMD package, OPR5911 from TT Electronics, a hemispherical lens of diameter 8 mm and focal length 5 mm mounted 2 mm above it and a cascade AC-coupled transimpedance and voltage amplifier stage. The RX schematic and the components used are shown in Fig. 6b and Table 2 respectively.

### 4.3 Tracking and Pointing

The Adafruit mini pan-tilt kit with standard "micro-servos" [17] (angular resolution of 0.5°) was chosen for the TP subsystem since the prototype was simply built for testing the principles set forth in the previous Sections using the QPD. An STM32F103C8T6, commonly known as the "blue pill", was used as the MCU. The MCU has a 12-bit ADC which was clocked at 50kSPS. Following the procedure described in Sect. 3.3 and Fig. 4, the MCU takes a 400-sample buffer from each QPD (corresponding to 8 ms), subtracts the mean, takes the square, finds the average of the buffer for each QPD for that 8 ms duration, which is then used for finding $x_{spot}/y_{spot}$. A simple proportional controller then tries to keep the spot on the QPD center. While <8 ms buffers gave faster TP response, values that approached the full wave periods of BFSK carriers created oscillating average values, hence oscillating $x_{spot}/y_{spot}$. Larger buffer sizes gave better stability and noise resilience but the MCU memory limited the size to 15 ms. Pan axis tracking performance results are shown in Fig. 7a and b, demonstrating this effect. Since the tilt axis performs in a similar manner (i.e. only gravity effects differ), only the pan axis results were shown. Performance for different buffer sizes under stationary and dynamic beam TP conditions are shown explicitly in the demo video [18].

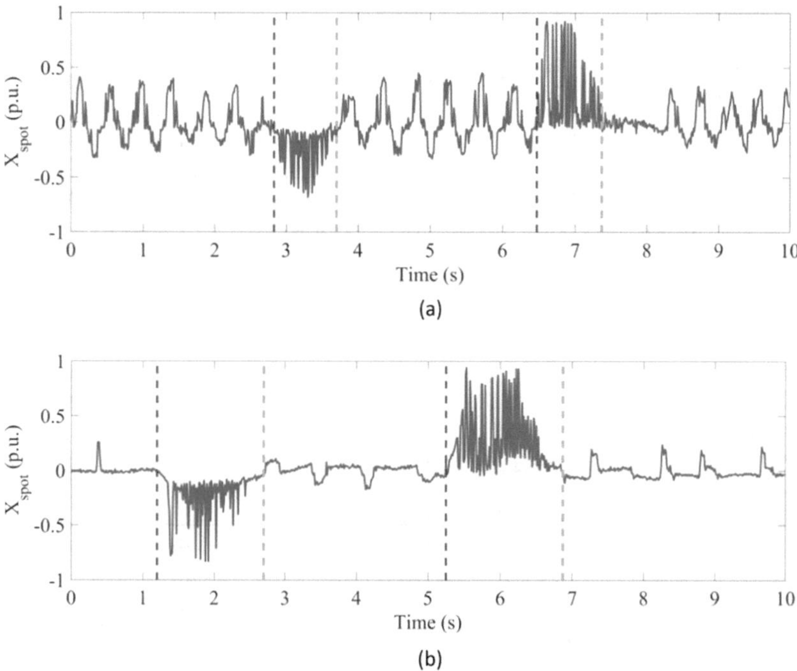


**Fig. 7.** Pan axis control performance with buffer sizes of 2 ms(a) and 10 ms(b) demonstrated via the QPD spot x-position in time. Red lines are where step changes in TX beam direction occur and green lines are where the controller settles. Same step change was applied in both cases. While settling time is $\sim$0.9 s in (a) and $\sim$1.8 s in (b), the steady state behaviour of (a) is very oscillatory since the buffer size is close to the full wave period of the TX VLC carrier waveform.

## 5 Conclusion

This paper presented a novel low-SWaP and low-cost transceiver (TRX) for physically secure visible light communication (VLC) in medium-range commercial UAV applications. To realize the design in low-SWaP, a single quad photodiode was used on the RX end both for VLC and tracking/pointing (TP) purposes. A proof-of-concept prototype using only off-the-shelf components was realized and demonstrated. While the basic prototype does not reflect the performance limits of the proposed TRX, it conveys the basic principle of the single-beam-single-detector VLC+TP.

**Acknowledgement.** The authors would like to thank Kıvanç Hedili from Koç University, Istanbul, Turkey for his contributions on the design of the RX subsystem.

## References

1. Zeng, Y., Zhang, R., Lim, T.J.: Wireless communications with unmanned aerial vehicles: opportunities and challenges. IEEE Commun. Mag. **54**(5), 36–42 (2016)
2. Chlestil, C., et al.: Reliable optical wireless links within UAV swarms. In: 2006 International Conference on Transparent Optical Networks, pp. 39–42. Nottingham (2006)
3. Uysal, M., Nouri, H.: Optical wireless communications — an emerging technology. In: 2014 16th International Conference on Transparent Optical Networks (ICTON), pp. 1–7. Graz (2014)
4. Kaymak, Y., et al.: A survey on acquisition, tracking, and pointing mechanisms for mobile free-space optical communications. IEEE Commun. Surv. & Tutor. **20**(2), 1104–1123 (2018)
5. Haan, H., Gerken, M., Tausendfreund, M.: Long-range laser communication terminals: technically interesting, commercially incalculable. In: 2012 8th International Symposium on Communication Systems, Networks & Digital Signal Processing (CSNDSP). Poznan (2012)
6. Jovicic, A., Li, J., Richardson, T.: Visible light communication: opportunities, challenges and the path to market. IEEE Commun. Mag. **51**(12), 26–32 (2013)
7. Ashok, A.: Position: DroneVLC: visible light communication for aerial vehicular networking. In: Proceedings of the 4th ACM Workshop on Visible Light Communication Systems (VLCS '17), pp. 29–30. ACM, New York, NY, USA (2017)
8. Amanor, D.N., Edmonson, W.W., Afghah, F.: Intersatellite communication system based on visible light. IEEE Trans. Aerosp. Electron. Syst. **54**(6), 2888–2899 (2018)
9. Bekmezci, I., Sahingoz, O.K., Temel, Ş.: Flying ad-hoc networks (FANETs): a survey. Ad Hoc Netw. **11**(3), 1254–1270 (2013)
10. Khan, M., Yuksel, M.: Autonomous alignment of free-space-optical links between UAVs. In: Proceedings of Hot Wireless '15, pp. 36–40. Paris, France (2015)
11. Multicomp 3 W High Power LED, White: http://www.farnell.com/datasheets/1678766.pdf
12. Carbonneau, R., Dubois, J., Harris, G.: An optical gun muzzle sensor to improve firing accuracy. In: Proceedings of SPIE 0661, Optical Testing and Metrology (1986)
13. Kazovsky, L.G.: Theory of tracking accuracy of laser systems. Opt. Eng. **22**(3), 223339 (1983)
14. Walther, F.G., Michael, S., Parenti, R.R., Taylor, J.A.: Air-to-ground lasercom system demonstration design overview and results summary. In: Proceedings of SPIE 7814, Free-Space Laser Communications X, p. 78140Y. 24 August 2010
15. Guelman, M., Kogan, A., Kazarian, A., Livne, A., Orenstein, M., Michalik, H.: Acquisition and pointing control for inter-satellite laser communications. IEEE Trans. Aerosp. Electron. Syst. **40**(4), 1239–1248 (2004)
16. Lens, PMMA, FWHM beam angle 5°: https://www.direnc.net/5-derece-lens
17. https://www.adafruit.com/product/1967. Accessed 4 Apr 2019
18. Demo video of the proof-of-concept prototype: https://youtu.be/wQPWQB3W_JM
19. Abualhoul, M.: Visible light and radio communication for cooperative autonomous driving: applied to vehicle convoy. PhD Thesis, MINES Paris Tech (2016)
20. Kahn, J.M., Barry, J.R.: Wireless infrared communications. Proc. IEEE **85**(2), 265–298 (1997)

# Validation of Novel System Identification Approach Based on Forced Oscillations Using Open-Loop Experiment

Rijad Sarić[1(✉)], Edhem Čustović[2], Dejan Jokić[1], and Željko Jurić[3]

[1] Faculty of Engineering and Natural Sciences, International Burch University, Sarajevo, Bosnia and Herzegovina
rijad.saric@stu.ibu.edu.ba, dejan.jokic@ibu.edu.ba

[2] School of Engineering and Mathematical Sciences, La Trobe University, Melbourne, Australia
e.custovic@latrobe.edu.au

[3] Faculty of Electrical Engineering, University of Sarajevo, Sarajevo, Bosnia and Herzegovina
zjuric@etf.unsa.ba

**Abstract.** The parametric identification is the primary consideration in developing a sophisticated automated control system. However, most tuning and system identification methods require the use of non-standard equipment such as relay which could cause a significant error and in turn affect the accuracy of the entire industrial process. The novel approach to the system identification in closed-loop feedback is similar to old Ziegler-Nichols (ZN) experiment, but it does not include any additional equipment while identifying the points in three quadrants in the Nyquist diagram. After applying this method to identify one complex object i.e. servomechanism with only damped oscillations, it is necessary to validate the correctness of the obtained model. Conducting the laboratory experiment in the open-loop loop represents the best way of verifying stated approach, since the transfer function including the Nyquist diagram of the model may provide enough data for further analysis. The verification laboratory experiments confirmed applicability together with the effectiveness of the new method in considerably less idealized conditions compared to computer-based simulations.

**Keywords:** Object identification · Open-loop · Ziegler-Nichols · Nyquist diagram

## 1 Introduction

In General, any unknown plant may be identified by unique transfer function applying certain experimental methods that could be divided into two categories such as open-loop and closed-loop feedback methods. Open-loop feedback methods refer to time-response where an object is entirely isolated from the control loop and usually the object response is scrutinized by bringing specifically selected signals. These methods are only suitable for simple asymptotically stable objects including short time


S. Avdaković et al. (Eds.): IAT 2019, LNNS 83, pp. 365–375, 2020.
https://doi.org/10.1007/978-3-030-24986-1_29

performance. Contrary to that, frequency response methods are based on determining significant number of points in the Nyquist diagram and such identification usually gives more information about the object itself even though it needs a great amount of time. Hence the frequency response methods are mainly used as a part of closed-loop feedback where there is no interruption of the control loop. In closed-loop feedback, an object is simply identified considering the behavior of the entire control loop supplied with the controller capable of providing stable operation together with motion of the system through the feedback. In fact, all the frequency based closed-loop methods assume full linearity of the object, which leads to difficult implementation in the presence of non-linear objects. Also, parametric type of identification where the shape of non-linear operator F normally containing derivative-integral characteristic, as in expression y(t) = F[x(t)], is known. However, it is necessary to experimentally determine unknown parameters available in F [1–3].

The oldest identification method performed in the closed-loop feedback experiment allowing approximate estimation of fundamental dynamic parameters of a simplified object is widely known as the Ziegler-Nichols (ZN) [4]. This method is mostly using for tuning PI/PID regulators by detecting one point in the Nyquist diagram. The primary disadvantage of the ZN method represents assumption of the non-standard object during the modeling process as well as lack of useful information acquired during the experiment that in the most cases requires unavoidable sustained high-amplitude undamped oscillations. Therefore, ZN could provide non-optimal tuning results, especially for the time-delay dominant or right half-plane finite zero processes [5]. However, ZN is widely used for relatively simple industrial control processes. An early extension of the ZN method is proposed by Åström et al. [6] in 1984 replacing the P-type controller with a relay that limits the amplitude of occurred oscillations. This method yields the same data as ordinary ZN containing the additional errors caused by the introduced non-linear element e.g. relay.

Advanced identification processes have a goal of reaching full knowledge of transfer function in its approximate form. This often requires the use of more sophisticated closed-loop methods as well as longer experimental time, while conducting the experiment necessary for providing a detailed model of object. The four most common closed-loop methods with the assumption of stable objects are Relay and Hysteresis by Åström et al. [6], Two Channel Relay (TCR) by Friman et al. [7], Auto Tune variation (ATV) by Li et al. [8], and lastly ATV+ method proposed by Scali et al. [9]. TCR method consists of two relays and integrator (dual channel relay) that replace a controller during the experiment. A representative number of points in the third quadrant of the Nyquist diagram could be obtained using TCR method. Likewise, ATV method improves the performance of parametric identification with previously known eventual pure time delay using a relay and variable delay line. On the other hand, the ATV+ method could provide the estimation of an unknown time delay using a trial-and-error procedure. The main weakness of all stated methods is unpreventable non-linearity due to the utilization of relays and additional non-standard equipment within the control loop. They may cause considerable inaccuracies, even up to 25%. Moreover, it is noticeable that all mentioned methods identify the significant number of points only in the third quadrant of the Nyquist diagram. However, the estimation of accurate process gain is almost impossible in the absence of useful points in the fourth quadrant of the Nyquist curve. There is also a novel method proposed by Juric et al. [10, 11] in 2007, that allows identification of points in the Nyquist diagram in all three quadrants, without additional equipment. The novel method is applied by Saric et al. [12] to demonstrate compensated PID regulation of ambient temperature in the closed-loop feedback.

This paper represents the open-loop verification of the novel approach to the system identification in the closed-loop feedback based on forced oscillations. The introductory part gives a brief overview of the system identification including most relevant closed-loop feedback methods. Section 2 describes the general idea under the novel approach to system identification. The methods used to conduct an experiment in open-loop feedback are described in the Sect. 3, whereas obtained results are demonstrated and summarized in the Sects. 4 and 5.

## 2 Theoretical Background

A novel approach to the frequency-response closed-loop identification is like classical ZN experiment, replacing the P-type controller with a PI, PID or another more complex controller, but without using any additional equipment. Primarily, the proposed approach allows obtaining a larger number of points in the second, third and fourth quadrant of the Nyquist diagram measuring the frequency of deliberately caused undamped oscillations, or alternatively, bringing damped oscillations supported by adequate frequency and damping ratio measurement. In the second case, adding limiter in the control loop could restrict the amplitude of damped oscillations, without introducing any errors. This method provides good results if the controller is already in use, even though tuning is not satisfactory. In short, such generalization of ZN works especially well for objects containing zeros in the right half-plane [11] as well as objects with a long-time delay [13] that are in most cases difficult to identify using conventional methods. As in frequency based open-loop, methods the experimental time is the same applying suggested approach. The block diagram of the new approach to the closed-loop system identification is depicted in Fig. 1.

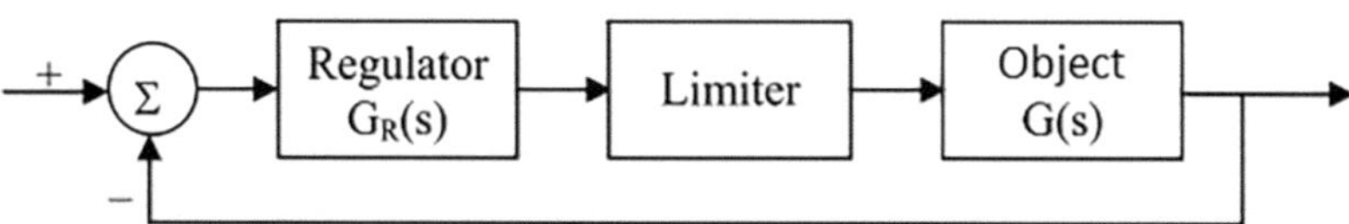


**Fig. 1.** The block diagram used for the novel identification approach in the closed-loop feedback

Here, the general idea of the method is recalled in [1, 10, 11]. It is assumed that the controller is described by transfer function $G_R(s, \Lambda)$ where $\Lambda$ is vector of tunable parameters and that there exists at least one vector $\Lambda = \Lambda^*$ making the control loop stable. Suppose that G(s) is the object transfer function. Then, the overall system transfer function could be expressed as (1)

$$W(s) = \frac{G(s)G_R(s, \Lambda)}{1 + G(s)G_R(s, \Lambda)} \tag{1}$$

At first, suppose that undamped oscillations are allowed. Starting from $\Lambda = \Lambda^*$, it is necessary to bring the system at the edge of the stability, by varying Λ. If sustained oscillations with frequency $\omega = \omega_0$ enable controller settings $\Lambda = \Lambda_0$, then $W(s, \Lambda_0)$ has one pole $s = j\omega_0$, which implies that

$$G(j\omega_0) = \frac{-1}{G_R(j\omega_0, \Lambda_0)} \tag{2}$$

Thus, the outcome $(\omega_0, \Lambda_0)$ of the stated experiment with undamped oscillations uniquely identifies a point in the Nyquist diagram. It may be easily proved that if (2) is satisfied for one setting $\Lambda = \Lambda_0$, then there exist an infinite set of pairs $(\omega_{k0}, \Lambda_{k0})$ that satisfies (2) as well. For instance, with an ideal PID regulator, all points in the third quadrant of the Nyquist diagram could be identified. In a similar way, the experiment with damped oscillations is performed when undamped oscillations are not allowed. Let assume that controller of settings $(\Lambda = \Lambda_0)$ produces damped oscillations with pseudo-frequency $\omega_0$ as well as damping factor $\sigma_0$, Then $W(s, \Lambda_0)$ has a pole $s = -\sigma_0 + j\omega_0$. Hence,

$$G(-\sigma_0 + j\omega_0) = \frac{-1}{G_R(\sigma_0 + j\omega_0, \Lambda_0)} \tag{3}$$

It is noticeable that although (3) does not identify a point in the Nyquist diagram, a collection of triplets $(\omega_{k0}, \sigma_0, \Lambda_{k0})$ could also be used for the parametric identification [14]. If it is necessary to obtain a point of the Nyquist diagram, it can be obtained by using extrapolation from results of two different experiments, i.e. from two different triplets [1].

## 3 Methodology

After identification of the object in the closed-loop feedback applying the novel approach the transfer function obtained is described as

$$G(s) = \frac{0.1744}{0.06985s + 1} \tag{4}$$

It is needed to evaluate the validity of the identified object which is one servo motor mechanism as well as model obtained in closed-loop experiment. Principally, it is possible to derive the exact transfer function of the motor (neglecting the nonlinear effects), but this is not very useful, because it requires knowledge of incentive flux, moment of inertia, and coefficient of viscose friction, which are very difficult to measure. Hence, the method of recording a representative number of points in the AF diagram in the open-loop feedback using already tested harmonic function is applied to verify the novel identification approach in the closed-loop feedback. The block diagram of the entire open-loop system is shown in Fig. 2 where drive circuit and data acquisition system are the same as in closed-loop experiment applied to servomechanism composed of two DC motors mutually connected by a pulley. The first propulsion motor provides rotation speed at the output roughly proportional to input voltage, whereas the second motor is used as tacho-generator (speed sensor) to measure the rate of angular speed at the output of the first motor. Additionally, the presence of the second motor could significantly increase the moment of inertia of object during the identification process. Assuming linearity of the magnetic circuit, this object could be considered as second-order linear model. However, during the experiment, a serious saturation nonlinearity is caused by the magnetic circuit of the first motor.

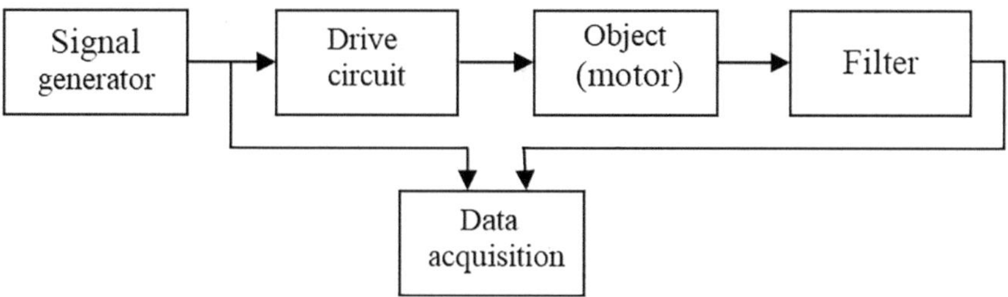


**Fig. 2.** The block diagram for performing open-loop parametric identification experiment

The filter used in the open-loop experiment is required because of the data concerning object speed generated at the output of tacho-generator are very noisy. The applied filter is described with a first-order transfer function as in (5)

$$GF(s) = \frac{1}{0.1s + 1} \tag{5}$$

During the conductance of the open-loop experiment, a common issue appears since DC motor which is used as the object of identification had a dead-zone in the voltage range between −2 V and 2 V. This issue is easily solved by increasing the offset of the input signal in order to avoid access to dead-zone. The offset of the input signal is increased for +5 V upward, so that input signal has mean value +5 V together with 3 V amplitude. Thus, the signal at the output of the signal generator is not a pure sinusoid, but the addition of a sinusoidal signal of 3 V amplitude and +5 V constant voltage. Experiments are performed by always generating the signal of the different frequencies in the range from 0.2 Hz to 10 Hz and the same amplitude. For each of the frequencies in the specified range the input/output diagram is recorded. The output of the object during harmonic (sinusoidal) excitation should be the same signal of harmonic frequency, but the different phase and amplitude. The difference in the phase and amplitude of input/output signal allows identification of one point in the Nyquist diagram for each frequency that matches the frequency available in the stated range. Furthermore, the postprocessing of collected input/output signals is not performed. The amplitude of the transfer function |G(jω)| is determined by dividing the roughly measured amplitude (graphical representation) of the output signal with the 3V fixed amplitude of the input signal. Phase $\phi(\omega) = \arg G(j\omega)$ is measured using also roughly estimated time span Δt between input and output signal while crossing the mean value ($\phi(\omega) = \omega\,\Delta t$). Obviously, it is expected to have certain imprecision during the reading, however, the goal is to simulate the reading of value in real industrial conditions which do not include any type of postprocessing. Thus, only graphically registered signals are available in the considered situation. The experimental research shows that responses deviate from the pure sinusoidal pattern, especially in the lower frequencies. For instance, Fig. 3 represents the signal excitation and response during the frequency of f = 0.2 Hz. An harmonic (non-sinusoidal) nature of the signal proves the presence of nonlinearity in the considered object. Interestingly, non-sinusoidal effects significantly decrease at higher frequencies as noticeable in Fig. 4. A similar experiment with motor maybe found in [15].

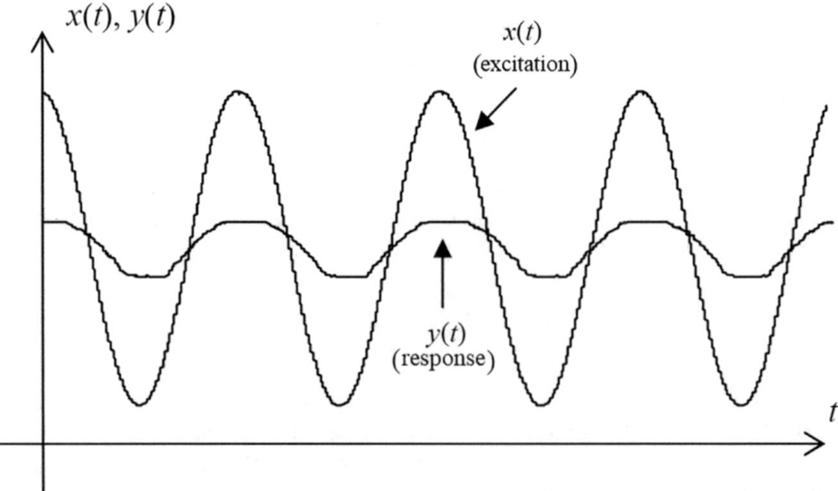


**Fig. 3.** Signal excitation and response during the open-loop experiment (f = 0.2 Hz)

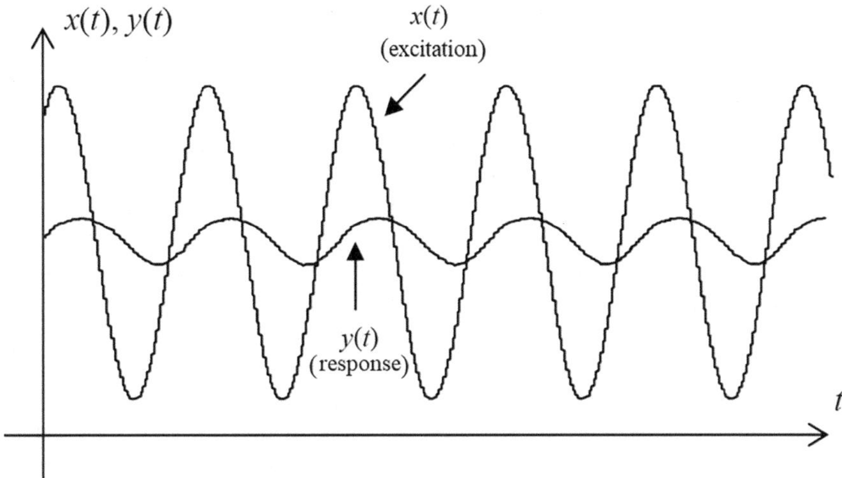


**Fig. 4.** Signal excitation and response during the open-loop experiment (f = 0.9 Hz)

## 4 Results and Analysis

The results achieved during the measurement process are represented in Table 1. Modulo of transfer function |G(jω)| is calculated as the quotient of input/output voltage amplitude, while the argument of the transfer function argG(jω) is described as the phase difference between input/output signal. Based on measured data it is possible to calculate transfer function G(jω) at a certain frequency whose real and imaginary part U(ω) and V(ω) represent the coordinates of one point in the Nyquist diagram.

**Table 1.** Results obtained during the parameter measurements including calculated coordinates for drawing the adequate Nyquist diagram

| Experiment | f (ω/2π) | \|G(jω)\| | arg G(jω) | U(ω) | V(ω) |
|---|---|---|---|---|---|
| I | 0.2 | 0.1750 | −10.8° | 0.1719 | −0.0328 |
| II | 0.3 | 0.1717 | −21.6° | 0.1596 | −0.0632 |
| III | 0.4 | 0.1683 | −29.5° | 0.1465 | −0.0829 |
| IV | 0.5 | 0.1667 | −32.4° | 0.1407 | −0.0893 |
| V | 0.6 | 0.1600 | −41.0° | 0.1208 | −0.1050 |
| VI | 0.9 | 0.1467 | −59.0° | 0.0756 | −0.1257 |
| VII | 2 | 0.0833 | −93.6° | −0.0052 | −0.0831 |
| VIII | 10 | 0.0017 | −180° | −0.0017 | 0 |

According to Table 1, it is possible to draw Nyquist diagram of the object in a broader sense which includes introduced filter as well. The distribution of points available in the Nyquist diagram in Fig. 5(a) suggests the second-order object. This is evident because of the first-order filter used as a part of control contour. The point (VIII) that is extremely close to the coordinate start (0) is ZN point though. Hence, it means that a small part of the Nyquist diagram exists in the second quadrant making possibility to form the third-order object based on detected points. This may lead to the second-order transfer function in case of object in a broader sense. The experiment of parametric identification is performed by taking different assumptions for a few poles (N) and zeros (M), even though it is clear that M = 0 and N = 2 or 3. Different characteristic parameters such as quality factor (V), condition number (CN), mean square error (MSE) and stability for different assumed M and N are taken into the consideration. The main goal is to prove that formal parametric identification procedure generates the same results as the closed-loop experiment. The best possible values for parametric identification are obtained when M = 0 and N = 2 or N = 3. Figure 5(b) illustrated the Nyquist diagram of the model obtained assuming M = 0 and N = 2, together with experimentally recorded points. Here, it may be seen that matching between the results of parametric identification and distribution of points is relatively good.

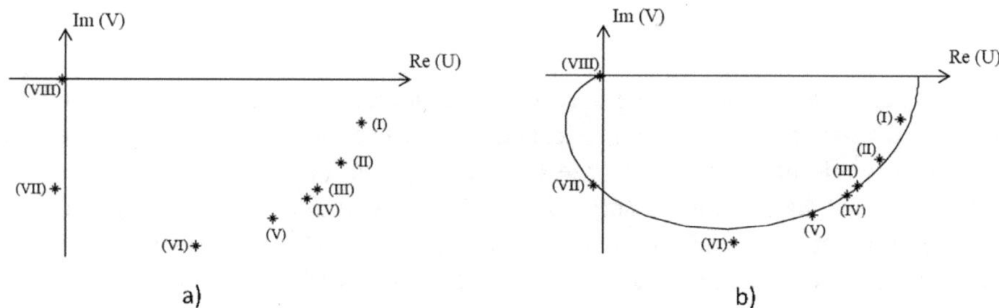


**Fig. 5.** (a) Nyquist diagram of experimentally recorded points during open-loop experiment (b) Nyquist diagram of the model obtained based on identification during open-loop experiment

Parametric identification procedure assuming N = 2 and M = 0 yields coefficient values $a_2 = 0.04612$, $a_1 = 1.0444$ and $a_0 = 5.4908$, therefore obtained transfer function is described as

$$G(s) = \frac{0.1821}{0.06972s + 1} \cdot \frac{1}{0.1205s + 1} \tag{6}$$

The second part in (6) originates from the filter dynamic, whereas the first part is the clear object in narrow sense. The better information about the object in the narrow sense may be obtained if we fully eliminate the transfer function of the filter. The effect of the filter is removed by dividing experimentally obtained data concerning transfer function $G(j\omega)$ available the Table 1 with values of filter transfer function $G_F(j\omega)$ at appropriate frequencies. Hence, the data regarding the object in the narrow sense is obtained and we could apply procedure of parametric identification when N = 1 resulting in coefficient values $a_1 = 0.4642$ and $a_0 = 5.5127$. Thus, the new transfer function of the object in the narrow sense is derived as

$$G(s) = \frac{0.1814}{0.08421s + 1} \quad (7)$$

The Nyquist diagram of the model in (7) together with experimentally recorded points of the object in the narrow sense obtained after filter elimination is represented in Fig. 6.

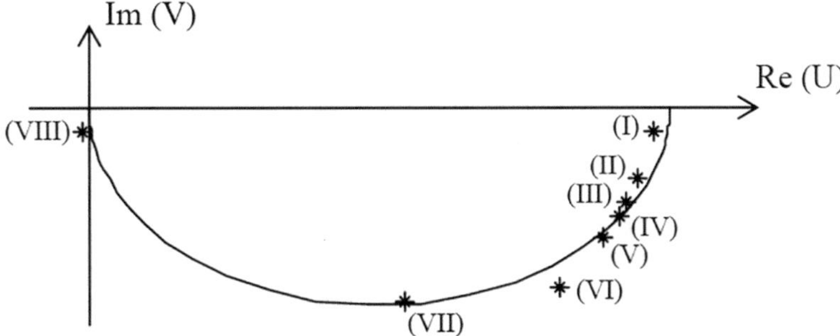


**Fig. 6.** Nyquist diagram of the model and experimentally recorded point after eliminating the impact of added filter

The Nyquist diagram represented in Fig. 6 suggests that modelling of the second-order object in the narrow sense is possible as well. Looking closely at experimentally obtained point (VIII) it may be seen its location in the third quadrant instead of fourth. If the procedure of parametric identification is applied on the experimentally recorded points as in Fig. 6, we would obtain a model with parameters $a_2 = 0.001044$, $a_1 = 0.4643$ and $a_0 = 5.534$. Therefore, the transfer function in the factorized form is

$$G(s) = \frac{0.1814}{(0.08159s + 1)(0.0002313s + 1)} \quad (8)$$

In the obtained model we have two time constants, where the first one is about 35.28 times greater than second. If we assume that first time constant originates from mechanical, while the second has electrical subsystem presented in the considered motor, it may be understood that electric subsystem is much faster than mechanical leading to dominant first-order object behaviour. Similar simulations and experiments are performed in [16, 17], however, in such an experiment the PI controller was also part of the controlled loop.

## 5 Conclusion

In summary, a novel system identification method proposed by Juric et al. [10, 11] which is an extension to the Ziegler-Nichols method is verified by performing an experiment in the open-loop. The open-loop experiment succeeds in detecting a DC motor as a second-order object, which was not a case during the closed-loop experiment. The reasonable explanation would be that poles of the introduced filter are the most dominant poles in the entire system. The pole of the dynamic of an electrical subsystem seems to be practically negligible compared to a set of filter poles. In this paper, it is demonstrated that the method is applicable in real use, under the conditions which are far from ideal. Finally, the next level of testing may be an application of the novel method in some real industrial processes and conduct sophisticated research which could provide profitability of this method.

## References

1. Sidorov, D.N.: Modelling of non-linear non-stationary dynamic systems by volterra series approach (VSA). J. Ind. Math. **3**, 182–194 (2000)
2. Ljung, L., Söderström, T.: Theory and Practice of Recursive Identification, 2nd edn. MIT Press, Cambridge, Massachusetts (1985)
3. Pierre, D.A.: Optimization Theory with Applications. Dover Publications Inc., New York (1986)
4. Ziegler, J.G., Nichols, N.B.: Optimal settings for automatic controllers. Trans. ASME **64**, 759–768 (1942)
5. Åström, K.J., Hägglund, T.: PID Controllers, 2nd ed. Instrument Society of America, North Carolina (1995)
6. Åström, K.J., Hägglund, T.: Automatic tuning of simple regulators with specifications on phase and amplitude margins. Automatica **20**, 645–651 (1984)
7. Friman, M., Waller, K.V.: A two-channel relay for autotuning. Ind. Eng. Chem. Res. **36**(7), 2662–2671 (1997)
8. Li, W., Eskinat, E., Luyben, W.L.: An improved autotune identification method. Ind. Eng. Chem. Res. **30**(7), 1530–1541 (1991)
9. Scali, C., Marchetti, G., Semino, D.: Relay with additional delay for identification and autotuning of completely unknown processes. Ind. Eng. Chem. Res. **38**(5), 1987–1997 (1999)
10. Jurić, Ž., Peruničić, B.: A method for closed-loop identification of Nyquist curve in three quadrants. Electr. Eng. **89**(3), 251–261 (2007)
11. Jurić, Ž., Peruničić, B.: A method for parametric closed-loop identification of plants with finite zeros. In: IEEE Mediterranean Conference on Control & Automation MED'04, pp. 1050–1055. Kuşadasi, Turkey, June 2004
12. Sarić, R., Čustović, E., Jurić, Ž.: Compensated PID regulation of ambient temperature using new approach to the system identification based on forced oscillations. In: 2018 5th International Conference on Control, Decision and Information Technologies (CoDIT), pp. 581–586. Thessaloniki (2018). https://doi.org/10.1109/codit.2018.8394811
13. Jurić, Ž., Peruničić, B.: A method for closed-loop identification of plants with unknown delay. In: IFAC TDS'04 Conference. Leuven, Belgium, August 2004, art. no. 156

14. Jurić, Ž., Peruničić, B.: A new method for the closed-loop identification based on the enforced oscillations. In: IASTED International Conference on Modelling, Identification and Control, MIC 2004, 23, pp. 85–91. Grindelwald, Switzerland, February 2004
15. Ishak, N., Abdullah, N.I., Rahiman, M.H.F., Samad, A.M., Adnan, R.: Model identification and controller design for servomotor. In: 2010 6th International Colloquium on Signal Processing & Its Applications, pp. 1–4. Mallaca City (2010). https://doi.org/10.1109/cspa.2010.5545294
16. Byeon, J., Kim, J.-S., Chun, D., Sung, S.W., Lee, J.: Third quadrant Nyquist point for the autotuning of PI controllers. In: 2009 ICCAS-SICE, pp. 3283–3286. Fukuoka (2009)
17. Shanshiashvili, B., Bolkvadzame, G.: Identification and modeling of the nonlinear dynamic open loop systems. In: 2012 IV International Conference “Problems of Cybernetics and Informatics” (PCI), pp. 1–4. Baku (2012). https://doi.org/10.1109/icpci.2012.6486373

# Edge Computing Framework for Wearable Sensor-Based Human Activity Recognition

Semir Salkic[1(✉)], Baris Can Ustundag[2], Tarik Uzunovic[1], and Edin Golubovic[3]

[1] Faculty of Electrical Engineering, University of Sarajevo, Sarajevo, Bosnia and Herzegovina
semir.salkic1@gmail.com, tuzunovic@etf.unsa.ba

[2] Department of Computer Engineering, Istanbul Technical University, Istanbul, Turkey
ustundag16@itu.edu.tr

[3] Inovatink, Istanbul, Turkey
edin@inovatink.com

**Abstract.** Human activity recognition is done based on the observation and analysis of human behavior to understand the performed activity. With the emergence of battery powered, low cost and embedded wearable sensors, it became possible to study human activity in various real-world scenarios. Together with the development in data collection, novel machine learning based modeling approaches show huge promise in modeling human activities accurately. Edge computing framework, that is capable of executing human activity recognition models at the edge of the network, is presented in this paper. Framework architecture and its implementation on a single board computer are presented. The framework allows the implementation of various machine learning models for human activity recognition in a standardized manner. The framework is demonstrated experimentally.

**Keywords:** Human activity recognition · Edge computing · Wearable sensors · Machine learning · IoT · Neural networks

## 1 Introduction

The popularity of wearable sensor technology is rising as standardized and easily accessible computing platforms are finding their place in multidisciplinary studies of human activity in various applications. Utilization of wearable devices is empowering human activity recognition applications in terms of comprehensiveness of data acquisition and meticulous measurements of human motion.

Systems including wearable sensors for human activity recognition mostly find its applications in health care, sports, rehabilitation, and human-robot collaboration [15].


S. Avdaković et al. (Eds.): IAT 2019, LNNS 83, pp. 376–387, 2020.
https://doi.org/10.1007/978-3-030-24986-1_30

We are witnessing specialized studies in virtual, augmented and mixed reality and physical activity [1], where we are still grasping utter importance of collected data. Exploiting these facts is a relatively simple task to implement with the help of data science. Most of the data collected through wearable devices have a simple flow which consists of data preprocessing, segmentation, feature extraction and modeling for a particular task (classification, prediction, detection). These approaches often require dedicated infrastructure for computing [2]. The development of various services based on artificial intelligence algorithms in the field of IoT opened doors for cloud-based platforms [3]. However, issues with cloud-based approach arise in the applications where data is collected by devices with limited bandwidth and power budget and in applications where latency is an operational issue. Due to these limitations on the application level, the concept of edge computing is prospering [4].

Edge computing refers to the enabling technologies allowing computation to be performed at the edge of the network, on downstream data on behalf of cloud services and upstream data on behalf of IoT services [4,5]. Edge computing approach, used in different implementations, is used to achieve significant results in regards to improved communication latency. Importance of leveraging edge power in the IoT domain is shown through everlasting problems, such as video transmission latency. This problem can be solved by the division of subchunks of video by end nodes [6]. Another example which requires notable resources, such as pool trainer, shape drawing helper, when combined with wearable, provided interesting results with improved overall response time in comparison with cloud-based platforms [7]. With improved latency, we are aware that the edge design of IoT networks, when paired with machine learning tasks, performs well. Layering and clustering nature of edge servers is enabling these benefits [8]. This approach results in a hybrid network of devices, which can be paired with cloud resources for optimal quality of service [9]. For most of IoT use cases, Edge-Cloud combination seems to be a satisfying solution which combines the best of both worlds in unified concept. Control of computing resources is done with the usage of various different architectures and innovative "smart" algorithms [10].

Together with the popularization of edge computing, the need for innovative, simpler, scalable and standardized design solutions is imminent. Application of edge-computing concept to human activity recognition using wearable sensors is still relatively new [2,11]. We have previously demonstrated that there is need to quickly process sensor data from wearable devices and make local decisions based on the analysis in applications such as sport coaching [12], rehabilitation [13] and human-robot collaboration in industrial environments [14]. In this research, we build on top of our previous work to present a comprehensive and novel edge computing system for human activity recognition using wearable sensors. The main focus of this paper is to present edge computing framework architecture and its implementation on a single board computer. The framework allows the implementation of various machine learning models for human activity recognition in an easy and standardized manner.

The rest of the paper is organized as follows; Sect. 2 introduces the proposed framework, in Sect. 3 details about implementation of framework on the single board computer are given, the application of framework in the human activity recognition scenario is presented in Sect. 4 and experimental results are provided. Finally, Sect. 5 concludes the paper.

## 2 Edge Computing Framework

The architecture of the edge computing system is given in Fig. 1. The edge computing process starts with data collection from the environment. The environment can be seen as tasks and activity performed by a human in a given application. The signals describing human activity are measured by wearable sensors. Data from wearable sensors is sent to the data processing hardware using wireless communication. Once received the data enters a *data processing pipeline*. The pipeline consists of following elements; data preprocessing, computation model, metric accumulation, feedback and external system interface.

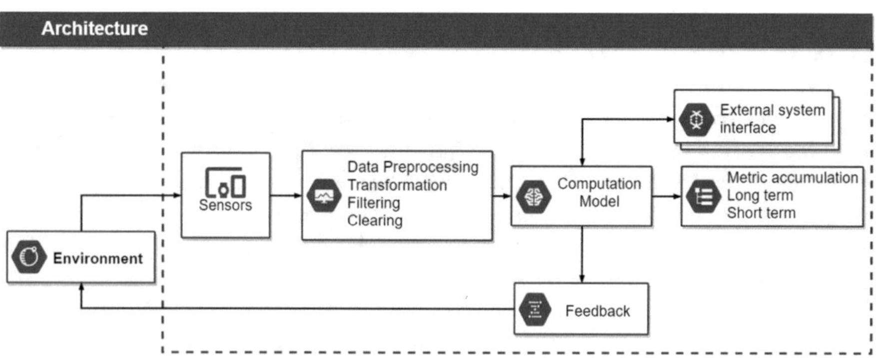


**Fig. 1.** Edge computing architecture

In data preprocessing element, data is transformed, filtered and cleaned of any erroneous instances. Essentially, the preprocessing element prepares the data in the format that is suitable to be used in the computation model. If machine learning models are employed as computation models then data segmentation and feature extraction are facilitated in this element too.

Computation model element of the data processing pipeline is a central part of human activity recognition. This model can include any combination of algorithms that are responsible for detection, classification or prediction of human activities in the given application. Computation models are developed based on application requirements. In our previous studies we have successfully developed models for basketball coaching [12], rehabilitation [13] and human-robot collaboration in industrial environments [14]. In the context of this paper, we exclusively consider models that are trained using machine learning methods.

Outputs of computation models are stored as metrics in the metric accumulation element. These metrics are intended for further use by other decision-making systems or as an input back to computation models.

Feedback element of the data processing pipeline is responsible for supplying information back to human based on the results of the computation model. This element can be very useful in certain applications where humans need to be aware of how well they are performing a task.

External system interface provides means of accessing configuration, control parameters, and data from the external system. Edge computing framework presented in this paper may be a part of a larger system as we previously demonstrated in [14], external system interface element is essential for successful integration. In the case of Edge-Cloud utilization, cloud communication is handled by this element.

The edge computing framework developed in this paper utilizes described architecture to deliver system for both, data collection for training and for utilization of computation models. Namely, the architecture from Fig. 1 can be used to collect data about human activities by taking out the computation model block and directly storing preprocessed data in the local database.

## 3 Implementation of Edge Computing Framework

Edge computing framework, described in the previous section, has been implemented as a software suite making use open source technologies and utilizing open source hardware. The framework is realized on Raspberry Pi 3 (RPi3). A detailed overview of the implementation is given in Fig. 2. The implementation consists of server backend, a wearable sensor communication component, metric accumulation components and components responsible for interface with external systems.

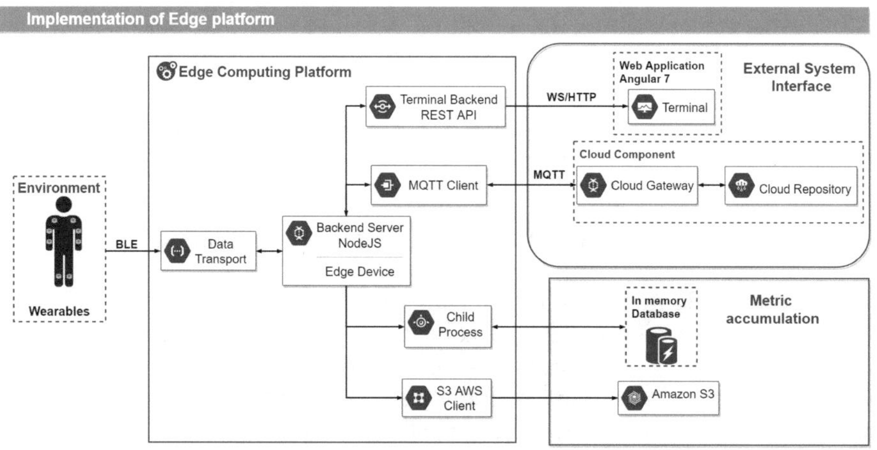


**Fig. 2.** Overview of edge computing implementation

Framework implementation is centrally coordinated by server backend implemented in NodeJS. Server coordinates the data flow between different system components. The server is used as a standard REST API interface which is also acting as a delegate mechanism for child processes. Data preprocessing, computation model, communication with sensors are all implemented as child processes written in Python programming language. Basically, these processes are scripts callable and continuously executable as background processes.

Two-way communication with the wearable sensor is coordinated by the data transport component. Data transport component is implemented in Python programming language making use of native Bluetooth on RPi3. Data received by transport component is sent to server backend via dedicated child process.

Metric accumulation is implemented using both local and remote databases. The local database is implemented as a MySQL database to locally store training data and long term metrics. Database operations are done by the backend server through a dedicated child process. Remote storage is realized on Amazon S3 through dedicated communication client. Through the client, the cloud-based storage is enabled. So far detailed Amazon billing and subscription in this platform iteration is not established.

The edge device is capable of sharing its internal states, results of computation models and locally stored metrics through an external system interface. In the current implementation, the external system interface is implemented in the form of communication with a dedicated Cloud system and web application that is accessible through a web browser. The edge device is able to exchange data with Cloud using MQTT client. Overview of data flow and user feedback is implemented through a custom web application.

Edge computing framework for human activity recognition is implemented as a generic approach with two main modes in mind: *training* and *model utilization*. Either mode can be easily achieved through configuration. From an implementation point of view, the difference between these two modes is that during *training* the preprocessed data is stored as metric in a local database and then extracted to 3rd party model training framework of choice. In *model utilization* mode, the preprocessed data is passed to the model and obtained outputs of the model are stored as metrics and used in other components of the system. This approach enables a generic foundation for wearable sensors utilization in various human activity recognition applications.

## 4 Application to Human Activity Recognition

To demonstrate the applicability of the implemented edge-computing framework, dataset previously collected for basketball exercise recognition [12] is utilized. Software program for data preprocessing and feature extraction was developed in Python programming language. Extracted features were used to train an Artificial Neural Network (ANN). The trained ANN model parameters were saved and utilized for the development of a software program that represents a computation model in an edge computing device (RPi3). Model training was done on the computing platform separated from edge device due to computing power constraints. Model is uploaded to the edge computing device and applicability has been demonstrated. Additionally, results demonstrating the processor and memory utilization of RPI3 during the model execution are presented.

### 4.1 Dataset

The dataset was collected for the purposes of basketball exercise recognition during basketball training. Ultimately, the idea is to track long term progress of basketball players in relation to the type and intensity of exercises they are performing in training sessions. During data collection, wearable motion sensor was placed on the dominant hand of trainee and trainee was instructed to perform 6 different types of basketball exercises in repetitions while spending ∼30 seconds on each exercise. The exercises were performed by 4 trainees. Sensor signals were received by the edge computing device and it was labeled and stored. Type of exercises performed and their labels are shown in Table 1. In this layout 6 classes are assigned to the exercises. Having this in mind, multiclass classification is implemented.

**Table 1.** Exercise types and corresponding labels

| Label | Exercise type |
|---|---|
| 1 | Forward-Backward Dribbling |
| 2 | Left-Right Dribbling |
| 3 | Regular Dribbling |
| 4 | Two Hands Dribbling |
| 5 | Shooting |
| 6 | Layup |

### 4.2 Data Preprocessing and Feature Extraction

For the model training purposes, accelerometer (3 axes) and gyroscope (3 axes) signals are filtered using a low-pass filter with 20 Hz cut-off frequency. The signals are normalized to sensors range so that values fall between $-1$ and 1, where signals are segmented into 3 s "windows" with 50% overlap. In total 11 statistical features are then calculated for each window. Calculated features are *maximum, minimum, mean, median, variance, skewness, 25th percentile, 75th percentile, root mean square, mean absolute deviation, mean crossing rate*. The training dataset is formed such that rows are formed from the calculation of features for each of 3 $s$ windows for 6 signals. This results in a dataset with 66 columns (6 signals $\times$ 11 features) and class label for each of the exercises from Table 1.

Preprocessing and feature extraction element of the pipeline is also implemented as a part of the edge-computing framework inside RPi3. For this purpose, a script in Python programming language is written to receive accelerometer and gyroscope signals at the sampling rate of 50 Hz. Same operations as in the training phase (filtering - normalization – 3 s window segmentation - feature extraction), are applied to the received signals to obtain the feature vector. This feature vector is then further passed to the computation model to obtain classification result every ~1.5 s (3 s window with 50% overlap).

### 4.3 Computation Model Development

Model for classification of basketball exercises was implemented using Artificial Neural Network (ANN). The reasons behind choosing ANN model for classification purposes is because ANNs can be trained in a straight forward fashion, their models can be made relatively small and model utilization can be done efficiently on platforms such as RPi3 without compromising speed and computing resources. The developed ANNs were trained offline using Python language, Jupyter Notebooks, Keras library, and Scikit-learn library. The developed model details are given in Table 2.

The developed model has been serialized and uploaded to edge computing device together with a script written in Python language responsible for utilizing model by supplying feature vector to the model and storing the model output as a metric.

**Table 2.** ANN classifier implementation details

| Parameter | Value |
|---|---|
| Architecture | $66 \times 99 \times 6$ |
| Activation function (hidden) | ReLU |
| Activation function (output) | Sigmoid |
| Loss function | Categorical cross-entropy |
| Optimizer | Adam |
| Number of training epochs | 130 |
| Training batch size | 40 |
| Weight initialization | Random uniform |

In Table 2 detailed overview of ANN classifier setup is given. The architecture specification indicates that an overall layer configuration is divided into 3 layers. The input layer is composed from 66 inputs, where we extract features from measurements. Hidden layer contains 99 neurons, where the third - output layer contains 6 neurons. In this case activation function of choice when it comes to hidden layer is **reLU** (*rectified Linear Unit*). The output layer is using a sigmoid activation function. In the majority of cases, the common practice is to include sigmoid function in the output layer, because basic non - linearity of this function is enabling us to observe the small changes in the output to come at the correct value of the input [16]. Generally, cross - entropy loss (popularly called log loss) is used to measure the performance of classification whose probability values changes from 0 to 1. The principle is based in the increment of cross - entropy loss when predicted probability diverges from the actual referent label. Having said that, categorical cross-entropy has been the best candidate for mandatory loss function. In this implementation where only one result can be correct (presented problem), this function is the most common setting for classification problems. Optimizers are often used to apply tested algorithms for better and faster ANN training. In this case, we are using Adam optimizer (Adaptive Moment Estimation). This algorithm is an extension to widely used stochastic gradient descent that has recently seen broader adoption for deep learning applications. This algorithm uses square gradients to scale the learning rate. This optimizer is an adaptive learning rate method, which practically means that this optimizer computes individual learning rates for different parameters.

### 4.4 Experimental Results - Model Validation

The trained model is validated using 5-fold cross-validation and performance metrics given in Table 3. Confusion matrices that represent the prediction results distribution over classes after 5-fold cross-validation are given in Fig. 3.

The average exercise recognition accuracy of the model is 88%. Figure 3 provides more clarity about the recognition capabilities of the trained model. The confusion matrix shows the correctly predicted instances in the main diagonal. All other elements of the matrix represent the misclassified data instances. It is obvious that the trained model cannot make a clear distinction between two similar dribble types (Forward-Backward vs Left-Right). These dribbles in terms of movements differ in the change of perpendicular directions respectively. Misclassification of the specified movements relies on the fact that certain labeled actions are practically the same movements with minor deviations in terms of sensor data which are limited by current hardware scope of sensing (differences can be viewed in slight changes between acceleration and gyroscope axes). Further improvements can be made by the introduction of direction information in the form of Euler angles obtained from the combination of acceleration and gyroscope signals [12]. Any further improvement of the model precision is out of the scope of this paper.

### 4.5 Experimental Results - Computing Resource Utilization

The developed model has been tested on the RPi3 by transferring the model description and execution scripts from the training environment to the device. The trained model consists of ANN description file that stores the network architecture and ANN model weights file. The processor and memory utilization are

**Table 3.** Classifier performance metrics

| | Precision | Recall | f1-Score |
|---|---|---|---|
| Forward-Backward Dribbling | 0.74 | 0.80 | 0.77 |
| Left-Right Dribbling | 0.76 | 0.69 | 0.72 |
| Regular Dribbling | 0.97 | 0.99 | 0.98 |
| Two Hands Dribbling | 0.98 | 0.99 | 0.99 |
| Shooting | 0.91 | 0.89 | 0.90 |
| Layup | 0.91 | 0.91 | 0.91 |
| Average | **0.88** | **0.88** | **0.88** |

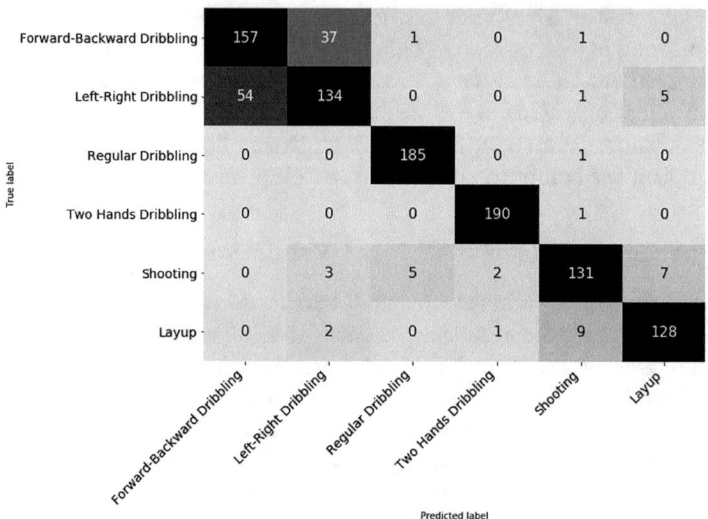


**Fig. 3.** Confusion matrix of classification results

the highest during the model loading period as demonstrated by Fig. 4a for first two seconds (0 s–2 s). After the model is loaded, it starts with execution and produces new result every 1.5 s as discussed in the previous previously. In reference to Fig. 4a, after loading of the model (2 s–7.5 s), spikes of CPU utilization can be noticed in regular intervals of 1.5 s. In the other graph in Fig. 4 the summary of processor and memory utilization for 2.5 min is given. The processor is utilized 28% on average with peak utilization at 77.8% while memory is utilized 72% on average with peak utilization at 74.1%.

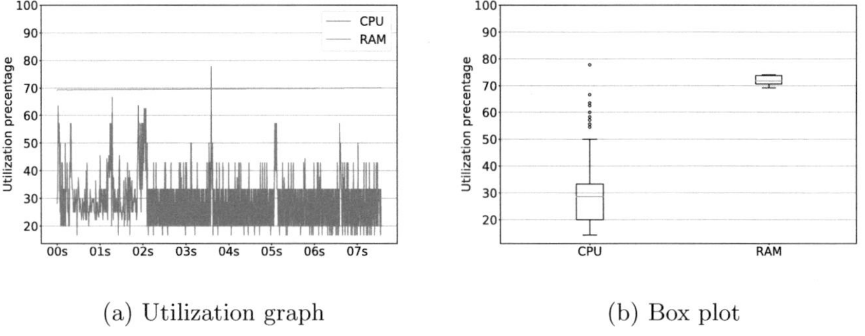


(a) Utilization graph

(b) Box plot

**Fig. 4.** RPi3 processor and memory utilization

# 5 Conclusion

In this work edge computing framework for human activity recognition based on the wearable sensors is presented. In studied human activity recognition scenarios, subjects perform tasks in the environment while data about their activity is recorded by wearable motion sensors. Data is sent to computing hardware via Bluetooth (or any other wireless connectivity). Received sensor data is processed to recognize the type of activity subject is performing and recognition result is stored locally, remotely in the cloud and exposed to the external systems through a dedicated interface. In the envisioned scenario, human activity recognition is being done by previously trained machine learning algorithms. The proposed system can be used for data collection necessary for the training of the machine learning models and for trained model execution.

Benefits of proposed edge computing framework for human activity recognition are most eminent given the application context. Namely, the applications that are part of the broader IoT picture where results of human activity recognition are consumed through cloud-based services, an edge computing framework offers advantages of improved latency and improved bandwidth utilization since data transmitted to the cloud has lower dimensionality (abstracted). Benefits of the proposed framework are also related to the system reliability; when the proposed system is part of a larger system, as previously demonstrated in [14], using the cloud for reporting metrics rather than human activity recognition increases the overall reliability of the system. In the cases of temporary loss of connectivity, the framework offers a seamless update of the metrics in the cloud after connectivity is restored.

Experimental results verify that framework can be used on rather inexpensive and computationally constrained hardware such as Raspberry Pi 3 to run ANN models for human activity recognition. Our future work will consist of extensive tests to figure out the system limitations in terms of execution speed, memory utilization, and power consumption. These test will be accompanied by the exploration of different human activity recognition applications.

**Acknowledgements.** The authors would like to thank the Ministry of Civil Affairs of Bosnia and Herzegovina for the financial support provided for this study.

Authors acknowledge Inovatink company for providing technical, logistic and financial support for this study.

Authors would also wish to express gratitude to Bosnia and Herzegovina Future Foundation (BHFF) for providing financial support in the form of a scholarship to lead author of this study.

## References

1. Gavrilova, M.L., Wang, Y., Ahmed, F., Paul, P.P.: Kinect sensor gesture and activity recognition: new applications for consumer cognitive systems. IEEE Consum. Electron. Mag. **7**(1), 88–94 (2018)
2. Greco, L., Ritrovato, P., Xhafa, F.: An edge-stream computing infrastructure for real-time analysis of wearable sensors data. Future Gener. Comput. Syst. **93**, 515–528 (2019)
3. Al-Turjman, F.: Artificial Intelligence in IoT. Springer (2019)
4. Shi, W., Cao, J., Zhang, Q., Li, Y., Xu, L.: Edge computing: vision and challenges. IEEE Internet Things J. **3**(5), 637–646 (2016)
5. Ren, J., Guo, H., Xu, C., Zhang, Y.: Serving at the edge: a scalable IoT architecture based on transparent computing. IEEE Netw. **31**(5), 96–105 (2017)
6. Long, C., Cao, Y., Jiang, T., Zhang, Q.: Edge computing framework for cooperative video processing in multimedia IoT systems. IEEE Trans. Multimedia **20**(5), 1126–1139 (2018)
7. Chen, Z., Klatzky, R., Siewiorek, D., Satyanarayanan, M., Hu, W., Wang, J., Pillai, P.: An empirical study of latency in an emerging class of edge computing applications for wearable cognitive assistance. In: Proceedings of the Second ACM/IEEE Symposium on Edge Computing - SEC 2017 (2017)
8. Li, H., Ota, K., Dong, M.: Learning IoT in edge: deep learning for the Internet of Things with edge computing. IEEE Netw. **32**(1), 96–101 (2018)
9. Pu, L., Chen, X., Xu, J., Fu, X.: D2D fogging: an energy-efficient and incentive-aware task offloading framework via network-assisted D2D collaboration. IEEE J. Sel. Areas Commun. **34**(12), 3887–3901 (2016)
10. Chen, X., Shi, Q., Yang, L., Xu, J.: ThriftyEdge: resource-efficient edge computing for intelligent IoT applications. IEEE Netw. 32(1), 61-65 (2018)
11. Chen, M., Ma, Y., Li, Y., Wu, D., Zhang, Y., Youn, C.-H.: Wearable 2.0: enabling human-cloud integration in next generation healthcare systems. IEEE Commun. Mag. 55(1), 54-61 (2017)
12. Acikmese, Y., Ustundag, B.C., Golubovic, E.: Towards an artificial training expert system for basketball. In: 10th International IEEE Conference on Electrical and Electronics Engineering (ELECO), pp. 1300–1304 (2017)
13. Sahinovic, S., Dzebo, A., Ustundag, B.C., Golubovic, E., Uzunovic, T.: An open and extensible data acquisition and processing platform for rehabilitation applications. Lecture Notes in Networks and Systems (2018)
14. Uzunovic, T., Golubovic, E., Tucakovic, Z., Acikmese, Y., Sabanovic, A.: Task-based control and human activity recognition for human-robot collaboration. In: IECON - 44th Annual Conference of the IEEE Industrial Electronics Society (2018)

15. Curic, M., Kevric, J.: Posture activity prediction using Microsoft Azure. In: Hadzikadic, M., Avdakovic, S. (eds.) Advanced Technologies, Systems, and Applications II. IAT 2017. Lecture Notes in Networks and Systems, vol. 28. Springer, Cham (2018)
16. Ittiyavirah, S., Allwyn Jones, S., Siddarth, P.: Analysis of different activation functions using backpropagation neural networks. J. Theor. Appl. Inf. Technol. **47**, 1344–1348 (2013)

# The Impact of Predictor Variables for Detection of Diabetes Mellitus Type-2 for Pima Indians

Maida Kriještorac(✉), Alma Halilović, and Jasmin Kevric

Faculty of Engineering and Natural Sciences,
Department of Electrical and Electronics Engineering,
International Burch University, Ilidža, Bosnia and Herzegovina
maida.Krijestorac@stu.ibu.edu.ba

**Abstract.** Diabetes Mellitus type-2 is one of the diseases of a modern age treated as a serious illness due to its symptoms in later stages, consequences if left untreated and its complexity in terms of detection, diagnosis, and prognosis widely spread among the Pima Indian population. The process of detecting the diabetes will require analysis of the data, processing, extracting portions of data into a set for training, testing and validation sets. Then applying several different machine learning algorithms, train a model, check the performance of the trained model and iterate with other algorithms until we find the most performant for our type of data. The goal of this research is to investigate which algorithm gives best results in terms of detecting the existing disease as well as predicting the possibility of getting one in the future, based on the diagnostic measurements of the patient. For this matter, MATLAB software will be used.

## 1 Introduction

This research will show and describe how data related to diabetes can be used to classify if a person has diabetes or not. Specifically, the focus is on how machine learning can be used to classify diseases such as diabetes. In the last couple of years, the commonness of diabetes has remarkably increased over the world. That has in turn effected on the many events related to with many complications, such as heart failure. In medicine, diagnosis of this health condition is a difficult task, thus there is a need for the development of right machine learning techniques to improve diagnosis of the disease in earlier stages. The subject of this article is the detection of Pima Indian Diabetes using machine learning techniques. Pima Indians females, a group of Native Americans living in Arizona, were investigated on the major factors of the disease and data information collected and conducted is used in this research. Diabetes belongs to the group of metabolic diseases where people have high blood sugar levels and whose body does not produce insulin or use it properly. If untreated properly and in right time, this condition represents an increased risk for developing kidney disease, blood vessel damage, blindness and play a part in heart disease development. In 21st century diabetes has become a global public health problem [1, 2]. Type-2 diabetes is often diagnosed with middle aged to elderly people, and one of the goals of this research is to


S. Avdaković et al. (Eds.): IAT 2019, LNNS 83, pp. 388–405, 2020.
https://doi.org/10.1007/978-3-030-24986-1_31

develop medical diagnostic decision support systems that can help middle aged and elderly people in the self-diagnostic process at home.

The type of dataset and problem is a classic supervised binary classification. Detection of diabetes type 2 using machine learning techniques, applying different types of algorithms is done in order to make better analysis of data information and developing relations between different data set attributes. This is used to provide better understanding on how this data can develop a smart machine which will be able to make a proper diagnosis while some clinical symptoms are not developed yet and while there is still no need for doctor intervention. This should lead to the reduction of insulin; hence machine will be able to detect disease in the early stages and make a warning so that patient can take a proper action such as loss of weight. This is one of major factors for the development of the disease and respectively an increase in psychical activity to prevent further spread of the disease and further declination which often, untreated. In this research we will attempt to classify the presence of diabetes based on relevant covariates.

Increased amount of body fat in the abdominal area is one of the common things that appears in patients suffering diabetes type 2, as well as lack of physical activity, older age, females what had prior glucose intolerance, a larger number of pregnancies, pedigree function. One data set cannot be used for all ethical or racial subgroups since there is a distinction in the frequent occurrence in different ethical or racial subgroups [3].

We examined and investigated literature review of articles published (from 2000 to 2018) in which researchers examined different machine learning algorithm types in order to make a relevant analysis of the data dependency and correlation with disease occurrence. In the research paper conducted by Ivan Contreras and Josep Vehi [4], in the last 30 years, from 1987 until 2017, increased rapidly, where a major increase started with the beginning of the 21st century (2000s) with the development of new technologies. In the range from 2000 to 2017, there is a significant exponential increase from nearly 900 articles published in 2000 to 10 300 articles published in 2017 on this topic. There is also an increase in the number of health care companies that are using these artificial machine learning algorithms to aid and assist them in overall health care when it comes to recognition and diagnosis of diabetes in patients, which is possible due to the development of new technologies and greater availability of data nowadays [5].

In the article published in 2014 [6], researchers investigated the performance of different neural network such as: probabilistic neural network, naive Bayes and decision tree, in the detection and classification of the disease. They gathered information from 6647 objects, at least 20 years old, to develop models that are based on the 21 common risk factors. Dealing with imbalanced database they found that naive Bayes and decision tree techniques are most optimal classifiers with the highest accuracy achieved when it comes to diabetes classification and detection.

Article published in 2010 [7], analyses different options for early diabetes diagnosis and provides deep analysis of critical issues of the algorithms used. It presents most accurate algorithms used in diabetes detection, as well as the strengths and weaknesses of these algorithms. Measuring accuracy, specificity, and sensitivity in 30 different algorithms from 16 different authors, in this article different methods advantages and disadvantages are presented. Data analysis using logistic regression method by Shaker [8] on 8 variables, on the Pima Indian Diabetes database was applied with the accuracy

79.17%. Then, when 4 of variables that are with leats importance were deleted, and the method is done with remaining 4 “more significant” variables, the accuracy obtained in this case was 80.21%. Considering the clustering technique, K-nearest neighbour (k-NN) algorithm applied gives accuracy in the range from 71–78%, while it is shown that better accuracy can be achieved using a hybrid model of k-NN and C4.5 algorithm, reaching value of 92.32%. Least square support vector machine (LS-SVM) method reach an accuracy of 78.21% with 10-fold cross validation, while smooth support vector machine (SSVM) technique gave 76.73% of accuracy in this research. Researches also replaced SSVM default-plus function with new multiple knot spline (MKS) smoothing function and in such way new accuracy value obtained reached 93.2% [9].

Jhaldiyal and Mishra, in the article [10] did a study where they compared two methods on this subject using a Pima Indians Diabetes database. In the first method, the database is lowered in dimensions by using Principal Component Analysis (PCA), and then such data are trained with Reduce Error Pruning Tree (REP Tree). In the second method database is also reduced in dimensions by PCA, then trained using Support Vector Machine (SVM). By using Matlab GUI, the obtained results showed that second method (PCA with SVM) obtained better accuracy of 96.66%, while first method accuracy was 78.93%.

In research conducted by Pradeep Kandhasamy and Balamurali [11], on this topic of diabetes mellitus-type 2 detections using Pima Indians Diabetes database, they investigated 4 classifier algorithms: J48, KNN (for 3 different k values: k = 1, k = 2, and k = 3), SVM and Random Forest technique, in 2 cases, and obtain results in terms of accuracy, sensitivity, and specificity. In the first case they applied these 4 algorithms to non-processed database with noisy data, while in second data was processed and modified. The result obtained shown that there is a major increase in accuracy, sensitivity and specificity in the second case where data was pre-processed before training performed, thus, demonstrating a need for data processing in order for achieving more accurate and results. Here, KNN with k = 1, and Random Forest technique gave the best results.

Kordos et al. [12] it their experiments on the observed database showed that KNN with properly adjusted parameters such as k, distance measure and weighing functions performs on average best with accuracy of 77%, while it has a zero training time and the test time which is significantly reduced by prior reference vector selection which is done only once. Purnami et al. [9] used smooth support vector machines (SSVM) technique to detect the diabetes with observed database and managed to obtain accuracy of 76.73%. Ster and Dobnikar [13] in for their research used KNN technique for detection of Pima Indian diabetes and obtained results with accuracy of 71.90% while it was tested with 10-fold cross-validation which is shown to be efficient for small data samples, without tuning the parametre, and showed in their comparison study that neural network give best results in terms of classification accuracy. Jahangeer et al. [14] investigated and made a comparison of different classifiers KNN with K = 1, KNN with K = 3, SVM, and Decision Tree-CART using different performance metrics and best results accuracies obtained are 64.60%, 68.79%, 74.21% and 76.60%, respectively, while data division ratio is 70%–30% for training and testing. In research they conducted they used Pima Indian Diabetes database and, by changing certain parameters, they have demonstrated that parameter tunning play an important role and that it could significantly improve the algorithms for diabetes detection.

Polat et al. [15] in their research by using GDA and LS-SVM algorithm with 10-fold cross-validation obtained accuracy of 78.20%. Angeline Christobel et al. [16] examined the performance of the classification algorithm KNN using a Pima Indians Diabetes database while their goal was to improve the accuracy of the KNN algorithm model. Since Pima Indian Diabetes database contains missing values, Angeline Christobel et al. by processing data and by imputation, scaling and normalization succeeded to get highest accuracy of 74.74% using 2-fold cross-validation. Accuracy of 71.84% is obtained by using 10-fold cross-validation and 73.59% using 5-fold cross-validation KNN algorithm model. Shital et al. [17] created an automated system which uses an SVM algorithm model, using RBF kernel function and 10-fold cross validation in the training dataset. In first case data is split at 80%–20%, while in second in 60% - 40% for train and test. In the first case, the accuracy obtained is 90.2%, while in second is 89.0% and obtained accuracy of 75.3% and 74.5%, respectively. In the article published by Barale, Shirke [18], data split in 70–30% for training and testing and classification techniques SVM and Decision Tree is used, and accuracies obtained are 79.39% and 76.10%, respectively. Using an SVM classification model, get the highest accuracy to be 80.22%, when 500 data used to train, having optimal parameter $\sigma = 1.5$, RBF kernel function. Hashi et al. [19] using wrapper did a feature selection and reduced features by 37.5% and 50% and in a such way increased accuracies from 74.48% (KNN) and 77.17% (SVM) to 81.17% and 87.01%, respectively. Using KNN algorithm with $K = 11$ and 10-fold cross-validation, by using wrapper feature selection Pregnancies, Glucose, BMI, DiabetesPedigreeFunction and Age are chosen as features than give best result when combined together.

The main contribution of the paper is implementation various Diabetes Mellitus type-2 classification strategies based on machine learning techniques. Furthermore, the paper studies different feature sets and examines significance of various features on the classification accuracy. The results are extensively compared the various classification methods reported in literature.

## 2 Materials and Methods

### 2.1 Pima Indian Diabetes Database

This dataset is acquired from the Kaggle of Machine Learning Databases and it was chosen from a bigger dataset held by the National Institutes of Diabetes and Digestive and Kidney Diseases [21]. All patients in this database are Pima Indian women, at least 21 years old living in or near Arizona, USA. The binary response variable takes the values "0"- that is a negative test for diabetes, or "1" – that is a positive test for diabetes. There are 268 (34.9%) cases in class "1" and 500 (65.1%) cases in class "0" and all examples have 8 input attributes (from X1 to X8) and 1 output attribute (Y). We can observe that the dataset contains 768 rows and 9 columns. 'Outcome' is the column which we are going to classify, which says if the patient is diabetic or not. Table 1 shows the attributes of this dataset [20].

**Table 1.** Pima Indian diabetes database

| № | Pima Indian diabetes database | | |
|---|---|---|---|
| | *Predictor variables abbreviation* | *Predictor variables* | *Type* |
| X1 | PREGNANT | Numbers of time pregnant | Numeric |
| X2 | GTT | Plasma Glucose | Numeric |
| X3 | BP | Blood pressure | Numeric |
| X4 | SKIN | Triceps skin fold thickness | Numeric |
| X5 | INSULIN | Serum insulin | Numeric |
| X6 | BMI | Body mass Index | Numeric |
| X7 | DPF | Diabetes pedigree function | Numeric |
| X8 | AGE | Age of patient | Numeric |
| Y | DIABETES | Diabetes diagnose results (“positive or negative”) | Nominal |

The age ranges from 21 to 80 years old, while the number of pregnancies is from 0 to 17. For this we need to apply a proper transformation [21]. We observed that the number of time pregnant and age attributes are integers. The population is generally young, less than 50 years old and some attributes where a zero value exist seem to be errors in the data. Upon examining the distribution of class values, we noticed that there are 500 negative instances (65.1%) and 258 positive instances (34.9%) (Table 2).

**Table 2.** Feature selection

| No | | Pima Indian diabetes | | | |
|---|---|---|---|---|---|
| | | *Statistics for attributes* | | | |
| | | *Min* | *Max* | *Mean* | *Stv. Dev.* |
| 1. | PREGNANT | 0 | 17 | 3.845 | 3.37 |
| 2. | GTT | 0 | 199 | 120.89 | 31.973 |
| 3. | BP | 0 | 122 | 69.105 | 19.356 |
| 4. | SKIN | 0 | 99 | 20.54 | 15.95 |
| 5. | INSULIN | 0 | 846 | 79.799 | 115.24 |
| 6. | BMI | 0 | 67.1 | 31.99 | 7.88 |
| 7. | DPF | 0.08 | 2.42 | 0.47 | 0.33 |
| 8. | AGE | 21 | 81 | 33.24 | 11.76 |

## 2.2 Data Analysis and Transformation

The data from the database was conducted in 1990. It contains a multivariate type of data: attribute types are Integer, Real numbers. 500 attributes from the database don’t have diabetes, while the rest, 268 of them, have. The data is saved as CSV file to be able to work with it later in other software. Since there are not column names by which the type of data is recognized inside in the original database, we added it in the program. After inspecting, the data was visualized. The correlation matrix is used to show the correlation of the attributes – range: −1 to 1, if the value is closer to 1 it means better correlation of two attributes. In order to achieve a higher precision and understand data better, the data is visualized graphically for each attribute in different corresponding range. Some values on the graphs were empty, thus for some samples there are null values for some attributes, that is some portion of the data in the database is missing. In this case were the database is incomplete, since machine learning techniques don't show very good results with null values, the transformation and cleaning of the data are done. Due to a large number of such null values, the cleaning of the data by just deleting them would cause a loss of a too much data. Considering the Pregnancies feature, the transformation of the data is not needed since it is possible for a person to not have any pregnancy in a record and be diagnosed with the diabetes type 2.

## 2.3 Classification Algorithms

Classification is the process of identifying a new observation category set on the basis of training set of data that contains examination whose category is known used to define the type of an object or class according to its features [22]. Several machine learning algorithms – classifiers - are tested, and ones with satisfying results are going to be presented in this article. To find the optimal classifier, the complete dataset with different combinations will be used with five-fold cross validation.

### 2.3.1 Logistic Regression

The classification algorithm establishes a model that can map data items in a given category, based on the existing data. It was used to extract significant data items from the model or to classify the tendency of data. The predictive values of the classification problem can only be 0 or 1, so we may set a critical point. This means that dependent variable should be a continuous function of the odds of events. Logistic regression has two category problem. The main purpose is to classify whether one person is diabetic or not, which is a typical binary-classification problem [23].

### 2.3.2 Decision Trees

Decision Tree is a supervised machine learning algorithm used to solve classification problems used in this research work for the detection of target class using decision rule taken from the prior data. It uses nodes and internodes for the detection and classification. Root nodes classify the instances with different features. In every stage Decision tree chooses each node by evaluating the highest information gain among all the features. Decision trees control the method of decision in a flow chart-like structure. Root nodes can have two or more branches while the leaf nodes represent a classification [24]. Decision Trees techniques used for this paper are Fine Tree, Medium Tree, and Coarse Tree. The main difference of these three methods is in the number of splits. For the Fine Tree method there are many levels to make many fine distinctions between classes where maximum number of splits is 100. In the case of Medium Tree method there are a medium number of leaves for finer distinctions between classes where maximum number of splits is 20. And, in Coarse Tree method there are few leaves to make coarse distinctions between classes where maximum number of splits for this method is 4.

### 2.3.3 Support Vector Machines – SVM

Support Vector Machines (SVM) Models - differentiate between the categories by separating the classes data points of one the class from those of the other class. Parameters that are to be selected for the model training in the SVM classification method are Kernel function and scale, bow constraint level, and multiclass model. The linear SVM algorithm builds a model that assigns new examples to one category or the other, while non-linear SVM methods use Kernel functions to train the model. SVM model types used for this paper examination are: Linear SVM, Quadratic SVM, Cubic SVM, Fine Gaussian SVM, MediumGaussian SVM, and Coarse Gaussian SVM. The difference of these SVM methods is in the separation of the data by mapping it to space of higher dimension using different Kernel functions. The idea of Linear SVM is to linearly separate data by mapping it to space of higher dimension using the linear Kernel function. Considering SVM methods that separate data using the Gaussian Kernel functions, we can distinguish between Fine Gaussian SVM with Kernel scale of 0.71, Medium Gaussian SVM with Kernel scale of 2.8, and Coarse Gaussian SVM with Kernel scale of 11. Quadratic SVM and Cubic SVM belong to the group of SVM methods using polynomial Kernel functions, quadratic and cubic that determines similarities of the input data, as well as it uses the combination of data for detection with different margins of separation [25, 26].

### 2.3.4 K-Nearest Neighbors - KNN

In K-Nearest Neighbors (KNN) models 'k' is the number of nearest neighborhood which are to be considered while classifying input variable, thus using local neighborhood to classify the outcome. The k-nearest neighbors are identified and majority class is assigned to input variable affect the output. Several parameters are to be determined when using this classification model for training the model, such as: number of neighbors, distance metric, distance weight. There are three types of distances in KNN model – Euclidean, Mahnattan and Correlation – and by changing it, there is a change in how the data is classified. Different model types of KNN algorithm that are examined in this paper are: Fine KNN, Medium KNN, Coarse KNN, Cosine KNN, Cubic KNN, and Weighted KNN. Different KNN models use different number of neighbours, distance metric and distance weight. Considering Fine KNN model training, number of neighbours is set to be 1, Euclidean distance metric, equal distance weight. For a Medium KNN number of neighbours is 10, with same Euclidean distance metric and equal distance weight. Coarse KNN is used with 100 neighbours, Euclidean distance and equal distance weight. Cosine KNN used 10 neighbours, Cosine distance metric and equal distance weight. For a Cubic KNN number of neighbours is set to 10, distance metric is Minkowski (cubic), and equal distance weight. In case of Weighted KNN, number of neighbours is set to be 10, with Euclidean distance and Squared inverse distance weight [25].

# 3 Result and Discussion

The following classification methods presented in the paper are used for model fitting done in MATLAB software version R2108b using the Statistics and Machine Learning Toolbox version 11.4. Data was previously processed, normalized, and used in MATLAB to compare results with the unprocessed, raw, data using k-fold cross validation for different machine learning classification methods.

## 3.1 Logistic Regression

Analysing the unprocessed raw database with the Logistic Regression, the accuracy obtained when all 8 features of the database used is 76.8%, while the highest accuracy with this model is obtained for only two features - Glucose and BMI - 77.1%. Combination of features for lowest accuracies is combination of SkinThickness or BloodPressure with Pregnancies, Insulin and Age features (64.8%). Using all features for normalized database model training, the accuracy obtained is 77.9%, while the lowest accuracy is with Insulin and Age features (65.4%) (Table 3).

**Table 3.** 5-fold CV

| No. | Logistic regression | | |
|---|---|---|---|
| | Accuracy | | |
| | *Processed data* | *Processes and normalized* | |
| All feature | 76.8% | All feature | 77.9% |
| **X2,X6** | **77.1%** | **X1-X8** | **74.0%** |
| **X2,X6,X7** | **77.1%** | X5,X8 | 65.4% |
| X2,X6,X8 | 76.8% | X7,X8 | 65.8% |
| X2,X3 | 75.0% | X2,X6,X7,X8 | 77.2% |
| X2,X7 | 74.7% | | |
| X2,X8 | 73.7% | | |
| X1,X3,X5,X8 | 64.8% | | |
| X5,X8 | 65.0% | | |
| X7,X8 | 65.2% | | |
| X4,X5 | 65.5% | | |

From results for both processed and unprocessed data, we can observe that Glucose and BMI play the major role in achieving high accuracy in all cases, thus for very good results all other features can be omitted and model trained only with Glucose and BMI feature data. Also, different combinations of Insulin, Age, SkinThickness and DiabetesPedigreeFunction give the lowest accuracies for this model type, thus can be omitted when training the model so higher accuracy is achieved. Using Logistic Regression method it can be seen that the accuracy of unprocessed data has been just slightly lower than processed data accuracy, thus it can be concluded that for Logistic

Regression method, with this particular database, unprocessed data can be successfully used for disease diagnosis.

### 3.2 Decision Tree: Fine Tree

For the unprocessed database Decision Tree – Fine tree classification method has been used, 5 folds Cross-validation, maximum number of splits 100, and Gini's diversity index split criterion. In cases where all features were used in the model trained, before PCA, the accuracy obtained is 70.6%. The higher accuracy of 70.7% is obtained when only Pregnancies and Glucose used in model training. Some of the lowest accuracies obtained were when BloodPressure and Diabetes PedigreeFunction used and it was 58.6%.

Using Decision Tree - Fine Tree - classification method with normalized data the highest accuracy achieved when all features used in model training were 73.2%. Among various cases with different combinations of features used in model train the value of accuracy closest to the highest one is 71.7% when Pregnancies, Glucose, Insulin and BMI used. Combination of features with lower accuracy is Pregnancies and DiabetesPedigreeFunction (60.1%) (Table 4).

**Table 4.** 5-fold CV

| No. | Decision tree-fine tree | | |
|---|---|---|---|
| | *Accuracy* | | |
| | *Processed data* | *Processes and normalized* | |
| All feature | 70.6% | All feature | 73.3% |
| **X1,X2** | **70.7%** | **X1,X2,X5,X6** | **71.7%** |
| X2,X8 | 69.4% | X2,X3,X5,X6,X7 | 70.9% |
| X2,X4 | 69.3% | X1,X7 | 60.1% |
| **X2,X5,X7,X8** | **69.3%** | **X1,X4** | **60.9%** |
| X3,X7 | 58.6% | X3,X6 | 61.1% |
| X4,X7 | 58.6% | X1,X3 | 61.5% |

### 3.3 Decision Tree: Medium Tree

For the unprocessed database Decision Tree – Medium Tree – classification is used 5 folds cross-validation with maximum number of splits 20, Gini's diversity index split criterion. The highest accuracy of 75.3% was achieved when all features used in the model train. And lowest accuracy values are with BloodPressure and Age features (61.5%).

While using the normalized data in the same Decision Tree – Medium - classification method, the highest accuracy achieved is 77.1% with all features used. Lowest accuracy values are obtained from the combination of SkinThickness and DiabetesPedigreeFunction (63.0%) (Table 5).

**Table 5.** 5-fold CV

| No. | Decision tree-medium tree | | | |
|---|---|---|---|---|
| | Accuracy | | | |
| | *Processed data* | *Processes and normalized* | | |
| All feature | 75.3% | All feature | 77.1% |
| **X2,X4** | **74.5%** | **X2,X4** | **74.3%** |
| X2,X6 | 73.7% | X1,X2,X6 | 73.7% |
| X5,X6,X7 | 73.4% | X3,X5,X7,X8 | 73.6% |
| X2,X7 | 72.7% | X2,X3 | 72.9% |
| X2,X8 | 73.8% | X5,X6,X7,X8 | 72.6% |
| X1,X2,X4 | 72.5% | X3,X7 | 63% |
| X3,X8 | 61.5% | X4,X6 | 63.5% |
| X3,X4 | 62.6% | X6,X7 | 63.7% |

### 3.4 Decision Tree: Coarse Tree

Using raw, unprocessed data with Decision Tree - Coarse Tree - classification method training model with all features gives an accuracy of 74.6% for 5 folds cross-validation, maximum number of splits equals 4, Gini's diversity index split criterion. The results obtained are very similar as in previous decision tree methods where Glucose feature in combination with other features gives the highest accuracy. The best result identical to the one when all features used is combination of only Glucose and BMI feature (74.6%). Lowest accuracies were with features: BloodPressure, Insulin and DiabetesPedigreeFunction (62.6%).

While using data from normalized database, the higher accuracy obtained is 75% when all features used, which is not not much higher in value than the accuracy of unprocessed data. The behaviour while training the model with certain features is same in both cases. Glucose and BMI used alone for both databases give the same accuracy as when all features used, thus it can be concluded that other features can be omitted when the accuracy is concerned. Lower values of accuracy obtained when BloodPressure and SkinThickness combined (62.4%).

In case where Maximum deviance reduction split criterion used for unprocessed data with Decision Tree - Coarse Tree - classification method, 5 fold cross-validation, and maximum number of splits 4, the accuracy obtained was 74.7% which is slightly higher than the accuracy in the first case where Gini's diversity index split criterion used (74.6%). In all other cases when Maximum deviance reduction or Twoing rule split criteria used for both processed and unprocessed data the accuracies were lower than in cases when Gini's diversity index split criterion used (Table 6).

**Table 6.** 5-fold CV

| No. | Decision tree-coarse tree | | |
|---|---|---|---|
| | Accuracy | | |
| | *Processed data* | *Processes and normalized* | |
| All feature | 74.6% | All feature | 75.0% |
| **X2,X6** | **74.6%** | **X1,X3,X4,X5,X7,X8** | **70.3%** |
| X1,X2,X6 | 74.6% | X3,X4 | 62.4% |
| X1,X2,X6,X7 | 74.6% | X3,X5,X7 | 64.1% |
| X3,X4,X5 | 74.6% | | |
| X1,X2,X6,X7,X8 | 74.6% | | |
| X1,X2,X6,X7,X5,X8 | 74.6% | | |
| X1X2,X6,X7,X8,X4 | 74.6% | | |
| X3,X5,X7 | 62.6% | | |
| X5,X7 | 63.4% | | |
| X3,X8 | 63.7% | | |

### 3.5 Support Vector Machines (SVM) Models

In Fine Gaussian SVM Kernel scale was 0.71, in Medium Gaussian SVM it was 2.8, and in Coarse Gaussian SVM 11. From the table it can be observed that the Fine Gaussian SVM method obtained the lowest accuracy for both unprocessed and processed data, while highest accuracy is achieved using Coarse Gaussian SVM for unprocessed data, and Linear SVM for processing data (Table 7).

**Table 7.** Results of model fitting done with SVM method in MATLAB

| SVM Method | Raw data accuracy | Normalized data accuracy |
|---|---|---|
| **Linear SVM** | 77.0% | **77.4%** |
| Quadratic SVM | 75.5% | 76.4% |
| Cubic SVM | 71.9% | 74.3% |
| Fine Gaussian SVM | 65.0% | 66.1% |
| Medium Gaussian SVM | 76.4% | 77.2% |
| Coarse Gaussian SVM | **77.5%** | 77.2% |

### 3.6 Linear Support Vector Machines - SVM

The linear SVM model is used for examining the unprocessed diabetes database with linear Kernel function, box constraint level equals 1, multiclass method is One-vs-One, standardized data: true. All features used for model training give accuracy of 77.0%. Both high and low accuracies obtained by this method are very similar to the high and low accuracies obtained by Logistic Regression Method, thus from the high accuracies best combination of features is Glucose, BMI, DiabetesPedigreeFunction and Age (76.6%). When it comes to the highest accuracies from the combination of only two features the best result is obtained when Glucose and BMI used (76.2%). Lowest accuracy obtained is 64.5% with BloodPressure, SkinThickness, Insulin, Diabetes PedigreeFunction and Age. When it comes to Linear SVM and normalized diabetes database, highest accuracy with all features is 77.7%. Lowest accuracy comes from DiabetesPedigreeFunction and Age (64.9%) (Table 8).

**Table 8.** 5-fold CV

| No. | Linear SVM | | |
|---|---|---|---|
| | Accuracy | | |
| | *Processed data* | *Processes and normalized* | |
| All feature | 77.0% | All feature | 77.7% |
| X2,X6,X7,X8 | 76.6% | X1,X3,X4,X5,X6,X7,X8 | 69.9% |
| X1,X2,X7,X8 | 75.4% | **X2,X6,X7,X8** | **77.6%** |
| X2,X5,X7,X8 | 74.2% | **X2,X4,X6,X7,X8** | **77.6%** |
| X2,X6 | 76.2% | X2,X6 | 76.3% |
| X2,X3 | 74.5% | X1,X7,X8 | 76.8% |
| X2,X5 | 74.3% | X7,X8 | 64.9% |
| X3,X4,X5,X7,X8 | 64.5% | X5,X8 | 65.2% |

### 3.7 KNN Models

There are several KNN model types that were examined in order to compare the accuracies of KNN models with this database. The parameters used for Fine KNN model training, number of neighbors is set to be 1, Euclidean distance metric, equal distance weight. For a Medium KNN number of neighbors is 10, with same Euclidean distance metric and equal distance weight. Coarse KNN is used with 100 neighbors, Euclidean distance and equal distance weight. Cosine KNN used 10 neighbors, Cosine distance metric and equal distance weight. Of the Cubic KNN number of neighbors is set to 10, distance metric is Minkowski (cubic), and equal distance weight. In case of Weighted KNN used 10 neighbors, Euclidean distance, and Squared inverse distance weight (Table 9).

**Table 9.** Results of model fitting done with KNN models in MATLAB

| KNN model | Rawdata accuracy | Normalized data accuracy |
|---|---|---|
| Fine KNN | 71.0% | 69.0% |
| Medium KNN | 74.1% | 75.4% |
| Coarse KNN | 74.0% | 75.4% |
| **Cosine KNN** | **76.0%** | **76.4%** |
| Cubic KNN | 74.6% | 75.5% |
| Weighted KNN | 75.0% | 75.4% |

### 3.8 Cosine KNN Model

Among different KNN model types examined, the best accuracy result is obtained by using a Cosine KNN model with the accuracy of 76% with raw data used, and 76.40% with normalized database. Number of neighbors is set to be 10 and equal distance weight. The results obtained for different combinations for this KNN model is presented in Table 10.

**Table 10.** 5-fold CV

| No. | Cosine KNN | | |
|---|---|---|---|
| | Accuracy | | |
| | *Processed data* | *Processes and Normalized* | |
| All feature | 76.0% | All feature | 76.4% |
| **X2, X5, X6, X7, X8** | **76.4%** | **X2, X5, X6, X7, X8** | **77.2%** |
| X2, X3, X5, X6, X7, X8 | 76.0% | X1, X2, X3, X5, X6, X7, | 75.8% |
| X2, X5, X6, X8 | 75.9% | X2, X5, X6, X8 | 75.7% |
| X1, X2, X5, X6, X7, X8 | 75.9% | X2, X5, X6, X8 | 75.5% |
| X2, X6 | 74.0% | X2, X6 | 73.4% |
| X2, X8 | 72.3% | X2, X7 | 74.0% |
| X3, X4, X5 | 63.0% | X3, X4, X5 | 63.0% |
| X3, X5 | 62.5% | X1, X4 | 61.9% |
| X4, X5 | 59.9% | X3, X5 | 61.1% |

Best accuracy results with raw database of 76.4% is obtained with the combination of five features - Glucose, Insulin, BMI, DiabetesPedigreeFunction and Age. This accuracy is higher for 0.4% than accuracy obtained when all features included, thus concluding that Pregnancies, BloodPressure and SkinThickness can be omitted from the raw database for better results. The reason to omit these features is, also, because when BloodPressure feature added to combination of five features with highest an accuracy, the accuracy decreases by 0.4%, thus having accuracy of 76%. Highest accuracy obtained by only two features used – Glucose and BMI – is 74%, while the lowest is 59.9% with the combination of SkinThickness and Insulin. In case when normalized database used, the accuracy result is similar for same combinations of features, where best result is obtained when the combination of Glucose, Insulin, BMI, DiabetesPedigreeFunction and Age used (77.2%), while lowest accuracy is obtained when the combination of BloodPressure and Insulin features used (61.1%).

### 3.9 Discussion

Polat et al. [15] using machine learning GDA and LS-SVM algorithm with 10-fold cross-validation managed to obtain accuracy of 78.20%, which is higher than all accuracies obtained by SVM algorithm models used in this paper with 5-fold cross-validation. Angeline Christobel et al. [16] examined the performance of the classification algorithm KNN using a Pima Indians Diabetes database. The aim of their research was to improve the accuracy of the KNN algorithm model. As stated previously in this research paper, Pima Indian Diabetes database containing missing values, thus Angeline Christobel et al. by processing data and by imputation, scaling and normalization managed to obtain best possible accuracy for KNN algorithm. Highest accuracy obtained is 74.74% with imputed, scaled and normalized data, using 2-fold cross-validation. Accuracy of 71.84% is obtained by using 10-fold cross-validation and 73.59% using 5-fold cross-validation KNN algorithm model, thus, it can be concluded that k-fold cross validation in all cases when $k < 10$ is better that when $k = 10$ considering KNN algorithm model. Shital et al. [17] created an automated system which uses an SVM algorithm model, using RBF kernel function and 10-fold cross validation in the training dataset. In first case data is split at 80%–20%, while in second in 60%–40% for train and test. In the first case, the accuracy obtained is 90.2%, while in second is 89.0%. Using Decision Tree technique in the same research, for the first case data division accuracy achieved is 75.3%, and the second case is 74.5%. In the article published by Barale, Shirke [18], data split in 70–30% for training and testing and classification techniques SVM and Decision Tree is used, and accuracies obtained are 79.39% and 76.10%, respectively. Using an SVM classification model, get the highest accuracy to be 80.22%, when 500 data used to train, having optimal parameter $\sigma = 1.5$, RBF kernel function. In the article *An Empirical Comparison of Supervised Classifiers for Diabetic Diagnosis* written by Jahangeer et al. [14], examined classification techniques SVM, KNN ($K = 1$ and $K = 3$) and Decision Tree-CART and obtained accuracies of 76.60%, 64.60%, 68.79% and 74.21%, respectively. The data division ratio is 70%–30% for training and testing. KNN algorithm, it can be observed that results for accuracy are lowest in this paper comparing to all listed in the Table 11 for the KNN method. The reason for this could be that the authors of the article did not do the processing, scaling and normalization of data as well as missing values, and since working with Pima Indian Diabetes database, obtained significantly lower accuracies. Accuracies, considering KNN, greatly depend on the parameters used in terms of distance metric, distance, weight as well as a number of neighbors.

Hashi et al. [19] did a feature selection using wrapper to obtain optimal features where they reduce features by 37.5% and 50% and in a such way increased accuracies from 74.48% (KNN) and 77.17% (SVM) to 81.17% and 87.01%, respectively. The KNN algorithm used $K = 11$ and by using wrapper feature selection Pregnancies, Glucose, BMI, DiabetesPedigreeFunction and Age are selected, and 10 fold cross-validation used. While in this paper Cosine KNN had the highest accuracy when performed with in combination of Glucose, Insulin, BMI, DiabetesPedigreeFunction and Age (77.2%). In case when SVM classification algorithm used, highest accuracy achieved, 81.17%, is done with same optimum features using Linear kernel function with 10-fold cross-validation.

Different performance of the algorithms is due to the difference in parameters used in articles, as well as processing and normalization of data, and feature selection. Great impact, also, have the data division ratio for training and testing, when considering the difference in the presented algorithm classification techniques. Another reason why the same algorithm of certain classification technique gives significantly different results is because the algorithm used in different program does not have the same structure of the

**Table 11.** Comparison of results in field of classification techniques for detection of diabetes using Pima Indian Diabetes database

| Algorithm | Accuracy | Reference |
|---|---|---|
| Logistic Regression | 79.17% | Shanker et al. [8] |
| Decision Tree | 76.10% | Barale, Shirke [18] |
| Decision Tree | 75.30% | Shital et al. [17] |
| Decision Tree-CART | 74.21% | Jahangeer et al. [14] |
| Lienar SVM | 87.01% | Hashi et al. [19] |
| SSVM | 76.73% | Purnami et al. [9] |
| SSVM (with MKS function) | 93.2% | Purnami et al. [9] |
| GDA and LS-SVM | 78.20% | Polat et al. [15] |
| SVM | 76.60% | Jahangeer et al. [14] |
| SVM | 79.32% | Barale, Shirke [18] |
| SVM | 80.22% | Abdillah, Suwarno [26] |
| KNN | 77.00% | Kordos et al. [12] |
| KNN | 71.90% | Ster, Dobnikar [13] |
| KNN | 74.74% | Angeline Christobel et al. [16] |
| KNN, K = 1 | 64.60% | Jahangeer et al. [14] |
| KNN, K = 3 | 68.79% | Jahangeer et al. [14] |
| KNN | 81.17% | Hashi et al. [19] |
| Logistic Regression | 77.90% | This paper |
| Decision Tree-Medium | 77.10% | This paper |
| Linear SVM | 77.40% | This paper |
| Cosine KNN | 76.40% | This paper |

algorithm code. Also, parameter tuning has a major role in machine learning, and further improvement for all classification techniques used can be made by more detailed analysis of parameters of each technique in order to achieve best possible results.

From the results obtained by using Logistic Regression model it can be concluded that Glucose and BMI play the major role in obtaining high accuracy in all cases, thus for very good results all other features can be omitted and model trained only with Glucose and BMI feature data. While, it has been shown that Insulin, Age, SkinThickness and DiabetesPedigreeFunction give the lowest accuracies for this model and can be omitted in order to achieve better accuracy. By using Logistic Regression method it can be observed from the results that the accuracy of unprocessed data has been just slightly lower than processed data accuracy, thus, we can state that for this method with this database unprocessed data can be successfully used for diabetes detection.

All types of Decision Tree model used in this research paper are used with 5-fold cross-validation, Gini's diversity index split criterion and different number of splits. Highest accuracy is obtained by using Decision Tree - Medium Tree classification model when all features from the database used for training the model, while

BloodPressure and Age features shown the worst results in terms of accuracy in case when unprocessed database used, and combination of SkinThickness and DiabetesPedigreeFunction shown worst result when normalized database used. It is shown that for the all Decision Tree models the lowest accuracy is obtained when BloodPressure, DiabetesPedigreeFunction, SkinThickness used in various combinations for model training. Observing results obtained from Decision Tree – Coarse Tree model, it can be seen that by using only Glucose and BMI feature it is possible to achieve equally high accuracy as it is achieved when all features used, and very high accuracy is achieved when only these two features used in all other models. Thus, it can be concluded that only these two features are sufficient for achieving high accuracy with observed unprocessed and processed Pima Indian diabetes database. When Maximum deviance reduction split criterion used for unprocessed data with Decision Tree - Coarse Tree classification method, 5-fold cross-validation, and maximum number of splits 4, the accuracy obtained was slightly higher than the accuracy in the first case where Gini's diversity index split criterion used. While in all other cases when Maximum deviance reduction or Twoing rule split criteria used for both processed and unprocessed data the accuracies were lower than in cases when Gini's diversity index split criterion used.

Considering SVM model, from the results provided, it can be seen that Fine Gaussian SVM model with Kernel scale of 0.71 obtained accuracy higher than Coarse Gaussian SVM and Linear SVM. It can be also seen that Fine Gaussian SVM method obtained the lowest accuracy for both unprocessed and processed data, while highest accuracy is achieved using Coarse Gaussian SVM for unprocessed data, and Linear SVM for processing data. As it was case in previously discussed classification models, Linear SVM model obtained very high accuracy when only Glucose and BMI feature used, and, since it is just slightly lower that accuracy obtained when all features used in model training, it can be concluded that these two features are sufficient for model training. Similarly as in previously discussed models, features whose combination gives lowest accuracy are BloodPressure, SkinThickness, DiabetesPedigreeFunction, Insulin and Age.

Several different KNN models with different parameters of distance metric and number of neighbors and the best accuracy among them is achieved in case when Cosine KNN model used where difference in accuracy is 0.4%. Best accuracy results with raw database of is obtained with the combination of five features - Glucose, Insulin, BMI, DiabetesPedigreeFunction and Age. This accuracy is higher for 0.4% than accuracy obtained when all features included, thus concluding that Pregnancies, BloodPressure and SkinThickness can be omitted from the raw database for better results. The reason to omit these features is, also, because when BloodPressure feature added to combination of five features which give with highest accuracy, the accuracy decreases by 0.4%. Combination of Glucose and BMI gives very high accuracy; thus, it can be said that in all classification models analyzed in this research paper these two features manage to obtain high accuracy which would be sufficient for diabetes detection from data obtained from Pima Indian diabetes database. SkinThickness, BloodPressure and Insulin are features that give the worst results in terms of accuracy, thus in most of the cases with different models analyzed these features could be removed and still obtain very high accuracy (Table 12).

**Table 12.** Summary of algorithms with best results obtained from this research in field of classification techniques for detection of diabetes using Pima Indian diabetes database

| Algorithm | Features | Percentage |
|---|---|---|
| Logistic regression | X2,X6,X7 | 77.9% |
| Decision tree-medium tree | X2,X4 | 77.1% |
| Linear SVM | X2,X4,X6,X7,X8 | 77.4% |
| Cosine KNN | X2,X5,X6,X7,X8 | 76.4% |

## 4 Conclusion

Diabetes Mellitus type-2 is a severe and complicated illness for both diagnoses as well as treatment, and it has a great impact on the patient's life as it disrupts it heavily if left undiagnosed and untreated. Nowadays, this topic is very popular due to the fact that there is an increase in overall population as well as an increase in the development of the new technologies that which aim to improve detection and diagnosis of the disease. The aim of this research paper was to investigate and show how machine learning with different classification methods can be used in diabetes diagnostic field to improve health care. For the purpose of this research paper results from different software tools for machine learning are examined and the results are presented. This process required data analysis through inspection and visualization, then transformation and normalization. Next, different algorithms were applied in MATLAB and accuracies of different classification method are examined. Throughout this paper the relationships of different features are presented all together with the accuracies of the most and least relevant features. Investigating several different classification methods, it has been found that same classification method give somewhat different results when model fitting performed in different software machine learning tools in different programs. Also, it was expected that in all cases and iterations of all classification models unprocessed, raw, data will give somewhat lower accuracy, it didn't happen.

The future work in the field of diagnosis, detection and prognosis of the diabetes Mellitus type-2 should be in the direction of further investigating machine techniques with the aim of creating more complex solutions with the proper parameters to acquire the best possible results in terms of model accuracy. As well, it is needed to find a better way of integrating these techniques into the healthcare diagnostic system for developing countries and areas with increased population suffering from this illness.

## References

1. American Diabetes Association: Diagnosis & classification of diabetes mellitus. Diabetes Care **37**(Supplement 1), S81–S90 (2014)
2. American Diabetes Association: Reports of the experts committee on the diagnosis and classification of diabetes mellitus. Diabetes Care **23**, S4–19 (2001)
3. National Diabetes Information clearinghouse (NDIC): http://diabetes.niddk.nih.gov
4. World Health Organization (WHO): Diabetes. 30 October 2018. http://www.who.int/mediacentre/factsheets/fs312/en/

5. Contreras, I., Vehi, J.: Artificial intelligence for diabetes management and decision support: literature review. J. Med. Internet Res. **20**, e10775 (2018)
6. CB Insights: This is how artificial intelligence is transforming diabetes care management. https://www.cbinsights.com/research/?s=diabetes+
7. Ramezankhani, A., Pournik, O., Shahrabi, J., Azizi, F., Hadaegh, F., Khalili, D.: The impact of oversampling with SMOTE on the performance of 3 classifiers in prediction of type 2 diabetes. SAGE J. **36**(1), 37–144 (2016)
8. Shankaracharya, D.O., Samanta, S., Vidyarthi, A.S.: Computational intelligence in early diabetes diagnosis: a review. Rev. Diabet. Stud. **7**, 252–262 (2010)
9. Shanker, M.S.: Using neural networks to predict the onset of diabetes mellitus. J. Chem. Inf. Comput. Sci. **36**, 35–41 (1996)
10. Purnami, S.W., Embong, A., Zain, J.M.: A new smooth support vector machine and its applications in diabetes disease diagnosis. J. Comput. Sci. **5**, 1006–1011 (2009)
11. Jhaldiyal, T., Mishra, P.K.: Analysis and prediction of diabetes mellitus using PCA, REP SVM. Int. J. Eng. Tech. Res. (IJETR) **2**(8) (2014) ISSN: 2321-0869
12. Kandhasamy, J.P., Balamurali, S.: Performance analysis of classifier models to predict diabetes mellitus. Procedia Comput. Sci. **47**, 45–51 (2015)
13. Kordos, M., Blachnik, M., Strzempa, D.: Do we need whatever more than k-NN? In: Proceedings of the 10th International Conference on Artificial Intelligence and Soft Computing, Part I, pp. 414–421. Springer-Verlag, Berlin (2010)
14. Ster, B., Dobnikar, A.: Neural networks in medical diagnosis: comparison with other methods. In: Proceedings of the International Conference on Engineering Applications with Neural Networks, pp. 427–430. London (1996)
15. Jahangeer, S., Zaman, M., Ahmed, M., Ashraf, M.: An empirical comparison of supervised classifiers for diabetic diagnosis. Int. J. Adv. Res. Comput. Sci. 8(1), 311–315 (2017)
16. Polat, K., Gunes, S., Arslan, A.: A cascade learning system for classification of diabetes disease: generalized discriminant analysis and least square support vector machine. Exp. Syst. Appl. **34**, 482–487 (2008)
17. Christobel, Y.A., Sivaprakasam, P.: Improving the performance of k-nearest neighbor algorithm for the classification of diabetes dataset with missing values IJCET **3**(3), 155–167 (2012)
18. Shital, T., Madan, S., Pranjali, C., Swati, S.: SVM based diabetic classification and hospital recommendation. Int. J. Comput. Appl. **167**(1), 40–43 (2017)
19. Barale, M.S., Shirke, D.T.: Cascaded modeling for PIMA Indian diabetes data. Int. J. Comput. Appl. **139**(11), 1–4 (2016)
20. Hashi, E.K., Zaman, S.U., Hasan, R.: Developing diabetes disease classification model using sequential forward selection algorithm. Int. J. Comput. Appl. **180**(5), 1–6 (2017)
21. https://www.kaggle.com/uciml/pima-indians-diabetesdatabase
22. Jayalakshmi, T., Santhakumaran, A.: Statistical normalization and back propagation for classification. Int. J. Comput. Theory Eng. **3**(1), 1793–8201 (2011)
23. Shantakumar, B.P., Kumarasawamy, S.: Predictive data mining for medical diagnosis of heart disease **1**(2), 161–176 (2009)
24. http://archive.ics.uci.edu/ml/datasets/Pima+Indians+Diabetes
25. Nakano, T., Nukala, B.T., Zupancic, S., Rodriguez, A., Lie, D.Y.C., Lopez, J., Nguyen, T. Q.: Gaits classification of normal vs. patients by wireless gait sensor and support vector machine (SVM) classifier. In: IEEE ICIS 2016, Okayama, Japan, June 2016
26. Abdillah, A.A., Suwarno, S.: Diagnosis of diabetes disease using support vector machines with kernel radial basis function. In: Conference on International Conference on Mathematics, Its Applications, and Mathematics Education (ICMAME) (2015)

# Prediction of Power Output for Combined Cycle Power Plant Using Random Decision Tree Algorithms and ANFIS

Lejla Bandić(✉), Mehrija Hasičić, and Jasmin Kevrić

International Burch University, Sarajevo, Bosnia and Herzegovina
lejla.bandic@ibu.edu.ba

**Abstract.** This paper presents methods for prediction of the power output of the combined cycle power plant (CCPP) with a full load. A dataset comprising 9568 samples include measurements of ambient temperature (AT), atmospheric pressure (AP), relative humidity (RH), exhaust steam pressure, i.e. vacuum (V) and power output of the CCPP (EP). The research was done two folded: using all features and the reduced set of features. Random Forest, Random Tree, and Adaptive Neuro Fuzzy Inference System (ANFIS) were used for regression. The performance of the methods studied in both folds showed that the best obtained results are gained using Random Forest. Results obtained on all features showed (Root Means Square Error) RMSE of 3.0271 MW, while feature selection leads to the RMSE of 3.0527 MW and Correlation coefficient (CC) of 0.9843, both obtained on 90% Percentage split.

**Keywords:** Random Forest (RF) · Random Tree · ANFIS · Power prediction · Power plant

## 1 Introduction

Considering that there is an increasing demand for electricity, many power plants with gas turbine derivatives have been founded all around the world [1]. Prediction of the power output of the gas turbine highly affects its reliability and sustainability and thus has become the subject to numerous research studies.

Gas turbine power generation primarily depends on the ambient conditions including ambient temperature, atmospheric pressure, and relative humidity. In addition, exhaust steam pressure, i.e. vacuum has an influence on the power generation.

In the past years, the effects of ambient conditions were studied together with artificial intelligence methods in order to predict electrical power. Some of them include artificial neural networks (ANN) and regression methods [1–10].

In [1], the impact of the ambient conditions including pressure and temperature, relative humidity and wind velocity on the power generation of the power plant was analyzed using the ANN model.

In [5–7], researches have considered ANN to model combined cycle power plant (CCPP), while [9] represents the prediction of electricity energy consumption using decision tree and neural networks.


S. Avdaković et al. (Eds.): IAT 2019, LNNS 83, pp. 406–416, 2020.
https://doi.org/10.1007/978-3-030-24986-1_32

Recently, researchers from [7] analyzed the application and experimentation of various options effects on feed-foreword (ANN) that was used to obtain regression model for prediction of electrical output power (EP) of CCPP based on 4 inputs. The study was performed on the same dataset that is used in this paper. Obtained results are compared with target values of output for the two groups: train and validation group and the test group. Comparison showed that results are very close to target outputs for both groups. Train and validation group had root-mean-squared error (RMSE) of 4.2358 MW while test group had 4.3223 MW as RMSE. The standard deviations of the error for these two groups are almost equal (4.2429 MW for train and validation group and 4.2825 for test group). Correlation factor of predicted and measured value they obtained was in average 0.88.

In [3], the prediction models of power generation of CCPP are described. Its aim was to find the best subset of the dataset and the best regression method that will give the highest prediction accuracy. The study was done on the same dataset which is used in this paper. As there are four features, all 15 combinations of the reduced sets (no feature selection algorithms used) were applied to 15 regression methods found in WEKA software [11]. Results showed that the best accuracy was obtained when all of the four features are included together with Bagging REP tree regression model for prediction, with a mean absolute error of 2.818 MW and RMSE of 3.787 MW within a range from 420 MW to 495 MW.

This paper represents the prediction of the power output of the combined cycle power plant with a full load using methods that were not used in [3], namely Random Forest and Random Tree models on all features. Furthermore, the experiment is repeated with selected features with mentioned algorithms and ANFIS. ANFIS is the most effective on the dataset with a small number of features, desirably two to three [12]. Using attribute selection methods in Weka (Cfs Subset Evaluation, Principal Component Analysis, and Relief Attribute Evaluation), we obtained that highest impact on result have AT, V and AP and used these three in the second fold of this research in order to obtain comparable results to the first fold and [3]. All computations are done in WEKA software while ANFIS was implemented in MATLAB.

## 2 Dataset

A dataset comprising 9568 samples collected from a Combined Cycle Power Plant over 6 years (2006–2011) and includes measurements of ambient temperature (AT), atmospheric pressure (AP), relative humidity (RH), exhaust steam pressure, i.e. vacuum (V) and power output of the CCPP (EP). The EP is measured in mega watt with the range of 420.26–495.76 MW. Above mentioned measurements are collected when CCPP is set to work with a full load over 674 different days. As explained in [13] a CCPP is composed of gas turbines (GT), steam turbines (ST) and heat recovery steam generators. Electrical energy in CCPP is generated by gas and steam turbines that are combined in one cycle and then transferred from one to another turbine. While collected Vacuum has an effect on a Steam turbine, another three variables effect on the GT performance. Information about the location of the power plant used for the data collection is not known. More detailed combined cycle power plant description is given in [3]. Table 1 represents the features of the dataset used and their units together with statistics of the dataset.

**Table 1.** Dataset features and statistics.

| # | Data features | Units | Min | Max | Mean | StdDev. |
|---|---|---|---|---|---|---|
| 1 | Ambient temperature (AT) | C | 1.81 | 37.11 | 19.65 | 7.45 |
| 2 | Vacuum (V) | cm Hg | 25.36 | 81.56 | 54.31 | 12.71 |
| 3 | Atmospheric pressure (AP) | mbar | 993.89 | 1033.30 | 1013.26 | 5.94 |
| 4 | Relative humidity (0.4 kV) | % | 25.56 | 100.16 | 73.31 | 14.6 |
| 5 | Full load electrical power output (EP) | MW | 420.26 | 495.76 | 454.37 | 17.07 |

Cross-correlation of the parameter is presented in Table 2, while in Fig. 1 the scatter plot of the data is illustrated. In Table 2 the linear relations strength between the variables can be observed. Input-output relations observed in the table show that AT and V have strong negative correlation with EP while AP and RH have weak positive correlation. Observed correlation show that RH has very low impact on the output while AT has the strongest but negative relationship with output. Moreover, multi-collinearity can be observed since input variables AT and V have strong positive correlation.

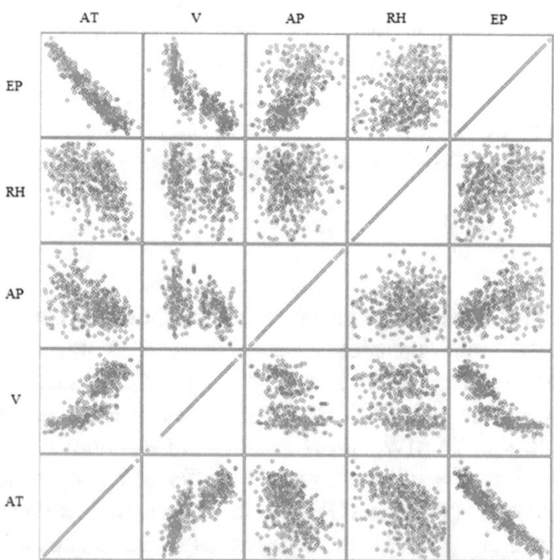


**Fig. 1.** Pairwise scatter plot of parameters.

**Table 2.** Correlation matrix

| | AT | V | AP | RH | EP |
|---|---|---|---|---|---|
| AT | 1 | 0.84 | −0.51 | −0.54 | −0.95 |
| V | 0.84 | 1 | −0.41 | −0.31 | −0.87 |
| AP | −0.51 | −0.41 | 1 | 0.10 | 0.52 |
| RH | −0.54 | −0.31 | 0.10 | 1 | 0.39 |
| EP | −0.95 | −0.87 | 0.52 | 0.39 | 1 |

# 3 Methods

## 3.1 Random Forest

Random forest is a powerful machine-learning method that was proposed by Breiman [14]. It is used for classification, regression, and unsupervised learning and it can be applied in many fields. Random forest is a decision tree algorithm that uses bagging to improve its results.

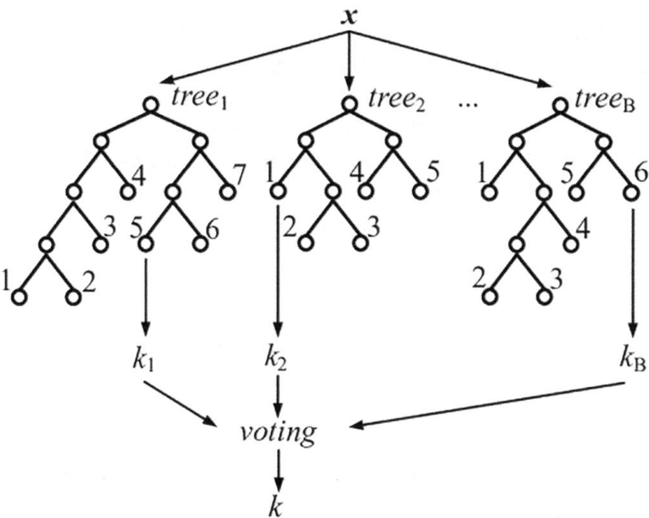


**Fig. 2.** A general Random Forest architecture [15]

The basic idea of a decision tree is to iteratively partition data into boxes using simple rules that minimize the error at each split (node). Each node is split using the best split among all variables, in standard trees. Instead of searching for the best feature while splitting a node, random forest searches for the best feature among a random subset of features. This process creates a wide diversity, which generally results in a better model [16]. A general Random Forest architecture is shown in Fig. 2. Each decision tree will generate a prediction for a given input, and the overall average of all predictions will be the result of voting.

There are only two parameters: the number of variables in the accidental subset and number of trees. Performances of RF are not sensitive to their values.

### 3.2 Random Tree

A random tree is a decision tree method that uses a tree-like graph or model of decisions. It is an ensemble learning algorithm that generates many individual learners. The algorithm can deal with both classification and regression problems.

Random trees represent the combination of two existing algorithms: single model trees are combined with random forest ideas. Model trees are decision trees where every single leaf holds a linear model which is optimized for the local subspace described by this leaf. Random trees use accidental forest process for split selection which then induces logically balanced trees where one global setup of ridge value works on all leaves. That leads to the simplification of the optimization process [17].

### 3.3 ANFIS

It is the kind of neural networks that is based on the Takagi - Sugeno fuzzy inference system, using a hybrid learning algorithm to identify the membership function parameters of single-output. A given set of input/output relationship is modeled by training the parameters of FIS membership function using a combination of back-propagation gradient descent methods and least-squares [18].

In MATLAB, the ANFIS represents the fuzzy logic toolbox for creation of fuzzy inference system using membership functions in order to track the given input/output data [19].

### 3.4 Feature Selection Algorithms

*Correlation-Based Feature Selection (CFS) Subset Evaluator*
CFS is a simple filter algorithm that lists feature subsets using a correlation based heuristic evaluation function. The correlation function leans toward the subset containing features that are mainly in correlation with class and uncorrelated to each other. Insignificant features should be neglected since they will be in very low correlation with the class, while redundant functions should be shown since they will be in high correlation with one or more of the remaining functions. The extent to which class is predicted in areas of the instance space which is not already predicted by other features effects the acceptance of a feature [20].

*Principal Component Analysis (PCA)*
The process of analyzing data in which observations are described by few inter-correlated quantitively dependent variables represents a technique known as PCA. Its goal is to extract a set of new, orthogonal variables called principal components out of the information from the data and to show the pattern of similarity in observations and variables as points on a map. Cross-validation techniques such as the bootstrap and the

jackknife can be used to evaluate the quality of the PCA model. To handle qualitative variables, PCA can be generalized as correspondence analysis (CA), while with heterogeneous sets of variables it may be generalized as multiple factor analysis (MFA). The eigen-decomposition of positive semidefinite matrices and the singular value decomposition (SVD) of rectangular matrices define the PCA mathematically [21].

*Relief Attribute Evaluator*
Feature selection algorithm inspired by instance-based learning is known as Relief Attribute Evaluator. It detects features that are statistically relevant to the target concept using training data S, sample size *m* and a threshold of relevancy τ which encodes a relevance threshold (0 ≤ t ≤ 1). The scale of every feature is assumed to be either nominal or numerical [22].

## 4 Results and Discussion

In order to compare the performance of the regression models, two performance metrics are used. The root mean square error (RMSE) and the mean absolute error (MAE) were used in the first fold:

$$RMSE = \sqrt{\frac{1}{n}\sum\nolimits_{i=1}^{n}\left(X_i - \underline{X_i}\right)} \tag{1.1}$$

$$MAE = \frac{1}{n}\sum\nolimits_{i=1}^{n}\left|X_i - \underline{X_i}\right| \tag{1.2}$$

Where $X_i$ is the actual measurement of the power output EP, $\underline{X_i}$ is the predicted value of the power output EP obtained by the algorithms used in this study that are described in sections above and $n$ is the number of the measurements which is defined as the number of samples in dataset, 9568.

Third metric used in this study was a correlation coefficient (CC) which is used to summarize the strength of the linear relationship between two data samples. CC is calculated as the covariance of the two variables divided by the product of the standard deviation of each data sample. Equations for both covariance and CC are:

$$cov(X_i, \underline{X_i}) = \frac{\sum_{i=1}^{n}((X_i - X_i') * (\underline{X_i} - \underline{X_i}'))}{n-1} \tag{1.3}$$

$$CC = \frac{cov(X_i, \underline{X_i})}{stdv(X_i) * stdv(\underline{X_i})} \tag{1.4}$$

where $X_i'$ and $\underline{X_i}'$ are mean values of $X_i$ and $\underline{X_i}$ variables, while their standard deviations are stdv($X_i$) and stdv($\underline{X_i}$) respectively. Value of the CC goes from −1 to +1 which represents the limits of correlation from a full negative correlation to a full positive correlation. A value of 0 means no correlation. Closer the value is to |1| the stronger relationship between real and predicted value is, +1 would represent positivie and −1 negative linear relationship.

In order to obtain the results and see the performance of the proposed regression models, five different cases were performed. The first case represents usage of 50% of instances of the dataset for training and 50% for testing. For the second case, 66% of instances were used for training and 34% for testing. The third case represents usage of 90% of instances for training and 10% for testing. For the fourth case, 10-fold cross validation (CV) was used. For the fifth case, 20-fold CV was applied.

### 4.1 The First fold

In Table 3, results using Random Forest are shown. The best performance was obtained using the third case, i.e. when 90% of instances are used for training and 10% for testing, with RMSE of 3.0271 MW and MAE of 2.2554 MW with CC of 0.946.

**Table 3.** Results of Random Forest algorithm on all attributes

| Random Forest | | | |
|---|---|---|---|
| | RMSE [MW] | MAE [MW] | CC |
| 50% train, 50% test | 3.4628 | 2.5008 | 0.9793 |
| 66% train, 34% test | 3.3031 | 2.3639 | 0.9815 |
| 90% train, 10% test | 3.0271 | 2.2554 | 0.9846 |
| 10-CV | 3.2522 | 2.2947 | 0.9817 |
| 20-CV | 3.228 | 2.2724 | 0.982 |

**Table 4.** Results of Random Tree algorithm on all attributes

| Random Tree | | | |
|---|---|---|---|
| | RMSE [MW] | MAE [MW] | CC |
| 50% train, 50% test | 4.852 | 3.4144 | 0.9597 |
| 66% train, 34% test | 4.6435 | 3.1938 | 0.9636 |
| 90% train, 10% test | 4.1891 | 2.9417 | 0.9704 |
| 10-CV | 4.4631 | 2.9931 | 0.9658 |
| 20-CV | 4.4515 | 3.0035 | 0.9661 |

Table 4 represents the results obtained using the Random Tree. The third case showed the best performance once more. RMSE for the Random Tree is 4.1891 MW, MAE is 2.9417 MW and CC is 0.9704. Comparing the results obtained, it can be concluded that the Random Forest has better performance than Random Tree algorithm.

## 4.2 The Second Fold

Using all three previously listed attribute selection methods we obtained that highest impact on result have AT, V and AP, whereas RH seems to be neglectable. We used these three selected attributes in the second fold of this research in order to obtain comparable results to the first fold and [3]. Listed in Tables 5 and 6, we may observe the results obtained using Random Forest and Random Tree algorithms, respectively, on feature selected data. From the obtained results, Random Forest gave the best results using 90% percentage split with RMSE of 3.0527 MW, MAE of 2.2575 MW and CC of 0.9843, while Random Tree and ANFIS gave significantly lower results.

The best result obtained with Random Tree is obtained with 90% Percentage split with RMSE of 4.0976 MW and MAE of 2.8615 MW and CC 0.9716.

**Table 5.** Results of Random Forest algorithm on selected attributes

| Random Forest | | | |
|---|---|---|---|
| | RMSE [MW] | MAE [MW] | CC |
| 50% Percentage split | 3.5134 | 2.5356 | 0.9786 |
| 66% Percentage split | 3.3428 | 2.4059 | 0.981 |
| 90% Percentage split | 3.0527 | 2.2575 | 0.9843 |
| 10-CV | 3.3037 | 2.3222 | 0.9811 |
| 20-CV | 3.2898 | 2.3042 | 0.9812 |

**Table 6.** Results of Random Tree algorithm on selected attributes

| Random Tree | | | |
|---|---|---|---|
| | RMSE [MW] | MAE [MW] | CC |
| 50% Percentage split | 4.6874 | 3.1782 | 0.9624 |
| 66% Percentage split | 4.542 | 3.0741 | 0.9649 |
| 90% Percentage split | 4.0976 | 2.8615 | 0.9716 |
| 10-CV | 4.5925 | 3.0696 | 0.9638 |
| 20-CV | 4.5595 | 3.0364 | 0.9643 |

In Table 7, results obtained with ANFIS on feature selected data using trapezoidal (trapmf), triangular (trimf), generalized bell shaped (gbellmf) and gaussian (gaussmf) membership functions are presented. We have used two and three membership functions (MF) on each feature and compared the results in Table 7. As an optimization method, the default hybrid method, which combines the least squares estimation with backpropagation, was used. Initial Sugeno-type FIS for ANFIS training using a grid partition on the data and linear output membership function were used for generating initial FIS. Data used in ANFIS was divided into train and test using 90% percentage split.

Obtained RMSE is approximated value on 100 epochs and is significantly higher when compared to either Random Forest or Random Tree.

**Table 7.** Results obtained with ANFIS on feature selected data

| ANFIS | | | | |
|---|---|---|---|---|
| MF | Numberf of MF | RMSE [MW] | MAE [MW] | CC |
| trapmf | 3 | 4.6143 | 3.5599 | 0.9625 |
| trapmf | 2 | 4.6049 | 3.5957 | 0.9626 |
| trimf | 3 | 4.1722 | 3.2870 | 0.9691 |
| trimf | 2 | 4.7885 | 3.7669 | 0.9602 |
| gbell | 3 | 4.2569 | 3.3287 | 0.9679 |
| gbell | 2 | 4.5636 | 3.5601 | 0.9632 |
| gaussmf | 3 | 4.3750 | 3.3838 | 0.9662 |
| gaussmf | 2 | 4.6946 | 3.6547 | 0.9614 |

### 4.3 Comparison with Previous Work

The best results obtained in [3] and our work are shown in Table 8. According to [3], bagging REP tree (BREP) gave the best results, when compared with other 14 regression methods. For the results shown in Table 5, BREP was performed with 5 × 2 cross-validation and using all features, the same as the number of features used in this paper in the first fold.

Comparison of the mean performances of used methods indicates that the random forest method is the best regression method for this study considering both folds. Furthermore, attributes selected in the second fold of this research are sufficient for EP prediction. The best obtained result is still the one that used all attributes on the Random Forest algorithm, i.e. first fold of this study.

**Table 8.** Comparison table

| | RMSE [MW] | MAE [MW] | CC |
|---|---|---|---|
| BREP – all features [3] | 3.787 | 2.818 | – |
| ANN – all features [7] | 4.3223 | – | 0.8810 |
| RF – all features [our work] | 3.0271 | 2.2554 | 0.9846 |
| BREP – selected features (AT, V, AP) [3] | 3.855 | – | – |
| ANN – selected features (AT, V, AP) [7] | – | – | 0.94 |
| RF - selected features (AT, V, AP) [our work] | 3.0527 | 2.2575 | 0.9843 |
| ANFIS - selected features (AT, V, AP) [our work] | 4.1722 | 3.2870 | 0.9691 |

## 5 Conclusion

In this paper, random forest, random tree and ANFIS models were developed to predict the electrical power output of the combined cycle power plant with a full load. The algorithms and solutions are implemented in the Weka environment and MATLAB.

It has been shown that presented methodology can be efficiently used for the prediction of the power of the CCPP. Moreover, the RF model yielded better results than the Random Tree and other methods used in research on the same dataset with RMSE of 3.0271 MW on full data set and RMSE of 3.0527 MW on the attribute selected data set. ANFIS and ANN (on the full dataset) yielded in poorest results (RMSE of 4.1722 MW and 4.3223 MW, respectively) when compared to obtained results on full and attribute selected data set in both [3, 7] and our research. Results obtained in [3] using attribute selected data lead to the RMSE of 3.855 MW which showed to be poorer performance than the one we obtained with Random Forrest (both on full data set and attribute selected data set), but better when compared to Random Tree algorithm and ANFIS. As for the computation time, ANFIS showed to be the slowest with the computation time of 20s while other methods took in average 5 s to compute the results, when worked on Intel(R) Core(TM) i3-3110 M CPU @ 2.40 GHz with 4 Gb of RAM memory.

Future work and research can be linked to the prediction of the power output of power plant using improved regression methods and their combinations. Also, more detailed analysis of the impact of different parameters on the power output of the CCPP can be performed.

## References


1. Lee, J.J., Kang, D.W., Kim, T.S.: Development of a gas turbine performance analysis program and its application. Energy **36**(8), 5274–5285 (2011)
2. Kaya, H., Tufekci, P., Gurgen, F.S.: Local and global learning methods for prediction of power of a combined gas and steam turbine. In: International Conference on Emerging Trends in Computer and Electronics Engineering (ICETCEE) (2012)
3. Tufekci, P.: Prediction of full load electrical power output of a base load operated combined cycle power plant using machine learning methods. Int. J. Electr. Power Energy Syst. **60**, 12–140 (2014)
4. Yari, M., Shoorehdeli, M.A.: V94.2 gas turbine identification using neural network. In: First RSI/ISM International Conference on Robotics and Mechatronics (ICRoM), IEEE Xplore (2013)
5. Kumar, A., Srivastava, A., Banerjee, A., Goel, A.: Performance based anomaly detection analysis of a gas turbine engine by artificial neural network approach. In: Proceedings of Annual Conference of the Prognostics and Health Management Society (2012)
6. Samani, A.D.: Combined cycle power plant with indirect dry cooling tower forecasting using artificial neural network. Decis. Sci. Lett. **7**(2), 131–142 (2018)
7. Elfaki, E., Hassan, A.H.A.: Prediction of electrical output power of combined cycle power plant using regression ANN model. Int. J. Comput. Sci. Control. Eng. **6**(2), 9–21 (2018)
8. Tayarani-Bathaie, S.S., Sadough Vanini, Z.N., Khorasan, K.: Dynamic neural network-based fault diagnosis of gas turbine engines. Neurocomputing **125**, 153–165 (2014)
9. Tso, G.K.F., Yau, K.K.W.: Predicting electricity energy consumption: a comparison of regression analysis, decision tree and neural networks. Energy **32**(9), 1761–1768 (2007)
10. Rahnama, M., Ghorbani, H., Montazeri, A.: Nonlinear identification of a gas turbine system in transient operation mode using neural network. In: 4th Conference on Thermal Power Plants (CTPP), IEEE Xplore (2012)

11. Machine Learning Group at the University of Waikato. https://www.cs.waikato.ac.nz/ml/weka/
12. Patel, K., Shah, H., Kher, R.: Face recognition using 2DPCA and ANFIS classifier. Adv. Intell. Syst. Comput. **336**, 1–12 (2015)
13. Tufekci, P., Kaya, H.: Combined Cycle Power Plant Dataset (CCPP) (2014). https://archive.ics.uci.edu/ml/datasets/Combined+Cycle+Power+Plant
14. Breiman, L.: Random forests. Mach. Learn. **45**, 5–32 (2001)
15. Gelzinis, A., Verikas, A., Vaiciukynas, E., Bacauskiene, M., Minelga, J., Hallander, M., Uloza, V., Padervinskis, E.: Exploring sustained phonation recorded with acoustic and contact microphones to screen for laryngeal disorders. In: IEEE Symposium on Computational Intelligence in Healthcare and e-Health (CICARE) (2014)
16. Liaw, A., Wiener, M.: Classification and regression by random forest. R. News **2**, 18–22 (2002)
17. Kalmegh, S.: Analysis of WEKA data mining algorithm REPTree, simple cart and randomtree for classification of Indian news. IJISET Int. J. Innov. Sci. Eng. Technol. **2**(2), 438–446 (2015)
18. Neuro-Adaptive Learning and ANFIS. https://www.mathworks.com/help/fuzzy/neuro-adaptive-learning-and-anfis.html
19. Moler, C.: MATrix LABoratory (MATLAB) R2018a, MathWorks. Massachusetts, United States (2018)
20. Hall, M.A.: Correlation-based feature selection for machine learning. PhD Thesis, University of Waikato, New Zealand (1999)
21. Abdi, H., Williams, L.J.: Principal component analysis. Wiley Interdiscip. Rev.: Comput. Stat. **2**, 433–459 (2010)
22. Kira, K., Rendell, L.A.: A practical approach to feature selection. In: Ninth International Workshop on Machine Learning, pp. 249–256 (1992)

# Artificially Intelligent Assistant for Basketball Coaching

Yasin Acikmese[1], Baris Can Ustundag[2], Tarik Uzunovic[3](✉), and Edin Golubovic[4]

[1] Department of Industrial Engineering, Galatasaray University, Istanbul, Turkey
yasin.acikmese@ogr.gsu.edu.tr

[2] Department of Computer Engineering, Istanbul Technical University, Istanbul, Turkey
ustundag16@itu.edu.tr

[3] Faculty of Electrical Engineering, University of Sarajevo, Sarajevo, Bosnia and Herzegovina
tuzunovic@etf.unsa.ba

[4] Inovatink, Istanbul, Turkey
edin@inovatink.com

**Abstract.** Technological advancements in wearable sensors, machine learning and Internet of Things (IoT) is opening new perspectives on the understanding of physiological, biomechanical, and psychological mechanisms of human movement in sport disciplines. Utilization of technology and field expertise allows the development of artificially intelligent (AI) assistants for sport coaching. This paper considers the development of AI assistant for basketball coaching. The assistant architecture and design is explained in detail. The details about the development of knowledge model for basketball exercise recognition are supplied and experimental verification of proposed method is done.

**Keywords:** AI assistant · Wearable · Machine learning · IoT · Basketball · Kinesiology · Neural Networks

## 1 Introduction

The advent of IoT, wearable sensors and machine learning allows for holistic study of kinesiology (physiological, biomechanical, and psychological mechanisms of human movement). These technologies improve the study of human movement in the context of rehabilitation, physical therapy, strength and conditioning training, sport, exercise and workplace activity. Promising benefits of faster progress and improvement through well measured, processed and analyzed movement, dictate the advancement of research in the application of IoT, wearable sensors and machine learning to the kinesiology studies.

In kinesiology studies, wearable sensors can be used to quantify performed movement and provide insight into performance. Machine learning algorithms


S. Avdaković et al. (Eds.): IAT 2019, LNNS 83, pp. 417–427, 2020.
https://doi.org/10.1007/978-3-030-24986-1_33

offer possibility to both quantify the movement and to provide reasoning assistance to the field experts. Using IoT, large scale databases and cloud based data analysis and knowledge generation tools can be explored. Ideally, the utilization of these technology for kinesiology applications would lead to truly autonomous artificially intelligent (AI) assistant.

The advancement in AI and human-computer interaction research opened up new areas of application for AI assistants in education [1], tele-rehabilitation [2], robotics [3], homes [4], smart phones [5] and entertainment [6]. AI assistants are also utilized in sports coaching with an aim to assist players during technical training. Authors in [7] introduce a smart assistant for professional volleyball training. The assistant utilizes combination of machine learning, statistical player data, wearable sensors and cameras to address controlling exercise effort and fatigue levels (technical-tactical) and exercise quality training. AI assistants are used for encouraging of physical activity [8,9]. In [8] AI assistant consists of an activity monitor that tracks user's location, a portal that provides an overview of the user's behavior and a reasoning engine built around computational model of behavior change. In [9] developed system for personalized physical activity coaching uses machine learning algorithms to automatize coaching based on the information about the probability of meeting daily physical activity goal.

Work on automated assessment of tennis swings to improve performance and safety is done in the context of qualitative coaching diagnostics in [10]. Authors demonstrate the possibility of automation of qualitative analysis of human motion in context of various tennis training exercises. Similarly authors in [11] develop coach for golf training using wearable sensor and machine learning algorithm. The system is used to assist the player in developing golf skills.

In this paper, the architecture, design and implementation of artificially intelligent assistant for basketball coaching is presented. This work builds on our previous research [12]. Developed AI assistant utilizes wearable sensors, machine learning and IoT technologies to build a complete solution. This paper has two major results. First, it presents in detail the software architecture necessary to build AI assistant for basketball coaching. Second, it builds on that architecture to demonstrate it's use in basketball training type recognition by utilizing neural network trained knowledge model. The rest of this paper is organized as follows; next section gives the overview of the assistant architecture and implementation. Section 3 gives details about developed knowledge model. Numerical results are presented in Sect. 4 and discussion is done in Sect. 5. Section 6 concludes the paper.

## 2 Design of AI Assistant for Basketball Coaching

### 2.1 System Architecture

The architecture of artificially intelligent assistant for basketball coaching is given in Fig. 1. Assistant perceives environment, the motion of trainee in particular, through wearable *sensors*. Data obtained through sensors is filtered to

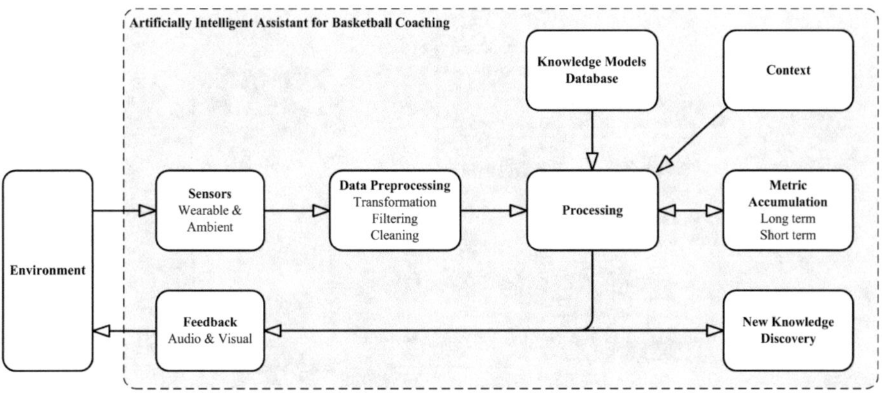


**Fig. 1.** Artificially intelligent assistant for basketball coaching architecture

remove noise and transformed to proper format by *data preprocessing* architectural block. Data is converted into knowledge and actionable intelligence by *processing* block. *Processing* block employs knowledge models previously developed and stored in *knowledge models database*. Assistant decides which models to use for data processing depending on the *context* of the training. Gathering of short and long term training metrics is done in training *metric accumulation* block. *Knowledge models* make use of training metrics to produce useful results. *Feedback* block provides audio and visual feedback to the trainees. The accommodation of new knowledge is done through *new knowledge discovery* element.

AI Assistant for basketball coaching makes use of knowledge models to rate the basketball training exercises and to monitor the progress of trainee. The knowledge models are developed under the assumption of certain context. For a model to provide correct results the context must be known. In this work the basketball coaching context has been divided into levels. The contextual map of developed assistant is shown in Fig. 2. The developed assistant is used for sports application of basketball coaching. The improvement in performance of basketball player is reached through physical, technical, tactical and psychological training. At this stage of progress of presented research the Level 3 of context must be supplied by user. The knowledge models are developed for assistant to be autonomous below this level, i.e. models must cover the autonomous progress monitoring and feedback generation for contextual levels 4, 5 and 6.

### 2.2 System Implementation

AI assistant for basketball coaching is implemented through three technological elements; wearable device, hub and cloud service. Wearable device sends motion data to the processing hub. Hub communicates with the wearable device, stores data and prepares it for further processing. The hub runs knowledge models and feedback algorithms. Hub is Internet connected device capable of storing data and providing processing offload from cloud in distributed manner. Cloud

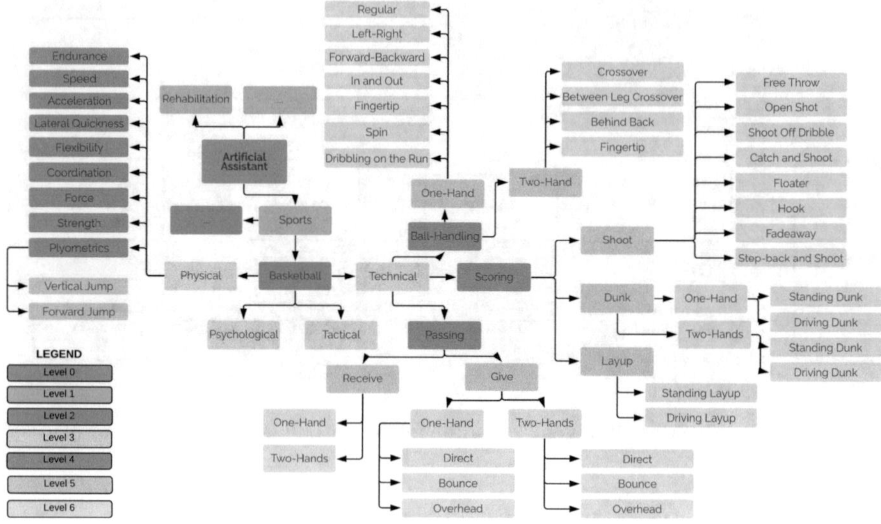


**Fig. 2.** Artificial assistant contextual map

service offers data storage, long term monitoring capabilities and utilization of large scale databases and cloud based data analysis tools. The implementation of the system is shown in Fig. 3.

**Wearable Device.** The operation of AI assistant for basketball coaching depends on the information about trainee's activity during training. Necessary information can be obtained through utilization of motion sensors inside the basketball, video capturing devices recording player during the training or wearable motion sensors. The sensor of choice depends on the design requirements of the

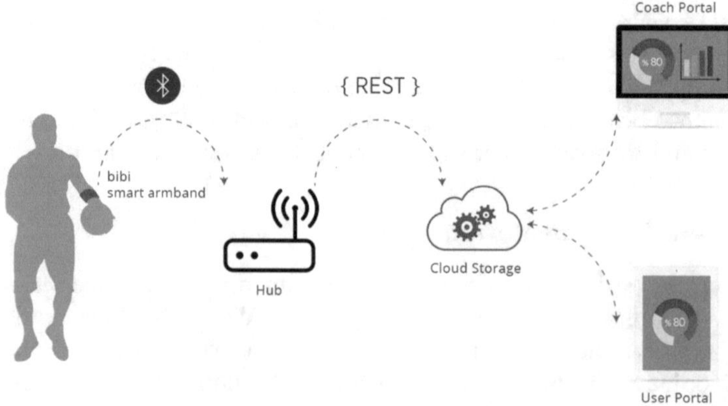


**Fig. 3.** System implementation

overall system. For the case of work presented in this paper following requirements are considered; sensor must be cost effective and easily implemented; next to the standard motion parameters such as position, velocity and acceleration, the force that hand exerts on the ball should be easily measured or observed; trainee should be able to utilize sensor both indoors and outdoors under different lighting conditions; trainee activity during rest/recovery period should be recorded. The best fit for these requirements would be the wearable device with motion sensor.

For the purposes of this work wearable device incorporating motion sensor was implemented. The wearable device is shown in Fig. 4. Sensor, microcontroller for signal acquisition and Bluetooth Low Energy (BLE) communication are coupled together in battery powered wearable device [15] to provide untethered and trainee friendly way of data collection [12]. Wearable device is capable of recording acceleration using a 3-axis accelerometer, angular velocity using 3-axis gyroscope and orientation in the form of quaternions. Recorded data is sent to hub through Bluetooth Low Energy (BLE) communication protocol. The device is intended to be worn on the arm as an armband. However, it might be placed anywhere on the body using a different size strap. The processor reads signals from sensor at a rate of 100 Hz.

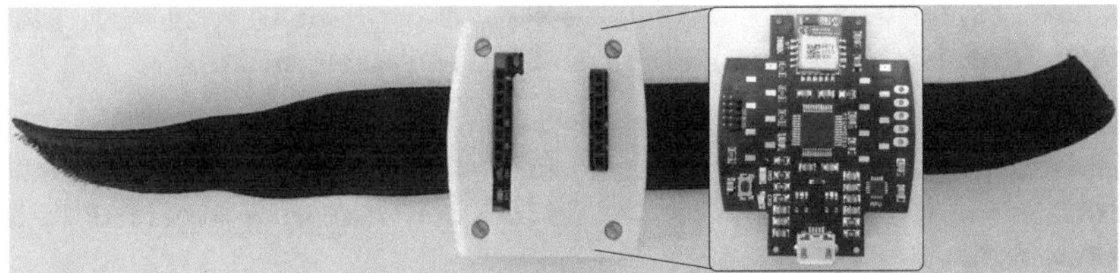

**Fig. 4.** Wearable device

## 3 Knowledge Model for Recognition of Training Exercises

In this section, knowledge model development for basketball exercise recognition is described in detail. Presented knowledge model is developed to autonomously understand the type of technical training that is being perform and records progress and gives feedback based on the perceived context. Such approach shows that given enough of empirical data, any desired context in basketball coaching can be perceived.

### 3.1 Model Development

The main steps in the development of knowledge model are (1) *data collection* to form learning dataset, (2) *preprocessing* of raw sensor data, (3) *segmentation* and *feature extraction*, (4) *modelling* and (5) *model verification*.

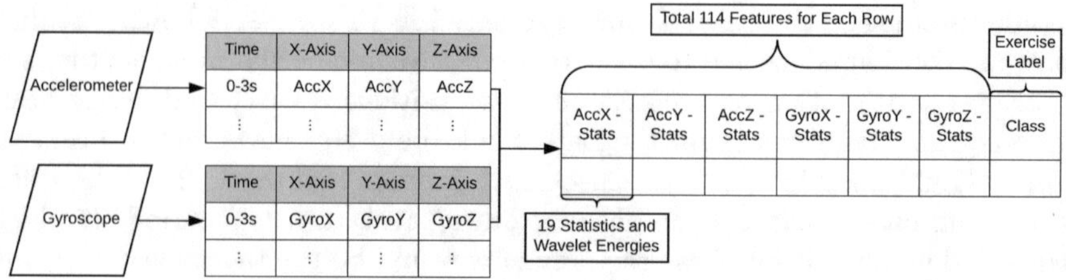


**Fig. 5.** Data segmentation and feature extraction process

**Dataset Collection.** Dataset used in this paper was formed using motion data collected during training exercises performed by 5 amateur basketball players. Single wearable sensor was worn by players on their dominant arm between wrist and elbow. Each player performed 6 different exercises (Table 1) in three different attempts. Each exercise was performed for 30 s. Acceleration and gyroscope signals were recorded and each exercise was labeled with appropriate exercise class.

**Data Preprocessing.** After forming learning dataset [12] the data is preprocessed. Acceleration and gyroscope signals are noisy and need to be filtered. To achieve replicability of models, sensor data is scaled to range $\pm 1$. Moving Average Filter is used to eliminate noise from sensor signals.

**Segmentation.** Raw sensor signals form timeseries segmented using *Sliding Window* method, to enable consistent feature extraction from raw data [16]. In his work 3s windows with 50% overlap factors were employed to extract features from raw data.

**Feature Extraction.** Human activity recognition systems with accelerometer and gyroscope sensors engage statistical and, frequency-domain features [13,14,16] to represent raw data. The feature extraction process contains the identification of description variables that include all relevant sensor information. In this work the abstraction of raw data has been done with *Statistical Calculations* and *Discrete Wavelet Transformation* [14,17]. Statistical features distinguish and represent raw data with high consistency. On the other hand the energies of 6 Daubechies (1 mother wavelet and 5 scaling functions) Discrete Wavelets are used as features [17,18]. In addition, *Hjorth Parameters* (mobility and complexity) are used [19]. The feature vector is formed for 3 s of each of sensor data and learning instance is formed using combination of feature vector and labeled exercise types. This process is depicted in Fig. 5.

**Modelling.** Training type recognition process includes defining a target set of exercises, collecting sensor data, and assigning sensor data to correct exercises. Most of activity recognition research focus on the use of statistical machine learning techniques [16] to get a grip on raw data of human activities. In this work, *Supervised Artificial Neural Networks* are used to classify different basketball exercises throughout the context levels.

The approach to the development of recognition model is done in cascaded fashion. Decisions are done for small number of classes and multiple machine learning classifiers are developed. There are 6 different types of exercises as it can be seen on Table 1 and all of them belong to technical training. Four of them are Ball-Handling exercises (One-Hand Ball-Handling) and 2 of them are scoring exercises (1 Open Shot and 1 Standing Layup). In recognition process, *Supervised Artificial Neural Networks (Multilayer Feed-Forward Neural Network or Multilayer Perceptron)* classification algorithm [20] runs for every context level. One ANN is trained to recognize either the exercise is ball handling or scoring (binary classifier). Scoring is further recognized as shooting or layup (binary classifier). Ball handling is further recognized as forward backward, left right, regular or two hands dribble (multi-class classifier).

**Table 1.** Exercise types

| Level 4 context | Level 5 & 6 context |
|---|---|
| Ball handling | Forward backward dribble |
| | Left - right dribble |
| | Regular dribble |
| | Two - hands dribble |
| Scoring | Shooting |
| | Layup |

**Model Verification.** Above described model is developed on relatively small dataset. The evaluation of such models is done by utilization of random selection when forming data partitions for learning and testing. Multiple repetition of the evaluation process using different randomly selected training and testing sets, as well as the averaging of the error estimation obtained, can provide satisfactory results. Cross-validation is based on this principle, with the corresponding substitution of the training set and the test set in each iteration. In this work results are validated using 10-fold cross-validation method [21].

### 3.2 Model Utilization in Real-World Training Scenario

In the context of basketball training player wears motion sensor on his/her dominant arm and performs exercises with the ball. Referring to the architecture from Fig. 1, data obtained from sensors is received by hub in real-time and data is preprocessed to form the same feature vector. Feature vector is supplied as the input to developed classification model. Model classifies the current training exercise and records the results in the long term metric accumulation block. Recorded metrics are compared against the current training plan supplied by the context block and feedback is supplied to the user. The training plan can simply consist of number of repetitions for each technical exercise with respect to predefined rules in regard to the age, physical characteristics, level of profession, prefered in-game position, etc. Training plan, in the context of developed AI basketball coach, serves as a set-point for accumulated metrics. Feedback is generated in the form of motivational suggestions.

## 4 Results

### 4.1 Features

Using domain knowledge and review of relevant machine learning and signal processing literature, a set of features is obtained to be used for model development [12]. Features are listed in Table 2. These features are extracted from raw data of all signals, total of 114 features create feature vector.

**Table 2.** Features list

| | | | | |
|---|---|---|---|---|
| Max | Skewness | Wavelet C energy | Wavelet D4 energy | Variance |
| Min | 25th percentile | Wavelet D1 energy | Wavelet D5 energy | Mean abs. dev. |
| RMS | 75th percentile | Wavelet D2 energy | Hjorth complexity | Mean cross. rate |
| Mean | Median | Wavelet D3 energy | Hjorth mobility | |

### 4.2 Exercise Recognition

Exercise recognition process is designed as pipeline of 3 classifiers. First classifier classifies the type of technical training as *Ball-Handling vs. Scoring.* Then, 2 different classifiers are used to differentiate sub-classes. Recognition using first classifier is done using only 2 neurons in one hidden layer of ANN. In Shooting and Layup exercise recognition, ANN of 24 neurons in one hidden layer is used. There are totally 132 instances, 67 Shooting and 65 Layup. Figure 6 shows the perfect classification capability of trained ANN in both *Scoring vs. Ball-Handling* and *Scoring.* In Ball-Handling type recognition process, *Forward-Backward Dribbling* and *Left-Right Dribbling* instances are confused due to the similarity between exercises. To recognize different *Ball-Handling* movements, neurons in hidden layer are set as 60 to reach satisfactory results. Total number of instances is 346, and *Forward-Backward, Left-Right, Regular and Two-Hands Dribbling* have 88, 88, 85, 85 observations in respectively.

The metrics that are used in validation process to evaluate performance of the model may change in different studies. However, the widely-used performance benchmarks in activity recognition research community are *Accuracy, Precision, Recall, F-score and Confusion Matrix* [16]. The average accuracy is 94% and worst performance is for *Left-Right Dribbling* which is 86%. The results are summarized in Table 3.

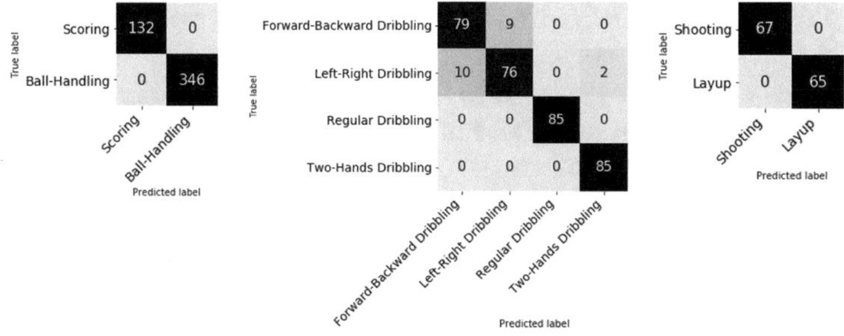


**Fig. 6.** Confusion matrices

**Table 3.** Metric results of ball-handling classification

| Classes/Metrics | Accuracy | Precision | Recall | F1-Score | Support |
|---|---|---|---|---|---|
| Forward-backward dribbling | 90% | 0.89 | 0.90 | 0.89 | 88 |
| Left-right dribbling | 86% | 0.89 | 0.86 | 0.88 | 88 |
| Regular dribbling | 100% | 1.00 | 1.00 | 1.00 | 85 |
| Two-hands dribbling | 100% | 0.98 | 1.00 | 0.99 | 85 |
| **Average/Total** | 94% | 0.94 | 0.94 | 0.94 | 346 |

## 5 Discussion

*t-Distributed Stochastic Neighbor Embedding (t-SNE)* [22] algorithm was used to understand data distribution of exercise type observations, which reduces dimensionality to 2D or 3D for better representation. Data distributions can be seen in Fig. 7. As it is seen, both *Ball-Handling vs. Scoring* and *Scoring* type data distributions can be distinguishable in 2D representation. However, *Forward-Backward* and *Left-Right Dribbling* observations are not easily distinguishable in 2D. Developed ANNs also show poor performance with these two exercise types. These results can be predictable considering the similarity between exercises, movements closely resemble each other.

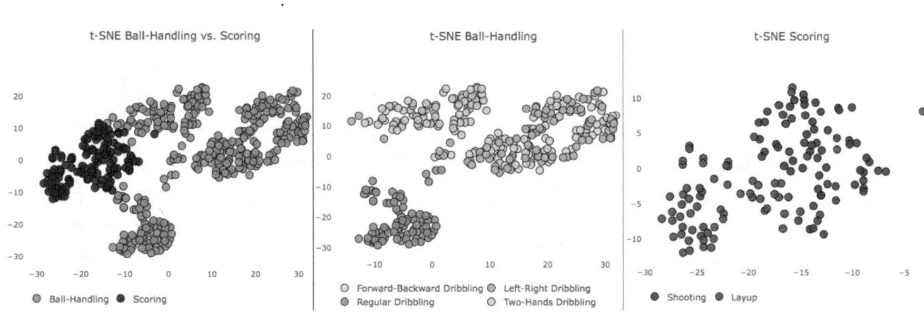


**Fig. 7.** t-SNE 2D representation of data distribution

The performance of trained ANNs is compared against the real labels using widely-used method of *Stratified K-fold Cross-Validation* [21] (Table 3). Average of validation metrics are calculated to understand general performance of model. Stratification ensures that every fold is a good representation of whole data. The verification process shows that the developed model is useful for the task in mind.

## 6 Conclusion

Training sessions in basketball are prepared, evaluated and monitored by professional coaches. Coaches are intensely involved in this process and have a huge influence on the performance quality of trainees. Coach's involvement in the training process, ultimately, decides the competition outcome. The convergence of technology for trainee monitoring and context processing with field experience opens up new horizons in basketball coaching. In this work artificially intelligent assistant is presented for basketball coaching. Once provided with certain amount of context developed assistant can monitor athletes, understand finer levels of supplied context (e.g. what exercise is being performed), evaluate the performance of athletes under certain context and provide necessary feedback. The AI assistant operates on the basis of knowledge models that can be developed for different training scenarios. The knowledge models allow embedding of the field experience into analytic, computational intelligence or artificial intelligence models. Presented knowledge model is developed to autonomously understand the type of technical exercise that is being performed and to record progress and give feedback based on the perceived context. Such approach shows that given enough of empirical data, any desired context in basketball coaching can be perceived. The ultimate goal of this research is to come up with fully autonomous assistant that is able to understand and act upon most general context of basketball training. The future work will consist of development of knowledge models for estimation of quality of performed exercises and further developments towards level 3 contextual training types.

## References

1. Gregg, D.G.: E-learning agents. Learn. Organ. **14**(4), 300–312 (2007)
2. Solana, J., Caceres, C., Garcia-Molina, A., Chausa, P., Opisso, E., Roig-Rovira, T., Gomez, E.J.: Intelligent therapy assistant (ITA) for cognitive rehabilitation in patients with acquired brain injury. BMC Med. Inf. Decis. Making **14**(1), 58 (2014)
3. Dragone, M., Saunders, J., Dautenhahn, K.: On the integration of adaptive and interactive robotic smart spaces. Paladyn J. Behav. Rob. **6**(1), 165–179 (2015)
4. Beetz, M., Jain, D., Mosenlechner, L., Tenorth, M., Kunze, L., Blodow, N., Pangercic, D.: Cognition-enabled autonomous robot control for the realization of home chore task intelligence. Proc. IEEE **100**(8), 2454–2471 (2012)
5. Biundo, S., Bercher, P., Geier, T., Müller, F., Schattenberg, B.: Advanced user assistance based on AI planning. Cogn. Syst. Res. **12**(3–4), 219–236 (2011)

6. Bercher, P., Richter, F., Hornle, T., Geier, T., Holler, D., Behnke, G., Nothdurft, F., Honold, F., Minker, W., Weber, M., Biundo, S.: A planning-based assistance system for setting up a home theater. In: Proceedings of the 29th National Conference on Artificial Intelligence (AAAI), pp. 4264–4265 (2015)
7. Vales-Alonso, J., Chaves-Dieguez, D., Lepez-Matencio, P., Alcaraz, J.J., Parrado-García, F.J., Gonzalez-Castano, F.J.: SAETA: a smart coaching assistant for professional volleyball training. IEEE Trans. Syst. Man Cybern.: Syst. **45**(8), 1138–1150 (2015)
8. Klein, M.C., Manzoor, A., Middelweerd, A., Mollee, J.S., te Velde, S.J.: Encouraging physical activity via a personalized mobile system. IEEE Internet Comput. **19**(4), 20–27 (2015)
9. Dijkhuis, T.B., Blaauw, F.J., van Ittersum, M.W., Velthuijsen, H., Aiello, M.: Personalized physical activity coaching: a machine learning approach. Sensors **18**(2), 623 (2018)
10. Bacic, B., Hume, P.: Computational intelligence for qualitative coaching diagnostics: automated assessment of tennis swings to improve performance and safety. arXiv preprint arXiv:1711.09562 (2017)
11. Ghasemzadeh, H., Loseu, V., Jafari, R.: Wearable coach for sport training: a quantitative model to evaluate wrist-rotation in golf. J. Ambient Intell. Smart Environ. **1**(2), 173–184 (2009)
12. Acikmese, Y., Ustundag, B.C., Golubovic, E.: Towards an artificial training expert system for basketball. In: 10th International IEEE Conference on Electrical and Electronics Engineering (ELECO), pp. 1300–1304 (2017)
13. Yang, C.C., Hsu, Y.L.: A review of accelerometry-based wearable motion detectors for physical activity monitoring. Sensors **10**(8), 7772–7788 (2010)
14. Ren, X., Ding, W., Crouter, S.E., Mu, Y., Xie, R.: Activity recognition and intensity estimation in youth from accelerometer data aided by machine learning. Appl. Intell. **45**(2), 512–529 (2016)
15. Wearable Sensor v2, Inovatink. https://github.com/inovatink/ws-hardware
16. Lara, O.D., Labrador, M.A.: A survey on human activity recognition using wearable sensors. IEEE Commun. Surv. Tutorials **15**(3), 1192–1209 (2013)
17. Rioul, O., Vetterli, M.: Wavelets and signal processing. IEEE Sig. Process. Mag. **8**(4), 14–38 (1991)
18. Guo, L., Rivero, D., Seoane, J.A., Pazos, A.: Classification of EEG signals using relative wavelet energy and artificial neural networks. In: Proceedings of the First ACM/SIGEVO Summit on Genetic and Evolutionary Computation (ACM), pp. 177–184 (2009)
19. Khan, M., Ahamed, S.I., Rahman, M., Smith, R.O.: A feature extraction method for realtime human activity recognition on cell phones. In: Proceedings of 3rd International Symposium on Quality of Life Technology (isQoLT 2011) (2011)
20. Russell, S.J., Norvig, P.: Artificial Intelligence: A Modern Approach. Pearson Education Limited, Malaysia (2016)
21. Kohavi, R.: A study of cross validation and bootstrap for accuracy estimation and model selection. In: International Joint Conference on Artificial Intelligence (IJCAI) (1995)
22. Maaten, L.V.D., Hinton, G.: Visualizing data using t-SNE. J. Mach. Learn. Res. **9**, 2579–2605 (2008)

# Subjective and Objective QoE Measurement for H.265/HEVC Video Streaming over LTE

Jasmina Baraković Husić[1(✉)], Sabina Baraković[2], and Irma Osmanović[3]

[1] Faculty of Electrical Engineering, University of Sarajevo, Sarajevo, Bosnia and Herzegovina
jasmina.barakovic@etf.unsa.ba

[2] American University in Bosnia and Herzegovina, Sarajevo, Bosnia and Herzegovina

[3] Systech, d.o.o. Sarajevo, Sarajevo, Bosnia and Herzegovina

**Abstract.** Mobile video traffic is the fastest growing segment of mobile data driven by proliferation of devices with larger screens and higher resolution, as well as increased network performance achieved through long term evolution (LTE) deployments. Emerging video formats and applications, such as H.265/high efficiency video coding (HEVC) will increase the video traffic consumption while improving the user quality of experience (QoE). Although QoE is affected by many influence factors (IFs), this paper focuses on media-related system IFs and their impact on subjective and objective QoE metrics for H.265/HEVC video streaming. The aim is to examine the impact of media-related system IFs on QoE for video streaming and compare the results of subjective and objective QoE measurements. Results obtained from experimental study are used to analyze the relationship between subjective and objective QoE metrics for H.265/HEVC video streaming.

## 1 Introduction

The potential advantages of the fifth generation (5G) networks are creating high expectations in the market, such as better speed and coverage, better battery life, high network reliability, and guaranteed quality [1]. Making these expectations for 5G reality will have significant implications for video traffic consumption. Mobile video traffic is predicted to grow by around 35% annually through 2024 to account for 74% of all mobile data traffic [2]. An emerging trend of increasing video usage is the main cause of the growth in mobile data traffic. Both streaming and sharing video is consumed anywhere and anytime. This trend is forecasted to continue since video is a part of most online content. Furthermore, evolving devices with larger screens and higher resolutions, as well as increased network performance achieved through long term evolution (LTE) deployments, seem to be important drivers for video traffic growth [3]. Additionally, emerging video formats and applications, such as streaming high-quality video and augmented/virtual reality, will increase the video traffic consumption while enhancing the user quality of experience (QoE) [4].


S. Avdaković et al. (Eds.): IAT 2019, LNNS 83, pp. 428–441, 2020.
https://doi.org/10.1007/978-3-030-24986-1_34

In order to manage QoE with success, one has to identify which characteristics affect the perceptual dimensions of QoE and QoE for specific service and understand how. This means it is necessary to understand influence factors (IFs) and multiple dimensions of human quality perception, and how they affect QoE for a given service [5].

Currently adopted classification groups IFs into three categories [6]: human, context, and system IFs. Many research studies addressing different IFs that affect and describe the user QoE focus on system IFs. They are related to properties and characteristics that determine the technically produced quality of an application or service. System IFs can be further divided into four sub-groups, i.e., content-, media-, network-, and device-related system IFs. Most analyzed of all system IFs are content- and media-related system IFs. Therefore, this paper focuses on media-related system IFs whose optimization enables the same level of quality with significant savings in network infrastructure.

Many research studies have considered media-related system IFs and their impact on the perceived quality as commonly analyzed perceptual dimension of QoE. Perceived quality of a given service can be measured using subjective and objective quality assessment methods. Subjective assessment methods require human observers to evaluate the video quality. The scores are then analyzed to determine the objective indicators for video quality. Although subjective evaluation is the most reliable way of quality assessment, it is expensive and usually too slow method to be useful in real-world applications. So objective video quality assessment is used to measure the impact of technical factors on the perceived quality of the given service [7].

This paper aims to investigate the impact of media-related system IFs on QoE, and to compare the results of subjective and objective QoE measurements for H.265/high efficiency video coding (HEVC) video streaming over LTE. Primary reason for that aim is to increase the awareness and understanding of this matter and indicate how interested stakeholders may use such findings in theory and practice.

The rest of the paper is organized as follows. Section 2 gives brief information about the most frequently analyzed media-related system IFs and their impact on subjective and objective quality metrics. Section 3 describes the conducted experimental study, which includes subjective and objective measurements of QoE for H.265/HEVC video streaming over LTE. The comparison of subjective and objective video quality assessment results is presented in Sect. 4, while conclusions are given in Sect. 5.

## 2 Related Works

This section gives a brief overview of recent research activities directed to media-related system IFs and their impact on subjective and objective quality metrics of QoE for video streaming.

Media-related system IFs refer to media configuration parameters. The importance of media-related system IFs and their impact on subjective and objective quality metrics has been emphasized by several research studies. Based on literature review summarized in Table 1, the following media-related system IFs have been identified as the most frequently analyzed in research studies: frame rate [9–15, 17–20, 22, 24–27], bitrate [8–12, 14–18, 20, 21, 23], resolution [8, 13, 14, 16–20, 22, 24–26], codec type [14, 16, 19–21, 23], sampling rate [12, 23, 27], and media synchronization [13].

*Frame rate* refers to the number of frames displayed by a viewing device in one second. Decreasing the frame rate decreases the perceived quality [12, 17, 19, 22, 23]. On the other side, increasing the frame rate increases the perceived quality only to certain degree.

*Bitrate* is defined as number of transmitted or processed bits per unit of time. Higher bitrate values indicates higher degree of perceived quality [13, 22], but also requires higher resolution and more network resources which can be problem when bandwidth is limited [12, 28].

*Resolution* represents the number of particular pixels in each dimension that can be displayed and perceived by the end user. Higher resolution usually means higher perceived quality [9, 17–19, 23].

*Codec type* affects the perceived quality. A variety of video coding standards have been analyzed in this context, such as H.264/advanced video coding (AVC) or H.265/HEVC. Although H.265/HEVC is the most effective video compression standard [29], it has received not nearly as much attention as its predecessor H.264/AVC in terms of perceived quality.

*Sampling rate* affects the perceived quality in a way that larger set of samples means better quality. However, each sample requires resources and processing time, so it is necessary to pay attention on the price-quality ratio to find their optimal relationship [23, 28, 30].

*Media synchronization* refers to the synchronization of lips and sound in video with audio content [31] or other human senses like smell, touch, etc. [32]. The perceived quality of multiple media is the same as their product, which indicates the significance of media synchronization.

Analysis of aforementioned media-related system IFs has been conducted based on 20 papers that consider their impact on perceived quality. It can be noticed that perceived quality has been analyzed in terms of frame rate (75%), bitrate (65%), resolution (60%), codec type (35%), sampling rate (15%), and media synchronization (5%). (Note: distribution exceeds 100% as multiple areas may have been studied in a single study.) Most of analyzed papers have neglected to consider mutual impact of media-related system IFs on perceived video quality due to complexity that such study incurs. These papers mainly analyzed the individual impact of media-related system IFs on perceived video quality. This motivated us to focus on investigating the mutual impact of most frequently analyzed media-related system IFs on QoE.

In order to measure the perceived quality of video streaming service, subjective and objective methods for video quality assessment can be used. Subjective quality assessment is the basic method of evaluating perceived quality in term of mean opinion score (MOS). However, it is sometimes more convenient to estimate perceived quality based on objective methods and associated quality estimation models, such as peak signal-to-noise ratio (PSNR), structural similarity (SSIM), and video quality metric (VQM). Subjective quality assessment requires more resources and effort, because it includes human viewers. On the other side, objective quality assessment using appropriate quality estimation models is much cheaper, but its precision depends on the accuracy of these models.

**Table 1.** Meta-analytical review of media-related SIFs and video quality assessment.

| | | References | N | P |
|---|---|---|---|---|
| Media-related SIFs | Frame rate | 9, 10, 11, 12, 13, 14, 15, 17, 19, 20, 22, 24, 25, 26, 27 | 15 | 75% |
| | Bitrate | 8, 9, 10, 11, 12, 14, 15, 16, 17, 18, 20, 21, 23 | 13 | 70% |
| | Resolution | 8, 13, 14, 16, 17, 18, 19, 20, 22, 24, 25, 26 | 12 | 65% |
| | Codec | 14, 16, 18, 19, 20, 21, 23 | 7 | 35% |
| | Sampling rate | 12, 24, 27 | 3 | 15% |
| | Media synchronization | 13 | 1 | 5% |
| Quality assessment | Subjective | 11, 12, 13, 15, 16, 17, 18, 24, 27 | 9 | 45% |
| | Objective | 8, 14, 21, 22, 23, 25, 26 | 7 | 35% |
| | Both | 9, 10, 19, 20 | 4 | 20% |
| Quality measures | MOS | 11, 12, 15, 16, 17, 18, 24, 27 | 8 | 40% |
| | PSNR | 8, 9, 10, 14, 19, 20, 21, 22, 23, 26 | 10 | 50% |
| | SSIM | 14, 21, 22, 25, 26 | 5 | 25% |
| | VQM | 21, 25, 26 | 3 | 15% |

Legend: MOS (Mean Opinion Score); N (Number); SIF (System Influence Factors); SSIM (Structural SIMilarity); P (Percentage); PSNR (Peak Signal to Noise Ratio); VQM (Video Quality Metric).

That is way most of research studies consider perceived video quality using quality assessment methods in the following ratio: subjective (45%), objective (35%), and both of them (20%). In addition, this serves us as a motivation to investigate the relationship between subjective and objective measures of perceived quality for H.265/HEVC video streaming.

## 3 Experimental Study

After summarizing the related work in Sect. 2, the aim of the paper can be defined as follows: (1) examine the impact of media-related SIFs on the subjective and objective QoE metrics for H.265/HEVC video streaming; (2) compare the results of subjective and objective QoE measurements.

Table 3 shows that the perceived video quality has been commonly analyzed in terms of bitrate, resolution, and codec. Accordingly, these media-related SIFs have been chosen to be manipulated in order to investigate their impact on subjective and objective QoE measurements. Table 2 shows that change of ffmpeg-specific parameters [33], i.e., resolution, coding tree unit (CTU), and constant rate factor (CRF), causes a change of selected media-related SIFs. In order to obtain the data necessary to investigate their impact on subjective and objective QoE measures, we have conducted experimental study.

**Phase I** included three steps: (1) preparation of reference and test video sequences, (2) configuration of LTE emulation environment, and (3) reconstruction of test video sequences.

Preparation of reference and test video sequences was based on San Francisco Cable Car Stock video clip. Original video clip was characterized by frame rate of 30 fps, resolution of 1920 × 1080 pixels, and duration of 39 s. Reference video sequence was prepared after removing audio content from original video clip that was shortened to 20 s. Test video sequences were created by the change of ffmpeg-specific parameters, i.e., resolution (858 × 480, 1280 × 720, 1280 × 960), CTU (16, 32, 64), and CRF (18, 28, 38). Table 2 contains definition of ffmpeg-specific parameters and their relation to media-related SIFs. 27 video sequences were prepared based on the variation of ffmpeg-specific parameters.

Configuration of LTE emulation environment was performed in LTE/EPC network simulator (LENA) using EvalVid framework [34]. LTE emulation environment is shown in Fig. 1 and contains user equipment (UE), eNodeB in LTE radio access network, packet data network – gateway (P-GW) in evolved packet core (EPC) network, and video server. Configuration parameters of connections between UE and eNodeB, and between P-GW and video server are summarized in Table 3. These parameters correspond to the macro base station [35], while carrier frequencies correspond to the European bandwidth [36].

Reconstruction of test video sequences was done after they were transmitted over LTE emulated network environment using etmp4 tool from EvalVid framework.

**Table 2.** Relationship between ffmpeg-parameters and selected media-related SIFs [4].

| ffmpeg parameter | Definition of ffmpeg parameter | Chosen ffmeg parameter values | Selected media-related SIFs |
|---|---|---|---|
| CTU | Maximal size of coding unit | 16, 32, 64 | Coding |
| CRF | Define the level of video quality by setting bitrate, compression and sampling rate to the corresponding values depending on the video and the defined CRF value | 18, 28, 38 | Bitrate, compression, sampling rate |
| WxH resolution | Video size in two dimensions | 858 × 480, 1280 × 720, 1280 × 960 | Resolution |

Legend: CRF (Constant Rate Factor); CTU (Coding Three Unit); SIFs (System Influence Factors).

**Table 3.** LTE network configuration parameters [4].

| | Parameter | Values |
|---|---|---|
| Radio link | Carrier frequency | Downlink 2110 MHz<br>Uplink 1920 MHz |
| | Base station power | 46 dBm (4 W) |
| | Resource blocks | 100 RB (20 MHz) |
| | User moving speed | 3 km/h |
| p2p link | Bandwidth | 80 Mbps |
| | MTU | 1500 B |
| | Delay | 0.01 s |
| | Distance base station-user | 100 m |

Legend: MTU (Maximal Transfer Unit); p2p (point-to-point).

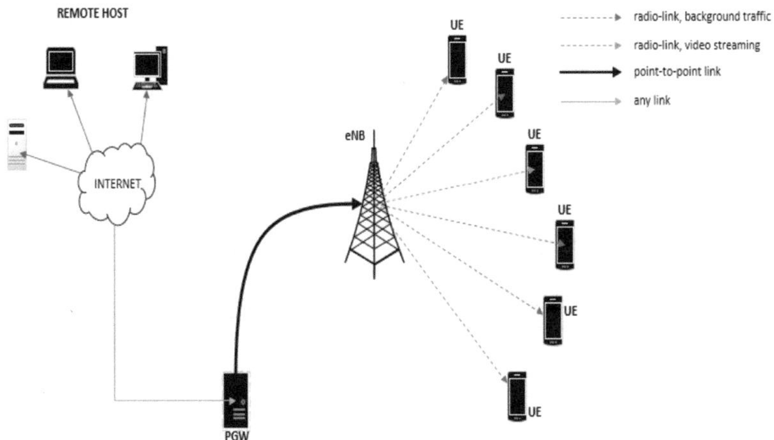


**Fig. 1.** LTE emulation environment [4].

**Phase II** involved the measurement of objective and subjective QoE metrics, i.e., PSNR, SSIM, and MOS.

Objective QoE measures were collected using full reference (FR) models as described in the recommendation ITU-R B.500-11. PSNR and SSIM were calculated with *psnr* tool using *yuv* files of the original and reconstructed videos.

Subjective QoE measures were performed using constant presentation and passive modality as described in the recommendation ITU-T G.1011. Test video sequences were displayed in no-reference (NR) mode, where only reconstructed video sequences were shown to viewers.

50 viewers took part in the experimental study. Experiments were carried out at home (36%), work (6%), or café (58%). Collected demographic data related to age, gender, educational level, and prior experience describe the group that approached the questioning, as shown in Fig. 2.

All viewers watched 27 video sequences using the Samsung Galaxy S4 mobile phone with 1920 × 1080 screen size. Then they were asked to express their opinion regarding perceived video quality.

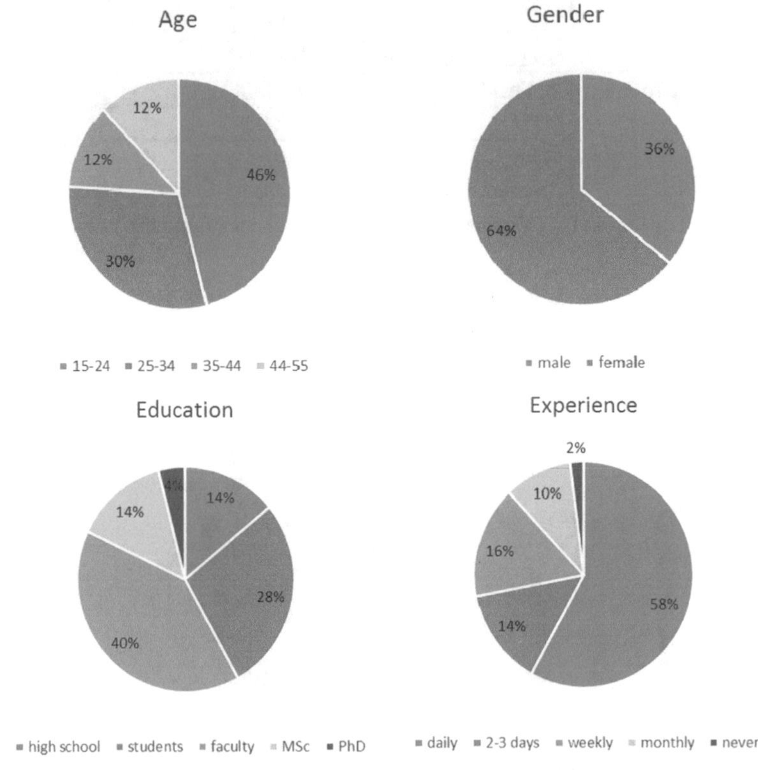


**Fig. 2.** Demographic data: (a) Age; (b) Gender; (c) Education; (d) Experience.

Electronic evaluation questionnaire was used to collect subjective QoE measures. It contained three parts: (1) the part that was completed at the beginning of the experiment, (2) part that covered information related to viewer's personal data and prior experience with video streaming service, and (3) part that deals with the viewer's ratings of the statement related to overall QoE when using video streaming service.

The experiment procedure lasted about 30 min and included the following steps [37]: (1) introduction and clarification of the experiment tasks that need to be performed by the examinee (5 min), (2) examinee training (5 min), and (3) testing and rating of video sequences (20 min). Video sequences have been displayed in the landscape mode and in the same order to all examinees. All examinees were asked not to think about their feelings during evaluation, but to be intuitive.

## 4 Results and Discussion

This section discusses the results of measurements of objective and subjective QoE metrics for H.265/HEVC video streaming over LTE. PSNR and SSIM are used as objective QoE metrics that mainly stem from the image quality domain. MOS is used as subjective QoE metric that is determined from the average of the individual ratings of video content.

Firstly, objective video quality assessment was performed to simulate the opinions of humans using computational models (phase I). Then, subjective quality assessment was conducted to calibrate objective quality assessment (phase II). Correlation between objective and subjective quality assessment was used to determine the accuracy of objective quality assessment.

Video quality assessment was performed to show dependency of objective QoE metrics on the media-related system IFs in the context of H.265/HEVC video streaming. The results are organized in figures showing the dependency of PSNR and SSIM on ffmpeg-specific parameters (i.e., resolution, CTU, CRF). The PSNR values are in range [25.50, 38.63] dB, while the SSIM values are in the interval [0.79, 0.96].

Figure 3 presents the dependency of objective and subjective QoE metrics (i.e., PSNR, SSIM, and MOS) on CTU and CRF for the corresponding resolutions. Results show that the increase of resolution causes the decrease of PSNR and MOS, and the increase of SSIM. This result is expected since higher resolutions require more network resources that has been kept constant during emulation. In addition, PSNR and MOS decreases, while SSIM increases when ffmpeg-specific parameters (i.e., CTU and CRF) increase since macroblocks are higher in this case and contains less data for recovering transmitted video. However, in order to draw non-misleading conclusions, especially for MOS, it must be taken into the account that it depends not only on media-related system IFs, but also human and context IFs [38].

Table 4 show the correlation coefficients between ffmpeg-specific parameters (i.e., resolution, CTU, CRF) and PSNR. Correlation coefficients indicates that there are weak negative correlation between PSNR and two ffmpeg-specific parameters (i.e., resolution and CTU), whereas there exist strong negative correlation between PSNR and CRF.

Table 5 shows the correlation coefficients between ffmpeg-specific parameters and SSIM. Correlation analysis indicates that there are weak positive relationship between all ffmpeg-specific parameters and SSIM.

Table 6 shows the correlation coefficients between ffmpeg-specific parameters and MOS. There exist weak negative relationship between MOS and all ffmpeg-specific parameters.

In addition, the correlation analysis was conducted in order to find the relationship between objective and subjective QoE metrics. The coefficient of correlation was calculated to give the true direction of the correlation between the objective and subjective QoE metrics. The obtained value of 0.4718 indicates a moderate positive linear relationship between PSNR and MOS. Similarly, the value of 0.4561 shows a moderate positive linear relationship between SSIM and MOS.

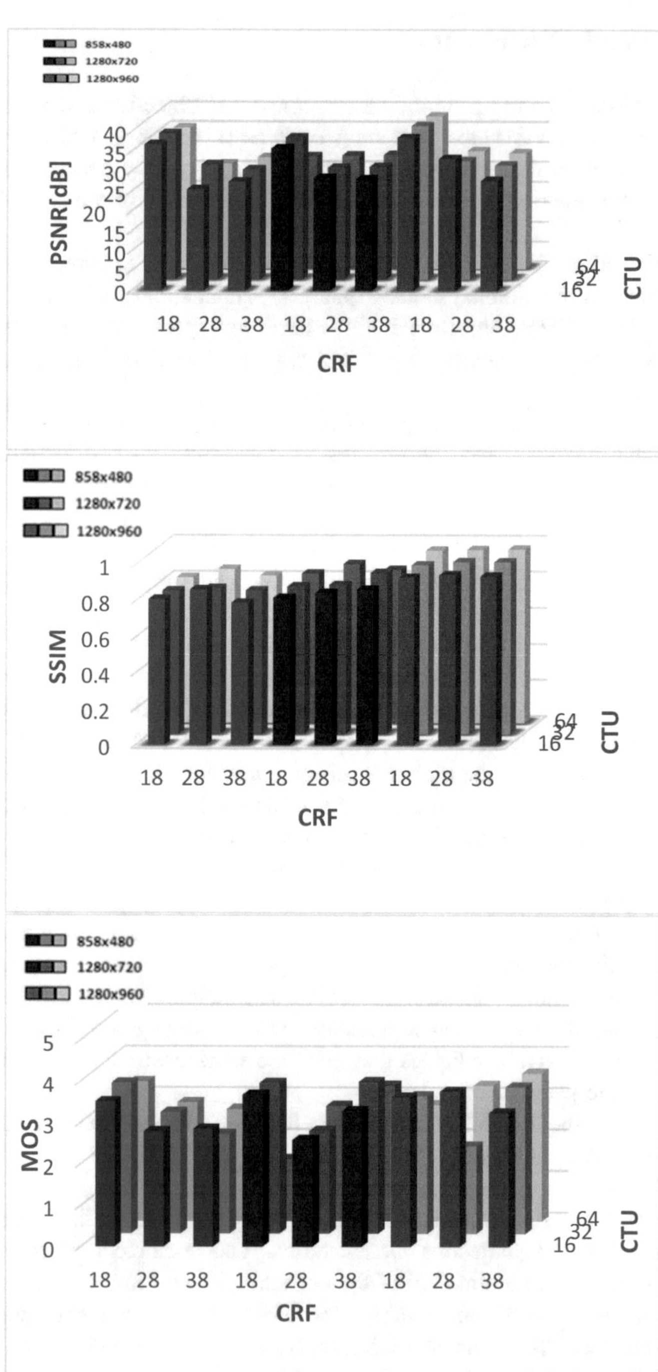


**Fig. 3.** Dependency of objective QoE metrics on ffmpeg-specific parameters: (a) PSNR; (b) SSIM; (c) MOS.

**Table 4.** Pearson correlation coefficients between ffmpeg parameters and PSNR.

| | PSNR | CTU | CRF | Resolution |
|---|---|---|---|---|
| PSNR | 1 | | | |
| CTU | −0.12036 | 1 | | |
| CRF | −0.76745 | 0 | 1 | |
| Resolution | −0.10713 | 0.98198 | 0 | 1 |

Legend: CRF (Constant Rate Factor); CTU (Coding Three Unit); PSNR (Peak to Noise Signal Ratio).

**Table 5.** Pearson correlation coefficients between ffmpeg parameters and SSIM.

| | SSIM | CTU | CRF | Resolution |
|---|---|---|---|---|
| SSIM | 1 | | | |
| CTU | 0.10536 | 1 | | |
| CRF | 0.12414 | 0 | 1 | |
| Resolution | 0.09545 | 0.98198 | 0 | 1 |

Legend: CRF (Constant Rate Factor); CTU (Coding Three Unit); SSIM (Structural SIMilarity).

**Table 6.** Pearson correlation coefficients between ffmpeg parameters and MOS.

| | MOS | CTU | CRF | Resolution |
|---|---|---|---|---|
| MOS | 1 | | | |
| CTU | −0.26697 | 1 | | |
| CRF | −0.04412 | 0 | 1 | |
| Resolution | −0.26980 | 0.98198 | 0 | 1 |

Legend: CRF (Constant Rate Factor); CTU (Coding Three Unit); MOS (Mean Opinion Score).

Figure 4 validates the correlation between subjective MOS and objective PSNR and SSIM values. This means that objective video quality assessment can provide similar results as an alternatively time consuming subjective quality assessment. For example, if objective PSNR values are higher than 33 dB and SSIM values are higher than 0.9, then subjective MOS value is higher than 3.5 that is acceptable from end-users perspective [38].

However, relationship between subjective and objective QoE metrics cannot be described as linear since there are many IFs that affect video quality assessment. Although objective video quality assessment has many advantages, it is often criticized due its weak or moderate correlation with subjective perceived quality since it does not take into account many human and context IFs, as well as other system IFs. QoE is multidimensional concept, so video quality assessment should be based on consideration of as many as possible different IFs in order to avoid misleading conclusions.

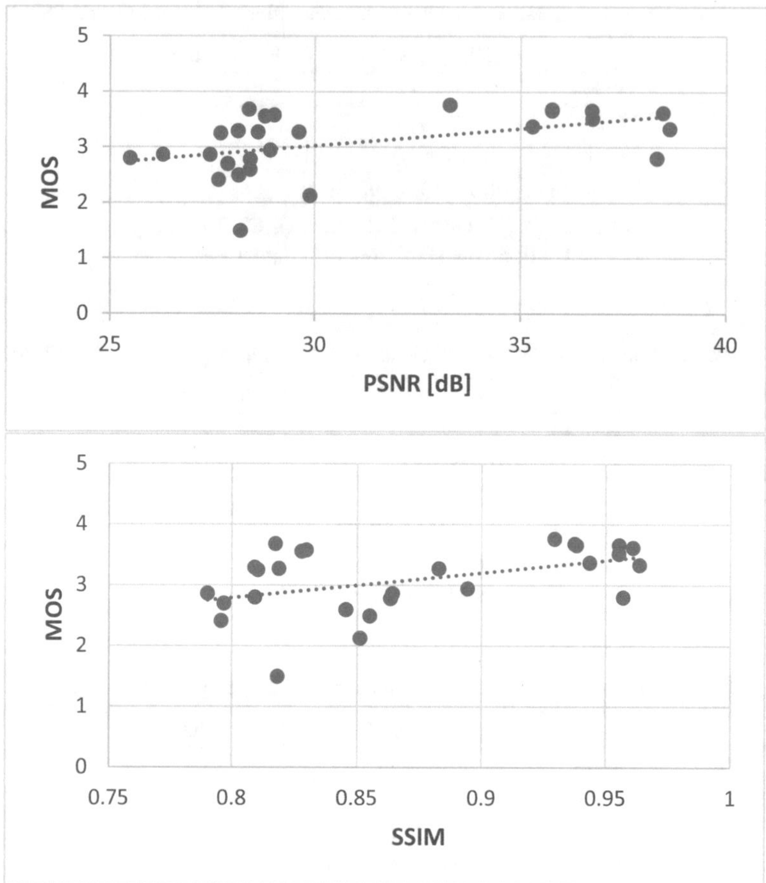


**Fig. 4.** Relationship between objective and subjective QoE metrics: (a) PSNR vs. MOS; (b) SSIM vs. MOS.

## 5 Conclusion

Media-related system IFs and their mutual impact on QoE for video streaming have been investigated to a limited extent. Consequently, the aim of this paper was to examine the impact of selected media-related system IFs on objective and subjective QoE metrics for video streaming and to compare the results in order to find out the relation between them.

In order to reach this aim, we have performed a brief meta-analysis of related works in domain of media-related system IFs and their influence on subjective and objective QoE measures for video streaming. Additionally, we have conducted experimental study in order to obtain data necessary for the analysis of the influence of media-related system IFs on QoE for a given service. After that, the collected data are analyzed in order to determine the relationship between subjective and objective QoE metrics for video streaming.

Results obtained from this experimental study provide several benefits to the interested stakeholders. Understanding the media-related SIFs and their impact on subjective and objective QoE metrics can increase the awareness and understanding of this matter and improve QoE for video streaming in general. Service providers can identify the expected range of QoE based on these media-related SIFs. Furthermore, the developers of media platforms for video streaming may exploit the obtained results to adapt media clients and maximize the perceived quality. Additionally, the QoE research community may use these results to analyze perceptual effects of media-related system IFs in the context of video streaming.

However, performed experimental study has certain limitations, which need to be overcome in the future research activities. Since QoE is a multidimensional concept, one have to include context and human IFs in addition to system IFs when addressing QoE for video streaming in order to avoid misleading conclusions. Furthermore, the QoE for video streaming is affected by many other system IFs in addition to three specific we have investigated that need to be addressed in future works. Also, more participants need to be included in the subjective QoE evaluation in order not to have deceptive conclusions.

## References


1. Golic, D., Barakovic Husic, J., Barakovic, S.: Impact of mobile ad-hoc network parameters on quality of experience for video streaming. In: Proceedings of the 17th International Symposium INFOTEH-JAHORINA 2018, Bosnia and Herzegovina (2018)
2. Cerwall, P., et al.: Ericsson Mobility Report (2018)
3. Begluk, T., Barakovic Husic, J., Barakovic, S.: Machine-learning-based QoE prediction for video streaming over LTE network. In: Proceedings of the 17th International Symposium INFOTEH-JAHORINA 2018, Bosnia and Herzegovina (2018)
4. Osmanovic, I., Barakovic Husic, J., Barakovic, S.: Impact of media-related SIFs on QoE for H.265/HEVC video streaming. J. Commun. Softw. Syst. **14**(2), 157–170 (2018)
5. Baraković, S., Skorin-Kapov, L.: Survey and challenges of QoE management issues in wireless networks. J. Comput. Netw. Commun., Article ID 165146 (2013)
6. Le Callet, P., Möller, S., Perkis, A.: Qualinet White Paper on Definitions of Quality of Experience. European Network on Quality of Experience in Multimedia Systems and Services (COST Action IC 1003) Version 1.2, Lausanne, Switzerland (2013)
7. Opozda, S., Sochan, A.: The survey of subjective and objective methods for quality assessment of 2D and 3D images. Theor. Appl. Inform. **26**, 39–67 (2014)
8. Nightingale, J., Wang, Q., Grecos, C., Goma, S.: The impact of network impairment on quality of experience (QoE) in H.265/HEVC video streaming. IEEE Trans. Consum. Electron. **60**(2), 242–250 (2014)
9. Aqil, A., Atya, A.O.F., Krishnamurthy, S.V., Papageorgiou, G.: Streaming lower quality video over LTE: how much energy can you save? In: Proceedings of the IEEE 23rd International Conference on Network Protocols (ICNP), San Francisco, CA, USA, 10–13 November 2015
10. Khan, A., Sun, L., Jammeh, E., Ifeachor, E.: Quality of experience-driven adaptation scheme for video applications over wireless networks. IET Commun. **4**(11), 1337–1347 (2010)

11. Khan, A., Sun, L., Ifeachor, E., Fajardo, J.O., Liberal, F., Koumaras, H.: Video quality prediction models based on video content dynamics for H. 264 video over UMTS networks. Int. J. Digit. Multimed. Broadcast. **2010**, 17 (2010)
12. Vakili, A., Grégorie, J.C.: QoE management for video conferencing applications. Comput. Netw. **57**, 1726–1738 (2013)
13. Alberti, C., et al.: Automated QoE evaluation of dynamic adaptive streaming over HTTP. In: Proceedings of the 5th International Workshop on Quality of Multimedia Experience (QoMEX), Klagenfurt am Wörthersee, pp. 58–63, Austria, 3–5 July 2013
14. Hammerschmidt, D., Wöllner, C.: The influence of image compression rate. In: Proceedings of International Conference on the Multimodal Experience of Music (ICMEM), Sheffield, UK, 23–25 March 2015
15. Vranješ, D., Žagar, D., Nemčić, O.: Comparison of objective quality assessment methods for scalable video coding. In: Proceedings of the 54th International Symposium ELMAR-2012, pp. 19–22, Zadar, Croatia, 12–14 September 2012
16. Silvar, I., Sužnjević, M., Skorin-Kapov, L.: The impact of video encoding parameters and game type on QoE for cloud gaming: a case study using the steam platform. In: Proceedings of the 7th International Workshop on Quality of Multimedia Experience (QoMEX), 2015, pp. 1–6, Pylos-Nestoras, Greece, 26–29 May 2015
17. Joskowicz, J., Ardao, J.: A general parametric model for perceptual video quality estimation. In: Proceedings of the International Workshop Technical Committee on Communications Quality and Reliability (CQR), 2010, pp. 1–6, Vancouver, BC, Canada, 8–10 June 2010
18. Joskowicz, J., Ardao, J.: Combining the effects of frame rate, bit rate, display size and video content in a parametric video quality model. In: Proceedings of the 6th Latin America Networking Conference, 2011, pp. 4–11, Quito, Ecuador, 12–13 October 2011
19. Joskowicz, J., Ardao, J.C.L., Sotelo, R.: Quantitative modeling of the impact of video content in the ITU-T G. 1070 video quality estimation function. Informática Na Educ.: Teor. Prática **14**(2) (2011)
20. Ohm, J., Sullivan, G.J., Schwarz, H., Tan, T.K., Weigand, T.: Comparison of the coding efficiency of video coding standards—including high efficiency video coding (HEVC). IEEE Trans. Circuits Syst. Video Technol. **22**, 1669–1684 (2012)
21. Řeřábek, M., Hanhart, P., Korshunov, P., Ebrahimi, T.: Quality evaluation of HEVC and VP9 video compression in real-time applications. In: Proceedings of the 7th International Workshop on Quality of Multimedia Experience (QoMEX), 2015, Pylos-Nestoras, Greece, pp. 1–6, 26–29 May 2015
22. Uhrina, M., Frnda, J., Sevcik, L., Vaculik, M.: Impact of H. 264/AVC and H. 265/HEVC compression standards on the video quality for 4K resolution. Adv. Electr. Electron. Eng. **12**, 368–376 (2014)
23. McDonagh, P., Vallati, C., Pande, A., Mohapatra, P., Perry, P., Mingozzi, E.: Quality-oriented scalable video delivery using H. 264 SVC on an LTE network. In: Proceedings of the 14th International Symposium on Wireless Personal Multimedia Communications (WPMC), 2011, Brest, France, 3–7 October 2011
24. Seeling, P., Reisslein, M.: Video traffic characteristics of modern encoding standards: H. 264/AVC with SVC and MVC extensions and H. 265/HEVC. Sci. World J. **2014**, 16 (2014)
25. Nasiri, R.M., Wang, J., Rehman, A., Wang, S., Wang, Z.: Perceptual quality assessment of high frame rate video. In: Proceedings of the 17th International Workshop on Multimedia Signal Processing (MMSP), 2015, Xiamen, China, pp. 1–6, 19–21 October 2015
26. Zinner, T., Hohlfeld, O., Abboud, O., Hossfeld, T.: Impact of frame rate and resolution on objective QoE metrics. In: Proceedings of the 2nd International Workshop on Quality of Multimedia Experience (QoMEX), 2010, Trondheim, Norway, pp. 29–34, 21–23 July 2010

27. Songi, W., Tjondronegoro, W.D.: Acceptability-based QoE models for mobile video. IEEE Trans. Multimedia **16**, 738–750 (2014)
28. Al-Nuaim, H., Nouf, A.: A user perceived quality assessment of lossy compressed images. Int. J. Comput. Graph. **2**, 23–36 (2011)
29. Grois, D., Marpe, D., Mulayoff, A., Itzhaky, B., Hadar, O.: Performance comparison of h. 265/mpeg-hevc, vp9, and h. 264/mpeg-avc encoders. In: Proceedings of Picture Coding Symposium (PCS), San Jose, CA, USA (2013)
30. Yang, W., et al.: Perceptual quality of video with periodic frame rate and quantization variation-subjective studies and analytical modeling. In: IEEE International Workshop on Multimedia Signal Processing, Xiamen (2014)
31. Shahab, D.U., Bulterman, D.: Synchronization techniques in distributed multimedia presentation. In: The Fourth International Conferences on Advances in Multimedia (MMEDIA), pp. 1–9 (2012)
32. Murray, N., et al.: Subjective evaluation of olfactory and visual media synchronization. In: Proceedings of the 4th ACM Multimedia Systems Conference, pp. 162–171 (2013)
33. FFmpeg development team. ffmpeg Documentation (2019, March). https://ffmpeg.org/ffmpeg.html
34. NS-3 development team (2017). https://www.nsnam.org/docs/manual/ns-3-manual.pdf
35. Shahab, S.N., Abdulkafi, A.A., Zainun, A.R.: Assessment of area energy efficiency of LTE macro base stations in different environments. J. Telecommun. Inf. Technol., 59 (2015)
36. Sauter, M.: Long term evolution. In: From GSM to LTE: An Introduction to Mobile Networks and Mobile Broadband. Wiley, West Sussex, UK (2010)
37. Baraković, S., Skorin-Kapov, L.: Multidimensional Modelling of quality of experience for mobile web browsing. Comput. Hum. Behav. **50**, 314–332 (2015)
38. Streijl, R.C.: Mean opinion score (MOS) revisited: methods and applications, limitations and alternatives. Multimedia Syst. **22**, 213–227 (2014)

# Examination of Digital Forensics Software Tools Performance: Open or Not?

Andrea Dizdarević[1], Sabina Baraković[1,2](✉), and Jasmina Baraković Husić[2]

[1] American University in Bosnia and Herzegovina, Sarajevo, Bosnia and Herzegovina
barakovic.sabina@gmail.com
[2] University of Sarajevo, Sarajevo, Bosnia and Herzegovina

**Abstract.** The performance evaluation in terms of digital forensics tools and software can be characterized as challenging research area due to constant development of technology in the digital world and rise of various manners in which it can be utilized for illegal purposes. There are many developed tools and software for digital forensics, some of them available for a license, and some of them free of charge. However, given that some practitioners from this field argue for commercial while others for open-source software, the reliability of the digital evidence which is collected, analyzed, and presented by both is constantly questioned. Motivated by the dilemma which tool or software for extracting digital evidence to use, we have conducted the review of the existing studies which directed us towards the examination of the performance of two different types of digital forensics tools: open-source (Linux Autopsy Sleuth Kit) and commercial (Magnet Axiom). The results of the research showed that the open-source digital forensics tool has better performance in comparison to the commercial one. In addition to this conclusion which can be useful for further investigations and research in both practical digital forensics and academic community, we also provide open issues to be addressed in the future.

## 1 Introduction

Throughout the last decades the Internet, communication networks, and information systems have experienced a phenomenal growth and thereby completely changed the lives of most people in various ways, i.e., how they communicate, obtain and exchange information, entertain, do business, take care of their health and environment, learn, govern, etc. [1]. In other words, computers and networks have realized users' requirements and expectations in terms of accessing a range of services anytime, anywhere, via any device, or network [2].

However, having so many data transmissions via computers and networks led to alteration in one very important field. That field is doing crime and everything linked to it. For example, processing electronic traces and evidence, i.e., digital forensics, has become the vital part of the legal process. One can claim that digital forensics has grown from a relatively obscure tradecraft to an important and integral part of the investigations [3]. For the evidence extracted by using digital forensics to be accepted


S. Avdaković et al. (Eds.): IAT 2019, LNNS 83, pp. 442–451, 2020.
https://doi.org/10.1007/978-3-030-24986-1_35

by courts, accepted processes and procedures need to be acquired and demonstrated, which is similar to the issues related to traditional forensics investigations. The admissibility of the digital evidence can be guaranteed if underlying techniques and methods are scientifically validated and recognized [4].

With the growth of digital forensics, the number of tools and software for conducting it has risen as well. Now, we have various commercial and open-source digital forensics tools on which the investigators depend when gathering, analyzing, investigating and guaranteeing the accuracy of the digital evidence. However, given that some practitioners from this field argue for commercial while others for open-source software, the reliability of the digital evidence, which is collected, analyzed, and presented by both, is constantly questioned.

Having in mind the increased importance of computer and network security nowadays and the expanding trends in cybercrime arena, it is important for the professionals in this field to understand the technology used in digital forensics. Moreover, it is very useful for them to solve the dilemma in relation to the performance of both, the commercial and open-source tools they may use in their investigations. The performance evaluation in terms of digital forensics tools and software can be characterized as challenging research area due to constant development of technology in the digital world and rise of various manners in which it can be utilized for illegal purposes. In other words, it needs to be stresses that current digital forensics tools are struggling to keep pace with the growing complexity and rapid evolution of technology in digital environment [5].

Motivated by this and aiming to help and ease the work to the practitioners in field of digital forensics, we have conducted the experiment investigating performance of two different digital forensics tools, i.e., open-source and commercial, and presented the results in this paper. In other words, the objective of this paper is given in its title: should one use open or commercial software in digital forensics domain. The obtained results may contribute to more accurate investigations, together with various other benefits.

The paper is structured as follows: After the introduction, Sect. 2 gives the background, while Sect. 3 describes used methodology which includes the hypothesis, research model, and experiment design. Section 4 provides the results analysis and discussion, while the Sect. 4 concludes the paper and points future research directions.

## 2 Background

Digital forensics software is a tool that assists digital investigators to acquire or locate a potential digital evidence. As already stated, one can differentiate between commercial and open-source software. The main difference between them is that open-source software allows open access to the source code, whereas commercial software does not allow the public to access its code free of charge. Prominent representatives of the open-source software tools are: SIFT Workstation [6], Linux Autopsy Sleuth Kit [7], and Data dumper [8]. The counterpart commercial software tools are: Magnet Axiom [9], FTK [10], Encase [11], Cellebrites UFED Physical Analyzer [12], and Win Hex Tool [13].

There are many arguments and misconceptions of both open-source and commercial software. Detailed analysis of the existing literature concerned with this topic is out of the scope of this paper, but the conclusions made there in relation to positive and negative sides of both types of software are discussed. Therefore, after the analysis of above listed tools and conclusion of the related studies one can conclude that the commercial software tools have better user experience design than open-source ones. The commercial tools also have an edge over their open-source counterparts due to their superior speed and accuracy during data extraction and analysis. Also, the former type is easier to learn to use because of their interface. Authors that analyzed the open-source tools considered them viable and better, and stress that these tools have more functional and useful options for handling sensitive information. Also, they have flexibility of working with a command line console or a Graphical User Interface (GUI) based application, logging capabilities, better fault tolerance, and tend to be more reliable since they operate using a Transmission Control Protocol (TCP) port.

In regard to reliability, there are opposite stances. Without the license, each and every commercial tool can be disputed in court and that is why needs to be validated. These software tools, unlike Linux, operate on an operating system which makes them easy target for disruptions during the analysis, while other programs are running in the background. However, the fact that they are commercial increases the perception of reliability given that you have the company guaranteeing for the security and accuracy of the results. On the other hand, source code made available to the public in case of open-source software tools attracts attackers to search and exploit vulnerabilities to achieve their goals, while on the other hand, having source code published allows, but not necessarily includes, the evaluation, security alert, and improvement by other developers.

The validity of digital forensics software must be fully assessed before the evidence is treated as admissible. Software reliability is satisfied in case when software faults do not cause a failure during a specified exposure period in a specified environment. An unreliable digital forensics software leads to untrustworthy results and may jeopardize the whole investigation.

## 3 Methodology

In order to contribute to the discussion of performance and reliability in terms of open-source and commercial software tools, we formulated the following starting hypothesis representing the baseline of this paper.

**H: Commercial software tool for digital forensics has better performance than the open-source software tool.**

In order to test the abovementioned hypothesis and reach the main objective of this research, i.e., empirically verify the performance of commercial and open-source software tool for digital forensics, we selected Magnet Axiom for the former, and Linux Autopsy Sleuth Kit for the latter tool for the experiment. The essential element of this research is the execution of a series of test scenarios using the selected tools based on the defined test requirements. Utilizing test scenarios based on a set of pre-defined requirements to verify disk-imaging tools is a common and recommended practice according to a variety of digital forensic tool studies [14–17]. The proposed research includes four phases: (i) description of the research environment, (ii) description of the research steps for every digital forensics software as well as the parameters, (iii) execution of tests and documentation of their results along with the reports, and (iv) analysis of test results and conclusion.

The research environment in this experiment was a laptop on which the analysis and the process of experimenting took place has the following characteristics:

- Windows edition licensed version Windows 10 Pro;
- Processor AMD E2-9000e RADEON R2 with 4 COMPUTE CORES 2C + 2G with 1.50 GHz;
- 64bit operating system, x64 based processor, memory 8 GB with 7.46 usable memory.

For better results and a wider area of analysis and testing the performances and capabilities of the two software tools, we have considered six different files and formats. We wanted to present some of the most used files that even the common digital user is fully aware of and knows how to use them. According to [18], those are: .jpg, . doc, .pdf, .xls, .rar, and .zip.

We have then converted information we created for the purpose of examination into all six file formats named above, after which they have been transferred to a Universal Serial Bus (USB). The USB contains 4 GB of memory (3.74 GB to be exact), and the files have taken up 1.5 GB in total. The file system is FAT32 which is an important information due to the fact that every report should contain the exact file system of any memory device it is analyzing. After we have collected all the information about the USB device as well, we then formatted that USB device in order to have all information on it to be deleted. With this, our data collection process is complete. Afterwards we created a disk image and started the process of analysis of the two software digital forensic tools.

Many research groups, government sections, and organizations have attempted to build standardized frameworks for digital investigation. However, a globally recognized investigation framework is yet to be established. A standardized scientific approach for digital investigation must be built to provide the foundation or common practice for digital investigation to identify any misconduct and malpractice. The current generation of digital forensics tools has certain limitations. It is not efficient to process investigation data at a single workstation, considering the limited capacity of the data storage device today. More powerful computers must be used to process a large amount of data efficiently. Otherwise, a new data acquisition approach must be used to cope with the ever increasing data capacity. Many digital forensic tools are still yet to be verified and validated before they can be used as forensic tools in the field. The National Institute of Standards and Technology (NIST) provides a guideline that discusses procedures for preservation, acquisition, examination, analysis, and reporting of digital evidence. Author in [19], claims that the guide is meant to be used by law enforcement, incident responders, and other types of investigators. It addresses common circumstances that may be encountered by organizational security staff. The program addresses one key problem of the industry and legal community. This problem is that there is no standard or credible test to validate the accuracy and completeness of the result extracted by disk imaging tools. The test results are able to assist the forensic software vendors to improve their tools and provide best practice reference to support the results produced by those tools for presentation in the court. The primary studies of NIST [15, 20], and Computer Forensics Tool Testing (CFTT) [21] program present the testing of disk imaging tools. The studies initiated by NIST have a direct link to the proposed research because the approach taken has been widely recognized and acknowledged by the scientific and legal community. NIST is also one of the few research organizations dedicated to digital forensic tool testing. The following parameters for the assessment of the two digital forensic tools have been set in accordance with the NIST methodology recommendations (which base the parameters on the tools preset basic functionalities):

- Availability of software;
- Ability of retrieving deleted data;
- Cost of software;
- Authorized and used by legal agencies;
- Difficulty installing the forensic software.

For this experimental research which includes testing the digital forensic tools, we will be using the following set of steps as previous authors [5, 22, 23] in order to gain the results:

- Determining and entering the case information;
- Imputing the digital evidence and starting the analysis process;
- Examining the digital evidence that is retrieved;
- Analyzing and determining the report of the digital forensic.

## 4 Results Analysis and Discussion

The results of the testing of the selected digital forensics software tools are provided in Tables 1 and 2. They can be divided based on two bases. Firstly, we have grouping based on the type of software, i.e., Linux Autopsy Sleuth Kit and Magnet Axiom. Second grouping is based on type of results, i.e., success retrieval given in Table 1 and parameters given in Table 2. All results are obtained in a process that is described in the following text.

In case of open-source software tool, after the analysis process we had a list of retrieved files. Other than the deleted and recovered files, there are other information like: when the files were accessed and created, their size and name, file system of the USB which was analysed, system volume information, and even the size of the metadata that each file contains. On the left side of the Autopsy dashboard, we also have extra options that we can use for file analysis or research like directory seek, file name search in case we have too many files but know the name of the one we want to retrieve, and an option for just deleted files to be shown. In this dashboard, the files that are blue are the ones that Autopsy can fully retrieve in their original size and format and the redo ones marked as deleted that it succeeded in retrieving. In our case, all of the six major files can also be fully retrieved.

**Table 1.** Results of open-source and commercial software tool analysis by retrieval success

| Files on USB | Linux Autopsy Sleuth Kit (open-source) | Magnet Axiom (commercial) |
|---|---|---|
| | Success in retrieval? | |
| .jpg | Yes | Yes |
| .pdf | Yes | Yes |
| .doc | Yes | Yes |
| .rar | Yes | No |
| .zip | Yes | No |
| .xls | Yes | Yes |
| Total | 6 | 4 |

**Table 2.** Results of open-source and commercial software tool analysis by parameters.........

| Parameters | Linux Autopsy Sleuth Kit (open-source) | Magnet Axiom (commercial) |
|---|---|---|
| Accuracy of extraction? | High | Medium |
| Speed of extraction? | High | High |
| Quality of report? | High | Medium |
| Hash values of files and USB? | High | Medium |
| Calculating metadata? | Yes | No |
| Availability of software? | Very | If paid |
| Ability of retrieving deleted data? | High | Medium |
| Cost of software? | None | High |
| Authorized and used by legal agencies? | Yes | Yes |
| Difficulty installing the forensic software? | Some | None |

In second case, once we opened the Magnet Axiom software and requested the option for a new case analysis on the opening dashboard, we were requested to file out some information about the case details that can later be very important like who the digital forensic examiner is, what is the date, source track of the folders in which the files will be, etc. After that, we are selecting the type of the image file that we are creating for the analysis, in this case a raw image type format. The software has scanned and found the evidence source from a USB and we continued to the processing details in which we had options like adding keywords to the search for better results, options for searching archives and mobile backups and adding hash values (both of which we have chosen), categorizing pictures and videos automatically in case of larger number of files and adding computer artefact. Continuing the process, we have reached an important phase in the analysis that is the choosing of the artefacts on the device that the search will be based on. It is noticed that Magnet Axiom had a lot of options of different artefacts and we have selected all of them presuming that we do not know what we are looking for. If we do not know what types of data we are looking for, it will be best to select all of those that Magnet Axiom is suggesting.

After conducting the testing of both digital forensics software tools in accordance with the pre-defined procedure and methodology, we can conclude that the open-source software tool had better accuracy in the testing of digital forensics software tools in comparison to the commercial one. This is concluded based on the fact that the open-source tool managed to retrieve all files on the digital evidence providing thereby more information in less time, while the commercial tool was not successful in retrieving all files.

Magnet Axiom tool failed to recognize compressed files, i.e.,.zip and .rar, although it successfully managed to find the remaining four file types and offer their full retrieval, but without some of the options that Linux Autopsy Sleuth Kit offered (meta-data and has files). In our case, these results mean that not only did the open-source software have better performance and results but that we have reached a conclusion that the commercial software we were testing, Magnet Axiom, does not have the technical capabilities of retrieving, analysing or collecting compressed files which was not known or stated in the specifications of the software. In practical sense, this information is valuable to digital forensic examinators and digital forensic community. These results lead to the conclusion that our hypothesis stating that the commercial digital forensics software tool has better performance in comparison to the open-source one is rejected. This rejection is the consequence of the former tool not being able to retrieve as many files as the latter, therefore proving for this examination that the open-source tool has better performance.

Also, the rejection has its grounds in parameter analysis given in Table 2, which shows that open-source tool outperforms the commercial one. Going further into discussion of our results, we refer to the authors in [24] which are one of the few that have executed similar testing and research in field of digital forensics software, but for mobile context. The authors have used NIST methodology and the same open-source software tools for digital forensics which allowed us to compare results following the same parameters as the authors. The author named and followed parameters characterized as “high” and “medium” based on the highest and lowest performance of the two testing software, using “medium” when the software had only satisfactory performance. However, opposite to our results, authors in [24] concluded that the commercial software has a slight advantage over the open-source one in mobile environment. Parameter by parameter comparison of our results and the results obtained in [24] is as follows:

- **Accuracy of extraction**: The authors in [24] has found the commercial and open-source software to have the same accuracy, while we concluded that the open-source software had better accuracy;
- **Speed of extraction**: The authors in [24] claimed open-source software had better speed of extraction and we concluded that both of the software had high speed of extraction;
- **Quality of report**: The authors in [24] had similar conclusion to ours in quality of report, concluding that open-source software has more detailed and stronger quality of the report;

- **Hash values of files and USB**: The authors in [24] had the same conclusion as we did, i.e., the hash values are in a higher level in open-source software;
- **Calculating metadata**: Both our analyses in comparison showed that open-source software calculates metadata while commercial software does not;
- **Availability of software**: Both our analyses and the compared work agree;
- **Ability of retrieving deleted data**: The authors in [24] concluded that commercial software had better deleted data retrieving than open-source software, and we concluded that open-source software has better performance;
- **Cost of software**: Both authors commented on the price of commercial software and concluded that open-source software is more available and more affordable;
- **Authorized and used by legal agencies**: We have concluded through the research that both software types are authorized and used in legal agencies as well has the authors in [24];
- **Difficulty installing the forensic software**: The authors in [24] commented on the immense difficulty of installing and running Linux Autopsy Sleuth Kit, while the results of the thesis indicate that there were none while installing the commercial software.

After analyzing and discussing the obtained results, which led us to the conclusion that according to the analyzed parameters and retrieval success open-source software tool for digital forensics is better than the commercial one and our hypothesis is rejected, we provide several limitations of our study that should be overcome in future research. Firstly, it needs to be noted that no research challenges are related to the software that were tested. One of the technical difficulties we had is related to the installation of the virtual box and Linux that was to operate on it, since both software had to be tested in the same testing environment, meaning that it is on the same laptop. Another challenge we experienced during this testing is also related to the virtual machine, where its stability was in question. Namely, it was tested a few times before the experiment took place, but in those first attempts it managed to crash two times before the installation was even complete, which imposed some additional adjustments to it before the installation was complete.

## 5 Conclusion

The main aim of this paper was to contribute to the dilemma of which software tool for the digital forensics analysis has better performance and is suggested for utilization in the investigation process: open-source or commercial. After conducting the research and the review of research studies' conclusions, as well as the experiment in which we tested the representatives of both tools, Linux Autopsy Sleuth Kit and Magnet Axiom, obtained results are in favor of open-source software tool. Namely, in addition to better retrieval success in comparison to commercial tool, open-source software has better user graphic experience, better accuracy and reliability, and allows multiple working stations at the same time. On the other hand, the process of installation is more difficult with open-source, while the commercial disrupts more easily. In addition, both software are legal, accepted in court, and used by the law enforcement agencies. Therefore, based on the analysis of the results and discussion, the question posed in the title of this paper gets its answer: open.

This paper enriches the body of knowledge of testing digital forensics tools and can be valuable to other researchers in law enforcement agencies, industry, and academia, as well as to other interested stakeholders that work on improving digital forensics tools.

Open issues that need to be addressed in the field of digital forensics by future research include building a systematic and scientifically proven methodology to validate the functions of the digital forensics' tools. What has been achieved by CFTT program and other researchers can be used as a stepping-stone to build a comprehensive testing framework. The framework must be automated, tool-independent, and future-proof. Along with that there must be continued academic and technical work on new high technology software that will incorporate the technology improvements into search and analysis or retrieval of data. Also, different test scenarios with different hardware types can be imposed on the test cases to create a more complete testing framework.

## References

1. Baraković, S., Baraković Husić, J.: We have problems for solutions: the state of cybersecurity in Bosnia and Herzegovina. Inf. Secur.: Int. J. **32**, 131–154 (2015)
2. Baraković, S., Kurtović, E., Božanović, O., Mirojević, A., Ljevaković, S., Jokić, A., Peranović, M., Baraković Husić, J.: Security issues in wireless networks: an overview. In: 11th International Symposium on Telecommunications (BIHTEL 2016) (2016)
3. Garfinkel, S.L.: Digital forensics research: the next 10 years. Digit. Investig. **7**, 64–73 (2010)
4. Erbacher, R.B.: Validation for digital forensics. In: 7th International Conference on Information Technology: New Generations (2010)
5. Ayers, D.: A second generation computer forensic analysis system. Digit. Investig. **6**, 34–42 (2009)
6. SIFT Workstation. https://digital-forensics.sans.org/community/downloads. Accessed March 2019
7. Autopsy Sleuth Kit. https://www.sleuthkit.org/autopsy/. Accessed March 2019
8. Data Dumper. https://perldoc.perl.org/Data/Dumper.html. Accessed March 2019
9. Magnet Axiom. https://www.magnetforensics.com/products/magnet-axiom/. Accessed March 2019
10. Forensic Toolkit (FTK). https://accessdata.com/products-services/forensic-toolkit-ftk. Accessed March 2019
11. EnCase Forensics. https://www.guidancesoftware.com/encase-forensic Accessed March 2019
12. Cellebrite UFED Ultimate. https://www.cellebrite.com/en/products/ufed-ultimate/. Accessed March 2019
13. WinHex: Computer Forensics & Data Recovery Software, Hex Editor & Disk Editor. https://www.x-ways.net/winhex/. Accessed March 2019
14. Wilsdon, T., Slay, J.: Validation of forensic computing software utilising black box testing techniques. In: Australian Digital Forensics Conference (2006)
15. NIST, Digital Data Acquisition Tool Specification (v4.0). Technical Report (2004)

16. Guo, Y., Slay, J.: Computer forensic functions testing: media preparation, write protection and verification. J. Digit. Forensics Secur. Law **5**(2), 5–20 (2010)
17. SWGDE/SWGIT, Guidelines & Recommendation for Training in Digital & Multimedia Evidence (2009)
18. Computer Hope. https://www.computerhope.com/issues/ch001789.htm. Accessed March 2019
19. Olivier, M.: On a scientific theory of digital forensics. In: IFIP International Conference on Digital Dorensics (2016)
20. NIST, Digital Data Aqusition Tool Test Assertions and Test Plan (v1.0). Technical Report (2005)
21. NIST, Computer Forensics Tool Testing (CFTT). http://www.cftt.nist.gov. Accessed March 2019
22. Bellin, K., Creutzburg, R.: Concept of a master course for IT and media forensics part II: android forensics. In: 9th International Conference of IT Security Incident Management and IT Forensics (2015)
23. Siddique, A., Alam, M.A., Chaudhary, O.: A proposed structured digital investigation and documentation model (DIDM). Int. J. Adv. Res. Comput. Sci. **8**(7) (2010)
24. Padmanabhan, R., Lobo, K., Ghelani, M., Sujan, D., Shirole, M.: Comparative analysis of commercial and open source mobile device forensics tools. In: 9th International Conference on Contemporary Computing (IC3) (2016)

# Intelligent Web Application for Search of Restaurants and Their Services

Arnela Gutlić(✉) and Edin Mujčić

Faculty of Technical Engineering, University of Bihac, Bihac, Bosnia and Herzegovina
arnelagutlic@gmail.com, edin.mujcic@unbi.ba

**Abstract.** Nowadays, we can say with certainty that we live in a world of busy lifestyle. For this reason, there is an increasing need for a simpler and more accessible way of collecting required and desired information. One of the ways to accomplish this need is to have easy access to all the desired data from one website. One of these websites is the application implemented in this paper. The intelligent web application for search of restaurants and their services has a purpose to help domestic population and tourists in search of the restaurants according to different criteria. There are four different search criteria in the designed web application: search by the current user location, search according to the key parts of the city near the restaurants, search according to the parts of the city and search for the restaurants with the best reviews and ratings. Search results contain all necessary information about restaurants. These results are including contact and main information about the restaurant, menu, user reviews and ratings. Also, the web application for search of restaurants and their services give possibility for users to write reviews and ratings for the visited restaurant. Processing user reviews and ratings is performed using fuzzy logic. This intelligent web application for search of restaurants and their services is developed for Bihac, but it can be used in any city with the adjustment of restaurants information. Designed web application currently consists information for thirty restaurants located in Bihac. Also, work analysis of the designed web application is performed and explained in this paper.

**Keywords:** Intelligent web application · PHP · Fuzzy logic · Restaurants · Reviews and ratings of restaurants

## 1 Introduction

Today's fast lifestyle requires an equally fast way to access all the necessary information. People want quick search, easily displayed search results, and the ability to collect all the necessary information from one place.

As a result of great progress in all areas of human life today, there is a growing number of data and information people have. Accordingly, it is necessary to group all the data in the right way, and to ensure the correct and easy way to search for them. Also, the unstoppable progress of technology increasingly allows the domination of the


S. Avdaković et al. (Eds.): IAT 2019, LNNS 83, pp. 452–469, 2020.
https://doi.org/10.1007/978-3-030-24986-1_36

virtual world and the Internet, which allows people quick access to the information they need [1]. Various web applications that meet user requirements have been developed. Rapid growth and development of web applications need to be thanked for the fact that they are available at anytime from anywhere, both computers and mobile phones [2].

Web application that can respond very well to all the challenges of today must have a synchronized working mode of all the tools used in the development, like front-end tools [3]: HTML, CSS, JavaScript [4], back-end tool [5], in this case PHP, and all other tools that are relevant to the application.

As restaurants are an important factor in tourism, it is necessary to find the right ways to adjust their functions and offer to today's way of life [6]. There are many different search applications for restaurants and their services. World's most famous platforms to search for restaurants are: Yelp, Foursquare and TripAdvisor- Great competition for these applications are user reviews on Google Maps. The basic service of these applications is the search for a restaurant. In addition to the basic role, these applications provide some additional services, such as restaurant reservations, food orders and delivery, personalized recommendations based on search history, location sharing, and more, all for the purpose of making a richer user experience [7–9]. All search applications also contain a section related to writing reviews of places found, which greatly helps future users [10, 11].

This paper describes an intelligent web application for searching of restaurants in Bihac [12]. This application is one of the first applications of this type in Bosnia and Herzegovina. Intelligent web application "Restaurants in Bihac" is written in PHP programming language. The main goal is to enable quick and easy search of the restaurants of the city of Bihac, according to one of the offered search methods. There are four searching methods, adjusted to all user groups, so each user can use any of the offered ones. Users can view all information about selected restaurant from one web location, and also, they can write reviews about restaurant they visited. Processing of user reviews and ratings is performed using fuzzy logic, and that is what makes the application stand out from all the others. Compared to classic logic, which offers two possibilities, either true or inaccurate, black or white, fuzzy logic offers a much wider range of possibilities, and it can be said that the fuzzy logic sees everything as a shade of gray. For this reason, the use of the fuzzy logic to calculate the average restaurant rating takes precedence over the use of classical logic [13].

## 2 Development of Intelligent Web Application for Search of Restaurants and Their Services

The main goal of the intelligent web application "Restaurants in Bihac" is to make the search of restaurants easier for tourists, visitors, but also by the inhabitants of Bihac. For this purpose, four ways of searching are defined:

1. Search for restaurants near the current location of the user
2. Search for restaurants by parts of the city

3. Search according to the key objects of the city near the restaurants
4. Search for restaurants with the best ratings and reviews.

The development of an intelligent web application included the following: designing the overall look and content of the application, using front-end tools, HTML, CSS and JavaScript.

The primary goal of using front-end development in this application was to provide users that all information are displayed in an easily readable and relevant format, once they open the website. For this reason, it was necessary to ensure that the website is properly displayed in various devices, different operating systems, and different browsers, which required careful planning and programming. The main task was to properly manage what users first see in their browser, so there is the look, the general impression of the website, and the last design of the site.

The next step in the process of developing the application was database modeling, and programming the complete application functionality, using a back-end tool PHP. The development of this application on the server-side controls what is happening "behind the scenes" of the intelligent web application. Also, the communication between the database and the browser has been achieved.

Back-end development has made it possible for a static intelligent web application which was designed using just front-end tools, to become dynamic, in which the content is constantly updated and changed. Updating, modifying and monitoring functionalities are some of the basic and very important responsibilities, at the same time.

The next step is programming the part related to statistics and average restaurant ratings, using the principle of fuzzy logic. This intelligent web application uses fuzzy logic to calculate the average restaurant rating. For that purpose, evaluations of certain parameters from user reviews are used. All the steps of fuzzy inference have been implemented and programmed using object-oriented PHP in order to ensure the correct assessment, accurate and relevant data.

The application has two types of users: users without privileges - guests and administrator. There are different user interfaces for each of the user types.

### 2.1 User Interface of the Intelligent Web Application for Search of Restaurants and Their Services for the User Without Privileges - Guest

User – "guest" is any person who visits our application. The home page of the application is shown on the Fig. 1 with the beginning of the page and the search form.

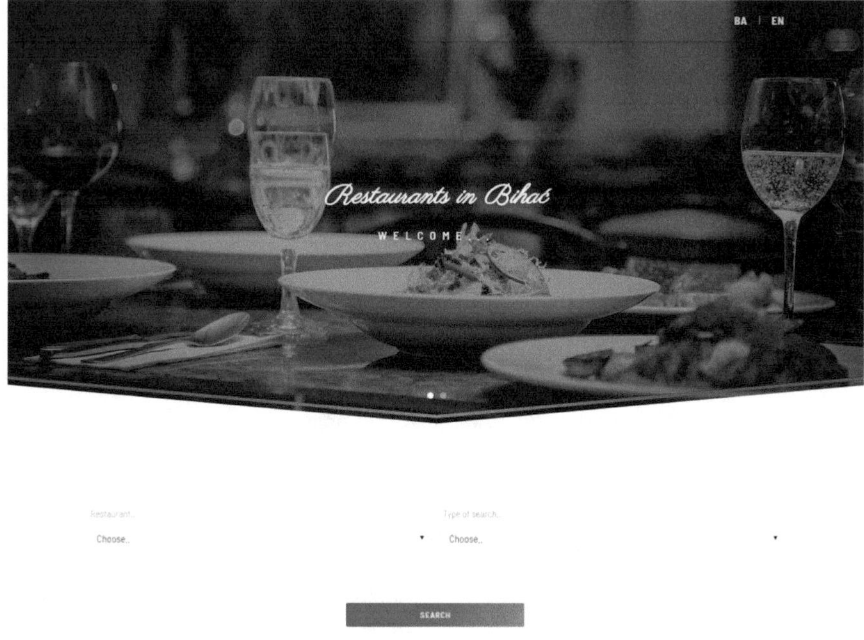


**Fig. 1.** Top of the home page of the application

Guest can view the section of the home page that contains basic information about the application, and the way it works. In addition, the map of the city, in this case Bihac, with marked parts of the city and restaurants, is also shown, for easier navigation and finding certain data. To add a map with all those features, it was necessary to use Places Library whose functions were to enable our application to search for places (defined in API as restaurants) contained within a defined area of Bihac. The Places service is separate from the main Maps JavaScript API code, but in our application and in this section, they cooperate to ensure full service. This section is very important for users, because basic principles of functioning and all the information they need to know about it are written. This section with all those information is shown on Fig. 2.

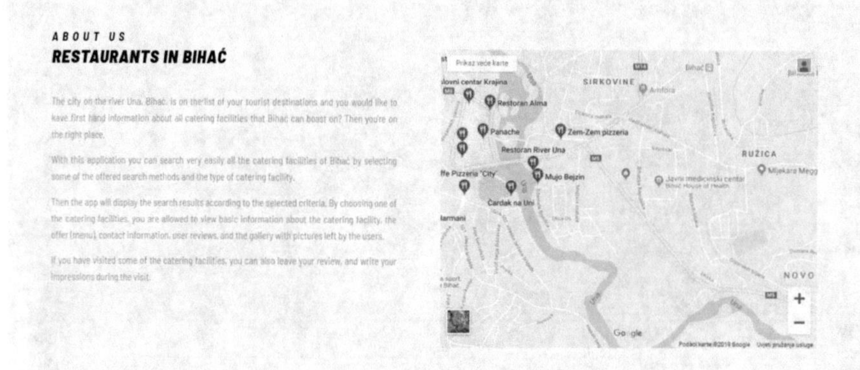


**Fig. 2.** Section of the home page with important information

The statistics section contains information about the number of visits to the application, the number of restaurants from the database, and the number of written reviews. Each time a user visits an app, a database field that contains the number of app visits is updated, and the users see this information as a counter, from 0 to the number of visits. Similarly, this section performs counting restaurants from the database and written reviews of it and these data are also displayed using a counter. Figure 3 shows the statistics section of the home page of the application.

**Fig. 3.** Statistics section

Application reviews, written earlier by previous users, are publicly available and contain a rating, commentary, and review author. Reviews of the application are added together with restaurant reviews, with star rating system from 1 to 5 stars. Star rating system used in this paper works through JavaScript code, but the information about rating, comment, and review author are added to the database using PHP. Reviews of the application are shown on Fig. 4.

**Fig. 4.** Reviews about the application

Finally, the home page of the application also contains a search form, earlier shown in Fig. 1, and it is the most important part of the application. The search form contains two fields: a field for selecting a type of restaurant, and a field to select one of the four offered search modes. Those fields, with their options, are shown on the Fig. 5.

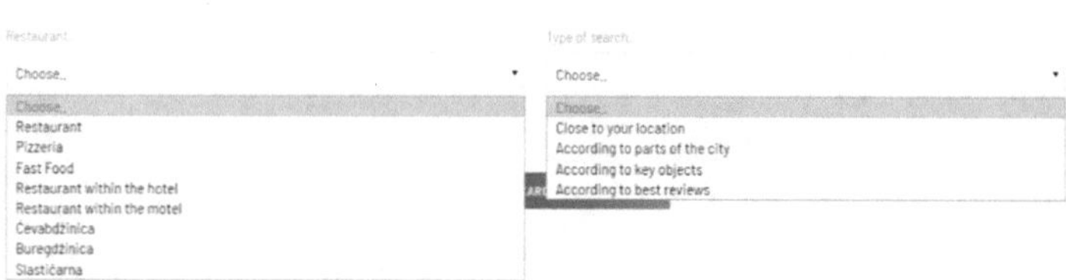


**Fig. 5.** Search form of the application

Users first select the type of restaurant. This field contains a list of restaurants from the database. Type of restaurant can be: restaurant, pizzeria, fast food, etc. This type of information is specific for different cities, and it's necessary to connect this field with all the other information specific for certain city.

After choosing the type of restaurant, users select one of four offered search modes explained below.

Searching for a restaurant near the current location of the user first requires permission from the user to collect information about his current location in order to show him adequate search results. This function is enabled thanks to, again Places Library in combination with HTML5 Geolocation browser's feature. Also, there is an indispensable need to use JavaScript API for Google Maps. The results contain a map with the indicated current location of the user, thanks to HTML5 Geolocation features and functions, and the marked restaurants of the selected type in its vicinity, thanks to Places Library, which enabled search of restaurants of the selected type within 500 meters of the user's current location. All the used parameters are defined through the JavaScript code. Also, search results contain the list of found restaurants, with an average rating, and it is also shown.

In case users choose a search method according to the parts of the city, the list of parts of the city from the database is also displayed. Similarly, in the case users choose a search method according to the key objects of the city, the list of key objects of the city from the database is also displayed. These data are entered by the administrator and they are specific to each city. These two search modes are enabled since restaurants also contain information about the part of the city they are in, as well as information about the key object nearby.

Searching for the best restaurants and reviews means the top 5 best-rated restaurants of the selected type. Average rating of restaurants, obtained by fuzzy logic, is saved into the database, and with this search mode, application takes the best rated ones. All of these search results can be sorted by name and according to the best ratings.

Users can view the offer of the selected restaurant after entering their e-mail. Entering the e-mail is necessary to let them write a review about the selected restaurant after visiting. This gives users some privileges, which gives them the ability to view a larger number of information related to the selected restaurant. After the user has selected a restaurant and successfully entered the email, a new page with information about the selected restaurant opens.

**Fig. 6.** The beginning of the restaurant page

The beginning of the page contains the restaurant text, along with the average rating obtained using the fuzzy logic. The average rating is displayed numerically but also by using the star rating system implemented within the application. The star rating system is associated with fuzzy estimation, so it shows the result with the number of stars. Figure 6 shows this part of the page with restaurant information, and average rating for certain restaurant.

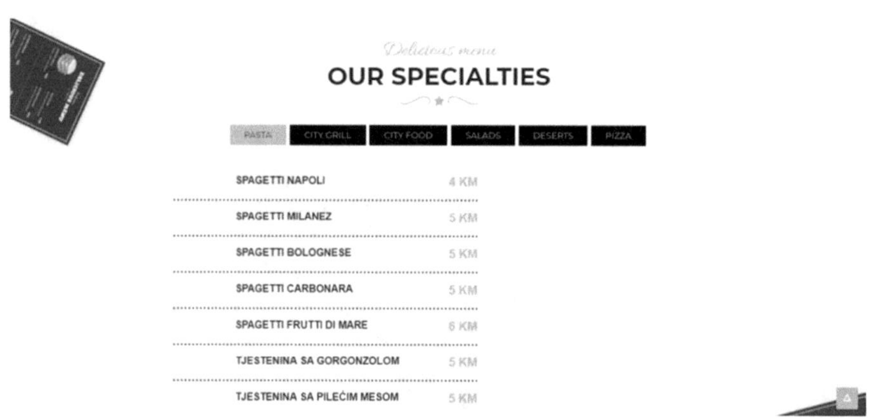


**Fig. 7.** Menu for selected restaurant

The menu is sorted according to the category specific for the selected restaurant, and menu items have the following information: name of the meal, price, and a description, if available. These menu categories are from the database, entered by the administrator, and are related to the selected restaurant and all its data. Menu with specific categories is shown on the Fig. 7.

**Fig. 8.** Section that allows writing reviews

For the purpose of providing better and more quality services and restaurant offers, customers are able to express their opinions and impressions. A section that allows user to write review, contains short text about how writing reviews work. It's said that if users want to write a review, by pressing the "Write Review" button, they will receive, to the e-mail address they have previously entered, a link through which they can write a review. The e-mail sending option is programmed with the help of PHP Mailer, a library that allows sending e-mails in the PHP programming language. The PHP Mailer has enabled sending mail without a local mail server, with adding recipient's e-mail, as e-mail user previously added, editing an e-mail lookup using HTML and CSS by setting up email formats as HTML. Section that allows users to write a review is shown on Fig. 8.

**Fig. 9.** Reviews section about selected restaurant

All of the users reviews are available in the "Reviews" section. Added privileges to the user allow him to read written reviews for the selected restaurant. Reviews include the title, comment, rating and author reviews, and those information are the ones that users write within the review form, so it's also a very important part of the application. Figure 9 shows restaurant reviews.

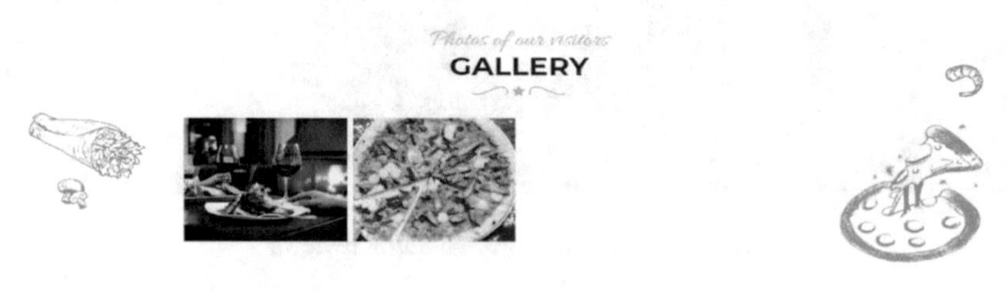


**Fig. 10.** Gallery with user pictures for selected restaurant

Gallery of pictures added by a review form, are optional when writing a review, but if user has some photo of visiting selected restaurant, he can share it with other users, and that photo will appear in this section. Hovering with mouse on every picture from gallery also shows information about author of the photo, and a short comment from review about restaurant. All pictures from gallery for certain restaurant are shown on Fig. 10.

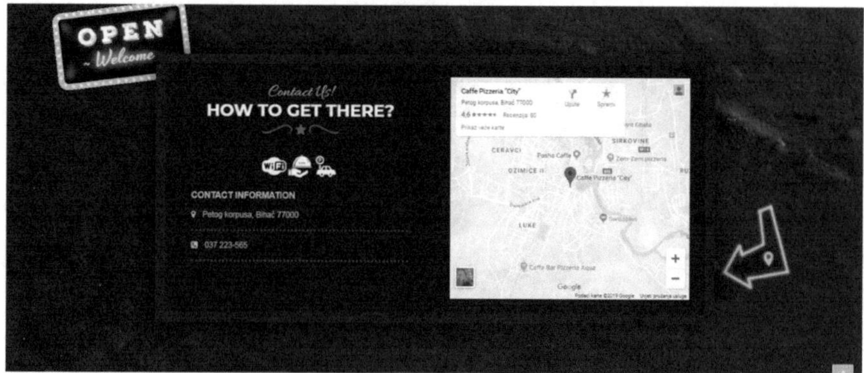


**Fig. 11.** Contact section at the bottom of page

A section with contact information, including additional restaurant services, and a map with the exact location of the restaurant, are at the bottom of the page, and those information are very important, because they can help users finding restaurant. Contact section is shown on Fig. 11.

If users decide to visit the selected restaurant and want to have a chance to write a review about it, they must press the "Write a review" button. Then a link to a page that allows them to write review is sent to the email address previously entered. The advantage of sending a link to an e-mail is that it does not require users to write a review at the time they are on the site of the restaurant. They can write it anytime, it is enough to visit the link from the e-mail. Also, writing reviews is optional and users do not have to do it, even when the link is sent to their e-mail. This is better for users, because it doesn't force a user to write a review, and the time of writing a review, if he wants, is when he wants, so it's fully customized to the user.

After the user opened an e-mail and visited the link, a page for writing the review is displayed. Page for writing a review is shown on Fig. 12.

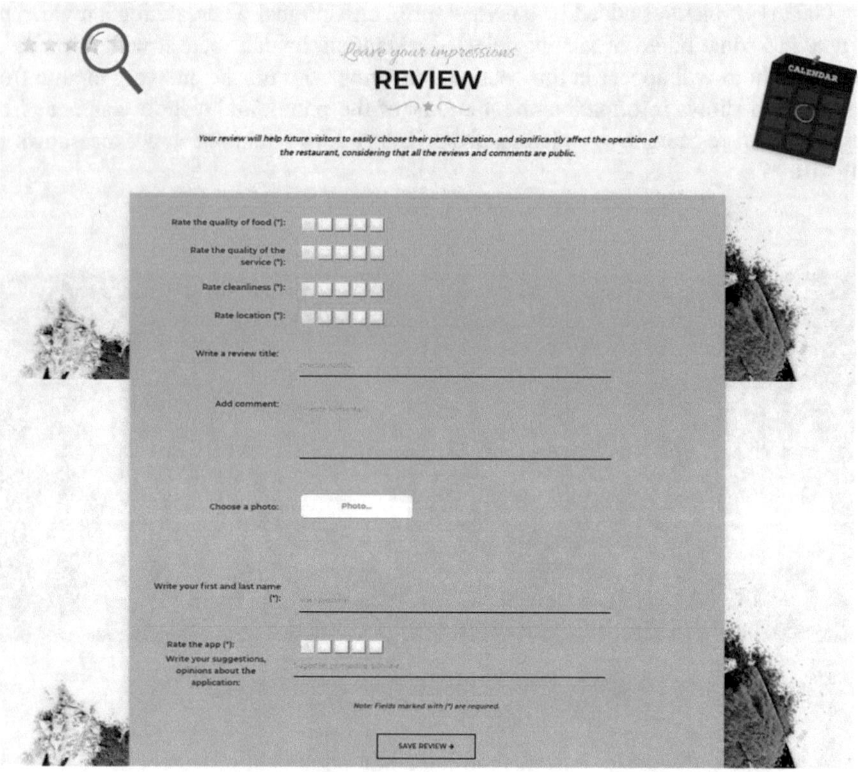


**Fig. 12.** Web page for writing a review

The review form includes mandatory and optional fields. Four parameters, food quality, quality of service, cleanliness and location are evaluated. These are mandatory fields and are rated by stars from 1 to 5. Also, the app is rated by stars from 1 to 5, and this option rating with stars is enabled thanks to implemented star rating system within the application. The required field is also the field for entering name and surname, which is linked to the previously entered e-mail and saved to the database. Other fields are optional and these are the title of the review, comment of restaurant review, app comment and image recorded during a visit to the restaurant. When adding reviews, the entered data is stored in two tables, a table of restaurant review data from form, and the date and time of writing review, with user and restaurant information. Identification of users and restaurants is enabled by id. User data (name and surname), if not previously entered, if the user first writes a review with the given e-mail address, are stored in the user table. Another table is affected by this activity. Those user ratings of food, service, cleanliness and location, at the moment of adding a review are forwarded to a fuzzy evaluation system that computes an average rating for a selected restaurant, including new values of parameters, and updates the average rating and saves changes in a separate table. The average rating as a result of fuzzy method of inference is therefore constantly updated by adding new reviews to the selected restaurant.

From the review page, it is possible to return to the selected restaurant page, also with a reminder of the email, and thus enable further review of the offer of that restaurant or all other restaurants from the search results according to the given criteria as well as a new search.

In addition to the above, the application is available in Bosnian and English. At any time, it is possible to change the language of the page by selecting the options offered at the top of the page, BA - for the Bosnian language and EN - for the English language.

### 2.2 User Interface of the Intelligent Web Application for Search of Restaurants and Their Services for the Administrator

The administrator is the person responsible for managing the complete application and its content and, the application the contains only one administrator. The administrator's user interface is the CMS (Content Management System), which allows classification, organization, and other forms of editing the content of this website. The great role of the CMS in this application is that it enables simple modification and input of content on the page using a graphical environment instead of programming. User interface for the administrator is shown on Fig. 13.

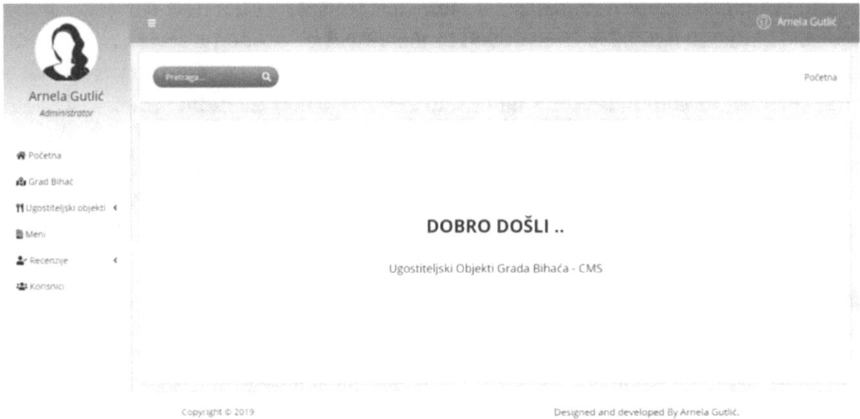


**Fig. 13.** User interface for administrator

The administrator has the following options: he can edit his own data, such as name, surname, e-mail, user name, password, and photo. All administrator information is visible only to him, in order to reduce the possibility of data misuse. Other options in the administrator's authority are displayed in the sidebar, such as the "Bihac City" option that allows administrator to enter data related, in this case, to the city of Bihac. These are the city's name, types of restaurants, parts of the city, key objects, and map of the city. This is very important, because this type of data are specific for each city, and precisely this information completely determines the city for which the application is customized. Types of restaurant, parts of the city, key objects are stored in different tables in database, and they are connected to the main table, "city", that contains information about city name and location (map).

The "Restaurants" option contains a list of added restaurants. Each of them can be reviewed in more detail, showing all available data, such as restaurant information, contact information, map with the exact location of restaurant, additional restaurant services, menu, and menu categories. All those data are stored in different tables in database, but they are connected, and when administrator edit one item, all items that includes edited one, are updated.

The "Menu" option offers a general overview of the menu of all restaurants, and the possibility of adding new ones. By adding a new menu item, it's necessary to connect it with a specific restaurant, and then with categories of menu of that restaurant. In this way it is possible to correctly add and display the content, and also enable all options related to these. All the above options offer the ability to view added items, delete and edit items, and add new ones.

The "Reviews" option contains two sub-options, a list of restaurant reviews, along with a gallery of images added together with a review, and a list of the app reviews. Those information contain just a table with review content, and administrator just can read it. This is a part where editing or deleting is not possible and all reviews, whether positive or negative, are publicly available to all users.

The "Contacts" option represents a table that consists of the user's e-mail, and the name and surname that have been added to the e-mail by writing a review. Some fields will have just e-mail, for those users who just reviewed restaurant offer, but did not write a review, as adding a user's name and last name is done at that time. Using PHP to complete those actions was very good, because this process of CRUD (create, read, update, delete) works very well to meet all the requirements. Complete control content is the responsibility of the administrator, and of his activities depends application's functionality.

### 2.3 Use of Fuzzy Logic to Calculate an Average Restaurant Rating

The processing of user reviews and ratings is performed using fuzzy logic. The parameters used to evaluate are food, services, cleanliness, and location, and they represent input variables of the system. The output variable is the average grade obtained using the Mamdani method of inference.

In order to get the correct results in both cases, it was necessary to program all the steps of the fuzzy inference procedure using PHP, where each step represents a particular class and its functionality. Therefore, in this application the principles of object oriented programming have been used, so that a fuzzy evaluation would be possible. Figures 14 and 15 show comparisons of the definition of the variable "Food", along with its parameters and membership function, using the application and the Matlab Fuzzy Toolbox.

```
//Funkcije pripadnosti lingvističke varijable Hrana
    $Hranaprip1 = new pripadnostTrokutasta(0,1.75,1);
    $Hranaprip2 = new pripadnostTrokutasta(1.25,2.75,2);
    $Hranaprip3 = new pripadnostTrokutasta(2.25,3.75,3);
    $Hranaprip4 = new pripadnostTrokutasta(3.25,4.75,4);
    $Hranaprip5 = new pripadnostTrokutasta(4.25,5,4.75);
    //5 lingvističkih oznaka
    $MBHrana = new LingvistickaOznaka('1',0,1.75,$Hranaprip1);
    $BHrana = new LingvistickaOznaka('2',1.25,2.75,$Hranaprip2);
    $NHrana = new LingvistickaOznaka('3',2.25,3.75,$Hranaprip3);
    $AHrana = new LingvistickaOznaka('4',3.25,4.75,$Hranaprip4);
    $MAHrana = new LingvistickaOznaka('5',4.25,5,$Hranaprip5);
    $Hrana=new LingvistickaVarijabla('Hrana',0,5,array($MBHrana,$BHrana,$NHrana,$AHrana,$MAHrana),1);
    // kraj deklaracije varijable Hrana
```

**Fig. 14.** Variable declaration using application and PHP

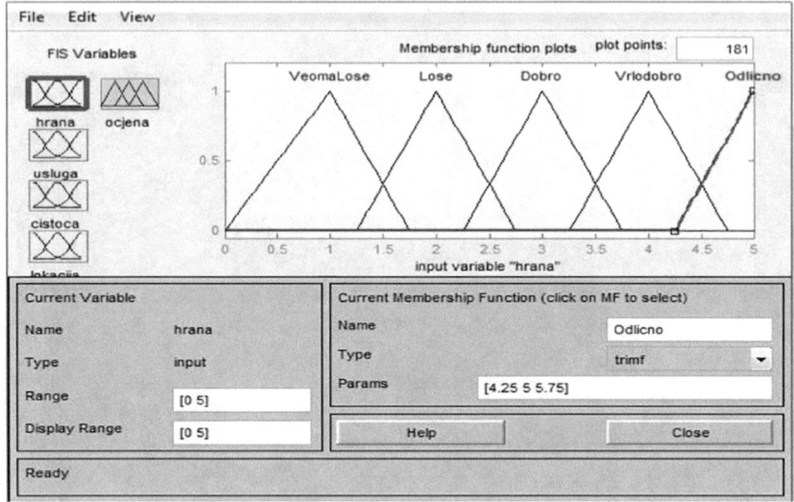


**Fig. 15.** Variable declaration using Matlab Fuzzy Toolbox

The Mamdani method of inference has its own parameters and rules that are also programmed using an object-oriented PHP. In general, the implementation of the fuzzy logic principles in any programming language is not simple and requires first a good knowledge of fuzzy logic, elements and steps in the inference process, and then detailed analysis and use through the selected programming language. In the application, the principles of fuzzy logic are presented through PHP programming language, as the PHP is being used as a back-end tool of the application, and as such, it has fully met all requirements and made it possible to create all the required functions and elements necessary for fuzzy calculation of average rating.

## 3 Analysis of the Developed Web Application

This section describes the analysis of a developed web application for search of restaurants and their services. Therefore, the user visited the application and the home page of the application is showed. After reviewing everything that it offers, the user searches, choosing the type of restaurant first, and then the search method that is the most appropriate for him. For example, the user selected the type of restaurant “pizzeria” and searched for the restaurants near the current location of the user. The application then requested the permission to collect information about the current user’s location, and after that showed adequate results. Search results consist of a map, with the indicated location of the user (“Location found”), and marked restaurants of the selected type near the user (restaurant icon). In addition, the search results also contain a list of found restaurants. These results can be searched or sorted by name and according to the best ratings and are shown on Fig. 16.

**Fig. 16.** Search results

The user selected one of the restaurants and pressed the “Show offer” button. Application then requires an e-mail entry. After a successful e-mail entry, the user has been forwarded to the page with data of the selected restaurant, with all the necessary information. On the page there is a map with the exact location of the restaurant, which helps the user to find the place.

User likes the restaurant offer and he decided to consider the option of writing a review, as he plans to visit the selected restaurant. He pressed the “Write a Review” button at the section of the site that allows writing a review, shown on Fig. 17, and got a link on the e-mail address he had previously entered, through which he can write a review, if and when he wanted to.

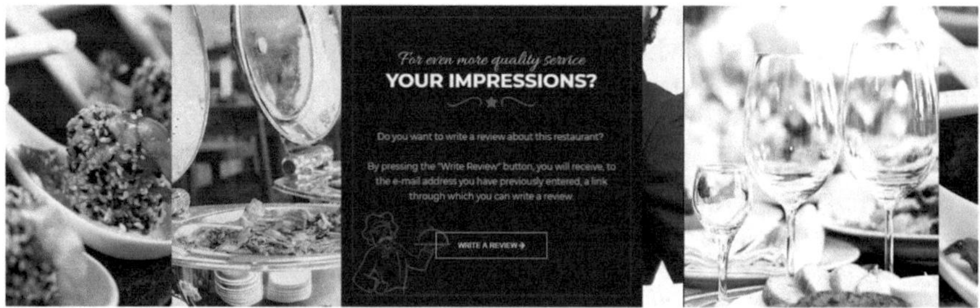


**Fig. 17.** Section of the site that allows writing a review

After visiting the restaurant, the user wants to write a review about the same, and opens his email containing the link that will allow him to do that. E-mail with the link is shown on Fig. 18.

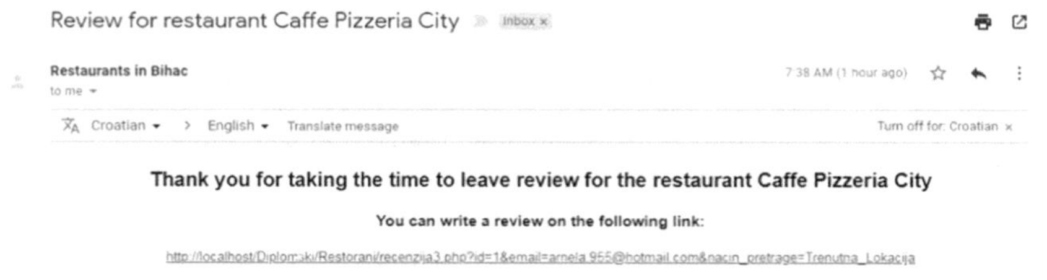


**Fig. 18.** E-mail with the link for writing a review

A link from an email opens a page that has a form to write a review, which contains the above-mentioned mandatory and optional fields.

**Fig. 19.** Alert notification about the successfully added review

After the user has filled in the necessary information, he will receive an alert informing him that the review has been added, as shown in Fig. 19, and he can continue reviewing the restaurant page, offer of other restaurants, or perform a new search.

## 3.1 Comparison of the Results for the Average Rating of Restaurant Obtained Through Fuzzy Logic

In this section the results obtained through the application and the identical system implemented using the Matlab Fuzzy Toolbox are displayed.

The results of the average rating of the given restaurant obtained using the application are shown in Fig. 20. The used data are the ratings from the user reviews stored in the database. The same data was also used for testing through Matlab.

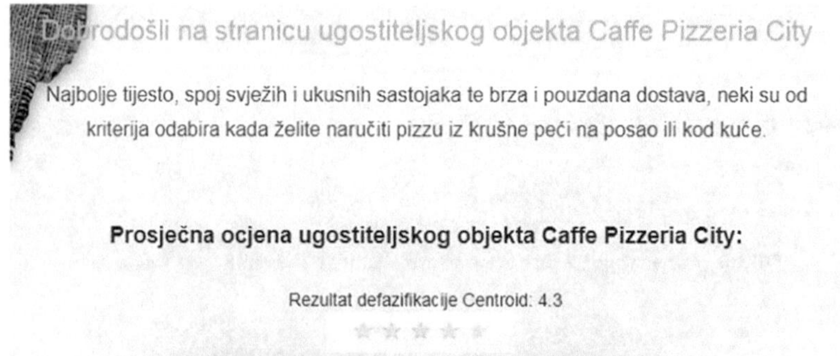


**Fig. 20.** Result of rating for restaurant with fuzzy logic through application and PHP

The results of the average rating of the given restaurant obtained using the Matlab Fuzzy Toolbox are shown in Fig. 21. They were introduced using the Matlab GUI for easier input, testing and displaying the results.

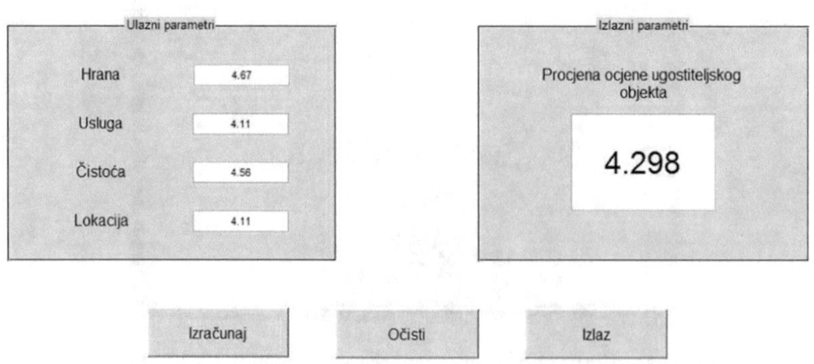


**Fig. 21.** Result of rating for restaurant with fuzzy logic through Matlab Fuzzy Toolbox

The average grade for the selected restaurant, obtained this way, using fuzzy logic, is much more relevant and it gives better results, since it offers a wider range of parameters necessary for the evaluation.

## 4 Conclusion

Thanks to the results from the analysis of the work, it can be said that the developed web application correctly performs all defined and specified functions. The application is fully applicable to any city, since all relevant data are entered using the CMS. The analysis and planning of the application before the development phase made it possible to identify potential problems or obstacles that could arise, and thus found a solution for them. Consequently, it was possible to properly program the part related to the average restaurant rating using the fuzzy logic, and this part is what sets the application apart from all the similar ones for searching restaurants and their services that already exist in Bosnia and Herzegovina.

## References


1. http://www.pewinternet.org/data-trend/internet-use/internet-use-over-time/
2. Mandava, A., Solomon, A.: A review and analysis of technologies for developing web applications. In ITERA (2012)
3. Crockford, D.: JavaScript: The Good Parts. O'Reilly Media Inc., Sebastopol (2008)
4. Aquino, C., Gandee, T.: Front-End Web Development: The Big Nerd Ranch Guide. Pearson Technology Group, Indianapolis, IN (2016)
5. Zandstra, M.: PHP Objects, Patterns, and Practice, 2nd edn. Apress, Berkeley, CA (2008)
6. Tousif, O., Maisha, M., Shahreen, P.: Adaptive food suggestion engine by fuzzy logic. In: 016 IEEE/ACIS 15th International Conference on Computer and Information Science (ICIS). IEEE (2016)
7. https://en.wikipedia.org/wiki/Yelp
8. https://en.wikipedia.org/wiki/Foursquare_City_Guide
9. https://en.wikipedia.org/wiki/TripAdvisor
10. Hajas, P., Gutierrez, L., Mukkai, S.: Analysis of Yelp Reviews (2014). https://arxiv.org/pdf/1407.1443v1.pdf
11. Agryzkov, T., Marti, P., Tortosa, L.: Measuring urban activities using Foursquare data and network analysis: a case study of Murcia (Spain). Int. J. Geogr. Inf. Sci. **31**(1), 100–121 (2016)
12. Mujiya, M., Rizal, L., Pradita, Y.: Evaluating service quality of Korean restaurants: a fuzzy analytic hierarchy approach. Ind. Eng. Manag. Syst. **15**(1), 77–91 (2016)
13. Bojadziev, G., Bojadziev, M.: Fuzzy Sets, Fuzzy Logic, Applications. World Scientific, London (1995)

# Design and Experimental Analysis of the Smart Traffic Control System Based on PLC

Amel Toroman([✉]) and Edin Mujčić

Faculty of Technical Engineering, University of Bihac, Bihac, Bosnia and Herzegovina
{amel.toroman, edin.mujcic}@unbi.ba

**Abstract.** Nowadays, traffic is a big problem in modern world. The most frequent traffic jams occur at intersection controlled by traffic control. These traffic control are usually based on PLCs that control traffic without any feedback information on traffic density. Today, there is a need to use intelligent traffic control system that can monitoring traffic density at intersections. In this way load at intersections are significantly reduced. In this paper is proposed the smart traffic control system based on PLC. Designed system aims to give traffic lights a certain 'intelligence'. This system uses motion sensors to monitoring the current state of the intersection with regard to the load. Based on that, designed system make decisions about the duration length of the red and green light. The purpose of designed system is to reduce waiting for pedestrians. For the smart traffic control system realization, the model of traffic intersection is created. The model is based on the real traffic control system that is applied in Bihac, Bosnia and Herzegovina. Siemens PLC is used for the designing smart traffic control system. The SCADA system is designed for control and monitoring. After designed of smart traffic control system the experimental analysis is performed. The results of the experimental analysis are shown and thoroughly analyzed.

**Keywords:** Smart traffic control · Traffic lights · Programmable Logic Controller – PLC · PLC siemens SIMATIC S7-300 · WINCC SCADA

## 1 Introduction

Today, modern plants are almost completely unimaginable without the existence of an automatic control systems, the use of computers and electrical devices, sensors and systems that enable real-time monitoring and control.

The most important part of the modern plant is the automatic control system, and the main part of the above mentioned system is PLC (Programmable Logic Controller), which also represents the center of the system. PLC, based on the received input signals from the input devices, according to a particular program, generates the output signals to control the output devices.

PLC is today the most widely used automation devices in various industrial plants. The main reasons for this are: reliability and speed in harsh operating conditions, simple connections for input and output signals, modularity for different tasks, and simple and easy to understand programming. Each PLC device has a larger or smaller


S. Avdaković et al. (Eds.): IAT 2019, LNNS 83, pp. 470–480, 2020.
https://doi.org/10.1007/978-3-030-24986-1_37

number of digital inputs/outputs to which the components of the control circuit can be connected, and here are the most important actuators and sensors [1].

All PLC, regardless of size, have the same hardware structure, similar to other computer systems, adapted to an industrial environment, which has the same basic features [2], such as CPU (Central Processing Unit), Program and data memory, Communication part, Power supply network, Input (digital, analog), Output (digital, analog), and Extension part.

In addition to the basic elements of PLC mentioned above, there are also special purpose modules, such as: visual control modules, PID control modules, radio frequency modules, and others.

In order to automate the general processes, i.e. collect data from sensors and instruments located at distant stations, as well as their transmission and display at the central station for the purpose of supervision and control, SCADA (Supervisory Control And Data Acquisition) systems are used.

Under the SCADA system, the most commonly understood is a central system that monitors and controls the entire technological process from a certain distance. SCADA performs visualization of data and presents them in a form that is suitable for the operation of the operator. If necessary, it sends the control signals back to the PLC that execute them [3].

The SCADA system is especially suitable for processes that work 24 h a day and require continuous monitoring and control, such as traffic regulation or traffic lights.

Traffic lights are the most appropriate way of controlling traffic at the intersection. But, we are witnessing that they are not able to control traffic effectively when a certain side of the road has higher traffic than other side of the road [4].

Given that in Bosnia and Herzegovina in the implementation of the control system does not exist smart traffic lights, except the set buttons for pedestrians. All the traffic lights are in advance of the time set, or programmed for the usual amount of traffic at the crossroads where the traffic lights are set up. The motivation for the work was to offer a solution for more optimal regulation of traffic, and for this purpose has been created a smart traffic control system, shown in this paper.

In order to solve this problem, in this paper an intelligent traffic lights was designed, and an analysis of his work was performed. The model of the real intersection in Bihać, Bosnia and Herzegovina, is used as a model of intersection.

For the practical realization of the work, are used: PLC Siemens, SIMATIC S7-300 with CPU 314C-2 PN/DP, PIR (Pyroelectric Infrared Radial) sensors, and SCADA system was implemented for control and monitoring of the entire system.

## 2 Overview of Traffic Control Systems

Although the railway used the first forms of signaling that associate at the traffic light, the first traffic light came in 1886, in London. The traffic light resembled old railway signal boards, it had two signal bars (red, which was a sign that movement was stoped and green, which meant that movement was allowed), and the police officer managed the bars. So, it can not be say that the first traffic light, which also worked on gas, was an independent mechanical invention. American invention is and modern electrical signaling. The first green and red lamps were installed in 1914 in Cleyeland. Third, the yellow light first time is set up in 1918 at a traffic light in New York. In Europe (England) for the first time a tricolor signal was set up in 1926, while in 1939 the first traffic light was implemented in Serbia (Belgrade), then in 1941 in Croatia (Rijeka) and in 1964 in Bosnia and Herzegovina (Sarajevo) [5].

Further, during the eighties, a new lighting technology was introduced, known as light emitting diode, i.e. LED (Light Emitting Diode) [6].

The United States (Oregon State) has implemented a project replacing traffic lights with new LED design in 1993, while during 2012, in Bosnia and Herzegovina (Tuzla), a project for replacing existing traffic lights with energy efficient LED traffic lights was implemented. Nowadays, other forms of energy are used to start a traffic light, so in Russia (Moscow) began the first traffic light using solar energy.

As we can see from the historical overview, in the last 70 years, several innovations have been introduced on the original traffic control concept, aimed at redundancy in increasing safety, more efficient and more economical reflectors, etc.

That's right, with the advancement of technology in the design of microprocessors, in the late 1960s, led to a revolution in control systems. An idea emerging about the creation of an electronic microprocessor control device that could easily be reprogrammed in the event of a change in the control tasks. The first such devices, which were named Programmable Logic Controllers, or abbreviated PLC, was created [2].

By using PLC and sensors, it is possible to achieve a more efficient traffic control.

Nowadays, in Europe and in the world, in the last few years there has been the implementation of a smart traffic light control system.

Namely, in 2016, the Netherlands (s-Hertogenbosch) implemented a smart traffic light control system, more specifically a traffic light that adapts to each individual, which aims to reduce crowds in traffic. That is, reduce the waiting time of participants in traffic, whether they drive a car, bicyclists or pedestrians. This intuitive system uses complicated algorithms and detection zones to determine the speed at which someone approaches intersection. Estimates its direction of movement, and ultimately calculates information and switch on the green light as soon as the crossing is safe. If there is only one vehicle waiting at the traffic light, it happens that the system leaves the green light for only 4 s, which is enough to cross. The system proved to be extremely safe, and each intersection is covered by a large number of cameras and all data is stored in the database [7]. This is done in case it is sometimes necessary in the court, but also for the analysis and further improvement of the system itself.

Also, the United Kingdom (London) has implemented a smart traffic lights control system that determine the interval based on the number of pedestrians waiting for the green light. The software is based on the evolution of technology which, depending on the traffic density, regulates the interval of traffic lights for vehicles. This system, instead of a car, "counts" the pedestrians in front of the pedestrian crossing and depending on their number, extends or shortens the duration of the green light. This system is currently being developed by Peek, Siemens and the British TRL [8].

At the end of 2017, a system of "smart traffic lights" was implemented in Serbia (Belgrade), which aims to increase the current percentage (17%) of the use of trams in relation to other public transport. Namely, unlike the bus, they are slow, because they stand at each traffic light, stops, and it takes time for passengers to get in and out of the vehicle. The solution for accelerating the tram is to set up "smart traffic lights" with sensors, which would open up at the tram station and thus accelerate the flow of the vehicle [7].

Over the coming years, using the traffic light will try to achieve increased energy efficiency. Namely, in Vienna a project for equipping a traffic light with special sensors for monitoring the weather and environment was launched. The aim is for all traffic lights in Vienna to be equipped with high quality sensors. Where is the basis of collected data can be discovered urban islands of heat or through smart control of traffic flows can improve air quality in the city, since that with the braking and accelerating frees significantly more harmful substances than when traffic flows normaly. In the first phase, sensors for measuring the temperature and humidity of the air will first be set up, and later, nitrogen monoxide, sulfur dioxide and noise will be measured [9].

## 3 Designing a Smart Traffic Control System Using PLC Siemens SIMATIC S7-300

In this part of the paper, the designing of smart traffic control system is described, and for this purpose, in addition to sensors and actuators, PLC is used as a central control element.

The reason for this implementation was that it is in context of current urban traffic, necessary to implement signaling systems that will provide the best traffic conditions [10].

Thus, practical realization is a signaling system that is real-time adjustable controlled by PLC, which consists of basic components, such as: PIR sensors, PLC Siemens Simatic S7-300, MPI-PROFIBUS DP interface, computer, as well the object of the control-intersection (Fig. 1) [11].

In Fig. 1, the block diagram of the control of intersection is shown. At the PLC's input, there are PIR sensors designed to identify pedestrians and shorten the duration of the red signaling light during the waiting performed by pedestrians, or prolonging the duration of the green light. The PLC communication with the computer is carried out through the MPI-PROFIBUS DP interface, where further, the computer communicates with the end-system, i.e. intersection [11].

### 3.1 Designing a Crossroads Model with Traffic Lights

The central control element for practical implementation was used by PLC the Siemens company, SIMATIC S7-300 with the CPU 314C-2 PN /DP.

The CPU 314C-2 PN /DP, shown in Fig. 2 is equipped with a microprocessor, extensive memory, flexible expansion capability, multi-point interface (MPI), PROFIBUS DP interface, Ethernet interface, and integrated inputs and outputs [11].

Input and output modules are used for data collection and execution, so PLC with CPU 314C-2 PN /DP has 24 digital and 5 analog inputs, and 16 digital and 2 analog outputs. The aforementioned model does not have external memory, except the attached Micro Memory Card, and the working memory size is 129 kB, while the external power supply is 24 V DC [12].

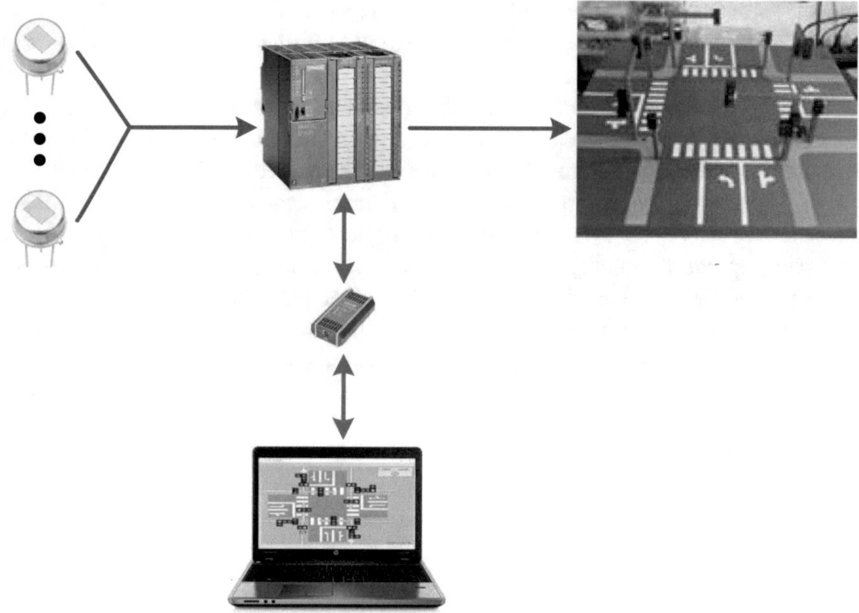

**Fig. 1.** Block diagram of the intersection [11]

**Fig. 2.** PLC SIMATIC S7-300 CPU 314C-2 PN/DP [11]

For programming the PLC SIMATIC S7-300 CPU 314C-2 PN /DP uses basic programming languages located within the Simatic Manager STEP7 V5.5 pro gramme package, such as: ladder diagram (LAD - Ladder logic), diagrams function blocks (FBD - Function Block Diagram), as well as command lists (STL - Statement List). There are also additional options such as SCL (Structured Control Language) and GRAPH.

As can be seen from the shown of the complete realization of the traffic light model, shown in Fig. 3, for the realization of the model, 8 traffic lights were needed for the cars, 8 pedestrian traffic lights and 4 smaller traffic lights with sign for turning. LEDs in red, yellow and green were used for the bulbs.

For the programming of the work of traffic lights, the ladder programming language of the Simatic Manager STEP7 package was used. And for the graphical part, through

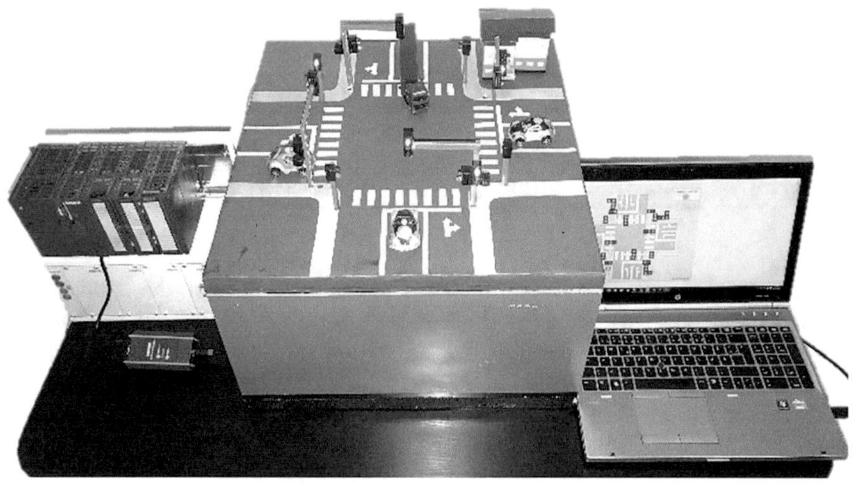

**Fig. 3.** Controlling intelligent traffic lights using a PLC [11]

which surveillance and traffic control is performed, WinCC Explorer, or SCADA system, was used.

As can be seen in Fig. 3, the overall control of the traffic lights is done by PLC controller. Its communication with the computer is performed through the MPI port on the PLC, while the DP port is used to connect the PLC to the PROFIBUS network. On a computer using WinCC SCADA performs overall surveillance and control of the intersection.

## 4 Overview and Analysis of Experimental Results

In this part of the paper, analysis of an intelligent traffic lights was performed. An intersection in Bihać, Bosnia and Herzegovina, is used as the model intersection.

Activating the PIR sensor changes the traffic light mode. On the side where the PIR sensor is activated, the duration of the red light for pedestrians is shortened, or if the green pedestrian light is currently green, the duration of the green light is prolonged.

In the experimental analysis, four cases were considered: the activation of the pedestrian sensor during the red light, the activation of the pedestrian sensor while the green light for them [11], the activation of two sensors on the orthogonal sides of the intersection, and simultaneous activation three or more sensors.

If the sensor is activated during the red light, waiting of pedestrians at a pedestrian crossing, or the duration of the red light will be shortened, depending on the pedestrian's arrival at the pedestrian crossing. Namely, the red light lasts for a maximum of 65 s, which will be spend witing by the pedestrian if it does not activate the sensor. If the pedestrian activates the sensor, waiting times for pedestrians will be realized according to Table 1.

**Table. 1.** Activating the sensor during the duration of the red light at the pedestrian traffic light

| Duration of the red light [s] | Waiting for pedestrians [s] | Time of shortening [s] |
|---|---|---|
| 65 - 60 - 55 | 50 - 45 - 40 | 15 |
| 50 - 45 | 40 - 35 | 10 |
| 40 | 35 | 5 |
| 35 - 0 | 35 - 0 | 0 |

If the sensor is activated during the duration of the green light, the duration of the green light will be extended, depending on the pedestrian's arrival at the pedestrian crossing. Namely, the green light lasts a maximum of 15.5 s, which will be available to the pedestrian if the sensor is not activated. If the pedestrian activates the sensor, the estimated time for crossing pedestrians through a pedestrian crossing will be realized according to Table 2.

**Table. 2.** Activating the sensor during the duration of the green light at the pedestrian traffic light

| Duration of the green light [s] | Crossing of pedestrian [s] | Time of extension [s] |
|---|---|---|
| 15.5 - 9.9 | 15.5 - 5.6 | 0 |
| 10 - 0 | 15.5 - 10 | 10 |

If two sensors are activated on the orthogonal sides of the intersection (road 1 - sensor 1, road 2 - sensor 2), there are several cases:

(a) If sensor 1 is activated before sensor 2 is activated, as shown in Fig. 4, the priority will be tracked with sensor 1, and depending on whether the sensor is activated for a red or green light, the actions shown in Tables 1 and 2 will be performed.
Figure 4 illustrates the work of the traffic light when activating the sensors on the orthogonal sides of the intersection. Analized is the case from Fig. 4(a), or activating the sensor 1 before activating the sensor 2 during the duration of the red light at the pedestrian traffic light.
As can be seen from Fig. 5, if on the orthogonal sides for the duration of the red light at the pedestrian traffic light, sensor 1 is activated before the activation of the sensor 2, priority will be given to the first activated sensor (sensor 1). So that the waiting for the pedestrians will be shortened depending on the duration of the red light, i.e. depending on the time of their arrival.
Figure 6 illustrates the work of the traffic light when activating the sensors on the orthogonal sides of the intersection. Analized is the case from Fig. 4(b), or activating the sensor 1 before activating the sensor 2 during the duration of the green light at the pedestrian traffic light.
As can be seen from Fig. 6, if on the orthogonal sides for the duration of the green light at the pedestrian traffic light, sensor 1 is activated before the activation of the sensor 2, priority will be given to the first activated sensor (sensor 1). So that the

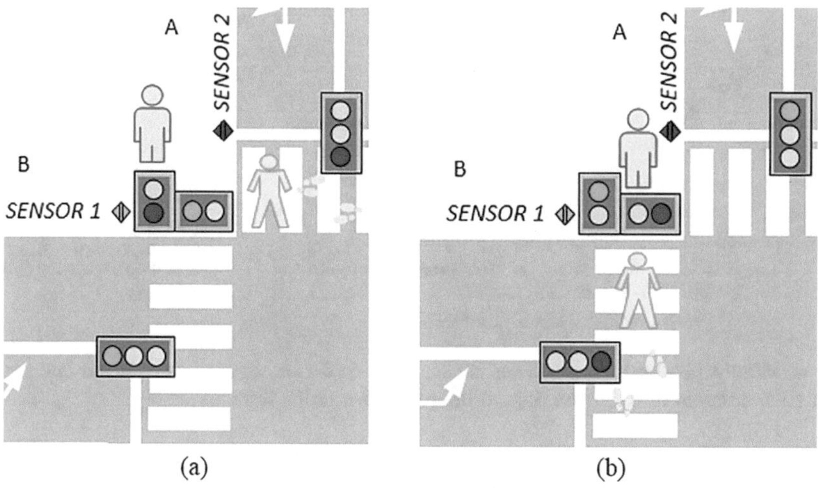


**Fig. 4.** The case of activating the sensor 1 before activating the sensor 2: (a) the red light at the pedestrian traffic light; (b) the green light at the pedestrian traffic light

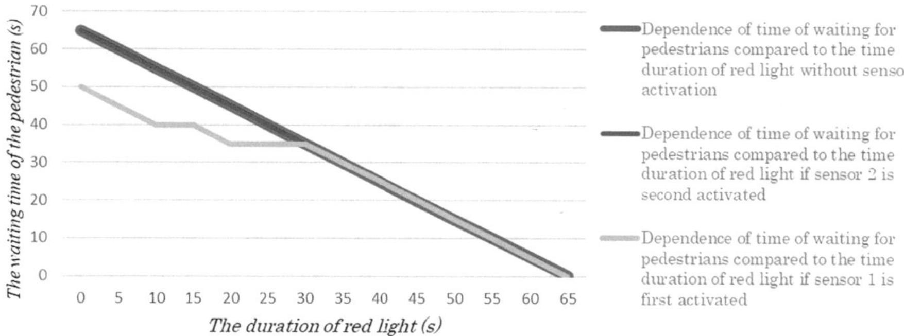


**Fig. 5.** Results obtained by activating the sensor 1 on the orthogonal sides of the intersection during the duration of the red light at the pedestrian traffic light

time for the crossing of pedestrian will be extended depending on the duration of the green light, i.e. depending on the time of their arrival.

(b) If sensor 2 is activated before sensor 1 is activated, the priority will be tracked with sensor 2, and depending on whether the sensor is activated for a red or green light, the actions shown in Tables 1 and 2 will be performed, and presented in case a).

(c) If both sensors are simultaneously activated, as shown in Fig. 7, which is a rare case, the pre-conditions will be kept.

Figure 7 illustrates the work of the traffic light when activating the sensors on the orthogonal sides of the intersection. Analized is the case from Fig. 7(a), or simultaneously activating both sensors during the duration of the red light at the pedestrian traffic light.

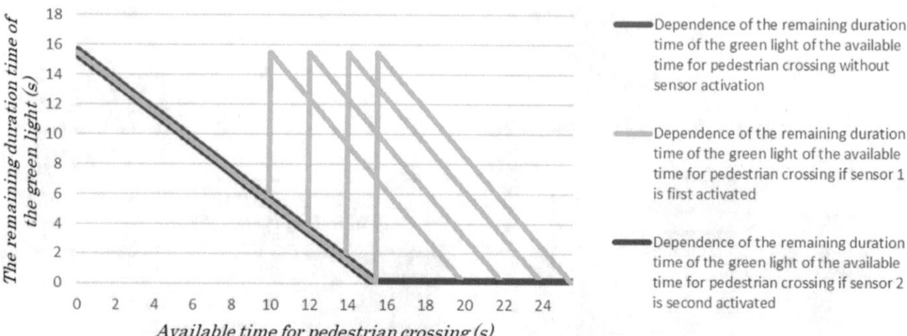


**Fig. 6.** Results obtained by activating the sensor 1 on the orthogonal sides of the intersec-tion during the duration of the green light at the pedestrian traffic light

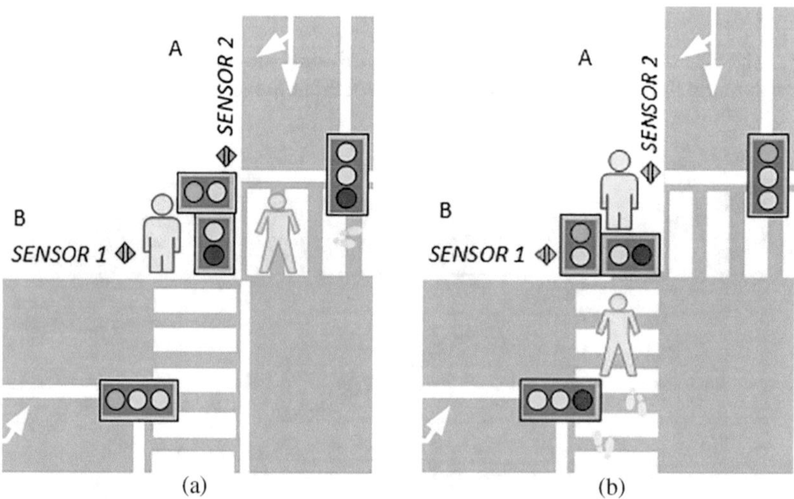


**Fig. 7.** The case of simultaneous activation of both sensors: (a) the red light at the pedestrian traffic light; (b) the green light at the pedestrian traffic light

As can be seen from Fig. 8, if on orthogonal sides of the intersection during the red light at the pedestrian traffic light, both sensors are activated simultaneously, priority will not be given to one activated sensor, that is, at the traffic lights the previous states will be retained.

Figure 9 illustrates the work of the traffic light when activating the sensors on the orthogonal sides of the intersection. Analized is the case from Fig. 7(b), or simultaneously activating both sensors during the duration of the green light at the pedestrian traffic light.

As can be seen from Fig. 9, if on orthogonal sides of the intersection during the green light at the pedestrian traffic light, both sensors are activated simultaneously,

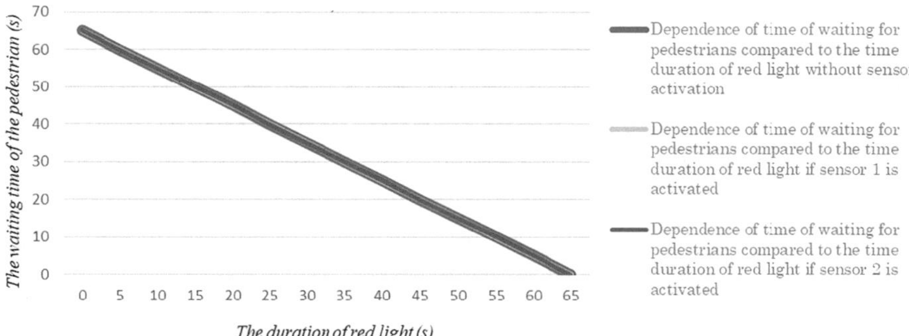


**Fig. 8.** Results obtained by simultaneously activating both sensors during the red light at the pedestrian traffic light

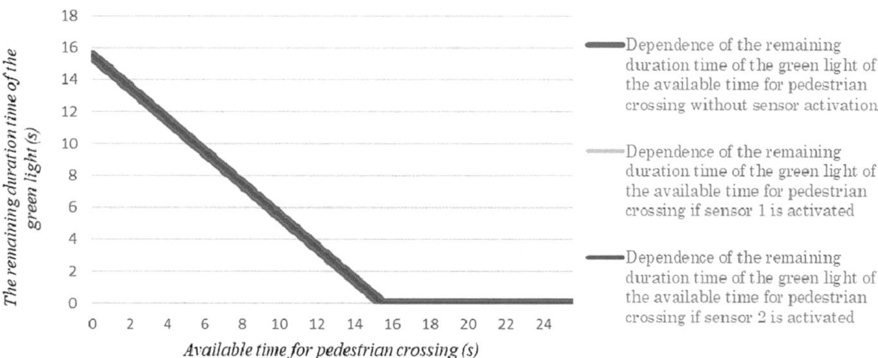


**Fig. 9.** Results obtained by simultaneously activating both sensors during the green light at the pedestrian traffic light

priority will not be given to one activated sensor. That is, at the traffic lights the previous states will be retained.

A very rare case is the simultaneous activating of three or more sensors, but if it happens, the traffic lights will not change, that is, on the traffic lights, the previous conditions will be retained.

## 5 Conclusion

The emergence of programmable logic controllers (PLC) is one of the milestones in the development of technology and one of the greatest technical achievements that marked the 20th century.

Today, automation in the industry has become unthinkable without PLC. As stated in the paper, and as demonstrated by the analysis of the results obtained from testing, PLC provides many possibilities and improvements in control.

In the practical realization of the work, it turned out that with the use of PLC significant improvements in the control are made by the use of sensors, which completes the automation itself, and in a more efficient way solves the control problem.

With the demonstration of intelligent traffic lights control, PIR sensors were used, PIR sensors were placed at pedestrian traffic lights. The goal was a using sensors to reduce the time spent waiting for crossing the pedestrian crossing, depending on the time of arrival of the pedestrians during the duration of the red light, or prolong the duration of the green light after a certain time, which would allow prolongation of the time for their crossing.

Experimental results showed that in places where the PIR sensors were used, a significant reduction in the waiting time of the pedestrian was achieved, or an extension of the time for crossing at the pedestrian crossing, depending on their arrival.

Recommendations for further research and possibilities of upgrading this work are the use of sensors in the car tape, regulation of daily /night traffic frequency (traffic of cars and pedestrians), as well as the use of additional sensors for recognizing certain sound frequencies, which would allow giving advantages vehicles with the right of priority (ambulances, firefighters and police vehicles).

## References

1. Siemens PLC Simatic S7-300, Elektronski fakultet, Katedra za automatiku, Univerzitet u Nišu. Dostupno: http://automatika.elfak.ni.ac.rs/82-icecation/100-plc-reklama.html
2. Application of PLC in mechatronics, Department of Energy Electronics and Converters, Faculty of Technical Sciences, Novi Sad. http://www.keep.ftn.uns.ac.rs/predmeti/masinski_4g_PrimenaPLC_u_meh/Industrijs%20kaInformatika.pdf. Accessed May 2018
3. SIMATIC HMI, WinCC SCADA, SIEMENS: http://w3.siemenscom/mcms/automation/en/human-machine-interface/Pages/Default.aspx. Accessed July 2018
4. Srivastava, M.D., Prerna, S.S., Sharma, S., Tyagi, U.: Smart traffic control system using PLC and SCADA. Int. J. Innov. Res. Sci., Eng. Technol. **1**(2), 169–172 (2012)
5. History of Traffic Signs: https://www.educentar.net/Vijest/10411/Povijest-prometnih-znakova. Accessed Apr 2019
6. Osorio, R.M.R., Ramón, M.C., Otero, M.Á.F., Navarrete, L.C.: Smart control system for LEDs traffic-lights based on PLC. In: Proceedings of the 6th WSEAS International Conference on Power Systems. Lisbon, Portugal, 22–24 September 2006
7. Intelligent traffic lights in the Netherlands: http://www.cgauto.me/zanimljivosti/item/3060-inteligentni-semafori-u-holandiji. Accessed Mar 2019
8. Intelligent traffic lights in the Netherlands: http://emedjimurje.rtl.hr/automobilizam/novi-pametni-semafori. Accessed Mar 2018
9. Toroman, A.: Primjena PLC-a Siemens SIMATIC S7-300 za upravljanje inteligentnim semaforima. Master's thesis. Technical Faculty, Bihac (2018)
10. Barz, C., Todea, C., Latinović, T., Preradović, D.M., Deaconu, S., Berdie, A.: Intelligent traffic control system using PLC. In: International Conference on Innovative Ideas in Science (IIS2015)
11. Toroman, A., Mujčić, E.: Application of industrial PLC for controlling intelligent traffic lights. In: 25th Telecommunications Forum TELFOR 2017, IEEE. Belgrade, Serbia (2017)
12. SIEMENS (6ES7314-6EH04-0AB0): Data sheet, Siemens AG, Germany, Editon 08/03/2017

# Remote Monitoring and Control System for Greenhouse Based on IoT

Una Drakulić(✉) and Edin Mujčić

Faculty of Technical Engineering,
University of Bihac, Bihac, Bosnia and Herzegovina
una.94@hotmail.com, edin.mujcic@unbi.ba

**Abstract.** Nowadays, the need for greenhouse production is growing more and more. This is affected by global climate change with high temperature oscillations, as well as an increasing number of natural disasters that can partly or completely destroy production on open fields. In addition, greenhouse production provides constant yields (fresh fruits, vegetables etc.) during the whole year. However, if the greenhouse is not well designed the yields from greenhouse production are declining considerably. This is the case with most greenhouses in B&H. Therefore, it is necessary to design the system that will maintain optimum conditions in the greenhouse with minimum energy consumption. In this paper is designed the greenhouse (greenhouse model size 120 × 60 cm). The microclimatic conditions in the greenhouse model are approximately the same as the real greenhouse conditions. Several types of plants in the greenhouse model are planted. The system is designed to control all important parameters in the greenhouse with minimal energy consumption. To control the parameters inside the greenhouse, the remote controller with touch 3.5 inch display is designed. This way is enabled for all users who are allowed access to review, analyze and make certain conclusions about the work of the greenhouse. Also, all relevant information about microclimate conditions inside the greenhouse model are sent to the cloud (IoT). In this way, remote access to all relevant data on microclimatic conditions in the greenhouse model is enabled, as well as analysis of the same.

**Keywords:** Greenhouse · Internet of Things · NodeMCU ESP8266 · Remote control · Energy consumption · Microcontroller

## 1 Introduction

Production in a protected area is an important and fast growing component of the agricultural industry [1]. In B&H, this production is still in development, but it can be noticed that more and more populations have the need for this type of production.

Protected areas are usually made of glass. In recent times, a greenhouse plastic film for the archive greenhouse effect is often used [2]. In this case, the growth of plants is affected by: the thickness of the greenhouse plastic film, the permeability of the amount of light and the possibility of heat retention.


S. Avdaković et al. (Eds.): IAT 2019, LNNS 83, pp. 481–495, 2020.
https://doi.org/10.1007/978-3-030-24986-1_38

The permeability of the amount of light is essential for plant growth. Clear greenhouse plastic film is used for growth lower plants because they require direct light.

While the diffusion greenhouse plastic film is used for the growth of higher plants. For these plants, scattered light is suitable. Diffuse greenhouse plastic film reduces sunburn and the temperature inside the greenhouse. In addition to these there are anti-condensing greenhouse plastic films to prevent condensation inside the greenhouse [3].

In the greenhouse, essential parameters can be controlled for creating optimal conditions for plant growth such as: temperature, lighting, airing, watering, etc.

In the opinion of experts, primarily economists and agronomists, it is necessary to construct and equip one or more greenhouses of at least 1000 $m^2$ of arable land, in which intensively produce fruit, vegetables and /or flowers all year round for economically cost-effective production [4].

For easier production, the greenhouse can be automated. In this way, the user of day-to-day duties is released, which are crucial for maintaining optimal conditions in the greenhouse. In automated greenhouse, the user can, regardless of his location, use a computer or some other smart device to control the lighting, temperature, soil moisture and air quality [5].

All relevant information about the microclimate conditions in the greenhouse are sent to the cloud (IoT). The reason for this is easier data availability and analysis of the same [6].

The greenhouse allows us to control lighting in greenhouse (duration and type of lighting). In addition to the sunlight, alternative light sources can be used to provide ideal conditions for plant growth [7]. One of these alternative light sources is LED lighting. The use of LED lighting in greenhouses was first presented in USA 1991. LED lighting has low energy consumption and gives light frequencies that are suitable for growing "useful" plants in the greenhouse [8]. Also, studies is shown how insects could be detected and expel from the greenhouse by using LED lighting [9].

In this paper an automated greenhouse system is proposed for control of all relevant microclimate conditions in the greenhouse.

## 2 Design and Implementation Remote Monitoring and Control System for Greenhouse Based on IoT

In this part of paper is described design and implementation remote monitoring and control system for greenhouse based on IoT. For easier implementation of the projected system, the model of greenhouse is made. Model of greenhouse is made from wood and special transparent foil. For the construction metal batten is used. Model of the greenhouse is shown in Fig. 1.

The greenhouse model consists of four separate parts for planting different plants. In every part is maintain different level of soil moisture. Based on this, the growth of plants is monitored and analyzed.

**Fig. 1.** Model of the greenhouse

## 2.1 Automated Control Greenhouse System

Automated control greenhouse system is designed to control microclimatic conditions in the greenhouse model. The functional block diagram of projected system is shown in Fig. 2.

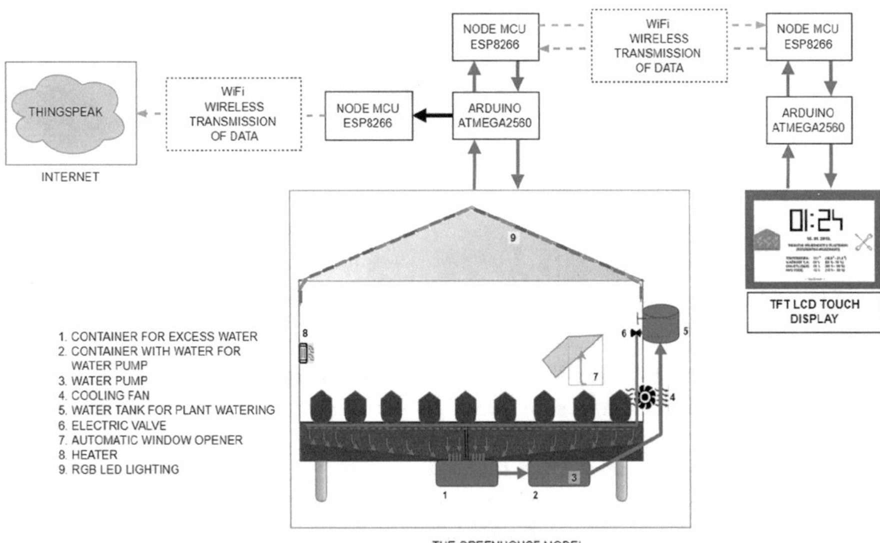


**Fig. 2.** Functional block diagram of the projected system

In Fig. 2 is shown that the projected system is very complex and consists from several independent subsystems. These subsystems are:

1. Watering subsystem in the greenhouse,
2. Lighting subsystem in the greenhouse,
3. Temperature control subsystem in the greenhouse,
4. Air quality subsystem in the greenhouse,
5. Remote monitoring and control subsystem for greenhouse and
6. IoT subsystem for greenhouse using ThingSpeak cloud.

Each subsystem is described below in this paper.

### 2.1.1 Watering Subsystem in the Greenhouse Model

In Fig. 3 is shown functional block diagram for the soil moisture and water level measurement. Soil moisture measurement is performed for each part of the greenhouse model separately. In water tank are placed two water level sensors, to detect low and high water level.

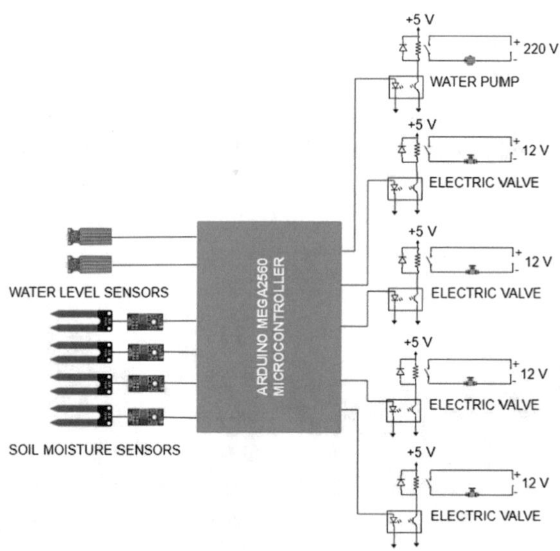


**Fig. 3.** Functional block diagram for the soil moisture and water level measurement

The watering subsystem in the greenhouse model is realized using water pump to fill the water tank. Water tank is placed at higher altitude than the greenhouse model bottom. Depending on which water lever sensor is active water pump is turn on or off. On the water tank are connected four water pipes. Every water pipe has electric valve to turn on or off the watering. Depending on soil moisture sensors, that are separate for each part for planting, electric valves are turn on or off. If the soil moisture sensor reads value below the desired value which the user had entered, electric valve in that part for planting is turn on. When soil moisture sensor match desired value, electric valve is turn off.

In Fig. 4 is shown functional block diagram for watering subsystem in the greenhouse model.

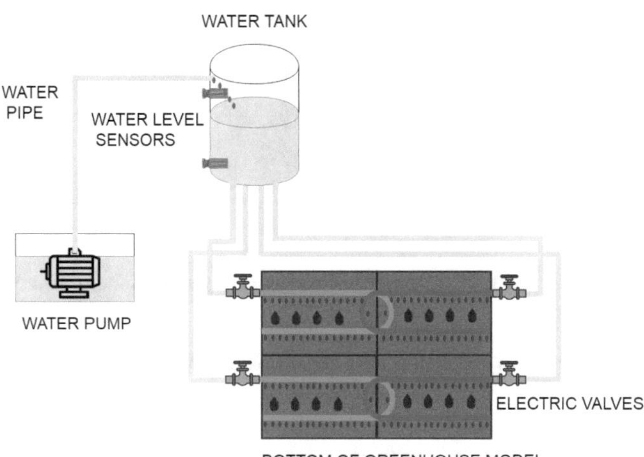


**Fig. 4.** Functional block diagram for the watering subsystem in the greenhouse model

### 2.1.2 Lighting Subsystem in the Greenhouse Model

Lighting subsystem in the greenhouse model is realized using the RGB LED strips. For the greenhouse model is used red, blue and green LED color. These colors are used because of their affect to the plant growth. Lighting subsystem in the greenhouse model is realized using photoresistor, Arduino board with ATMega2560 microcontroller and transistor BD137.

Arduino is microcontroller board with Atmel microcontroller ATMega2560 and additional parts needed to operate the microcontroller and external connectors for easy operation and connecting to external elements [10, 11].

The level of light in the greenhouse model depends of the level of the day light. RGB LED light is brighter if day light is darker. In this way plants in the greenhouse gets required light day and night. In Fig. 5 is shown the functional block diagram of the lighting subsystem in the greenhouse model.

Each type of the plants reacts differently depending on the color of the light it absorbs. The source of energy in plants is light. The process of converting light into food is photosynthesis. The color of light that plants absorbs has influence on the amount of energy that plant provides. The reason is that in the light there are different wavelengths. Depending on if wavelengths are short or long, light provides different level of energy.

The light color that provides the most energy is on ultrapurple end of the spectrum of colors. Purple lights have short wavelengths which means higher energy. At the end of the spectrum of colors is red light. Red light has long wavelengths which means

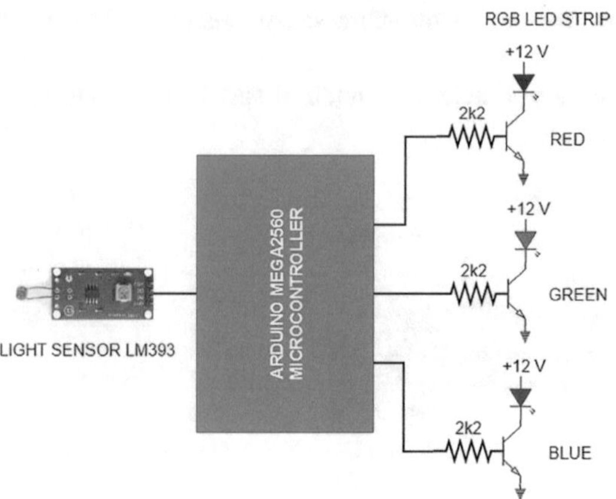


**Fig. 5.** Functional block diagram of the lighting subsystem in the greenhouse model

lower energy. The green light has a sooting affect on plants because it is the color of their pigment of chlorophyll. Because of that green light do not affect to energy level in plants [11].

Different light colors help plants to achieve different goals. Blue light helps to stimulate vegetative growth and leaves growth. Red light combined with blue light allows early flowering of plants [12]. Knowing that different color of light can affect on how plant grows people can make the most from plants.

Advanced LED technology allows people to control the strength and the color of light. Lighting subsystem is realized in the greenhouse model to encourage fast growth and higher yields.

### 2.1.3 Temperature Control Subsystem in the Greenhouse Model

Temperature control subsystem in the greenhouse model is used to control temperature in the greenhouse model. Temperature control subsystem consist of: temperature sensor DS18B20, ATMega2560 microcontroller, two optocouplers with relay and heater. In the case when the temperature in the greenhouse model is low the heater is turn on. The heater will be turn on until the temperature grow to desired value. In Fig. 6 is shown temperature control subsystem in the greenhouse model.

Temperature control subsystem also can periodically (with manual user speed) open the window and trigger the fan.

### 2.1.4 Air Quality Subsystem in the Greenhouse

Air quality subsystem in the greenhouse is designed to track air quality in the greenhouse model. Air quality subsystem consist of: MQ2 gas sensor, ATMega2560 microcontoller, cooling fan and servo motor. In case air quality is lower than the desired air quality cooling fan is turn on for airing. Also, in the greenhouse model is implemented a window for airing which is open by servo motor. The greenhouse model is designed to have a possibility of adding $CO_2$ to increase the productivity of plants. In Fig. 7 is shown functional block diagram for air quality subsystem.

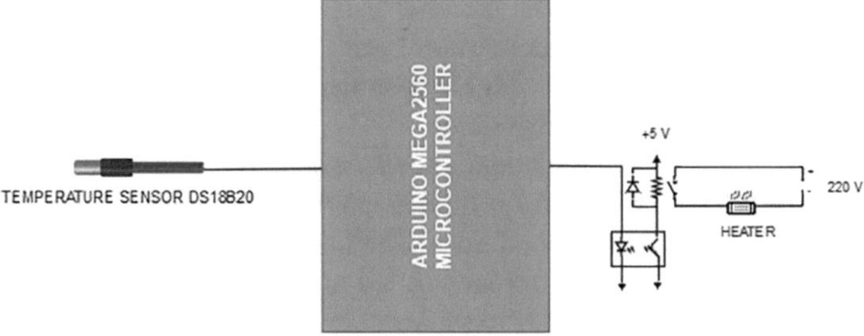


**Fig. 6.** Functional block diagram for temperature control subsystem

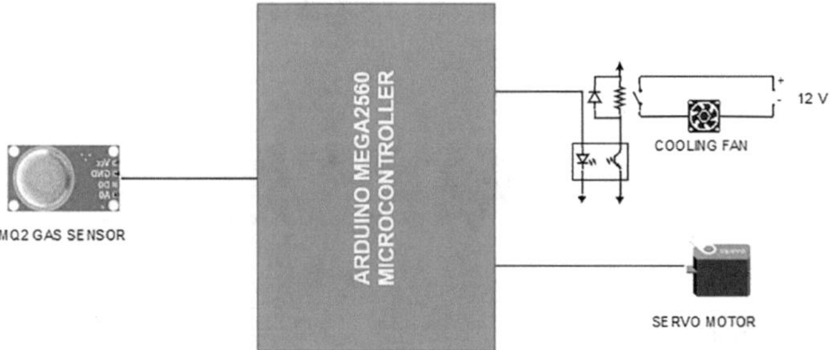


**Fig. 7.** Functional block diagram for air quality subsystem

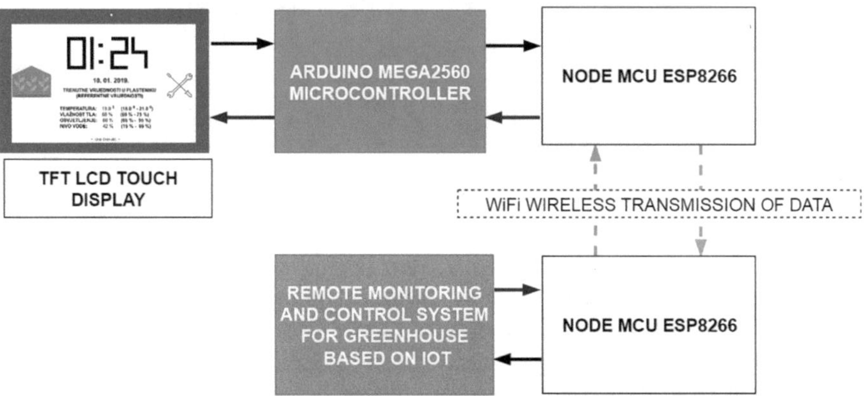


**Fig. 8.** Functional block diagram for the remote monitoring and control subsystem

### 2.1.5 Remote Monitoring and Control Subsystem for Greenhouse

Remote monitoring and control subsystem for greenhouse is designed for remote monitoring and control the greenhouse model. Designed subsystem consists of: screentouch displej 3.5 inch and Arduino board with ATMega 2560 microcontroller which is connected to Node MCU esp8266. This part of remote monitoring and control subsystem is on client side. The second part of remote monitoring and control subsystem is on server side. The second part consists of Arduino board with ATMega2560 microcontroller which control the entire designed greenhouse system and regulate all the parameters. In Fig. 8 is shown the functional block diagram for remote monitoring and control subsystem.

In this way, the user can track and set all desired parameters in the greenhouse model. The distance of the user from the greenhouse model can be several hundred meters depending on the current conditions. The main part of this subsystem is remote controller which is realized using 3.5 inch tft lcd touch display. This remote controller must be programmed and for that is used Arduino IDE environment.

Automation of the greenhouse means that all the work that needs to be done in the greenhouse is done by devices. In order to enable automation of the greenhouse, remote controller is made. Remote controller allows user to be away from the greenhouse. The user has access to all relevant data and can set the desired parameters in the greenhouse. Remote controller is shown in Fig. 9.

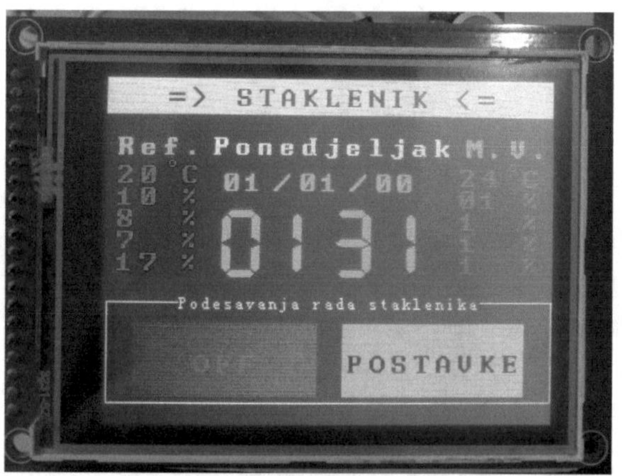


**Fig. 9.** Remote monitoring and control device for the greenhouse system

In Fig. 10 we can see that the first window display time and date, desired values (left) and measured values (right). At the bottom of the display are two buttons. The first button is for turn on or off the designed greenhouse system and the second for settings. When button for settings is pressed new window display (see Fig. 10).

In order to protect the user and his greenhouse system, it is implemented a password to proceed other settings. If the entered password is incorrect, the message: 'Lozinka je netacna!' (engl. The password is incorrect!) is showing and a new password entry is required. If the entered password is correct, the message: 'Dobrodosli!' (engl. Wellcome!) is shown and new window is displayed (see Fig. 11).

In Fig. 11 is shown main menu for settings. The user has tri buttons to choose:

(1) Time and data (in Fig. 11 'DATUM I VRIJEME')
(2) Settings in the greenhouse model (in Fig. 11 'PODESAVANJE Staklenika')
(3) Main window display (in Fig. 11 'POCETNA')

**Fig. 10.** Window display for password

**Fig. 11.** Main menu for settings

If the button 'Time and date' is pressed, window in Fig. 12 is displayed.

After desired values are entered, all data are saved in EEPROM memory in microcontroller ATMega2560.

In Fig. 13 is shown window display for button 'Settings in the greenhouse model'.

All entered data, using Wi-Fi wireless network, are sent to Arduino board with ATMega2560 microcontroller in the greenhouse model.

### 2.2 IoT Subsystem for Greenhouse Using ThingSpeak Cloud

In Fig. 14 is shown IoT subsystem for greenhouse. This system is realized using Node MCU esp8266 module which is connected to Arduino board with ATMega2560 microcontroller in the greenhouse model. Node MCU esp8266 model, using Wi-Fi wireless network, sends received data to the cloud. For this purpose is used ThingSpeak platform.

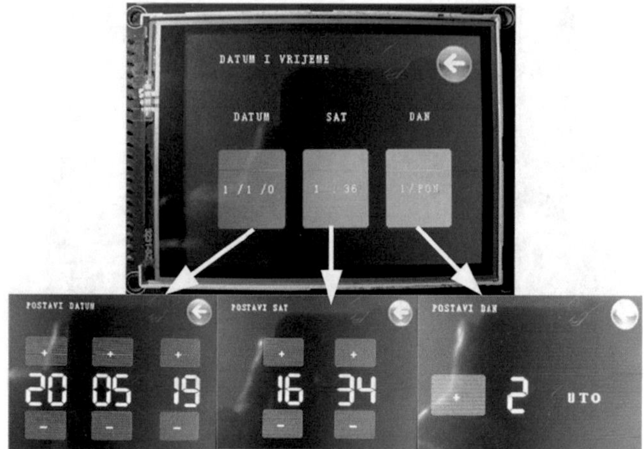


**Fig. 12.** Time and date settings

ThingSpeak is the open IoT platform with MATLAB analytics [13].

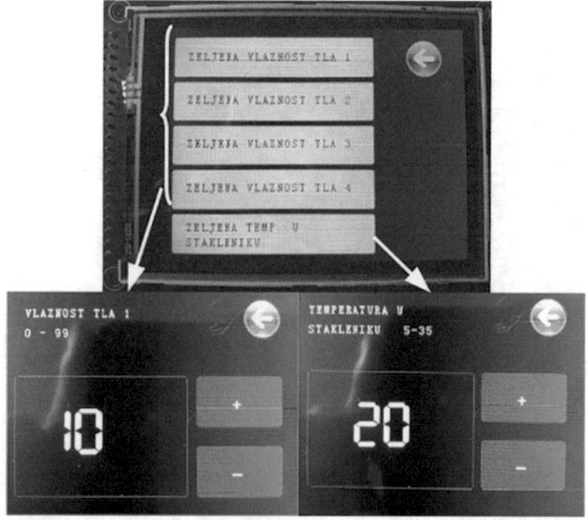


**Fig. 13.** Settings in the greenhouse model

In this way is enabled data monitoring in the greenhouse model for a longer period of time as well as overview of all relevant data from any location in the world that has access to the Internet.

Because our system is based on the use of wireless Wi-Fi network, it is important that there are no physical gaps that would significantly reduce the power of the Wi-Fi signal [14].

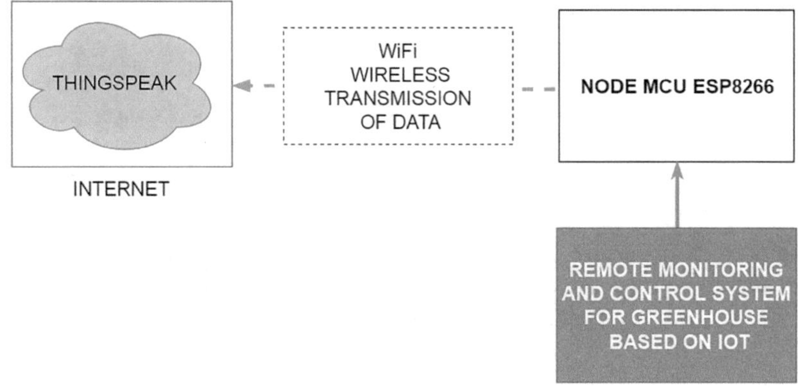


**Fig. 14.** Functional block diagram for the IoT subsystem

In Fig. 15 is shown appearance of window when user access to cloud to see data.

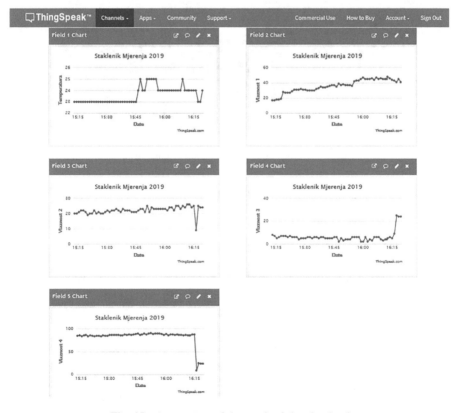

**Fig. 15.** Appearance of the received data in cloud

# 3 Final Remote Monitoring and Control System for Greenhouse Using IoT

After connecting all the parts we get the final system which is shown in Fig. 16.

**Fig. 16.** Appearance of the final designed system

Four types of plants are planted in the greenhouse model (two types of flowers, strawberries and peppers).

All parts of the projected system, except the actuators and sensors, are located below the greenhouse. This is shown in Fig. 17.

## 3.1 Experimental Analysis of the Designed System

Remote monitoring and control system for greenhouse is automated system. This means that system consists of sensors. Based on the values that sensors read actuators will work. Using remote controller the user sets desired values for different senosors.

For example, if user sets desired temperature in the greenhouse model 21 °C it will maintain that temperature all time. If temperature sensor DS18B20 reads lower temperature the heater in the greenhouse model will turn on until the temperature rise up to desired temperature. If temperature sensor DS18B20 reads higher temperature the cooling fan will turn on and the window will open until the temperature drop down to desired temperature.

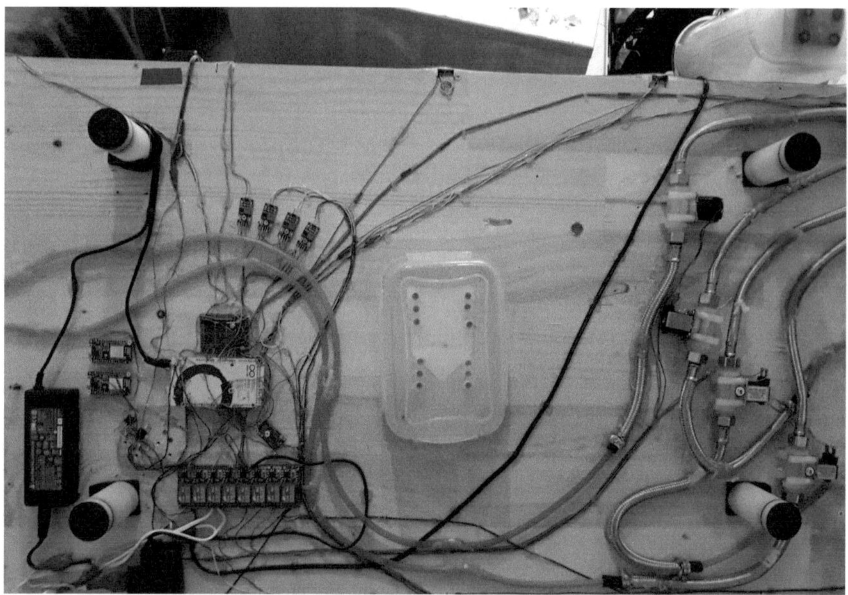

**Fig. 17.** Appearance of the final designed system from bottom side

**Fig. 18.** Appearance of the watering part and the airing part in the greenhouse model

When user sets desired soil moisture in the greenhouse model 60% it will maintain that moisture. If soil moisture sensor reads lower moisture then water pump will turn on. Water pump is turn on to fill water tank in which are two water level sensors, for low water level and high water level. When lower level sensor is active water pump is turn on and if higher water level sensor is active water pump is turn off.

In Fig. 18 is shown watering part and the airing part in the greenhouse model.

Lighting subsystem in the greenhouse model is realized so plants always have needed amount of light. If the day light is normal, i.e. bright, RGB LED lights are turn off. When day light is getting lower, RGB LED lights are turning on. If it is dark outside, RGB LED lights are turn on maximum.

In Fig. 19 is shown realization of lighting subsystem in the greenhouse model.

**Fig. 19.** Appearance of the lighting subsystem in the greenhouse model

## 4 Conclusion

In this paper is described design and implementation remote monitoring and control system for greenhouse based on IoT. Based on the experimental analysis we can see that the projected system is working properly. We can also conclude that the projected system can, with minimal changes, be applied to the real system. Possible future upgrade of the projected system is adding a $CO_2$ production system. $CO_2$ enrichment in greenhouses allows crops to meet there photosynthesis potential. Enriching the air with $CO_2$ can be done by means of the combustion of natural gas or with liquid $CO_2$.

On the basis of the obtained results, apply (with minor changes primarily to the actuator) automated greenhouse system for controlling all microclimate conditions in a greenhouse of larger dimensions and apply a knowledge flow on the influence of certain factors on plant growth.

## References


1. Latake, P.T., Pawar, P., Ranveer, A.C.: The greenhouse effect and its impacts on environment. Int. J. Innov. Res. Creat. Technol. **1**(3), IJIRCT1201068 (2015). ISSN: 2454-5988
2. Link: https://www.colic.hr/2017/09/25/folije-za-plastenike/. 3 April 2019
3. Parađiković, N., Kraljičak, Ž.: Zaštićeni prostori – plastnici i staklenici. Osijek 2008 (4 April 2019)
4. Shah, N.P., Bhatt, P.P.: Greenhouse automation and monitoring system design and implementation. Int. J. Adv. Comput. Res. Comput. Sci. **8**(9) (2017). ISSN: 0976-5679
5. Ignatius, R.W., Martin, T.S., Bula, R.J., Morrow, R.C., Tibbitis, T.W.: Method and apparatus for irradiation of plants using optoelectronic devices. US Patent 5,012,609, 7 May 1991 (4 April 2019)
6. Mantere, T., Hajru, T., Valisuo, P., Alander, J.: Using the LED lighting in the greenhouses – a pre-study (4 April 2019)
7. Chen, T.Y., Chu, C.C., Henneberry, T.J., Umeda, K.: Monitoring and trapping insects on poinsettia with yellow sticky card traps equipped with light-emitting diodes. HortTechnology **14**(3), 337–341 (2004)
8. Mujčić, E., Drakulić, U., Škrgić, M.: Advertising LED system using PIC18F4550 microcontroller and LED lighting. In: International Symposium on Innovative and Interdisciplinary Applications of Advanced Technologies IAT, Neum, B&H (2016)
9. Link: https://sensing.konicaminolta.us/blog/can-colored-lights-affect-how-plants-grow/. Can colored lights affects how plants grow? (01.05.2019)
10. Mujčić, E., Drakulić, U., Hodži, M.: Using microcontrollers and android for control matrix display. In: 25th International Electrotechnical and Computer Science Conference ERK 2016 (IEEE), Portoroz (2016)
11. Mujčić, E., Drakulić, U., Merisa, Š.: Closed loop temperature control using MATLAB@Simulink, Real-Time Toolbox and PIC18F452 microcontroller. In: Internacionalni simpozij o inovativnim i interdisciplinarnim aplikacijama savremenih tehnologija (IAT) (BHAAAS), Neum (2016)
12. Link: https://www.researchgate.net/publication/266465755_Use_of_LED_lights_in_Plant_Tissue_Culture_and_Greenhouse_Industry, Use of LED lights in plant tissue culture and greenhouse industry (01.05.2019)
13. Link: https://thingspeak.com/ (01.05.2019)
14. Mujčić, E., Pajazetović, S.: Deformacija signala zbog prisustva fizičkih prepreka u Wi-Fi bežičnim mrežama. In: XV International Scientific Professinonal Symposium INFOTEH JAHORINA, Jahorina (2016)

# Analysis of Optical Fibers Characteristics Due to Different Influences

Anis Maslo[1(✉)], Mujo Hodzic[1], Aljo Mujcic[2], and Nermin Goran[1]

[1] BH Telecom Bosnia and Herzegovina, Sarajevo, Bosnia and Herzegovina
{anis.maslo,mujo.hodziv,nermin.goran}@bhtelecom.ba
[2] Faculty of Electrical Engineering, Tuzla, Bosnia and Herzegovina
aljo.mujcic@untz.ba

**Abstract.** This paper is based on practical experience collected during many years of the intervention maintenance on optical cables. Analyzed data are relating to the practical measurement and statistical data which considers interference on optical cables. In addition, a comparative analysis of the cable characteristics was made. Tested cables were laid recently into ground and some of them even 16 years ago.

## 1 Introduction

Regardless of which cables are considered (multimode, monomodal), beside the attenuation, they have various types of fiber characteristics degradation. The attenuation of the fiber is a consequence of the fiber structure, and the degradation is the consequence of the external factors (vandalism, weather conditions, effects of laser-equipment …) [1].

The lifetime of the optical cable/fiber is not clearly calculated value and it is influenced by several factors that determine it. These factors can be grouped into the following categories: impacts on changes in fiber optic parameters during the installation of optical cables, the purpose or place of use of optical cables, the influence of the environment in which the cables are used. All three of these influences may more or less affect weakening of the optical cable [2].

## 2 Types of Degradation on Optical Cable

Regarding the influence of the change of optical fiber parameters when installing fiber optic cables, then the stretching forces of optical cables were in focus. On this occasion, the structure of the protective tubule - a tube or the protective layer (varnish) of the optical fiber is changing. Respectively, there is a disruption of the surface structure of the optical fiber thereby creating the possibility of penetration and the negative effect of water on the sheath of the optical fiber core.

Striking higher than the permissible strain affects the fiber optic structure and inserts attenuation into the transmission or permanently damages the fiber to the limit of interrupting the signal transfer capability. Stretching is characteristic for all types of


S. Avdaković et al. (Eds.): IAT 2019, LNNS 83, pp. 496–503, 2020.
https://doi.org/10.1007/978-3-030-24986-1_39

optical network access (OAN) installations: airline optical cables, ground optical cables with direct laying in the trench, and cables that are pulled into the underground duct (cable) system [2, 3]. Macro and micro bending of optical cables can also create macro and micro attenuation on optical fibers during installation. Such attenuation (macro and micro bending) on optical fibers is more characteristic for underground OANs, and appears especially during the insufflation of optical fiber cables into cable ducts.

If the (listed) damage during testing (after installation) does not exceed a certain size, then it is not noticed [2]. However, during the exploitation (during the life cycle) of the optical cable, there is a possibility that these small degradations will turn into macro degradation which becomes the total degradation. In this case, this could affect the signal attenuation in the optical fiber during transmission.

Purpose or place of using an optical cable/fiber is essentially determined the change of parameters of optical fiber, or its working life. If optical fiber is used in access networks with low power TX laser emissions, it is anticipated that the life cycle of the fiber should last two to three times longer than the life cycle of the copper pair. It is considered to be approximately 35 years. However, the optical fiber cable is often used in urban transmission networks and in the transfer between cities using wave multiplexing (Dense wavelength division multiplexing-DWDM). In this case, high power TX lasers have been used and then the heating of the optical fiber occurs. Greater power of impulses changes the structure or parameters of optical cable during time. The paper provides a review of the heating effect due to the transfer of greater power to the life cycle of the fiber. It has been clearly shown that this heating shortens the life cycle of the optical fiber [2, 4].

The physical location of the installed optical cable significantly influences life cycle of optical fiber. Based on previous maintenance experience of fiber optical cables (about 15 years old), their life cycle is determined by the amount of construction work around the optical cable route and the number of cable damages. Cable damages on the optical cable are the most common: the destruction of underground optical cables (cable window), cutting aerial cable and underground fiber optical cables and a like. Construction works performed on optical cable routes often cause the optical fiber to cut. After several such cases happen on one part of the route, optic cables must be replaced [5].

Thus, this type of impact on optical cables (physical cutting) can be shown in the percentages:

- the number of breaks due to vandalism is cca 10 [%]
- the number of breaks due to construction works is 60 [%].

In addition, the work of the operator during manipulation with cables affects the lifetime of optical cables. In [6] the number of problems caused by the work of the operator during manipulation with cables is presented.

In practice, the problems of the operator's optical cables occur even if the life cycle of the optical cables is interrupted due to the limit in their capacity. This is the result of bad planning. Fifteen years ago, the practice of some operators was to lay optical cables with capacity of 24 fibers. These optical cables initially connected the remote subscriber unit (RSU) with the aggregation networks. Over time, between these two points there was a larger number of users who demanded their direct connection with the main switch center. Therefore, the initial capacity of optical cable became insufficient and the main optical cables with 24 fibers are replaced by cables of larger capacity.

In addition to various mentioned problems, the most important factor for determining the life cycle is the change in fiber optical parameters which are determined by the quality of protective optical connectors. The primarily aim was protection from water penetration. It is possible to solve water penetration on the fiber optic using the coupler and opening tubes through which the water can be horizontally expanded through the optical cable. Since in the protective optical coupler the fibers are not protected except with the thin layer of lacquer, there is direct vulnerability to the surface structure of the optical fibers due to water existence. Over time, this protective coating (lacquer) is degraded also in geographic areas with low temperatures in which the ice can occur, which may break the optical fiber in the coupler [6, 7]. According to research, connecting optical fibers which have a longer period of exploitation (over ten years) under presence of water is difficult because the fibers become fragile and the straight cut cannot be made on the cutter needed for the preparation of optical splices. In order to analyze the change of parameters of optical fiber we made some practical measurements. The results of these measurements are compared to the results obtained for more than 16 years ago.

## 3 Practical Experience and Results Measurement

For the needs of doing this research, we have observed a cable of capacity 24 SM (monomodal fibers) according to the ITU T G652C standard designed for installation in the cable duct pipeline. The cable cross section is shown in Fig. 1.

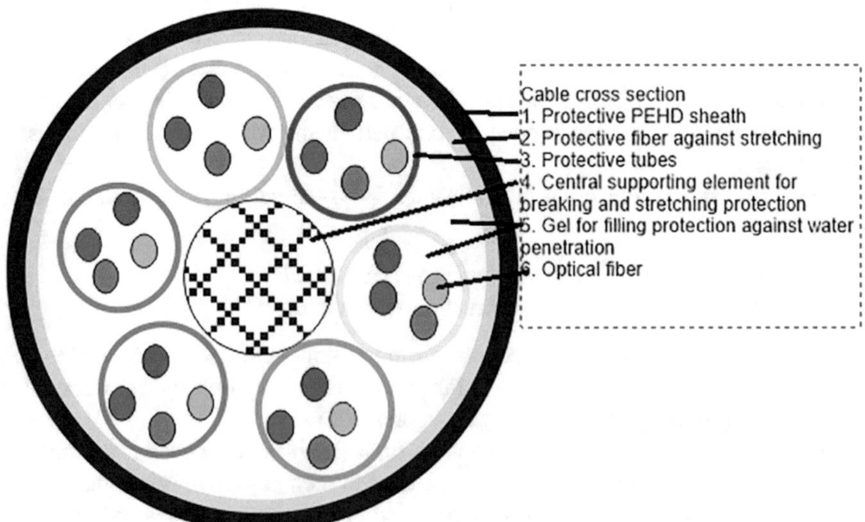


**Fig. 1.** Cross-section of an optical fiber 24-fiber capacity [7]

In the Fig. 1 it can be noticed that the stretching and breaking of the cable is made with the fibers placed under the main coat and with the central element. Water penetration protection is performed by means of the gel between the tubes and the alongside [7].

Figure 2 illustrates an optical fiber damaging diagram expressed in percentages for a wider urban area in BiH (approximately 600,000 inhabitants).

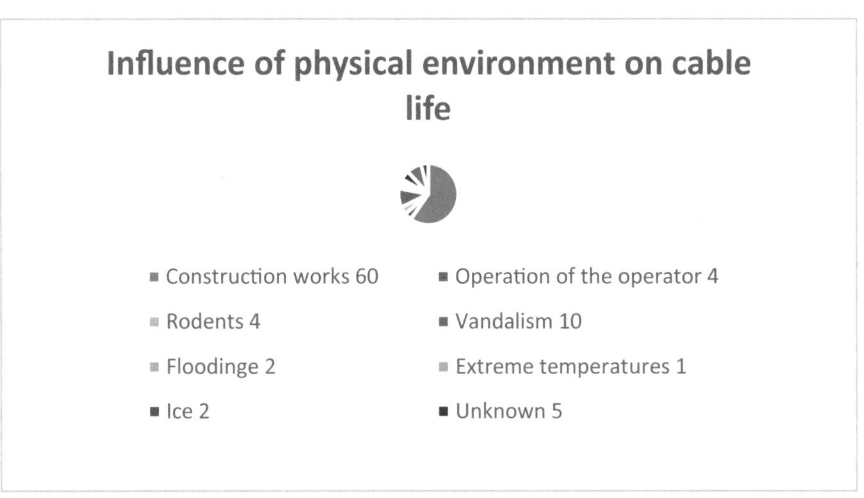


**Fig. 2.** The impact of the external environment on the life cycle of optical cables

The number of breaks due to vandalism is approximately 10%, and the number of breaks due to construction works is 60%. The operator's manipulation with the cables also affects the life cycle of optical cables. The number of breaks due to this effect is cca 4%.

Practical measurements were made with the instrument OTDR(Optical Time Domain Reflectometer) EXFO FTB 7400E, character, ± 0.03 dB /dB, up to 256 000 sampling points per cable track. The instrument has Dead Zone Accuracy of 0.8 m and Detection of Dead Zone of 4 m. Fiber Pitch was tested at 1310 nm. Dynamic range was up to 42 dB which is used for long-term optical fiber testing [3, 8].

Practical measurements have been done on the length of optical cable of 33028 m (Gorazde-Hrenovica). When it was built 15 years ago, this optical cable had 18 connectors. Three of the connections were made with the aim of extending the fiber optical cable to the base station or the remote cabinet (Fig. 3).

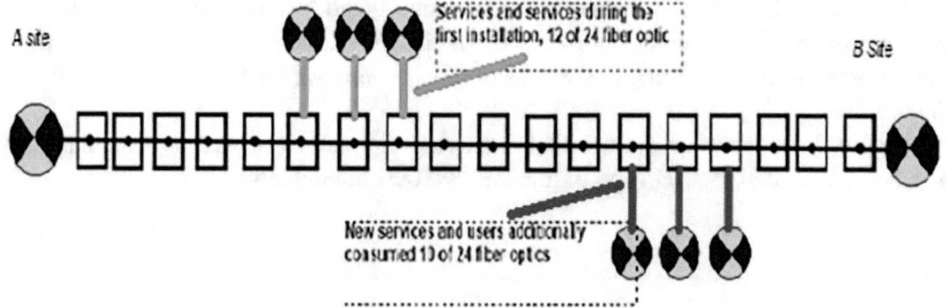


**Fig. 3.** Optical cable path where measurements were made

During the exploitation (16 years), new 6 extensions have been made, three of them due to cable breaking generated by construction work, and the others due to joining of new users. After 16 years of exploitation of the fiber optic cables in urban and suburban part two damages were made due to construction works. There was no damage in the part of the optical cable route through the rural area.

Measurements of OTDR on the analyzed cable are shown in Figs. 4 and 5.

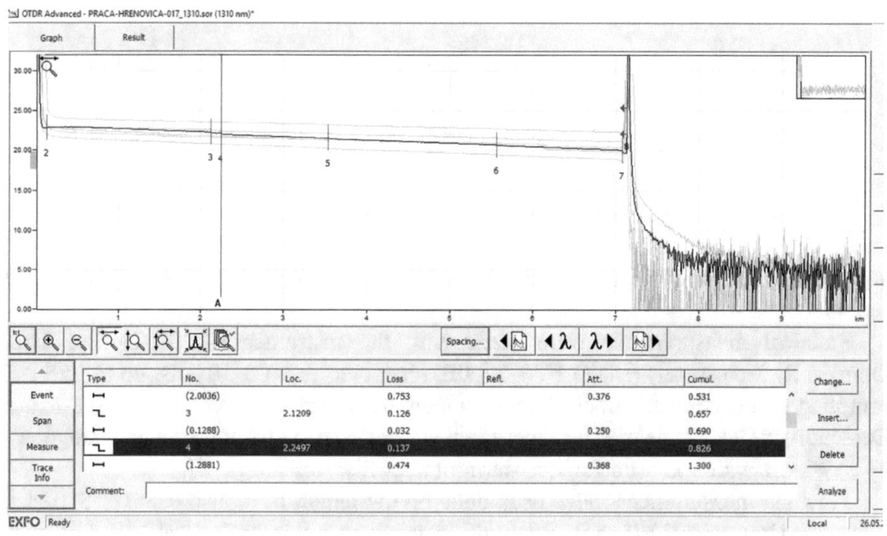


**Fig. 4.** Measure the OTDR of the whole optical fiber cable

Analysis Results -- PRAGOR13.17

| Feature #/Type | Location (km) | Event-Event (dB) | Event-Event (dB/Km) | Loss (dB) | Refl (dB) |
|---|---|---|---|---|---|
| 1/R | 3.5640 | 1.48 | 0.416 | 0.06 | -73.20 |
| 2/N | 7.0422 | 1.37 | 0.394 | 0.30 | |
| 3/E | 7.1035 | ?? | ?? | >3.00 | -39.62 |

Overall (End-to-End) Loss: 3.15 dB

```
Primary Trace: PRAGOR13.17
         Date: 10/02/03              Range:        32 km
         Time: 10:18 AM         Resolution:     2.000 m
 Product Type: K2320           Pulse Width:       250 ns
  Opt. Module: K2320-25              Index:  1.467000
   Fiber Type: Singlemode       WaveLength:      1310 nm
FAS Thresholds:                Horz. Shift:    0.0000 km
         Loss:    0.05 dB      Vert. Shift:      3.05 dB
  Reflectance:  -60.00 dB No. Averages:     12544
  Fiber Break:    3.00 dB
  Backscatter:  -80.00         Trace Type: TS(TS)
  Trace Flags: Analysis, Smooth
         ORL:  <31.65 dB [A] [0.00,32.75]
```

*measurements 2003.*

Analysis Results -- PRACA-HRENOVICA-017_1310.sor

| Feature #/Type | Location (km) | Event-Event (dB) | Event-Event (dB/Km) | Loss (dB) | Refl (dB) |
|---|---|---|---|---|---|
| 1/N | 0.1173 | 0.16 | 1.366 | -0.25 | |
| 2/N | 2.1209 | 0.81 | 0.406 | 0.13 | |
| 3/N | 2.2497 | 0.02 | 0.179 | 0.14 | |
| 4/N | 3.5378 | 0.48 | 0.373 | 0.06 | |
| 5/N | 5.5631 | 0.76 | 0.376 | 0.06 | |
| 6/N | 7.0692 | 0.60 | 0.401 | 0.36 | |
| 7/E | 7.1317 | 0.02 | 0.347 | >3.00 | -13.37 |

Overall (End-to-End) Loss: 3.35 dB

```
Primary Trace: PRACA-HRENOVICA-017_1310.sor
         Date: 05/23/19              Range:        10 km
         Time: 12:56 PM         Resolution:     1.250 m
 Product Type:                 Pulse Width:       100 ns
  Opt. Module:                       Index:  1.469170
   Fiber Type: Singlemode       WaveLength:      1310 nm
FAS Thresholds:                Horz. Shift:    0.0000 km
         Loss:    0.02 dB      Vert. Shift:      0.00 dB
  Reflectance:   -6.46 dB No. Averages:       493
  Fiber Break:    3.00 dB
  Backscatter:  -79.40         Trace Type: SR4731
  Trace Flags: Analysis
         ORL:   50.28 dB [O] [0.00,7.13]
```

*measurements 2019.*

**Fig. 5.** Comparison of measurement results from 2003 and 2019 on 17 fibers from side B to side A at wavelength 1310 nm

Figure 5 shows the comparison of the original measurement during the 2003 optical cable installation and the OTDR track measurement in 2019. At the damaged cable, some disorders were repaired in a way that the reserve cable was withdrawn with the aim of making a continuation of the cable [9]. In the second case, a new piece of 150-metre-long cable was added and two new sequences were made. These two new sequences have caused additional attenuation as seen in Figs. 4 and 5. From the analysis of the results of measurements (Figs. 4 and 5) for the wavelength of 1310 nm, we can conclude the following:

- additional cable extensions have introduced a new 0.26 dB attenuation on the observed optical fiber,
- certain fiber optic connections (for example, a connection at 5631 m away from the B side) have degraded the characteristics of the optical fiber cable due to the presence of water and possibly ice in the winter.

Hence, aging of optical fibers and the influence of water and temperature made changes in the parameters in the way that they have increased weakening of optical fibers.

## 4 Conclusion

It is very common around the world that most damages to optical cables are the result of construction works. Telecom operators cannot influence this. They may affect the interference of their equipment and their improper maintenance and try to decrease number of these cases.

The results of the OTDR measurement on the optical cable made in 2019 are worse than the measurements made in the year 2003. This is the consequence of degradation due to aging, new joints and mechanical strains.

On the other hand, power measurements in 2019 are slightly better than in 2003. This is due to using of better and more precise instruments during measurements, better measuring techniques and better connecting cables.

Experience has shown that a good measurer, with great probability, can estimate and discover the type of interference in the cable whether it is break of the cable, stretching or excessive bending.

## References

1. Maslo, A., Hodzic, M., et al.: Last Mile at FTTH Networks: Challenges in Building Part of the Optical Network From the Distribution Point to the Users in Bosnia and Herzegovina. BHAAS-ISICT, Jahorina, BiH (2018)
2. ITU-T Recommendation L.12 SERIES L: Construction, installation and protection of cables and other elements of outside plant
3. Hodzic, M., Skaljo, E., Mujcic, A., Suljanovic, N.: Transmission of Two Optical Signals Through the Fibber in Opposite Directions Using PLC Splitters—Practical Measurements Chapter. In: BHAAS-ISICT Simposium, Teslic 2017, BiH 8, p. 776. Springer (2017)
4. Thermal effects in high power CW fiber laser Marc-André Lapointe*, Stephane Chatigny, Michel Piché, Michael Cain-Skaff, Jean-Noël Maran, CorActive High-Tech, 2700 Jean-Perrin, suite 121, Québec, Canada, G2C 1S9

5. https://www.linkedin.com/pulse/impact-water-moisture-mechanism-optical-fiber-cable-using-mohiuddin
6. The Fiber Fuse - from a curious effect to acritical issue: a 25th year retrospective Raman Kashyap 1, 2. *1 Fabulas and the Advanced Photonics Concepts Laboratory, Department of Physics Engineering, École Polytechnique de Montréal, 2900 Édouard-Montpetit, Qc, Montreal H3T 1J4, Canada. 2 Department of Electrical Engineering, PolyGrames, École Polytechnique de Montréal, 2900 Édouard-Montpetit, Qc, Montreal H3T 1J4, Canada
7. Van Vickle, P.: Optical Fiber Cable Design & Reliability. Sumitomo Electric Lightwave (2005)
8. NECA/FOA 10/2016 Installing and testing fiber optic an American National Standards, USA (2016)
9. FTTH Handbook Edition D&O Committe Revision date: 16/02/2016
10. Impact of wateeron fiber optic cable © Datwyler Cabling Solution (2014)

# Internal Stress in Water - Water UV Varnish

Esed Azemović, Ibrahim Busuladžić(✉), and Izet Horman

Faculty of Mechanical Engineering, University of Sarajevo, Sarajevo, Bosnia and Herzegovina
busuladzic@mef.unsa.ba

**Abstract.** This paper specified the research results of internal stress with films of water and UV water varnish formed on a solid base. Internal stress occurring by the change of volume in the hardening and adhesion between film and substrate. The hardening process of the studied water-based varnish is achieved by UV radiation. The console method was used to determine the internal strain. This method is based on the principle that one end of the thin elastic narrow panel is fixed. The influence of the film thickness as well as internal stress kinetics has been controlled. In the kinetics of the internal stresses on investigated water and UV water varnish, it is possible to notice the period in which we have an intense evaporation of the volatile components and the first occurrence of the internal stresses, the time of the internal stresses to reach the maximum, the period of maximum retention stress at the achieved level.

**Keywords:** UV hardening · Internal stress · Water varnish · Cantilever method · *Film thickness*

## 1 Introduction

All film-forming cover materials, applied to wood or any other substrate have the task of protecting the substrate from various mechanical damage and mitigating the influence of external parameters on the substrate (temperatures, humidity of the surrounding air etc.) or to increase the esthetic properties of the substrate or product. Applying the water or any other covering material on the substrate the film forming process begins. In this process, the UV water varnish causes evaporation of the water and coupling the molecules, caused by the accelerated drying process using ultra violet air [1]. This evaporation and the chemical reaction of coupling the molecules cause binding of the film, which is simultaneously bonded to the adhesive forces for the substrate.

The result of these changes is the appearance of internal stresses in the film. Internal stresses in the film can cause separation of surface varnish and cracking on the substrate.

By studying the laws of creating, developing and relaxing internal stresses in films, it is necessary to develop the varnish recipe and the rules for applying varnishes to obtain a film with minimal internal stresses [2].

Summing up so far, we come to the conclusion that the internal stresses have a negative phenomenon because it tends to lower the level of mechanical characteristics of the film [3]. Many authors have investigated the internal stresses in the cover


S. Avdaković et al. (Eds.): IAT 2019, LNNS 83, pp. 504–512, 2020.
https://doi.org/10.1007/978-3-030-24986-1_40

materials applying various methods. Sanzarovski and others [4] have investigated the kinetic consonance method and the intensity of internal stresses in nitrocellulose varnish of various thicknesses.

The main task of this paper is to present the influence of film thickness on the occurrence and development of internal stresses in overlaid films on surface treatment of wood.

## 2 Theory of Internal Stress

Physically speaking, the cause of the formation of internal stresses is the unevenness in the size of plate changes in the film-substrate system, if the film and the substrate, for simplicity are conceived as panels.

All water-based materials used for surface protection are subject to volumetric changes during the hardening period on the substrate. In the film of material applied to a solid substrate, the adhesion is only possible on the substrate, while changing the film dimension in parallel with the substrate obstructs adhesion of the film on substrate.

Changing the volume of the film results in internal stresses in the film itself that is exposed to stretching and at the border of the film - substrate occur shear stress [5] (Fig. 1).

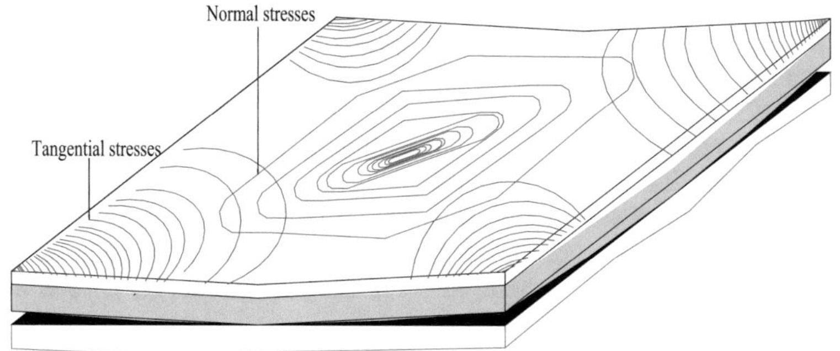


**Fig. 1.** Stress distribution in hardened paint applied to a rigid substrate

Internal stresses are particularly pronounced on dimensionally unstable substrates, such as wood and wooden products. Internal stresses in the filaments are a negative phenomenon which must be reduced to a permissible limit that does not cause separation or cracking of the varnish on the substrate.

The total internal stress at the film of wood treatment is defined by the expression (1) [5]:

$$\sigma = \frac{E_1 S^3}{6Rt(S+t)(1-\mu_1)} + \frac{E_2(S+t)}{2R(1-\mu_2)} \tag{1}$$

S - plate thickness [mm],
T - film thickness [mm],
E1 - modulus of elasticity of substrate [MPa],
E2 - elasticity modulus of film [MPa],
$\mu 1$ - Poisson Coefficient for Substrate,
$\mu 2$ - Poisson coefficient for film,
R - Radius of curvature [mm]

During the hardening of polymeric film, the internal stresses are accompanied by a well-known sequence of changes, characteristic for all macromolecular systems, respectively we distinguish three periods in their development. At period I, the film is in liquid state, there are an intense loss of volatile components and the first occurrence of internal stresses in the film. In the period II, internal stresses and their intense growth to maximum value occur. At the end of the III period, the film's maximum stresses are maintained [6]. These are the final stresses that are significant for assessing the durability of the surface of finishing film. It is important to note that in proportion to the increased evaporation of volatile components, the thickness of the film is reduced.

## 3 Materials

In the last few years, large changes are taking place in the production and application of surface treatment for wood. More attention is devoted to systems that are less polluting the environment and require less energy. In this regard, great research and development efforts are made, even by the manufacturer of materials for surface treatment, as well as the supplier of technological equipment and users of these materials.

For experimental analysis of internal stresses in this work, water and UV water varnishes were used. Water-based varnishes, or as they are referred to as water-repellent are dispersions or solvents of resin in water, in the presence of a certain amount of organic solvents. The basic characteristics of water and UV water materials are given in Table 1.

**Table 1.** Characteristics of water and UV water materials

| Property | | Property | UV |
|---|---|---|---|
| Basic | Acrylic resin | Basic | Acrylic resin |
| Specific gravity [g/cm$^3$] | 1,01 | Specific gravitay [g/cm$^3$] | 1,04 |
| Dry substance content [%] | 34,4 | Dry substance content [%] | 35,3 |
| Viscosity [mPa.s] | 750 | Viscosity [mPa.s] | 1230 |

## 4 Application and Hardening of Varnish on Substrate

In the surface treatment of wood, there are various processing stages used to form the cover or the film on the substrate. The basic three stages of processing used for the film forming are:

- phase of varnish application,
- hardening of the varnish,
- varnish drying stage.

Materials for the surface treatment of wood are liquid, semi-solid or paste-like. Surface finishing materials are liquid, semi-solid or paste-like. Application of these materials to treated surface is carried out by various methods, using different devices, arrangements, hand tools and apparatus.

### 4.1 Water Varnishes

To apply water-based varnish to the substrate, airless mode dispersion - manual application is used. It is called airless, the dispersion of liquid material by a considerable static pressure under which the fluid material is rapidly released into the atmosphere. The air is opposed to the liquid circulation in the atmosphere. When the forces of confrontation overcome the force of cohesion of the fluid itself, their dispersion appears.

The drying conditions of the finishing material surface, consist of transferring liquid-to-hard coatings, which is also referred to hardening. Increasing the temperature, increases the elasticity of vapor constituents, which accelerates vaporization and the increase of temperature accelerates the process of chemical reactions, characteristic for the surface treatment of wood.

The drying process on the console, takes place at ambient conditions, at temperature 21°C and relative humidity of 51%. The disadvantage of this process is longer drying time, reducing the productivity, the demand for higher drying capacity and others.

The water-drying process lasts 48[h]. This drying method is used for larger grid construction, chair assemblies and other wood products. This method also monitors the internal stresses kinetics of water-based varnish.

### 4.2 UV Water Varnish

The application of UV water-based varnish on the substrate is electronically based at airless dispersion method. It is dispersing the material by means of a considerable static pressure, by which the fluid material is released at high speed into the atmosphere. Air circulation is a matter of movement of liquid in the atmosphere. When the forces of confrontation overcome the force of cohesion of the same liquidity, its dispersion occurs. By electronically applying varnish we increase productivity, save material and steam vapor constituents generated by dispersion are retained by the air stream, without the possibility of spreading them in the unwanted direction (Fig. 2).

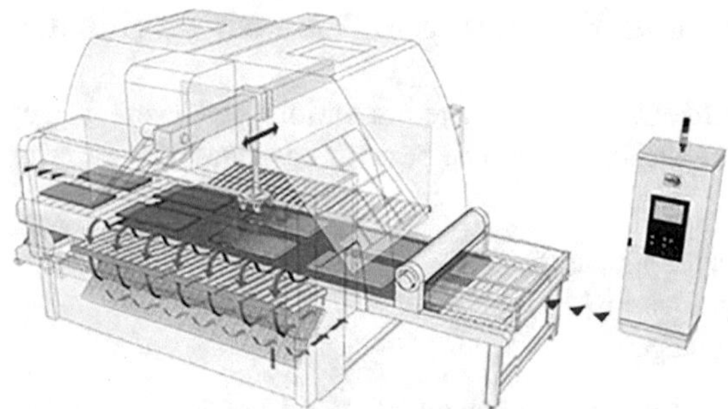

**Fig. 2.** Automatic self-priming cabin with pneumatic dispersion [www.cefla.com]

With application of the aqueous varnish on the surface the drying process begins (hardening). In practice there are various methods of drying, but in this research a combined drying method has been used. The complete drying process takes place in a closed UV chamber. Drying process takes place in the chamber with the convection and radiation (Fig. 3).

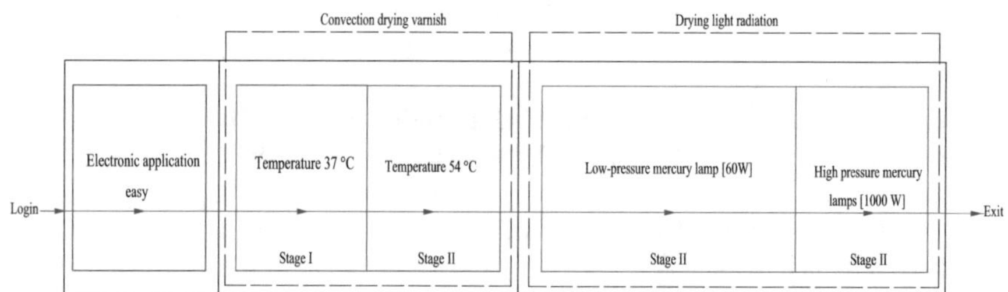


**Fig. 3.** Application and hardening scheme of UV water varnish

## 5 Testing Method

The console method was used for experimental internal stress analysis [7]. The principle of this method is that on one side of thin, narrow and long sheet of elastic material, whose characteristics are known, the film of material for the surface treatment is deposed. During the hardening of the film, due to changes in its dimensions and the adhesion forces to the substrate, the bending of the newly formed biplane occurs. Since its length is considerably larger than the width, the plate is folded to the cylindrical surface. Measuring the curvature radius (R), deflection of the panel (f) or deflection of the panel console (h), which is clamped at one end, as shown in Fig. 4, the internal stresses in the film of the applied material can be determined.

S - substrate thickness; t - film thickness; L - length of substrate with film; R - radius curve; f - deflection; h - deflection of the console.

Between the size of R, f and h, there is a dependency given by expression:

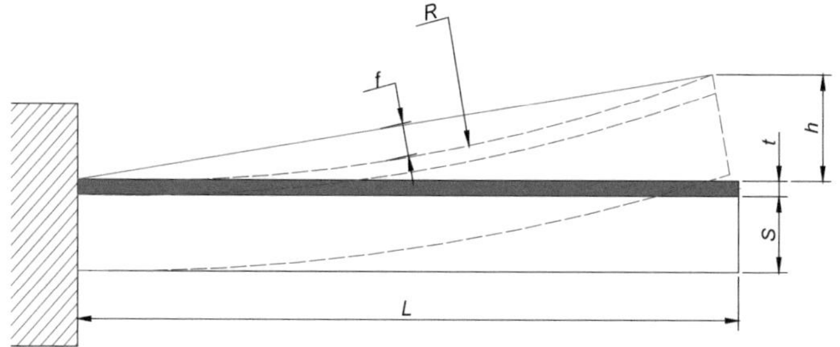


**Fig. 4.** Schematic representation of the console method

$$R = \frac{L^2}{8f} = \frac{L^2}{2h} \tag{2}$$

where:

L is plate length and h is deflection of the console.

On the basis of the relation (1) and the expression (2), internal stresses in the polymer material film can be determined by means of the Eq. (3):

$$\sigma = \frac{hE_1S^3}{3L^2t(S+t)(1-\mu_1)} + \frac{hE_2(S+t)}{L^2(1-\mu_2)} \tag{3}$$

For the console method for testing the polymeric material films, on surface treatment of wood, the term (3) is used without the second member, because $E_1 >> E_2$ and d>>t. The second member of the expression is less than 1% so it can be neglected.

As a base for testing this method, metal is most often used. For this internal stress testing in the film, aluminum has been used, with basic features:

- Elasticity modulus: $7\times10^4$ [MPa],
- Poisson coefficient: 0.346,
- Dimensions of the substrate: S-100; L-13; t-0,18 [mm].

## 6 Results and Discussions

In this experimental analysis, the results of internal stresses in water and UV water-based varnish were obtained. These varnishes were applied twice on the substrate by an electronically-dispersed method in airless atmosphere obtaining dry films of different thicknesses:

(a) water varnish t = 30 [μm] and t = 60 [μm]
(b) UV water varnish t = 20 [μm] and t = 50 [μm]

### 6.1 Water Varnish

In this test, the total time for drying process of the film was 48[h]. During this drying process, at different time intervals 0, 0.5, 1, 24 and 48[h], the displacement of the console were measured (see Figs. 5 and 6).

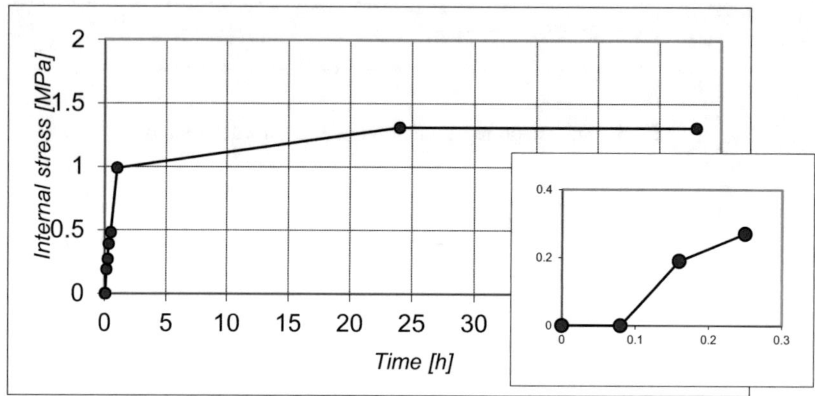


**Fig. 5.** Cumulative internal stresses in water film with thickness t = 60 μm

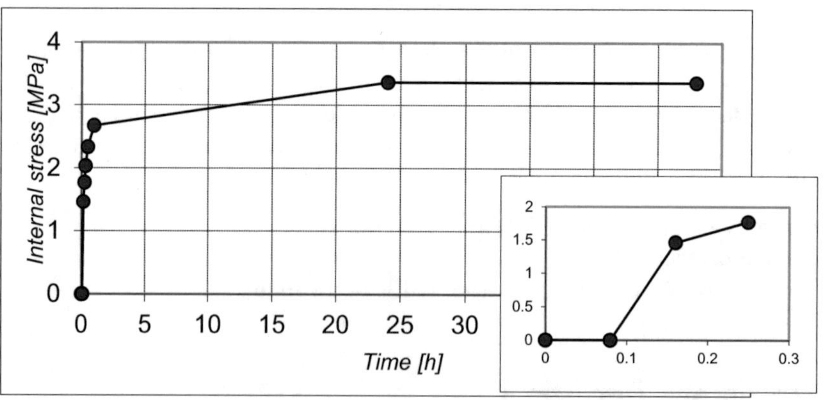


**Fig. 6.** Cumulative internal stresses in water film with thickness t = 30 μm

For a water film having dry film thickness t = 60[μm], the maximum stress is σs max = 1,314 [MPa], and the mean vertical displacement of the console is hs = 1.31 [mm]. For the thickness of the dry film t = 30[μm] the maximum stress is σs max = 3,36 [MPa], and the mean vertical displacement of the console is hs = 1.56 [mm]. We can clearly conclude that films of smaller thickness have greater internal stresses.

## 6.2 UV Water Varnish

In this test, the total drying process of the film was 1[h]. During this drying process, at different time intervals 0, 0.13, 0.3 and 0.5[h] the console displacements were measured (Figs. 7, 8).

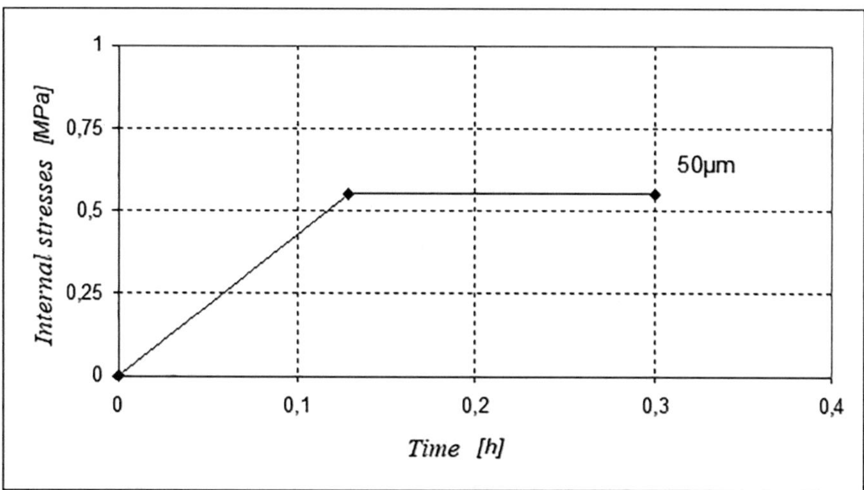


**Fig. 7.** Average value of internal stresses of UV water-based varnish, with film thickness of 50 μm

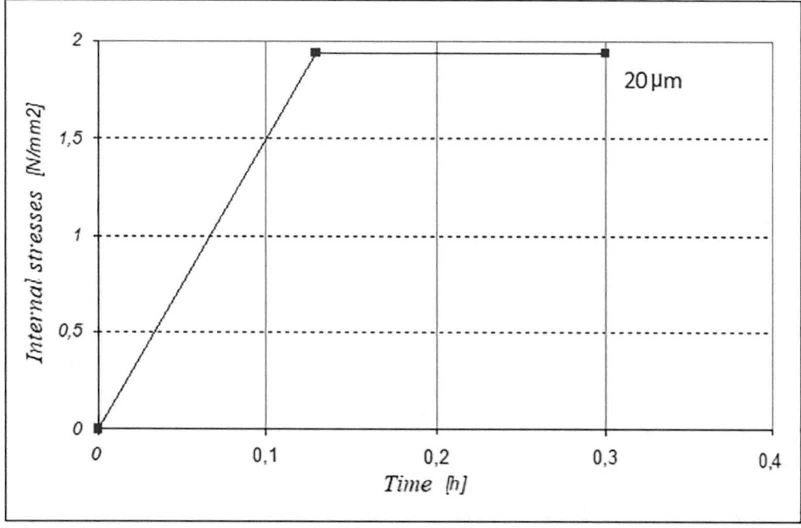


**Fig. 8.** Average value of internal stresses of UV water-based varnish, with film thickness of 20 μm

For a UV water film with a dry film thickness t = 50[μm], the maximum stresses are σsmax = 0.558[MPa] and the mean vertical console displacement is hs = 0.38 [mm]. For the dry film thickness t = 20[μm] the maximum stresses are σsmax = 1.94 [MPa] and the vertical center displacement of the console is hs = 0.475[mm].

We can clearly conclude that films of smaller thickness have greater stress on the film.

## 7 Conclusion

Based on the theory and experimental analysis, presented in this paper, we can conclude that internal stresses in films produce negative phenomenon on material. In order to reduce the internal stresses in the film, we must act on influencing parameters that can reduce stress. The basic task of this paper is to analyze one of these parameters - the thickness of dry film and the kinetics of stress development.

The obtained results clearly show that films with water varnish and UV water varnish, respectively with thickness t = 30[μm] and t = 20[μm] have higher internal stresses, or contrary to that, thicker films t = 60[μm] and t = 50[μm] have lower internal stress.

The illustrated horizontal stress diagrams in water and UV water-based varnish confirms that theory of internal stress kinetics: the period in which we have an intense evaporation of volatile components and the first occurrence of internal stresses, the period of internal stresses growth to reach maximum and the period of maximum stress retention at the achieved level.

## References

1. Axelsen, S.T., et al.: Topcoat flaking - a mechanism study. Party 1; Laboratory testing internal stress, mechanical properties, adhesion and ageing. Corrosion, March 16–20, New Orleans, USA, p. 14 (2008)
2. Grigore, E., Ruset, C., Short., Hoeft, D., Dong, H., Li, X.Y., Bellt, T.: In situ investigation of the internal stress within the nc-Ti2 N/nc –TiN nanocomposite coatings produced by a combined magnetron sputtering and ion implantation method. Surf. Coat.S Technol. **200**(1–4), 744–747, ISSN 0257-8972 (2005)
3. Jaić, M., Živadin, R.: Površinska obrada drveta – svojstva materijala, kvalitet obrade. SITZAMS, Beograd, Srbija (1993)
4. Sanzarovski, A.T.: Metodi opredelenija mehaničeskih i adgezionih svoistv polimerih pokriti, Moskva (1974)
5. Volinsky, A.A., Moody, N.R., Gerberich, W.W.: Interfacial toughness measurements for thin films on substrates. Act Mater. **50**(30), 441–466 (2002)
6. Gerald Stoney, G.: The stress of metallic films deposited by electrolysis. In: Proceedings of the Royal Society of London. Series A, Containing Papers of a Mathematical and Physical Character, vol. 82, no. 553, pp. 172–175 (1999)
7. Vela, J.B., Adhihetty, I.S., Junker, K., Volinsky, A.A.: Mechanical Properties and fracture toughness of Organe – silicate Glass (OSG) low-k dielectric thim films for microelectronic application. Int. J. Fract. **119**(4), 487–499 (2003)

# Energy Consumption Regression Curve Analysis for Large Educational Building

Armin Teskeredžić[1(✉)], Rejhana Blažević[1], and Marko Kovačević[2]

[1] Faculty of Mechanical Engineering, University of Sarajevo, Sarajevo, Bosnia and Herzegovina
teskeredzic@mef.unsa.ba

[2] Petrolinvest d.d. Sarajevo, Sarajevo, Bosnia and Herzegovina

**Abstract.** Energy management in buildings is strongly promoted in the current Energy Efficiency Directive (2012/27/EU). The main goal of the energy management introduction is an optimal energy use either for heating or cooling. This paper shows how the data from an advanced building energy data acquisition system, implemented at the Mechanical Engineering Faculty building in Sarajevo, can be used to determine the baseline energy consumption for heating. The key parameters are measured on hourly basis during the whole calendar year. Raw data collected from different sensors, in the form of the simple ASCII files, are processed in advanced MS Excel tool, specially designed for this purpose. This paper discusses the applicability of the linear regression models and demonstrates how to construct the daily regression curve based on available hourly measured data. After the regression curve is established, the relationship between energy consumption for heating and outside temperature is validated against measured values in the whole heating season. It is also demonstrated how the hourly measured data can be used to analyze the occupants behavior. The results are discussed and the further steps are proposed.

## 1 Introduction

Systematic energy management in heating requires estimation of the current energy consumption and evaluation of the thermal comfort conditions in the occupied space [1–4]. Based on the measurement results the mathematical relationship between the energy consumption and outside temperature or the temperature difference is established as a baseline for any further use. If the treated spaces are overheated, the new targeted energy consumption should be set. This is to be achieved either by interventions at the heating system or by introducing and promoting the changes in occupants behavior [5].

In case that the current energy consumption is optimal, the energy manager has the task to keep consumption at the previously prescribed level, which in other terms means that the energy consumption should be as close as possible to the established and targeted regression curve [6, 7].

In this work simple linear regression is used, defining the linear dependency of the energy consumption versus outside temperature. Mathematically this implies the treatment of the heat transfer within the building as a quasi-steady-state process in


S. Avdaković et al. (Eds.): IAT 2019, LNNS 83, pp. 513–524, 2020.
https://doi.org/10.1007/978-3-030-24986-1_41

which the thermal inertia terms are neglected in governing equations [8]. In case that the heating system works for 24 h per day, this approach can be used with higher level of confidence, since the transient terms coming from the thermal inertia of the outside walls only and are not dominant in the equations.

However, in case of the intermittently heated buildings with large periods of the system in stand-by, the transient behavior of the inner walls can play a role of the heat sources or sinks, depending on the fact that the heating system is ON or OFF. These transient effects influence the thermal balance of the treated rooms and introduce additional complexity and non-linearity in the mathematical treatment of equations which are usually solved by using numerical methods [8].

Subject of the current study is the Mechanical Engineering Faculty building in Sarajevo, with 5200 $m^2$ of heated space, including classrooms and the teaching staff offices. Building has a centralized heating unit with 400 kW condensing boiler, two-pipe distribution system with 20 vertical branches supplying the radiators without thermostatic valves. Weather compensated control is implemented which makes the system suitable for the linear regression between the energy consumption and outside temperature. Heating schedule is 10 h of heating per day during the working week and only few hours of heating on weekends and holidays where operator makes a decision for how long the system will run. Operator can also manually change the heating schedule any time and can switch the system off.

## 2 Mathematical Model and Physical Assumptions

In order to estimate the energy consumption for heating in observed building a linear regression model is used which assumes the linear dependency between the energy consumption and outside temperature. Targeted mathematical expression will have the simple linear form:

$$E = kt_{ok} + n \tag{1}$$

where E is the energy consumption, $k$ and $n$ are the slope and intercept, respectively. Time interval for the proposed curve can be set in hours, days, weeks or months and the slope and intercept will change accordingly.

For $N$ discrete points or in our case measurements, the linear relationship is defined so that the variables $k$ and $n$ are determined by forcing the sum of the square errors $e_i$ to minimum.

$$\sum\nolimits_{i=1}^{N} (E_i - n - kt_{oki})^2 = S(n,k) \rightarrow minimum! \tag{2}$$

where $E_i$ is the measured energy consumption in prescribed time interval $i$, $t_{oki}$ is the outside temperature in time interval, while $N$ is the number of available measurements.

Partial derivations of function $S(n,k)$ per $n$ and $k$ are set to zero, providing the extreme of function while the second derivatives are definitely positive providing the minimum of the function. After algebraic manipulations the unknown values of $k$ and $n$ are determined as:

$$k = \frac{\sum_{i=1}^{N} t_{oki}E_i - N\bar{t}_{ok}\bar{E}}{\sum_{i=1}^{N} t_{oki}^2 - N\bar{t}_{ok}} \tag{3}$$

$$n = \bar{E} - k\bar{t}_{ok} \tag{4}$$

Where $\bar{E}$ and $\bar{t}_{ok}$ are average energy consumption and average outside temperature, respectively, both for N measurements within prescribed time interval.

The quality of correlation is estimated by the correlation coefficient $r$ such as:

$$r = \frac{N\sum_{i=1}^{N} t_{oki}E_i - (\sum_{i=1}^{N} t_{oki})(\sum_{i=1}^{N} E_i)}{\sqrt{N(\sum_{i=1}^{N} t_{oki}^2) - (\sum_{i=1}^{N} t_{oki})^2}\sqrt{N(\sum_{i=1}^{N} E_i^2) - (\sum_{i=1}^{N} E_i)^2}} \tag{5}$$

Where the squared value of $r$ or $R^2$ is the coefficient of determination and is used in the present study to judge the quality of the correlation as a key parameter.

Since the heat transfer mechanism for the intermittently heated buildings is strongly transient in this particular case (10 h of heating per day), the hourly values are less appropriate for construction of the regression curves. Transient effects as already explained introduce the non-linear dependencies between chosen variables. Therefore, the time interval in days is used for determination of regression curves. It is recognized that the aforementioned transient problems will be muted on a daily basis, especially after the heating system comes close to the steady-state operation.

Transformation of the hourly values is done for all measured variables; temperatures are represented with average values over the calculation period, cumulative energy consumption as a sum of consumed energy and the calculation period is determined when the heating system is ON.

Based on the average values of outside temperatures and cumulative energy consumption per days, a single or multiple regression curves can be constructed. The following cases can lead to the construction of multiple regression curves, such as:

- Operator can override the standard boiler control schedule and turn off the boiler any time by changing the number of hours when the heating system is ON,
- Boiler safety control can also override the standard schedule by introducing additional firing of the boiler during the periods with extremely low temperatures and this can also happen during the night when the boiler is usually OFF or in the setback mode,
- Special treatment for situations where boiler works only one or two hours with full capacity which practically excludes any dependency between the energy consumption and the outside temperature.

In the next subsection of this study the regression curve will be constructed and some of the mentioned assumptions will be treated.

## 3 Experimental Setup and Data Analysis

An advanced system for data measurement and collection is applied in the building of Mechanical Engineering Faculty in Sarajevo. The readings from the sensors are sent to the central database and the current day measurements are available next morning in the form of the simple ASCII file. The measurement system consists of sensors for measuring indoor (8 rooms) and outdoor temperature, hot water supply and return temperature, calorimeters, water meters, gas meters, electric meters, and data collection and transmission devices.

The outdoor temperature sensor is located on the first floor of the northern wall of the building (Fig. 1 left). It is enclosed in the distribution box and is thus protected from external disturbances (e.g. wind and/or sun).

Indoor temperature sensors (Fig. 1. right) are placed in six classrooms and two offices for teaching staff and they read current room temperature and humidity.

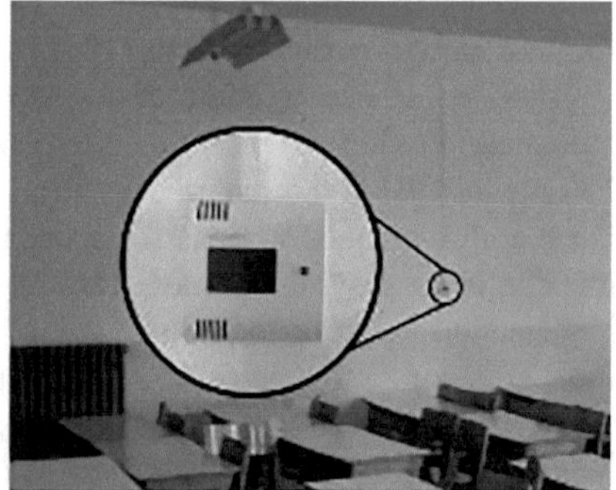

**Fig. 1.** Sensor position: outdoor temperature sensor (left) and indoor temperature sensor (right)

Every year more than 100,000 individual measurement data is available for the analysis, which requires appropriate automated software. In this work an advanced MS Excel tool is specially designed for the purpose of the reading, sorting and analysis of data. It is flexible in a sense that it can either read huge set of data for the whole season or can import data for a single day. The tool processes the data automatically, fills in the separate hourly and daily sheets and creates the graphical representation of the data sets which helps the operator to quickly find deficiencies. Deficiencies can either be non-physical values in measurement quantities which reflect the problems with sensors or data loggers, or it can be a problem in heating system settings such as the manual override of the regular control settings or can also be the problems with occupants behavior such as leaving the window open during the night or using electrical heaters for re-heating of the rooms.

Data sorting part which deals with calculation of daily values based on the hourly measurements simplifies the whole process and put the operator into the position to focus on energy consumption and not on data manipulation which could be time consuming.

## 4 Results and Discussion

For the creation of the regression curve in the heating season 2016/2017 the period from October 1st 2016 until April 1st 2017 is taken. Hourly values are automatically read into the Hourly sheet and the background calculations are introduced to convert and sort the data into the Daily sheet, where every single row of data contains the daily values, either averaged or cumulative as explained earlier in the text.

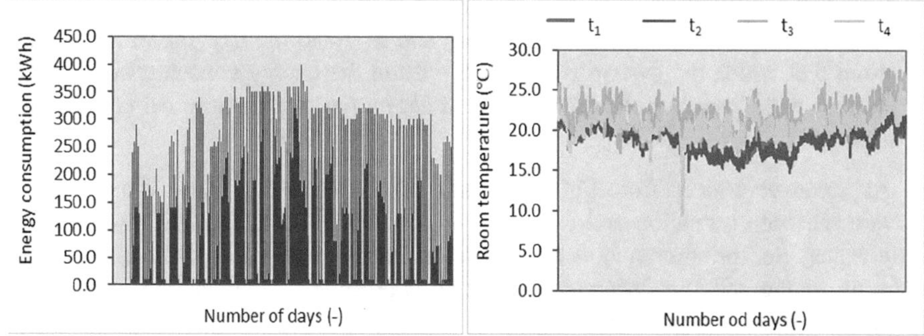


**Fig. 2.** Hourly energy consumption (left) and temperatures in 4 characteristic rooms (right).

Figure 2 shows the diagrams of hourly measured overall energy consumption (left) and temperatures (right) in four observed rooms for the whole heating season. One may note that orange line (Fig. 2 right), which represents the temperature in the room 4 has the minimum value during one or more heating days and this fact is elaborated at the end of this subsection.

After the data on hourly basis are converted into daily values, the result is 187 days of heating coupled with all variables set as cumulative or average values, as previously explained. The results from the read and integrate procedure is given in Fig. 3.

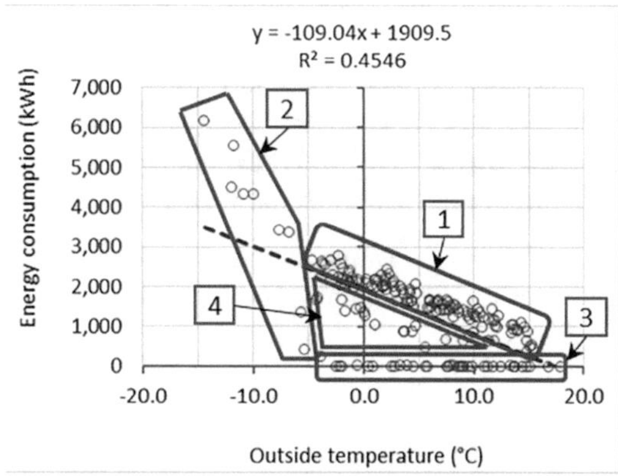


**Fig. 3.** Different characteristic zones in energy consumption versus outside temperature curve

Figure 3 shows very weak correlation between energy consumption and outside temperature, represented by the coefficient of determination $R^2$. In Fig. 3 one may also notice four different zones enclosed into four different red shapes, which can be explained as follows:

- **Zone 1** at which the linear relationship between the energy consumption and outside temperature is visible,
- **Zone 2** at which a different slope and interception values may be recognized. This zone is characteristic with extremely low outside temperatures where boiler control settings overrides the standard control schedule in order to prevent the system from intensive sub-cooling of the heated space and to avoid the freezing in the system,
- **Zone 3** at which the system is switched off and the energy consumption is zero,
- **Zone 4** with 'flying' points for which an additional explanation and correction is needed.

As it can be noticed from Fig. 3, the row data without any intervention does not provide reliable correlation and cannot be treated as relevant, even if we work with daily values. So, the strategy how to resolve the existence of 4 different zones and how to come to the reliable dependency between the energy consumption and outside temperature is implemented as follows:

- First the zone 3 (zero consumption) is avoided by the fact that the relevant data are filtered for the situations where the heating system is ON, which is defined by a minimum of 20 kWh energy consumption during one hour. It can be said that even when the system is off, the natural circulation within the system may drive the minor flow and can result in 'parasite' energy consumption, sometimes at the level of up to 10 kWh.
- Second, the data are sorted for outside temperature larger than −6 °C, as it was recognized that this is the temperature at which the boiler control overrides the manual settings. Note that these data for the outside temperatures less than −6 °C will be treated separately in order to get the specific slope and intersection values for these extreme situations.
- Zone 4 with flying points is recognized as the zone at which the number of operational hours was different from the standard settings, including the manual manipulation of the boiler control by the operator. For this zone, the correction factor depending from the number of hours of operation is introduced, with intention to normalize energy consumption values for these non-standard operating hours. Graphically it means that the corrected energy consumption in these cases should bring the 'flying' points closer to the regression curve.

By applying all assumptions explained above, it can be seen how the implementation of the strategy increases the coefficient of determination, which is also demonstrated graphically in Fig. 4.

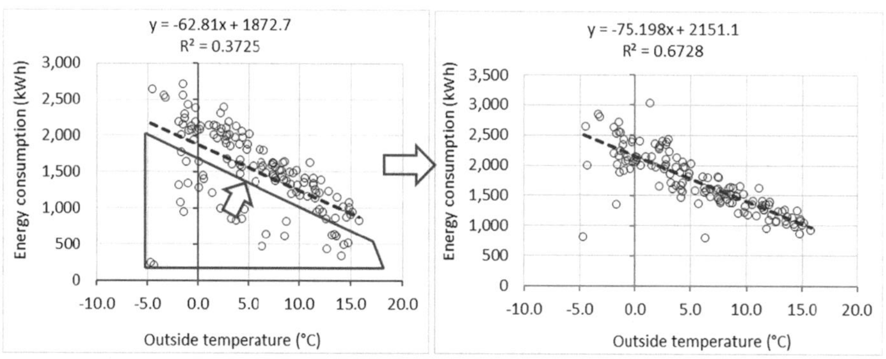


**Fig. 4.** Regression curves for energy consumption E (left) and corrected energy consumption E* (right), when heating system is ON

At the right hand side of Fig. 4 the corrected value of energy consumption is used E*, which takes into account the different operating hours of the system. As the correction was applied the coefficient of determination improved immediately.

In Fig. 5 left, the existence of three remaining 'flying' points can be recognized. After the deeper analysis of the data it is realized that the boiler operator manually switched off the boiler, letting the system run only for 3 h or less per day. In that case any extrapolation of the values would lead to additional error and energy consumption versus outside temperature is impossible to find. This is also visible from the graph in Fig. 5 (left) since the same energy consumption is recorded for outside temperatures of 6 °C and −5 °C which physically makes no sense. Therefore, the removal of these points was justified and the coefficient of determination gets the final value of $R^2$ = 0,8074 (see Fig. 5 right).

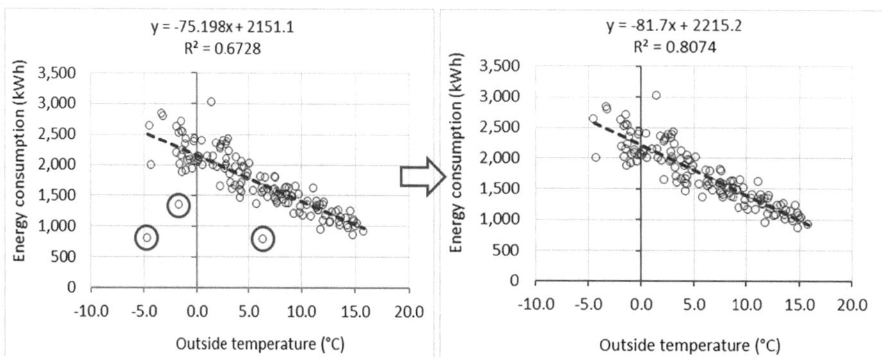


**Fig. 5.** Corrected energy consumption E* with remaining 'flying' points (left) and with their removal (right)

The result of the regression analysis is the equation given in the form:

$$E_I^* = -81.7 * t_{ok} + 2215.2(kWh) \qquad (6)$$

where $E_I^*$ is corrected energy consumption. It should be said that this equation is valid for Zone 1 (see Fig. 3) where the heating system is ON and where outside temperature is higher than −6 °C.

After the regression line is defined for Zone 1, the same procedure is applied for the outside temperatures lower than −6 °C, which corresponds to the Zone 2 presented in Fig. 3. The resulting equation for Zone 2 is:

$$E_{II}^* = -777.73 * t_{ok} - 4469.3(kWh) \qquad (7)$$

where $E_{II}^*$ is corrected energy consumption for outside temperature lower than −6 °C. Note that $t_{ok}$ is average daily outside temperature.

Results of the analysis established two regression curves, which are jointly presented in Fig. 6. Blue line in the Fig. 6 shows the regression curve for daily $t_{ok}$ larger than −6 °C, while the red line presents the regression curve for outside daily temperatures less than −6 °C.

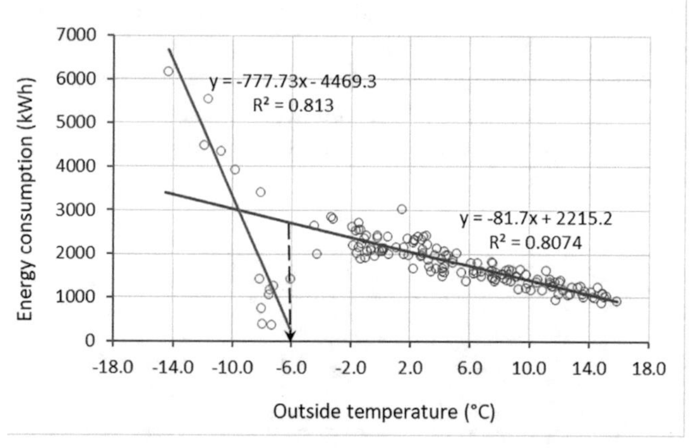


**Fig. 6.** Cumulative regression diagram for the analyzed heating season

Figure 6 shows two regression curves defined for the whole heating season. Within the red zone, despite the extremely low temperatures, there is a small data set with lower energy consumption. After deeper analysis of the data, it is concluded that the operator of the system manually changed the standard system settings for these days and this is also taken into consideration when the final form of the second regression curve was constructed.

It is noticed that the reference number of hours for the blue line was set to 10 h, while for the red line (extremely low outside temperatures) the reference hours was set to 4 h.

The final form of the corrected energy consumption is written in the compact form with help of the delta operators, such as:

$$E^{*} = \delta_1 * \left[(-81.7 * t_{ok} + 2215.2) * \left(1 - \delta_2 + \frac{\delta_2 * n_h}{10}\right)\right] + (1 - \delta_1) * \left[(-777.73 * t_{ok} - 4469.3) * \left(1 - \delta_3 + \frac{\delta_3 * n_h}{4}\right)\right], (kWh) \quad (8)$$

where $E^{*}$ is corrected energy consumption, $t_{ok}$ is outside temperature, $n_h$ is number of system working hours, and three introduced delta operators have the following values:

$$\delta_1 = \begin{cases} 0; t_{ok} < -6°C \\ 1; t_{ok} \geq -6°C \end{cases};$$

$$\delta_2 = \begin{cases} 0; n_h \geq 10h \\ 1; n_h < 10h \end{cases};$$

$$\delta_3 = \begin{cases} 0; n_h \geq 4h \\ 1; n_h < 4h \end{cases}$$

After the regression curve is defined, energy consumption is calculated by applying the values and coefficients from the Eq. (1.) for the whole heating season 2016/2017. As an input parameter a daily average value of temperature $t_{ok}$ is used and as a result daily energy consumption is calculated. The comparison of measured and calculated energy consumption result by the regression curve is shown in Fig. 7.

Figure 7 (top) shows the comparison of calculated and measured daily energy consumption and the agreement is very good. The difference of the cumulative calculated and measured energy consumption for the whole heating season is less than 1%, which can be treated as an excellent agreement. The peak in energy consumption which is clearly visible at the graph comes with extremely low outside temperatures in this period. This is the part of the graph where regression for extremely low temperatures (lower than −6 °C) is used.

Figure 7 (middle) shows the number of hours per day when the heating system was ON and it is clear that the system worked for 24 h for extremely low outside temperatures. It can also be seen that the real operating hours in many days differ from the prescribed heating schedule which is in line with our assumption that the energy consumption should be normalized with number of heating hours per day.

The lowest diagram in Fig. 7 (down) shows the daily values of the room 4 temperatures for the whole heating season from which it can be seen that the temperature is maintained between minimum 19 °C and maximum 23 °C, highlighted with black lines in the graph. There are three exceptions, where the measured daily temperature is below prescribed minimum and the most drastic one is during the 44$^{th}$ day of heating.

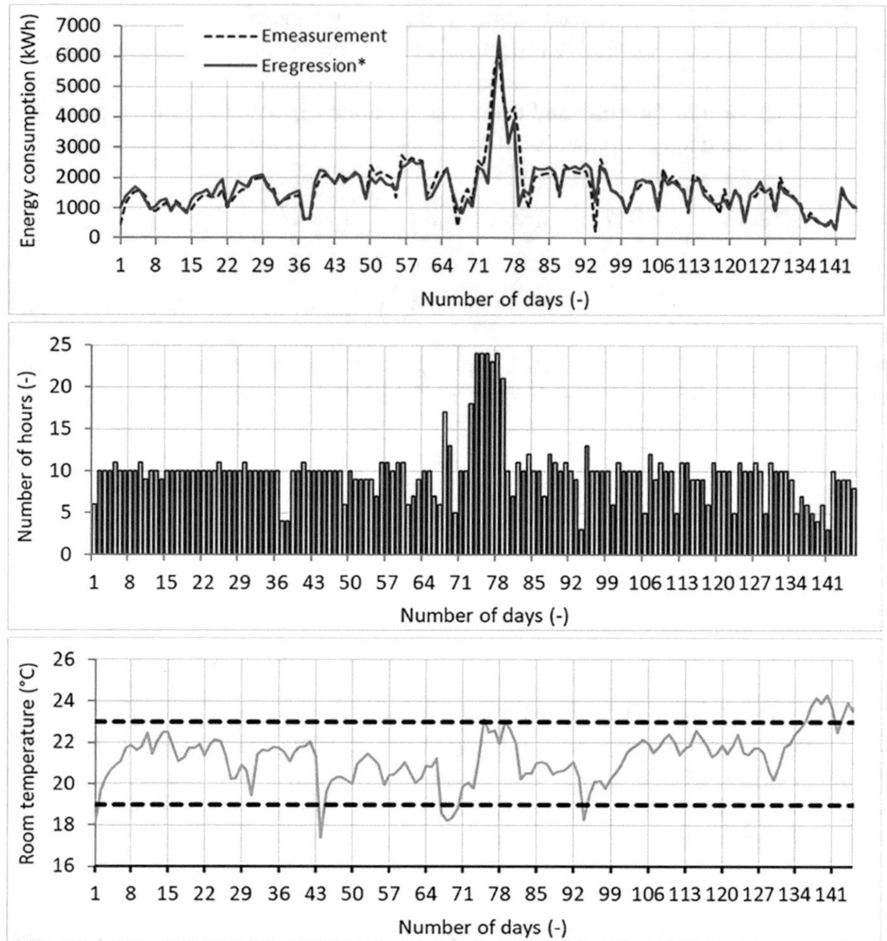


**Fig. 7.** Comparison of measured and calculated values of energy consumption (top), number of hours per day when the heating system was ON (middle) and room daily values of temperature in the heating season 2016/2017 (down)

Presented procedure offers a simple way to analyses what was wrong in this case. The data which corresponds to the 44$^{th}$ day of heating in the days sheet was recognized by the date December the 5$^{th}$ 2016 and is further processed in the hourly sheet. For the demonstration purposes four consecutive days are chosen from 3$^{rd}$ of December until 6$^{th}$ of December.

On the left hand side in Fig. 8 the sudden temperature drop can be recognized with the minimum value of 9,5 °C. Figure 8 (right) presents the water supply and return temperatures and it is visible that the heating system was off in the observed period. Only reasonable explanation is that the user left the window open and therefore the air temperature in this room follows the drop in the outside air temperature (temperature gradients are the same). Sometime in the early morning the housekeeper recognized the

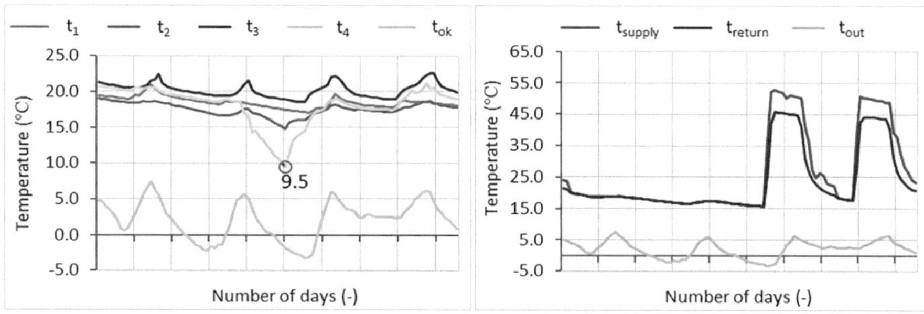


**Fig. 8.** Temperature profile from 3rd of December until 6th of December (left) and the water supply temperature, return temperature and outside temperature (right)

problem and closed the window before the heating system started. Due to the accumulated heat in the inner walls, as soon as the window was closed the room gained some heat and the air temperature started to grow gradually.

Another observation is that the temperature in the observed room 4 did not reach the comfort level during the whole next day despite the fact that the heating system was ON. Temperature in the room 4 was the whole day below 20 °C, while the desired temperature in the room was reached only after the second day of heating. This clearly shows how the irresponsible behavior of the occupants can have influence on the thermal comfort in the building. It also demonstrates the possibilities of the systematic energy management introduced in buildings and offers many other types of analysis as well.

## 5 Conclusion

Construction of the daily regression curve of energy consumption for heating versus outside temperature is demonstrated on a practical example of large educational building. Calculated values are compared with measurements and very good agreement is reached on the daily basis, with annual energy consumption differences between calculated and measured values of less than 1%. It was shown how the huge amount of data, more than 100,000 measured values per heating season, can be analyzed, filtered and used in a systematic way.

Paper also deals with occupants behavior issue and on a practical example demonstrates how the open window influenced the thermal comfort conditions in the observed room. Developed advanced MS Excel tool offers also other types of analysis targeted on the heating system settings and heating schedule, which will be demonstrated in some other publication. Despite the fact that the presented procedure brings many benefits in terms of the optimal energy use it also requires educated personnel with deep understanding of both; physics of the problem and the equipment control settings.

## References

1. Xia, X., Zhang, J., Cass, W.: Energy management of commercial buildings–a case study from POET perspective of energy efficiency. J. Energy South. Afr. (2012)
2. Harris, D.J.: A Guide to Energy Management in Buildings, 2nd edn. Routledge, Taylor & Francis Group (2017)
3. Dragicevic, S., Peulic, A., Bjekic, M., Krneta, R.: Measurement and simulation of energy use in school building. Acta Polytechnica Hungarica (2013)
4. Turner, W.C., Doty, S.: Energy Management Handbook, 6th edn. CRC Press, Taylor & Francis Group (2007)
5. Vilnis, V.: Energy Management Principles and Practice, BSI (British Standards Institution) (2009)
6. Morvay, Z.K., Gvozdenac, D.D.: Applied Industrial Energy and Environmental Management. IEEE Press, Wiley (2008)
7. Interreg Central Europe: Step-by-step procedures' handbook for EnMS in public buildings, European Union (2014–2020)
8. Teskeredzic, A., Blazevic, R.: Transient radiator room heating – mathematical model and solution algorithm, buildings (2018)

# Categorization of Educational Institutions in Sarajevo Canton According to TABULA Approach

Sandra Martinović, Armin Teskeredžić(✉), and Rasim Bajramović

Faculty of Mechanical Engineering, University of Sarajevo, Sarajevo, Bosnia and Herzegovina
teskeredzic@mef.unsa.ba

**Abstract.** In this paper, the categorization of educational institutions in Sarajevo Canton, Bosnia and Herzegovina, has been proposed. TABULA approach [1] is used for classification of buildings according to the climate zone, period of construction, type of institution and specific architectural characteristics of buildings. The matrix of buildings is developed and the representative samples for each category were determined by using multiple regression analysis. It is demonstrated how the typical representative for schools with the sloped roof has been chosen out of detailed energy audit reports. Calculation of energy need, delivered and primary energy, as well as the $CO_2$ emission factors, for all representative buildings was done according to the TABULA methodology. Current study introduces a systematic approach within the public buildings stock, which can be used for proposal of the cost-optimal energy efficiency measures in the future.

**Keywords:** Building typology · Educational institutions · TABULA methodology · Regression analysis

## 1 Introduction

Building typologies can help for better understanding of the energy characteristics of a group of buildings. That was one of the reasons for developing the structure of European building typologies, focused on residential sector, within the framework of the European international research project TABULA [1].

TABULA project is harmonized with the Directive 2010/31/EU [2], which is obligatory for European Union member states (EU) and as well for Bosnia and Herzegovina as a member of Energy Community. Each participating country in the TABULA project has defined criteria for residential buildings classification and for every type of building a representative building has been selected. This was also done for residential buildings in Bosnia and Herzegovina [3].

In this paper, the categorization of public buildings, applied to educational institutions in Sarajevo Canton, using the TABULA concept has been proposed [4]. Educational institutions have been chosen as an example because they represent the largest energy consumer in the public buildings sector in Sarajevo Canton with more


S. Avdaković et al. (Eds.): IAT 2019, LNNS 83, pp. 525–534, 2020.
https://doi.org/10.1007/978-3-030-24986-1_42

than 46% share [5] of the sector's heat consumption. Above that, educational buildings are the most common in the total public buildings stock with 38.5% share based on the heated area.

Figure 1 shows the share of the energy consumption for heating for different types of public buildings in Sarajevo Canton.

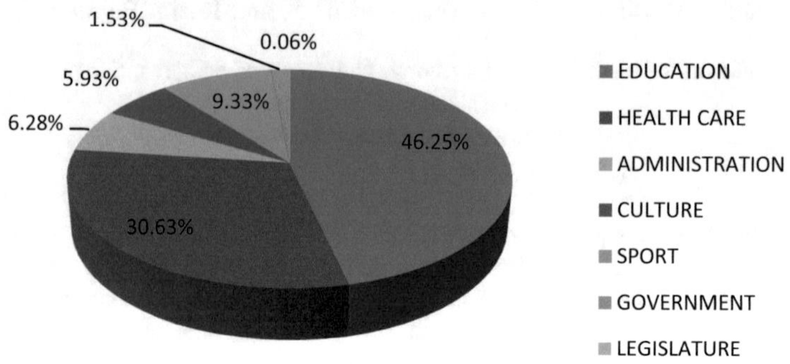


**Fig. 1.** Structure of heat consumption in public buildings in Sarajevo Canton

## 2 The Development of Educational Buildings Inventory Based on TABULA Concept

The classification of educational institutions (school and faculty buildings) in Sarajevo Canton was done by taking into the account following parameters:

- period of construction,
- type of institution,
- architectural characteristics of buildings.

Note that the climate zone influence was excluded from the list of parameters, since all buildings belong to the same zone. Based on the period of construction, educational institutions were grouped in the following groups:

- buildings constructed before 1940,
- buildings constructed from 1940 to 1970,
- buildings constructed from 1970 to 1987,
- buildings constructed from 1987 to 2009,
- buildings constructed in accordance with valid Regulation [6], that's been in line with European standards and obligatory since January 2011.

The periodization of educational buildings, specified above, was conditioned with the socio – historical context, construction technology and applied standards and regulations defining the field of thermal protection of buildings.

For the first two periods of construction is characteristic that there were no regulations defining the field of thermal protection of buildings.

The first regulation on thermal protection of buildings in Bosnia and Herzegovina was enacted in 1970, as the Rulebook on Technical Measures and Conditions for Thermal Protection of Buildings, in which the highest values of the heat transfer coefficient for certain construction elements, were defined. In spite of the mentioned Regulation, the heat losses of buildings from this period are very high [4]. In 1987 the new regulation [7] was enacted, which raised the criteria for thermal protection of the buildings. For this reason, the buildings built after 1987 are not large energy consumers compared to buildings from the previous periods. In 2009 the Regulation [6], in line with European standards, was put in place. The preliminary analysis of existing data suggests that the renovation activities to increase energy efficiency in buildings should be targeted to buildings built before 1987. In this study only these buildings will be analyzed.

Based on the type, educational institutions are grouped in two groups:

- primary and secondary schools and
- faculties.

In terms of the specific architectural characteristics, the division is made for:

- buildings with sloped roof (Type I) and
- buildings with flat roof (Type II).

Proposed assumptions resulted in three characteristic construction periods with three different types of buildings. Having in mind that primary and/or secondary schools with flat roofs (Type II) were not built in the period before 1940, the final number of typical buildings was 8. In the next section, linear multiple regression model is used in order to choose the typical representative of the chosen type and construction period.

## 3 The Regression Analysis of Influential Parameters on Heat Losses

In this chapter, the linear multiple regression model is used to determine the functional dependence of the selected dependent variable $y_i$ on the predicted independent variable $x_i$ in the form [8]

$$y_i = \theta_1 f_1(x_i) + \theta_2 f_2(x_i) + \ldots + \theta_m f_m(x_i) + \epsilon_i \quad i = 1, 2, \ldots; n \quad j = 1, 2, \ldots; m,\ n \geq m \tag{1}$$

where $f_1(x_i), f_2(x_i), \ldots$ are $m$ different, pre-known element functions of $x_i$.

It is often $f_1(x_i) = 1$ for every $i$, $\theta_j$ are unknown parameters and $\epsilon_i$ are experimental errors.

The presented regression model has been applied to one category of educational institutions, schools, constructed from 1940 to 1970, and according to specific architectural characteristics they belong to the group of buildings with sloped roof. Using this model, the influence of selected independent variables on the dependent variable, the specific energy need for heating ($y_i$) was examined. Independent variables are geometric characteristics of buildings such as useable floor area ($p_i$) and building shape factor ($q_i$). Building shape factor is the ratio between external envelope area and the inner volume of a building and it can have a significant impact on the energy consumption [9].

Beside these selected independent variables, other parameters also have influence on energy need for heating, but they have not been taken into account in regression analysis, because standardized values of these parameters, defined by BAS EN ISO 13790 were used for calculation of energy need for heating. For example, the internal heat gains gained from occupants, according to BAS EN ISO 13790 for school buildings are 7 W/m$^2$ and the number of working hours is 14 h, 5 days a week.

Starting with the general form of regression model (1), the functional dependence has been assumed here [4]:

$$y_i = \theta_1 + \theta_2 p_i + \theta_3 q_i + \theta_4 p_i q_i \tag{2}$$

where $i$ represents the number of buildings within a certain class, defined by parameters explained in Sect. 2.

In the Eq. (2) the values of the parameters $\theta_j$ need to be determined. Therefore, a system of equations for the sample of 28 schools (current condition), for which detailed energy audits have been done, has been formed. For each school net surface of the heated space was measured, a building shape factor and specific energy need for heating, based on EN ISO 13790, were calculated.

The energy need for heating is calculated as given by Eq. (3):

$$Q_{H,nd} = Q_{ht} - \eta_{H,gn} Q_{H,gn} \tag{3}$$

where $Q_{ht}$ is the total heat transfer for the heating mode, determined in accordance with Eq. (4), $Q_{H,gn}$ is the total heat gains for the heating mode determined in accordance with Eq. (5) and $\eta_{H,gn}$ is the dimensionless gain utilization factor.

$$Q_{ht} = Q_{ht,tr} + Q_{ht,ve} \tag{4}$$

where $Q_{ht,tr}$ is the total heat transfer by transmission during the heating season and $Q_{ht,ve}$ is the total heat transfer by ventilation during the heating season.

$$Q_{H,gn} = Q_{Sol} + Q_{int} \tag{5}$$

where $Q_{Sol}$ is the total solar heat load during the heating season and $Q_{int}$ are the total internal heat gains during the heating season.

Also, adequate climatic data for the heating season, all the necessary building characteristics, indoor design temperature, number of heating hours have been used in calculation.

Thus, 28 sets of data $y_i$, $p_i$, $q_i$, were obtained, shown in Table 1.

**Table 1.** The sample of 28 schools constructed from 1940 to 1970 with sloped roof

| $y_i$ (kWh/m$^2$a) | $p_i$ (m$^2$) | $q_i$ (m$^{-1}$) | $y_i$ (kWh/m$^2$a) | $p_i$ (m$^2$) | $q_i$ (m$^{-1}$) |
|---|---|---|---|---|---|
| 153 | 1665 | 0,50 | 153 | 1117 | 0,50 |
| 221 | 3148 | 0,58 | 146 | 2683 | 0,47 |
| 161 | 3794 | 0,47 | 116 | 5515 | 0,42 |
| 152 | 1356 | 0,83 | 161 | 3146 | 0,41 |
| 187 | 3415 | 0,42 | 143 | 900 | 0,52 |
| 168 | 1794 | 0,43 | 152 | 2000 | 0,50 |
| 168 | 897 | 0,51 | 168 | 1254 | 0,49 |
| 142 | 684 | 0,63 | 221 | 2193 | 0,56 |
| 155 | 2808 | 0,29 | 171 | 3350 | 0,43 |
| 152 | 2972 | 0,56 | 169 | 1523 | 0,57 |
| 167 | 1171 | 0,50 | 159 | 3631 | 0,43 |
| 172 | 3240 | 0,51 | 150 | 2745 | 0,33 |
| 179 | 2084 | 0,66 | 172 | 1870 | 0,58 |
| 167 | 1295 | 0,42 | 168 | 4350 | 0,48 |

From the linear regression model (2) can be seen that there are 4 unknown parameters ($\theta_j$). To calculate these parameters the program written in C++ was used. The results are shown in Table 2.

**Table 2.** The values of parameter $\theta_j$ t - values

| $\theta_1$ | $t_1$ | $\theta_2$ | $t_2$ | $\theta_3$ | $t_3$ | $\theta_4$ | $t_4$ |
|---|---|---|---|---|---|---|---|
| 288,6 | 6,12 | −0,069 | 3,4 | −256,13 | 2,8 | 0,15 | 3,46 |

In the regression model the number of degrees of freedom is: number of equations – number of unknown parameters, i.e. 28 − 4 = 24. To test the significance of parameters, t – distribution was used. For analyzed sample, the critical value is $t_{cr}$ = 2,064 for 95% confidence interval [10]. For calculated parameters $\theta_j$ t – values obtained for certain parameters are given in Table 2. All parameters $\theta_j$ are equally significant for 95% confidence interval ($t > t_{cr}$), so the presumed function given by (2) is justified and can be written in the final form:

$$y = 288,60 - 0,069p - 256,13q + 0,15pq \tag{6}$$

The dependence of parameter $y_i$ from $p$ and $q$ is shown in Fig. 2.

Figure 2 shows that in the area of larger useable floor area the building shape factor has bigger influence and the energy demand increases faster.

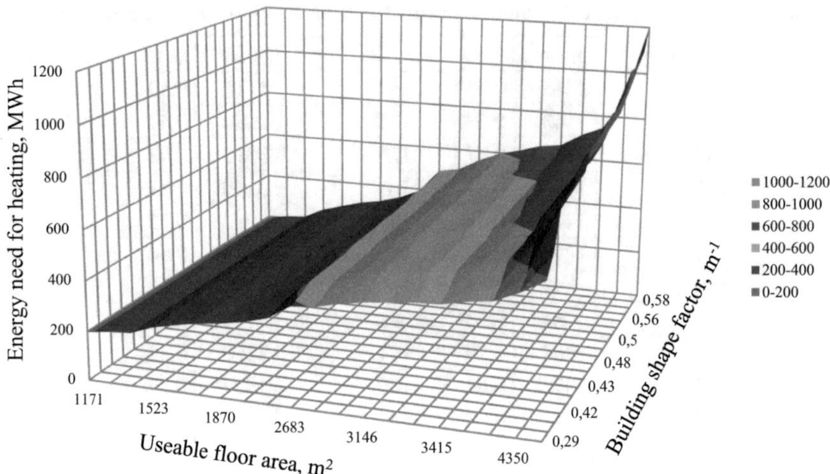


**Fig. 2.** Dependence of building shape factor and usable floor area on energy need for heating

## 4 The Selection of a Representative Building for Different Categories of Buildings

The dependence of building shape factor and useable floor area on energy need for heating (6) has a multivariable normal distribution [4] and dependent variables values $p$ and $q$ for typical building were calculated as the arithmetic mean of the values of the variables within the analyzed class.

Therefore,

$$p_m = \frac{\sum_1^n p_i}{n} \quad , \quad q_m = \frac{\sum_1^n q_i}{n} \tag{7}$$

where $n$ represents the number of buildings within a certain class.

On the diagram $p$, $q$, Fig. 3, the pairs of independent variables for 28 school buildings are shown.

In Fig. 3 two subsets can be seen within which the highest concentration of analyzed buildings is. Within both subsets, the building shape factor varies between 0,4 and 0,6 $m^{-1}$, but useable floor area significantly varies in different ranges. In order to determine the subset from which a representative building of the group is going to be chosen, for the analysis of useable floor area parameter, on the top of the existing detailed energy audits for 28 buildings, additional available data for schools were taken into account [5]. By not taking into account buildings with the extreme values of building shape factor, as well as buildings with an extremely small useable floor area ($p < 1000\ m^2$), within the extended set, the mean values of the useable floor and the mean value of the building shape factor were obtained.

Calculation for the extended data set resulted in mean value of useable floor area of $p_m = 2973\ m^2$ and the mean value of the building shape factor of $q_m = 0{,}47\ m^{-1}$. For

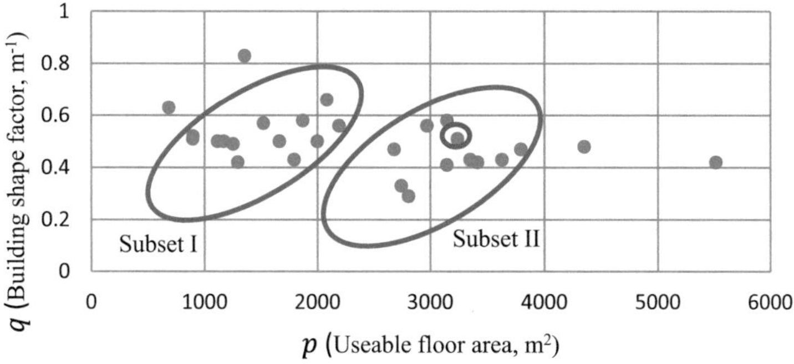


**Fig. 3.** The sample of 28 buildings with defined useable floor area and building shape factor

the representative building of the group, the actual building that has the closest values for $p_m$ and $q_m$ to those calculated, has been chosen and that is the building from the subset II, with the useable floor area of 3240 $m^2$ and with building shape factor of 0,51 $m^{-1}$, highlighted in a red circle in Fig. 3. The building code is KS.N.EHS.T1.02. Gen (Fig. 4).

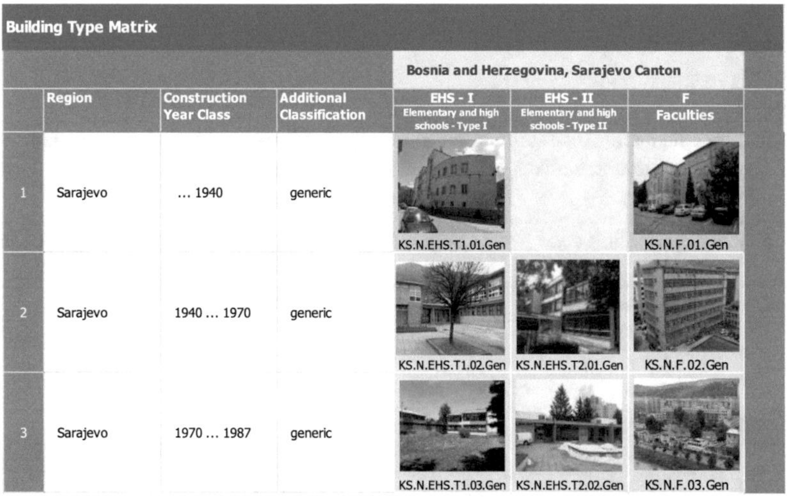

**Building Type Matrix**

| | | | | Bosnia and Herzegovina, Sarajevo Canton | | |
|---|---|---|---|---|---|---|
| | Region | Construction Year Class | Additional Classification | EHS - I Elementary and high schools - Type I | EHS - II Elementary and high schools - Type II | F Faculties |
| 1 | Sarajevo | ... 1940 | generic | KS.N.EHS.T1.01.Gen | | KS.N.F.01.Gen |
| 2 | Sarajevo | 1940 ... 1970 | generic | KS.N.EHS.T1.02.Gen | KS.N.EHS.T2.01.Gen | KS.N.F.02.Gen |
| 3 | Sarajevo | 1970 ... 1987 | generic | KS.N.EHS.T1.03.Gen | KS.N.EHS.T2.02.Gen | KS.N.F.03.Gen |

**Fig. 4.** Building type matrix applied to educational buildings in Sarajevo Canton

The same methodology has been used to choose representative buildings within all other classes. As a result, a typology matrix of educational buildings in Sarajevo Canton, based on TABULA methodology, was created which can be seen in Fig. 4.

## 5 The Calculation of Energy Demand for the Representative Buildings

The calculation of specific energy need for heating, delivered and primary energy for typical representatives, based on TABULA methodology, was done using TABULA excel calculator [1]. Energy need for heating was calculated as given by Eq. (3).

The delivered energy is calculated as given by Eq. (8)

$$q_{del,h,i} = \alpha_{nd,h,i} q_{g,h,out} e_{g,h,i} \tag{8}$$

where $\alpha_{nd,h,i}$ is fraction of heat generator *i*, $q_{g,h,out}$ is heat output of heat generator *i* and $e_{g,h,i}$ is heat generation expenditure factor of heat generator *i*.

The annual primary energy demand was calculated as given by Eq. (9)

$$q_{p,nonren,h} = \sum_i f_{p,nonren,h,i} q_{del,h,i} + f_{p,nonren,aux} q_{del,h,aux} \tag{9}$$

where $f_{p,nonren,h,i}$ is primary energy factor of the energy ware used by heat generator *i* of the heating system, $f_{p,nonren,aux}$ is primary energy factor of electricity used for auxiliary devices and $q_{del,h,aux}$ is the annual auxiliary energy use of heat generator *i* of the heating system.

The results are presented in Table 3.

**Table 3.** The calculation results of specific energy need for heating, delivered and primary energy for typical representatives of educational buildings in Sarajevo Canton, based on TABULA methodology

| Construction period | Building code | $E_{need}$ (heating), $kWh/m^2$ | $E_{del}$, (heating), $kWh/m^2$ | $E_{prim}$ (heating), $kWh/m^2$ |
|---|---|---|---|---|
| do 1940 | KS.N.EHS. T1.01.Gen | 130,2 | 170,9 | 188,0 |
| do 1940 | KS.N.F.01. Gen | 117,6 | 140,2 | 154,2 |
| 1940–1970 | KS.N.EHS. T1.02.Gen | 16 4,6 | 213,9 | 235,3 |
| 1940–1970 | KS.N.EHS. T2.01.Gen | 173,4 | 203,3 | 223,6 |
| 1940–1970 | KS.N.F.02. Gen | 120,1 | 158,2 | 174,0 |
| 1970–1987 | KS.N.EHS. T1.03.Gen | 188,7 | 244,0 | 268,4 |
| 1970–1987 | KS.N.EHS. T2.02.Gen | 208,3 | 242,7 | 266,9 |
| 1970–1987 | KS.N.F.03. Gen | 159,1 | 207,0 | 227,7 |

The defined building types, for which specific delivered and primary energy for heating have been calculated, can serve as a base for calculating the cost of energy efficiency measures applied [4].

On the top of the existing average number per different building, the data on specific energy use for the components of the envelope elements is calculated. The results of calculation for the representative building with the building code KS.N.EHS. T1.02.Gen are presented in Fig. 5.

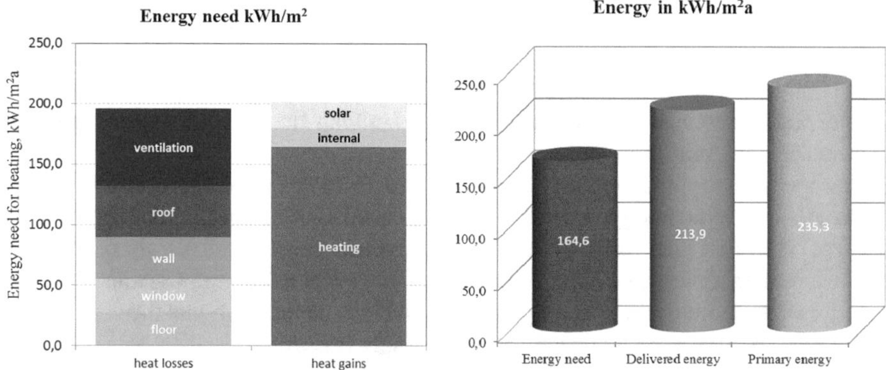


**Fig. 5.** Energy need decomposed per different envelope elements and net energy need for heating (left) and energy need, delivered energy and primary energy (right) all for the building with the building code KS.N.EHS.T1.02.Gen

Figure 5 (left) shows the graphical representation of the calculated energy need, decomposed per building envelope elements and also shows the solar, internal and net heating gains from the heating system. At the right hand side of the same figure, presented graphs show the energy need for heating, delivered and primary energy, which tells us about the efficiency of the heating system and about the fuel which is used. Decomposition of the energy need per envelope elements can clearly show which parts of the envelope contribute the most in overall heat losses.

# 6 Conclusion

In this paper, the categorization of educational buildings in Sarajevo Canton, based on the TABULA concept, was carried out by taking into the account following parameters: period of construction, type of institution and architectural characteristics of buildings.

The selection of representative building for every category was carried out using the linear regression analysis.

The functional dependence of specific energy need on useable floor area and building shape factor has been assumed.

On the sample of 28 buildings, constructed from 1940 to 1970, with sloped roof has been proven that assumed functional dependence has a multivariable normal distribution and that the chosen parameters are significant for 95% confidence interval, using t-distribution. This means that the assumed functional dependence of the influential parameters is correct.

The influence of building shape factor and useable floor area on energy need for heating has been examined and it has been shown that the influence of building shape factor is significant when useable floor area has a higher value.

Since the functional dependence for the selected dependent variable, energy need for heating, has normal distribution, the typical representatives for certain categories of buildings could have been chosen by using the mean value of the useable floor area and of the building shape factor and it was explained in detail for one category of buildings.

For the representative building, the actual building that has the closest values to those calculated, has been chosen.

As a result, a typology matrix of educational buildings in Sarajevo Canton, based on TABULA methodology, was created. The calculation of specific energy need for heating, delivered and primary energy for typical representatives, based on the TABULA methodology, was done using TABULA excel calculator. Calculated energy need for every building was decomposed per envelope elements so that the contribution of the heat losses can be analyzed for every single element.

The methodology presented in this paper can be the starting point for developing the TABULA methodology for the whole public buildings sector in Bosnia and Herzegovina and other countries.

## References


1. www.building-typology.eu
2. Energy performance of buildings directive 2010/31/EU
3. Arnautović-Aksić, D., Burazor, M., Delalić, N., Gajić, D., Gvero, P., Kadrić, Dž., Salihović, E., et al.: Typology of residential buildings in Bosnia and Herzegovina. Faculty of Architecture, Sarajevo (2016)
4. Martinović, S.: Development of methodology for cost – optimal solutions of energy efficiency projects, implemented on individual and typical buildings, based on TABULA concept. Ph.D. thesis, University of Sarajevo, Sarajevo (2019)
5. HVAC DESIGN d.o.o.: Akcioni plan energetski održivog energetskog razvoja Sarajeva (Seap). REGEA, Sarajevo (2011)
6. The regulations on technical requirements for thermal insulation of buildings and rational energy use, SN FBiH, 49/09
7. JUS U.J5.600: Thermal equipment in construction work, technical requirements for the design and construction of buildings. Savezni zavod za standardizaciju, Beograd (1988)
8. Bajramović, R.: Osnovi linearnog programiranja. Mašinski fakultet, Sarajevo (2001)
9. Matić, M.: Energija i arhitektura. Školska knjiga, Zagreb (1988)
10. Huntsberger, D.V., Billingsley, P.: Elements of Statistical Inference. Allyn and Bacon Inc, Boston (1975)

# Energy Transition of Power Utility JP Elektroprivreda BiH Through Upgrading and Retrofit of the Generation Portfolio Supported by Horizon 2020 Projects

Anes Kazagić(✉), Ajla Merzić, Dino Trešnjo, and Mustafa Music

Department of Strategic Development, Power Utility Elektroprivreda of Bosnia and Herzegovina, Sarajevo, Bosnia and Herzegovina
a.kazagic@epbih.ba

**Abstract.** Today, many power utilities are found at crossroads, facing challenges of decarbonisation and the imperative of competitiveness on the market, often considered as contradictory requirements for their further development. In this regard, a particular challenge is placed in front of utilities which generation portfolio is based on conventional resources, dominantly coal for electricity and thermal energy production.

On a concrete example of the generation portfolio development of Power Utility Elektroprivreda of Bosnia and Herzegovina – Sarajevo (EPBiH), in this work so far realised actions within the transitional process and its further continuation are presented. Effects of improvements of the existing thermal power plants (TPPs), i.e. TPP Tuzla and TPP Kakanj through the reconstruction/modernization process, as well as further planned measures for increasing energy efficiency, reducing CO2 emissions and reducing pollutant emissions are shown. Particular attention is given to the possibilities of extending the cogenerative production of EPBiH's TPPs, as well as the upgrading process of existing and construction of new district heating systems, thus including the integration of renewable energy sources. Further, the paper focuses on the application of biomass technologies in TPPs, through full and partial conversion to biomass (co-firing technology), shown on examples of EPBiH's opt-out thermal units.

The proposed measures are also processed as illustrative examples within three Horizon 2020 projects, namely CoolHeating, Upgrade DH and BioFit, funded by the European Commission and in which EPBiH participates with its European partners. Transfer of knowledge and technology, mastering the know-how and implementing the proposed measures is a significant contribution and stimulation of the ongoing transition process in EPBiH's power system in a responsible and sustainable way and can serve as a demonstration example to other power utilities/countries facing similar challenges.


S. Avdaković et al. (Eds.): IAT 2019, LNNS 83, pp. 535–548, 2020.
https://doi.org/10.1007/978-3-030-24986-1_43

# 1 Introduction

Due to increased concerns regarding rising greenhouse gas (GHG) emissions and other negative impacts on the environment caused by actions within the energy sector, the search for clean and efficient energy sources has been intensified. In doing so, energy producers are searching for sustainable ways to reconcile requirements facing and develop their own transition paths, depending on specific circumstances in which they operate, primarily relating to available resources, regulations, price policies, etc. By these processes the electricity and heating sector have been affected.

In the electricity power sector governments are designing energy and environmental policies to minimize their exposure to volatile international fossil fuel prices and reduce carbon emissions in the energy sector. Appropriate directives, e.g. [1], strategies, e.g. [2, 3] and guidelines have been issued, in order to promote the use of renewable energy sources (RES) and lead to their proper integration. In addition, conclusions of the UN Climate Change Conference held in Paris in December 2015 emphasize the transition towards a clean economy, suspending dangerous climate change [4].

According to the EU Strategy on Heating and Cooling (2016), the contribution of district heating (DH) in the EU accounts for 9% and is mainly driven by fossil fuels such as gas and coal. District heating networks present a high potential for the transition of the heat sector, both technically and organisationally. They allow integration of renewable energies, improve the overall energy efficiency, as well as facilitate sector coupling (coupling between heating, electricity and mobility) [5].

Countries with a dominantly coal-based generation portfolio are facing particular challenges in this regard. The illustrative generation portfolio example that has been considered with in this paper is the system of Power Utility Elektroprivreda of Bosnia and Herzegovina – Sarajevo (EPBiH). Within its generation portfolio, the production of electrical and thermal energy is carried out, so it is of great importance to conceive its further development in such a way that both types of useful energy forms are produced sustainably.

# 2 Methodology

In order to meet challenges placed before conventional power systems and contribute towards a sustainable development, in this paper methodologies used within three Horizon 2020 projects, CoolHeating, Upgrade DH and BioFit, funded by the European Commission in which EPBiH participates with its European partners have been presented and applied.

The focus of this analysis has been on the thermal power sector and all the measures applied are mapped to the concrete example of EPBiH's generation portfolio and incorporated into the overall transitional process of this system. Performed actions on the modernization of thermal power units and increase in energy efficiency are addressed in this paper, too.

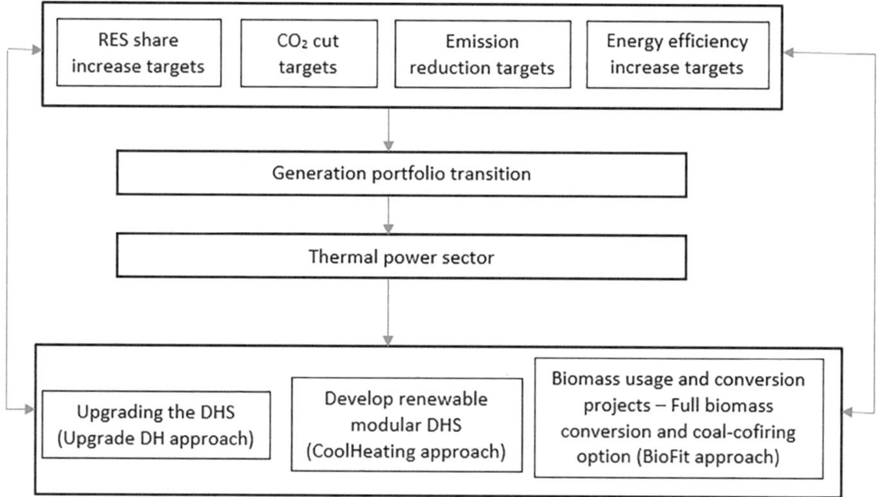


**Fig. 1.** Methodological approach of EPBiH's thermal power generation transition

The overall methodology is presented in Fig. 1.

## 2.1 EPBiH's Generation Portfolio Structure and Development Targets

The generation portfolio of the considered system consists of two coal-fired TPPs, three cascade related hydro power plants (HPPs) and a small number of small HPPs. In addition to the electricity produced within these units, during the heating season, cogenerative blocks are also engaged in the production of thermal energy. By this, from TPP Tuzla cities Tuzla and Lukavac are supplied by heat energy and from TPP Kakanj the city of Kakanj. There are plans to extend the heat supply to the city of Sarajevo and Zenica from TPP Kakanj and to Živinice city from TPP Tuzla. On a yearly basis, approx. 7.500 $GWh_e$ and 400 $GWh_t$ are produced. The installed capacity ratio (GW) between fossil fuels and RES based facilities is 69/31%, whereby the generation ratio (GWh) depends on the hydrology and approx. accounts for 79/21% in favor of fossil fired units.

EPBiH's business policy is in line with the EU energy policy and promotes security of supply, competitiveness and sustainability in economic, environmental and social aspects. Development targets are based on the following principles:

- energy independence - usage of domestic resources for electricity and heat generation
- economic growth - competitive energy prices and more jobs
- environmentally acceptable business
- consumer satisfaction and protection.

Few papers published so far [4, 6, 7] treated EPBiH's sustainable generation portfolio development, as a typical example of a conventional power system, mostly based on fossil fuels. By an approach proposed in [7] addressing EPBiH's generation portfolio optimization as function of specific RES and decarbonisation targets resulted in three scenarios as shown in Fig. 2.

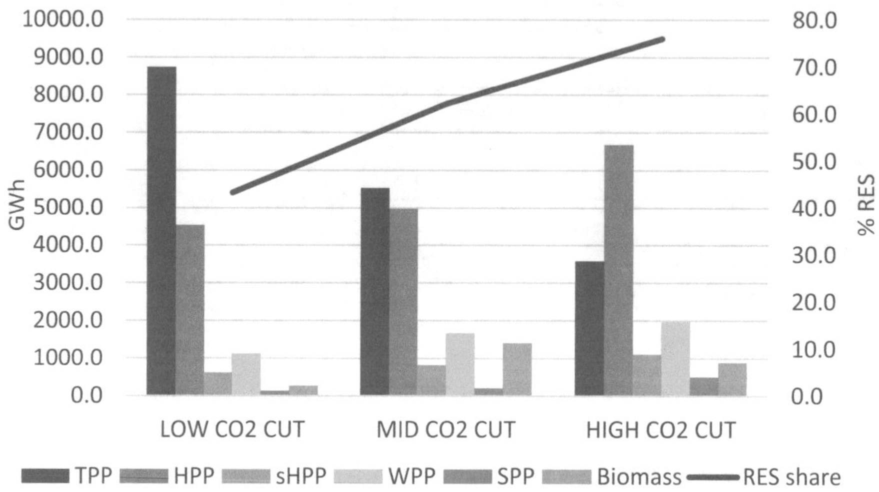


**Fig. 2.** EPBiH's generation portfolio optimization as function of specific RES and decarbonisation targets

The preferable option is the "High $CO_2$ cut" scenario which requires drastic measures for these effects to be reached. In addition to the intensive construction of RES-based facilities, the shutting down of inefficient thermal power units according to the National Emission Reduction Plan, this scenario includes the expansion of cogeneration and increase in energy efficiency.

Nevertheless, this paper couples the power and thermal sector and includes upgrading measures in the heating domain, as an inevitable segment of the transitional process of the company's product portfolio. This is specified and explained in more details in the chapters below and also reflects the specific contribution of this paper and the analysis provided within.

## 2.2 Increase of Energy Efficiency by the Modernization of Power Units

### 2.2.1 Energy Efficiency Requirements for Large Power Utilities

Energy efficiency is the central objective both for Energy 2020 and Energy roadmap 2050 – it is a key factor in achieving long-term energy and climate goals. Efficiency, including in electricity use, must become a profitable business itself, leading to a robust internal market for energy-saving techniques and practices and commercial opportunities internationally. Special attention should be given to the sectors with the largest potential to make energy efficiency gains. Beside the existing building stock and transport sector, industry sector, and particularly power industry, is in the focus. As set in EU strategic documents, as well as in new energy efficiency Directive 2012/27/EU, the industry sector needs to incorporate energy efficiency objectives and energy technology innovation into its business model. The ETS contributes significantly to doing so for larger companies, but there is need for a wider use of other instruments, including energy audits and energy management systems in smaller companies and supporting mechanisms for SMEs. Efficiency benchmarking can indicate to companies where they stand in efficiency terms in comparison with their competitors.

For power industry, energy efficiency in both production as well as distribution should become an essential criterion for the authorization of generation capacities, and efforts are needed to substantially increase the uptake of high-efficiency cogeneration, district heating and cooling. Distribution and supply companies (retailers) are required to secure documented energy savings among their customers, using means such as third-party energy services, dedicated instruments such as 'white certificates', public benefit charges or equivalent and speeding up the introduction of innovative tools such as 'smart meters' which should be consumer-oriented and user-friendly so that they provide real benefits for consumers.

### 2.2.2 Energy Efficiency and $CO_2$ Emissions - Current Situation

EPBiH supplies electricity to near 750.000 consumers in Bosnia and Herzegovina, via its distribution network operated by EPBiH distribution company organized in five regional distributive parts. Furthermore, EPBiH exports about 20% of electricity. Annual production of heat energy, generated in cogeneration power units of TPP Tuzla and TPP Kakanj, is app. 400 GWh. The heat is supplied over long-distance district heating systems to the consumers in the city of Tuzla, Lukavac and Kakanj. In the past ten years, by different measures that have being undertaken in generation and distribution sectors of the company, energy efficiency in EPBiH increased by 30% compared to the 1990 level. In generation sector, by appropriate measures like decommissioning old power units (4 × 32 MW in TPP Kakanj and 2 × 32 MW in TPP Tuzla) because of exhausted life time and low efficiency of these installations, as well as modernizations of all other existing coal-based power units in period between 2002 and 2012, total net efficiency of the EPBiH's power plants increased from 24% up to 31%, which is an increasing of app. 30%. For the same period, for comparable electricity production to the 1990 level, $CO_2$ emission was reduced from 9.500.000 t/a (1990) to the current 6.500.000 t/a.

The major measures for improvement of energy efficiency in coal-based power plants that have been implemented through the modernization campaign in the last ten years, can be summarized in increasing boiler efficiency through modernization of the boiler design and replacing the combustion system, replacing LP and HP steam turbine parts by a modern design, modernization of generator and improvements of cooling towers. After last modernization performed on TPP Tuzla Unit 6-225 MW (the modernized unit commissioned in January 2013), average heat consumption rate of EPBiH's TPPs reduced at 11.722 kJ/kWh (avr. net efficiency in condensing regime of 30.71%, i.e. not incl. co-generation effects).

### 2.3 Retrofitting the Generation Portfolio by CHP Expansion and Upgrading DHS with RES

The planned increase in cogeneration production within EPBiH's product portfolio by increase of heat supply to Tuzla, Lukavac and Kakanj, but especially by plans to extend the heat supply to cities of Sarajevo, Zenica and Živinice will increase the energy efficiency of the thermal units and positively affect environmental and social issues in these areas.

An important step in the improvement of the district heating sector is also the implementation of small, modular, renewable district heating systems, as done within the project CoolHeating (https://www.coolheating.eu/en/). Core activities, besides techno-economical assessments, included measures to stimulate the interest of communities and citizens to set-up renewable district heating systems as well as the capacity building about financing and business models. The outcome was the initiation of new small renewable district heating and cooling grids in 5 target communities up to the investment stage [8]. The lighthouse project developed in Bosnia and Herzegovina was a district heating system for Visoko Municipality which showed complementary to the planned heat supply from the combined heat and power facility (CHP) Kakanj [9]. The fundamental idea of this district heating system was to use local fuel or heat resources that would otherwise be wasted, in order to satisfy local customer demands for heating, by using a heat distribution network of pipes as a local market place. Existing heating systems in Visoko are mainly individual and currently dominated by coal as the cheapest energy source on the market, therefore they should be upgraded or new networks created, using solid biofuel and solar and geothermal energy technologies.

Applying the developed methodology within the CoolHeating project which started with stakeholder identification and involvement, followed by a consumer survey in order to identify the expected connection rate and the identification of available heat resources for the district heating system, the specialized software tool EnergyPRO [10] has been used for the forecast of the demand and determination of the capacities of the production units, as well as the optimization of the operating mode itself. For feasibility assessments, the CoolHeating economic tool has been applied. District heating concept based on excess heat from CHP Kakanj and RES integrated facilities would help in meeting the rising urban energy needs, improve efficiency, increase energy security, reduce greenhouse gas emissions and pollutants and significantly improve local air quality in Municipality of Visoko. It is foreseen that the elaborated district heating system for Visoko Municipality will have a long-term impact on the further development of small modular renewable heating grids in Bosnia and Herzegovina, as one of the target countries within the CoolHeating project.

On the other hand, as proposed by the Upgrade DH project (https://www.upgrade-dh.eu/en/home/), many district heating systems could be upgraded by improvements of the heat consumption (substations and consumer connections, heat demand), heat distribution and heat generation. Core actions could involve lowering of leakage rates and heat losses, reducing operation temperatures, adapting piping dimensions and hydraulics, introducing modern IT-based management systems and options for user control [5]. The heat generation could be improved by the integration of RES and waste heat. Upgrading processes must go hand-in-hand with predictions of future heat demand as well as with efficiency measures on the end use of heat.

The overall objective of the Upgrade DH project, funded by the EU's Horizon2020 program, is to improve the performance of inefficient district heating networks in Europe by supporting selected demonstration cases for upgrading, which can be replicated in Europe. Specific objectives of the Upgrade DH project are as follows:

- to initiate the DH upgrading process for eight district heating systems in Europe up to the investment stage
- to save, by initiating the eight demo cases, more than 190 GWh/a primary energy and of 77,000 t $CO_2$ equiv. greenhouse gas emissions
- to increase the share of waste/residual heat (currently 7% in the demo cases) by more than 6% and the share of renewable heat (currently 28% in the demo cases) by more than 20% in the eight demo cases and beyond.
- to replicate the proposed upgrading solutions across Europe
- to develop concrete regional/national action plans for the retrofitting of inefficient district heating networks by including the results of the retrofitting approaches.

The considered case study from BiH in this regard is the district heating system of Tuzla city supplied by heat from CHP Tuzla. The district heating system Tuzla will be upgraded in a way to propose measures to solve identified hydraulic problems, conceptualize heating solutions based on RES, namely utilizing solar energy, waste and biomass, as well as introducing energy efficiency measures and reducing the primary energy demand thereby.

The proposed methodology includes the identification and elaboration of best practice examples with implemented upgrading measures which can serve as replication basis; the identification of stakeholders and their early involvement in the upgrading process; formation of local working groups and interviews with all relevant participants in it; application of software tools and instruments to assess possible upgrading measures, and perform feasibility checks as well. All this can be applied to other district heating systems aiming for upgrade and sustainable development.

## 2.4 Retrofit of the Generation Portfolio by Biomass Co-firing

Bioenergy is an essential form of renewable energy, providing an estimated 60% of EU's renewable energy production in 2017 [11]. In the future, bioenergy will remain important; in its 2017 Roadmap [12] the International Energy Agency (IEA) notes that bioenergy plays an essential role in its 2DS (2 °C Scenario), providing almost 20% of the global cumulative $CO_2$ emission savings by 2060. Bioenergy is a complex and sometimes controversial topic. There is an increasing understanding that only bioenergy that is supplied and used in a sustainable manner has a place in a low carbon energy future.

Modern bioenergy takes on many forms. Relatively straight-forward applications, such as space heating by combustion of wood are implemented alongside biogas production through anaerobic digestion and production of transport fuels. Spurred by innovation, technologies are becoming more advanced and diverse, leading to the production of variety of advanced transport fuels (first and second-generation bioethanol, biodiesel and bio-kerosene), intermediate bioenergy carriers and high-efficiency, low carbon emission production of power, heating and cooling [13]. Various ways of reducing $CO_2$ from coal-based power generation are currently in certain phases of research, development and demonstration. Many of them involve biomass co-firing. Co-firing coal with biomass can be carried out directly (in the same combustion chamber), indirectly (after pre-treatment), in parallel (separate combustion), and completely (full conversion to biomass).

In the segment of technologies and the processes of using biomass, EPBiH has recognized its possibilities even in 2011 and started a pilot project on coal and waste woody biomass co-firing TPP Kakanj. In this chapter RD&D (Research, Development and Demonstration) activities are presented carried out by EPBiH and Faculty of Mechanical Engineering of University in Sarajevo, to develop and implement solutions of retrofitting large thermal power plants of EPBiH with biomass. The target is achieving a sustainable power production in TPPs in a long-term view.

The experimental results in laboratory and trial run in TPP Kakanj Unit 5 do show that there is reasonable expectation that tested coal/woody biomass/Miscanthus blends could be successfully co-fired in real operating conditions of PC Large Boiler, not producing any serious ash-related problems. Low or low-to-medium slagging propensity was noticed for all co-firing cases, at usual PC process temperature 1250 °C. The increase of the unburnt carbon content (UBC) in the ash deposits collected in lab-scale furnace when increasing co-firing rate from 0.0 to 0.3 is only minor. However, in the slag collected at the bottom of the furnace, the UBC is increased for 0.15 and 0.2 biomass co-firing which implies that attention has to be paid to the combustion organization when co-firing above 15% of biomass at similar process conditions as used here, in order to keep combustion efficiency at an acceptable level.

In regard with emissions issue, it can be concluded that the results from performed biomass co-firing tests give some ground for optimism for achieving additional benefits, in particular for the co-firing with Miscanthus.

The results suggest that woody sawdust and Miscanthus co-firing with Bosnian low rank coal types shows promising effects at higher ratios for pulverized combustion, [14].

Further investigations in this field are foreseen within the BioFit project activities (https://www.biofit-h2020.eu/) where EPBiH will investigate full biomass repowering of one unit in TPP Kakanj and biomass cofiring in one unit in TPP Tuzla (Fig. 3).

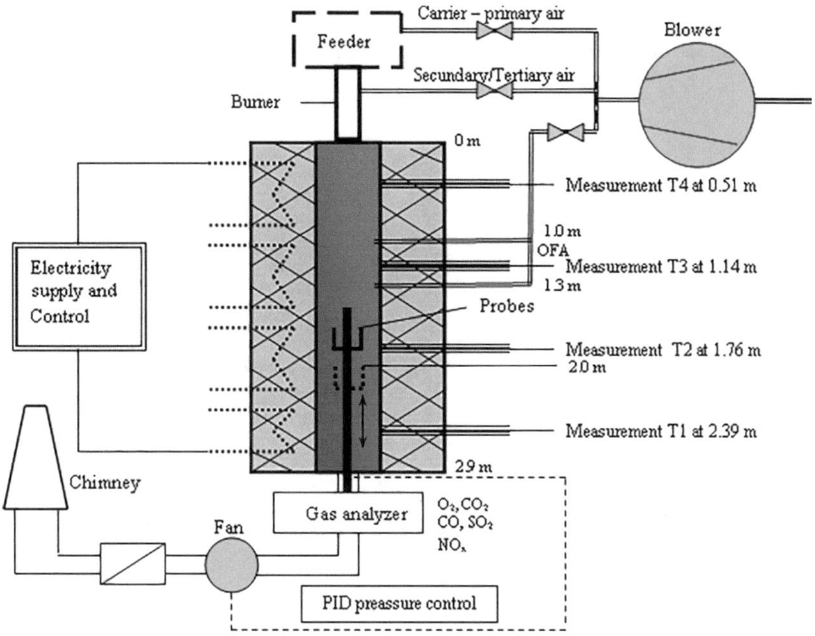


**Fig. 3.** Biomass cofiring RD&D activities of EPBiH – lab-scale test furnace, schematic layout, [14]

## 3 Results and Discussion

Following the proposed methodology and considering improvements in different product portfolio domains, obtained results are given below. In each sub-section, the expected results of the projects and activities under consideration are given, and the direct impact on the further transition of EPBiH's production portfolio is estimated, as a contribution to sustainable development in general.

### 3.1 Expected Energy Efficiency Increase

Specific heat consumption for existing units of TPP Tuzla and TPP Kakanj is approximately at the level of unit consumption for similar production units of the older generation from the 70-ies of the last century worldwide, with average specific heat consumption from 11.500 to 12.000 kJ/kWh, and i.e. efficiency of 30 to 31.5%. This level of energy efficiency of existing thermal units, along with the measures of planned maintenance and investments, will remain until their decommissioning, with proportionate increase in energy efficiency with the planned entry of new more efficient thermal power units in the company's production portfolio.

Although EPBiH has achieved significant progress in raising energy efficiency in recent years, it still lags behind in performance compared to developed countries, primarily due to the obsolescence of the existing production facilities and associated technologies. In this sense EPBiH faces challenges in introducing efficient and clean technologies, in the form of power plants using RES & thermal units with clean coal technologies, in order to further improve energy efficiency and reduce emissions (Fig. 4).

**Fig. 4.** Projection of net efficiency of EPBiH existing thermal power units

According to one of the expected scenarios, with the planned decommissioning all existing coal-based power plants and entry into operation of two high-efficient CHP replacement units in period 2023–2035, it is expected to continue with the improvement of energy efficiency of the production from thermal facilities from the current 30.5% to 34.9% in 2020, and 40% by 2035. Thereby, $CO_2$ emissions will be reduced drastically by 2050, see Fig. 5.

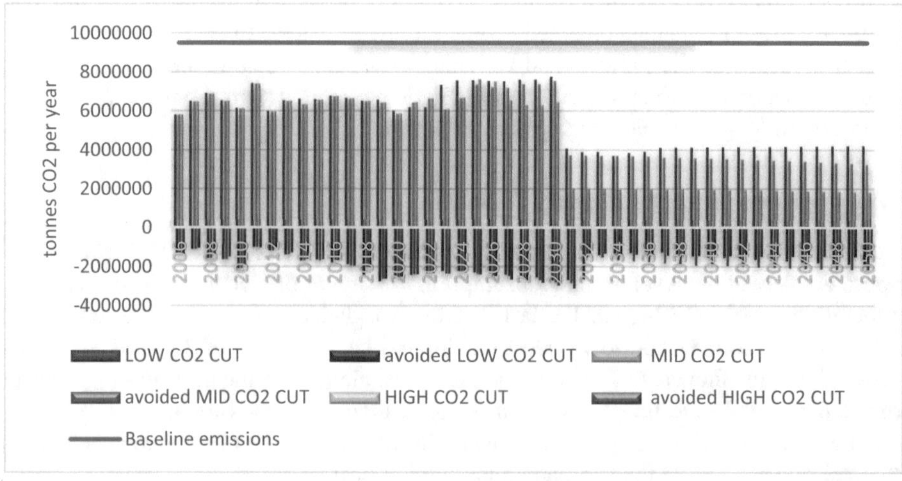


**Fig. 5.** $CO_2$ cut scenarios until 2050

According to this scenario, the specific $CO_2$ emission from EPBiH thermal power units would be reduced below 800 kg/MWh from the current 870 kg/MWh by 2030, see Fig. 6.

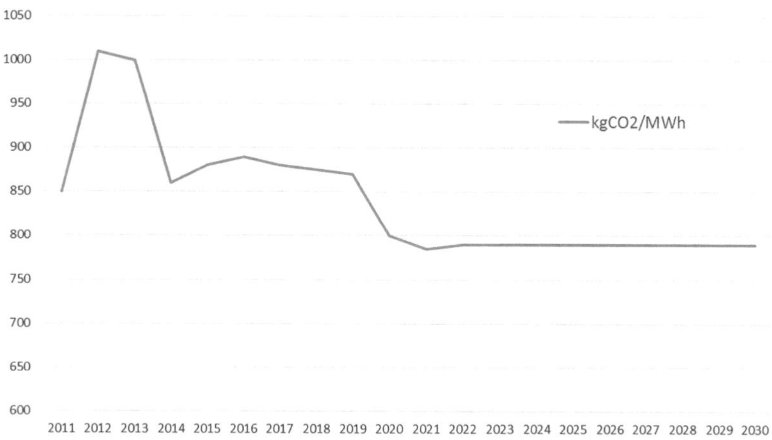


**Fig. 6.** Projection of network emission coefficient (the net rate of $CO_2$) until 2030

## 3.2 Expected Impacts of CHP Expansion and DHS Retrofit and Upgrade

District heating uptake has grown with the increasing need for cleaner and more efficient energy supply. This has resulted in a rising number of new developments signing up to a district heating scheme, typically powered by CHP or another solution based on RES.

### 3.2.1 DHS Based on RES Modular Solutions – CoolHeating Concept

According to the set CoolHeating project work program, the expected impacts were to increase the share of renewable electricity, heating and cooling in the final energy consumption, to reduce the time taken to authorize the construction of renewable energy plants and related infrastructure and to develop better policy, regulatory, market support and financing frameworks.

CoolHeating addressed all expected impacts in Municipality of Visoko, Fig. 7, and in addition to that, clear improvement of economic and environmental indicators have been obtained. The realization of the project will result in savings of around 500,000 € of fuel costs per year, and 24 € operational costs per produced MWh of heat energy. Perhaps a clearer picture of the justification of such project is given by environmental indicators, where emissions savings amount: 64,464.6 t of $CO_2$, 118.1 t of SOx, 561.6 t of NOx, 626.3 t of PM(10) and 623.7 t of PM(2,5) during the DHS lifetime of 20 years.

**Fig. 7.** Visoko Municipality CoolHeating solution: Map of locations for installation of RES production units; thermal storage, photovoltaics and main pipes from the DH grid

### 3.2.2 Increase of DH Energy Efficiency and CHP and RES Hybridization – Upgrade DH Concept

The aim of the UpgradeDH project is to achieve primary energy and GHG emission savings triggered by the proposed actions, to increased share of waste heat and RES of heat and to scale the replicability potential of the proposed solutions. By implementing various upgrade measures in the DHS of Tuzla, savings of 90 GWh/a primary energy demand and 25,000 $tCO_2$equiv/a of GHG emissions would be achieved, and the share of heat energy production based on RES would be increased to 20%, Fig. 8.

**Fig. 8.** Potential solar thermal collector fields (green) in DHS Tuzla

The figures do not yet include impacts triggered by the Upgrade DH Bosnian replication cases. Therefore the expected impact will be higher than the calculated data.

### 3.3 Expected Impacts of the Generation Portfolio Retrofit

The use of biomass may mitigate greenhouse gas emissions by displacing the use of coal as fossil fuel for electricity generation. Within the BioFit project several impacts are expected to lead to market uptake of bioenergy retrofits in Bosnia and Herzegovina, and consequently to more renewable bioenergy production. BioFit is expected to lead to more bioenergy due to the expected implementation of the concrete proposals, and because of the many other activities that will – more indirectly - facilitate bioenergy retrofits.

EPBiH will investigate full biomass repowering of one unit (118 MWe + 150 MTth) in TPP Kakanj which is equivalent generation of 7.7 PJ renewable energy per year. The unit currently uses local brown coal and, apart from power, also supplies heat to the local district heating network. Two pathways will be considered: wood pellets and thermally treated wood pellets.

Along with the full biomass conversion, EPBiH will investigate biomass cofiring in the existing brown coal Unit 5 (200 MWe) in TPP Tuzla which would be equivalent to generation of renewable energy of 1.7 PJ/year. Local biomass sources (e.g. sawdust, agricultural residues) will be considered in order to substitute up to 30% of the thermal input from coal.

## 4 Conclusion

Today, power utilities worldwide are facing challenges of decarbonisation and competitiveness on the market, often considered as contradictory requirements for their further development. In this regard, a particular challenge is placed in front of utilities which generation portfolio is based on conventional resources, dominantly coal for electricity and heat production.

On a concrete example of the generation portfolio development of Power Utility Elektroprivreda of Bosnia and Herzegovina – Sarajevo (EPBiH), a typical large power utility of South-east Europe, actions of the transitional process are presented and elaborated in this paper. Transitional measures can be divided in several pillars; RES implementation, energy efficiency measures, conversion of CHP to biomass-co-firing, as well as co-generation expansion and DH sector transition into $4^{th}$ generation of DHS.

The proposed measures have been developed through three Horizon 2020 projects, namely CoolHeating, Upgrade DH and BioFit, funded by the European Commission and in which EPBiH power utility participates with its European partners. Transfer of knowledge and technology, mastering the know-how and implementing the proposed measures, is a significant contribution to the ongoing transition process in EPBiH to achieve a sustainable power system, and can serve as a demonstration example to other power utilities/countries facing similar challenges.

**Acknowledgements.** The authors would like to thank the Coolheating, Upgrade DH and BioFit partners for their contributions in the project. The authors would like to thank the European Commission and the EASME for the support of the three Horizon 2020 projects.

CoolHeating has received funding from the European Union's Horizon 2020 research and innovation program under grant agreement No. 691679.

Upgrade DH has received funding from the European Union's Horizon 2020 research and innovation program under grant agreement No. 785014.

BioFit has received funding from the European Union's Horizon 2020 research and innovation program under grant agreement No. 817999.

Disclaimer: The sole responsibility for the content of this paper lies with the authors. It does not necessarily reflect the opinion of the European Union. Neither the EASME nor the European Commission are responsible for any use that may be made of the information contained therein.

## References

1. European Parliament and the Council of the European Union, Directive 2009/28/EC on the Promotion of the Use of Energy from Renewable Sources (2009)
2. European Commission, Energy 2020, European Union, Brussels, Belgium (2011)
3. European Climate Foundation, Energy Roadmap 2050, European Union, Brussels, Belgium (2010)
4. Merzic, A., Music, M., Haznadar, Z.: Conceptualizing sustainable development of conventional power systems in developing countries - a contribution towards low carbon future. Energy **126**, 112–123 (2017)
5. Mergner, R., et al.: Upgrading the performance of district heating networks in Europe. In: 27th European Biomass Conference & Exhibition, Lisbon, Portugal, 27–30 May 2019
6. Kazagic, A., Merzic, A., Redzic, E., Music, M.: Power utility generation portfolio optimization as function of specific RES and decarbonisation targets – EPBiH case study. Appl. Energy **135**, 694–703 (2014)
7. Kazagic, A., Music, M., Redzic, E., Merzic, A.: Transition of a conventional power utility to achieve 2050 RES and carbon cut targets—EPBIH case study, advanced technologies, systems, and applications II, Springer publishing, January 2018
8. Rutz, D., Doczekal, C., Zweiler, R., Hofmeister, M., Laurberg Jensen, L.: Small modular renewable heating and cooling grids. In: Handbook, WIP Renewable Energies, Munich, Germany (2017)
9. Kazagic, A., Merzic, A., Redzic, E., Tresnjo, D.: Optimization of modular district heating solution based on CHP and RES - demonstration case of the municipality of Visoko. Energy (2019). https://doi.org/10.1016/j.energy.2019.05.132
10. EnergyPRO User Guide. EMD International A/S, 2013 Aalborg.S
11. http://www.europeanbioenergyday.eu/
12. IEA Technology Roadmap Delivering Sustainable Bioenergy (2017)
13. Reumerman, P., et al.: Bioenergy retrofits for Europe's industry – the biofit project horizon 2020. In: 27th European Biomass Conference & Exhibition, Lisbon, Portugal, 27–30 May 2019
14. Kazagic, A., Hodzic, N., Metovic, S.: Co-combustion of low-rank coal with woody biomass and miscanthus: an experimental study. Energies **11**(3), 601 (2018)

# Konjic District Heating System Sustainability Parameters

Haris Lulić[1(✉)], Emir Sirbubalo[1], Milovan Gutović[2], Galib Šenderović[2], Kerim Šenderović[2], and Adnan Đugum[3]

[1] Faculty of Mechanical Engineering, University of Sarajevo, Sarajevo, Bosnia and Herzegovina
lulic@mef.unsa.ba
[2] HVAC Design Doo, Group for Design and Engineering, Sarajevo, Bosnia and Herzegovina
[3] IGT - Research and Development Center of Gas Technology, Sarajevo, Bosnia and Herzegovina

**Abstract.** Development of district heating systems (DHS) is one of the solutions for rational consumption of energy and environmental protection. Many old DHS in Bosnia and Herzegovina are not operational due to obsolete and unusable heat generation systems although the distribution network exists. The existing DHS in Konjic municipality is presented in this work, along with the climate characteristics of that area. Estimation of the heat demand accounting for the possible extension of the system is given. Detailed routes of the main and branch pipelines are determined according to the identified heating zones. Sustainability parameters, that enable comparative analysis with other similar DHS, are calculated and presented.

## 1 Introduction

District heating systems are economical and ecologically acceptable way of generation and distribution of heating energy to end users. Centralized generation of heat makes optimization of that process and better control of pollutant emissions possible. During the process of transformation of primary energy to the energy delivered to end users, a significant part is lost, mainly as heat. It also has to be emphasized that some 60% of the final energy is used for space heating. Climate change challenges and necessity for sustainable and economic development require a change in planning and design of DHS. Many countries actively promote and support district heating as an important component in national strategic energy plans. Fossil fuels are gradually being replaced with renewable energy sources due to increased awareness of rational energy use and more strict regulations regarding the environmental protection. A property of DHS is invariability in respect of the energy source. This property makes possible utilization of local and renewable resources and waste heat from industrial processes, what represents a certain advantage over the decentralized systems. However, besides those advantages, DHS is often neglected by the policy makers as an efficient way of providing the necessary demand for space heating. The reason might be high investment costs, which are, in this case, usually financed from the local community budget. One


S. Avdaković et al. (Eds.): IAT 2019, LNNS 83, pp. 549–562, 2020.
https://doi.org/10.1007/978-3-030-24986-1_44

of the possible solutions is co-financing by the end users, in the amount of the investment they would have in case of constructing their own heating system.

In order to provide sufficient data to the decision makers, various sustainability parameters have been proposed. They make possible comparison of various sizes and types of DHS, providing guidelines in developing new and assessing existing DHS. Talebi et al. [1] proposed a new a categorization of DHSs based on their geographical location, scale, heat density, and end-user demand. Rezaie et al. [2] provided an overview of district energy systems in Canada, from technical, economic and environmental perspectives. Wolff et al. [3] made a study of the DHSs in Germany, providing the thorough analysis of their current state, and analysing potentials and risks for future development. Within the study, sustainability parameters of DHSs in Germany are provided. Nussbaumer et al. [4] analysed and compared DHSs in Austria, Denmark, Finland, Germany, and Switzerland covering a total of 800 district heating systems based on characteristic the liner heat density and the connection load. Although the line heat density is confirmed to be an important parameter, the study showed that the heat losses are distributed over a range of more than a factor of three at a given linear heat density. They have shown that additional parameters, such as pipe diameter, insulation class, network layout, the ratio between the operation hours of the DHS and the full load hours, etc. also influence the heat distribution losses. Otepka in [5] analysed best bio-energy practices and successful business models from Finland and Austria, in comparison to Poland, Romania and Slovakia, providing the local stakeholders with the basis for making decisions about developing the biomass powered DHS in those region. Zinko et al. [6] concluded that the areas with low heat demand density and line heat density can be economically heated with DHS. Reidhav et al. [7] analysed profitability of sparse district heating in Sweden, and identified line heat density and heat sold per house as the main influencing factors. Determining the accurate heat demand is essential for design of DHS. Swan et al. [8] provided the details of two approaches, top-down, where the energy consumption is regressed from historic aggregate energy values, and bottom-up approach, which extrapolates the estimated energy consumption of a representative set of individual houses to regional and national levels.

Konjic is one of the municipalities in Bosnia and Herzegovina which had DHS with coal as the energy source, with developed hot and warm water primary and secondary network and tertiary piping in the buildings connected to DHS. That system is not operational at the moment Heat demand of Konjic is determined through bottom-up approach. In addition to that, the heat map of Konjic is presented, which clearly illustrate the directions and dynamics regarding the DHS expansion. Sustainability parameters of the expansion of that DHS are determined and presented in this work. These sustainability parameters enable comparison of Konjic DHS with other systems, and provide a basis for scientifically supported technical and economic evaluation of it.

## 2 Description of Climate Conditions and Existing DHS in Konjic

Region around Konjic belongs both to the Mediterranean climatic area, as well as to the area of Bosnia continental climate of high karst, with influences of both climates being felt in the area. Mediterranean climate is influencing the area along the valley of river Neretva, while the mountainous ranges provide influences of continental climate. It is generally accepted that the Mediterranean climate influences are felt up to 450 m altitude, while the continental climatic factors prevail in higher areas. One of the most important characteristics of the area is especially high maximal air temperature in vegetative periods. This confirms strong influence of the Mediterranean climate, especially during summer season. Highest maximal absolute temperatures are registered during summer months June, July and August. Temperature and solar radiation monthly diurnal averages for Konjic are shown in Fig. 1.

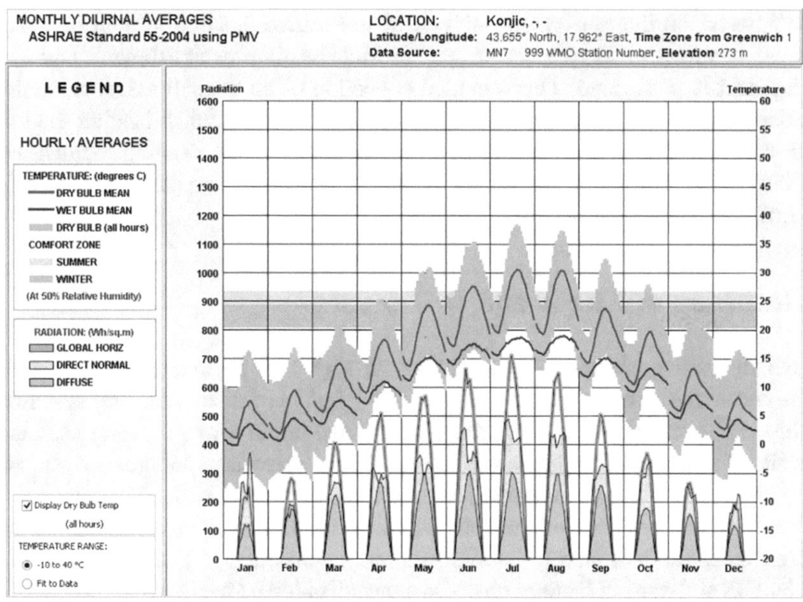


**Fig. 1.** Temperature and solar radiation monthly diurnal averages.

Average temperature decreases by 0.6°C per 100 m altitude, thereby proportionally increasing the possibility of frosting. The warmest month is July. The absolute maximal temperatures in July and August may reach beyond 40°C. The coldest month is January, with absolute minimal temperatures very rarely below −10°C. Relative humidity of an area is directly connected to the air temperature and cloudiness of the area, and follows an opposite trend to the one of air temperature during a calendar year. In Konjic area, average annual humidity is 79%, with minimal average monthly value of 74% in July, and maximal monthly average value of 89% in December. Konjic area is characterized by south-eastern and north-western winds, while wind in other directions appears significantly less often as a result of diurnal wind cycles. During a year, wind is significant for about 30% of the time. Average wind intensity is about 3° of the Beaufort scale.

Parts of the urban area of Konjic have had a functioning district heating system until 2013. The existing Konjic DHS was constructed between 1982 and 1984. The boiler plant supplied process and space heating energy for industry, as well as district heating of parts of urban area of the town. According to the original design documentation, 3 steam boilers of total nominal capacity of 57t/h of superheated steam at 12barg and 220°C were installed. About 11.5 MW of the capacity was dedicated to district heating via central steam to hot water heat exchanger station at hot water temperatures of 130/90°C. Pipe network consists of 2350 m of primary hot water distribution lines and 2300 m of secondary warm water networks. Pipe network has been constructed in 1980-ties, consisting of steel inner pipe, glass wool insulation layer and aluminum sheet covering. Actual peak heat load required by the system during several heating seasons before interruption of operations is estimated in range of 6.7-9 MW. Based on the performed analysis of the district heating system in Konjic, a technical assessment is performed on district heating establishment and further development is performed. This included collection of all available documentation on the existing system and detailed town survey in order to establish heating load of the urban area. The survey included walk-thorough audits of the existing thermal substations and boiler houses of significant capacity, as well as inspection of the existing stated of the distribution network.

## 3 Heat Map of Konjic and DHS Zones

In order to distinguish areas of the town with high heating load density, which represents one of the most important parameters for design of district heating systems heat map has been created. Figure 2 represents heat map of the town, showing heating load per unit of building footprint area calculated by creating polygons over official orthophoto maps of the town from 2008 available at 1:2500 scale. The map has been produced by interpolation of the total building heating load data normalized by its footprint area, and assigned to building footprint centroid points automatically calculated by GIS software. It thereby provides a visualization of heating load per town area, as required for design of district heating systems.

It can be seen from the map that the town may be classified to the following categories with respect to heating load density:

- Favorable areas with high heating load densities over multiple buildings, where heating load reaches 80 kW/m$^2$ of footprint area. These are clearly areas which are most appropriate for connection to district heating system.
- High individual heating loads, which is characteristic for schools, hospital, and some other public or commercial buildings with high heating load, which are either positioned in more remote town areas, or surrounded by buildings with low heating load.

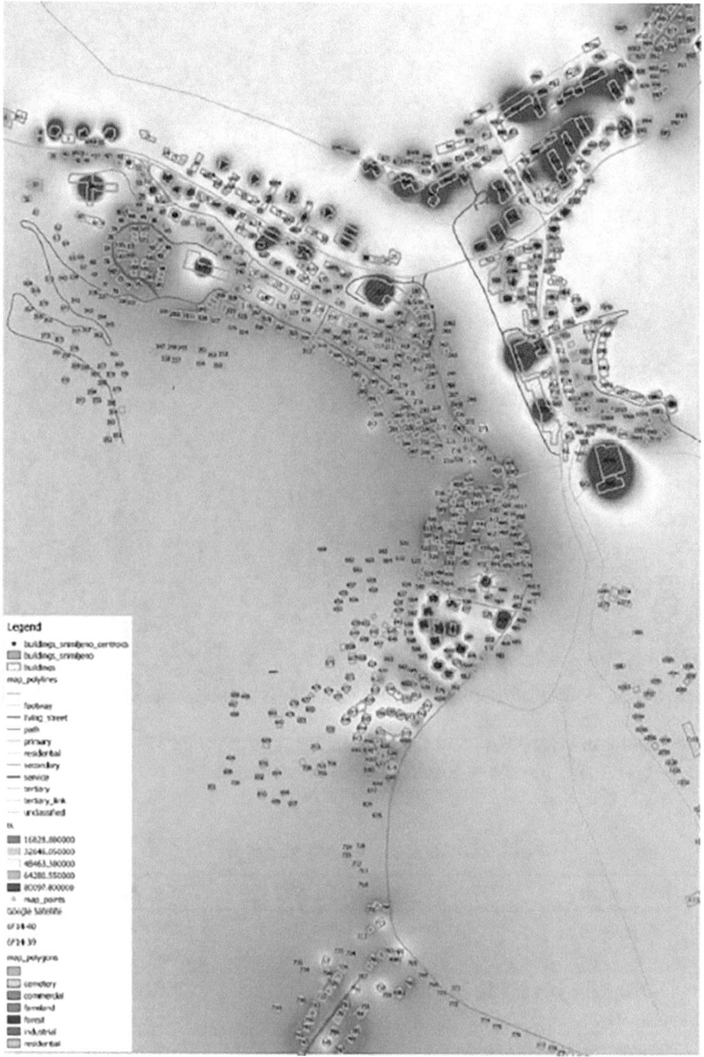


**Fig. 2.** Heat map of Konjic

- Unfavorable areas with significantly lower heating load densities, for which specific investment cost into local distribution network would be high. These are areas consisting mainly of individual family houses or small commercial areas, and it is justified to connect parts of these areas to the district heating system only if their location is very close to district heating network designed for the above two categories of buildings.

Town area respect to district heating system is divided into 10 zones, as shown in the Fig. 3 and Table 1.

**Fig. 3.** Town division with respect to district heating system zones (solid lines district heating zone boundaries, dashed line town administrative boundaries)

**Table 1.** District heating zones in Konjic

| Number | Zone name | Number | Zone name |
|---|---|---|---|
| 1 | Orasje | 6 | Tresanica 1 and 2 (residential area) |
| 2 | Druga osnovna skola (primary school) | 7 | Drustveni dom (community center) |
| 3 | Srednja mjesovita skola (secondary school) | 8 | Prva osnovna skola (primary school) |
| 4 | Robna kuca (shopping mall) | 9 | Hospital |
| 5 | Ustalica basta (residential area) | 10 | Radava (residential area) |

## 4 Project Development Phases for DHS in Konjic

The performed analysis has been used to identify four Project development phases with respect to implementation priorities. The Project development phases may be described as follows

Phase 1 – This phase of district heating system development in Konjic is planned to encompass measures and investments necessary in order to achieve the following major goals

- Commissioning of the existing infrastructure after necessary refurbishment measures
- Establishing biomass heating system in order to improve local economy and employment

This phase of the Project aims at protecting existing infrastructure from further degradation due to interruption of operation and lack of maintenance, as well as partially recovering investments made in infrastructure in recent years. Service area for the Phase 1 of district heating development is given on the Fig. 4.

**Fig. 4.** Service area for Phase 1 of district heating system development in Konjic

As shown in Table 2, total expected heating load for the Phase 1 of the Project is 6.66 MW. Considering heating load duration curve as representative consumption dynamics of the town, two boilers have been selected for the boiler house

- Wood chip fueled boiler of total nominal capacity of 4 MW,
- Light fuel oil (LFO) fueled boiler of total nominal capacity of 4 MW.

**Table 2.** Summary of technical data on the project development phases

| Phase | Boiler house capacity (MW) | Substation capacity (MW) | Number of connected buildings | Gross surface area of connected buildings (m$^2$) | Heating load of connected buildings (MW) |
|---|---|---|---|---|---|
| Phase 1 | 8 | 14.9 | 72 | 90949 | 6.66 |
| Phase 2 | 8 | 14.9 | 103 | 134410 | 9.26 |
| Phase 3 | 12 | 18.9 | 186 | 200374 | 13.25 |
| Phase 4 | 16 | 22.9 | – | – | 17.25 |

Wood chip boiler is planned to be used as a baseload boiler, covering more than 90% of heating energy demands in a heating season. Light fuel oil boiler selected here is used for three purposes (1) peak load boiler for normal operation, (2) backup boiler in case there is wood chip supply interruption, and (3) low load boiler which is more efficient at the lowest allowable part loads than corresponding wood chip boiler. New heating plant is planned to be constructed at the existing plant location in this phase of the Project. High-temperature primary water network is sized for flow and return temperature levels of 140/75°C. Secondary medium-temperature distribution network is planned at supply and return temperature levels of 90/70°C for compatibility with the existing indoor heating installation. These design temperatures are also suitable for most of the future extensions of secondary distribution networks, due to the fact that original installations have been found during town survey to be sufficiently oversized.

Phase 2 – The major goal of the Phase 2 of the Project development for district heating system in Konjic is to maximize utilization of heating generation and distribution capacities of the system installed in Phase 1 of the development. This is planned to be achieved by partial replacement and extension of the existing secondary medium-temperature water network, and connection of new consumers with appropriate heating load density to DHS. New buildings connected to the system are mainly multi-apartment residential buildings. Total planned service area of the Phase 2 of district heating system development is presented in the Fig. 5. Additionally, the existing boiler houses (described in Phase 1) in zones covered by the district heating system should enable peak load servicing if the need arises, so that installation of additional generation capacities in the central boiler house is not necessary at this stage of Project development.

Phase 3 – The major goal of the Phase 3 of the Project development for district heating system in Konjic is to extend service area of the system to the left side of river Neretva in order to enable connection of areas with very high heating load density to the central heating system. This is planned to be achieved by extension of central boiler plant and primary high-temperature distribution network over the river, construction of new secondary medium-temperature distribution network in Zone 10 of the system, and connection of building with significant heating load density consisting of mainly multi-

**Fig. 5.** Service area for Phase 2 of district heating system development in Konjic

apartment residential units. Total planned service area of the Phase 3 of district heating system development is presented in the Fig. 6. It is planned that the boiler plant is extended by additional wood chip fired boiler of nominal capacity of 4 MW of the same technical characteristics described in Phase 1 of project development.

New boiler nominal capacity has been selected by considering evidence from previous studies and existing district heating systems in BiH which show that usually optimistic assumptions related to consumer base do not materialize in actual system operation. Additionally, the existing capacity of larger boiler houses throughout the town may contribute in case installed capacity of central boiler plant is not sufficient during peak load periods. Based on the assumptions, total nominal capacity of central boiler house of 12 MW is undersized by 1.25 MW based on estimations of connected load. Regarding fuel selection of additional boiler, wood chips have been selected due to the fact that they represent local fuel and contribute to generally established goals of supporting local employment and economy, environmental protection and increased utilization of renewable, environmentally friendly heating fuels. Having in mind that wood chips are local fuel source, security of fuel supply would be increased and sensitivity of district heating system to variable fossil fuel prices would be reduced. It is worth noting that technical conceptual design of boiler house is compatible with selection of boilers powered by light-fuel oil or natural gas. Fuel selection for additional boiler is also significant for overall fuel efficiency of the boiler plant. It is generally recommended that biomass based district heating plants are designed as combination of base load-serving biomass boiler sized at about 30–40% of total nominal heating load, and fossil-fueled peak load boiler covering 60–70% of peak load.

**Fig. 6.** Service area for Phase 3 of district heating system development in Konjic

In this arrangement biomass boiler operates close to its full capacity for very long period of the heating season which is operational point where such boilers reach their peak efficiencies. Lighter fossil fuel powered boilers operate on as-needed basis, have higher part-load efficiencies compared to biomass boilers, and are capable of faster start-up and shut-down times.

Phase 4 – Phase 4 of the Project development includes extension of the district heating system to Zone 7 – Drustveni dom (community center). This area is characterized by low specific heating load and hilly terrain, both of which are very unfavorable for district heating system development and operation. Implementation of this phase of the Project is entirely dependent on preferences of the Beneficiary and the future district heating company, as well as real operational performance parameters (both technical and financial) of the system after the Phase 3 of the Project development has been implemented. Summary of technical data on the Project development phases is shown in the following table.

## 5 Forest Fuel Wood Biomass Potential in Konjic

The data on fuel wood from the forestry management institutions in FBiH indicates that annual consumption of fuel wood in FBiH is about 136 ktoe [9]. Data on forestry in the Municipality of Konjic has been taken from the local forest management company (Sumarija Prenj Konjic) representatives. The following data were obtained

- Total available quantity of round wood to a future district heating company would be between 1500 and 2000 $m^3$ per month
- Total available wood residues are between 4000 and 5000 $m^3$ annually
- Wood waste from primary wood processing facilities is estimated to 500 $m^3$
- In another, significantly smaller, forest management company (Sumarstvo Ljuta) there is 5–10% of the above stated quantities available for the future district heating company

The biomass as the primary fuel for DHS in Konjic will substitute the fuel wood currently used in individual stoves at significantly lower efficiency.

## 6 Cost Estimates and Energy Parameters Related to the Project Development Phases

Process of Konjic DHS reconstruction is planned in four phases. It has to be emphasized that the Phase 4 assumes connection of all objects not encompassed by the first three phases, meaning that even the areas with low heat load density will be connected. For that reason, the economic aspects of the phase 4 are not considered in this work. Estimation of the investment costs for the first three phases of the DHS reconstruction, as well as necessary energy consumption for heating and energy sources are given in Table 3.

**Table 3.** Cost estimates and energy parameters related to the project development phases

| Phase | Investment cost estimate* (€) | Energy generated by wood chips boiler (MWh/a) | Energy generated by LFO boiler (MWh/a) | Wood chips consumption (t/a) | Light fuel oil consumption (l/a) |
|---|---|---|---|---|---|
| Phase 1 | 2142500.00 | 13392 | 499 | 6696 | 53261 |
| Phase 2 | 157500.00 | 16673 | 2098 | 8337 | 240508 |
| Phase 3 | 1165000.00 | 26762 | 734 | 15381 | 84150 |
| OVERAL | 3465000.00 | | | | |

*Without VAT

## 7 Sustainability Parameters of Konjic District Heating System

Basic parameters for comparison of different district heating systems, based on the engaged power, total annual heat demand, surface of the area covered by the system and length of the pipelines, are:

- heat demand density - HDD [kWh/m$^2$-a]

The heat demand density is the ratio between annual heat demand and area. This indicator gives information about suitability of a city for district heating.

- line power density: LPD [kW/m]

This indicator is estimated by dividing the total heating load of connected buildings with the total length of pipe network

- line heat density: LHD [kWh/m-a]

The line heat density is ratio between total annual heat demand and the total length of the DH network.

- specific investment cost for heating plant: SICp [€/KW]

Specific investment cost for heating plant includes the overall investment costs of the heating plant (boiler house and storage with all installations without land and development costs) referred to as the sum of nominal capacity ("power" in kW) of all heat generators.

- specific investment cost for heat distribution network: SICn [€/MWh-a]

Specific investment costs of heat network consist of the overall investment costs of heat network without transmission stations (substations), referred to the annual heat amount, which is consumed through the network without losses in the distribution network.

Sustainability parameters for the Konjic district heating system for each phases of implementation are given in Table 4.

**Table 4.** Sustainability parameters for the Konjic district heating system according to phases of implementation

| Phase | HDD (kWh/year m$^2$) | LPD (kW/m) | LHD (kWh/year m) | SICp (€/KW)* | SICn (€/MWh-a) |
|---|---|---|---|---|---|
| Phase 1 | 76.9 | 1.86 | 3879.3 | 268 | 154 |
| Phase 2 | 103.9 | 2.03 | 4114.8 | 288 | 123 |
| Phase 3 | 98.0 | 2.10 | 4586.3 | 289 | 126 |
| Phase 4 | 66.5 | 2.11 | 4262.0 | | |
| TOTAL | 89.5 | 2.02 | 4210.6 | | |

*Partial reconstruction

Overall heat demand density after all phases implementation for Konjic district heating system is 89.5 [kWh/a-m$^2$]. Various lowest values of the heat demand density as the limit for economic viability, and mainly do not exceed 70 [kWh/m$^2$-a]. According to [6] even systems with heat demand density and line heat density as low as 10 [kWh/m$^2$-a] and 300 [kWh/m-a] respectively, can be justified. Line power density and line heat density are calculated according to the length of the hot water pipeline are

2.02 [kW/m] and 4210.6 [kWh/m-a] respectively. According to [3] line power density as low as 1.5 [kW/m] represent efficiency criterion of the district heating systems.

According to [7], the economic and environmental threshold for the LHD of different networks is 1MWh/m for DHSs with biomass heat source. As it can be seen in Table 4 SICp and SICn parameters are significantly below the average Austrian values which are 519 €/KW and 214 €/MWh-a, respectively [11].

## 8 Conclusion

Regarding high investment costs of design and construction of DHS, sustainability parameters play significant role in financial and economic analysis. Different countries use different values sustainability parameters during the decision making process of the investments in such systems. One of the deciding factors for construction of a DHS is also favorable impact regarding environmental protection.

Construction of the DHS in Konjic is planned in four phases, and sustainability parameters are calculated for the each of the phases, apart from specific investment cost parameters of the phase 4, whose economy aspects are not analyzed. Sustainability parameters for the first three phases are within the ranges proposed by various authorities. The overall sustainability parameters for all four phases together are still within the proposed ranges. This analysis clearly showed that reconstruction of DHS in Konjic is justified not only from environmental, but also from economic point of view.

**Acknowledgements.** This work was done within the Project " Provision of technical assistance for 3 high-scale energy efficiency projects – Western Balkans, Bosnia and Herzegovina, Konjic, Fojnica and Brcko" financed by the European Union EC/BIH/TEN/13/029 according to the contract no: 2014/348-406 closed with company HVAC Design doo, Sarajevo.

## References

1. Talebi, B., Mirzaei, P.A., Bastani, A., Haghighat, F.: A review of district heating systems: modeling and optimization. Front. Built Environ. **2** (2016). https://doi.org/10.3389/fbuil.2016.00022
2. Rezaie, B., Rosen, M.A.: District heating and cooling: review of technology and potential enhancements. Appl. Energy **93**, 2–10 (2012). https://doi.org/10.1016/j.apenergy.2011.04.020
3. Wolff, D., Jagnow, K.: Überlegungen zu Einsatzgrenzen und zur Gestaltung einer zukünftigen Fern- und Nahwärmeversorgung [Online] (2011)
4. Nussbaumer, T., Thalmann, S.: Status report on district heating systems in IEA countries, IEA Bioenergy Task 32, Swiss Federal Office of Energy, and Verenum, Zürich (2014)
5. Otepka, P.: Guidebook on Local Bioenergy Supply Based on Woody Biomass (2013) ISBN: 978-1-938681-99-8
6. Zinko, H. et al.: District Heating Distribution in Areas with Low Heat Demand Density, International Energy Agency (2008)
7. Reidhav, C., Werner, S.: Profitability of sparse district heating. Appl. Energy **85**(9), 867–877 (2008). https://doi.org/10.1016/j.apenergy.2008.01.006

8. Swan, L.G., Ugursal, V.I.: Modeling of end-use energy consumption in the residential sector: A review of modeling techniques. Renew. Sustain. Energy Rev. **13**(8), 1819–1835 (2009). https://doi.org/10.1016/j.rser.2008.09.033
9. Informacija o gospodarenju šumama u Federaciji BiH u 2013. godini i planovima gospodarenja šumama za 2014. godinu, Federal ministry of Agriculture, Water Management and Forestry
10. Biomass Consumption Survey for Energy Purposes in the Energy Community, Bosnia & Herzegovina, National Report, source: https://www.energy-community.org/pls/portal/docs/1378189.PDF
11. ÖKL-Arbeitskreis Energie., ÖKL-Merkblatt Nr.67 – Planung von Biomasseheizwerken und Nahwärmenetzen, 2. Auflage, Wien: Österreichisches Kuratorium für Landtechnik und Landentwicklung (2009)
12. Sokolov, J.J.: Toplifikacija i toplotne mreže, Građevinska knjiga Beograd (1985)

# Collecting Geospatial Data Using Unmanned Aerial Photogrammetric System

Admir Mulahusić[1], Nedim Tuno[1(✉)], Jusuf Topoljak[1], Faruk Čengić[2], and Seat-Yakup Kurtović[1]

[1] Faculty of Civil Engineering, Department of Geodesy and Geoinformatics, University of Sarajevo, Sarajevo, Bosnia and Herzegovina
{admir_mulahusic, nedim_tuno, jusuf.topoljak}@gf.unsa.ba, seat.yakup@gmail.com

[2] Geobiro, Konjic, Bosnia and Herzegovina
faruk.cengic@gmail.com

**Abstract.** The development of unmanned aerial vehicles was improved applying of aerial photogrammetry in the process of collecting geospatial data. Photogrammetry has been defined by the American Society for Photogrammetry and Remote Sensing (ASPRS) as the art, science, and technology of obtaining reliable information about physical objects and the environment through processes of recording, measuring and interpreting photographic images and patterns of recorded radiant electromagnetic energy and other phenomena. The primary purpose of unmanned aerial vehicles was for military use. The unmanned aerial vehicles are made from different materials and they have to be of very good quality. Some parts and construction are entrusted to the mechanical engineers. Electronic parts of the unmanned aerial vehicles are entrusted to the electrical engineers. In the field of aerial photogrammetry, geodesy is "involved" through the implementation of the aerial photogrammetric system and GNSS technology. Practical procedures of aerial photogrammetric geodetic survey, with analysis of the obtained results are shown in this paper. Area of analysis results of aerial photogrammetric geodetic survey is area around the Faculty of Civil Engineering and Faculty of Architecture (University of Sarajevo). This research examines the possibility of using unmanned aerial vehicles for the purpose of surveying the medium size area of complex urban surface.

**Keywords:** UAV · Photogrammetry · Camera · DTM · DSM

## 1 Introduction

Photogrammetry is traditionally the most appropriate method for the collecting data about different objects [1]. Aerial photogrammetry, as a geodetic survey technique, is a part of photogrammetry, but it is not for commercial use in small areas of geodetic survey [2]. The most important characteristic of aerial photogrammetry is the realization of geodetic survey when measurement camera is not on the ground. Carriers of the aerial photogrammetry cameras were balloons, zeppelins, airplanes, and today's very successful carriers are unmanned aerial vehicles (UAV). Unmanned aerial vehicles are vehicles capable of performing continuous flight without pilots [3]. There are


S. Avdaković et al. (Eds.): IAT 2019, LNNS 83, pp. 563–575, 2020.
https://doi.org/10.1007/978-3-030-24986-1_45

navigation devices built in the unmanned aerial vehicle. These vehicles are very successful for a geodetic survey. Result of geodetic survey is a digital geodetic plan or a digital orthophoto plan. As it was mentioned in the abstract, this research examines the possibility of using unmanned aerial vehicles for the purpose of surveying the medium size area of complex urban surface.

## 2 Unmanned Aerial Vehicle SurveyDron01

First officially documented use of unmanned aerial vehicle for civilian use was in 1979 [4]. The fundamental parts of unmanned aerial vehicle are:

- High quality construction,
- Navigation system,
- Software solutions and
- Measurement camera.

Construction of unmanned aerial vehicle is made from the parts of different materials. Mechanical engineers are responsible for making a construction of unmanned aerial vehicle. Some parts of unmanned aerial vehicles need an electrical energy to work and electrical engineers participate in its design. During the development of unmanned aerial vehicles, there were different versions of them. They were produced in micro and mini versions. Some of them are made in the form of balloons, zeppelin, airplanes, helicopters and various combinations of these structures [5].

Unmanned aerial vehicles can be divided in two groups [4]:

- Unmanned aerial vehicles made by fixed wings and
- Unmanned aerial vehicles made by rotating wings.

There is no unique division of unmanned aerial vehicles. There are several ways to sort them out: purpose, flight height, area of flight, constructive characteristic of unmanned aerial vehicles [2].

It is necessary to navigate unmanned aerial vehicle in order to successfully finish geodetic survey. Unmanned aerial vehicle made in Bosnia and Herzegovina is a product of company GeoWILD from Sarajevo. The name of this unmanned aerial vehicle is SurveyDron01 (Fig. 1). It is the unmanned aerial vehicle made by rotor wings [6].

**Fig. 1.** SurveyDron01 [7]

SurveyDron01 was accompanied with appropriate Aerial image processing software solution 3Dsurvey (Ver.: 2.1.2014). Its main Technical data, Functionalities, Advance Functionalities, Hardware and Software Requirements are given in [8] (Tables 1 and 2).

**Table 1.** Technical characteristic of unmanned aerial vehicle SurveyDrone01

| Category | Technical characteristics |
|---|---|
| Length/Width | 960 mm |
| Height | 430 mm |
| Material | Carbon fiber |
| Empty weight | 3400 g |
| Maximum takeoff weight | 6500 g |
| Recommended payload | 700 g |
| Maximum payload | 2500 g |
| Maximum speed | 40 km/h |
| Rate of climb | 10 m/s |
| Maximum altitude | Up to 1000 m |
| Flight duration | Up to 30 min |
| Navigation | GNSS receiver, gyroscope, accelerometer, barometer |
| Controlling of unmanned aerial vehicle | Remote controlling |
| Power supply | Li-Po battery |
| Number of blades | 6 |

**Table 2.** Working characteristic of unmanned aerial vehicle SurveyDron01

| Atmospheric parameters | Working characteristics |
|---|---|
| Temperature | From −20 °C up to +40 °C |
| Wind tolerance | Usable images up to 10 m/s of wind |
| Radius of flight | Minimum 500 m |
| Fog | Maximum 90% |

## 3 Geodetic Survey of Area of the Faculty of Civil Engineering and the Faculty of Architecture

Testing area is around 7.5 ha. It is situated in the center of Sarajevo (wider location of the Faculty of Civil Engineering in Sarajevo), as it is shown in Fig. 2.

**Fig. 2.** Testing area

For the purpose of preparing geodetic survey, aerial photo signals of ground control points were prepared (Fig. 3). For the photo signalization of ground control points appropriate signals in the shape of square were used. Dimensions of signals were 50 cm × 50 cm with the black circle in the centre of square. Radius of circle is 35 cm. In the centre of circle is a cavity (exact place of ground point mark).

**Fig. 3.** Signal of ground control point

Eleven points were stabilized in the research area. When planning and placing the control points positions, it is especially important to keep in mind that control points are geometrically properly arranged, i.e. they have to cover the entire range of the test area (Fig. 4).

**Fig. 4.** Correct layout of control points (left), improper layout of control points (right) in 3D survey software solution

Coordinates of all ground control points were calculated. In this case, coordinates were determined by Topcon GR5 GNSS receiver and by using the RTK (Real-Time Kinematic) with the addition of using FBiHPOS system. FBiHPOS is the positioning system for the Federation of Bosnia and Herzegovina. It consists of networked GNSS reference stations, and its purpose is to provide real time positioning with accuracy of 1 to 2 cm. Authorized users have continuous access to the network. FBiHPOS system use the latest GNSS technology available and track all operational GNSS signals like GPS-NAVSTAR (L1, L2, L2C) and GLONASS (L1, L2). Calibration of aerial camera and a plan of the unmanned aerial vehicle flight were made prior to an aerial survey. Next step was the realization of geodetic survey. As it was mentioned before, there were eleven control points in the test area. The layout of the control points is shown in Fig. 5.

The flight plan is performed in the Mission Planner software (Fig. 6). To work in this software, internet connection was necessary. As a background Google Earth Map was used.

**Fig. 5.** Signaled ground control points (red circles)

The Area of Interest (AOI) is marked by a polygon. The options that are essential for the flight itself are selected, namely: flight speed, flight altitude, flight direction, photo overlay, camera type. The setting of these options affects the required accuracy, terrain configuration, and surface area. Based on these criteria, the most optimal settings were selected. After adjusting the settings, the program is computing all important flight plan statistics: drone path coordinates, surface (area), flight time, number of images, ground resolution, number of strips, etc.

After finishing geodetic survey, first step was the calculation of orientation parameters. Creating of Digital Terrain Model (DTM) is a very important step in creating a digital orthophoto plan. Digital Terrain Model (DTM) represents terrain area without any objects and plants [5]. Nowadays, it is possible to use digital terrain model (DTM) for all kinds of terrain analysis. Digital terrain model was created using appropriate application.

**Fig. 6.** Flight plan layout in Mission Planner software

Unfortunately, many geodetic and other geospatial experts are mistaken when they want to define precisely the terms of Digital Terrain Model (DTM) and Digital Surface Model (DSM). The following figures (Figs. 7 and 8) show the difference between the mentioned.

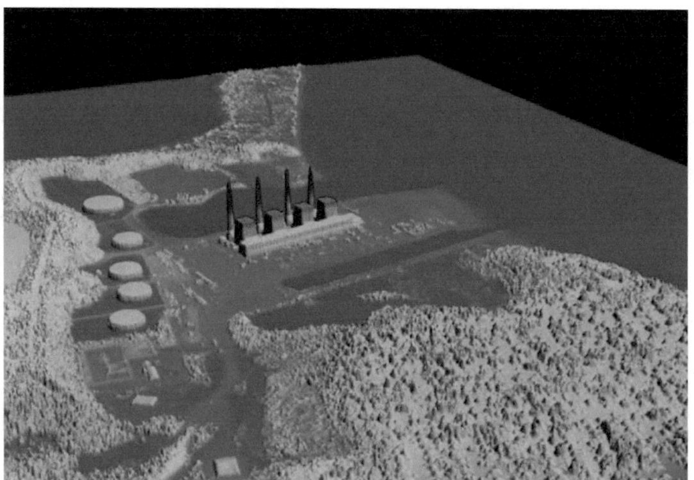

**Fig. 7.** Example of Digital Surface Model [9]

As a written and simple explanation of the mentioned terms, it should be noted that:

- The Digital Surface Model (DSM) contains elevations of natural terrain features including objects on it (Fig. 7), i.e. vegetation and cultural features such as buildings, bridges, and power lines [10].

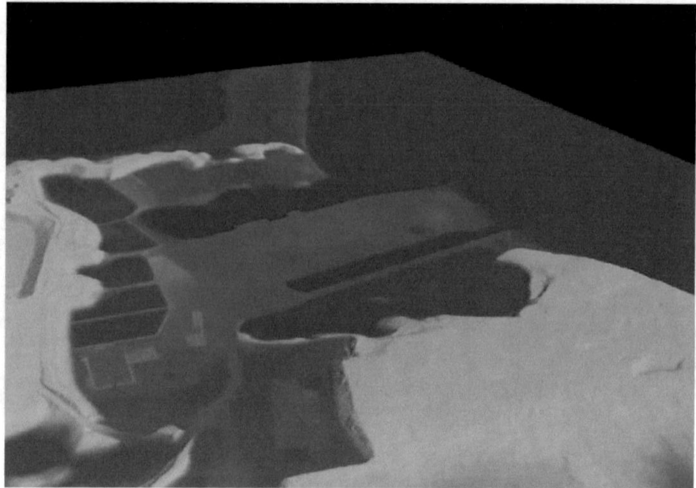

**Fig. 8.** Example of Digital Terrain Model [9]

- The Digital Terrain Model (DTM) is an ordered set of sampled data points that represent the spatial distribution of various types of information on the terrain (both topographic and non-topographic) (Fig. 8), e.g. elevation, slope, slope form, rivers, ridge lines, break lines, etc. It usually represents the elevation of "bare earth", i.e. the shape of terrain without any objects on it [10].

In this case, the digital terrain model is in the following figure (Fig. 9):

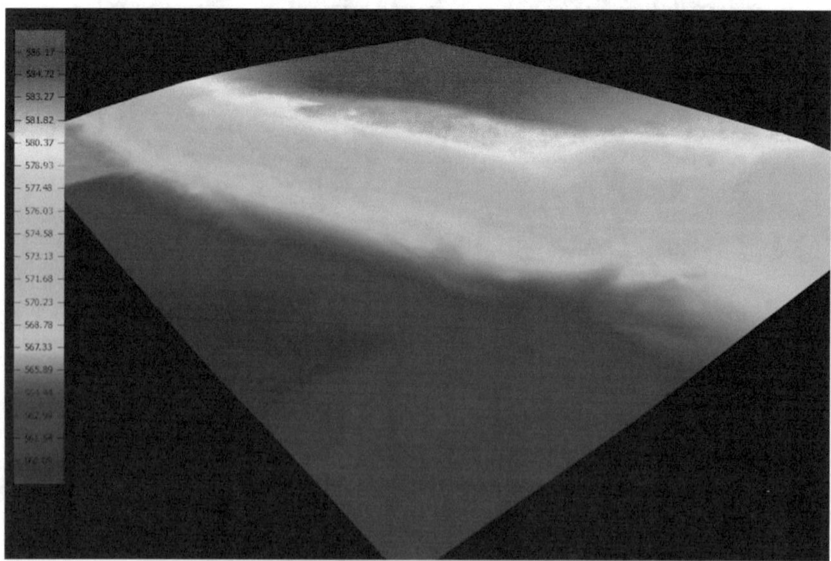


**Fig. 9.** Digital Terrain Model (DTM) of the research area

Some more important data regarding the Digital Terrain Model (DTM) of the research area are as follows:

- Minimum height: 558.64 m,
- Maximum height: 587.62 m,
- Number of triangles: 7616808.

Creation of digital orthophoto plan (DOF) is based on the creation of Digital Terrain Model. The digital orthophoto plan area around the Faculty of Civil Engineering and the Faculty of Architecture (Fig. 10) was created using appropriate application. The digital orthophoto plan is a digital image with orthogonal projection characteristics obtained from the aerial photogrammetric images [11, 12].

Pixel size in Digital orthophoto plan (Fig. 10) is 0.02 m. Size of Digital orthophoto plan is 429.60 m × 354.60 m (Fig. 10).

**Fig. 10.** Produced digital ortophoto plan

The option of creating a dense cloud of points (visually very attractive model) was also used and based on 1471680 points (Fig. 11).

**Fig. 11.** Dense point cloud of the research area

### 3.1 Accuracy of Digital Orthophoto Plan

Analysis of the accuracy of the orthophoto plan was done by using the coordinates of ground control points. The most important factors for the analysis accuracy of digital orthophoto plan (DOF) are:

- Flight height and characteristics of measurement of camera,
- Pixel size,
- The accuracy of Digital Terrain Model (DTM) and
- Camera position [11].

For the accuracy analysis, in addition of eleven control points, ten more points were used. Calculated were differences in coordinates that were obtained through GNSS process and coordinates obtained from DOF (Figs. 12, 13 and 14). Checking the height of the control point is done by displaying Point Picking from the point cloud. This option selects the closest point to the appropriate control point. The height of every control point is possible to determine using appropriate application. There is a possibility to made a simultaneous display of point clouds and orthophoto.

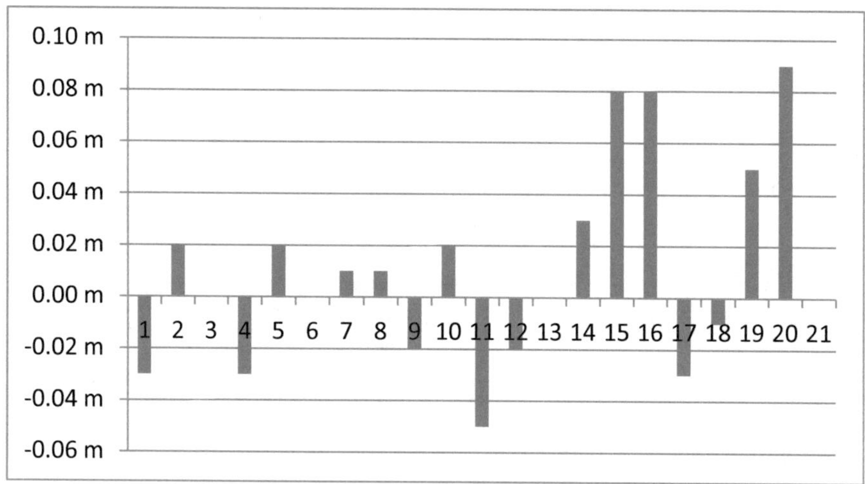


**Fig. 12.** Graphic representation of differences in *y* coordinates

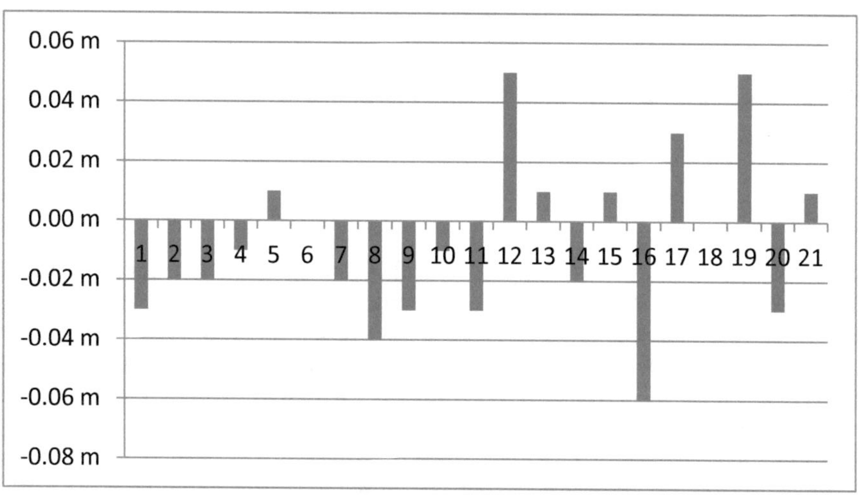


**Fig. 13.** Graphic representation of differences in *x* coordinates

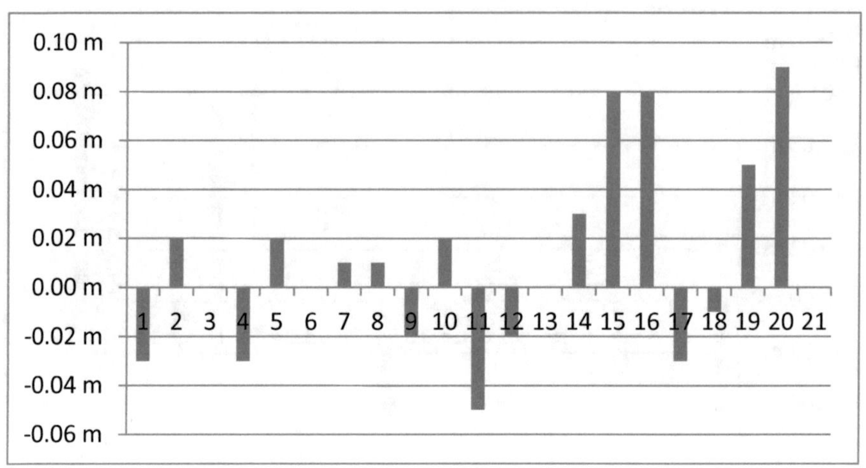


**Fig. 14.** Graphic representation of differences in $z$ coordinates

The maximum difference in $y$-axis is 5 cm, in $x$-axis is 6 cm and the maximum difference in height is 9 cm. Based on differences, root mean square positional error and root mean square height error were calculated:

- root mean square positional error: 3.1 cm,
- root mean square height error: 3.2 cm.

## 4 Conclusion

Among others, progress of geodesy in the field of photogrammetry is reflected through the use of unmanned aerial vehicles. Using unmanned aerial vehicles and applying the appropriate software applications are of great assistance for geodetic survey. Final result of geodetic survey using unmanned aerial vehicles is digital orthophoto plan. Digital orthophoto plans can be used for different purposes.

It is important to analyze accuracy of digital orthophoto plan (DOF). Unmanned aerial vehicle used for research in this paper is a product of company GeoWILD from Sarajevo. The official name of this unmanned aerial vehicle is *SurveyDron01*. Area of research is the area around the building of the Faculty of Architecture and the Faculty of Civil Engineering (University of Sarajevo).

Analysis of the accuracy of digital orthophoto plan implied the determination of the coordinates for 21 points (obtained by two methods). The first included the use of GNSS receiver. Second method used the assessment of coordinates from the digital orthophoto plan.

The analysis started with coordinate differences computation between positional coordinates and the heights of the corresponding points. After that, the standard deviation of coordinate residuals with regard to the mean values of coordinates was calculated.

The biggest coordinate difference for $y$ axis is 5 cm and for $x$ axis is 6 cm. The biggest difference in heights is 9 cm. Standard deviations show that the maximum positional accuracy is less than 2 cm, while the height accuracy of the model is less than 3 cm. Used method is good enough for many purposes (cadastral, planning, etc.) and it proved to have its future in many fields of human activities, where the above mentioned accuracy is sufficient.

It can be concluded that the use of unmanned aerial vehicles is an exceptionally perspective technology for geodetic survey. Obtaining digital orthophoto plan in a very short time period is the greatest advantage of using the unmanned aerial vehicles, but every user must be careful because the impacts of meteorological components (sunny and windy weather) on the results are very important.

## References


1. Mulahusić, A., Tuno, N., Topoljak, J., Balić, D.: Izdelava 3D-modela kompleksnega kulturno zgodovinskega spomenika z uporabo digitalne fotogrametricne postaje: Using an imaging station in making a 3D model of a complex cultural and historical heritage object. Geodetski vestnik **60**(1), 28–41. https://doi.org/10.15292/geodetski-vestnik.2016.01.28-41 (2016)
2. Jurić, V., Kolobarić, J., Kvesić, V., Bjeliš, D.: Bespilotne letjelice: Unmanned aerial vehicles. Geodetski glasnik **50**(47), 57–67 (2016)
3. Bento, M.F.: Unmanned aerial vehicles: an overview. Inside GNSS, pp. 54–61 (2008)
4. Eisenbeiss, H.: A Mini Unmanned Aerial Vehicle (UAV) system overview and image acquisition. In: International Workshop on "Processing and Visualization Using High Resolution Imagery". Pitsanulog, Thailand (2004)
5. Everaerts, J.: The Use of Unmanned Aerial Vehicles (UAVS) for Remote Sensing and Mapping, Remote Sensing and Earth Observation Processes Unit. Beijing: The International Archives of the Photogrammetry, Remote Sensing and Spatial Information Sciences, vol. XXXVII, Par B1 (2008)
6. Čengić, F.: Prikupljanje geoprostornih podataka pomoću bespilotnog aerofotogrametrijskog sistema. Sarajevo, MA thesis, Građevinski fakultet Univerziteta u Sarajevu, (2017)
7. Zedis. SurveyDrone01 (2014). http://www.zedis.info
8. 3Dsurvey. Aerial image processing software (2014). http://www.3dsurvey.si/upload/downloads/3dsurvey_specifications.pdf
9. Oštir, K., Mulahusić, A.: Daljinska istraživanja, Građevinski fakultet Univerziteta u Sarajevu, (2014)
10. Kokalj, Ž., Hesse, R., Mulahusić, A.: Vizuelizacija rasterskih podataka laserskog skeniranja iz zraka, ZRC SAZU, Inštitut za antropološke in prostorske študije - Založba ZRC (2018)
11. Mulahusić, A., Topoljak, J., Tuno, N.: Geodezija za građevinske inžinjere, Univerzitet u Zenici (2017)
12. Žilić, A.: Application of unmanned aerial vehicles in geodesy on the example of aerial photogrammetric system SenseFly Ebee : Primjena bespilotnih letjelica u geodeziji na primjeru aerofotogrametrijskog sistema SenseFly Ebee. Geodetski glasnik **49**(46), 18–27 (2015)

# Genetic Algorithms Applied to the Map Registration

Nedim Tuno(✉), Admir Mulahusić, Jusuf Topoljak, and Seat-Yakup Kurtović

Faculty of Civil Engineering, Department of Geodesy and Geoinformatics, University of Sarajevo, Sarajevo, Bosnia and Herzegovina
{nedim_tuno, admir_mulahusic, jusuf.topoljak}@gf.unsa.ba, seat.yakup@gmail.com

**Abstract.** Genetic Algorithm (GA) is an artificial intelligence procedure which efficiently searches a large space of possible solutions to find the best possible solution for the given problem. This paper is focused on the GA and its application in the problem of finding optimal parameters for map registration. It has been shown that genetic algorithms can solve very complex problems, such as geometric transformation of raster datasets. After determining the appropriate fitness function and tuning the GA parameters, the optimal transformation parameters were obtained. Finally, the historical topographic map was efficiently georeferenced to the state plane coordinate grid, i.e., the Bosnian-Herzegovinian state coordinate system.

**Keywords:** Genetic algorithm · Map registration · Transformation

## 1 Introduction

The registration of map data is an important issue in multi-source data integration, management and analysis for many geomatic applications. The quantitative use of scanned maps requires the geometric distortions to be corrected, or rectified, to desired map projection. Geometric correction, also known as georeferencing and map rectification, is necessary when the output products of image analysis are overlaid on a map or merged into a geographic database [1]. Determination of optimal transformation parameters is a complex optimization problem. Genetic algorithms (GA) offer an effective approach to solve optimization problems [2] including those in spatial sciences, as shown in [3, 4, 5 and 6].

This study is the continuation of the research conducted by [7, 8] in which the original methodologies based on GA were successfully applied in transformation models of various topographic maps. Aforementioned research has shown that the determination of optimal transformation parameters is a complex optimization problem. It has also been shown that GA is a general, but weak method. To be successful in a real application, it needs to incorporate knowledge of solved problem. The key questions that have to be answered in application are proper representation of individuals (to allow simple and efficient operators) and genetic operators [9]. This study sought to improve the solution obtained in previous research.


S. Avdaković et al. (Eds.): IAT 2019, LNNS 83, pp. 576–586, 2020.
https://doi.org/10.1007/978-3-030-24986-1_46

# 2 Materials and Methods

## 2.1 Genetic Algorithms

Genetic algorithms (GA) belong to probabilistic algorithms, combining elements of directed and stochastic search. At the beginning, a set of potential solutions (so-called initial population) is generated. Then an iteration process could begin. In any iteration, the potential solutions - members of the population (individuals) - are evaluated as to their suitability to solve the given problem (fitness). More fitting individuals are picked and transformed by genetic operators to form a new population. Iteration continues for some (usually given) number of generations and the best individual of the last generation (or, better, the best individual found during all the iteration process) is kept as solution. Here are usually two genetic operators: crossover and mutation. Crossover is a binary operator. It combines two individuals - 'parents' into two new individuals - 'children'. Mutation is unary and makes a small change in an individual. How are these operations done depends on the choice of representation and on the problem characteristics [9]. Fundamentals of genetic algorithms can be found in [4, 10 and 11].

## 2.2 Historical Topographic Map

The genetic algorithm and its application in transformation of cartographic documents will be examined using a section of the Special map of the Austro-Hungarian Monarchy at scale of 1:75 000 (map sheet XIX.30 Sarajevo) (Fig. 1).

**Fig. 1.** Scanned section of the historical topographic map

Following the 1878 Austro-Hungarian occupation of Bosnia and Herzegovina, the Austrian Military Headquarters issued a command for the Military Geographical Institute in Vienna (Germ. Wien: K.u.K. Militärgeographisches Institut - MGI) to carry out a survey of the entire territory of Bosnia and Herzegovina [12, 13]. Aforementioned map was created on the basis of the results of that survey and published in 1888.

### 2.3 Selection of Tie Points

In order to georeference the map, it is necessary to select the appropriate tie points [14, 15].

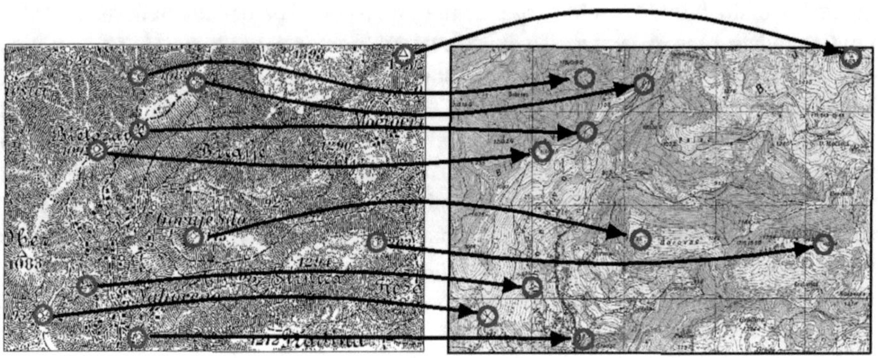

**Fig. 2.** Principle of georeferencing based on identified tie points on maps at scale of 1: 75 000 and 1: 25 000

By inspecting the contents of the historical map and modern topographic map at scale of 1: 25 000 (map sheet 525-2-2 Faletići), the content that was assumed to have not undergone any changes was identified. A total of 11 identical points were selected, which could be identified on both maps (Fig. 2). On the basis of these points, the georeferencing of the historical map can be made using the affine transformation method. Considering the number and disposition of tie points, as well as the character of map deformation, it is assumed that this transformation model provides a satisfactory solution.

## 3 Results and Discussion

### 3.1 Fitness Function Design

The research assumes that a genetic algorithm can be used efficiently to determine the 6 transformation parameters of affine transformation. To implement the genetic algorithm, the Genetic Algorithm Toolbox was used within the MATLAB program. To use the routines of the GA Toolbox, M-Files were created. Their purpose was to calculate functions that are optimized (fitness values). The fitness function is determined by the condition that the sum of the squared difference between the transformed coordinates $(x_i^{tr}, y_i^{tr})$ and the coordinates in the reference system $(x_i, y_i)$ is minimal, i.e.:

$$\sum_{i=1}^{m}((x_i - x_i^{tr.})^2 + (y_i - y_i^{tr.})^2) = \min, \tag{1}$$

where $m$ is the number of identical points.

Transformation parameters are determined separately for each coordinate axis. Accordingly, the minimum conditions (fitness functions) are:

$$(a_0 + a_1v_1 + a_2u_1 - y_1)^2 + (a_0 + a_1v_2 + a_2u_2 - y_2)^2 + \cdots + (a_0 + a_1v_m + a_2u_m - y_m)^2 = \min,$$
$$(b_0 + b_1v_1 + b_2u_1 - x_1)^2 + (b_0 + b_1v_2 + b_2u_2 - x_2)^2 + \cdots + (b_0 + b_1v_m + b_2u_m - x_m)^2 = \min. \quad (2)$$

where:
$u_i$, $v_i$ - the coordinates of points determined in the map at scale of 1: 75 000,
$y_i$, $x_i$ - the coordinates of points determined in the map at scale of 1: 25 000,
$a_i$, $b_i$ – unknown transformation parameters.
Considering the characteristics of the map to be transformed, a maximum transformation error can be defined as $\Delta_t$ = 25 m, which gives:

$$\sum_{i=1}^{m} ((x_i - x_i^{tr.})^2 + (y_i - y_i^{tr.})^2) \leq 6875. \quad (3)$$

According to the above condition, the maximum value of the fitness function (target value) is 3438:

$$(a_0 + a_1v_1 + a_2u_1 - y_1)^2 + (a_0 + a_1v_2 + a_2u_2 - y_2)^2 + \cdots + (a_0 + a_1v_m + a_2u_m - y_m)^2 \leq 3438.$$
$$(b_0 + b_1v_1 + b_2u_1 - x_1)^2 + (b_0 + b_1v_2 + b_2u_2 - x_2)^2 + \cdots + (b_0 + b_1v_m + b_2u_m - x_m)^2 \leq 3438. \quad (4)$$

The functions that are minimized are presented in the M-File with the following expressions:

**a. Function for Determining the Parameters $a_0$, $a_1$, $a_2$**

```
function f = f(x)
f = (1*x(1)+4865361.63137253*x(2)+6533579.59363197*x(3)-
6533567.33966717)^2+(1*x(1)+4865292.59301976*x(2)+6534214.39
224915*x(3)-
6534213.26640835)^2+(1*x(1)+4865596.10399485*x(2)+6536512.40
913142*x(3)-
6536512.96007978)^2+(1*x(1)+4864590.10576256*x(2)+6533127.38
615627*x(3)-
6533110.13750536)^2+(1*x(1)+4864790.8694435*x(2)+6533588.367
95998*x(3)-
6533583.80262393)^2+(1*x(1)+4863639.06309127*x(2)+6534197.43
491776*x(3)-
6534187.95940408)^2+(1*x(1)+4863469.09530365*x(2)+6532284.91
709677*x(3)-
6532315.38193546)^2+(1*x(1)+4862799.60908719*x(2)+6532489.31
541758*x(3)-
6532515.90812452)^2+(1*x(1)+4862532.81642243*x(2)+6533538.45
349389*x(3)-
6533526.65916133)^2+(1*x(1)+4863119.75182385*x(2)+6532992.04
675944*x(3)-
6533004.62303021)^2+(1*x(1)+4863539.46653521*x(2)+6536199.70
919874*x(3)-6536211.98147372)^2
```

**b. Function for Determining the Parameters $b_0$, $b_1$, $b_2$**

```
function g = g(x)
g = (1*x(1)+4865361.63137253*x(2)+6533579.59363197*x(3)-
4865362.9199416)^2+(1*x(1)+4865292.59301976*x(2)+6534214.392
24915*x(3)-
4865301.0288533)^2+(1*x(1)+4865596.10399485*x(2)+6536512.409
13142*x(3)-
4865588.39663433)^2+(1*x(1)+4864590.10576256*x(2)+6533127.38
615627*x(3)-
4864569.42852291)^2+(1*x(1)+4864790.8694435*x(2)+6533588.367
95998*x(3)-
4864789.34919787)^2+(1*x(1)+4863639.06309127*x(2)+6534197.43
491776*x(3)-
4863630.18869693)^2+(1*x(1)+4863469.09530365*x(2)+6532284.91
709677*x(3)-
4863510.02398756)^2+(1*x(1)+4862799.60908719*x(2)+6532489.31
541758*x(3)-
4862801.4411889)^2+(1*x(1)+4862532.81642243*x(2)+6533538.453
49389*x(3)-
4862531.91007752)^2+(1*x(1)+4863119.75182385*x(2)+6532992.04
675944*x(3)-
4863112.77198684)^2+(1*x(1)+4863539.46653521*x(2)+6536199.70
919874*x(3)-4863574.71386812)^2
```

## 3.2 Parameter Tuning in Genetic Algorithm

After defining the fitness function, the GA parameters were defined and adjusted. The permanent parameters and options of the graphic presentation are the following:

- Number of variables: 3
- Population type: Double Vector
- Creation function: Uniform
- Scaling function: Proportional
- Selection function: Stochastic uniform
- Initial penalty: 30
- Penalty factor: 100
- Function tolerance: 1E-018
- Plot: Best Fitness

In the first experiment to determine the parameters $a_0$, $a_1$, $a_2$, the initial algorithm settings were used:

- Population size: 20
- Reproduction:
  - elite count: 2,

- crossover fraction: 0.8
- Mutation: Gaussian
  - scale: 1.0
  - shrink: 1.0
- Crossover: Two point
- Stopping criteria: 200 generations

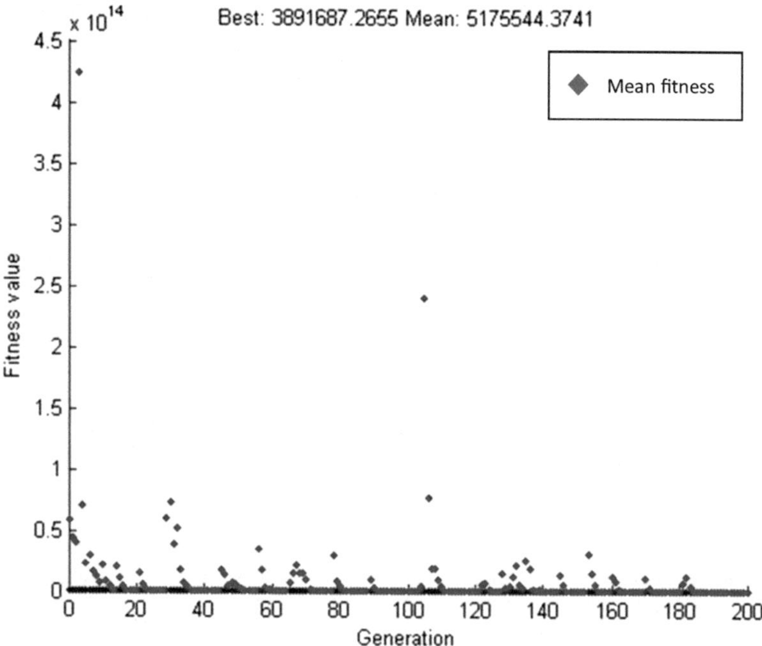


**Fig. 3.** Graphic representation of GA optimization in the first experiment (determination of parameters $a_0$, $a_1$ and $a_2$)

Figure 3 shows graphical presentation of GA optimization. After 200 generations, the best chromosome gene values were obtained, as shown in Table 1. The fitness function value was 3891687.3 (target value < 3438).

**Table 1.** Transformation parameters obtained after the first experiment

| Parameter | $a_0$ | $a_1$ | $a_2$ |
|---|---|---|---|
| Value | 11.74333 | 0.54052 | 0.59762 |

In the next experiment, the size of the population was changed to 80, and the number of generations was increased to 500. Since the value of the fitness (5276455.9) was even worse than in the first experiment, the settings of the genetic algorithm were corrected.

In the third experiment, the crossover function was changed to *Scattered*, and other parameters were retained. After 300 generations the process was stopped because the change in fitness in successive iterations became insignificant. A slightly better solution was obtained compared to the result of the first experiment.

In the fourth experiment, the previous parameters were used, and mutation was changed to *Adaptive feasible*. An even better solution was obtained.

In the fifth experiment, the settings from the fourth experiment were retained, and the size of the population increased to 100 and the crossover fraction was reduced to 0.42. The resulting fitness value was 2454.9. The optimal values of the parameters $b_0$, $b_1$, $b_2$ are determined on the basis of the same GA parameters as in the case of determining the parameters $a_0$, $a_1$, $a_2$. The values of the transformation parameters are given in Table 2.

**Table 2.** Transformation parameters obtained after the fifth experiment

| Parameter | $a_0$ | $a_1$ | $a_2$ | $b_0$ | $b_1$ | $b_2$ |
|---|---|---|---|---|---|---|
| Value | 3.1081307 | −0.0033325 | 1.0024807 | 5.3462539 | 0.9956374 | 0.0032474 |

### 3.3 Determination of Definitive Transformation Parameters

Based on the obtained transformation parameters, the transformed coordinates were calculated and the accuracy assessment was performed by comparing the transformed coordinates $(x_i^{tr}, y_i^{tr})$ and reference values ($x_i$, $y_i$) (Table 3).

**Table 3.** Transformed coordinates and the accuracy assessment of transformation.

| Point number | $y_i$ [m] | $x_i$ [m] | $y_i^{tr.}$ [m] | $x_i^{tr.}$ [m] | $r_y = y_i - y_i^{tr.}$ [m] | $r_x = x_i - x_i^{tr.}$ [m] | $r_p = \sqrt{r_y^2 + r_x^2}$ [m] |
|---|---|---|---|---|---|---|---|
| 1 | 6533567.3 | 4865362.9 | 6533576.9 | 4865358.7 | −9.6 | 4.2 | 10.4 |
| 2 | 6534213.3 | 4865301.0 | 6534213.5 | 4865292.1 | −0.2 | 9.0 | 9.0 |
| 3 | 6536513.0 | 4865588.4 | 6536516.2 | 4865601.7 | −3.2 | −13.3 | 13.7 |
| 4 | 6533110.1 | 4864569.4 | 6533126.1 | 4864589.1 | −16.0 | −19.7 | 25.4 |
| 5 | 6533583.8 | 4864789.4 | 6533587.6 | 4864790.5 | −3.8 | −1.1 | 4.0 |
| 6 | 6534188.0 | 4863630.2 | 6534202.0 | 4863645.7 | −14.0 | −15.5 | 20.9 |
| 7 | 6532315.4 | 4863510.0 | 6532285.3 | 4863470.2 | 30.1 | 39.8 | 49.9 |
| 8 | 6532515.9 | 4862801.4 | 6532492.4 | 4862804.3 | 23.5 | −2.9 | 23.7 |
| 9 | 6533526.7 | 4862531.9 | 6533545.1 | 4862542.1 | −18.4 | −10.2 | 21.1 |
| 10 | 6533004.6 | 4863112.8 | 6532995.4 | 4863124.7 | 9.3 | −12.0 | 15.1 |
| 11 | 6536212.0 | 4863574.7 | 6536209.6 | 4863553.0 | 2.4 | 21.7 | 21.8 |

The root mean square error (RMSE) resulting from the residue errors ($r_x$, $r_y$) is:

$$RMSE = \sqrt{\frac{\sum_{i=1}^{m}((x_i - x_i^{tr.})^2 + (y_i - y_i^{tr.})^2)}{m}} = 22.7 \text{ m}. \quad (5)$$

By observing the values in Table 3, it can be noticed that point 7 has a significantly higher deviation than other points. In order to improve the transformation model, this point was eliminated and the process of determining the transformation parameters was repeated.

The seventh point was excluded from the target's functions. The corrected values of the transformation parameters (Table 4) were determined with the same GA settings as in the 5th experiment. After 408 generations, the values of the parameters $a_0$, $a_1$, $a_2$, with the fitness value of 1730.83 were obtained (Fig. 4). After 710 generations a solution was obtained for $b_0$, $b_1$, $b_2$, with a fitness value of 2025.8 (Fig. 5).

**Table 4.** Definitive transformation parameters

| Parameter | $a_0$ | $a_1$ | $a_2$ | $b_0$ | $b_1$ | $b_2$ |
|---|---|---|---|---|---|---|
| Value | 3.99459661 | 0.000148652 | 0.99988866 | 6.21361601 | 0.98675268 | 0.009860727 |

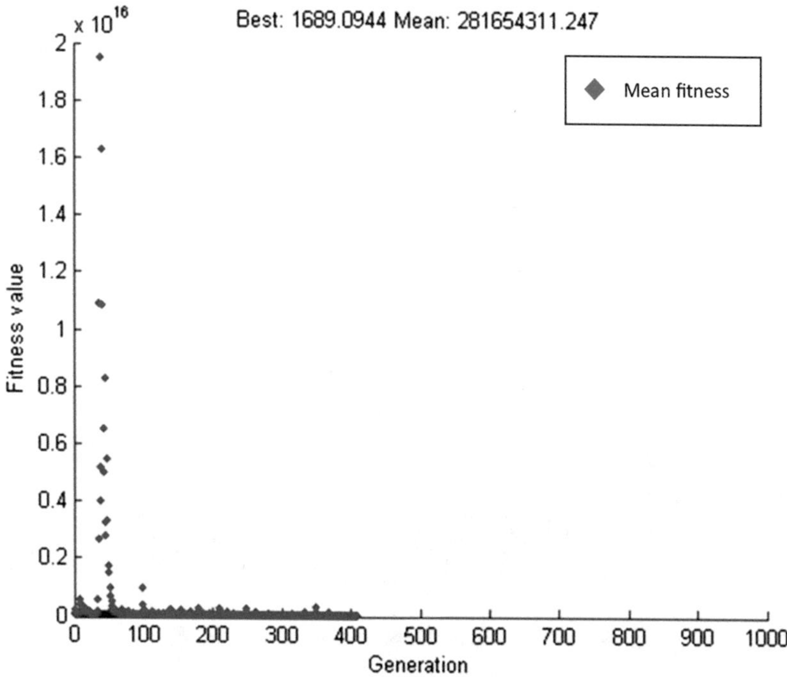


**Fig. 4.** Determination of definitive parameters $a_0$, $a_1$, $a_2$

On the basis of the obtained transformation parameters, definitive transformed coordinates were calculated and the accuracy assessment was performed by comparing the transformed coordinates ($x_i^{tr}, y_i^{tr}$) and reference values ($x_i$, $y_i$) (Table 5).

**Table 5.** Transformed coordinates and the accuracy assessment of transformation.

| Point number | $y_i$ [m] | $x_i$ [m] | $y_i^{tr.}$ [m] | $x_i^{tr.}$ [m] | $r_y = y_i - y_i^{tr.}$ [m] | $r_x = x_i - x_i^{tr.}$ [m] | $r_p = \sqrt{r_y^2 + r_x^2}$ [m] |
|---|---|---|---|---|---|---|---|
| 1 | 6533567.3 | 4865362.9 | 6533579.4 | 4865340.7 | −12.0 | 22.2 | 25.3 |
| 2 | 6534213.3 | 4865301.0 | 6534214.1 | 4865278.8 | −0.8 | 22.2 | 22.2 |
| 3 | 6536513.0 | 4865588.4 | 6536511.9 | 4865601.0 | 1.1 | −12.6 | 12.6 |
| 4 | 6533110.1 | 4864569.4 | 6533127.1 | 4864574.9 | −17.0 | −5.5 | 17.8 |
| 5 | 6533583.8 | 4864789.4 | 6533588.1 | 4864777.6 | −4.3 | 11.8 | 12.5 |
| 6 | 6534188.0 | 4863630.2 | 6534196.9 | 4863647.0 | −8.9 | −16.8 | 19.1 |
| 8 | 6532515.9 | 4862801.4 | 6532488.8 | 4862801.9 | 27.1 | −0.4 | 27.1 |
| 9 | 6533526.7 | 4862531.9 | 6533537.8 | 4862548.9 | −11.2 | −17.0 | 20.4 |
| 10 | 6533004.6 | 4863112.8 | 6532991.6 | 4863122.7 | 13.1 | −9.9 | 16.4 |
| 11 | 6536212.0 | 4863574.7 | 6536198.9 | 4863568.5 | 13.1 | 6.2 | 14.5 |

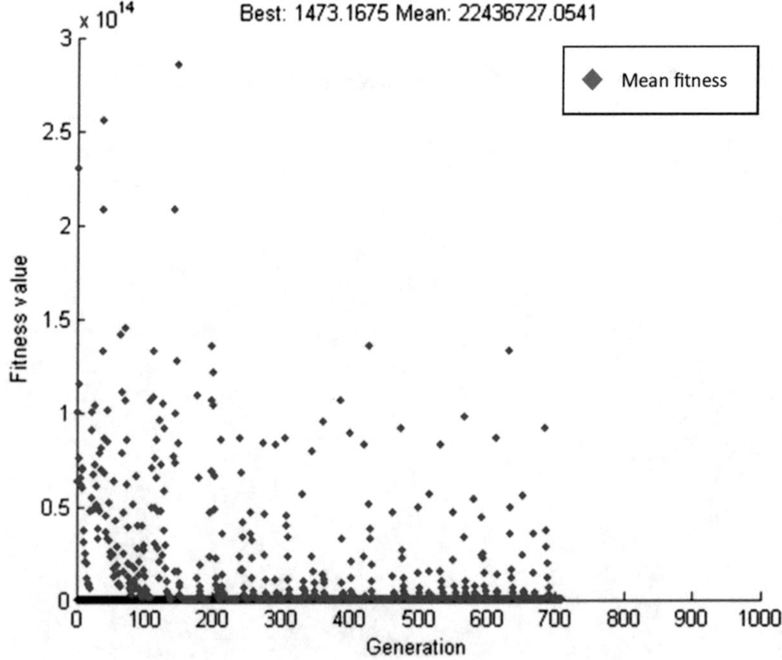


**Fig. 5.** Determination of definitive parameters $b_0$, $b_1$, $b_2$

The root mean square error (RMSE) resulting from the residue errors $(r_x, r_y)$ is:

$$RMSE = \sqrt{\frac{\sum_{i=1}^{m}((x_i - x_i^{tr.})^2 + (y_i - y_i^{tr.})^2)}{m}} = 19.4 \text{ m.} \tag{6}$$

It is evident that a somewhat better overall accuracy of transformation has been achieved, but it is much more important that the deviations on the individual points are uniform. This indicates the good quality of the identical points and that the transformation model is properly formed.

Based on the obtained parameters, the map section can be definitely georeferenced, and 3D visualized [16] using the appropriate GIS application (Fig. 6).

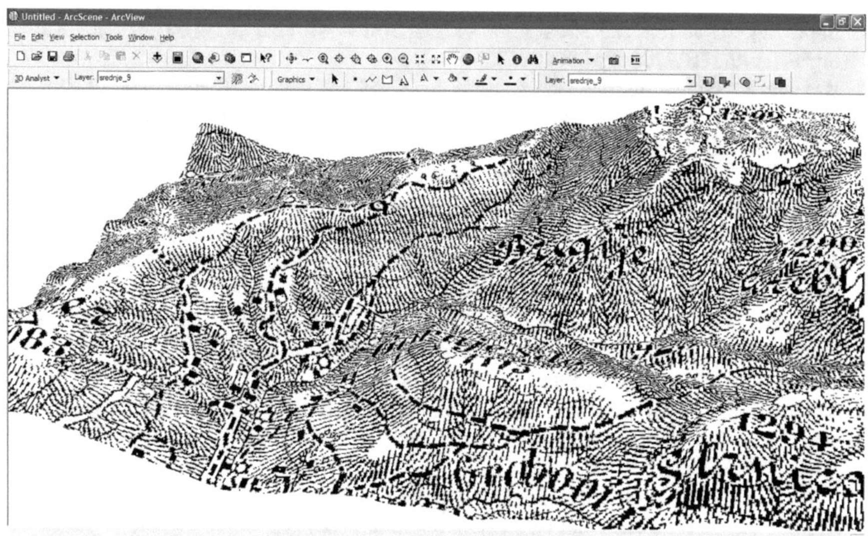


**Fig. 6.** Perspective DTM (Digital Terrain Model) view of the georeferenced map

# 4 Conclusion

This study confirms that the tie points play a key role in the final quality of the georeferencing results of the topographical map. The tie points can be selected with the use of different criteria and with different accuracy levels. The analysis of the topographical map's quality, evaluated on the basis of the remaining positional distortions, revealed that a satisfactory positional accuracy of the transformed contents of the map is achieved with an appropriate approach to transformation. A genetic algorithm can be applied in this manner, as a tool to search the entire range of possible solutions efficiently, in order to find the optimal transformation parameters. It is possible to solve very complex problems, but in order to achieve this, problems must be adapted appropriately according to the genetic algorithm. The methodology proposed for determining the transformation parameter gives very good results, but it is time-consuming (it requires a long-lasting parameter tuning in genetic algorithm).

## References


1. Nguyen, T.: Optimal ground control points for geometric correction using genetic algorithm with global accuracy. Eur. J. Remote Sens. **48**(1), 101–120 (2015). https://doi.org/10.5721/eujrs20154807
2. Haupt, R.L., Haupt, S.E.: Practical Genetic Algorithms. Wiley, Hoboken (2004)
3. Srinuandee, P., Satirapod, C.: Use of genetic algorithm and sliding windows for optimising ambiguity fixing rate in GPS kinematic positioning mode. Surv. Rev. **47**(340), 1–6 (2015). https://doi.org/10.1179/1752270614y.0000000088
4. Shnaidman, A., Shoshani, U., Doytsher, Y.: Genetic algorithms: a stochastic approach for improving the current cadastre accuracies. Surv. Rev. **44**, 325, 102–110 (2012). https://doi.org/10.1179/1752270611y.0000000012
5. Baselga, S.: Global optimization solution of robust estimation. J. Surv. Eng. **133**(3), 123–128 (2007). https://doi.org/10.1061/(asce)0733-9453(2007)133:3(123)
6. Coulot, D., Pollet, A., Collilieux, X., Berio, P.: Global optimization of core station networks for space geodesy: application to the referencing of the SLR EOP with respect to ITRF. J. Geod. **84**(1), 31–50. https://doi.org/10.1007/s00190-009-0342-1 (2009)
7. Tuno, N., Avdagić, Z., Ponjavić, M.: Georeferencing a raster data using genetic algorithm. In: 10th International Multidisciplinary Scientific GeoConference SGEM2011 Conference Proceedings, vol. 1, pp. 1003–1010 (2010)
8. Tuno, N., Mulahusić, A., Kozličić, M., Orešković, Z.: Border reconstruction of Bosnia and Herzegovina's access to the Adriatic Sea at Sutorina by consulting old maps. Kartografija i geoinformacije **16**, 26–55 (2011)
9. Kolingerová, I.: Genetic optimization of the triangulation weight. In: Proceedings of the 2nd International conference Computer Graphics and Artificial Intelligence, Limoges, pp. 23–34 (1998)
10. Gen, M., Cheng, R.: Genetic Algorithms & Engineering Design. Wiley, New York (1997)
11. Kramer, O.: Genetic Algorithm Essentials, Springer (2017)
12. Tuno, N., Topoljak, J., Mulahusić, A., Kozličić, M.: Cartographic depiction of religious buildings and cemeteries on cadastral maps created during the first cadastral survey of Bosnia and Herzegovina. Geoadria **20**(2), 175–214 (2015) https://doi.org/10.15291/geoadria.7
13. Topoljak, J., Lapaine, M., Tuno, N., Mulahusić, A.: Analiza vanjskih elemenata sadržaja katastarskih planova stare izmjere Bosne i Hercegovine. Geodetski list **71** (94) (1), 55–76 (2017)
14. Tuno, N., Mulahusić, A., Kogoj, D.: Improving the positional accuracy of digital cadastral maps through optimal geometric transformation. J. Surv. Eng. **143**(3), 1–12 (2017). http://dx.doi.org/10.1061/(ASCE)SU.1943-5428.0000217
15. Tuno, N.: Polinomska transformacija u georeferenciranju: Polynomial transformation in the georeferencing. Geodetski glasnik **39**, 38–46 (2007)
16. Borisov, M., Petrović, V.M., Vulić, M.: Vizuelizacija 3D modela geopodataka i njihova primjena : Visualisation of the 3D geodata models and their application. Geodetski glasnik **45**, 29–45 (2014)

# A Rule Based Events Correlation Algorithm for Process Mining

Almir Djedović(✉), Almir Karabegović, Emir Žunić, and Dino Alić

Faculty of Electrical Engineering, University of Sarajevo, 71000 Sarajevo, Bosnia and Herzegovina
almir.djedovic@infostudio.ba

**Abstract.** Process mining is a technique for extracting process models from event logs. Process mining can be used to discover, monitor and to improve real business processes by extracting knowledge from event logs available in process-aware information systems. This paper is concerned with the problem of grouping events in instances and the preparation of data for the process mining analysis. Often information systems do not store a unique identifier of the case instance, or errors happen in the system during the recording of events in the log files. To be able to analyze the process, it is necessary that events are grouped into case instances. The aim of the presented rule based algorithm is to find events belonging to the same case instance. Performances of the algorithm, for different sizes of log file events and different levels of errors within log files in the real process, have been analyzed.

## 1 Introduction

Business process models were traditionally done by domain experts, based on their experience and perceptions of organizations. This way of modeling is subjective and time-consuming. It often happens that process models acquired in this way, differ from processes which happen in reality.

Process mining techniques start from the recorded process data and therefore the main benefits of process mining relate to speed, completeness and correctness [1]. The first step of gathering the recorded data is a manual task and the results of process mining techniques can be tempered if no correct set of data is collected.

Some actions have to be taken before applying process mining. First, a search for the data in the IT systems should be performed. Most of today's information organizations store relevant data about the execution of processes in some structural form. For example, Workflow systems most often note the time of the beginning and end of the activity [2]. Enterprise Resource Planning systems such as SAP (Systems, Applications and Products) store all transactions, such as filling a form or change of documents [3–6]. The second step is structuring data, identifying single process steps (events) and group of process steps that belong to the same process execution (case instances), and converting this data to the format required by the process mining tool. In case there is data found in different sources, then one additional action is performed: merging the data into one log file. If the data is structured then different actions can be taken: using process mining techniques, the workflow of the process can be analyzed [3]. Also, the


S. Avdaković et al. (Eds.): IAT 2019, LNNS 83, pp. 587–605, 2020.
https://doi.org/10.1007/978-3-030-24986-1_47

simulation model can be built [1, 7]. The analysis of the resources' behaviour in the process can be performed [8], and the best allocation of the resource in the process can be found [9–12]. The bottlenecks in the process can be also analyzed [13], different systems can be integrated in order to improve processes [14, 15] and more, can be done. The prerequisite for analyses like this, and other similar analyses is that the events are grouped into case instances.

In this paper the problem of grouping events in the same case instance is addressed. This problem does not appear in systems where along with the events, also the information about the case instance is recorded. The problem appears in systems which do not have a unique identifier for grouping events into case instances. That can happen because of errors in log files, or because different systems are used for the business process or it is necessary to merge log files at a structured level in an inter-organizational process. In that case, different techniques for determining the connection between the events have to be used. This paper presents a rule based algorithm which has a task of connecting the most similar events. The algorithm calculates the similarity between all of the events, and then connects those events which have a similarity higher than the predefined threshold. The threshold value is not constant. It changes from the maximum value to the minimum. The algorithm can group events, even if errors are present in the log files.

This paper is organized as follows. The related work is discussed in more detail in Sect. 2. Section 3 has a more detailed explanation of the problem being solved, as well as the purpose of solving it. In Sect. 4 an implementation is given and the working principles of the proposed rule based algorithm are explained. Results acquired via experimentation are given in Sect. 5. Section 6 provides the conclusion of the paper.

## 2 Related Work

The problem of preprocessing data for the process mining analysis is already known in literature. Burattin and Vigo [16] decide whether sequences belong to the same case instance by comparing the corresponding attributes to the sequence. If the number of equal attribute values is higher than the predefined threshold, then it means that the sequences belong to the same case instance. One of the shortcomings of this approach is that the performances of the algorithm depend on the predefined threshold. If the predefined threshold is higher, then the algorithm will give a higher accuracy, but the searching time will also be high. On the other hand, if the predefined value is lower, the algorithm will form the case instances quicker, but the acquired results will be less accurate. The authors also used assumptions about data types (they excluded data about time, which are type timestamp from the analysis), assumptions about minimum and maximum length of the unique event indicator, a variation of its length, special characters which it does or does not include etc. All of the listed assumptions constrain the generalization of the approach.

Authors in [17] considered the many to many relation, allowing one event to belong to many case instances. Authors in [18] presented a new way of abstraction on the level of event classes. The events here are globally grouped according to classes based on rules. However, the presented method can only be used in specific cases. The approach as presented by [19] can work with the many to many relation using substantially dependent rules which exist between events. Paper [20] analyzed the appearance of events in order to find the laws related to events, i.e. determine the context in which the same events appear and where is assumed that events are semantically related. The context which appears in his paper is exclusively related to the events, which represents a large restriction for activities.

Papers [21] and [22] solve the same problem of forming case instances from events partially or under specific circumstances which restricts the generalization of the usage. Authors in paper [23] gave a detailed review of the literature from this field and they presented a new approach which enables gathering data which will be analyzed. The technique described in this paper requires modification of the source code of the application, which is not always possible. Some papers assume that a connection between a group of events and a group of activities already exists. Paper [24] presented the clustering of case instances with similar behavior, so that differences between clusters could be identified. This approach is interesting, but it has a shortcoming, which is that it can not identify events which always happen together. Also, some papers [25–27] tend to establish the behavior of a group of events via generalization, and the classification of the event flow aims to decrease the complexity of the process and enable its better understanding. None of the listed papers start with the assumption that defined activities in the process exists. Instead, they tend to group events which belong to the same activity, without referencing to the existing activities.

Paper [28] introduced a new technique for mapping events in an activity. The paper is an extension of their previous papers [29, 30]. Mapping is done using rules which are acquired from existing process models and event logs which are generated with the aim of connecting conceptual process models and execution data. The authors have used behavior and semantical knowledge. The presented approach proved itself robust to noise and precise. Recorded events in log files can be abstracted to recognizable activities. Authors in paper [31] presented an approach for the event abstracting using behaviour of activity in processes. They tried to improve the existing approaches which require the existence of a process model for solving this problem, and can not be used when noise exists in the files.

Paper [32] presented a new method for merging log files from different systems. The authors combined two algorithms: artificial immune algorithm and simulated annealing algorithm. The approach exploits the artificial immune algorithm from paper [33]. Aside from event attributes, for the function of connecting events, the authors included occurrence frequency between two execution sequence and temporal relations between two processes. They also implemented the algorithm as a plug-in in the ProM tool.

Authors in paper [34] presented an approach to deduce case identifiers for unlabeled event logs. The authors have addressed acyclic processes in this paper. In their next paper [35] have extended the approach so that it can be used for cyclic processes as well. For solving this problem the authors used the decision tree. The authors also, used statistical data about the execution duration of activity for determining case identifiers. Execution duration of activity is represented in interval and with an average value. But the time of execution of the activity in which end users are included, depends on the users, their work experience, their engagement in the process and so on.

Paper [36] presented the rule based approach for connecting input data from different sources for process mining analysis. Merging input data is done based on rules which the users define and merging is done at a structured data level. The users define which attributes are going to be compared and define the comparison operator. The advantage of this approach is that it enables the use of a-priori knowledge from the user. The paper has certain disadvantages. The first disadvantage is that only simple rules can be used, the use of complex rules is not enabled. Also, the rules are based only on the equals operator. While the other disadvantage is equality among all the rules. Differentiation of rules is not enabled in the paper, so all the rules have an equal impact on decision making. But we think that that this information, if available, can be of great importance in connecting events.

By analyzing state of the art it can be noted that very little to no attention at all was paid to connecting events in the case where errors in the log files are present. In real log files the probability of errors occurring is high. Errors can occur during data entry, during file conversion, current cessation of the system's work and so on. Errors cause the attributes of events to be incorrect or to not exist at all, which can lead to an incorrect connecting of events. In the end this will result in acquiring the wrong process model using process mining techniques. That's why we consider that analyzing the connecting of events in case of errors in the log files is very important aside from the problem of connecting events itself. The primary goal of this work is to present the rule based algorithm which can work with log files with errors. The algorithm needs to ensure a greater resistance to errors compared to the standard rule based algorithm.

## 3 Problem of Grouping Events into Case Instances

Event logs show the appearance of events in specific moments in time, where every event refers to a certain process and case instance. An example of a log file of a process event is given in Table 1. The log file is available on the page [37]. As mentioned previously, log files of events serve as a starting point for process mining analysis. A log file of events represents a group of case instances. A case instance is a sequence of activities which have been executed in the process. Every case instance has its own unique identifier. Event log files L can also be represented in the form of a group of pairs of events $l \in L$ where l has the form (1):

$$\langle ProcessInstanceID, WorkflowModelElement, \ldots \rangle \quad (1)$$

where ProcessInstanceID is the unique case instance identifier and Workflow ModelElement a unique activity identifier. Let's assume that $a_1, a_2, \ldots, a_n$ are unique activity sets in the process, and R is set of tuples $l \in L$, so that $R \subseteq L$, on set R the projection operator $\pi_{a_1,a_2,\ldots,a_n}$ is defined as a restriction of R on attributes $a_1, a_2, \ldots, a_n$. If it is assumed that the set of attributes is $A = \{a_1, a_2, \ldots, a_n\}$, then a projection on all attributes in A is denoted as $\pi_A(R)$. Similarly, for the given attribute a, the value of constant $c$ and the binary operator $\Delta$, an operator of selection $\sigma_{a\Delta c}(R)$ is defined representing a selection of tuples $l \in R$, where operator $\Delta$ connects attribute $a$ and value

$c$. For example, if $a$ = *activity* and $c$ = "*Amount*" and $\Delta$ the function of identity, then $\sigma_{a\Delta c}(R)$ is a set of all tuples $l$ having attribute *activity* as the value "*Amount*".

**Table 1.** Structure of repairExample.mxml log file

| Process instance | Workflow model element | Event type | Timestamp | Originator |
|---|---|---|---|---|
| 1 | Register | Complete | 1970-01 02T12:23:00.000 | System |
| | Analyze defect | Start | 1970-01-02T12:23:00.000 | Tester3 |
| | Analyze defect | Complete | 1970-01-02T12:30:00.000 | Tester3 |
| | Repair (complex) | Start | 1970-01-02T12:31:00.000 | SolverC1 |
| | Repair (complex) | Complete | 1970-01-02T12:49:00.000 | SolverC1 |
| | Test repair | Start | 1970-01-02T12:49:00.000 | Tester3 |

The case instance can be defined as a set of events having the same attribute ProcessInstanceID. If the case instance is denoted as PI then it can be defined using the relation (2):

$$PI = \{l | \pi_{ProcessInstanceID}(l) = const., l \in L\} \quad (2)$$

For performing the process mining analysis, it is necessary to have information about the case instance. This information is used for forming the process model, workflow analysis, analysis of performances, etc. In organizations, it is often the case that information systems are not process oriented, and business processes are covered by different applications, and the information about the unique case instance does not exist. This prevents the use of the process mining analysis, so it is necessary to group the events into case instances first. The events need to be grouped by the presented algorithm.

## 4 Rule Based Algorithm for Events Correlation

Let's assume that n is the given number of events and that each event is described with m attributes, i.e. (3):

$$l_i = \langle a_1, a_2, \ldots, a_m \rangle, \qquad i = 1, \ldots, n, \qquad n, m \in \mathbb{N} \quad (3)$$

For those events having a smaller number of attributes k, it is assumed that the other attributes are null, $a_j = null, j = k + 1, \ldots, n$. For two events i, j, and their two attributes p, q, the similarity function will be denoted $f_{p,q}^{i,j}$. It is assumed that i ≠ j, i.e. the same events will not be considered. The function $f_{p,q}^{i,j}$ can have two values {0, 1}. In case two events belong to the same case instance according to attributes p, q, then the value of the function $f_{p,q}^{i,j} = 1$, otherwise it is $f_{p,q}^{i,j} = 0$. The similarity of events according to all attributes is now defined with relation (4):

$$f^{i,j} = \sum_{p=1}^{n}\sum_{q=1}^{n} f_{p,q}^{i,j} \tag{4}$$

A larger value of the fitness function points to the higher probability that the events $i, j$ are connected, i.e. that they belong to the same case instance. Some attributes are more significant and can point to a higher similarity compared to others. In order to accept this differentiation weight coefficients $K_{p,q}$ are introduced, representing the significance of the similarity between the $p, q$ attributes. A higher value of coefficients points to a higher significance. Accordingly the relation (4) is transformed into relation (5):

$$f^{i,j} = \sum_{p=1}^{n}\sum_{q=1}^{n} K_{p,q} \cdot f_{p,q}^{i,j} \tag{5}$$

It is important to mention that most of the values $f_{p,q}^{i,j}$ are 0, because of the lower connectivity between different attributes, i.e. for $p \neq q$. If there is $N$ in log event files (the total number of attributes for all events is taken) the similarity matrix $F$ is defined:

$$F = \begin{bmatrix} f_{1,1} & f_{1,2} & \cdots & f_{1,n} \\ f_{2,1} & f_{2,2} & & f_{2,n} \\ & \vdots & \ddots & \vdots \\ f_{n,1} & f_{n,2} & \cdots & f_{n,n} \end{bmatrix}$$

Elements on the main diagonal represent the similarity of attributes with themselves. The events at the beginning can be identified by the fact that the user executing the activity is labeled as System and the event type is Complete i.e. $\pi_{Originator}(l_i) = System$ and $\pi_{EventType}(l_i) = complete$. For such events it is considered that the diagonal elements are equal $f_{i,i} = -1$. After the matrix of similarity $F$ is calculated, it is necessary to find the beginning elements, and then for every beginning element in the matrix to find events which are most similar to the observed event. In order to determine whether two events are from the same case instance a threshold T is defined. In case the value of the similarity function of two events is equal to the threshold then it is considered that those events belong to the same case instance, otherwise it is considered that they do not belong to the same case instance. The threshold should have an initial value equal to the maximum value of the similarity function. In case multiple similar events with equal value of the similarity function are found in the step where similarity between events and the event $l_i$ is calculated, the first event found is chosen.

If the algorithm does not find such two events in the first iteration, then the threshold is updated and its value is decreased by 1 and the search process is continued. In case if two events whose similarity functions are equal to the new value of the threshold are still not found, the threshold will be decreased by 1 again. In this way the threshold is decreasing to the minimum value $T_{min}$ and then the search process stops and all the selected events belong to the same case instance, and then the search process continues again from the event whose similarity function value is $-1$. The way the rule based algorithm works is explained in the following example. If there is a matrix of similarity $F$:

$$F = \begin{bmatrix} -1 & 3 & 1 & & 1 \\ 2 & 2 & 4 & \cdots & 0 \\ 1 & 1 & 2 & & 2 \\ & \vdots & & \ddots & \vdots \\ 2 & 1 & 3 & \cdots & 0 \\ -1 & 3 & 2 & & 1 \end{bmatrix}$$

Elements which have the value $-1$ mark the beginning of the process. The search starts moving at the rows and looking for the starting elements. A starting element is already found in the first row and first column.

$$F = \begin{bmatrix} -1 & 3 & 1 & & 1 \\ 2 & 2 & 4 & \cdots & 0 \\ 1 & 1 & 2 & & 2 \\ & \vdots & & \ddots & \vdots \\ 2 & 1 & 3 & \cdots & 0 \\ -1 & 3 & 2 & & 1 \end{bmatrix}$$

According to that, all the values from that row of the similarity function, i.e. the values $f_{1,j}, j = 1, \ldots, n$ will be considered. The first maximum value of the similarity function is in the second column $f_{1,2} = 3$.

$$F = \begin{bmatrix} -1 \rightarrow & 3 & 1 & & 1 \\ 2 & 2 & 4 & \cdots & 0 \\ 1 & 1 & 2 & & 2 \\ & \vdots & & \ddots & \vdots \\ 2 & 1 & 3 & \cdots & 0 \\ -1 & 3 & 2 & & 1 \end{bmatrix}$$

That means that that event is the next in the case instance and it is added to the case instance. A case instance which has two events $\{1, 2\}$ has already been analyzed. It is assumed that the value of the threshold $T$ fulfills the relation $f_{1,2} > T$. In case it is not, then the threshold will be decreased by 1 and it will be searched again in the first row. This process will be continued either until an event whose similarity function is higher than the threshold or until the threshold reaches its minimum value. Since it is found that the value of the similarity function is the largest in the second column, the algorithm continues in the second row and searches this row according to the same principle. The connection of the events can be shown graphically after the matrix of similarity is calculated. The matrix of similarity is calculated using real log file of credit requirement process and Fig. 1. shows the similarity of the starting event with the rest of the events. There have been 1545 events in total. Analysis of the figure shows that a peak exists whose value of the similarity function is equal to 3. The serial number of that event is 1084. This shows that in the analyzed log file of events, the first event is mostly connected with the 1084th event. Figure 2 shows that the function of the connection of the 1084th event with the rest of the events. It can be noticed that the values of the function of similarity between this event and the others are higher than the previous events'. The reason for that is that the previous event is one of the starting events in the process which means that not all of the attributes have been entered into the process yet. Because of the smaller number of attributes, the value of the function of similarity is also smaller. This also indicates that this approach can also be used for detecting the phase in the process in which the events currently are. Usually the group of attributes is smaller at the beginning of the process, and as the process progresses the number of attributes increases. The larger number of attributes is, the larger the value of the function of similarity is as well.

### 4.1 Memory Introduction

The process of searching for events with the highest similarity and their connection, which are described in the previous section, can run into the problem which is given in the following example. If there is a matrix of similarity:

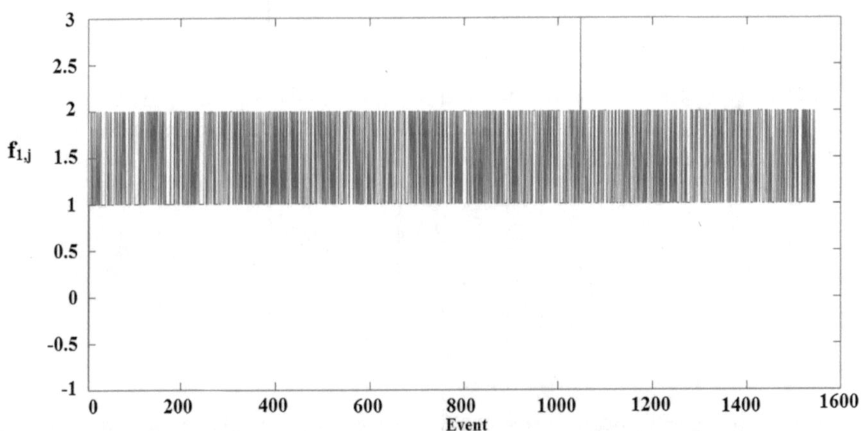


**Fig. 1.** Graphic representation of the similarity function of the 1st event

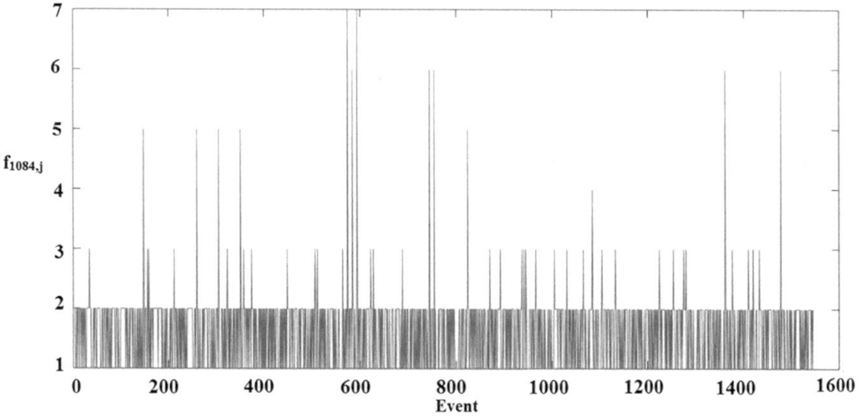


**Fig. 2.** Graphic representation of the similarity function of the 1084th event

$$F = \begin{bmatrix} -1 & 3 & 1 & & 1 \\ 2 & 2 & 4 & \cdots & 0 \\ 1 & 3 & 2 & & 2 \\ & \vdots & & \ddots & \vdots \\ 2 & 1 & 3 & \cdots & 0 \\ -1 & 3 & 2 & & 1 \end{bmatrix}$$

In the same way, the algorithm starts forming the instance from elements which are in the first row and first column, $f_{1,2} = 3$. It can be observed that the most similar event is located in the second column. Positioning to the second row, the most similar event is located in the third column, $f_{2,3} = 4$. After positioning to the third row, the algorithm finds that the most similar event is located in the second column $f_{3,2} = 3$ and it positions in the second row. The algorithm will position to the second row again and then position to the third and so on. The main problem is going into an infinite loop. In order to solve this problem, the algorithm has been designed so it is able to remember previous states. Adding the starting event to the instance $\{1\}$, the algorithm will remember that it was in the position $\{(1, 1)\}$, and after adding the second event to the instance $\{1, 2\}$, the algorithm will remember states $\{(1, 1), (1, 2)\}$. Before positioning to the next event, it is been checked whether it already exists among the memorized states. If it does not appear among the memorized states, then that event is picked. If it is among the memorized states, then the algorithm continues with the next most similar event which is also being checked in the same way. Introducing the memorization of states enables solving the problem of going into an infinite loop, but on the other hand causes increase of the execution time of the algorithm. If the log file that's being analyzed is larger, then more time will be needed for the algorithm to find a solution. In each next step the algorithm needs more time, because the matrix of the memorized states is larger. The pseudocode of the algorithm is given below.

**Pseudocode 1.** Pseudocode of rule based algorithm

```
// define all possible rules for L
defineRules(L)
// define the significance for each rule r_i ∈ R by determining coefficients // K_i
defineCoefficients(R)
// calculate the similarity between all events and form matrix F
foreach l_1 ∈ L do
  foreach l_2 ∈ L do
    // calculating similarity using predefined rules r_i
          formSimilarityMatrix(l_1, l_2)
  end
end
foreach l_1 ∈ L
  // finding most similar event to current event
  foreach l_2 ∈ L do
    find the most similar event to l_1
  end
  if (found) then
        T = T_max
    if (f == T) then
      // event found. Connect events and add it to memory
      connectEvents(l_2)
      // adding events to memory to avoid infinite loop
      addEventToMemory(l_2)
    else
      T = T − 1
        if (T = T_min or existsInMemory) then
          // go to the next event
                              continue;
        end if;
    end if;
  else
          // go to the next event
          continue;
  end if;
end;
```

## 4.2 Problem of Errors

In real systems it is expected that errors with the analyzed log files exist. An error can occur due to incorrect entry from the system, a user mistake whilst entering data, or due to preparation of data from different sources etc. Any error related to attributes directly affects the function of similarity and can cause the algorithm to incorrectly connect the events. This problem can not be fought with increasing the lower limit of the threshold of similarity. The increase of the lower limit of the threshold can lead to rejection of one part of the events which were connected earlier, but had the values of the similarity function close to the lower value of the threshold. One thing which is known with certainty is that attributes which change less in the process, as well as attributes which are automatically filled have a lower probability of an error occurring unlike attributes which change from step to step in the process, as well as attributes which are improperly entered from the user in the process. Knowing this and using the relation (4), different values of weight coefficients $K_{p,q}$ can be specified. For those attributes for which it is known that the probability of error occurrence is rare, it is necessary to choose different values of weight coefficients $K_{p,q}$. Attributes which are more frequently prone to errors need to have smaller values of weight coefficients. The larger value of weight coefficients indicates that the similarity between the corresponding attributes is more significant than the similarity of the other attributes. It is important not to specify larger values of weight coefficients. In that case only attributes with large values of weight coefficients determine the affiliation of events to the same instance, while the similarity of other attributes is being disregarded.

# 5 Experimental Results

The Java programming language and MATLAB environment have been used for the implementation of the rule based algorithm. Java has been used for the preparation of data, and all the calculations in the algorithm have been done in MATLAB. Data about a credit requirement process which is in the MXML format has been used for the analysis of this algorithm[1]. The dataset contains information about the time of the event, the participants in the event, type of the event (start or complete), as well as the name of the event. The data is gathered over a time period of over 6 months and the dataset contains over 150 000 events. The credit request is chosen because it represents one of the core processes in banks. The Java programming language is only used for extracting events into distinct text files, and MATLAB has been used for calculating the similarity of events. All the experiments were executed on machine with following characteristics: 4 core Intel Core i7 4700MQ (2,4 GHz), 4 GB RAM.

## 5.1 Discussion and Comparison of Results

A time analysis necessary for the execution of the algorithm was conducted for a different number of events. The execution time of the algorithm was measured in accordance with the number of events in the log file. The results are shown in Fig. 3. For small values of event numbers the execution time of the algorithm is negligible. The execution time of the algorithm is 11 s with a number of events equal to 600. With the increase of the number of events to 1200, the execution time of the algorithm

[1] https://data.4tu.nl/repository/uuid:35e6200e-ff3c-4019-af1e-c1ee8df19977

increases to about 45 s. The execution time of the algorithm increases because on one hand the number of calculations increases in the matrix of similarity, where the similarity between each two events is calculated. The size of the memory vector, which keeps the most similar events found, increases. The execution time of the algorithm can also be expressed by the size of the event's log file. The aforementioned analysis is shown in Fig. 4. For finding linked events, the given test parameters are: $K_{1,1} = K_{2,2} = K_{3,3} = 2$, $T_{min} = 5, T_{max} = 6$, and three rules are used. The rules are between each two events is calculated. The size of the memory vector, which keeps the most similar events found, increases. The execution time of the established in accordance with the values of the four primary attributes WorkflowModelElement, EventType, Timestamp and Originator.

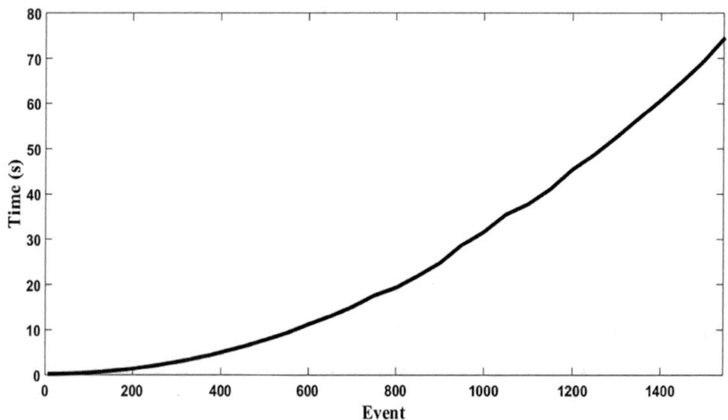


**Fig. 3.** Algorithm execution time for a different number of events

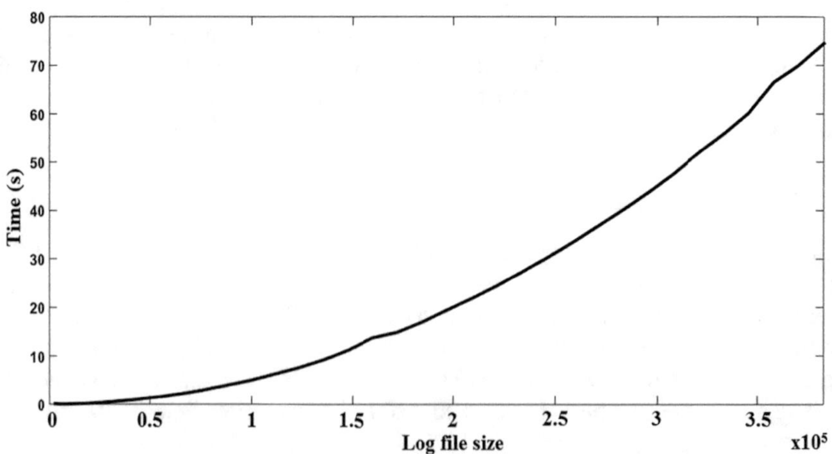


**Fig. 4.** Algorithm execution time for different size of log file

Rule No. 1:

$$f^{i,j}_{WorkflowModelElement,WorkflowModelElement} = \begin{cases} 1, & \pi_{EventType}(l_i) = start \wedge \pi_{EventType}(l_j) = complete \wedge \pi_{Originator}(l_i) = \\ & \pi_{Originator}(l_j) \wedge \pi_{WorkflowModelElement}(l_i) = \pi_{WorkflowModelElement}(l_j), \\ 0, & otherwise \end{cases}$$

Rule No. 2:

$$f^{i,j}_{EventType,EventType} = \begin{cases} 1, & \pi_{EventType}(l_i) \neq \pi_{EventType}(l_j), \\ 0, & \pi_{EventType}(l_i) = \pi_{EventType}(l_j) \end{cases}$$

Rule No. 3:

$$f^{i,j}_{Time,Time} = \begin{cases} 1, & \pi_{Timestamp}(l_i) = \pi_{Timestamp}(l_j) \\ & \wedge\ \pi_{WorkflowModelElement}(l_i) \neq \pi_{WorkflowModelElement}(l_j), \\ 1, & \pi_{Timestamp}(l_i) < \pi_{Timestamp}(l_j) \\ & \wedge\ \pi_{WorkflowModelElement}(l_i) = \pi_{WorkflowModelElement}(l_j) \\ 0, & otherwise \end{cases}$$

The first rule shows that two events are similar if they share the same label, completed by the same participant in the process and the first event has a label showing that it started, and the second has a label showing that it finished. By the second rule, events are considered to be the same if they are of a different type. By the third rule, events are considered to be the same if they have the same label and the second event is executed after the first, or if they have a different label, but their times of execution (of events) are the same. If more attributes describing the execution of the process exists, it is possible to form a larger set of rules. The bigger the set of rules is, the more resistant the algorithm will be to errors in log files, and at the same time the execution time of the algorithm will increase. In Fig. 5. it is shown how with the increase of rules, the execution time of the algorithm increases as well. For a small number of events that difference is negligible, but when the number of events reaches a value of 500 and 600 the data is incorrectly entered or that the data is not entered into the log files of events at take as many attributes as possible, i.e. rules which decrease the impact of errors on the result. Also, in cases like that it is important to pay attention to the selection of weight all. Errors can occur due to the aforementioned reasons: because of a problem with the differences in time can be noticed for a different amount of rules. With the increase of events, the differences increase as well. Errors can occur in real systems, such as that upper value of the threshold is supposed to be equal to the maximum value of similarity between events. In these cases, it is necessary to coefficients, as well as the smallest and largest value of the threshold. Generally, the two attributes can be $f^{i,j}_{p,q} \in \{0, 1\}$, is taken into account, then the largest value of the threshold is equal to:

$$T_{max} = \sum_{p=1}^{n} \sum_{q=1}^{n} K_{p,q} \tag{6}$$

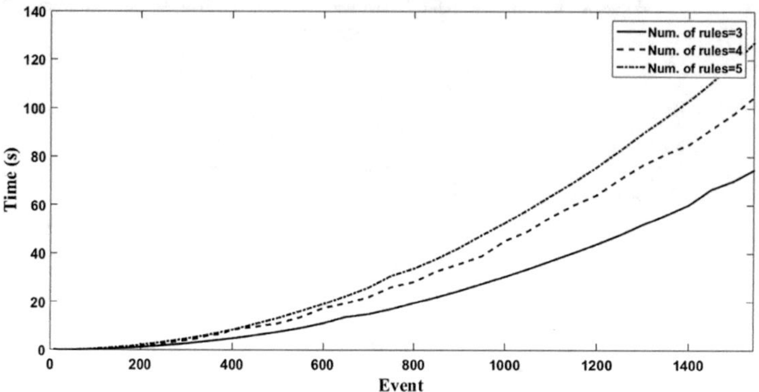


**Fig. 5.** Algorithm execution time for a different number of rules

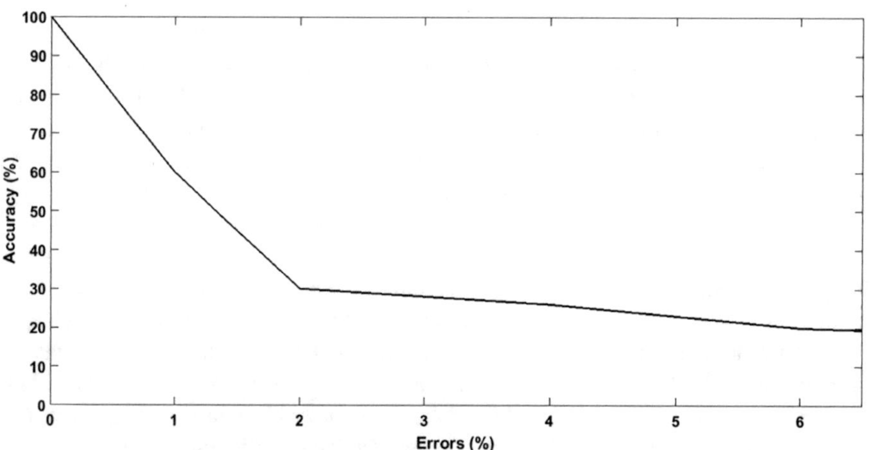


**Fig. 6.** Algorithm accuracy with errors in the log file with four attributes

It is important to be more cautious when choosing the smallest value of the threshold. If the value chosen is too high, then the possibility exists that, because of the errors, the algorithm will not recognize the events that belong to the same case instance. Small values of the lower threshold can cause that the wrong events get linked.

It is very important to know that the occurrence of all errors in the system is not equal. This information can be used to increase the accuracy of recognizing an event in the algorithm. The probability of an error occurrence in the system is larger in data that the user enters than in data which is automatically entered. Also, the probability of an error is larger in data whose entry is more frequent, meaning data which is more often entered, unlike data which is once entered in the form in the process and which can not be changed afterwards. For those attributes, it is necessary to define larger values of coefficients $K_{p,q}$, because their function of similarity has a higher impact in the connecting of events into case instances than events which are more susceptible to errors. The impact of errors on the accuracy of the algorithm is shown in Fig. 6. The parameters of the algorithm were: $K_{1,1} = K_{2,2} = K_{3,3} = 2, T_{min} = 5, T_{max} = 6$. Four main attributes have been included: WorkflowModelElement, EventType, Timestamp i Originator. Errors have been randomly allocated in the log file of events according to a uniform distribution. Given that the order of events has been known, this information has enabled to measure the accuracy of the algorithm in a way that the obtained results have been compared to the real order of events. The x-axis shows the percentage presence of errors in the log file with events. The figure shows that with the increase of errors, the accuracy of recognizing events from the same case instance decreases drastically. At a 1.5% presence of errors in the event's log file, the accuracy has decreased below 50%. The accuracy oscillates with the increase of the number of errors, the reason for that is because the errors are randomly generated and the layout of errors has an impact on the algorithm's execution. The results of the algorithm will not be the same when two errors happen in the same event and when two errors happen in different events. The first case is more dangerous than the second, because when an error happens in the log file of the event, the similarity of the other attributes that do not have the error might prevail so the similarity of the event is recognized regardless of the error.

It is very important to know the occurrence of all the errors in the system is not the same. This information can be used to increase the accuracy of recognizing events in the algorithm. The possibility of an error occurring in the system is higher in the data the end user enters rather than the data which is automatically entered. Also, the possibility of an error occurring is higher with data which is entered more frequently, compared to data which is entered once in the form in the process and later can not be changed. For those attributes it is necessary to define larger values of the coefficients, because their similarity function has a more significant impact on connecting events into case instances, than attributes which are less resistant to errors.

In accordance with that, two additional attributes in the process were observed. Their weights were $K_{4,4} = K_{5,5} = 4$, unlike the weights of other attributes $K_{1,1} = K_{2,2} = K_{3,3} = 2$. Also, the possibility of error occurrence on these attributes was 10 times lower than the possibility of error occurrence on other attributes. The results are shown in Fig. 7. The first and most important thing which can be noticed by comparing Figs. 6 and 7 is that the accuracy of the algorithm's execution has increased. The increase of the number of errors causes the accuracy of connecting events to decrease, but the decrease is reduced by increasing the number of attributes and specifying larger values of the weight coefficients. The maximum value of the threshold is chosen considering the relation (6), so the maximal value of the threshold is $T_{max} = 14$. The minimal value of the threshold is $T_{min} = 5$.

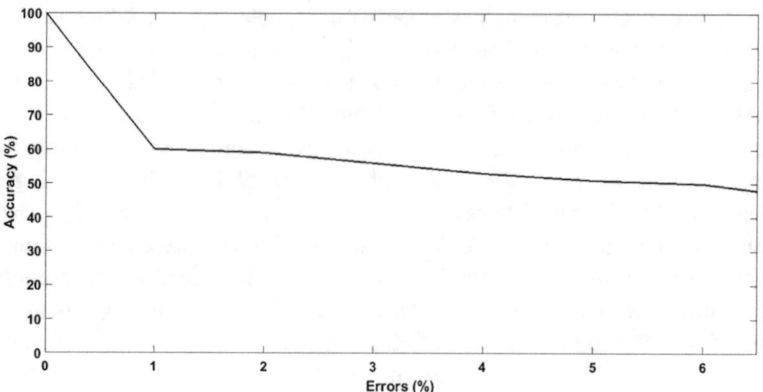


**Fig. 7.** Algorithm accuracy with errors in the log file with six attributes

The minimal value of the threshold is chosen so the similarity of all attributes can impact the linking of events. For example, if the value $T_{min} = 6$ is chosen, then for the first four attributes the following applies: $K_{1,1} + K_{2,2} + K_{3,3} = 6$. Which means even if the events are similar from the aspect of these attributes, that is not sufficient to consider two events to belong to the same case instance. However, as is shown in Fig. 6, in case when there are no errors, the given set of attributes is sufficient to link the events into a case instance. Because of that, if there are attributes whose similarity is sufficient to link events into a case instance, then their coefficients should be chosen so the following relation applies (7):

$$\sum_{i=1}^{m} K_{i,i} > T_{min} \tag{7}$$

The other attributes are there to increase the similarity in case errors appear in any of the first four attributes. The way their coefficients are supposed to be chosen is so that their single similarity is not enough for two events to be connected, otherwise they would already be in the first category of attributes (8):

$$K_{i,i} < T_{min}, \forall i = m + 1, \ldots, n \tag{8}$$

## 6 Conclusion

This paper presents a rule based algorithm for preprocessing data for the process mining analysis. A new way of comparing events has been defined and weight coefficients have been introduced. Weight coefficients determine the significance of the similarity of event attributes in the forming of case instances, because the similarity between all of the event attributes does not necessarily have to be of equal importance. In literature it is usually considered that there are no errors within the analyzed log files.

The case in which errors are present in the log files of events has also been observed. The effect of errors on the efficiency of the algorithm has been tested. Also a selection of parameters for the algorithm has been proposed so that the accuracy of the connection of events into case instances is increased. The algorithm has been tested on the log file of events of a real process, which has proven its practical usefulness.

In our future work we plan to expand the approach so it takes into account statistical behavior of the activity duration in the process during event grouping. It is considered that this will make the algorithm more resistant to errors. Also, we plan on enabling a parallel approach for connecting events which will enable the algorithm to work faster, which is especially significant for big and complex real log files.

## References

1. Rozinat, A., Mans, R.S., Song, M., van der Aals, W.M.P.: Discovering simulation models. Inf. Syst. **34**, 305–327 (2009). https://doi.org/10.1016/j.is.2008.09.002
2. van der Aalst, W.M.P., van Hee, K.M.: Workflow Management: Models, Methods and Systems. MIT Press, Cambridge (2004)
3. van der Aalst, W.M.P., van Dongen, B.F., Herbst, J., et al.: Workflow mining: a survey of issues and approaches. Data Knowl. Eng. **47**, 237–267 (2003)
4. Agrawal, R., Gunopulos, D., Leymann, F.: Mining process models from workflow logs. In: 6th International Conference on Extending Database Technology. LNCS, vol. 1337, pp. 467–483 (1998). https://doi.org/10.1007/bfb0101003
5. Grigori, D., Casati, F., Dayal, U., Sha, M.C.: Improving business process quality through exception understanding, prediction, and prevention. In: Proceedings of the 27th VLDB Conference, pp. 159–168 (2001)
6. Sayal, M., Casati, F., Dalay, U., Shan, M.C.: Business process cockpit. In: Proceedings of 28th International Conference on Very Large Data Bases (VLDB 2002), pp. 880–883 (2002)
7. Djedović, A., Žunić, E., Karabegović, A.: Model business process improvement by statistical analysis of the users' conduct in the process. In: 2016 International Multidisciplinary Conference on Computer and Energy Science, pp. 1–6 (2016)
8. Joy, J., Rajeev, S., Narayanan, V.: Particle swarm optimization for resource constrained-project scheduling problem with varying resource levels. Proc. Technol. **25**, 948–954 (2016)
9. Huang, H., Ma, H., Zhang, M.: An enhanced genetic algorithm for web service location-allocation. LNCS, vol. 8645, pp. 223–230 (2014)
10. Park, J., Seo, D., Hong, G., et al.: Human resource allocation in software project with practical considerations. Int. J. Softw. Eng. Knowl. Eng. **25**, 5–26 (2015). https://doi.org/10.1142/S021819401540001X
11. Djedović, A., Žunić, E., Avdagić, Z., Karabegović, A.: Optimization of business processes by automatic reallocation of resources using the genetic algorithm. In: XI International Symposium on Telecommunications – BIHTEL, pp. 1–7. IEEE (2016)
12. Tan, B., Ma, H., Zhang, M.: Optimization of location allocation of web services using a modified non-dominated sorting genetic algorithm. LNCS, vol. 9592, pp. 246–257 (2016)
13. van der Aalst, W.M.P., et al.: Process mining manifesto. In: Business Process Management Workshops. LNBIP, vol. 99, pp. 169–194 (2011)
14. Yimwadsana, B., Chaihirunkarn, C., Jaichoom, A., Thawornchak, A.: DocFlow: an integrated document workflow for business process management. Int. J. Digit. Inf. Wirel. Commun. (IJDIWC), 219–229 (2011)

15. Djedović, A., Žunić, E., Alić, D., Omanović, S., Karabegović, A.: Optimization of the business processes via automatic integration with the document management system. In: International Conference on Smart Systems and Technologies, pp. 117–122. IEEE (2016)
16. Burattin, A., Vigo, R.: A framework for semi-automated process instance discovery from decorative attributes. In: IEEE SSCI 2011: Symposium Series on Computational Intelligence - CIDM 2011: 2011 IEEE Symposium on Computational Intelligence and Data Mining, pp. 176–183 (2011)
17. Steinle, M., Aberer, K., et al.: Mapping moving landscapes by mining mountains of logs: novel techniques for dependency model generation 2000, pp. 1093–1102 (2000)
18. Günther, C.W., Rozinat, A., van Der Aalst, W.M.P.: Activity mining by global trace segmentation. LNBIP, pp. 128–139 (2010)
19. Li, J., Liu, D., Yang, B., Mining process models with duplicate tasks from workflow logs. LNCS, vol. 4537, pp. 396–407 (2007)
20. Bose, R.P.J.C., Verbeek, E.H.M.W., van Der Aalst, W.M.P.: Discovering hierarchical process models using ProM. In: CEUR Workshop Proceedings, pp. 33–40 (2011)
21. Walicki, M., Ferreira, D.R.: Mining sequences for patterns with non-repeating symbols. In: 2010 IEEE World Congress on Computational Intelligence, WCCI 2010 - 2010 IEEE Congress on Evolutionary Computation, CEC 2010 (2010)
22. Ferreira, D., Zacarias, M., Malheiros, M., Ferreira, P.: Approaching process mining with sequence clustering: experiments and findings. LNCS, vol. 4714, pp. 360–374 (2007). https://doi.org/10.1007/978-3-540-75183-0_26
23. Perez-Castillo, R., Weber, B., et al.: Generating event logs from non-process-aware systems enabling business process mining. Enterp. Inf. Syst. **5**, 301–335 (2011). https://doi.org/10.1080/17517575.2011.587545
24. Greco, G., Guzzo, A., Pontieri, L.: Mining taxonomies of process models. Data Knowl. Eng. **67**, 74–102 (2008). https://doi.org/10.1016/j.datak.2008.06.010
25. Polyvyanyy, A., Smirnov, S., Weske, M.: Process model abstraction: a slider approach. In: Proceedings of the 12th IEEE International Enterprise Distributed Object Computing Conference, EDOC 2008, pp. 325–331 (2008)
26. Bose, R.S.P.J.C., van Der Aalst, W.M.P.: Process diagnostics using trace alignment: opportunities, issues, and challenges. Inf. Syst. **37**, 117–141 (2012). https://doi.org/10.1016/j.is.2011.08.003
27. Fahland, D., van Der Aalst, W.M.P.: Simplifying discovered process models in a controlled manner. Inf. Syst. **38**, 585–605 (2013). https://doi.org/10.1016/j.is.2012.07.004
28. Baier, T., Di Ciccio, C., Mendling, J., Weske, M.: Matching events and activities by integrating behavioral aspects and label analysis. Softw. Syst. Model. 1–26 (2017)
29. Baier, T., Mendling, J., Weske, M.: Bridging abstraction layers in process mining. Inf. Syst. **46**, 123–139 (2014)
30. Baier, T., Solti, A., Mendling, J., Weske, M.: Matching of events and activities - an approach based on behavioral constraint satisfaction. In: SAC, pp. 1225–1230. ACM (2015)
31. Mannhardt, F., de Leoni, M., Reijers, H.A., van der Aalst, W.M.P., Toussaint, P.J.: From low-level events to activities - a pattern-based approach. LNCS, vol. 9850, pp. 125–141 (2016)
32. Xu, Y., Lin, Q., Zhao, M.Q.: Merging event logs for process mining with hybrid artificial immune algorithm. In: International Conference on Data Mining, pp. 10–16 (2016)
33. Burke, E.K., Kendall, G.: Search Methodologies-Introductory Tutorials in Optimization and Decision Support Techniques, 2nd edn. Springer, New York (2014)
34. Bayomie, D., Helal, I.M.A., Awad, A., Ezat, E., el Bastawissi, A.: Deducing case IDs for unlabeled event logs. In: BPI Workshop (2015)

35. Bayomie, D., Awad, A., Ezat, E.: Correlating unlabeled events from cyclic business processes execution. In: International Conference on Advanced Information Systems Engineering, pp. 274–289 (2016)
36. Claes, J., Poels, G.: Merging event logs for process mining: a rule based merging method and rule suggestion algorithm. Expert Syst. Appl. **41**, 7291–7306 (2014)
37. Process mining: Event logs. http://www.processmining.org/logs/start Accessed 10 Jan 2017

# Unification Algorithms

Mirna Udovicic(✉)

SSST, Sarajevo, Bosnia and Herzegovina
mirna.udovicic@ssst.edu.ba

**Abstract.** The name unification and the first formal investigation of this notion is due to J.A.Robinson. The notions of unification and most general unifier were independently reinvented by Knuth and Bendix. Various researchers have studied the problem further. Among other results, it was shown that linear time algorithms for unification exist. In this paper, the already existing unification algorithm is modified and improved. Also, an interesting implementation of a new algorithm is presented.

## 1 Introduction

The name unification and the first formal investigation of this notion is due to Robinson [3] in a year 1965. He introduced unification as the basic operation of his resolution principle, showed that unifiable terms have a most general unifier and described the algorithm for computing this unifier. Also, Herbrand investigated this topic in [4, 5]. The description by Herbrand of the unification algorithm for the general case is rather informal and there is no proof of correctness. A similar transformation-based algorithm was constructed by Martelli and Montanari [7].

The notions unification and most general unifier were independently reinvented by Knuth and Bendix [8] as a tool for testing term rewriting systems for local confluence by computing critical pairs.

Almost complete coverage of all the results obtained in unification theory can be seen in the overview article by Baader and Siekmann [2].

Syntactic unification of first-order terms was introduced by Post and Herbrand in the early part of this century. Various researchers have studied the problem further, including Martelli and Montanari [7], Paterson and Wegman [9]. Among other results, it was shown that linear time algorithms for unification exist (Martelli and Montanari [6], Paterson and Wegman [9]).

In this paper, the main result is improving the algorithm described in a Subsect. 2.3. An original algorithm is presented and it is proved that it has a linear complexity.

The algorithm is implemented in Mathematica. Also, the implementation contains the original part. Namely, in standard implementations in programming languages the terms are represented using either explicit pointer structures (as in C or Pascal) or built-in recursive data types (as in Lisp). In the implementation shown in this paper, the terms are presented in a form of arrays with three dimensions.


S. Avdaković et al. (Eds.): IAT 2019, LNNS 83, pp. 606–620, 2020.
https://doi.org/10.1007/978-3-030-24986-1_48

## 2 Syntactic Unification

In this section we review the major approaches to syntactic unification. Let us introduce the basic notions first, presented by Baader and Snyder in [1].

We assume the reader is familiar with the basic algebraic notions.

Terms in a first order language L are defined by induction:

1. Variables and constant symbols are terms
2. If $F \in Fnc_L$ has arity $n$ and $t_1$ , ... , $t_n$ are terms in a language L, then $F(t_1 ,..., t_n)$ is a term of a language L.
3. Any term of a language L can be constructed using rules 1 and 2 finite numbers of times.

We will denote terms by the letters $l,r,s,t,u$ and $v$; variables will be denoted by $w,x,y$ and $z$. The set of variables occuring in a term t will be denoted by Vars($t$).

A **substitution** is a mapping from a set of variables to a set of terms. The identity substitution is represented by Id.

The application of a substitution $\sigma$ to a term $t$, denoted $t\sigma$, is defined by induction on the structure of terms:

$$t\sigma := \begin{cases} x\sigma, & ako\ t = x \\ f(t_1\sigma, \ldots .t_n\sigma), & ako\ t = f(t_1, \ldots .t_n) \end{cases} .$$

In the second case of the definition, if $n$ =0 then $f$ is a constant symbol and $f\sigma = f$.

A **domain** of a substitution $\sigma$ is the following set of variables,

$$Dom(\sigma) := \{x | x\sigma \neq x\},$$

a **range** is the set of terms,

$$Ran(\sigma) := \bigcup_{x \in Dom(\sigma)} \{x\sigma\}.$$

A substitution can be represented explicitly as a function by a set of bindings of variables in its domain,

$$\{x_1 \rightarrow s_1, \ldots, x_n \rightarrow s_n\}.$$

A composition of two substitutions $\sigma$ and $\theta$ is written $\sigma\theta$ and is defined by

$$t\sigma\theta = (t\sigma)\theta.$$

An algorithm for constructing the composition $\sigma\theta$ of two substitutions represented as sets of bindings is given below:

1. Apply $\theta$ to every term in $Ran(\sigma)$ to obtain $\sigma_1$;
2. Remove from $\theta$ any binding $x \rightarrow t$, where $x \in Dom(\sigma)$, to obtain $\theta_1$;
3. Remove from $\sigma_1$ any trivial binding $x \rightarrow x$, to obtain $\sigma_2$; and

4. Take the union of the two sets of bindings $\sigma_2$ and $\theta_1$

Two substitutions are equal, denoted $\sigma = \theta$, if $x\sigma = x\theta$ for every variable $x$. We say that $\sigma$ is more general than θ if there exists an $\eta$ such that $\theta = \sigma\eta$.

A substitution $\sigma$ is a **unifier** of two terms $s$ and $t$ if $s\sigma = t\sigma$; it is a **most general unifier** (or **mgu**) if for every unifier $\theta$ of $s$ and $t$, $\sigma$ is more general than $\theta$.

## 2.1 Unification by Recursive Descent

The following algorithm is essentially the one first described by Robinson [3] and has been almost universally used in symbolic computation systems.

```
global σ: substitution;  { Initialized to Id }

Unify( s: term; t: term )
 begin
   if s is a variable then   { Instantiate variables }
      s := sσ;
   if t is a variable then
      t := tσ;
   if s is a variable and s = t then
      { Do nothing }
   else if s = f(s_1,...,s_n) and t = g(t_1,....,t_m) for n,m ≥ 0 then begin
      if f = g then
         for i := 1 to n do
            Unify(s_i,t_i);
      else Exit with failure   { Symbol clash }
   end
   else if s is not a variable then
      Unify(t,s);
   else if s occurs in t then
      Exit with failure;  { Occurs check }
   else σ := σ{s → t};
 end;
```

At the beginning of the algorithm, Id is represented by the empty binding set. We can see that in the end of the algorithm σ contains a result (a list of all bindings), if the unification is possible.

The previous algorithm can take exponential time and space. Clearly the problem is that a substitution may contain many duplicate copies of the same subterms. The solution to this problem is to develop a more subtle data structure for terms and a different method for applying substitutions.

## 2.2 Unification of a Term Dag, Recursive Descent

In this paper, we consider two approaches to speeding up the unification process. The first approach, which we adapt from Corbin and Bidoit, fixes the problem of duplication of subterms created by substitution by using a graph representation of terms which can share structure; this results in a quadratic algorithm. However, to develop an asymptotically faster algorithm, it is necessary to abandon the recursive descent approach and recast the problem of unification as the construction of a certain kind of equivalence relation on graphs. This second approach is due to Huet ([10]). Both approaches are presented in [1].

A term dag is a directed, acyclic graph whose nodes are labeled with function symbols, constants, or variables, whose outgoing edges from any node are ordered and where the outdegree of any node labelled with a symbol *f* is equal to the arity of *f*.

Note that variables have the outdegree 0. In such a graph, each node has a natural interpretation as a term, which means that a node *f(a,x)* is one labeled with *f* and having arcs to nodes *a* and *x*. The only difference between various dags representing a particular term is the amount of structure sharing among the subterms.

Assuming that names of symbols are strings of characters, it is possible to create a dag with unique, shared occurrences of terms in *O(n),* where *n* is a number of all characters in the string representation of the unification problem.

So, suppose we are given a term dag which represent two terms which need to be unified. Also, we are given an attribute *parents(t)* for every node *t*. It represents a list of all parents of a node *t* in a graph.

Now when we use the new term structure, a substitution does not duplicate subterms. However, the existing problem is that we might visit the same terms multiple times. So, the solution is to avoid already solved problems in a graph. The best solution is to create the union of the nodes *s* and *t* in a graph *Δ* which are already unified. So, let us denote a list of parents of a node s by

$$parents(s) = \{p_1, \ldots, p_n\};$$

The procedure is the following:

1. for every node $p_i$, we replace a subterm arc $p_i \to s$ by $p_i \to t$
2. let $parents(t) := parents(s) \cup parents(t)$;
3. let $parents(s)$: = Ø.

The previous steps share a structure for t and isolates a node s. We will denote by *Replace(Δ,s,t)* a new graph given from a graph *Δ* using previous three steps. The procedure *Replace(Δ,s,t)* is used in the unification algorithm.

Even with this modification of a term structure, the unification algorithm is mostly unchanged so there is no need to present it in this paper.

The correctness of the algorithm depends of the following lemma, which can be proved by induction.

**Lemma 1.** Suppose we are given a term structure *Δ* with nodes *x* and *t* such that there is no path from *t* to *x*.

- *Replace(Δ,x,t)* is an acyclic graph which contains the same nodes (with same notions) as *Δ*.
- Consider a distinguished node in *Δ* corresponding to the term *s* and let $s_1$ be the term corresponding to the same node in *Replace(Δ,x,t)*; then it holds:

– if $s = x$, then $s_1 = x$;
– otherwise, $s_1 = s\{x \rightarrow t\}$.

Now let us consider the complexity of the algorithm UnifyDag. Since each call to this function isolates a node, there can not be more than n calls in total (where $n$ is a number of symbols occurring in the original terms). Each call does a constant amount of work except for the occurs check (which traverses no more than $n$ nodes) and the moving of no more than $n$ pointers. Maintaining a list of parents costs $O(n)$ at each call. The original construction of a dag takes $O(n)$. This results in a time complexity of $O(n^2)$.

### 2.3 Unification of a Term Dag, an Almost Linear Algorithm

The unification closure approach to unification was first presented by Huet in [10] in a year 1976. It makes the following fundamental changes to the approach considered so far:

- instead of recursive calls to pairs of subterms which must be unified, we will construct an equivalence relation whose classes consist of these subterms
- substitution will be replaced by the union of equivalence classes
- the repeated calls to the occurs check will be replaced by a single pass through the graph

Now we will define a term and unification relation.

A **term relation** is an equivalence relation on terms and it is **homogeneous** if no equivalence class contains $f(\ldots)$ and $g(\ldots)$ with $f \neq g$; it is **acyclic** if no term is equivalent to a proper subterm of itself.

A **unification relation** is a homogeneous, acyclic term relation satisfying the unification axiom: for any $f$ and terms $s_i$ and $t_i$,

$$f(s_1, \ldots, s_n) \cong f(t_1, \ldots, t_n) \rightarrow s_1 \cong t_1 \wedge \cdots \wedge s_n \cong t_n.$$

The unification closure of $s$ and $t$, when it exists, is the least unification relation which makes $s$ and $t$ equivalent.

In the proof of the correctness of the algorithm the following lemma is used.

<u>**Lemma 2.**</u> If $s$ and $t$ are unifiable, then there exists a unification closure for $s$ and $t$.

It can be proved that constructing a unification relation on two terms corresponds to finding a mgu. It is necessary to introduce the following notions before presenting the algorithm.

For any term relation $\cong$, let a **schema function** be a function $\varsigma$ from equivalence classes to terms such that for any class $C$, it holds

1. $\varsigma(C) \in C$.
2. $\varsigma(C)$ is a variable only if $C$ consists entirely of variables

The term $\varsigma(C)$ will be called the **schema term** for $C$.

Note that schema functions are not unique.

For any unification closure $\cong$, we define $\sigma_{\cong}$ by:

$$x\sigma_{\cong} = \left\{ \begin{array}{c} y,\ ako\ \ \varsigma([x]) = y \\ f(s_1\sigma_{\cong}, \ldots, s_n\sigma_{\cong}),\ ako\ \ \varsigma([x]) = f(s_1, \ldots, s_n) \end{array} \right\}.$$

**Theorem 3.** Terms s and t are unifiable if and only if there is a unification closure for $s$ and $t$. In the affirmative case, $\sigma_{\cong}$ is an mgu for $s$ and $t$.

By the previous theorem it can be concluded that the first part of the efficient unification algorithm attempts to build a unification closure for two terms and the second part extract the mgu. The most efficient data structure represents classes as trees of class pointers with a class representative at the root.

To determine whether two terms are equivalent, it is only necessary to find the roots of the trees and check for identity; and to join two classes, one class is made a subtree of the others root. To reduce the height of the trees as much as possible, it should maintain a counter of the size of each class in the representative and when joining classes, make the smaller one a subtree of the larger.

The term dag for this approach needs

- class pointers
- a counter of the size of the class stored in the representative
- a pointer from each representative to the schema term for the class
- boolean flags visited and acyclic in each node used in cycle checking
- a pointer vars from each representative to a list of all variables in the class
- (used when generating solutions).

Note that at the beginning of the algorithm a representative is simply a node whose class pointer points to itself. Also, the term dag $\Delta$ for $s$ and $t$ is initialized to the identity relation, where each class contains a single term; thus for each node the class and schema pointers are initialized to point to the same node and the size is initialized to 1. The vars list is initialized to empty for non-variable nodes and to a singleton list for variable nodes.

Let us explain the main parts of the algorithm.

```
global Δ : termDag;  { Term structure for s and t }
global σ : list of bindings:= nil;  { A list σ contains a solution}

Unify( s: node; t: node )
  begin
    UnifClosure(s,t);
    FindSolution(s);
  end;
```

We can see that the unification algorithm Unify consists of two main procedures: UnifClosure and FindSolution. The first procedure UnifClosure forms all classes with its elements, if there is no a symbol clash. A procedure Find Solution uses the previous result and forms a result of the unification in a form of a list σ, if there are no cycles in a graph. So, in case there exist a symbol clash or a cycle in a graph, the two subterms can not be unified and the algorithm returns failure.

The correctness of this method depends on checking that it implements correctly a construction of an acyclic unification closure.

The important properties in the proof are the following ones:

- the equivalence is clearly homogenous
- equivalence classes are joined iff required by the unification axiom, hence the relation is least;
- FindSolution fails if and only if there is a cycle in a graph
- whenever a binding $\{x \rightarrow s\}$ is added to $\sigma$, all relevant bindings of variables in $s$ already occur in $\sigma$

At the beginning of the procedure UnifClosure created by the author Huet, an additional function Find for two terms $s$ and $t$ is called. This function is defined in a such way that it returns a representative of a class for a given term.

Note that at the beginning of the algorithm for each node $s$, a class pointer is initialized to point to that same node $s$. Since a procedure UnifClosure is recursive, in practical cases a function Find will not be called.

Complexity of the algorithm is $O(n\alpha(n))$, as, with the exception of Find, each function can be called at most $n$ times for terms with $n$ symbols and each call performs a constant amount of work. The dominating cost is therefore the calls to Find, which, as explained previously, can cost $O(n\alpha(n))$, where $\alpha(n)$ is the inverse of the Ackerman function.

The modified unification algorithm is presented below. Since a main change is that a new algorithm does not contain a function Find, the proof of the correctness of the algorithm is the same as the proof of the correctness of the original Huets algorithm. The complexity of the algorithm is improved now (the complexity is $O(n)$), which represent one of the results in this paper.

Note that we assume here that our given terms $s$ and $t$ do not share subterms. The reason is that a new algorithm is created in order to be implemented in Mathematica with a data structure presented in a form of three dimensional arrays.

The parts of the unification algorithm (the procedures and the function), are presented below.

```
  UnifClosure( s: node; t: node)
begin
  if s and t are the same node then
       { Do nothing}
  else begin
    if s = f(s_1,...,s_n) and t = g(t_1,...,t_m) for n,m ≥ 0
    then begin
      if f = g then begin
        Union(s,t);
        for i := 1 to n do
            UnifClosure(s_i, t_i);
      end
      else Exit with failure    { Symbol clash }
    end
    else Union(s,t);
  end;
end;
```

```
Union( s: node; t: node )
  begin
    if s is not a variable then begin
      size(s):= size(s) + size(t);
      vars(s):= concatenation of lists vars(s) and vars(t);
      class(t):= s;
    end
    else begin
      size(t):= size(t) + size(s);
      vars(t):= concatenation of lists vars(t) and vars(s);
      class(s):= t;
    end;
  end;
```

```
FindSolution ( s : node );
  begin
    t:= class(s);
    if  t = f(s_1,...,s_n)  for some n ≥ 0 then  begin
       for  i:= 1 to n  do
       FindSolution( s_i );
    end;
    for each  x ∈ vars(t)  do
      if  x ≠ t  then
      Add [x→s] to front of σ;
  end;
```

## 2.4 Analysis of the Already Implemented Algorithms and a New Algorithm

Let us give some basic definitions in order to represent the unification problem in terms of the equational unification. We present a solution of a problem as the solution of a system of equations. We first introduce the concept of a multiequation.

A multiequation is the generalization of an equation and it allows us to group together many terms which should be unified. To represent multiequations we use the notation $S = M$, where the left side S is a nonempty set of variables and the right side $M$ is a multiset of nonvariable terms. An example is:

$$\{x_1, x_2, x_3\} = (t_1, t_2).$$

The solution (**unifier**) of a multiequation is any substitution which makes all terms in the left and right sides identical.

Now let us compare the performance of the algorithm created by the authors A. Martelli and U.Montanari with that of two well-known algorithms: G.P.Huets and M.S. Paterson and M.N.Wegmans algorithm. Huets algorithm has an almost linear time complexity. Paterson and Wegmans algorithm is theoretically the best because it has a linear complexity.

As we have previously mentioned, we will make a description of all three algorithms using the same terminology (terms of the equational unification).

The algorithms deal with the sets of multiequations whose left-hand sides are disjoint and whose right-hand sides consist of only one term of depth equal one, that is, of the form $f(x_1, \ldots, x_n)$, where $x_1, \ldots, x_n$ are variables.

Furthermore, we have a set $S$ of the equations whose left and right sides are variables; for instance,

$$S\{x_1 = x_3\}.$$

A step of the algorithms consist is choosing an equation from $S$, merging the two corresponding multiequations and adding to $S$ the new equations obtained as the outcome of merging.

The three mentioned algorithms differ in a way they select the equation from $S$.

In Huets algorithm $S$ is presented in a form of a list; at every step, the first element of it is selected and the new equations are added at the end of the list. The algorithm stops when $S$ is empty and up to this point it has not yet checked the absence of cycles. Thus, there is a last step which checks whether the final multiequations are partially ordered, which is equivalent to acyclicity test if we use different terminology.

The source of the nonlinear behaviour of this algorithm is the same as for the algorithm created by A. Martelli and U. Montanari; that is, the access to the multiequations after they have been merged.

In order to avoid this, M.S.Paterson and M.N.Wegman choose to merge two multiequations only when their variables are no longer accessible. To select the multiequations to be merged, their algorithm "climbs" the partial ordering among multiequations until it finds a multiequation which is "on top"; thus the deduction of cycles is intrinsic in this algorithm.

Considering the algorithm of A. Martelli and U. Montanari, the equations which contain the pairs of variables to be unified are kept in multiterms and the merging is delayed until the corresponding multiequation is eliminated. An important advantage of this algorithm is that it may use terms of any depth. This fact entails a gain in efficiency, because it is certainly simpler to compute the common part and the frontier of deep terms than to merge multiequations step by step.

In order to compare the essential features of these three algorithms, we point out that they can stop either with success or failure for detection of a cycle or detection of a clash.

Let us denote by $P_m$, $P_c$ and $P_t$ the probabilities of stopping with one of these three events, respectively. In order to make a detail analysis of the algorithms, we consider three extreme cases:

(1) $P_m \gg P_c$, $P_t$ (very high probability of stopping with success): Paterson and Wegmans algorithm is asymptotically the best, because it has a linear complexity whereas the other two algorithms have a comparable nonlinear complexity.
However, in a typical application, such as a theorem prover, the unification algorithm is not used for unifying very large terms, but instead it is used a great

number of times for unifying rather small terms. In this case we can not exploit the asymtotically growing difference between linear and nonlinear algorithms. So, the computing times of the three algorithms will be comparable, depending on the efficiency of the implementation.

An experimental comparison of these three algorithms was carried out by Trum and Winterstein. The algorithms were implemented in the same language, PASCAL, with similar data structures and tried on five different classes of unifying test data. The result was that the algorithm created by A. Martelli and U. Montanari had the lowest running time for all test data. In fact, a previously mentioned algorithm is more efficient than Huets because it does not need a final acyclicity test. Also, it is more efficient than Paterson and Wegmans algorithm because it needs simpler data structures.

(2) $P_c \gg P_t \gg P_m$ (very high probability of detecting a cycle): Huets algorithm has a very poor performance in this case and Paterson and Wegmans algorithm is the best because it starts merging two multiequations only when it is sure that there are no cycles above them.

(3) $P_t \gg P_c \gg P_m$ (very high probability of detecting a clash): Huets algorithm is the best because, if it stops with a clash, it has not paid any overhead for cycle detection.

The new algorithm presented in this paper is very similar to the algorithm created by author G.P.Huet. A main difference between the two algorithms is that a Huets algorithm contains a function Find. Since it has the dominating cost $O(n\alpha(n))$, a new algorithm created without it is linear.

Considering the results when applying a new algorithm, it can be concluded that they are similar to the results when applying of the Huets algorithm. So, it is very efficient in cases where a symbol clash is detected and as well as in all cases when the two terms can be unified. Since the algorithm is created in order to be implemented in Mathematica in the way described in a Subsect. 2.5, its performance is good for the most of the cases.

There exist many implementations of the unification algorithms in programming languages like C and Pascal. It can be seen from the previous that the authors Martelli and Montanari implemented their unification algorithm in Pascal. All data structures used by their algorithm are dynamically created data structures connected through pointers. The examples of the definitions of these data types can be seen in [7].

## 2.5 Implementation

The algorithm is implemented in Mathematica and the approach to the representation of a data structure is completely new.

The main idea for implementing a new unification algorithm in Mathematica was that there are existing functions implemented in Mathematica which are convenient for data structures as term dags.

Namely, all terms and its subterms are presented in a form of three dimensional arrays. If some subterm does not exist, we write a number zero instead of that element.

At the beginning of a program the Input represent terms *s* and t. Using a function TreeForm implemented in Mathematica, they are transformed in a form of graph (new variables are called st and tt). Variables depth1 and depth2 contain a number of levels (depth) for each graph.

We form the arrrays a and b which contain all data for the beginning terms s and t.

The first counter of an array a represent a number of the level in a term graph. Since we have evaluated depth of a term s using a function Depth in Mathematica, we know the exact number of levels. So, there is no unnecessary data.

The second dimension is a serial number of a subterm in one determined level.

The third dimension is number 1,2,3 or 4.

1. If the third dimension is equal 1, then that element contains a value of a subterm. The line: a[1,1,1] = s means that an element at level 1 and number in that level equal 1 is the beginning term s.
   Also, the elements at level equal 2 (the first counter is 2) represent the subterms of s or equivalently the child nodes.
2. If the third dimension is equal 2, then that element contains a class pointer for that subterm.
   We know that in the beginning of the algorithm all elements point to itself; for example:
   $$\mathrm{a}[2,1,2] = \{2,1,2\}$$
3. If the third dimension is equal 3, then that element contains a number of elements in that class.
   We know that in the beginning of the algorithm all classes have a size equal 1; for example:
   $$\mathrm{a}[1,1,3] = 1$$
4. If the third counter is equal 4, then that element contains a list of variables for that subterm. We know that in the beginning of the algorithm if a subterm is a variable then a list contains that variable and if subterm is not a variable a list is empty; for example:
   $$\mathrm{a}[1,1,4] = \{\}$$
   $$\mathrm{a}[2,1,4] = \{\mathrm{x}\}.$$

The first part of the program which creates a term structure for a given term s is presented below.

```
s=f[x,g[a]];
t=f[g[y],g[y]];
p[a_]:=(If[MemberQ[{x,y,z},a],1,0]);
st=s//TreeForm;
tt=t//TreeForm;
depth1=Depth[st]-1;
depth2=Depth[tt]-1;
pom1=2^(depth1-1);
pom2=2^(depth2-1);

//We form 3 dimensional array a in the beginning, which contain data for a given term s
//first dimension is level in a graph, the second one is a serial number at that
//level and for every node there are 4 additional data: term in an algebraic
//form, pointer on a class representative, number of elements in a class
//and a list of variables

Array[a,{depth1,pom1,4}];
Array[b,{depth2,pom2,4}];
a[1,1,1]=s;
Array[broj1,depth1];
Do[broj1[i]=2^(i-1),{i,depth1}];
Do[broj1[i]=0,{i,depth1}];

//A function Formnivoa forms for a term s all elements of array a
//which have third dimension equal 1 in the beginning; means that
//all terms and its subterms are the elements of the array a
// If some element does not have left or right subterm, instead of empty place we write 0

 Formnivoa[nivo_]:=
Module[{s2,s3,l,pom},
Do[s2=a[nivo-1,i,1];
If[!(!(s2==0)&&!(Depth[s2]==1)),
Do[broj1[nivo]++; pom=broj1[nivo]; a[nivo,pom,1]=0,{2}]];
If[Depth[s2] ≥2, s3=s2//TreeForm;
l=Level[s3,{2}];
broj1[nivo]++; pom=broj1[nivo];
a[nivo,pom,1]=First[l];
l=Rest[l]; broj1[nivo]++;
pom=broj1[nivo];
If[l ≠{},a[nivo,pom,1]=First[l]];
If[l == {},a[nivo,pom,1]=0]],{i,broj[nivo-1]}]]

Do[Formnivoa[j],{j,2,depth1,1}];
```

Now let us explain a structure of a function Formnivoa (without loss of generality, a function Fornivob is the same).

We can see from the program that for every level of a term s a function creates all the elements of an array a at that level. Level is denoted by a variable j. The first value for j is equal 2 and a function Formnivoa creates all the elements at the level equal 2. Of course, at the level equal 1 there is only one element, a term s. So, a counter j takes values from 2 to depth1 (this can be seen in the program).

Note that when any level denoted by a variable i is created, there exist all necessary elements for creating the next level i + 1.

A function Formnivoa also uses functions Level, First and Rest which are the existing functions implemented in Mathematica.

We can point out that as a result of applying a function Formnivoa all elements of an array a with third dimension equal 1 are created.

After that step, it follows the part of a program which creates all other elements and is presented below.

```
Do[Do[If[a[i,j,1]=!=0,p1=p[a[i,j,1]]];
  a[i,j,2]={a,i,j};
  a[i,j,3]=1;
  If[p1==1,a[i,j,4]={a[i,j,1]}];
  If[p1==0,a[i,j,4]={}],{j,broj1[i]}],{i,depth1}]
```

A function Union is additional function called in a scope of a function UnifClosure (Union[n_,redbr_]). It makes the union of the two elements and uses the existing function Union implemented in Mathematica in order to make the union of two lists.

A function UnifClosure is the main function in a program. It compares pairs of terms at a certain level and makes the unions of pairs of the terms if the unification is possible. If there exist a symbol clash and unification is not possible, it returns failure. It also uses the existing function Head implemented in Mathematica, which returns a name of the main functional symbol of a given term; for example

Head[s]=f.

A function UnifClosure is given below.

```
UnifClosure[n_,redbr_]:=
          Module[{s1,t1},
                    s1=a[n,redbr,1];t1=b[n,redbr,1];
                    If[(s1=!=0)&&(t1=!=0)&&(s1=!=t1),
                    If[(p[s1]==0)&&(p[t1]==0),
Which[(Depth[s1]≥2)&&(Depth[t1]≥2)&&(Head[s1]===Head[t1]),
                              Spajanje[n,redbr],
!((Depth[s1]≥2)&&(Depth[t1]≥2)&&(Head[s1]===Head[t1])),
                              Print[failure]]];
                    If[!((p[s1]==0)&&(p[t1]==0)),
                         Spajanje[n,redbr]]]
Do[Do[UnifClosure[i,j],{j,broj[i]}],{i,depth1}];
```

At the end of a program, a result of the unification is presented in a list named Sigma, which is initialized to the empty list at the beginning.

The program was tested for the following pairs of terms:

$$\textbf{Input}: \ \mathrm{s} = \mathrm{f}[\mathrm{x}, \mathrm{g}[\mathrm{a}]] \ \text{ and } \ \mathrm{t} = \mathrm{f}[\mathrm{g}[\mathrm{y}], \mathrm{g}[\mathrm{y}]]$$

Since the terms s and t can be unified, the result was a list of bindings of variables:

$$\textbf{Output}:\ \text{Sigma}\ = \{\{y\}, a, \{x\}, g[y]\}$$

$$\textbf{Input}:\ s = f[f[f[z, y], a], g[x]]\ \text{ and }\ t = f[f[f[g[a], b], x], z]$$

Since the terms s and t can be unified, the result was a list of bindings of variables:

$$\textbf{Output}:\ \text{Sigma} = \{\{z\}, g[x], \{x\}, a, \{z\}, g[a], \{y\}, b\}$$

We can see from the previous result that the resulting substitution can be equivalently presented in the following way:

$$\{\{x\}, a, \{z\}, g[a], \{y\}, b\}$$

$$\textbf{Input}:\ s = f[f[f[f[z, y], a], g[x]], b]\ \text{and}\ t = f[f[f[f[g[a], b], x], z], y]$$

Since the terms s and t can be unified, the result was a list of bindings of variables:

$$\textbf{Output}:\ \text{Sigma}\ = \{\{y\}, b, \{z\}, g[x], \{x\}, a, \{z\}, g[a], \{y\}, b\}$$

We can see from the previous result that the resulting substitution can be equivalently presented in the following way:

$$\{\{x\}, a, \{z\}, g[a], \{y\}, b\}$$

The following test was for the terms which can not be unified:

$$\textbf{Input}:\ s = f[f[f[f[z, y], a], g[x]], b]\ \text{and}\ t = f[f[f[g[g[a], b], x], z], y]$$
$$\textbf{Output}:\ \text{failure}$$

## 3 Conclusion

It was already shown that linear time algorithms for the unification exist (Martelli and Montanari [6], Paterson and Wegman [9]). In this paper, the existing almost linear algorithm was modified and the complexity of a new algorithm is also *O(n)*. In a Subsect. 2.5 an interesting implementation of this algorithm was presented.

## References

1. Baader, F., Snyder, W.: Handbook of Automated Reasoning. In: Robinson, A., Voronkov, A. (eds.) Elsevier Science Publishers B.V. (2001)
2. Baader, F., Siekmann, J.H.: Unification Theory, Handbook of Logic in Artificial Intelligence and Logic Programming, pp. 41–125. Oxford University Press, Oxford, UK (1994)
3. Robinson, J.A.: A machine oriented logic based on the resolution principle. J. of the ACM **12**(1), 23–41 (1965)
4. Herbrand, J.: Investigations in proof theory: the properties of true propositions. In: van Heijenoort, J. (ed.) From Frege to Godel: A Source Book in Mathematical Logic, 1879–1931, pp. 525–581. Harvard University Press, Cambridge, MA (1967)
5. Herbrand, J.: Recherches sur la theorie de la demonstration. In: Goldfarb, W.D. (ed.) Logical Writings. Reidel, Dordrecht (1971)
6. Martelli, A., Montanari U.: Unification in linear time and space: A structured presentation, Technical Report B76-16, University of Pisa (1976)
7. Martelli, A., Montanari, U.: Am efficient unification algorithm. ACM Trans. Program. Lang. Syst. **4**(2), 258–282 (1982)
8. Knuth, D.E., Bendix, P.B.: Simple words problems in universal algebras. In: Leech, J. (ed.) Computational Problems in Abstract Algebra. Pergamon Press, Oxford (1970)
9. Paterson, M.S., Wegman, M.N.: Linear unification. J. Comput. Syst. Sci. **16**(2), 158–167 (1978)
10. Huet G.P.: Resolution d'equations dans les langages d'ordre 1,2,…,omega, These de doctorat d'etat, Universite Paris VII (1976)

# Static Based Classification of Malicious Software Using Machine Learning Methods

Ali Kutlay[1] and Kanita Karađuzović-Hadžiabdić[2(✉)]

[1] Zemana Ltd., Sarajevo, Bosnia and Herzegovina
alikutlay@hotmail.com
[2] Faculty of Engineering and Natural Sciences,
Computer Sciences and Engineering Program,
International University of Sarajevo, Sarajevo, Bosnia and Herzegovina
kanita@ius.edu.ba

**Abstract.** In this work, we perform classification of malicious software by evaluating the performance of six machine learning methods: Multilayer Perceptron Neural Network (MLP), Support Vector Machine (SVM), C4.5, CART, Random Forest and K-Nearest Neighbors (K-NN). The classification is performed using only structural information from portable executable file header that can be extracted from Win32 driver files. The best classification accuracy was achieved by the Random Forest method with 93.3% overall classification accuracy, followed by C4.5, CART, K-NN, SMV and MLP method with classification accuracy of 92.9% 92.5%, 91.6%, 77.7% and 89.0% respectively.

## 1 Introduction

Malicious software (i.e. malware) is a generic name for unwanted harmful software that is intended to cause malfunction or to interfere with the operation of a computer system or other devices infected by it. Examining, identifying and knowing about possible threats may help computer users to avoid harm that may be caused by malware. Malicious software can be designed to steal personal information such as login information, banking data, or encrypt personal files on a computer and request a ransom for the password key from the user. Other types of malware such as advertising software, browser hijackers, etc. are designed to monetize only the promotional content to the designers per click. Almost all malicious software has the ability to block security software. Furthermore, it can also update itself, download additional malware, or cause system security vulnerabilities to be compromised.

According to a survey conducted by Beaming Ltd., cybercrime is on the rise, with 2.9 million UK organizations being hit by a cyber-attack in 2016 at a cost of £29.1 billion [1]. Beaming Ltd.'s report found that phishing is the most common type of attack and affects almost 1.3 million organizations, followed by computer viruses and computer hijackings. According to Verizon's most recent data titled "2018 Data Breach Investigations Report", ransom attacks were the most prevalent variety of malware in 2017 [2]. While numbers speak for themselves, lack of strong prevention mechanisms cause numerous users to be victims of ever evolving cyber threats every day. As the severity of the case may very well be underestimated, such threats not solely target


S. Avdaković et al. (Eds.): IAT 2019, LNNS 83, pp. 621–628, 2020.
https://doi.org/10.1007/978-3-030-24986-1_49

home users, businesses also can become infected with malicious software such as ransomware, resulting in negative consequences as well as temporary or permanent loss of sensitive or proprietary info, disruption to regular operations, monetary losses incurred to revive systems and files, and potential damage to organization's reputation [3].

Malware classification can be performed using either static or dynamic malware analysis. Static malware analysis is based on static features that are extracted from binary code without the need of code execution. This is in contrast to dynamic malware analysis which is based on analyzing the system call sequences of malware while it is being executed. While there are many studies focused on working with one of two aforementioned approaches, there are also other studies conducted by combining these techniques and proposing a hybrid approach to deal with the malware classification problem.

Liu et al. [4] propose automatic malware classification and new malware detection using machine learning approach where gray scale image, opcode n-gram and import functions are used to extract malware features. The detection system applies the shared nearest neighbor clustering algorithm to discover new malware. The system achieved an accuracy of 98.9% to classify unknown malware and 86.7% to detect new malware. In a study conducted by Burnap et al. [5], authors took a dynamic analysis approach using continuous machine activity data (e.g. CPU use, RAM/SWAP use, Network I/O) to automatically distinguish between malicious and trusted portable executable software samples. In [6], Xu et al. proposed a novel malware classification method, Hybrid Analysis for Detection of Malware (HADM,) that focuses on Android malware classification. The method uses both static and dynamic features collected from Android applications and further uses deep neural networks (DNN) to improve the features by concatenating the original features with DNN learned features. The final classifier is built by using hierarchical Multiple Kernel Learning achieving 94.7% classification accuracy. Bounouh et al. [7] also adopted a hybrid approach combining both static and dynamic feature extraction techniques achieving 99.41% classification accuracy.

In this work we analyze the performance of several machine learning methods to detect malware by performing only static analysis of Win32 driver files. In our case static features are obtained from portable executable (PE) file header that can be very easily extracted from Win32 driver files. The advantage of this approach is that it provides a simple and yet fast malware classification achieving a reasonably high classification accuracy. Machine learning methods used in this work are: Multilayer Perceptron Neural Network, Support Vector Machine, C4.5, CART, Random Forest and K-Nearest Neighbors method.

## 2 Materials and Methods

### 2.1 Dataset

The dataset used in this work consists of 138,407 files: 107,025 clean and 31,381 infected Win32 driver files collected from Zemana AntiMalware [8] company. Zemana AntiMalware is a malware detection and removal software that protects computer users from malware, spyware, adware, ransomware, rootkits and bootkits. The decision to use Win32 driver files for malware detection and analysis is because a well-defined structural information, i.e. portable executable (PE) information can be extracted from Win32 driver files. Furthermore, such files are also a common target for malware creators. Since malware detection was performed by supervised machine learning methods, each file was labeled as infected or a clean file by processing it by Zemana AntiMalware cloud scan engine.

## 2.2 Feature Extraction and Dimensionality Reduction

After collectingWin32 infected and clean files, it was necessary to extract useful information from the obtained files, to be used as discriminative features as inputs into machine learning methods for malware classification. This was done by writing a disassembling tool in Go language. Go programming language was selected for feature extraction from Win32driver files as the language provides advanced library tools suitable for systems programming. Extracted information from Win32 disassembled driver files were initially stored in a plain text file in JSON format including file's MD5 unique identifier and a label indicating whether the file is infected or not. Initially 18 features presented in Table 1 were extracted by a disassembler tool. Afterwards, extracted information stored in JSON format was processed to create a dataset stored in SQL Server database. This data pre-processing involved removing noisy data such as broken strings as well as handling NULL attribute values. Since simply removing data containing NULL attributes would considerably reduce the size of the dataset, NULL values were replaced by computing the corresponding column mean. The final dataset obtained contained 138,406 instances; 107,025 clean instances and 31,381 infected instances corresponding to clean and infected files respectively. Feature evaluation and selection was performed next. This is one of the most important steps in classification problems. In general, feature evaluation and selection process help to improve the model performance. The process involves identification of attributes (i.e. features) that are most relevant to the predictive modeling problem.

Feature evaluation was performed using *Information Gain Attribute Evaluation* and *Correlation Attribute Evaluation* algorithms. This led to removal of redundant and irrelevant attributes from the feature set and at the same time the selection of most relevant features. Information Gain Attribute Evaluation algorithm computes the information gain value of an attribute with respect to the output class, (i.e. attribute's worth) in the range of 0–1 for each evaluated attribute. 0 indicates no information gain, and 1 maximum information gain. The attributes with low information gain value can be removed from the final feature set. Conversely, the attributes with high information gain value can be included in the final feature set. Correlation Attribute Evaluation algorithm evaluates attributes by computing the Pearson correlation between each attribute and the output class. The method selects only the attributes with moderate to high positive or negative correlation values, and neglects the remaining attributes.

Both techniques require the selection of minimum support threshold value. Threshold values of 0.111 and 0.030 were selected as minimum support threshold values for Information Gain Attribute Evaluation and Correlation Attribute Evaluation algorithms respectively. Both feature evaluation methods resulted in the same selection

**Table 1.** Initial feature set extracted from Win32 driver files

| Feature | Description |
|---|---|
| PeEntryPoint | Point at which the processor enters a program or a code fragment and begins executing |
| TextSecRawSize | Raw size value of the .text section contained in executable |
| DataSecRawSize | Raw size value of the .data section contained in executable |
| DataSecEntropy | Entropy of the .data section of the executable |
| TextSecEntropy | Entropy of the .text section of the executable |
| ContainedSections | Total number of contained sections in the executable |
| InitializedDataSize | Total size of the initialized variables in .data section of the executable |
| CodeSize | Total size of the code contained in executable |
| FileSize | Total size of the packed file |
| SubsystemVersion | Specifies the min. version of the subsystem on which the generated executable file can run |
| LinkedVersion | Version of the linker utility contained in executable |
| SigVerified | Flag to determine whether executable has a verified signature |
| OSVersion | Determines the oldest OS version in which executable can run |
| ImageVersion | Executable's image version |
| InternalName | Name of the driver DLL contained in executable |
| LegalCopyright | Legal copyright information of the executable |
| CompanyName | Name of the company for which executable was signed |
| FileVersionNum | Binary version number for the executable file |

of attributes, however differently ranked. Using these two attribute selection methods, the number of attributes was reduced from initial 18 attributes, to 11 attributes. Selected attributes are listed in Table 2, and were used as the final reduced feature set as inputs into evaluated classifiers.

**Table 2.** Final feature set used in malware classification

| **Feature** |
|---|
| PeEntryPoint |
| TextSecRawSize |
| DataSecEntropy |
| TextSecEntropy |
| ContainedSections |
| CodeSize |
| InitializedDataSize |
| FileSize |
| SubsystemVersion |
| DataSecRawSize |
| LinkedVersion |

### 2.3 Experiments and Methods

For malware classification, we evaluated the performance of six machine learning methods: Multilayer Perceptron Neural Network [9], Support Vector Machine [10], C4.5 [11], CART [12], Random Forest [13] and K-Nearest Neighbors [10]. The experiments were run on a well-known open-source machine learning software, Weka [14]. 70%–30% dataset split was used for training and testing sets respectively. We then performed an experiment to determine the optimal configuration parameters for each of the evaluated machine learning methods. The obtained parameters are shown in Table 3.

**Table 3.** Configuration parameters used for the evaluated machine learning methods

| Method | Optimal configuration |
|---|---|
| MLP | Hidden layers: 1; Learning rate: 0.2; Momentum: 0.2 |
| SMV | Kernel: Polynomial; Exponent Value: 0.1; Complexity: 0.1 |
| k-NN | Distance function: Manhattan; No of neighbors: 1 |
| CART | Batch size: 100; Confidence Factor: 0.25; No of Folds: 3; Pruning: true |
| C4.5 | Batch size: 100; Confidence Factor: 0.25; No of Folds: 3; Pruning: true |
| Random Forest | Bag size: 100%; No of Features: 0; No of iterations: 200 |

## 3 Results

The performance results of the malware classification were evaluated based on the overall accuracy, true positive rate (TPR), false positive rate (FPR), true negative rate (TNR) and false negative rate (FNR). TPR represents the rate of correctly classified infected data instances; FPR represents the rate of incorrectly classified infected data instances; TNR represents the rate of correctly classified clean data instances; FNR represents the rate of incorrectly classified clean data instances. Other than achieving high overall accuracy results, it is in particular important to have as high TPR and as low FPR results, followed by high TNR and low FNR. FP and TP rates represent the quality of infected file detection, whereas the TN and FN rates represent the quality of clean file detection.

The formulas for these metrics are defined as follows:

$$Accuracy = \frac{(TP + TN)}{(TP + TN + FP + FN)} \tag{3.1}$$

$$TPR = \frac{TP}{TP + FN} \tag{3.2}$$

$$FPR = \frac{FP}{FP + TN} \tag{3.3}$$

$$TNR = \frac{TN}{TN + FP} \tag{3.4}$$

$$FNR = \frac{FN}{FN + TP} \tag{3.5}$$

Table 4 shows the obtained classification results for each tested method.

**Table 4.** Classifier performance

| Classifier | Accuracy | TPR | FPR | TNR | FNR |
|---|---|---|---|---|---|
| MLP | 0.890 | 0.577 | 0.180 | 0.982 | 0.423 |
| SMV | 0.777 | 0.130 | **0.00** | **0.100** | 0.987 |
| k-NN | 0.916 | 0.772 | 0.430 | 0.957 | 0.228 |
| CART | 0.925 | 0.734 | 0.200 | 0.980 | 0.266 |
| C4.5 | 0.929 | 0.742 | 0.170 | 0.983 | 0.258 |
| Random Forest | **0.933** | **0.776** | 0.210 | 0.979 | **0.224** |

Results presented in Table 5 show that Random Forest method outperformed other classifiers achieving the best results in three out of five evaluated performance measures, i.e. the highest overall accuracy, 93.3%, and highest TPR, 77.6%, and the best (i.e. lowest) FNR, 22.4%.

C4.5 and CART algorithms also achieved high classification results, achieving very close and comparable results to that of Random Forest, i.e. 92.9%, and 92.5% overall accuracy results respectively, 74.2% and 73.4% TPR respectively, and 1.7% and 2.0 FPR respectively. Fourth best performing classifier was K-NN: 91.6% classification accuracy. A somewhat weaker classifier than Random Forest, C4.5, CART and k-NN classifiers was MLP classifier, with 89.0% overall accuracy, 57.7% TPR, 1.8% FPR, 98.2 TNR and 42.3% FNR. Even though SMV classifier had the best (i.e. lowest) FPR, 0.0%, and the best (i.e. highest) TNR,100%, it was the worst performing method in the remaining three evaluation criteria: it achieved the worst overall accuracy, 77.7% (a most important evaluation criteria), and the worst (lowest) TPR 1.3% (a second important evaluation criteria) as well as the worst FNR, 98.7%.

## 4 Conclusion

In this work we proposed a malware classification system composed of four phases: (1) data collection: collection of driver files from Zemana AntiMalware company; (2) data processing: feature extraction from malware database by a disassembler tool, data pre-processing and creation of a dataset; (3) feature evaluation and selection: evaluation and selection of most relevant features using Information Gain Attribute Evaluation and Correlation Attribute Evaluation methods to be used in the decision making phase; (4) decision making (classification): training machine learning methods using the training dataset, and malware classification on the unknown test dataset.

To perform classification, we analyzed six machine learning methods: Multilayer Perceptron Neural Network, Support Vector Machine, C4.5, CART, Random Forest and K-Nearest Neighbors. While many malware classification studies found in literature are aimed at detecting malware using a very large number of low level features such as binary images, op-code n-gram and function calls, where some methods have achieved higher results than the ones presented in this work, the main strength of the presented approach is a successful malware detection by using only structural PE information. This information can be easily extracted and analyzed from Win32 driver files. Furthermore, with the proposed approach it is not even necessary to analyze the whole file. The reason for this is because the malicious software can be detected by only reading the first and last 100 bytes of the file since the extracted features are only contained within this range. This is important since analysing the whole file can take considerably longer amount of time, especially when performing real-time classification.

The obtained results showed that from the tested machine learning methods Random Forest achieved the best overall accuracy results 93.3%, the best TPR 77.6% and best (i.e. lowest) FNR, 22.4%.

As part of the future work other then measuring the performance of machine learning methods using the results from the confusion matrix, we also plan to evaluate and measure the algorithm's execution time for the test dataset. This will allow possible identification of the machine learning method to be used in real-time analysis for malware classification. Further future work can be an extension of the proposed malware classification to apply clustering methods to generate different file clusters in an attempt to discover new unknown types of malware.

## References


1. White, R.: The cost of cyber security breaches: British business lost almost £30 billion in 2016 (2017) https://www.beaming.co.uk/press-releases/cyber-security-breaches-cost-businesses-30-billion
2. Verizon: Data breach investigations report (2018) https://www.verizonenterprise.com/resources/reports/rp_DBIR_2018_Report_execsummary_en_xg.pdf
3. Berkeley ISP: What is the possible impact of ransomware (2018). https://security.berkeley.edu/faq/ransomware/what-possible-impact-ransomware
4. Liu, L., Wang, B., Yu, B., Zhong, Q.: Automatic malware classification and new malware detection using machine learning. Front. Inf. Technol. Electron. Eng. **18**, 1336 (2017)
5. Burnap, P., French, R., Turner, F., Jones, K.: Malware classification using self organizing feature maps and machine activity data. Comput. Secur. **73**, 399–410 (2018)
6. Xu, L., Zhang, D., Jayasena, N., Cavazos, J.: HADM: hybrid analysis for detection of malware. In: Proceedings of SAI Intelligent Systems Conference (IntelliSys) (2016)
7. Bounouh, T., Zakaria, B., Al-Nemrat, A., Benzaid, C.: A scalable malware classification based on integrated static and dynamic features. In: Communications in Computer and Information Science book series (CCIS, vol. 630). Springer, Cham (2017)
8. https://www.zemana.com/
9. Haykin, S.: Neural Network: A Comprehensive Foundation. Prentice Hall, Upper Saddle River (1999)

10. Mitchell, T.: Machine Learning. McGraw Hill, New York (1997)
11. Deville, B.: Decision Trees for Business Intelligence and Data Mining: Using SAS Enterprise Miner. SAS Institute Inc, Cary, ISBN - 13:978-1-59047-567-6 (2006)
12. Breiman, L., Friedman, J.H., Olsen, R.A., Stone, C.J.: Classification and Regression Trees. Taylor & Francis, Wadsworth (1984)
13. Breiman L.: Random forests. Machine Learning **45**, 5–32, Kluwer Academic Publishers (2001)
14. Weka: Weka 3: data mining software in Java. Weka The University of Waikato (2018). https://www.cs.waikato.ac.nz/ml/weka

# Stochastic Efficiency Evaluation of the Lightning Protection System of Base Station

Adnan Mujezinović(✉), Nedis Dautbašić, Maja Muftić Dedović, and Zijad Bajramović

Faculty of Electrical Engineering, University of Sarajevo, Sarajevo, Bosnia and Herzegovina
adnan.Mujezinovic@etf.unsa.ba

**Abstract.** Base stations for mobile telephony are often exposed to the direct lightning strikes, which can lead to the significant damage of telecommunication equipment, measuring equipment and other electronic equipment within the base station. This paper describes methods of assessing the stochastic efficiency of the lightning protection systems of the mobile cellular base stations. In this paper Monte Carlo method is applied and electrogeometric model, which were used for the simulation of random natural of the lightning strikes. The presented methodology was applied on the example of high tower base station.

**Keywords:** Lightning discharge current · Lightning protection system (LPS) · Base station · Stochastic efficiency

## 1 Introduction

The rapid development of mobile telephony at the end of the 20th century required the accelerated construction of a large number of cellular base stations. Users' access to service at any time and in any place has conditioned that in addition to base stations in urban areas, they must be built on remote, elevated and difficult locations. Such objects, due to their position, are strongly exposed to lightning discharges. Malfunctions that would lead to interference in normal functioning or to a complete shutdown could cause, in addition to direct costs for operators, repairs or purchases of damaged equipment, and the indirect costs that are reflected in the termination of service delivery by users, then their dissatisfaction with the operator and possible termination of the contract for the provision of telecommunications services [1, 2].

The base stations of mobile telephony consist of antenna tower, power supply and telecommunication equipment containers and antenna (one or more) placed on the antenna tower. The container with the equipment and antennas must be placed within the protection zone of the antenna tower or separate lightning protection system (if it used) to avoid direct lightning strikes [3].


S. Avdaković et al. (Eds.): IAT 2019, LNNS 83, pp. 629–638, 2020.
https://doi.org/10.1007/978-3-030-24986-1_50

The consequences of direct lightning strikes can undermine the operational availability of mobile base stations as an elementary part of the network for providing telecommunication services to users. Therefore, it is necessary to carry out a detailed analysis of the efficiency of the lightning protection system of base stations of mobile telephony.

In this paper the Monte Carlo method and the electrogeometric model are used for the modeling of random nature of lightning strikes. In order to evaluate the efficiency of the lightning protection system, a large number of simulations were carried out and data on the number of direct lightning strikes in the component of the analyzed system were collected.

## 2 Lightning Parameters

Numerous experimental studies show that the statistical change in the peak of the lightning discharge current corresponds to log-normal distribution. Today, lightning discharging parameters data can be obtained using lightning location systems [4, 5]. Therefore, the cumulative probability that the peak of the atmospheric discharge current will be higher than the considered peak current $I$ can be determined from the following equation [6]:

$$P(I) = 0.5 \cdot \left(1 - \frac{2}{\sqrt{\pi}} \sum_{n=0}^{+\infty} \frac{(-1)^n u_0^{2n+1}}{n!(2n+1)}\right) \tag{1}$$

$$u_0 = \frac{\log(I) - \log(I_\mu)}{\sqrt{2}\sigma} \tag{2}$$

where:

$I$ - peak value of the lightning current,

$I_\mu$ - median of the lightning current peak,

$\sigma$ - standard deviation of the of the lightning current peak

The probability density and cumulative distribution of the peak currents of the lightning discharge was considered at $I_\mu = 31,1\,kA$ and $\sigma = 0.484$ are shown on Figs. 1 and 2.

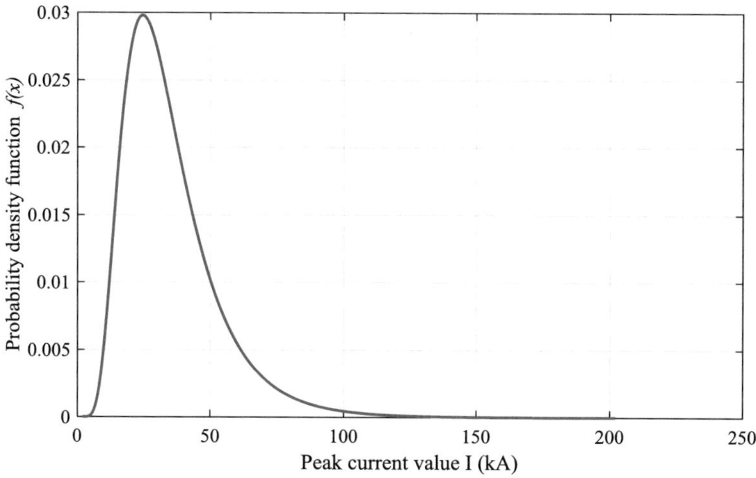


**Fig. 1.** The function of the cumulative probability of the peak of the lightning current

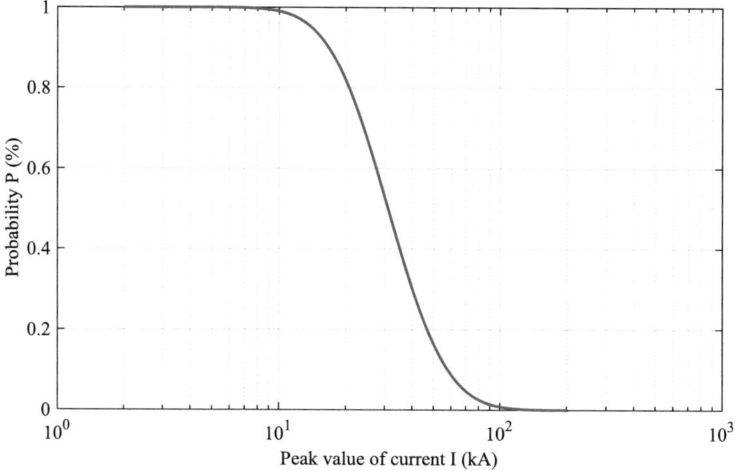


**Fig. 2.** Probability density function of peak values of the lightning current

The lightning current peak, at each simulated lightning discharge are stochastically selected from the considered interval $[I_{min}, I_{max}]$. The peak current values in the considered interval $[I_{min}, I_{max}]$ are divided into the corresponding number of classes of the same range, with the equal width range [7]:

$$\Delta I = \frac{\mathrm{I_{max}} - \mathrm{I_{min}}}{N_C} \tag{4}$$

where:

$N_C$ - number of classes.

The number of simulated lightning discharges per class can be determined using the following equation:

$$N^i = N \cdot \left[ \frac{P(I^i_{min}) - P(I^i_{max})}{P(I_{min}) - P(I_{max})} \right] \tag{5}$$

where:

$N$ - total number of simulated lightning discharges,

$N^i$ - total number of simulated lightning discharges of the $i$-th class,

$I^i_{min}$ - minimum peak value of the lightning current of the $i$-th class,

$I^i_{max}$ - maximum peak value of the lightning current of the $i$-th class

## 3 Stochastic Model

The spread of lightning discharges from the thunderous cloud is considered to moves stochastically in jumps. The length of each jump of lightning discharge is a function of the lightning current peak and can be determined using the following empirical equation:

$$R = \alpha I^{\beta} \tag{6}$$

where is:

$R$ - the striking distance expressed in [m],

$I$- the peak value of lightning current (leader) expressed in [kA],

$\alpha, \beta$- empirical constants whose values are proposed by many authors and are summarized in Table 1.

**Table 1.** Empirical constants values [8]

| Author | $\alpha$ | $\beta$ |
|---|---|---|
| Young | 27 | 0.32 |
| Love | 10 | 0.65 |
| Brown - Whitehead - CIGRE | 6.4 | 0.75 |
| IEEE - 1995 | 8 | 0.65 |

Within this paper for calculating the breakthrough distance, the empirical constants suggested by the author of Love were used. The calculation of the efficiency of the lightning protection system was carried out using the Monte Carlo method. As the input parameter for the calculation of the protection system efficiency, it is necessary to know the lightning current peak value. As the value of the lightning discharge current is not known, the same is determined based on the known distribution of peak values [9].

The first step in the calculation is to generate a series of random numbers in the range [0–1]. The size of this series is defined by class of the peak of the atmospheric discharge current. Then, based on the generated random numbers, the peak currents can be determined using the following equation:

$$I^i = I^i_{min} + \lambda_1\left(I^i_{max} - I^i_{min}\right) \tag{7}$$

where is:
$i$ - $i$-th class,
$\lambda_1$ - randomly generated number

Each lightning discharge starts at a randomly chosen point on the thunder cloud $T(x_0, y_0, z_0)$, where the coordinates $x_0$ and $y_0$ are randomly generated by the random number generator while the coordinate $z_0$ is maintained constant throughout all the simulations. If it is assumed that the lightning discharge leader was at point $G(x_G, y_G, z_G)$, then after a leap, the leader came to a new point $H(x_H, y_H, z_H)$ located on the semi-sphere as shown on Fig. 3 [10].

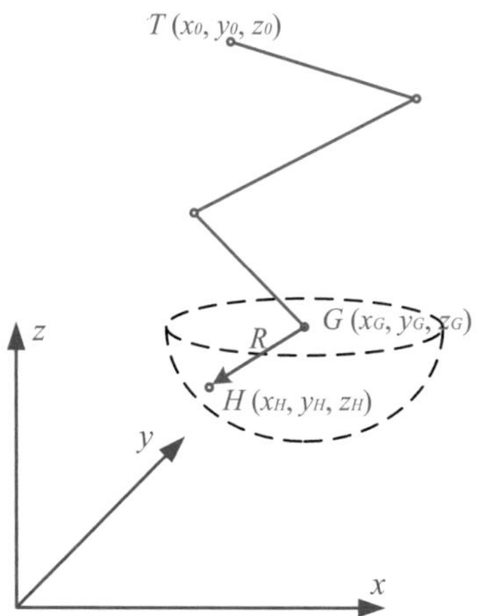


**Fig. 3.** Movement of the leader of the atmospheric discharge

The movement of atmospheric discharges is stochastic and jumpy, and hence the coordinates of the new point $H$ are determined in a random way using the following equations [7]:

$$x_H = x_G + R \cdot sin\vartheta \cdot cos\varphi \tag{8}$$

$$y_H = y_G + R \cdot sin\vartheta \cdot sin\varphi \tag{9}$$

$$z_H = z_G + R \cdot cos\vartheta \tag{10}$$

where is:

$\vartheta, \varphi$ - angles of the spherical coordinate system and can be determined using Eqs. (11) and (12):

$$\varphi = \lambda_2 \cdot 2\pi \tag{11}$$

$$\vartheta = (\lambda_3 + 1) \cdot \frac{\pi}{2} \tag{12}$$

where:

$\lambda_2, \lambda_3$ - randomly generated numbers from the interval [0–1].

In order to determine the movement of lightning, it is necessary to determine the shortest distance between the leader of the lightning discharge and other segments (the elements of the protected object or the lightning protection system). The shortest distance between the leader located at point $G(x_G, y_G, z_G)$ and other segments can be determined from the parameter $t_H$ given by the following equations [10]:

$$t_H = \frac{a(x_G - x_P) + b(y_G - y_P) + c(z_G - z_P)}{a^2 + b^2 + c^2} \tag{13}$$

$$a = x_K - x_P, b = y_K - y_P, c = z_K - z_P \tag{14}$$

where:

$P(x_P, y_P, z_P)$ - starting point of the segment,
$K(x_K, y_K, z_K)$ - endpoint of the segment

The shortest distance between the leader of the atmospheric discharge and the considered segment is under condition [11]: $t_H \in [0, 1]$.

Finally, stohastic efficiency of some lightning protection system can be calculated by using following equation [7]:

$$EC = \frac{\sum_{i=1}^{N_c} N_{SZ}^i}{\sum_{i=1}^{N_c} \left(N_{SZ}^i + N_{OB}^i\right)} \tag{15}$$

where:

$N_{SZ}^i$ - number of lightning strikes of $i$-th class to lightning protection system,
$N_{OB}^i$ - number of lightning strikes of $i$-th class to protected object

## 4 Result and Discussion

The previously presented mathematical model was used to calculate the stochastic efficiency of the base station LPS. The base station analyzed consists of an antenna tower with lightning rod at the top of tower of total height 42 m. At the base of the antenna tower container with telecommunication and power supply equipment is

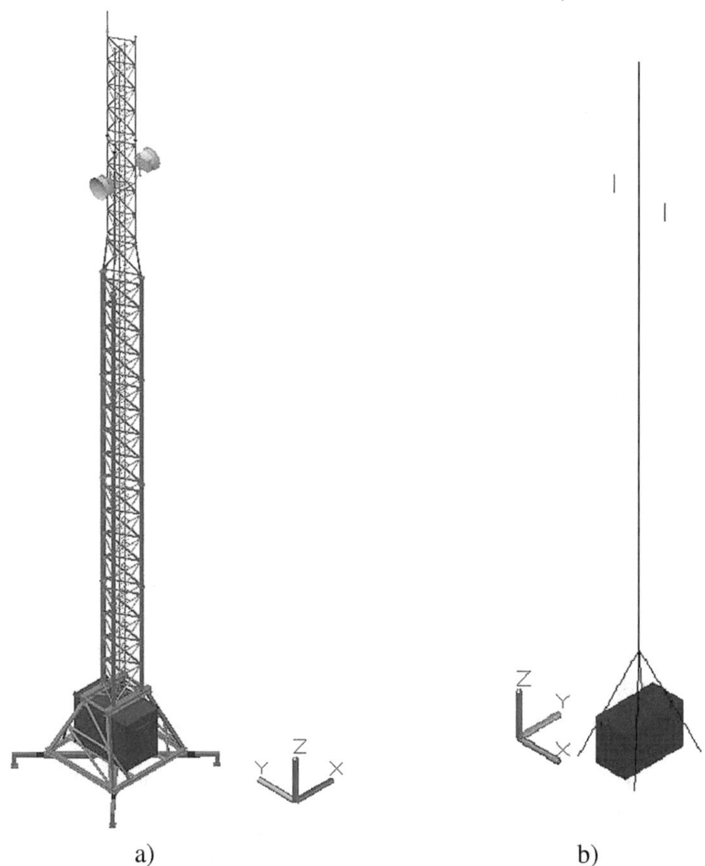


**Fig. 4.** Geometry of the analyzed base station (a) original geometry (b) approximate geometry [12, 13]

placed of a total area 4.1 × 4.2 m and height of 2.7 m. On the antenna tower two antennas are placed. Both antennas are set at the same height of 33.1 m from the ground surface. The total heights of both antennas are 1.18 m. The geometry of analyzed base station is shown in Fig. 4a. In order to accelerate the calculation, the base station components are approximated with the corresponding 1D straight line elements, as shown in Fig. 4b.

A total of 199981 simulations were performed, with peak values of lightning discharge divided into 40 classes with a range of 5 kA per class. The highest peak value of the simulated atmospheric discharge current was 202 kA, while the minimum peak value of simulated atmospheric discharge was 2 kA. The thunderstorm cloud was set at 1 km above the ground surface and the total analyzed surface area was 1 km$^2$. The histogram of distribution of peak values of lightning strikes by class is shown in Fig. 5.

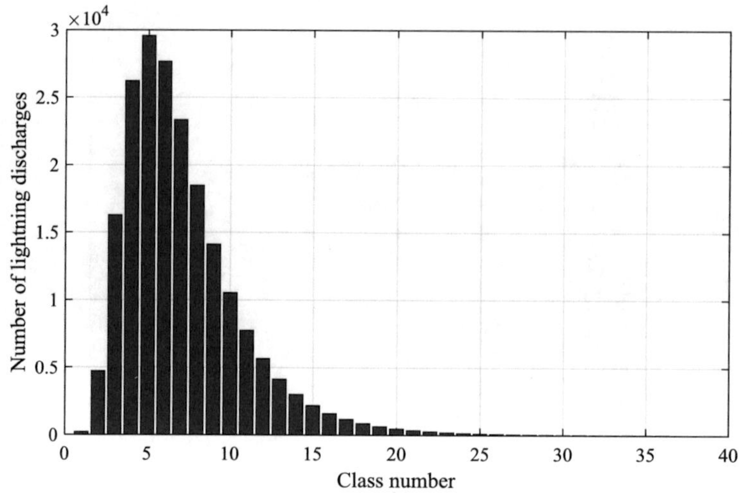


**Fig. 5.** Histogram of the distribution of the lightning current peak values by classes

On the histogram given on Fig. 5, a number of class was applied on abscise while the number of simulated lightning discharge was applied to the ordinate. From the previous histogram it is evident that the most of simulated lightning discharge belongs to class 5, i.e. the most simulated lightning discharge has a peak value ranging from 22 kA to 27 kA. The total number of simulated lightning discharges belonging to class 5 was 29576. The lowest number of simulated lightning discharges belongs to class 40, or the lowest number of simulated lightning discharges has a peak value ranging from 197 kA to 202 kA. The total number of simulated lightning discharges from class 40 was 2.

Figure 6 shows the histogram of the peak values of lightning currents that strikes to the base station tower i.e. LPS system of base station.

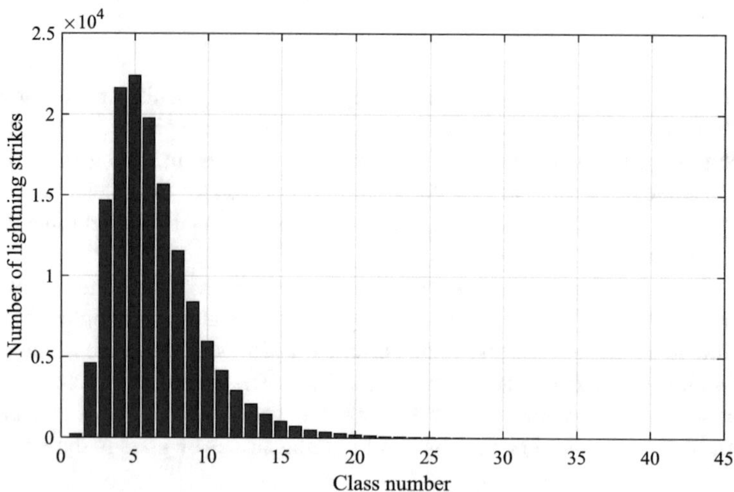


**Fig. 6.** Histogram of the lightning current peak values that strikes to the base station tower

The total number of lightning strikes to the antenna tower i.e. base station LPS is 139031, which is 69.52% of the total number of simulated lightning discharge. A high percentage of simulated lightning strikes to the antenna tower were expected because the antenna tower compared to other base station components is most prominent. Also, analyzes were made on the assumption that it was a lonely object, which additionally contributed to the high number of lightning strikes to the antenna tower.

On Fig. 7, the histogram of the peak values of the lightning currents which strikes to base station antennas is given.

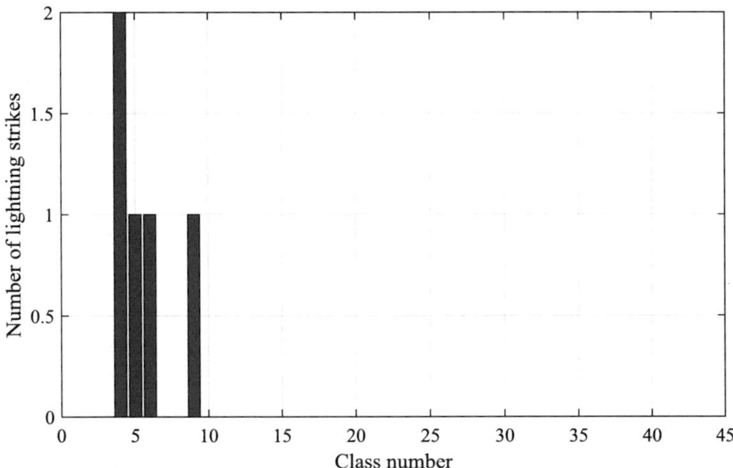


**Fig. 7.** Histogram of the lightning current peak values that strikes to the base station antennas

The total number of lightning strikes to base station antennas was 5, which is 0.0025% of the total number of simulated lightning discharges. From the histogram given on Fig. 7 it is noticeable that the highest value of lightning strikes to base station antennas was in class 4.

During the simulation, one simulated lightning strike the base station container. A small number of atmospheric discharges that hit the base station container were also expected with respect to the position of the container in relation to the base station antenna tower.

For the analyzed example, the stochastic efficiency EC was 0.99995. Based on the previous it can be concluded that the level of stochastic efficacy of the lightning protection system of the analyzed example is satisfactory.

## 5 Conclusion

The paper deals with problems of assessment of the stochastic efficiency of the base telecommunication stations and supporting equipment lightning protection system. For the purpose of modeling lightning discharges on the considered system, the discharge parameters are calculated. The behavior of the lightning peak current was treated with a stochastic model, whereby this behavior was treated as a probability density function of log-normal distribution. The spread of atmospheric discharges with a stochastic model has been worked out in a way that is also described by jumps, whereby an electro-geometric model with empirical constants suggested by the author Love was used to determine the length of each jump. Experimental testing of proposed lightning discharge models carried out in the developed simulation software in such a way that the discharge processes are initiated by a random number generator. Simulation tests on the selected base station facility, made up of a container with telecommunication and power supply equipment and an antenna tower with two antennas, showed a high degree of stochastic efficiency of the lightning protection system.

## References

1. Čaršimamović, A., Čaršimamović, S.: Protection of GSM base stations from lightning discharges. In: 9th Symposium BH K CIGRE, Neum (2009)
2. Mujezinović, A., Veladžić, N., Alihodžić, A., Haračić, S.: Determination of protection zone of lightning installation by computer. In: 12th Symposium BH K CIGRE, Neum (2015)
3. Milan, P.: Zaštita od groma i prenaponska zaštita baznih stanica mobilne telefonije, 16. Telekomunikacijski forum TELFOR, Srbija, Beograd (2008)
4. Uglešić, I., Milardić, V., Franc, B., Filipovic-Grčić, B., Horvat, J.: Prva iskustva sa sustavom za lociranje munja u hrvatskoj, 9. simpozij o sustavu vođenja EES-a HRO CIGRE, Hrvatska, Zadar (2010)
5. Franc, B., Uglešić, I., Filipovic-Grčić, B., Nuhanović, R., Tokić, A., Bajramović, Z.: Primjena sustava za lociranje munja u elektroenergetskim sustavima, 10. Savjetovanje bosanskohercegovačkog komiteta CIGRE, Sarajevo (2011)
6. Čaršimamović, S.: Lightning discharges (in Bosnian). Institut zaštite od požara i eksplozije, Sarajevo (1998)
7. Vujević, S., Sarajčević, P., Saračević, I.: Stochastic assessment of external LPS of structures. In: 29th International Conference on Lightning Protection, Sweden, Uppsala (2008)
8. Grujić, A., Stojković, Z.: Software tool for estimating the 3D lightning protection zone of high voltage substations. Int. J. Electr. Eng. Educ. **48**(3), 307–322 (July 2011)
9. Tokić, A., i ostali: Provjera ispravnosti primjenjih mjera zaštite odatmosferskih prenapona i udara groma na otvorenimrazvodnim postrojenjima 110 kV i 220 kV u HE na Neretvi, Studija rađena za JP Elektroprivreda BiH, April 2011
10. Vujević, S., Sarajčević, P., Lovrić, D.: Efficiency assessment of the external lightning protection system. In: 9th International Conference on Applied Electromagnetics, Serbia, Niš (2009)
11. Uglešić, I., Milardić, V.: Izabrana poglavlja tehnikevisokog napona. Fakultet elektrotehnike i računarstva, Zagreb (2007)
12. Bajramović, Z., Hadžialić, D., Mujezinović, A.: Efficiency assessment of the lightning protection system of mobile communications base stations. In: 13th Symposium BH K CIGRE, Neum (2011)
13. Hadžialć, D.: Protection of mobile phones base station from lightning discharges. MSc Thesis, Faculty of Electrical Engineering, University of Sarajevo, Sarajevo (2017)

# Author Index




S. Avdaković et al. (Eds.): IAT 2019, LNNS 83, pp. 639–640, 2020.
https://doi.org/10.1007/978-3-030-24986-1

Printed in the United States
By Bookmasters